W9-AGB-192

McGRAW-HILL YEARBOOK OF SCIENCE & TECHNOLOGY

2015

McGRAW-HILL YEARBOOK OF SCIENCE & TECHNOLOGY

2015

Comprehensive coverage of recent events and research as compiled by the staff of the McGraw-Hill Encyclopedia of Science & Technology

New York Chicago San Francisco Athens London Madrid

Mexico City Milan New Delhi Singapore Sydney Toronto

On the front cover
Wind turbines of the Burbo Bank Offshore Wind Farm in the Irish Sea.
(Courtesy of Siemens AG, Germany; photograph by Paul Langrock)

ISBN 978-0-07-183576-3
MHID 0-07-183576-8
ISSN 0076-2016

1 2 3 4 5 6 7 8 9 0 DOW/DOW 1 9 8 7 6 5 4

This book was printed on acid-free paper.

*It was set in Garamond Book and Neue Helvetica Black Condensed by
Aptara, New Delhi, India. The art was prepared by Aptara.
The book was printed and bound by RR Donnelley.*

Contents

Editorial and Production

John Rennie, Editorial Director

Jonathan Weil, Senior Staff Editor

David Blumel, Senior Staff Editor

Stefan Malmoli, Senior Staff Editor

Hilary Maybaum, Senior Online Editor

Charles Wagner, Manager, Digital Content

Renee Taylor, Editorial Coordinator

Pamela A. Pelton, Senior Production Supervisor

Frank Kotowski, Jr., Managing Editor

Consulting Editors

Prof. Vernon D. Barger. *Vilas Professor and Van Vleck Professor of Physics, University of Wisconsin-Madison.* CLASSICAL MECHANICS; ELEMENTARY PARTICLE PHYSICS; THEORETICAL PHYSICS.

Prof. Rahim F. Benekohal. *Department of Civil and Environmental Engineering, University of Illinois, Urbana-Champaign.* TRANSPORTATION ENGINEERING.

Dr. James A. Birchler. *Division of Biological Sciences, University of Missouri, Columbia.* GENETICS.

Dr. Marty K. Bradley. *The Boeing Company, Boeing Research and Technology, Huntington Beach, California.* AERONAUTICAL ENGINEERING AND PROPULSION.

Robert D. Briskman. *Technical Executive, Sirius XM Radio, New York.* TELECOMMUNICATIONS.

Dr. Luca Callegaro. *Electromagnetism Division, Istituto Nazionale di Ricerca Metrologica (INRIM), Torino, Italy.* ELECTRICITY AND ELECTROMAGNETISM.

Prof. Wai-Fah Chen. *Department of Civil and Environmental Engineering, University of Hawaii at Manoa, Honolulu, Hawaii.* CIVIL ENGINEERING.

Prof. Peter J. Davies. *Department of Plant Biology, Cornell University, Ithaca, New York.* PLANT PHYSIOLOGY.

Prof. Mohamed E. El-Hawary. *Associate Dean of Engineering, Dalhousie University, Halifax, Nova Scotia, Canada.* ELECTRICAL POWER ENGINEERING.

Dr. Gaithri A. Fernando. *Department of Psychology, California State University, Los Angeles.* CLINICAL PSYCHOLOGY.

Barry A. J. Fisher. *Director, Scientific Services Bureau, Los Angeles County Sheriff's Department (Retired), Los Angeles, California.* FORENSIC SCIENCE AND TECHNOLOGY.

Dr. Kenneth G. Foote. *Woods Hole Oceanographic Institution, Woods Hole, Massachusetts.* ACOUSTICS.

Dr. Margaret L. Fraiser. *Department of Geosciences, University of Wisconsin-Milwaukee.* INVERTEBRATE PALEONTOLOGY.

Dr. Richard L. Greenspan. *Retired; formerly, The Charles Stark Draper Laboratory, Cambridge, Massachusetts.* ACOUSTICS.

Prof. Joseph H. Hamilton. *Landon C. Garland Distinguished Professor of Physics, Department of Physics and Astronomy, Vanderbilt University, Nashville, Tennessee.* NUCLEAR PHYSICS.

Prof. Terry Harrison. *Department of Anthropology, Paleoanthropology Laboratory, New York University, New York.* ANTHROPOLOGY AND ARCHEOLOGY.

Dr. Ralph E. Hoffman. *Yale Psychiatric Institute, Yale University School of Medicine, New Haven, Connecticut.* PSYCHIATRY.

Dr. Beat Jeckelmann. *Chief Science Officer, Federal Office of Metrology (METAS), Bern-Wabern, Switzerland.* ELECTRICITY AND ELECTROMAGNETISM.

Dr. S. C. Jong. *Senior Staff Scientist and Program Director, Mycology and Protistology Program, American Type Culture Collection, Manassas, Virginia.* MYCOLOGY.

Prof. Robert E. Knowlton. *Department of Biological Sciences, George Washington University, Washington, DC.* INVERTEBRATE ZOOLOGY.

Prof. Chao-Jun Li. *Canada Research Chair in Green Chemistry, Department of Chemistry, McGill University, Montreal, Quebec, Canada.* ORGANIC CHEMISTRY.

Prof. Donald W. Linzey. *Department of Biology, Wytheville Community College, Wytheville, Virginia.* VERTEBRATE ZOOLOGY.

Dr. Dan Luss. *Cullen Professor of Engineering, Department of Chemical and Biomolecular Engineering, University of Houston, Texas.* CHEMICAL ENGINEERING.

Prof. Albert Marden. *School of Mathematics, University of Minnesota, Minneapolis.* MATHEMATICS.

Dr. Ramon A. Mata-Toledo. *Professor of Computer Science, James Madison University, Harrisonburg, Virginia.* COMPUTING.

Prof. Krzysztof Matyjaszewski. *J. C. Warner Professor of Natural Sciences, Department of Chemistry, Carnegie Mellon University, Pittsburgh, Pennsylvania.* POLYMER SCIENCE AND ENGINEERING.

Prof. Jay M. Pasachoff. *Director, Hopkins Observatory, and Field Memorial Professor of Astronomy, Williams College, Williamstown, Massachusetts.* ASTRONOMY.

Prof. Stanley Pau. *College of Optical Sciences, University of Arizona, Tucson.* ELECTROMAGNETIC RADIATION AND OPTICS.

Dr. Marcia M. Pierce. *Department of Biological Sciences, Eastern Kentucky University, Richmond.* MICROBIOLOGY.

Dr. Donald Platt. *Micro Aerospace Solutions, Inc., Melbourne, Florida.* SPACE TECHNOLOGY.

Dr. Kenneth P. H. Pritzker. *Professor, Laboratory Medicine and Pathobiology, and Surgery, University of Toronto, and Pathology and Laboratory Medicine, Mount Sinai Hospital, Toronto, Ontario, Canada.* MEDICINE AND PATHOLOGY.

Dr. John D. Protasiewicz. *Department of Chemistry, Case Western Reserve University, Cleveland, Ohio.* INORGANIC CHEMISTRY.

Dr. Roger M. Rowell. *Professor Emeritus, Department of Biological Systems Engineering, University of Wisconsin, Madison.* FORESTRY.

Dr. Thomas C. Royer. *Department of Ocean, Earth, and Atmospheric Sciences, Old Dominion University, Norfolk, Virginia.* OCEANOGRAPHY.

Prof. Ali M. Sadegh. *Director, Center for Advanced Engineering Design and Development, Department of Mechanical Engineering, The City College of the City University of New York.* MECHANICAL ENGINEERING.

Prof. Joseph A. Schetz. *Fred D. Durham Endowed Chair Professor of Aerospace & Ocean Engineering, Virginia Polytechnic Institute and State University, Blacksburg.* FLUID MECHANICS.

Dr. Alfred S. Schlachter. *Advanced Light Source, Lawrence Berkeley National Laboratory, Berkeley, California.* ATOMIC AND MOLECULAR PHYSICS.

Prof. Ivan K. Schuller. *Department of Physics, University of California, San Diego, La Jolla.* CONDENSED-MATTER PHYSICS.

Jonathan Slutsky. *Naval Surface Warfare Center, Carderock Division, West Bethesda, Maryland.* NAVAL ARCHITECTURE AND MARINE ENGINEERING.

Dr. Arthur A. Spector. *Department of Biochemistry, University of Iowa, Iowa City.* BIOCHEMISTRY.

Dr. Anthony P. Stanton. *Teaching Professor and Director, Graphic Media Management, Tepper School of Business, Carnegie Mellon University, Pittsburgh, Pennsylvania.* GRAPHIC ARTS AND PHOTOGRAPHY.

Dr. Michael R. Stark. *Department of Physiology, Brigham Young University, Provo, Utah.* DEVELOPMENTAL BIOLOGY.

Prof. John F. Timoney. *Maxwell H. Gluck Equine Research Center, Department of Veterinary Science, University of Kentucky, Lexington.* VETERINARY MEDICINE.

Dr. Timothy E. Valentine. *Director, Radiation Safety Information Computational Center, Leader, Nuclear Security Modeling Group, Reactor and Nuclear Systems Division, Oak Ridge National Laboratory, Oak Ridge, Tennessee.* NUCLEAR ENGINEERING.

Dr. Daniel A. Vallero. *Adjunct Professor of Engineering Ethics, Pratt School of Engineering, Duke University, Durham, North Carolina.* ENVIRONMENTAL ENGINEERING.

Dr. Bruce White. *Department of Physics, Applied Physics and Astronomy, Binghamton University, State University of New York.* PHYSICAL ELECTRONICS.

Prof. Mary Anne White. *Department of Chemistry, Dalhousie University, Halifax, Nova Scotia, Canada.* MATERIALS SCIENCE AND METALLURGICAL ENGINEERING.

Dr. Thomas A. Wikle. *Department of Geography, Oklahoma State University, Stillwater.* PHYSICAL GEOGRAPHY.

Article Titles and Authors

Preface

In Mark Twain's novel *A Connecticut Yankee in King Arthur's Court,* the eponymous New Englander travels 1300 years backward in time and dazzles the people of sixth-century England with his knowledge of machinery, plumbing, firearms, and astronomy. Twain, writing in the late 1880s, had observed the miraculous transformations that science and industrial technology had already wrought on society, even though both were relatively new enterprises. (The modern meaning of scientist as someone systematically seeking knowledge of the natural world was not introduced until 1834.) Now imagine how Twain or his contemporaries would have reacted if they had glimpsed the science of today's world, only a century and a quarter later. Would they have been any less thunderstruck than Camelot's peasants were by his hero?

The breakneck pace of scientific progress can leave any of us in the dust: More discoveries are reported in more journals than ever before. That is where the 2015 edition of the *McGraw-Hill Yearbook of Science & Technology* comes in. Since 1962 the volumes in this series have tracked key research and technology developments for the sake of students and professionals exploring disciplines other than their own. The *Yearbooks* do so by offering concise overviews on subjects from across the scientific spectrum, invited by a distinguished panel of consulting editors and written by international leaders in their fields.

In this edition, we report on recent astronomical findings about planets orbiting distant stars; a DNA-editing technique revolutionizing the field of genetic engineering; the detection of high-energy neutrinos from deep space inside a cubic kilometer of Antarctic ice; Middle East respiratory syndrome (MERS); the use of polymers for drug delivery; emerging energy technologies, including wind power innovations; the evolutionary history of sharks; quantum communications; results from the GRAIL, WISE, Gravity Probe B, and MESSENGER space missions; the effects of gut bacteria on obesity; the engineering and logistics that keep cruise ships afloat; the phenomenon of metamemory; Dutch elm disease; our earliest hominin ancestors and the origins of the anthropoids; the first known venomous crustacean; gene therapies against cancer; functional nanomaterials made from inorganic compounds; the research honored with a Nobel Prize in 2014; and more than 100 other astounding topics.

Each contribution to the *Yearbook* is the result of a well-informed collaboration. Our consulting editors, representing dozens of disciplines, select the topics with our editorial staff based on the present significance and potential applications of recent work. One or more authorities are then invited to write brief yet thorough articles that describe those new accomplishments and concepts. Through careful editing and extensive use of specially prepared graphics, McGraw-Hill Education strives to make every article as readily understandable as possible.

The *McGraw-Hill Yearbook of Science & Technology* is where librarians, students, teachers, researchers, journalists, the merely curious, and others turn to find the information they need. Its contents hold the seeds from which will grow the world of the twenty-first century—and of the centuries beyond that. Why wait for a time traveler to reveal them to you?

John Rennie
EDITORIAL DIRECTOR

McGRAW-HILL YEARBOOK OF SCIENCE & TECHNOLOGY

2015

A–Z

Accident-tolerant fuels

The safe, reliable, and economic operation of the nation's nuclear power reactor fleet has always been a top priority for the U.S. nuclear industry. Continual improvement of technology, including advanced materials and nuclear fuels, remains central to the industry's success. Decades of research and operation have produced steady technology advances and yielded an extensive base of data, experience, and knowledge on light-water reactor (LWR) fuel performance under both normal and accident conditions. Thanks to efforts by both the U.S. government and private industry, nuclear technologies have advanced over time to optimize economic operations in nuclear utilities while ensuring safety.

One of the missions of the U.S. Department of Energy's (DOE) Office of Nuclear Energy (NE) is to develop nuclear fuels and claddings with enhanced accident tolerance. Enhancing the accident tolerance of LWRs became a topic of serious discussion following the 2011 Great East Japan Earthquake, resulting tsunami, and subsequent damage to the Fukushima Daiichi nuclear power plant complex. As a result of direction from the U.S. Congress, the DOE-NE initiated accident-tolerant fuel (ATF) development as a primary component of the Fuel Cycle Research and Development (FCRD) Advanced Fuels Campaign (AFC).

Prior to the unfortunate events at Fukushima, the emphasis of advanced LWR fuel development was on improving nuclear fuel performance in terms of increased burnup for waste minimization, higher power density for power upgrades, and greater fuel reliability. Fukushima highlighted some undesirable performance characteristics of the standard fuel system during severe accidents, including accelerated hydrogen production under certain circumstances. Thus, enhanced fuel system behavior under design-basis accident (DBA) and severe-accident (beyond design-basis accident, BDBA) conditions became the primary focus for advanced fuels development. In addition, research continued on improving performance under normal operating conditions to ensure that proposed new fuels would be economically viable.

The goal of ATF development by the DOE-NE is to demonstrate performance with a lead test rod (LTR) or lead test assembly (LTA) irradiation in a commercial power reactor by 2022. An LTR refers to a full-length rod composed of candidate fuel and cladding material that would be irradiated in a commercial reactor (its intended use environment) to provide performance data necessary for eventual licensing of the fuel for use in a full core reload. Commercial irradiation of a full assembly to support licensing is referred to as an LTA, which could be a 9 × 9 or 10 × 10 array for a boiling-water reactor or a 14 × 14 to a 17 × 17 array for a pressurized-water reactor. ATF research and development activities are now being conducted at multiple national laboratories and universities and within the nuclear industry through the support of the DOE. Initial concept irradiations will be performed in a test reactor prior to LTR/LTA irradiation in a commercial reactor.

Current LWR fuel. The U.S. commercial LWR fleet—which provides 70% of the nation's low-carbon dioxide (CO_2)-emitting energy sources—produces electricity for consumers with a standard uranium dioxide (UO_2)–zirconium alloy fuel system. Decades of industry research and operational experience have produced an extensive database supporting the performance of LWR fuel during normal power operations and during postulated accident conditions. The nuclear power industry deploys design enhancements to the fuel system (typically small, incremental improvements) as they become available. U.S. boiling-water reactors (BWRs) currently utilize Zircaloy-2 cladding. Pressurized-water reactors (PWRs) previously used Zircaloy-4 but have transitioned to zirconium-niobium (Zr-Nb) cladding (M5® and Zirlo™), which has demonstrated better corrosion behavior.

ATF design goals and constraints. The overall goal of ATF development is to identify alternative fuel system technologies that would enhance the safety, competitiveness, and economics of the U.S. commercial power LWR fleet. An enhanced fuel system supports the sustainability of the U.S. LWR nuclear fleet by allowing it to continue to generate low-CO_2-emitting electrical power in the United States.

Fuels with enhanced accident tolerance will demonstrate enhanced "grace time" relative to the

Improved Reaction Kinetics with Steam
- decreased heat of oxidation
- lower oxidation rate
- reduced hydrogen production (or other combustible gases)
- reduced hydrogen embrittlement of cladding

Improved Fuel Properties
- lower fuel operating temperatures
- minimized cladding internal oxidation
- minimized fuel relocation/dispersion
- higher fuel melt temperature

Enhanced Tolerance to Loss of Active Core Cooling

Improved Cladding Properties
- resilience to clad fracture
- robust geometric stability
- thermal shock resistance
- higher cladding melt temperature
- minimized fuel - cladding interactions

Enhanced Retention of Fission Products
- gaseous fission products
- solid/liquid fission products

Fig. 1. Key considerations in establishing accident-tolerant fuel attributes.

current fuel system—meaning that they would be able to tolerate the loss of active cooling in the reactor core for considerably longer and at higher temperatures—while maintaining or improving fuel performance during normal operations. As summarized in **Fig. 1** and defined relative to the current state-of-the-art system, key requirements for advanced fuels are categorized by nuclear fuel performance, cladding performance, and adherence to overall system constraints.

Advanced nuclear fuel and cladding must maintain, at a minimum, the level of performance and safety margin associated with the current industry-standard fuel system. To warrant the additional costs associated with the development, licensing, and implementation of a new fuel technology, advanced fuel options must not only meet these performance levels but exceed them. In addition to providing significantly enhanced safety performance under off-normal conditions, advanced fuels should provide the possibility of power up-rates, increased fuel burnup, reduced numbers of assemblies per reload, and longer cycle lengths (for reduced outage frequency).

Any new fuel concept proposed for enhanced accident tolerance under rare events (BDBAs) must also be compliant with and evaluated against current design, operational, economic, and safety requirements. The constraints associated with commercial nuclear fuel development and deployment include (**Fig. 2**):

Economics: The fuel should maintain economic viability with respect to additional costs (for example, fabrication) and potential cost reductions realized through better performance (higher burnup for extended cycles and power upgrades) or increased safety margin.

Backward compatibility: The fuel must be compatible with existing fuel-handling equipment, fuel-rod or assembly geometry, and co-resident fuel in existing LWRs.

Operations: The fuel should maintain or extend plant operating cycles, reactor power output, and reactor control.

Safety: The fuel must meet or exceed current fuel-system performance under normal, operational transient, design-basis accident (DBA), and beyond-design-basis accident (BDBA) conditions.

Front end of the nuclear fuel cycle: The fuel must adhere to regulations and policies, for both the fuel fabrication facility and the operating plant, with respect to technical, regulatory, equipment, and fuel performance considerations.

Back end of the nuclear fuel cycle: The fuel cannot degrade the storage (whether wet or dry) and repository performance of the fuel (assuming a once-through fuel cycle). The fuel should also be compatible with the possibility of future transition to a closed fuel cycle.

ATF evaluation metrics. A common set of technical evaluation metrics is required to aid in the optimization and selection of candidate designs on a more quantitative basis. Because of the complex multiphysics behavior of nuclear fuel and the large set of performance requirements that must be met, evaluation metrics proposed by Idaho National Laboratory

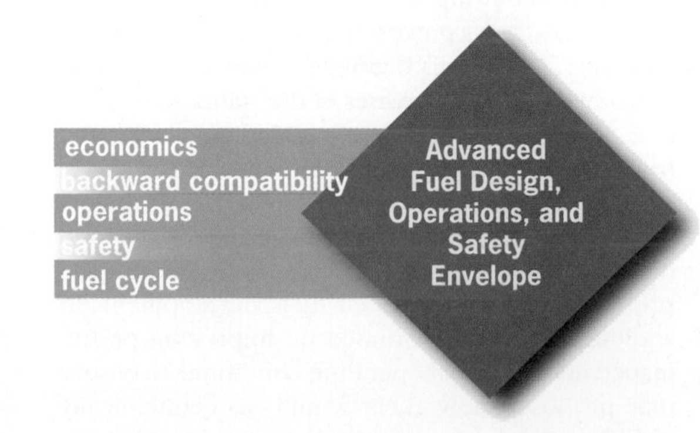

Fig. 2. Considerations that constrain new fuel designs.

(INL) in 2014 describe a technical methodology for concept evaluation rather than establish specific quantitative targets for each property or behavior. An independent expert panel review team will be established to review the ATF concepts under development. This team will be composed of technology experts selected for their knowledge of the technologies under review, including materials (metals and ceramics), neutronics, thermal-hydraulics, and severe accidents. Candidate designs will be evaluated for their relative benefits or vulnerabilities across relevant performance regimes, including (1) fabrication and manufacturability (including their potential for licensing), (2) normal operation and anticipated operational occurrences, (3) postulated accidents (DBAs), (4) severe accidents (BDBAs), and (5) used-fuel storage, transport, and disposition (including potential for future reprocessing).

The proposed technical evaluation methodology will result in a ranked, prioritized list of candidate ATF designs based on their estimated benefits and any remaining gaps or vulnerabilities that must be addressed via performance characterization or design modifications. The review panel may choose to develop two ranked lists: one for near-term technologies, fitting within the defined 10-year development window (to meet the 2022 deadline for LTR insertion in a commercial reactor), and a second for longer-term technologies that appear to offer a significant benefit eventually but that are unlikely to meet the desired development timeframe. The ranking can be used to select the most promising ATF design options for continued development.

ATF development activities. The ATF program is currently in its early R&D phases, supporting the investigation of a number of technologies that may improve fuel system response and behavior in accident conditions. The U.S. DOE program is sponsoring multiple teams to develop ATF concepts within national laboratories, universities, and the nuclear industry. These concepts offer both evolutionary and revolutionary changes to the current fuel system. **Figure 3** provides a notional schematic of potential performance improvements (depending on the specific LWR design and operating scenario) and relative development times for several proposed ATF concepts. The **table** provides a summary of the leading concepts currently under investigation in the United States.

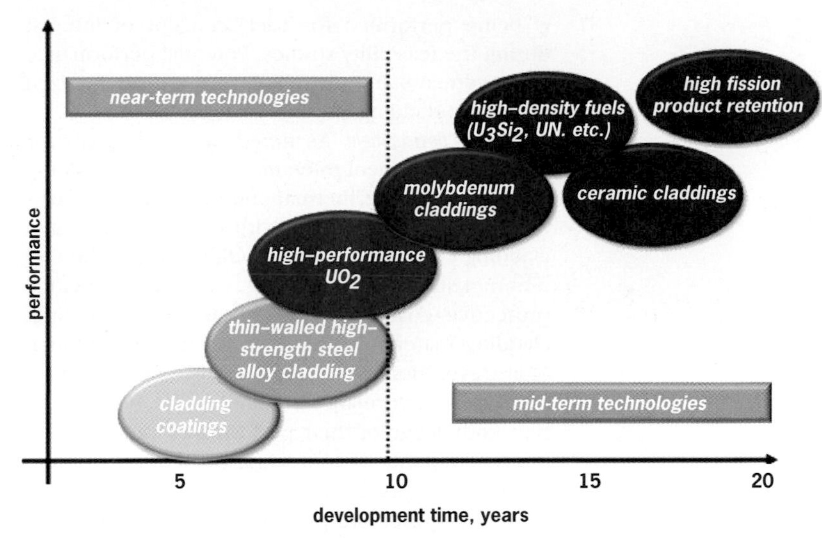

Fig. 3. Notional schematic of potential performance improvements and relative development times for several proposed ATF concepts. Specific performance enhancements will depend on the specific LWR design and operating scenario.

Each concept under development offers some potential performance benefits across the spectrum of normal operating conditions, anticipated transient conditions, and possible accident scenarios. However, additional data are necessary to fully characterize the performance of the proposed fuel and cladding materials, and some key challenges remain. A systematic analytical and experimental evaluation

Summary of major U.S. DOE–funded ATF development projects

Lead organization	Category—Major technology area
Oak Ridge National Laboratory (ORNL)	Fuel: Fully ceramic microencapsulated (FCM)-UO_2; FCM-UN Cladding: FeCrAl alloy; silicon carbide (SiC) concepts
Los Alamos National Laboratory (LANL)	Fuel: Enhanced UO_2; advanced composites
Electric Power Research Institute (EPRI) + LANL	Cladding: Advanced molybdenum alloys (multilayer design)
AREVA	Fuel: High-conductivity fuel (UO_2+Cr_2O_3, +SiC) Cladding: Coated Zr-alloys (protective materials, MAX phase)
Westinghouse Electric Company	Fuel: U_3Si_2, and UN +U_3Si_2 fuel Cladding: Coated Zr-alloy; SiC concepts
GE Global Research	Cladding: Advanced steel (ferritic/martensitic, including FeCrAl)
University of Illinois	Cladding: Modified Zr-based cladding (coating or modification of bulk cladding composition)
University of Tennessee	Cladding: Modified Zr-based cladding via ceramic coatings (MAX phase; multilayer ceramic coatings)

is being performed for each concept of interest during the feasibility studies. Potential performance enhancements and key challenges are summarized below for cladding and fuel materials of interest.

Cladding development. As noted in Fig. 1, goals for enhanced, accident-tolerant nuclear fuel cladding focus on cladding thermal and mechanical properties and reaction kinetics with steam. Evolutionary cladding concepts include modifications to the current metallic Zr-alloy cladding, which could include protective coatings or modifications to the bulk cladding material, or alternative metallic cladding. More revolutionary concepts offer the potential for significant performance enhancements, but the current knowledge of their performance under all the relevant regimes is more limited and will require longer development time.

Coatings offer the potential for improved steam reaction kinetics. Primary challenges facing coated cladding designs (pursued by AREVA, Westinghouse, the University of Illinois, and the University of Tennessee) include the ability to achieve good coating adherence, uniform thickness, and scratch resistance without altering the performance of the base material. These coatings must be shown to maintain adherence during fuel handling and loading and when subjected to thermal cycling in the harsh in-core environment. Many performance characteristics can be assessed in separate effects tests prior to irradiation, followed by in-pile tests (conducted in a reactor environment) to verify irradiation stability and post-irradiation performance.

Alternative metallic cladding concepts under investigation include advanced steels [Oak Ridge National Laboratory (ORNL), General Electric (GE) Global Research], refractory metal concepts [Electric Power Research Institute (EPRI)], and modification of the standard bulk Zr-alloy (University of Illinois). These concepts offer greater strength and reduced reactions with high-temperature steam, but they require additional development to fabricate long, thin-walled tubes and to weld endcap closures that will provide a hermetic seal. Some materials of interest, such as the advanced steel alloys (for example, FeCrAl), would increase parasitic neutron absorption relative to the current cladding, such that they must be coupled with either higher-enrichment fuel or fuel having a higher fissile density than standard UO_2. FeCrAl, however, has demonstrated significant reduction in steam reaction kinetics relative to Zr-based alloys. Advantages of the proposed molybdenum cladding include high melting point, excellent high-temperature strength and creep resistance, and good ductility at low temperatures; however, oxidation at high temperatures is a current concern. EPRI is therefore developing coated molybdenum cladding to improve high-temperature oxidation resistance. Cladding designs can and should be tested both out-of-pile (out of a reactor) and in-pile to characterize the complete material behavior.

Revolutionary cladding designs of interest include ceramic SiC cladding and hybrid metal–ceramic concepts. While offering enhanced structural integrity under both normal and accident conditions and reduced oxidation rate and heat of oxidation when in contact with high-temperature steam, ceramic cladding faces several challenges. Ceramics are brittle by nature. Hence, fully ceramic concepts generally consider multilayer designs that incorporate a monolithic layer to provide hermeticity and a composite layer for structural integrity (metallic–ceramic hybrid concepts rely on the metallic layer for hermeticity). The order and thickness of each of these layers can have a significant impact on the cladding mechanical, thermal, and neutronic performance. A key challenge for SiC cladding is the ability to fabricate long, thin-walled tubes and to provide an endcap seal that will maintain hermeticity under the intended operating conditions. (As previously noted, these requirements also apply to alternative metallic cladding concepts.) Additionally, these designs must be shown to withstand quench (which would be caused by core reflood to mitigate a loss-of-coolant accident) and to maintain sufficient thermal conductivity and strength after irradiation.

Fuel development. Key goals for enhanced, accident-tolerant fuel material focus on reduced internally stored energy, operating temperature that is low relative to the fuel melting temperature, and enhanced fission product retention. Fuel concepts under development range from modifications of the current UO_2 fuel to more drastic changes in the fuel composition. Many of the fuels under development require research to establish basic fabrication feasibility, followed by irradiation in a test reactor to demonstrate performance under in-core conditions. If initial irradiation proves successful, the ability to ramp up fabrication to commercial scale must also be demonstrated.

Modified UO_2 fuel would introduce additives to the fuel material to reduce fission-gas generation, improve load-following characteristics, increase uranium density, improve wash-out characteristics in the event of rod failure, and retain fission products in the fuel matrix. Chromia doping, for instance, offers the potential to lock up cesium in the fuel matrix, whereas SiC fibers could lock up iodine in the matrix while also enhancing fuel thermal conductivity (both additives are under development by the AREVA team via University of Florida collaboration). Los Alamos National Laboratory (LANL) research in ceramic fuels includes an evolutionary path based on UO_2 fuels, and a revolutionary path based on composite fuels. Fuel design seeks to improve fracture toughness and plasticity to relieve fuel-pellet cracking, enhance thermal conductivity (to reduce stored energy and internal fuel temperature), and enhance oxidation resistance to increase the coping time under accident conditions and to increase the stability of used fuels during wet and dry storage. A range of ceramic/ceramic candidate composite systems with at least one high-uranium-density phase are undergoing initial assessment using a combination of thermodynamic analysis, thermochemical and thermophysical property review, material compatibility screening experiments, and assembly-level neutronic analysis.

High-density fuels, such as metallic, nitride, and silicide fuels, offer the potential for higher thermal conductivity and higher fissile density than standard UO_2 fuel. The latter property can improve fuel efficiency and may be necessary to compensate for neutronic inefficiencies presented by some new cladding concepts—which would minimize the need to significantly increase fuel enrichment. U_3Si_2 would offer up to 17% increase in ^{235}U loading and a fivefold increase in thermal conductivity over standard UO_2, while "waterproofed" UN (using a silicide additive) would offer up to 35% increase in ^{235}U loading and tenfold increase in thermal conductivity. Development efforts for these fuels, led by Westinghouse in coordination with INL, LANL, and Texas A&M University, have demonstrated feasibility of their manufacture, but fuel performance must be demonstrated under in-core conditions.

More drastic changes to the current UO_2 fuel could include fully ceramic microencapsulated (FCM) fuel, currently under development at ORNL. Microencapsulated fuel is essentially particle fuel dispersed in a ceramic or metallic matrix. FCM offers a barrier to fission product release via the silicon carbide (SiC) shell of the tristructural-isotropic (TRISO) particles, which are compacted in a dense, high-conductivity, and radiation-resistant SiC matrix that serves as a secondary boundary to fission product release. Results to date for several fuel development hurdles, such as the ability to achieve a robust high-TRISO-density compact fuel, demonstration of the radiation stability of fuel constituents, and preliminary core analyses have been encouraging. Remaining challenges include fabrication of the uranium nitride (UN) kernel fuel required to achieve adequate fissile density, compacting scale-up, and full-core analysis to assess the impact of this fuel on reactor operations.

ATF irradiation testing. Concepts that have matured to the point of irradiation testing will be sent to INL for final assembly and integration into an irradiation capsule for further evaluation. An irradiation test plan has been drafted for ATF concepts. The plan progresses from feasibility experiments under normal operating conditions to integral demonstrations under accident conditions, providing data to support the LTR/LTA program and eventual qualification of an ATF concept. The initial irradiation series, referred to as ATF-1, will utilize drop-in capsules in the INL Advanced Test Reactor (ATR) to investigate the performance of a variety of proposed ATF concepts under normal LWR operating conditions. This test series is intended to investigate the irradiation behavior of new fuels and their interaction with the cladding under normal LWR conditions. **Figure 4** shows an ATF irradiation capsule schematic. Resulting data on fuel behavior and fuel–cladding interaction will inform selection of one or more promising concepts prior to more complex irradiation tests.

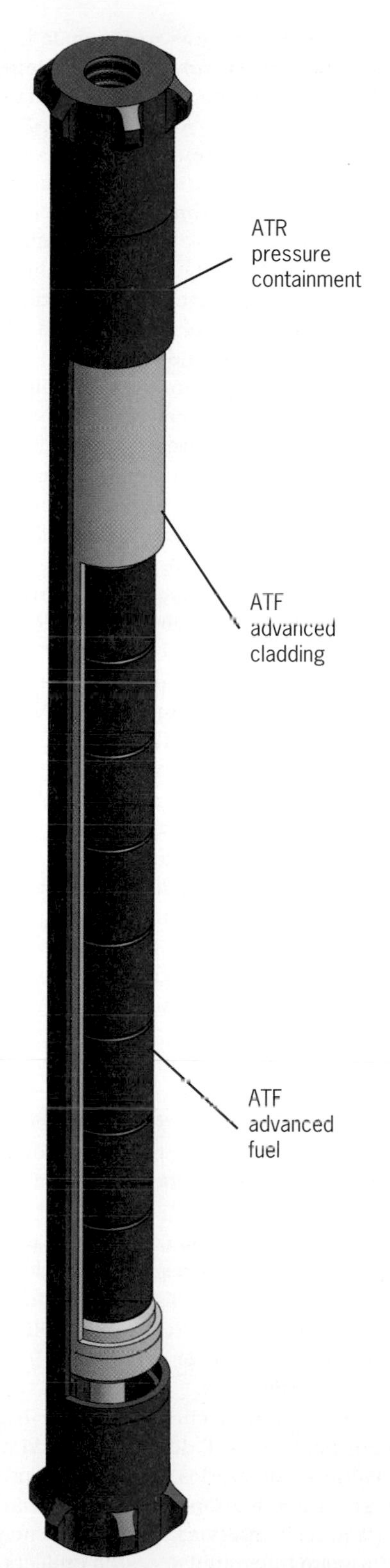

Fig. 4. ATF-1 capsule assembly schematic. (*Figure by INL, from S. Bragg-Sitton, Development of advanced accident-tolerant fuels for commercial LWRs, Nucl. News, 57(3):83–91, March 2014*)

Several ATF concepts began irradiation in the ATF-1 series in late summer 2014, with additional concepts ready for ATR insertion in early FY2015 (October 2014–September 2015).

Follow-on irradiation tests will include loop testing in the ATR (ATF-2) and transient testing (ATF-3 and -4). ATF-2 will use the pressurized-water loop in the INL ATR, such that the ATF fuel rods will be in direct contact with high-pressure water coolant with active chemistry control to mimic the conditions of PWR coolant. The most promising concept(s) would then proceed to ATF-3. ATF-3 would include transient testing of rods in the INL Transient Reactor Test (TREAT) facility to investigate performance under reactivity insertion accident scenarios. The final step in this test series would be transient testing of a subset of LTRs, possibly in TREAT, while LTRs or LTAs are irradiated in a commercial LWR.

Next steps. The ATF development effort adopts a three-phase approach to commercialization. The approximate timeframe for each phase is noted, where FY is the U.S. government fiscal year (October–September).

Phase 1: Feasibility assessment and preliminary selection of the most promising concepts. Fuel concepts are developed, tested, and evaluated. Feasibility assessments are performed to identify promising concepts. These include laboratory-scale experiments, such as fabrication, preliminary irradiation, and material-property measurements; fuel performance code updates; and analytical assessments of economic, operational, safety, fuel cycle, and environmental impacts (FY 2012–FY 2016).

Phase 2: Development and qualification. Prototypic fuel rodlets are irradiated in a test reactor at LWR conditions to provide the data required for the LTRs and LTAs. The fabrication process expands to industrial scale for LTRs and LTAs (FY 2016–FY 2022).

Phase 3: Commercialization. Partial-core (region-sized) reloads are tested to verify the performance observed for the LTRs and LTAs and to provide additional data for final licensing of the product. Commercial fabrication capabilities are established (FY 2022 and beyond).

Each development phase corresponds roughly to the technology readiness levels (TRLs) defined for nuclear fuel development. TRLs 1–3 correspond to the proof-of-concept stage (Phase 1), TRLs 4–6 to proof-of-principle (Phase 2), and TRLs 7–9 to proof-of-performance (Phase 3).

The United States is currently in the feasibility assessment phase of ATF development, with teams led by national laboratories, industry (via proposals made to a DOE Funding Opportunity Announcement on ATF), and academia (via DOE Nuclear Energy University Program Integrated Research Project proposals). Awards were made to industry and university teams in 2012 to support concept development; industry teams are currently preparing proposals for continued funding through the end of Phase 1. The AFC is now preparing for a formal expert panel review to be conducted in 2016 to select concepts that will proceed to Phase 2 development.

For background information *see* NUCLEAR FUEL CYCLE; NUCLEAR FUELS; NUCLEAR FUELS REPROCESSING; NUCLEAR POWER; NUCLEAR REACTOR in the McGraw-Hill Encyclopedia of Science & Technology.

Shannon M. Bragg-Sitton

Bibliography. S. Bragg-Sitton, Development of advanced accident-tolerant fuels for commercial LWRs, *Nucl. News*, 57(3):83–91, March 2014; *Light Water Reactor Accident Tolerant Fuel Performance Metrics*, INL/EXT-13-29957, February 2014; Special section on "accident tolerant fuels," *J. Nucl. Mater.*, 448:373–540, 2014.

Agile methods in software engineering

Agile methods have significantly changed the practice of software development since their inception around the turn of the twenty-first century. Traditional software development methods emphasize careful preparation and planning, followed by closely monitored development work culminating in the delivery of a finished product. In contrast, agile methods emphasize close collaboration with customers in creating small product increments in a very short time, providing opportunities to change product requirements quickly and easily (hence the term "agile"). This article surveys the rise of agile methods and characterizes them by describing Extreme Programming and Scrum, the two most influential agile methods.

Traditional methods. Digital computers were invented in the 1940s and were programmed by the engineers and mathematicians who created them. These early machines ran relatively short and simple programs, mostly to solve mathematical problems, so programming them was not much of a problem. Computer power and capability grew rapidly, and before long people were using computers to manage large databases and to perform complex computations in business, science, and engineering. These more demanding applications required much larger and more complicated programs, and by the late 1960s a software crisis had emerged. The software crisis was characterized by an inability to deliver correct and useful programs on time and within budget. Many programs were never delivered at all. This crisis precipitated the creation of the discipline of software engineering, whose goal was to figure out how to deliver high-quality software products (programs, data, documentation, and so forth) on time and within budget.

The approach taken by software engineers in response to the software crisis was modeled on the way that engineers in other disciplines developed products. These methods, which we call "traditional," emphasize planning both the product to be created and the process used to create it. Process planning includes estimating the time and money needed to create the product, scheduling the work to be done, and allocating human and other resources. Product planning includes carefully specifying product features, capabilities, and properties; the program components and structures to realize them; and the tests for confirming that the product behaves as it should. These methods also include data collection and analysis activities to see if the development process is going according to plan, and to use as a basis for better planning in later projects. Thus, traditional methods devote much effort to formulating, writing, checking, and modifying various plans, as well as to collecting and analyzing data.

The traditional approach succeeded in relieving the software crisis to some extent. The success rate for software development projects improved steadily (but slowly) for many years. Organizations using traditional methods often deliver excellent products on time and within budget. But there are still serious problems. Failure rates for projects using traditional methods remain unacceptably high. Even when projects succeed, customers are often disappointed because the development effort took so long (often years) and the product was not as useful as they had expected. Traditional methods make it very expensive to change product specifications during development, so it is difficult to alter specifications in response to evolving customer needs, changing market conditions, new technologies, and so forth. Developers sometimes find the culture of caution, order, control, and thorough documentation inherent in the traditional approach stifling and dehumanizing.

Agile methods. Toward the end of the 1990s, several prominent software developers began to advocate a different approach in direct opposition to traditional software development methods. These proposals, collectively called agile methods, emphasize the need to create software products quickly and to respond to product specification changes immediately. Agility is achieved by developing software products in small increments over a short period, typically a month or less. Product increments can be delivered very quickly. Developers do not plan their activities or the product in detail beyond the current increment, so it is easy to react to changing product specifications. Project overhead is further trimmed by avoiding written specifications as much as possible. Instead, developers work closely with customers to pin down product specifications verbally just as they are needed for implementation. Feedback, in the form of tests and of customer reviews of the current state of the product, is made as immediate as possible to catch problems quickly and with minimal overhead.

Agile methods have been used successfully on many projects, and there has been a steady migration from traditional to agile methods in the software development community. Although many organizations, particularly those creating systems with high reliability, safety, and security requirements, continue to use traditional methods, it appears that agile methods are well on their way to becoming the dominant software development approach.

The best-known agile methods are Scrum, Extreme Programming, the Crystal Methods, Lean Software Development, and the Dynamic System Development Method. Extreme Programming (XP) was initially the best-known agile method, but today Scrum is by far the preferred approach. However, continuous improvement is an agile principle. Thus, successful techniques pioneered by one approach are quickly adopted by the others, which results in close similarities among the various methods.

Extreme programming. Although XP includes values and principles, its greatest influence has been its development practices. The first version of XP (from 1999) listed 12 practices; the more recent version (from 2005) lists 13 primary practices and 11 corollary practices. The later practices overlap considerably with the earlier practices. Below are the 13 primary practices from the most recent version of XP, and most of these are used in other agile methods.

Sit together. Work with the entire team in a common space at least part of the time. This practice promotes communication.

Whole team. Form cross-functional teams consisting of people who together have all the skills necessary for the project to succeed. This is often further understood to mean that team members work across multiple disciplines.

Informative workspace. Adorn the workspace with "information radiators" that quickly convey the state of the project.

Energized work. Require people to work only as many hours as they can be effective. This practice, also called sustainable pace, is in contrast with the common practice in software development, which requires people to work 60 or more hours per week.

Pair programming. Write all code with two developers sharing a single computer (programmers traditionally work alone). This practice helps keep people on task, promotes brainstorming, helps people get past problems, spreads knowledge of the code through the team, and holds programmers to project coding standards.

Stories. Drive efforts using small units of customer functionality, called user stories, or just stories. A story is a brief description of some product feature or function like "Provide command history accessible with arrow keys," or "Display pie, bar, or line charts as the user directs." Stories are the agile version of software requirements, which in traditional approaches are usually expressed as detailed and exhaustive specifications collected in large software requirements specification documents. Stories are not elaborated until they are implemented, so they are easily changed. However, they are still an adequate basis for effort estimation and project planning. Using stories is one of the most widely adopted XP practices.

Weekly cycle. Start each week by reviewing the current state of the project, having the customer choose stories that can be implemented in a week, breaking the selected stories into tasks, and letting team members choose tasks. The rest of the week is for doing the tasks. This cycle of developing small product increments is common among agile methods, but other methods use longer increments and may handle the steps in the cycle differently.

Quarterly cycle. Plan work every 3 months. This larger planning horizon is an opportunity to identify and address larger problems, choose large stories or coherent groups of stories for implementation in the quarter (perhaps forming a release or a major part of a release), and examine how the project fits into the organization. Other agile methods may use a different time scale or focus on different issues, but all incorporate a larger-horizon planning activity.

Slack. Build extra tasks into a plan that can be dropped, if necessary, to ensure that all committed stories can be implemented. Some other agile methods emphasize the collection of tracking data to improve estimates as a way to meet commitments.

Ten-minute build. Compile all code and run all automatic tests in 10 minutes. The point of this practice is that building and testing processes that take too long will not be used.

Continuous integration. Combine and test all code written by different programmers frequently, at least every few hours. Traditional practice used to be to integrate and test code only after a lot of component development, but nowadays at least daily integration and testing has become standard in both agile and traditional methods.

Test-first programming. Write tests before writing code, then write code to pass the tests. This practice, usually called test-driven development, has also become common in both traditional and agile methods.

Incremental design. The program is "cleaned up" continuously as it is changed to improve its design. If this were not done, the program would become too convoluted as it evolved. This practice is not needed (as much) in traditional approaches, because the design is determined before the code is written.

Scrum. The heart of Scrum is the process it uses to plan and manage development increments. One person, the product owner, is responsible for this activity. The development team is a self-managing group of about five developers responsible for carrying out the work. All desired product features are listed in a prioritized list (of stories, usually) called the product backlog. At the start of a product development increment, called a sprint, the product owner and the team agree on a selection of features chosen from the product backlog for implementation in the sprint; this set of features is the sprint backlog. The team is then responsible for implementing the sprint backlog by the end of the sprint, typically in about 3 weeks.

The team consults with the product owner and customers during the sprint to clarify specifications and resolve issues, but neither customers nor the product owner are allowed to alter the sprint backlog. The team works with the product owner to help groom the product backlog during sprints, however. Grooming the backlog consists of adding, removing, or altering features, decomposing large features into smaller features suitable for implementation in a sprint, estimating implementation efforts, and prioritizing features. Grooming allows Scrum projects to respond to changed circumstances or customer preferences, new ideas, and so forth.

At the end of the sprint, the team delivers its results, even if all the features in the sprint backlog are not finished (unfinished features return to the product backlog). The team presents its work to the product owner and customers, who are thereby apprised of the current state of the product. The team also examines its own performance during the sprint looking for ways to improve. The result of the sprint is a deliverable product, so the results of each sprint can be released to customers as a new version of the product. Usually customers only receive new versions after several sprints, however. The next sprint usually begins immediately.

Scrum includes several practices besides those governing the sprinting process. For example, it defines the role of Scrum master as a Scrum expert who helps the team work effectively and tries to remove roadblocks; it recommends burn charts (a simple line chart that shows accomplishments against time) for tracking sprint and project progress; it recommends daily, short stand-up meetings (called Scrums) as a way for teams to keep in touch; and so on. However, Scrum accommodates other practices and techniques, and many XP practices in particular have become common among Scrum practitioners, including sitting together, having an informative workspace, energized work, pair programming, using stories, continuous integration, test-first programming, and incremental design.

Outlook. When agile methods were first proposed, many developers thought they were nothing more than an excuse to avoid the discipline and effort of traditional software engineering. Time has shown that agile methods require just as much effort and discipline as traditional methods, and that they can solve many of the problems that continue to plague traditional methods. Agile methods have matured and grown and are now just as sophisticated in their practices, techniques, and intellectual underpinnings as traditional methods. In fact, while innovation in traditional methods has lagged in the last decade, agile methods continue to evolve in exciting directions. Agile methods have contributed significantly to the growth of software engineering, and it appears that they will continue to be popular and important in the future.

For background information *see* COMPUTER PROGRAMMING; METHODS ENGINEERING; PROGRAMMING LANGUAGES; SOFTWARE; SOFTWARE ENGINEERING; SOFTWARE TESTING AND INSPECTION in the McGraw-Hill Encyclopedia of Science & Technology.

Christopher Fox

Bibliography. K. Beck, *Extreme Programming Explained*, 2d ed., Addison-Wesley, Upper Saddle River, NJ, 2005; R. Pressman, *Software Engineering: A Practitioner's Approach*, 7th ed., McGraw-Hill,

Boston, 2009; K. Rubin, *Essential Scrum: A Practical Guide to the Most Popular Agile Process*, Addison-Wesley, Upper Saddle River, NJ, 2012; K. Schwaber and M. Beedle, *Agile Software Development with Scrum*, Prentice Hall, Upper Saddle River, NJ, 2001.

Alkane metathesis

Alkane metathesis is a reaction that transforms an alkane (C_nH_{2n+2}) or an alkane mixture into lower and higher homologues. An example of alkane metathesis is the ideal reaction of propane, which would yield butane and ethane [the $(n+1)$ and $(n-1)$ homologues]; however, in practice, higher and lower homologues (hexanes, pentanes, and methane) are also formed [reactions (1)]. These reac-

$$C_nH_{n+2} \rightleftharpoons \sum_i C_{n-i}H_{2(n-i)+2} + \sum_i C_{n+i}H_{2(n+i)+2} \quad (1a)$$

$$\mathrm{R{\frown}} \rightleftharpoons \mathrm{R{\frown}{\smile}R} + C_2H_6 \quad (1b)$$

tions are typically thermoneutral [the change in the Gibbs energy (ΔG) is close to 0] and require a catalyst or a mixture of catalysts. While reaction (1*b*) is formally analogous to alkene and alkyne metathesis, where alkylidene ($R_2C{=}M$) and alkylidyne ($RC{\equiv}M$) groups are exchanged at a transition-metal center upon reaction with alkene and alkyne mixtures, the redistribution of alkyl fragments does not take place by cleavage of carbon-carbon sigma (σ) bonds with a metal-carbon σ-bond (σ-bond metathesis mechanisms), but rather involves alkylidene and alkene intermediates. The detailed mechanism depends on the catalyst system, which can be divided into two families: tandem systems involving two catalysts and single-component catalysts. Overall, tandem and single-component alkane metatheses are promising reactions that could enable the upgrading of hydrocarbons, in particular, converting natural-gas resources into valuable liquid fuels.

Tandem alkane metathesis. In tandem alkane metathesis, two catalysts are present: a dehydrogenation/hydrogenation catalyst and an olefin metathesis catalyst for alkene homologation. The alkane homologation process involves three steps: alkane dehydrogenation [reaction (2)], alkene metathesis to generate the alkene homologues [reaction (3)], and alkene hydrogenation to form the lower- and higher-molecular-weight alkanes [reaction (4)].

$$\mathrm{R{\frown}} \rightleftharpoons \mathrm{R{\frown}{=}} + H_2 \quad (2)$$

$$2\mathrm{R{\frown}{=}} \rightleftharpoons \mathrm{H_2C{=}CH_2} + \mathrm{R{\frown}{=}{\smile}R} \quad (3)$$

$$\mathrm{R{\frown}{=}{\smile}R} \xrightleftharpoons{H_2} \mathrm{R{\frown}{\smile}R} \quad (4)$$

The original system developed by Chevron in the 1970s was based on a mixture of classical heterogeneous catalysts: Pt/Al_2O_3 as the dehydrogenation/hydrogenation catalyst and WO_3/SiO_2 as the alkene metathesis catalyst. This process operates at relatively high temperature in order to favor dehydrogenation, a highly endothermic process. More recently (2006), a lower-temperature tandem alkane metathesis process was developed by combining tridentate (pincer) ligand complexes of iridium (homogeneous dehydrogenation catalysts) with alkene metathesis catalysts (**Scheme 1*a*** and *b*). Initial reports used homogeneous Mo- or W-alkylidene metathesis catalysts as well as heterogeneous Re_2O_7/Al_2O_3 (Scheme 1*b*). In all cases, this process leads to a distribution of linear alkane homologues and no methane, which is consistent with the proposed tandem process involving homologation via alkene metathesis (Scheme 1*c*). The lack of selectivity of this process is related to the formation of

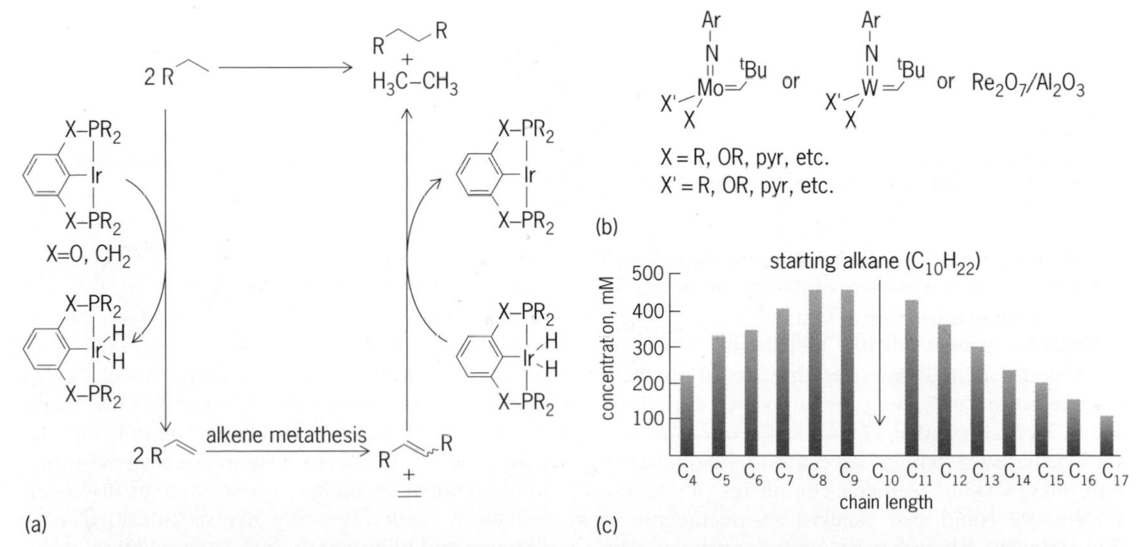

Scheme 1. Homogeneous alkane metathesis by (*a*) the pincer-Ir complex, where X = O, CH_2 and (*b*) a Mo-, W-alkene, or Re_2O_7/Al_2O_3 metathesis catalyst. (*c*) Distribution of linear alkane products.

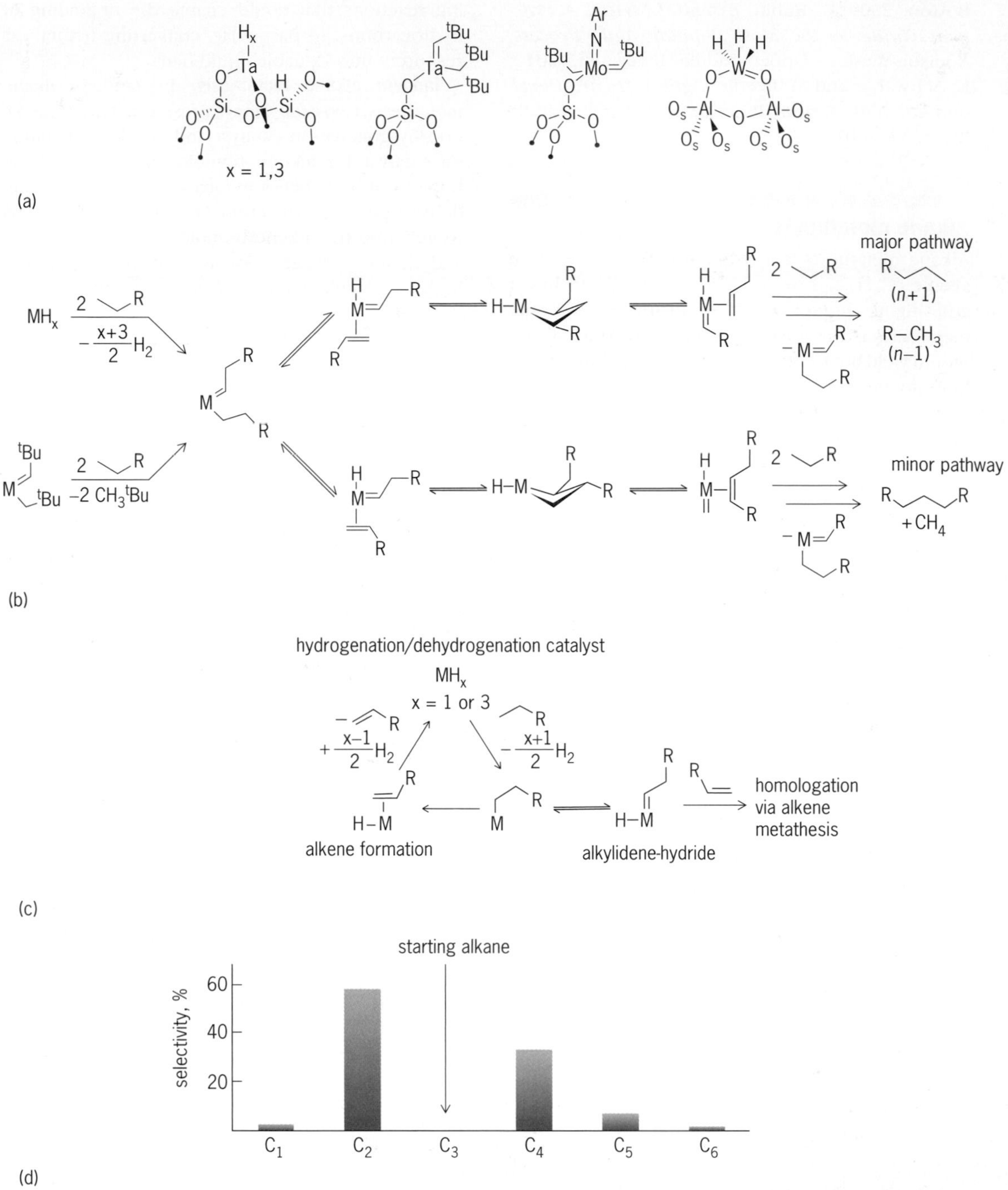

Scheme 2. (a) Structures of well-defined Ta-H and W-H heterogeneous alkane metathesis catalysts. (b) One-site catalytic reaction pathways. (c) Two-site catalysis. (d) Distribution of alkane products using the W-H/Al_2O_3 catalyst.

alkene isomers in the dehydrogenation step as well as the subsequent reactivity of the alkane products, which are similar to the reactants.

Single-component alkane metathesis. The first single-component alkane metathesis catalysts were discovered in 1997 using the supported early transition metal hydrides, Ta and W. Later, it was shown that well-defined supported high-oxidation-state alkyl alkylidene metal complexes of Ta, Mo, W, and Re could also catalyze alkane metathesis (**Scheme 2*a***). The proposed active species rely on metal alkylidene hydride intermediates, which are generated in situ from supported early transition metal hydrides or metal alkyl alkylidenes. To date, the most active and stable catalysts, which can be regenerated under H_2, are the alumina-supported tungsten hydrides. These systems, in particular those based on W, selectively transform an alkane (C_nH_{2n+2}) into its lower and higher homologues, favoring the $(n-1)$ and $(n+1)$ homologues, in contrast to the tandem alkane metathesis systems discussed previously, which typically give statistical mixtures of lower and higher homologues. In addition, these single-component systems convert linear alkanes

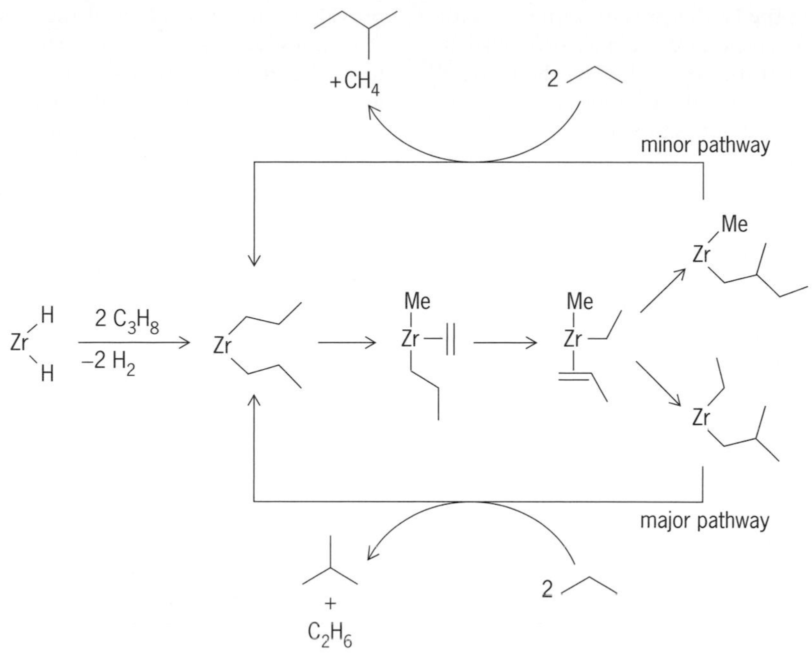

Scheme 3. Mechanism of supported Zr-H propane homologation.

into mainly linear alkanes, whereas branched alkanes give only branched alkanes.

While a detailed understanding of the mechanism has not yet been fully established, the single-component catalysts probably operate through different reaction intermediates from those in tandem metathesis (Scheme 2*b* and *c*). While the mechanism can involve one or two sites, the metal alkylidene hydride is generated through C-H bond activation, preferentially of the primary carbon of the alkyl chain. Subsequent β-H transfer forms alkenes that are metathesized by the alkylidene hydride. The favored head-to-tail metathesis generates the $n+1$ and $n-1$ homologues and accounts for the observed product selectivity. Note that this mechanism contrasts with the tandem processes discussed earlier, which require two independent catalysts and rely on the metathesis of all possible olefins.

Alkane homologation is also catalyzed by supported group IV metal hydride complexes, but this yields very different products because it takes place via β-alkyl transfer and chain transfer steps in addition to C-H activation, as shown in **Scheme 3**. With supported Zr-hydrides, propane is converted into mostly 2-methylpropane and ethane, although higher homologues (for example, 2-methylbutane) are also observed.

All these supported metal hydrides also catalyze the hydrogenolysis (cleavages in lower homologues with the addition of hydrogen) of alkanes. The group 5 and 6 metal hydrides catalyze the cross-metathesis of methane and higher alkanes ($CH_4 + C_3H_8 \rightarrow 2C_2H_6$), the cross-metathesis of alkanes and aromatics, and the nonoxidative coupling of methane into ethane and H_2 at high reaction temperature because of the highly endothermic nature of this reaction.

For background information *see* ALKANE; ALKENE; CATALYSIS; C-H ACTIVATION; DEHYDROGENATION; FREE ENERGY; HETEROGENEOUS CATALYSIS; HOMOGENEOUS CATALYSIS; HYDROGENATION; METAL HYDRIDES; TRANSITION ELEMENTS in the McGraw-Hill Encyclopedia of Science & Technology.

Matthew P. Conley; Christophe Copéret

Bibliography. J.-M. Basset et al., Metathesis of alkanes and related reactions, *Accounts Chem. Res.*, 43(2):323–334, 2010, DOI:10.1021/ar900203a; M. C. Haibach et al., Alkane metathesis by tandem alkane-dehydrogenation-olefin-metathesis catalysis and related chemistry, *Accounts Chem. Res.*, 45(6):947–958, 2012, DOI:10.1021/ar3000713; C. Copéret, C-H bond activation and organometallic intermediates on isolated metal centers on oxide surfaces, *Chem. Rev.*, 110(2):656–680, 2010, DOI:10.1021/cr900122p; F. Rascón and C. Copéret, Alkylidene and alkylidyne surface complexes: Precursors and intermediates in alkane conversion processes on supported single-site catalysts, *J. Organomet. Chem.*, 696(25):4121–5131, 2011, DOI:10.1016/j.jorganchem.2011.07.015.

Amborella genomics

Few fossils of early flowering plants (angiosperms; Magnoliophyta) have been found. The group appears suddenly in the fossil record and quickly rose to ecological dominance. As a result, Charles Darwin famously referred to the origin of angiosperms as an "abominable mystery." The oldest known angiosperm fossils date from 130–136 million years ago (MYA), but the age of the flowering plant group (clade) is thought to be much older, at least 160 MYA. The origin of the flowering plants prompted one of the Earth's greatest terrestrial radiations because angiosperms have since diversified to more than 350,000 species. Following their origin, the angiosperms became the ecologically dominant group before the end of the Cretaceous Period, and flowering plants now occupy nearly all terrestrial and many aquatic environments. Angiosperms are enormously important for economic reasons, providing the vast majority of human food and shelter. They are also a major contributor to global photosynthesis and carbon sequestration. Because of these factors, as well as issues ranging from crop improvement to the clarification of the processes underlying the assembly of entire ecosystems, a better understanding of angiosperm evolution and diversification is crucial. This background is essential for understanding the rationale for sequencing the entire nuclear genome of the flowering plant *Amborella*.

Plant genome sequencing. Dating to the *Arabidopsis* and rice genome sequencing projects (Arabidopsis Genome Initiative, 2000; International Rice Genome Sequencing Project, 2005), plant biologists have recognized the importance of a broad, comparative sequencing of plant genomes. Subsequently published genome sequences, including those of poplar, grape, papaya, maize, *Sorghum*, *Brachypodium*, soybean, *Medicago*, *Lotus*, *Mimulus*, and tomato, have provided an excellent sampling for comparative genomics within the angiosperms. However, these species all reside on just two branches, that is, the monocots and eudicots, within the angiosperm tree of life (**Fig. 1**). Hence, these sequenced genomes do not help to provide an understanding of the characteristics of the earliest angiosperms, or what researchers refer to as the most recent common ancestor of all angiosperms. Many key innovations of flowering plants [such as the origins of the flower and fruit, pollination systems, double fertilization, special large water-conducting cells (vessel elements), and diverse biochemical pathways], as well as many of the specific genes that regulate key developmental processes, first appeared among the basal angiosperm lineages. A better understanding of the evolutionary processes that shaped the genes and genomes in flowering plants requires a perspective that can be obtained only by the study of those lineages that branch from the base of the angiosperm tree of life (Fig. 1).

***Amborella* as an evolutionary reference.** In this regard, a pivotal flowering plant for comparative genome sequencing is *Amborella* (Fig. 1). *Amborella trichopoda*, the sole member of the Amborellaceae family, is a shrub known only from the island of New Caledonia. Phylogenetic analyses strongly support the placement of *Amborella* as sister to all other extant angiosperms. That is, *A. trichopoda* is the one surviving species of the earliest diverging lineages of the living flowering plants. To put this in perspective, this is similar to the position that the duck-billed platypus occupies in the mammalian tree of life, and it is the reason why the nuclear genome of the duck-billed platypus was sequenced—namely to provide a reference genome for mammals. Likewise, *Amborella* is a unique evolutionary reference for inferring the features of the

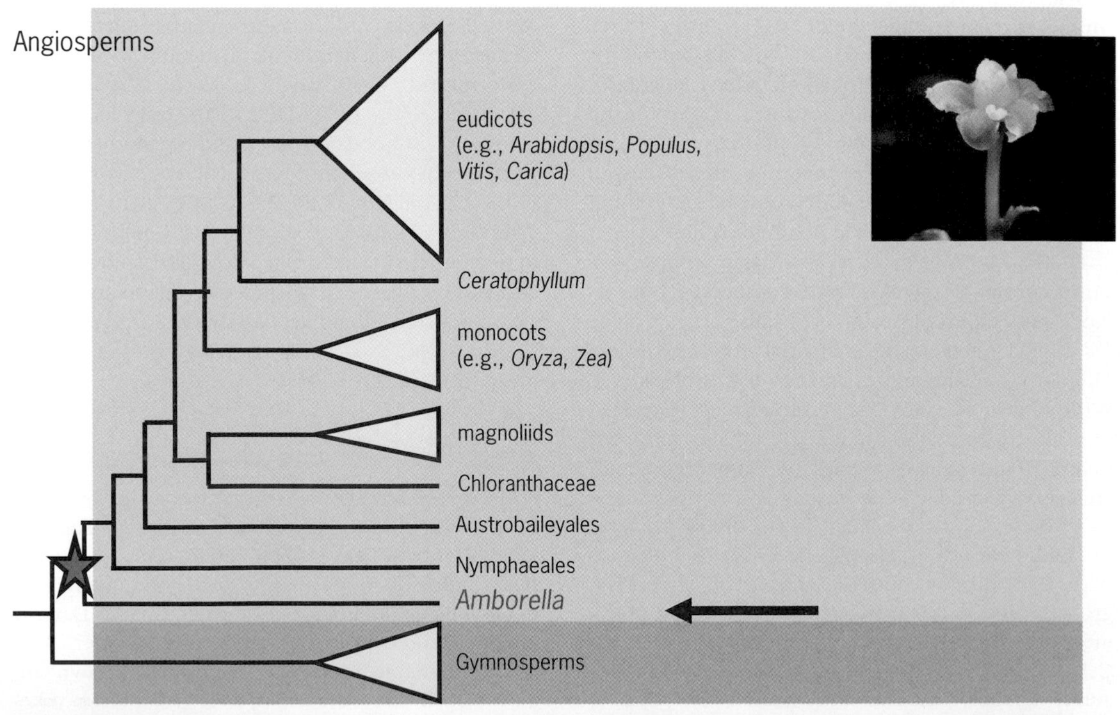

Fig. 1. Summary phylogenetic tree for angiosperms. *Amborella trichopoda* holds a pivotal position in the angiosperm tree of life (designated by the star and arrow). An image of an *Amborella* flower is also shown in the inset. (*Adapted from D. E. Soltis et al., The Amborella genome: An evolutionary reference for plant biology, Genome Biol., 9:402, 2008, DOI:10.1186/gb-2008-9-3-402*)

ancestral angiosperm. The phylogenetically pivotal position that *Amborella* occupies in the angiosperm tree of life facilitates an estimation of the ancestral characteristics of all angiosperm features from gene families to genome structure, and from physiology to morphology (for example, characteristics of the earliest flowers). Because of this, the genome of *Amborella* was sequenced to provide an evolutionary reference to which all other angiosperm genomes could be compared.

As hoped, the sequencing and assembly of the *Amborella* genome provided many new insights into the flowering plants. For example, the complete genome sequence of *Amborella* reveals that an ancient whole genome duplication (polyploidy) event likely took place in the common ancestor of all angiosperms. Polyploidy is known to be of great importance in the evolution of eukaryotes, particularly plants. The *Amborella* genome suggests that whole genome duplication or polyploidy may have been pivotal to the success of the angiosperms as a whole. This ancient polyploidy event created many new genes that are important in flowering and other angiosperm processes. Importantly, however, the *Amborella* genome (unlike all other sequenced angiosperm genomes) shows no evidence of additional more recent, lineage-specific genome duplications. This makes *Amborella* uniquely well suited to help interpret the genomic changes in all other angiosperms. As a result, the *Amborella* genome is an excellent reference genome for comparison with all other flowering plants.

Genomic analyses. As one example of the utility of the *Amborella* genome, a comparison of this genome to the sequenced genome of eudicots (this clade comprises approximately 75% of all flowering plants) confirms that another whole genome duplication event or events occurred in the common ancestor of most eudicot species (Fig. 1). This event near the origin of eudicots appears to be a paleohexaploid event—that is, a genome triplication event or two genome duplications close in evolutionary time (**Fig. 2**). In addition, genome comparisons also reveal the remarkable conservation of gene order (synteny) among the genomes of *Amborella* and other angiosperms (Fig. 2). That is, gene order has been remarkably conserved over the millions of years of angiosperm evolution. This conservation in gene order has enabled in turn the reconstruction of the ancestral gene arrangement in eudicots (a clade that contains many important crops).

The *Amborella* genome has also allowed the reconstruction of the ancestral angiosperm genome. An ancestral angiosperm gene set has been identified and shown to contain at least 14,000 protein-coding genes. Subsequent changes in gene content and genome structure across diverse flowering plant lineages thereafter affected the evolution of important crops and model species. *Amborella* genome analyses indicate that more than 1000 gene lineages were gained in the ancestral angiosperm relative to other seed plants. These gains include genes important in flowering, wood formation, and environmental responses. Taken together, the analysis

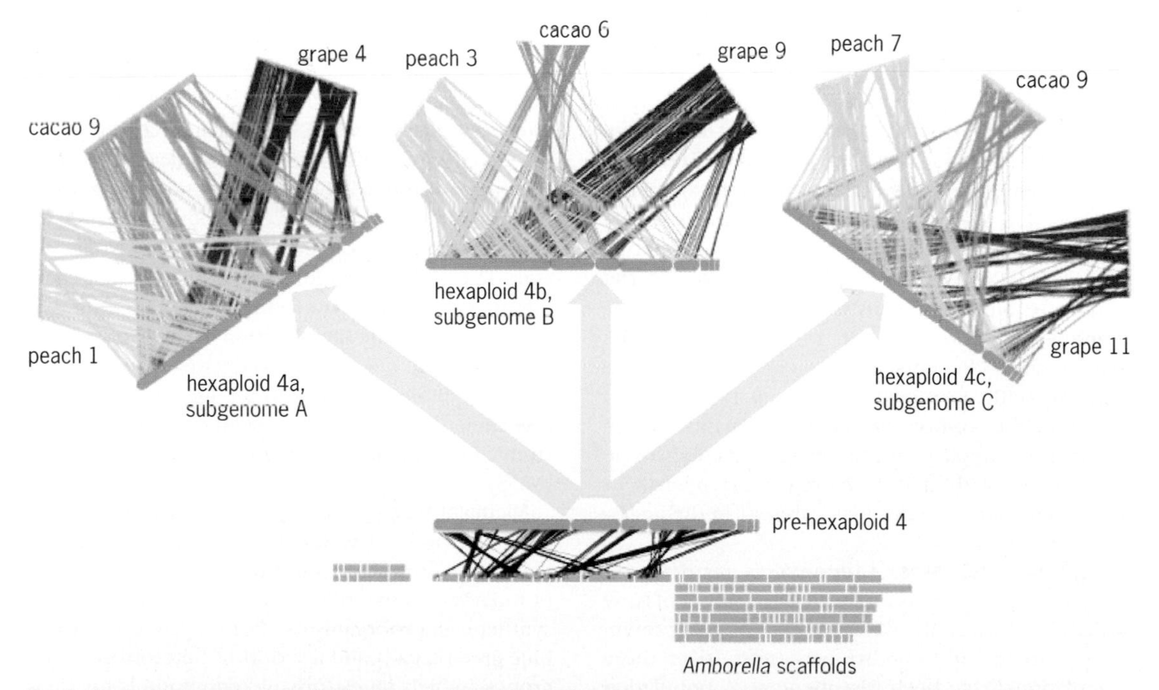

Fig. 2. Ancient hexaploidy in eudicots. Comparison of the *Amborella* genome to the sequenced genome of eudicots indicates an inferred triplication event in eudicots and the conservation of gene order (synteny) in eudicots. Shown are gene order alignments between one of seven ancestral core eudicot chromosomes and a subset of the *Amborella* scaffolds, as well as between the three post-hexaploidization copies of this chromosome, and the peach, cacao, and grape chromosomes descending from it. (*Adapted from Amborella Genome Project, The Amborella genome and the evolution of flowering plants, Science, 342(6165):1241089, 2013, DOI:10.1126/science.1241089*)

of the *Amborella* genome suggests that numerous gene families, gene family expansions, and certain floral gene and protein interactions first appeared in the ancestral angiosperm.

Whereas the origin of the flower may be partly explained by novel genes that first appeared with the origin of angiosperms, other floral genes (such as petal- and stamen-specific genes) seem to have appeared before the origin of angiosperms. More than 70% of the gene lineages with known roles in flowering (including genes involved in floral timing and initiation, meristem identity, and floral structure) were actually present in the most recent common ancestor of all extant seed plants. Other major gene components of the floral regulatory pathway were even older, with core components of the pathway present in the ancestral vascular plant. Together, these observations suggest that orthologs of most floral genes existed long before their specific roles were established in flowering, and these genes were later coopted to serve floral functions. In addition, after the origin of angiosperms, new genes originated or were recruited to refine or more narrowly distribute functions associated with flower development. This genetic pattern is consistent with the observation that the floral organ transcriptional program is highly canalized (constrained) in the large eudicot clade relative to the less organ-constrained (more flexible) transcriptomes of the earlier-diverging, basal angiosperm lineages.

Importantly, many novel genes that first arose in the angiosperms play no actual role in reproduction. For example, genes with specific functions in the formation of water-conducting vessels also first appeared in the early angiosperms. Gene families that expanded in the angiosperms based on the *Amborella* genome include those relevant to the importance of plant–herbivore coevolution in the diversification of angiosperms and insects, including those with expression elicited by herbivory (the consumption of living plant tissue by animals).

There are also other noteworthy aspects of the *Amborella* genome. Transposable elements (mobile pieces of DNA) have played a major role in the genome evolution of many flowering plants (for example, corn). However, in *Amborella*, all of the transposable elements are ancient and highly divergent, with no recent transposon radiations. In addition, the *Amborella* genome contains a large number of atypical lineage-specific 24-nucleotide microRNAs, with at least 26 regulatory microRNA families inferred to have been present in the ancestral angiosperm.

Population genomics. *Amborella* is restricted to the wet tropical forests on the isolated slopes of New Caledonia. Fewer than 15 populations are known. The genomes of 12 individuals from 10 of these populations have been resequenced. A population genomic analysis of these individuals from across the small native range of *Amborella* has revealed a geographic structure with important conservation implications. Specifically, the genetic variation among the *Amborella* populations indicates that there are four geographic clusters of populations located across the island of New Caledonia. This population genomic study also provided evidence for a genetic bottleneck event that occurred 300,000 to 400,000 years ago, followed by some recovery of genetic diversity thereafter.

For background information *see* BOTANY; GENE; GENETIC MAPPING; GENETICS; GENOMICS; HERBIVORY; MAGNOLIOPHYTA; PLANT EVOLUTION; PLANT KINGDOM; PLANT PHYLOGENY; POLYPLOIDY in the McGraw-Hill Encyclopedia of Science & Technology.

Douglas E. Soltis

Bibliography. Amborella Genome Project, The *Amborella* genome and the evolution of flowering plants, *Science*, 342(6165):1241089, 2013, DOI:10.1126/science.1241089; S. Chamala et al., Assembly and validation of the genome of the nonmodel basal angiosperm *Amborella*, *Science*, 342(6165):1516–1517, 2013, DOI:10.1126/science.1241130; D. E. Soltis et al., *Phylogeny and Evolution of the Angiosperms*, Sinauer, Sunderland, MA, 2005; D. E. Soltis et al., The *Amborella* genome: An evolutionary reference for plant biology, *Genome Biol.*, 9:402, 2008, DOI:10.1186/gb-2008-9-3-402.

Ancient microbial ecosystem

At first, it may appear paradoxical that we must study life in the context of modern Earth in order to understand the earliest life in the fossil record. The oldest part of Earth's history is preserved in rocks of the Archean time period, from about 4 to 2.5 billion (giga) years ago (GYA). Detailed research on the ancient marine sedimentary rocks exposed in the 3.48 billion-year-old Dresser Formation in the Pilbara area of Western Australia has revealed microfossils of individual cells of bacteria beautifully preserved in glasslike cherts (hard, dense sedimentary rock composed of fine-grained silica, most commonly quartz) and stromatolites (boulder-shaped architectural buildups created by microbiota from long ago). The Archean world of life, however, was vastly different in comparison to the world of life known today. Still, it is the investigation of microorganisms in comparable coastal environments of the present time that allows insight into the life and ecology of this ancient scenario. One avenue to follow is the study of microbially induced sedimentary structures (MISS).

Microbial mats. In the photic zone (the uppermost layer of a body of water that receives enough sunlight to permit the occurrence of photosynthesis) of today's oceans, trillions of cyanobacteria (photosynthetic microorganisms, formerly referred to as blue-green algae) and a world of heterotrophic microbes (which ingest organic compounds for their nutrition) form dense organic layers on top of the seafloor. The very coherent layers may have thicknesses of less than a millimeter, but they also can be meter-thick accumulations of biomass. Such

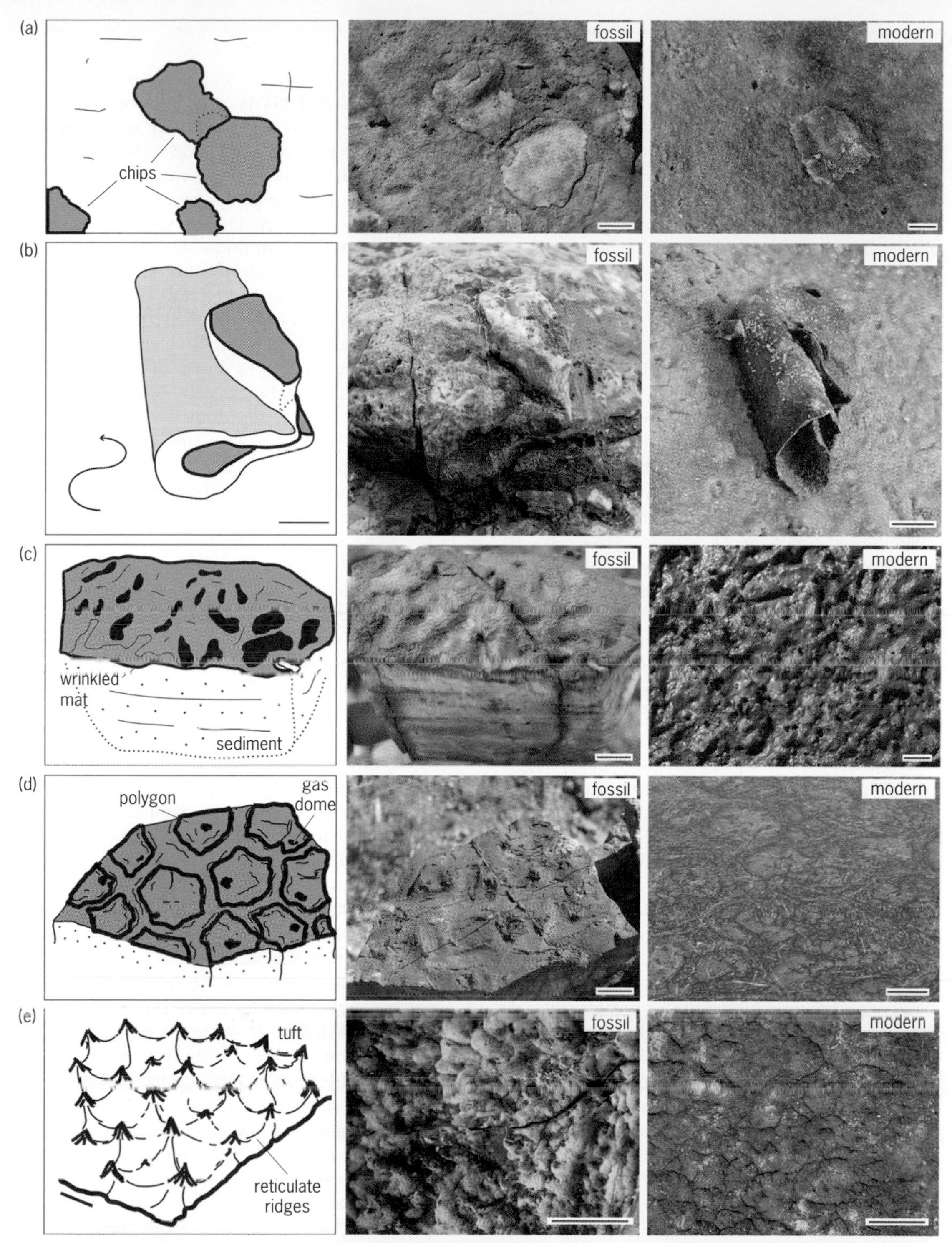

Microbially induced sedimentary structures (MISS) arise from the interaction of microbial mats with ocean floor sediments in coastal environments. Such MISS occur in the present time, but surprisingly also in the 3.48 billion-year-old sedimentary rocks from the Dresser Formation, Pilbara, Western Australia. The photos show examples for MISS from the fossil rocks of the Dresser Formation (*center*). For better visualization, the MISS are outlined in the sketches on the *left*. The photos on the *right* show some modern counterparts of the MISS. (*a*) Fragments of a microbial mat on an ancient seafloor recorded in the Dresser Formation rocks. In equivalent modern marine settings, such "chips" are common. Here, an example is shown from Portsmouth Island, North Carolina, United States. Scale bar: about 1 cm (0.4 in.). (*b*) Rolled up piece of microbial mat in the Dresser Formation rocks. In modern coastal environments, microbial mat fragments are rolled up in this fashion by water currents. The modern example is from Portsmouth Island, North Carolina, United States. Scale bar: about 1 cm (0.4 in.). (c) The wrinkled upper surface of this rock bed records an ancient microbial mat. In modern environments, such wrinkle structures are typical for microbial mats rich in extracellular polymeric substances. The modern example is from Mellum Island, Germany. Scale bar: about 1 cm (0.4 in.). (*d*) The surface of this sedimentary rock is arranged into polygons of several centimeters in diameter. Many polygons have a hole in their center. In modern coastal environments, such polygons form within microbial mats exposed to a seasonal climate. The climate is characterized by dry and hot summers, during which the microbial mat dries out and cracks appear, and wet winters, during which the cracks "heal." The structures are called polygonal oscillation cracks. Each individual polygon is separated from its neighbors by a 3- to 10-cm-wide (1.2- to 4-in.-wide) desiccation crack. The holes in each of the polygons are collapsed gas domes. The microbial mat rises during the summer because gases are produced and assemble beneath the microbial mat. The mat finally ruptures locally and the gases escape. The modern example for polygonal oscillation cracks is from the El Bibane sabkha (salt flat) in Tunisia. Scale bar: about 10 cm (4 in.). (e) The honeycomb pattern ridges are here exposed on a surface of a sedimentary rock in the Dresser Formation. In modern environments, such ridges arranged into a honeycomb pattern are typical for microbial mats developing in very shallow ponds close to the beach. The modern example for such ridges is from Portsmouth Island, North Carolina, United States. Scale bar: about 5 cm (2 in.).

layers resemble greenish colored carpets; hence, the term "microbial mats" was applied to them. In the present time, microbial mats occur along the coasts of all oceans, indeed forming one of the largest ecosystems on Earth. The coherent microbial mats withstand the erosive forces of waves and currents and thus protect their substrates from being washed away by the water's motion. In close-up view, microbial mats appear to be simply a collection of trillions of microbial organisms embedded in their slimy and adhesive secretions—the "extracellular polymeric substances." However, recent studies have revealed that, within a microbial mat, the microorganisms form a highly developed collective, within which the individual cells communicate with each other. In fact, microbe "decisions" are made by quorum sensing (a process whereby microbes sense their own population by measuring the levels of autoinducer signal molecules), and necessary tasks are executed to allow the optimal use of nutrients and other functions to promote the survival of the entire community. The microbes also control carbonate mineral precipitation, which leads to the preservation of many of the microbial mats in rocks. However, minerals are not precipitated by all microbial mats. The interaction of the microbial mat with its substrate, the seafloor, often causes simple traces in the sediment. Such traces are the microbially induced sedimentary structures (MISS). The MISS can become fossilized and can be detected in sedimentary rocks dating to all the ages of the Earth, from the Archean time period to the present. In contrast to stromatolites, the MISS are not buildups, and they result from the lack of carbonate minerals precipitated within the microbial mats.

Microbially induced sedimentary structures. So far, 17 main types of MISS have been described, all being biogenic structures of clearly defined morphology and dimension. Transitions between the individual structure species do not exist. Examples of MISS include microbial mat chips, roll-ups, wrinkle structures, and polygonal oscillation cracks (see **illustration**). These MISS are not distributed at random in a coastal area, but each species relates to a very specific zone. For example, thick, epibenthic microbial mats in the lower supratidal zone form gas domes or honeycomb-like patterns of ridges, whereas endobenthic microbial mats in the intertidal zone produce multidirected ripple marks.

The investigations on modern microbial mats and the MISS that they form have allowed the detection of similar MISS in the sedimentary rock records of equivalent coastal environments in Earth's history. For example, the Pleistocene rocks forming a fringe around the modern Mediterranean Sea have revealed a suite of MISS indicating a similar ecosystem of microbial mats as found today. More surprisingly, systematic follow-up studies have detected the same type of ecosystems in coastal settings preserved in rocks dating back to the Ordovician [480–440 million years ago (MYA)], Neoproterozoic (1 GYA–540 MYA), Meso-Archean (2.9 GYA), and Paleoarchean (3.2 GYA). Could it be that even Earth's oldest coastal sedimentary rocks in the Dresser Formation would include such MISS?

MISS and the Dresser Formation. MISS have been found to occur in the Pilbara area of Western Australia. The morphology of the individual MISS and their patterns of distribution show stunning similarities to those of modern MISS and may lead to the amazing conclusion that the microbial mat–forming microbiota have not changed over the past 3.5 billion years. The illustration documents the similarity in appearance of MISS in both modern coastal settings and the 3.48 billion-year-old Dresser Formation rocks. However, which microbes were around at that time? Were they the same as those present today? This question is difficult to answer. Indeed, the Dresser Formation MISS point toward the existence of "cyanobacteria" already at this early time in the Earth's history. The fossil MISS indicate an immense sediment stabilization by the coherent microbial mats during their lifetime. In modern environments, such a protection of the sediment against erosion by waves and currents is intermediated by enormous amounts of extracellular polymeric substances. Such a high amount of these mucilages can be produced only by the effective photoautotroph microbes, that is, the cyanobacteria. However, any conclusion that the ancient microbial mat–forming "cyanobacteria" were indeed the cyanobacteria in the modern sense must be cautioned. The genetic information of a group of microorganisms is not stable enough to survive billions of years without change. With confidence, however, we can say that the ancient microbial mats were complex microbial communities that resemble strongly those dominated by the cyanobacteria found in present-day coasts. With respect to the question of the origin of life, however, we know that life must have been present on Earth well before the rock record. Life was not only present in the early Archean time, but it had evolved already to a highly diverse and complex community—a process that would have taken a long time, in the range of hundreds of millions of years.

For background information *see* ARCHEAN; CYANOBACTERIA; ECOLOGY; ECOSYSTEM; FOSSIL; GEOLOGY; MARINE MICROBIOLOGY; MARINE SEDIMENTS; MICROBIAL ECOLOGY; PALEOECOLOGY; SEDIMENTOLOGY; STROMATOLITE in the McGraw-Hill Encyclopedia of Science & Technology. Nora Noffke

Bibliography. R. Buick and J. S. R. Dunlop, Evaporitic sediments of early Archaean age from the Warrawoona Group, North Pole, Western Australia, *Sedimentology*, 37:247–277, 1990, DOI:10.1111/j.1365-3091.1990.tb00958.x; A. W. Decho, R. L. Frey, and J. L. Ferry, Chemical challenges to bacterial AHL signaling in the environment, *Chem. Rev.*, 111:86–99, 2011, DOI:10.1021/cr100311q; C. Dupraz et al., Processes of carbonate precipitation in modern microbial mats, *Earth Sci. Rev.*, 96:141–162, 2009, DOI:10.1016/j.earscirev.2008.10.005; N. Noffke, *Geobiology: Microbial Mats in Sandy Deposits from the Archean Era to Today*,

Springer, New York, 2010; N. Noffke and S. Awramik, Stromatolites and MISS: Differences between relatives, *GSA Today*, 23:5–9, 2013, DOI:10.1130/GSATG187A.1; N. Noffke et al., Microbially induced sedimentary structures recording an ancient ecosystem in the ca. 3.48 billion-year-old Dresser Formation, Pilbara, Western Australia, *Astrobiology*, 13(12):1103–1124, 2013, DOI:10.1089/ast.2013.1030; P. Stoodley et al., Biofilms as complex differentiated communities, *Annu. Rev. Microbiol.*, 56:187–209, 2002, DOI:10.1146/annurev.micro.56.012302.160705; M. J. Van Kranendonk et al., Geological setting of Earth's oldest fossils in the ca. 3.5 Ga Dresser Formation, Pilbara Craton, Western Australia, *Precambrian Res.*, 67:93–124, 2008, DOI:10.1016/j.precamres.2008.07.003.

Aneuploidy in neurons

A human being develops from one cell at conception. This single cell will grow, duplicate its genome, and then divide into two cells. Each daughter cell will then repeat this process, and some 37 trillion cell divisions will occur before the single cell becomes an individual. An essential requirement of cell division is that the duplicated genome is evenly partitioned between each single cell's two progeny. The genome is organized into chromosomes, and humans have 23 pairs of chromosomes. When human cells divvy up, or segregate, their chromosomes evenly, each daughter cell will have 46 total chromosomes. Sometimes, cells missegregate chromosomes, and daughter cells have more or fewer than 46 chromosomes. Cells with a normal chromosome number are called euploid. In contrast, cells with an abnormal chromosome number are not euploid, so these are called aneuploid.

Mitosis. Mitosis is the term used to describe the cellular process of chromosome segregation (**Fig. 1**). Before a cell divides, it enters mitosis. Mitosis proceeds through four distinct phases. During the first phase, called prophase, the chromosomes condense into discrete units. During the final phase, called telophase, the cell divides into two cells. During the middle two phases, that is, metaphase followed by anaphase, the chromosomes are segregated; that is, the replicated chromosomes are divided and the two parts are separately distributed to the daughter cells. In metaphase, the chromosomes are aligned in the middle of the cell; then, in anaphase, the duplicate copies of each chromosome are pulled apart. Chromosome missegregation leading to aneuploidy happens when errors occur during mitosis (**Fig. 2**).

Aneuploid neurons. In 2001, researchers at the University of California, San Diego, discovered aneuploid stem cells in the developing mouse brain. They determined that chromosome missegregation occurred frequently in these stem cells, known as neural progenitor cells (NPCs). The aneuploid NPCs gave rise, in turn, to aneuploid neurons. Additional studies showed that aneuploid neurons formed appropriate connections with other neurons and participated in neural circuits. Although the initial studies focused on the cerebral cortex, follow-up studies also found aneuploid neurons in the human cortex and cerebellum, as well as in the zebrafish brain.

The loss or gain of an entire chromosome directly affects cellular function. An average chromosome contains approximately 1000 genes, and aneuploidy changes the copy number of all of these genes. This alteration, in turn, will change the protein levels in that cell. Moreover, the loss or gain of any chromosome can activate a cellular stress response that leads to additional differences in aneuploid cells relative to euploid cells. When aneuploidy occurs in one cell, that cell will be different from euploid cells in the population. Many aneuploid NPCs and neurons may die during brain development. However, some aneuploid neurons clearly survive and integrate into mature neural circuits.

Aneuploidy among neurons is significant because it shows that not every cell in an individual's brain has the same genome. The discovery of aneuploid neurons was the first direct evidence of genetic differences between neurons in the same brain. In the late 1990s, there was renewed interest in the possibility that somatic gene rearrangement, which was then thought to occur only in the immune system, might also occur in the brain. Aneuploid NPCs were discovered serendipitously while researchers were developing methods to look for somatic gene rearrangement in neurons. Subsequent studies found additional types of genome changes in single neurons; for example, mobile genetic elements "jump" in some neurons, but not others. Most recently, large subchromosomal changes [>1 megabase (Mb)] in DNA copy number have been reported in as many as 41% of human cortical neurons. In addition, other smaller copy number variations (CNVs) have been found to exist in as many as 25% of cells in one region of an individual's brain, but not in other regions of that brain. When different cells in an individual have different genomes, this is called somatic mosaicism. Collectively, these discoveries suggest that somatic mosaicism is a normal state for vertebrate brains.

Relation to disease and development. Somatic mosaicism has also been linked to neurological diseases. A neurodevelopmental disorder, hemimegencephaly, can occur when many brain cells acquire a mutant gene that leads to local cellular overgrowth. Other studies have linked increased levels of aneuploidy and chromosomal translocations to the neurodevelopmental disorder ataxia-telangiectasia, and increased levels of chromosome 1 aneuploidy have been associated with schizophrenia. Similarly, increased mobile element activity has been reported in the brains of patients with ataxia-telangiectasia, schizophrenia, and Rett syndrome (an autism spectrum disorder).

Most aneuploidies are incompatible with human development. However, trisomy (the presence in triplicate of a particular chromosome) for chromosome 21 is not lethal, although it leads to Down syndrome. Because the Alzheimer's disease risk gene, β-amyloid, is on chromosome 21, an interesting hypothesis has been developed regarding the different levels of chromosome 21 mosaicism in

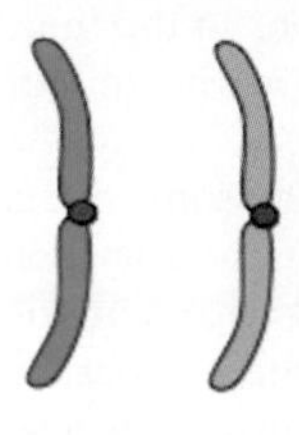

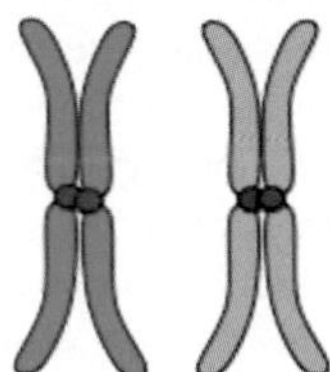

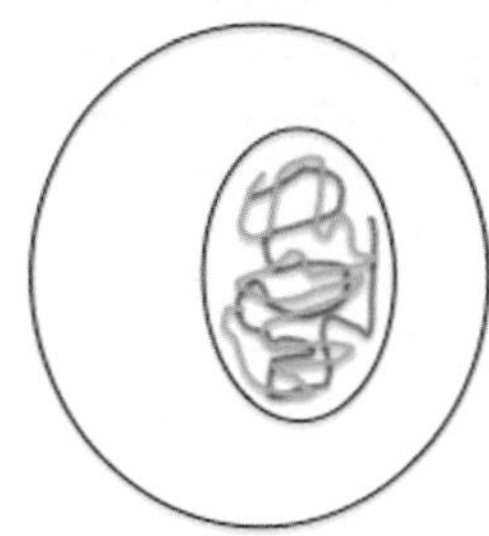

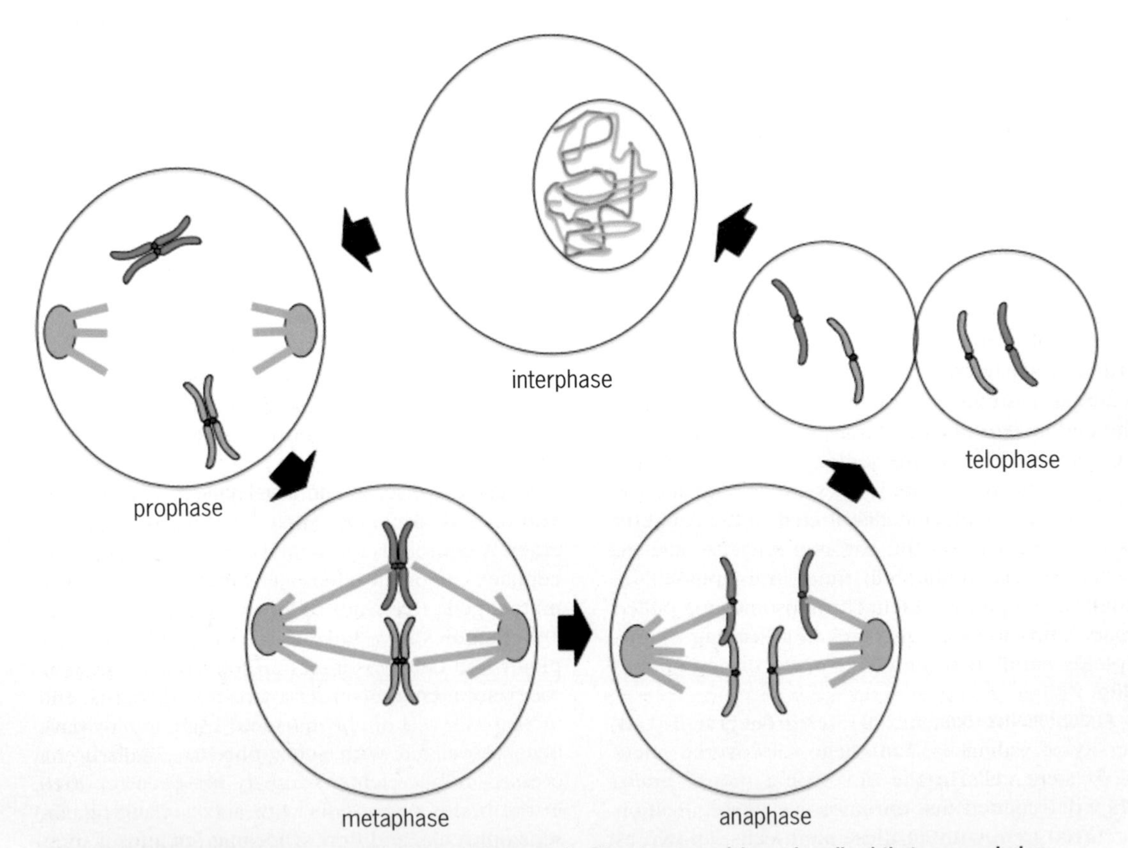

Fig. 1. Illustration of chromosome segregation during mitosis. The structures (shown in yellow) that appear during prophase are mitotic spindles. These are composed of centriole proteins at the poles, and microtubules that extend from the pole, capture the chromosomes, and then pull the chromosomes to each pole.

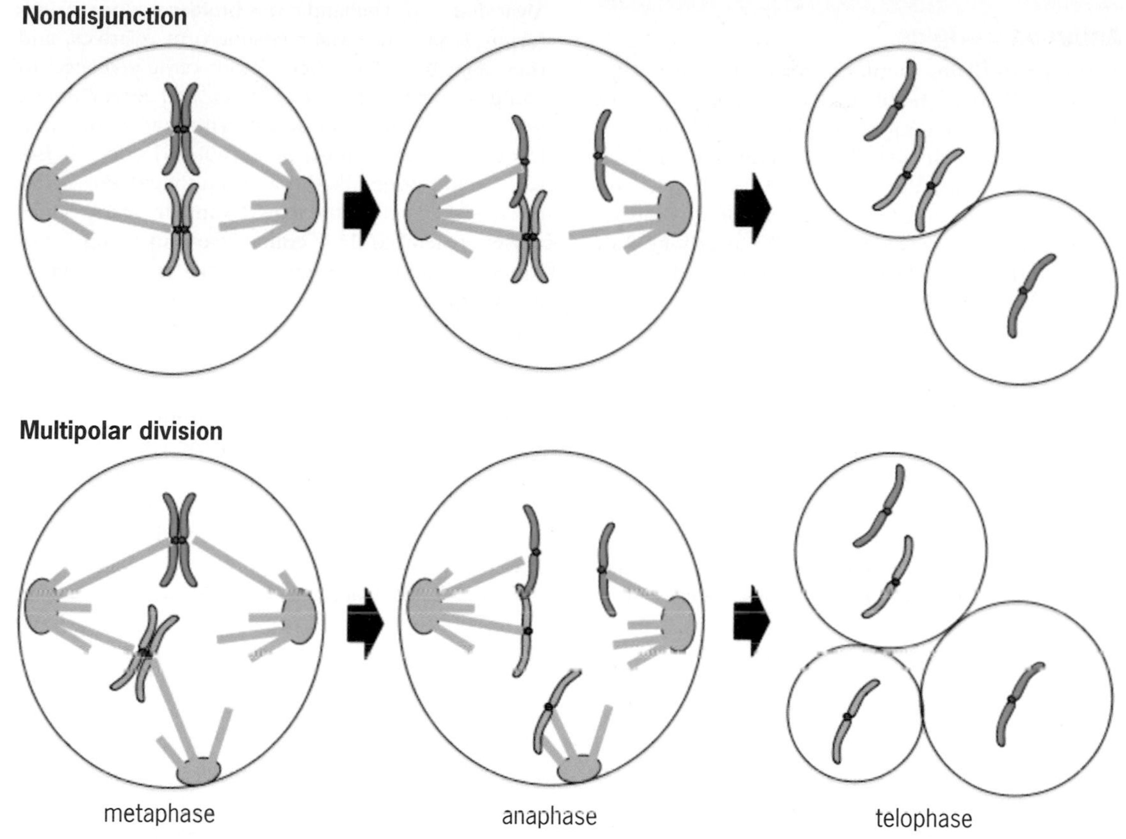

Fig. 2. Two examples of chromosome missegregation leading to aneuploidy.

different individuals. Aged individuals with Down syndrome invariably present with Alzheimer's disease. Because chromosome missegregation can often lead to euploid cells in trisomic individuals, it is possible that the severity of the Down syndrome phenotype is related to the number of euploid cells acquired during development. Moreover, it is possible that Alzheimer's disease in the normal population is related to the percentage of trisomy 21 neurons in an individual's brain. These associations are not yet certain, but they represent provocative hypotheses that should promote future studies.

Neuronal diversity. Diversity is a hallmark of the nervous system, and neurons come in innumerable types and shapes. This diversity is generally thought to be the product of genetic programs that dictate the orderly, combinatorial use (that is, expression) of specific genes. Gene expression signatures are one element of a neuron's identity, and somatic mosaicism adds an additional level of complexity to neuronal identity. Neuron-to-neuron somatic mutations can be small (for example, approximately 7-kilobase mobile element insertions) and affect single genes; they can be large (for example, >1 Mb CNVs) and affect tens of genes; or they can be chromosomal (for example, aneuploid) and affect thousands of genes. The ongoing development of novel approaches for single cell analysis is expected to provide insight into how somatic mosaicism contributes to phenotypic diversity among neurons in the same brain. Moreover, advances in the scientific understanding of psychiatric genetics will be promoted by knowing how somatic mosaicism affects the genetic risk for neuropsychiatric diseases.

For background information *see* ALZHEIMER'S DISEASE; BRAIN; CENTRAL NERVOUS SYSTEM; CHROMOSOME; CHROMOSOME ABERRATION; DEVELOPMENTAL GENETICS; DOWN SYNDROME; HUMAN GENETICS; MITOSIS; MOSAICISM; NERVOUS SYSTEM (VERTEBRATE); NEUROBIOLOGY; NEURON; POLYPLOIDY; STEM CELLS in the McGraw-Hill Encyclopedia of Science & Technology.

Michael J. McConnell; Fred H. Gage

Bibliography. D. M. Bushman and J. Chun, The genomically mosaic brain: Aneuploidy and more in neural diversity and disease, *Semin. Cell Dev. Biol.*, 24:357–369, 2013, DOI:10.1016/j.semcdb.2013.02.003; J. Chun, Choices, choices, choices, *Nat. Neurosci.*, 7:323–325, 2004, DOI:10.1038/nn0404-323; A. Poduri et al., Somatic mutation, genomic variation, and neurological disease, *Science*, 341:1237758, 2013, DOI:10.1126/science.1237758; H. Potter and L. N. Geller, Alzheimer's disease, Down's syndrome, and chromosome segregation, *Lancet*, 348:66, 1996, DOI:10.1016/S0140-6736(05)64399-1; T. Singer et al., LINE-1 retrotransposons: Mediators of somatic variation in neuronal genomes?, *Trends Neurosci.*, 33:345–354, 2010, DOI:10.1016/j.tins.2010.04.001; Y. C. Tang and A. Amon, Gene copy number alterations: A cost-benefit analysis, *Cell*, 152:394–405, 2013, DOI:10.1016/j.cell.2012.11.043.

Anthropoid origins

Anthropoids (living monkeys, apes, and humans, together with their fossil relatives) share a host of anatomical, behavioral, and genetic traits that are absent in other primates. These features include relatively large brains, complete bony eye sockets, and multiple changes to the genome known as short interspersed repetitive element (SINE) insertions. As a result, anthropoids have always been recognized as biologically distinctive. Anthropoids are sometimes called "higher primates" because of the wide gap separating them from other primates and their evolutionary proximity to humans. The study of anthropoid origins is compelling because it illuminates a distant phase of our evolutionary history that has previously been elusive.

Evolutionary context. Aside from anthropoids, living primates can be segregated into two major lineages or clades. The first of these includes lemurs, lorises, and bushbabies, known collectively as strepsirhines. Tarsiers, which are small, nocturnal primates inhabiting certain islands in Southeast Asia, are the only extant representatives of the other major group of primates. An impressive set of anatomical, physiological, and genetic features indicates that tarsiers are the closest living relatives of anthropoids. The recent discovery of a tiny and very primitive primate called *Archicebus achilles* in central China, represented by a nearly complete skeleton, has revealed that early relatives of tarsiers and anthropoids possessed a unique combination of features, with some resembling conditions seen in early anthropoids.

Early Asian anthropoids. Eocene fossil primates from Asia have enjoyed a long-standing role in efforts to comprehend anthropoid origins. Recent fossil discoveries have vastly improved our knowledge of these early Asian anthropoids, which can be sorted into at least two main groups. The first of these, known as eosimiiforms, includes the oldest and most primitive anthropoids discovered to date. Eosimiiforms were uniformly small and had yet to evolve many of the features that characterize living anthropoids. For example, *Eosimias* from the middle Eocene of China retains an extra cusp, known as the paraconid, on its lower molars, which was suppressed among later anthropoids. At the same time, *Eosimias* possesses an advanced, anthropoid-like chin region, in which its vertically oriented front teeth, or incisors, are implanted. Ankle bones of *Eosimias* show features that are highly characteristic of anthropoids, although in some ways these bones are transitional between those of early fossil relatives of tarsiers (known as omomyids) and later anthropoids. The recent discovery of *Afrasia*, an eosimiiform from the Late Middle Eocene of Myanmar (Burma) that closely resembles the North African eosimiiform *Afrotarsius*, signals an early biogeographic connection between the anthropoids of Asia and Africa.

A second group of early Asian anthropoids, called amphipithecids, is known from later Eocene sites in Myanmar and Thailand. Amphipithecids were uniformly larger than their eosimiiform relatives, and they appear to have been more closely related to living anthropoids. Their upper and lower cheek teeth were highly specialized, with low crowns that frequently bear rugose or crenulated enamel. Isolated ankle bones of the amphipithecid *Pondaungia* have been reported from Myanmar. These bones closely resemble their counterparts in living South American monkeys, suggesting that amphipithecids moved quadrupedally through the forest canopy in a fashion resembling some of the less specialized living New World monkeys.

The antiquity, diversity, and basal evolutionary position of early Asian anthropoids indicate that Asia was the ancestral homeland for the anthropoid lineage. This hypothesis is also supported by the fact that living and fossil tarsiers, which are the closest relatives of anthropoids, have never been found in Africa.

Early African anthropoids. The oldest African anthropoids discovered to date hail from Late middle Eocene or Late Eocene sites in Algeria, Libya, and Egypt. Both the absolute and relative ages of these North African sites remain contested, but all are substantially younger than the oldest Asian sites yielding anthropoid fossils. Of special note is the recently discovered anthropoid assemblage from the Dur At-Talah escarpment in southern Libya. This fauna contains three surprisingly divergent taxa of early anthropoids, including the eosimiiform *Afrotarsius*, the parapithecid *Biretia*, and the oligopithecid *Talahpithecus*. Oligopithecids are often reconstructed as close relatives of catarrhine anthropoids (Old World monkeys, apes, and humans). If so, the early occurrence of *Talahpithecus* at Dur At-Talah likely marks the first appearance of modern "crown" anthropoids in the fossil record.

The diversity of the oldest known African anthropoid assemblages indicates either that the earlier evolutionary history of African anthropoids is missing from the fossil record or that multiple anthropoids colonized Africa from Asia at roughly the same time.

Paleobiogeography. Africa was an island continent located south of its current position when the earliest African anthropoids made their debut there. Reflecting its geographic isolation, Africa hosted a highly endemic mammalian fauna prior to its collision with Eurasia. Exactly how and when the earliest anthropoids colonized Africa from Asia remain unknown, but these anthropoids apparently did so by crossing a substantial marine barrier known as the Tethys Sea, probably by rafting on "floating islands" of matted vegetation. The recent discovery of the eosimiiform *Afrasia djijidae* in Myanmar, which closely resembles species of *Afrotarsius* from North Africa, suggests that the colonization of Africa by early Asian anthropoids occurred during the latter part of the Eocene. At least two species of Asian anthropoids, an eosimiiform and a second taxon

Small eosimiid anthropoid, depicted atop a human hand. This restoration is based on fossil jaws and postcranial bones from a middle Eocene site in eastern China. (*Drawing by Kim Reed-Deemer*)

more closely related to living anthropoids, seemingly colonized Africa at roughly the same time.

Mosaic evolution. Paleontologists have long assumed that the numerous anatomical features distinguishing living anthropoids from other primates must have evolved consecutively, rather than all at once. Not only is the fossil record of early anthropoids now sufficient to show that this was indeed the case, but it also suggests that the earliest anthropoids were more primitive in certain respects than anyone could have guessed a few years ago.

Although living anthropoids tend to be considerably larger than other living primates, it now appears that the earliest members of the anthropoid lineage were actually smaller than the smallest living primate, that is, the mouse lemur *Microcebus myoxinus* from Madagascar. Tiny primate ankle bones much smaller than those of *Microcebus* have been identified at one of the sites in China that has yielded fossils of *Eosimias* and related taxa. Some of these primate ankle bones are morphologically very similar to those of *Eosimias* and clearly pertain to an early anthropoid, yet they belong to animals much smaller than *Eosimias*. Using a variety of mathematical regression techniques, the body mass of these diminutive Chinese anthropoids has been estimated at approximately 15 g (0.5 oz), or roughly the size of a small mouse (see **illustration**). That early anthropoids could have been this small was entirely unexpected. Because all mammals are endothermic, or warm-blooded, such a tiny body mass poses significant physiological challenges, implying an active lifestyle with a shrewlike metabolic rate.

For background information *see* ANIMAL EVOLUTION; APES; EOCENE; FOSSIL; FOSSIL APES; FOSSIL PRIMATES; METABOLISM; MONKEY; PALEOGEOGRAPHY; PALEONTOLOGY; PRIMATES in the McGraw-Hill Encyclopedia of Science & Technology.

K. Christopher Beard

Bibliography. K. C. Beard, *The Hunt for the Dawn Monkey: Unearthing the Origins of Monkeys, Apes, and Humans*, University of California Press, Berkeley, CA, 2004; K. C. Beard and J.-W. Wang, The eosimiid primates (Anthropoidea) of the Heti Formation, Yuanqu Basin, Shanxi and Henan Provinces, People's Republic of China, *J. Hum. Evol.*, 46:401–432, 2004, DOI:10.1016/j.jhevol.2004.01.002; Y. Chaimanee et al., Late middle Eocene primate from Myanmar and the initial anthropoid colonization of Africa, *Proc. Natl. Acad. Sci. USA*, 109:10293–10297, 2012, DOI:10.1073/pnas.1200644109; J.-J. Jaeger et al., Late middle Eocene epoch of Libya yields earliest known radiation of African anthropoids, *Nature*, 467:1095–1098, 2010, DOI:10.1038/nature09425; X.-J. Ni et al., The oldest known primate skeleton and early haplorhine evolution, *Nature*, 498:60–64, 2013, DOI:10.1038/nature12200; E. R. Seiffert, Early primate evolution in Afro-Arabia, *Evol. Anthropol.*, 21:239–253, 2012, DOI:10.1002/evan.21335; B. A. Williams et al., New perspectives on anthropoid origins, *Proc. Natl. Acad. Sci. USA*, 107:4797–4804, 2010, DOI:10.1073/pnas.0908320107.

Application and research of modular buildings

Modular buildings generally consist of multiple three-dimensional units (modules) that have been constructed and preassembled in a factory, with the trim work and the electrical, mechanical, and plumbing systems having been installed at the factory as well. These units are then shipped to the construction site for installation on permanent foundations.

The modules are prefabricated off-site under controlled conditions, using the same materials and codes as would be used for conventionally built facilities. Off-site construction ensures better management of the construction quality and earlier completion, as well as considerable savings in labor costs. Modular buildings are generally stronger than conventional buildings, as the independent modules need to withstand the rigors of transportation and craning onto their foundations.

Modular buildings can be divided into four categories: container buildings (**Fig. 1**),

Fig. 1. Container buildings.

Fig. 2. Corner-supported modules. (*Courtesy Kingspan Off-Site, Modular, and Yorkon, UK*)

Fig. 3. Four-sided modules and partially open-sided modules. (*Courtesy Kingspan Off-Site and Modular, UK*)

Fig. 4. Mixed modules. (*Courtesy Feilden Clegg Bradley and Rollalong, UK*)

corner-supported modules (**Fig. 2**), four-sided modules and partially open-sided modules (**Fig. 3**), and mixed modules (**Figs. 4–6**).

In container buildings (Fig. 1), remodeled shipping containers are used as construction modules. To be used for transportation, shipping containers must have considerable strength and stiffness, which are mainly provided by corrugated steel. As a result, containers use more steel than other types of modules. Container modules are mainly used when maritime transportation is required.

Corner-supported modules (Fig. 2) are designed to provide fully open sides by transferring the loads to the corner posts. The walls in each module are usually treated as nonstructural elements, with deep edge beams being used to sustain the loads. It is easy to directly connect these modules to create a large open-plan space. The stability of corner-supported modules is generally provided by bracing systems.

Four-sided modules and partially open-sided modules (Fig. 3) transfer loads continuously through longitudinal composite walls, which are usually made of light-gage steel joists, thermal insulating material, and decorative material. Therefore, sufficient load-bearing capacity and building function are guaranteed.

There are several types of mixed modules (Figs. 4–6), including "hybrid" or mixed modular and panel systems. Mixed modules optimize the application of three- and two-dimensional (3D and 2D) components in terms of both the amount of space provided and the manufacturing costs. These modular units are used for highly serviced areas, such as bathrooms, wall panels, and floor cassettes for a more flexible open space. Figure 4 shows a recent project of mixed modules in Fulham, UK.

Fig. 5. Twenty-five-story modular building in Wolverhampton, England. (*Courtesy R. M. Lawson, UK*)

Fig. 6. Non-load-bearing modular system. (*Courtesy Yorkon and Joule Consulting Engineers, UK*)

Modular concrete-core systems and non-load-bearing modules are two other types of mixed modules. Modular concrete-core systems are designed to be supported by a primary structure—for example, when modules are clustered around a core that provides stability. This type of system is suitable for high-rise buildings, such as the 25-story building shown in Fig. 5.

Non-load-bearing module systems are similar in form to fully modular units, but they lack resistance to external loads. Generally, they are supported directly by a separate structure. A typical non-load-bearing module system is shown in Fig. 6.

Advantages of modular buildings. Compared to conventional buildings, modular buildings have the following advantages.

Shorter construction time. Time savings is one of the most substantial benefits of modular buildings. Their use reduces the risk of delays for a given project.

Financial savings. The use of prefabrication techniques in modular buildings allows cost savings at every stage of the production chain, such as material savings at the procurement stage and labor savings at the construction stage; the cost reduction may be up to 10% for the overall project costs and 25% for on-site labor costs. In addition, there may be savings resulting from efficient installation and standardized design processes.

Reduced site disruption. Prefabrication can enhance on-site safety by reducing the exposure of workers to inclement weather, heights, hazardous operations, and on-site working time. Prefabrication can also provide more working space to reduce the possibility of on-site accidents.

More consistent quality. Quality control is easier for factory-made products than it is for products made on-site. On-site assembly should ensure that products perform as designed. Meanwhile, prefabrication of modules can alleviate skilled labor shortages.

Reduced environmental impact. Careful quality control during the manufacturing process reduces construction waste through appropriate design and recycling measures. Negative environmental impact can then be alleviated by reducing on-site construction time, noise, and waste. A study by the European Community (EC) indicated the following benefits of modular construction: 50% reduction in water use for constructing a typical house, 50% reduction in quarried materials, and at least 50% reduction in energy consumption.

Applications of modular buildings. In Canada and Australia, modular buildings are widely used for office buildings and for dormitories in mining and forestry operations. Since these areas are usually far from urban areas, the modular buildings have inherent advantages compared to conventionally constructed buildings.

A typical timber modular building is shown in **Fig. 7**. The area of a single modular unit is 53.5 m^2.

Fig. 7. A typical timber modular building in Canada. (*http://www.northgateindustries.com/*)

Fig. 8. A timber modular dormitory in Canada. (*http://www.northgateindustries.com/*)

Fig. 9. A dormitory at the University of Manchester. (*Courtesy Ayrshire Framing, UK*)

Fig. 10. The Uxbridge Hotel in London. (*http://openarchitecturenetwork.org/projects/6683*)

This type of modular building is mainly used for office buildings, dormitories, or cafeterias in mining operations. **Figure 8** shows another type of dormitory for mining operations. It consists of 264 modules, and the total time for its manufacture and installation was 155 days.

In the United Kingdom, modular buildings have found widespread application in residential, hotel, and hospital markets. Off-site construction techniques are applied for about 15–20% of all new buildings constructed in the United Kingdom, of which 40% are modular buildings. A dormitory at the University of Manchester is shown in **Fig. 9**. Conventional composite structures were used for the ground floor and the underground parking garage. The modular portion of the building consists of 1400 modules for the upper seven floors. A 25-story modular building in Wolverhampton is shown in Fig. 5, where 824 modules are connected to the concrete core. The modules resist the vertical loads, and the concrete core sustains the lateral loads.

The Uxbridge Hotel built by Travelodge Hotels is shown in **Fig. 10**. It was the first container modular hotel built in Europe, and installing the 86 containers took only 20 days. The construction time was reduced by 40–60% compared to conventional construction, and the costs for the overall project were reduced by 10%.

Figure 11 shows the Keetwonen (Amsterdam student housing) in Holland, which is the largest container building system. This system includes 12 dormitory buildings, with a capacity of 1000 students. The project was finished in one year, with 20–25 containers being installed per day. The manufacturing cost of each container was €20,000.

In Japan, 50–80% of low-rise residential buildings are modular. A typical postcataclysm container building (built after the East Japan earthquake in 2011) is shown in **Fig. 12**. A single-family unit consists of two containers with an area of 30 m^2. The price of each unit is about $39,000.

Today, China dominates the container manufacturing market, as 95% of all containers used worldwide are made there. From 2008 to 2010, China International Marine Containers (Group) Co., Ltd. (CIMC) produced 20,000 building containers. These containers were used in many international construction projects, such as the British Travelodge container hotel (Fig. 10) and the largest container city, Keetwonen (Amsterdam student housing; Fig. 11).

Modular building research. Some research and design highlights follow, with detailed information available in the references.

Container buildings. The idea of container buildings was first proposed by John M. DiMartino in the United States. Recent research on container buildings has mainly concentrated on their architectural composition, not their structural performance. J. D. Smith investigated the development of container

Fig. 11. Keetwonen (Amsterdam student housing). (*http://www.tempohousing.com/projects/keetwonen.html*)

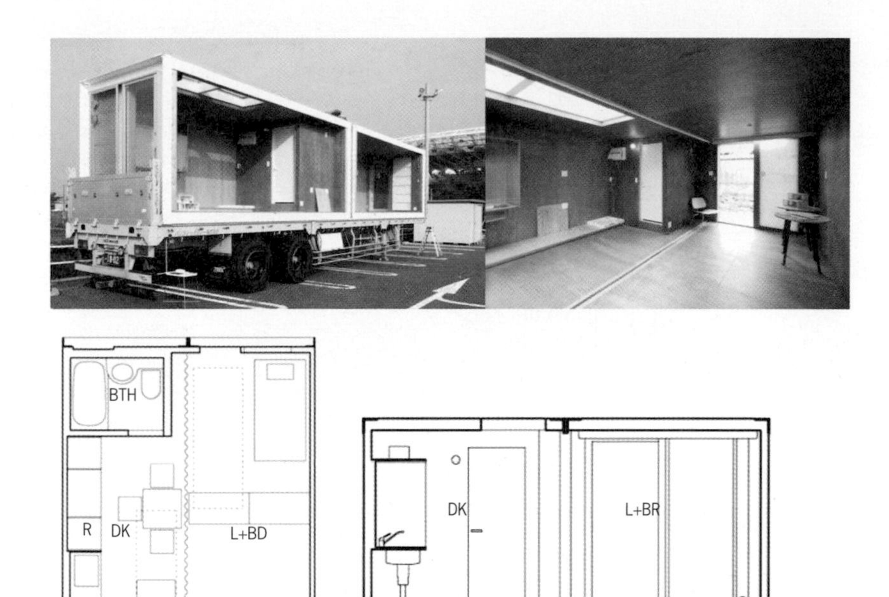

Fig. 12. A typical postcataclysm container building (one constructed after the East Japan earthquake). (*http://www.designboom.com/architecture/yasutaka-yoshimura-architects-ex-container/*)

buildings in the United Kingdom and emphasized that thermal insulation is the most important property.

Spanish architect Jure Kotnik suggested the use of container buildings to solve housing problems.

P. Sawyers reported on construction methods for single-family houses using two to four containers, as well as the principles for choosing foundations, designing spaces, and connecting modules.

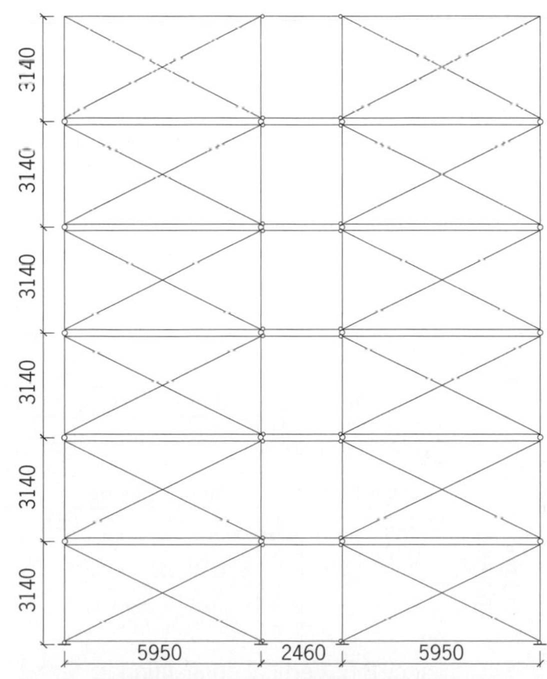

Fig. 13. Equivalent cross-braced corrugated panel for container building. (*Y Lu et al., Structural design technology of container buildings, Industrial Construction, 44(2):130–137, 2014*)

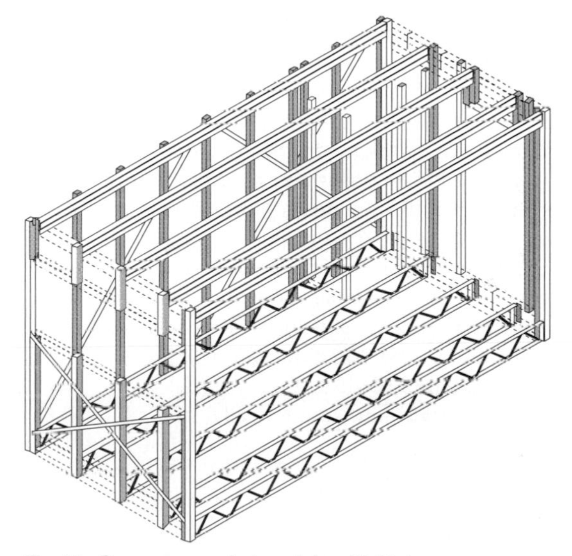

Fig. 14. Corner-supported modules. (*R. M. Lawson, Building design using modules, New Steel Construction, 15(9):28–29, 2007*)

Y. Lu treated the corrugated panels of containers as steel-plate shear walls. By determining the initial stiffness and yielding capacity of corrugated panels through finite element analysis, the corrugated panels were simplified into the equivalent of cross-braced frames (**Fig. 13**).

Corner-supported modules. The structural characteristics of corner-supported modules are the same as those for frame structures. **Figure 14** shows a typical corner-supported module. These modules provide great flexibility in the structural layout and the ability to create large open-plan spaces. However, corner-supported modules without other lateral-resisting systems can be used only for two-story buildings because of their poor stability. For two- to six-story

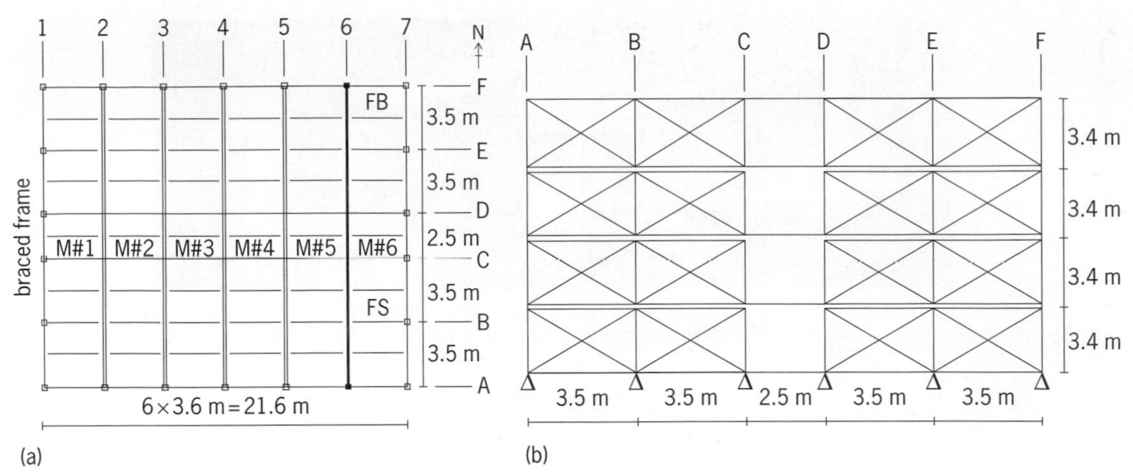

Fig. 15. The floor plan and the elevation of the modular building. (*a*) Floor plan. (*b*) Elevation, centerline 1 or 7. (*C. D. Annan, M. A. Youssef, and M. H. El Naggar, Experimental evaluation of the seismic performance of modular steel-braced frames, Engineering Structures, 31(7):1435–1446, 2009, DOI:10.1016/j.engstruct.2009.02.024*)

buildings, stability depends on separate bracing systems in the form of X-bracing in the separating walls. For higher buildings (more than six stories), mixed modules can be used, such as modular concrete-core systems and non-load-bearing module systems.

C. D. Annan studied the seismic behavior of a four-story corner-supported modular building (**Fig. 15**). The connections between the modules are illustrated in **Fig. 16**. Regular and modular braced frames were designed. Observations were as follows. (1) Both regular and modular frames experienced obvious nonlinear deformation when braces and columns yielded, respectively. (2) The energy dissipation capacity of the modular frame was better than that of the regular frame, but not significantly. (3) The modular frame had various internal force-distribution and force-transfer mechanisms, compared to the regular frame. In addition, Annan experimentally and numerically studied the seismic overstrength factor for modular buildings using pushover analysis.

Four-sided modules and partially open-sided modules. Four-sided modules and partially open-sided modules (Fig. 3) are commonly made of light-gage steel joists. R. M. Lawson and coworkers (2008) experimentally investigated the in-plane and torsional behavior of a pair of four-sided modules under unusual action. The test setup is illustrated in **Fig. 17**. The natural frequency was over 12 Hz for an imposed floor load of 0.5 kN/m^2, and the damping ratio was 3–5.6%. A deflection of 8.8 mm was measured for a pair of modules and 6.1 mm for a single module.

Outlook. Modular buildings are constructed off-site under controlled plant conditions. Their application can shorten the construction period, improve efficiency, save labor cost, and reduce environmental impact. These advantages make them particularly suitable for use in temporary or permanent residential buildings in China and elsewhere in the world. However, research on modular buildings is lacking. Therefore, comprehensive studies on modular buildings will not only improve modular construction

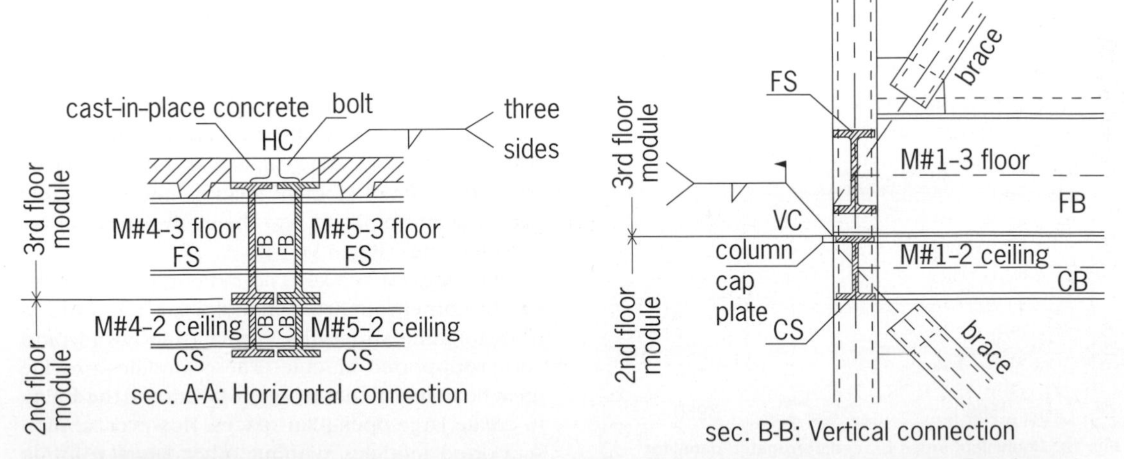

Fig. 16. Typical details for a multistory modular building. (*C. D. Annan, M. A. Youssef, and M. H. El Naggar, Experimental evaluation of the seismic performance of modular steel-braced frames, Engineering Structures, 31(7):1435–1446, 2009, DOI:10.1016/j.engstruct.2009.02.024*)

Fig. 17. Load test on a pair of four-sided modules. (*R. M. Lawson et al., Robustness of light steel frames and modular construction, Proceedings of the ICE: Structures and Buildings, 161(1):3–16, 2008, DOI:10.1680/stbu.2008.161.1.3*)

systems, but also extend the market for modular buildings.

For background information *see* BUILDINGS; CONCRETE; EARTHQUAKE ENGINEERING; FLOOR CONSTRUCTION; FOUNDATIONS; STRUCTURAL ANALYSIS; STRUCTURAL PLATE; STRUCTURAL STABILITY; STRUCTURAL STEEL; STRUCTURE (ENGINEERING); WALL CONSTRUCTION in the McGraw-Hill Encyclopedia of Science & Technology. Guo-Qiang Li; Ye Lu; Ke Cao

Bibliography. C. D. Annan, M. A. Youssef, and M. H. El Naggar, Experimental evaluation of the seismic performance of modular steel-braced frames, *Eng. Struct.*, 31(7):1435–1446, 2009, DOI:10.1016/j.engstruct.2009.02.024; C. D. Annan, M. A. Youssef, and M. H. El Naggar, Seismic overstrength in braced frames of modular steel buildings, *J. Earthquake Eng.*, 13(1):1–21, 2009, DOI:10.1080/13632460802212576; M. Ball, Chasing a snail: Innovation and housebuilding firms' strategies, *Hous. Stud.*, 14(1):9–22, 1999, DOI:10.1080/02673039982975; N. G. Blismas et al., Constraints to the use of off-site production on construction projects, *Architect. Eng. Des. Manag.*, 1(3):153–162, 2005, DOI:10.1080/17452007.2005.9684590; J. M. DiMartino, Modular container building system, U.S. Patent 4599829, 1983; D. M. Gann, Construction as a manufacturing process? Similarities and differences between industrialized housing and car production in Japan, *Construct. Manag. Econ.*, 14(5):437–450, 1996, DOI:10.1080/014461996373304; A. G. F. Gibb, Standardization and pre-assembly—distinguishing myth from reality using case study research, *Construct. Manag. Econ.*, 19(3):307–315, 2001, DOI:10.1080/01446190010020435; J. Kotnik, *Container Architecture*, Links Books, Barcelona, Spain, 2008; R. M. Lawson, Building design using modules, *New Steel Construct.*, 15(9):28–29, 2007; R. M. Lawson and R. G. Ogden, "Hybrid" light steel panel and modular systems, *Thin-Walled Structures*, 46:720–730, 2008, DOI:10.1016/j.tws.2008.01.042; R. M. Lawson et al., Developments in pre-fabricated systems in light steel and modular construction, *Struct. Eng.*, 83(6):28–35, 2005; R. M. Lawson et al., Pre-fabricated systems in housing using light steel and modular construction, *Int. J. Steel Struct.*, 5(5):477–483, 2005; R. M. Lawson et al., Robustness of light steel frames and modular construction, *Proceedings of the ICE: Structures and Buildings*, 161(1):3–16, 2008, DOI:10.1680/stbu.2008.161.1.3; N. Lu, *Investigation of Designers' and General Contractors' Perceptions of Offsite Construction Techniques in the United States Construction Industry*, Clemson University, Clemson, SC, 2007; Y. Lu et al., Structural design technology of container buildings, *Industrial Construction*, 44(2):130–137, 2014; P. Sawyers, *Intermodal Shipping Container Small Steel Buildings*, 2d ed., Paul Sawyers Publications, 2008; J. D. Smith, *Shipping Containers as Building Components*, University of Brighton, Brighton, UK, 2006; C. B. Tatum, J. A. Vanegas, and J. M. Williams, *Constructability Improvement Using Prefabrication, Preassembly and Modularization*, Construction Industry Institute, Austin, TX, 1987.

Application of software-defined radio

This article describes the ways that software-defined radio (SDR) has been practically applied. The aim of an SDR, like that of any radio, is to transmit and receive data or voice-based information by means of wireless communication. In addition, an SDR must include unavoidable analog components, such as an antenna to allow radio-frequency coupling with free space and radio-frequency amplifiers for signal strengthening. Such an SDR may be simply defined as shown in **Fig. 1**.

Evolution of SDR toward an all-digital, reconfigurable radio. Signal and data processing are necessary for information transfer. The capabilities for digital signal processing have increased substantially over the last several years, making SDR more and more an all-digital radio. This development has been made possible through technology advances in processing devices such as digital signal processors (DSPs), field-programmable gate arrays (FPGAs), and analog-to-digital and digital-to-analog converters (ADCs and DACs). These advances in hardware components have been paired with software component evolution that has provided designers and developers with increasingly effective and user-friendly development tools. Among the software advances have been development systems that allow software design and coding in the environment of the real-time operating system (RTOS), which, in turn, offers developers a direct link to the relevant application program interface (API) libraries, including efficient debugging features. The adoption of higher-level programming languages, among them C and C++, was also supported by formal description languages such as the Unified Modeling Language (UML), directed at a diffused approach known as model-based design, which, in turn, is based on visual modeling

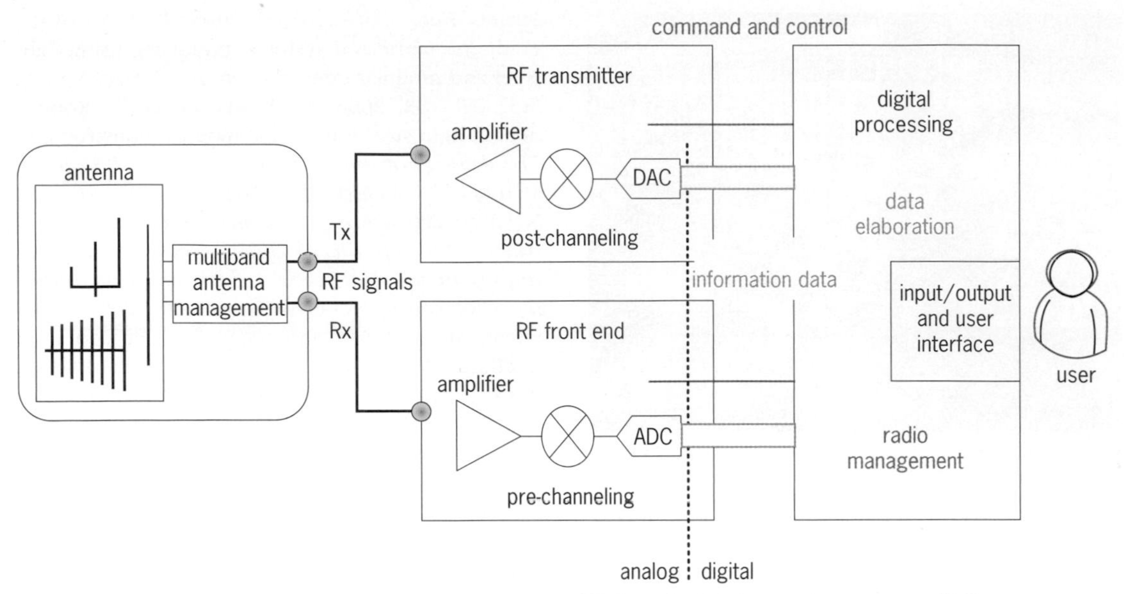

Fig. 1. General schematic diagram of a software-defined radio (SDR). DAC, digital-to-analog converter; ADC, analog-to-digital converter; Tx, transmission; Rx, reception; RF, radio frequency.

techniques. This software development has transformed digital radio from a fixed to a reconfigurable design. SDR designers currently devote their efforts to exploiting these new software capabilities to make the SDR a many-purpose device, able to perform a wide set of network-based capabilities. So, within the limits imposed by the previously mentioned analog components, SDR has evolved from a simple fixed-design digital radio to the present reconfigurable software-defined-based architecture (**Fig. 2**).

The "software-defined" concept can be further exploited. A conventional digital radio relies on an RTOS. An RTOS offers the capability of accessing platform resources in the form of specific APIs. Software developers have to know the platform-specific features, because their software has to execute specific calls to RTOS-specific APIs necessary for hardware management. For example, this occurs in the implementations of file managers or input/output of data on specific buses for board interconnections. This level of hardware dependency is still retained in more advanced SDRs, but a specific user's environment has been developing and deploying over the past decade, which has an SDR architecture with a higher flexibility level.

In the United States, the military has been managing a national procurement program, the Joint Tactical Radio System (JTRS), with the purpose of fielding a new generation of military radios having a common software architecture known as the Software Communications Architecture (SCA). The JTRS

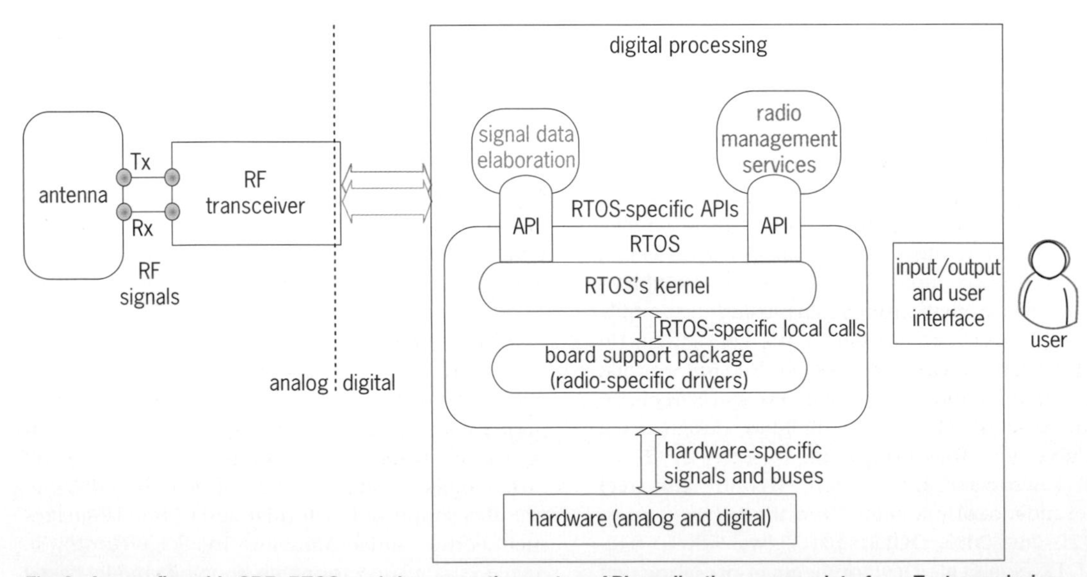

Fig. 2. A reconfigurable SDR. RTOS, real-time operating system; API, application program interface; Tx, transmission; Rx, reception; RF, radio frequency.

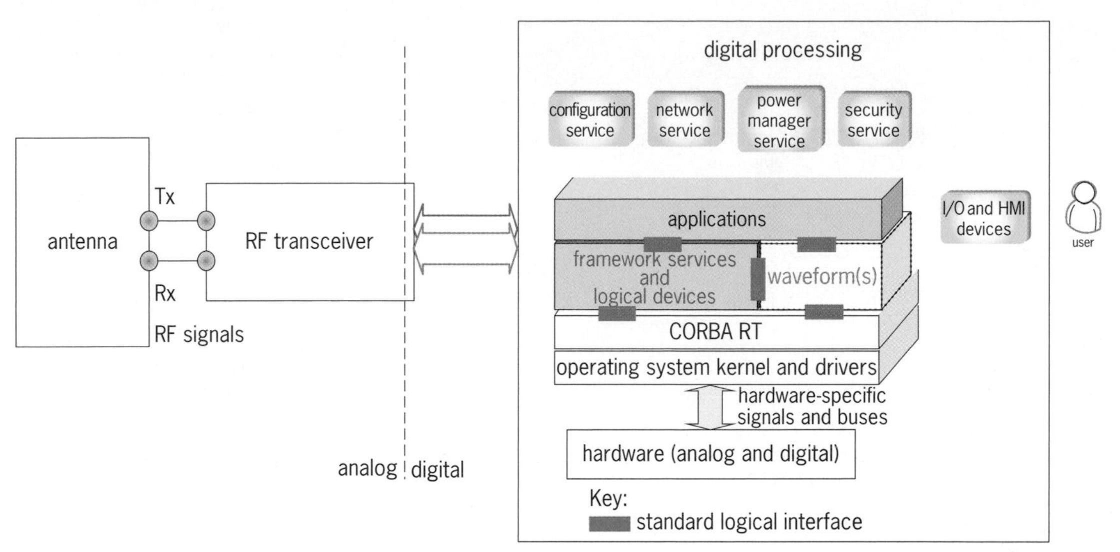

Fig. 3. A Software Communications Architecture (SCA)-based SDR. CORBA RT, Real-time Common Object Request Broker Architecture; I/O, input/output; HMI, human-machine interface; Tx, transmission; Rx, reception; RF, radio frequency.

aims to replace some 25 legacy types of radios with a limited set of platforms and relevant waveform-based functionalities. The JTRS's managers came to the decision to adopt the SCA in order to pragmatically face the problem of software portability. This program has been making some relevant changes in military radio communications, both technology- and market-based, and in both cases due to the new software architecture. A typical model made of platform/waveform-integrated solutions is shown in **Fig. 3**.

The adoption of an SCA-like architecture featuring real-time middleware and APIs of the related framework has raised the SDR concept to near its maximum flexibility level. The Common Object Request Broker Architecture (CORBA) is employed as the middleware session protocol. Each developer can adopt standardized common interface definition language (IDL)-based APIs that enable interoperability and portability of logical interconnections between the multisource software environments. The limit to full software portability comes from the traditional digital technologies constraining the developers to adopt specific features of low- or high-level programming languages. There is an analogy between the SCA-like SDR architecture and the layers of the Open Systems Interconnect (OSI) model that is used in Internet Protocol (IP)-based networks (**Fig. 4**).

Recent developments and implementations. The SDR has been applied mainly on the commercial mobile market with specific architectures. Examples can be found in "SDR-like"-based base stations that are able to support multiple radio access technologies (RATs), including second-, third-, and fourth-generation (2G/3G/4G) mobile communications and multiple wireless access technologies, including GSM (Global System for Mobile Communications), UMTS (Universal Mobile Telecommunications System), CDMA2000 (Code-Division Multiple Access-2000), and LTE (Long-Term Evolution). The commercial mobile communication stakeholders are currently developing technology that would minimize radio-frequency shortages that could cause serious constraints in the short- or mid-term time period in some countries. White-space exploitation (using the free frequency ranges between television channels) and the licensed spectrum available in areas shared between primary and secondary users under specific conditions are already the targets of specific applications such as cognitive radio. In these applications, software implementing specific algorithms and protocols provides an SDR with an application layer performing dynamic spectrum management.

Public safety networks. Radio communication for public safety is an essential element in various operational scenarios and at different levels of the hierarchy of public safety organizations. First responders should be able to exchange information (that is, voice and data) in a timely manner to coordinate the relief efforts and to improve situational awareness of the environment (**Fig. 5**). The user applications specific to public safety and the previously described SDR features have been effectively integrating so as to pave the way for the most compelling evolution of current narrowband networks based on radio communications system standards, such as Terrestrial Trunked Radio (TETRA) in the European Union and Project 25 (P25 or APCO-25, after the Association of Public-Safety Communications Officials-International) in the United States.

Smart radio terminals. Development of smart radio terminals capable of hosting computer applications has begun. In 2012, both Motorola Mobility and Harris Corporation stated that they were developing their own broadband mission-critical handheld terminal that is compliant with the LTE standard. Both handheld terminals enable seamless interoperability among P25 networks and commercial cellular technology. Both the Motorola Mobility and Harris

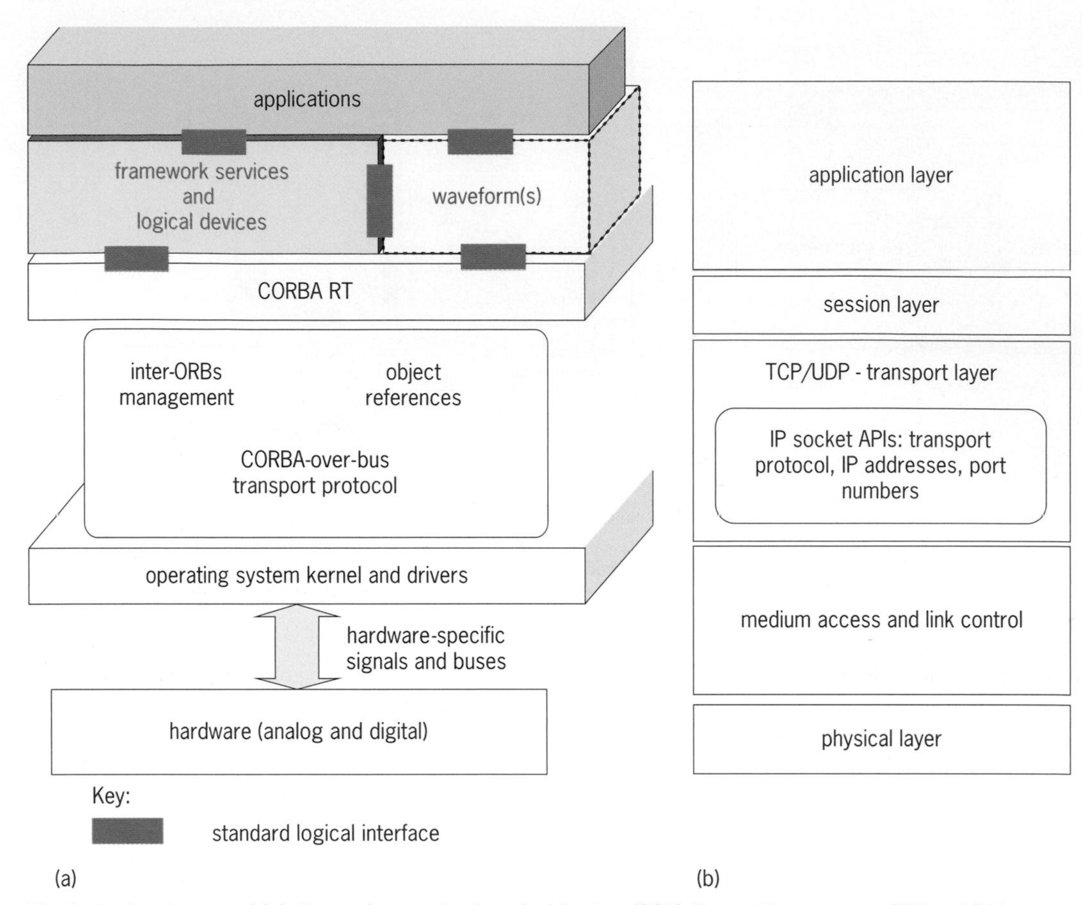

Fig. 4. Analogy between (*a*) Software Communications Architecture (SCA)–like architecture of an SDR and (*b*) layer stack in the Open Systems Interconnect (OSI) model, used in Internet Protocol (IP)–based networks. CORBA RT, Real-time Common Object Request Broker Architecture; ORB, object request broker; TCP/UDP, Transmission Control Protocol/User Datagram Protocol; API, application program interface.

terminals adopt the Android operating system, providing the handhelds with software architecture suitable for an effective integration of applications and services. An earlier development, carried out by the Communications Research Centre Canada (CRC), ported a complete APCO P25 waveform and SCA radio system into a small-form-factor Android device.

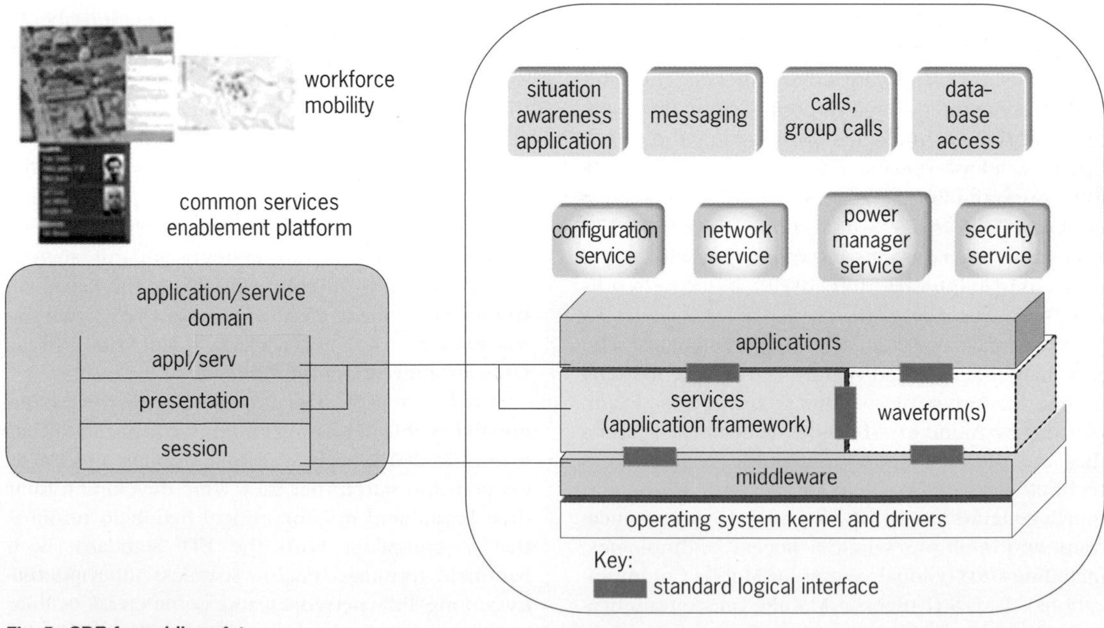

Fig. 5. SDR for public safety.

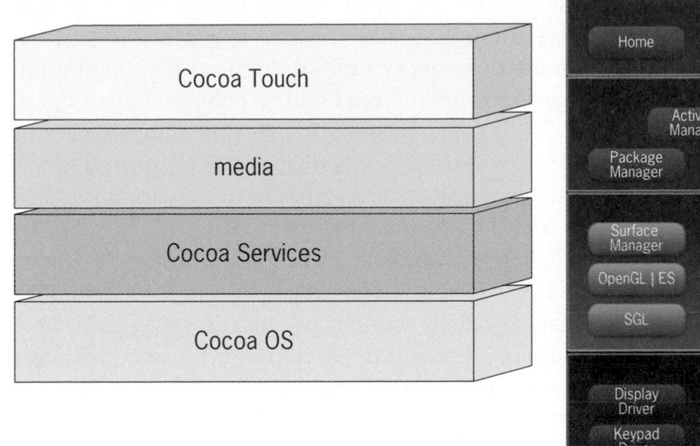

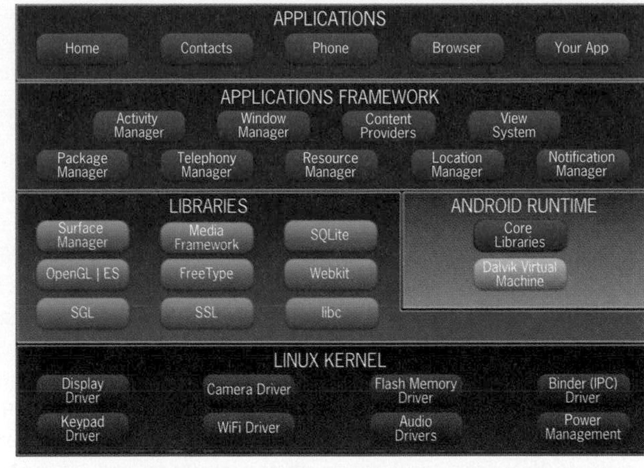

(a) (b)

Fig. 6. Operating system (OS) architectures of smartphones. (*a*) iOS. (*b*) Android OS.

Space communications. SDR also has found application in space communications. The first SDR-for-space architecture was implemented by NASA's Jet Propulsion Laboratory (JPL) for the UHF Electra-Lite SDR. In 2012, the Mars rover *Curiosity* landed on the red planet's surface equipped with an ultrahigh-frequency (UHF) SDR for communicating with the Mars orbiters, *Mars Odyssey* and *Mars Reconnaissance Orbiter*. NASA has also been defining and adopting a suitable specification known as the Space Telecommunications Radio System (STRS) Architecture Standard.

Mass market. A special deployment of SDR that has been underway since 2007 is the mass public implementation of the smartphone (**Fig. 6**). Users have the capability to install new applications allowing the devices to provide a GSM connection, audio and video players, car navigators, text editors, messaging, and Internet connections. Users can also update the operating systems of their terminals.

For background information *see* AMPLIFIER; ANALOG-TO-DIGITAL CONVERTER; DATA COMMUNICATIONS; DIGITAL-TO-ANALOG CONVERTER; FIELD-PROGRAMMABLE LOGIC ARRAYS; MOBILE COMMUNICATIONS; MODELING LANGUAGES; OPERATING SYSTEM; PROGRAMMING LANGUAGES; RADIO; RADIO RECEIVER; RADIO SPECTRUM ALLOCATION; RADIO TRANSMITTER; SIGNAL PROCESSING; SOFTWARE; SPACE COMMUNICATIONS; SPACE PROBE; WAVEFORM in the McGraw-Hill Encyclopedia of Science & Technology.

Fabrizio Vergari

Bibliography. F. Casalino, G. Middioni, and D. Paniscotti, Experience report on the use of CORBA as the sole middleware solution in SCA-based SDR environments, *Proceedings of the SDR '08 Technical Conference and Product Exposition*, Oct. 26–30, 2008, Washington, DC, SDR Forum, 2008; European Telecommunications Standards Institute, *Reconfigurable Radio Systems (RRS); Cognitive Pilot Channel (CPC)*, Tech. rep. ETSI TR 102 683 V1.1.1, ESTI, Sophia Antipolis, France, 2009; European Telecommunications Standards Institute, *Reconfigurable Radio Systems (RRS); Functional Architecture (FA) for the Management and Control of Reconfigurable Radio Systems*, Technical report ETSI TR 102 682 V1.1.1, ESTI, Sophia Antipolis, France, 2009; J. Mitola III and G. Q. Maguire, Jr., Cognitive radio: Making software radios more personal, *IEEE Pers. Commun. Mag.*, 6(4):13–18, August 1999, DOI:10.1109/98.788210; R. C. Reinhart et al., *Space Telecommunications Radio System (STRS) Architecture Standard*, Release 1.02.1, rev. ed., NASA Glenn Research Center, Cleveland, OH, 2012; E. Satorius et al., *The Electra Radio*, in J. Hamkins and M. K. Simon (eds.), *Autonomous Software Defined Radio Receivers for Deep Space Application*, pp. 19–43, Wiley, Hoboken, NJ, 2006; F. Vergari, Software-defined radio: Finding its use in public safety, *IEEE Veh. Tech. Mag.*, 8(2):71–82, June 2013, DOI:10.1109/MVT.2013.2252292; F. Vergari and G. Baldini, Reconfigurable radio systems for public safety: New generation public safety ICT, *Proceedings of the European Conference on Communications Technologies and Software Defined Radio*, June 22–24, 2011, Brussels, Belgium, Wireless Innovation Forum, 2011.

Artificial spinal disc replacement

Chronic back and neck pain (chronic spine pain) is a significant and growing problem that afflicts an estimated 15% of the population. Chronic spine pain, as opposed to transient spine pain, is characterized by its extended duration (longer than 6 weeks) and its recurrence. It is the second most common reason for visiting the doctor, right behind the common cold. It is also considered to be one of the most disabling conditions in the world and its associated costs make up the bulk of workers' compensation costs in the United States.

Traditional approaches to pain relief for back and neck pain include both surgical and nonsurgical alternatives. Nonsurgical alternatives are usually the first choice for both clinicians and patients and include physical therapy, ergonomic adjustments, heat treatments, electrical stimulation, chiropracture, acupuncture, and pharmaceutical pain management. Unfortunately, these alternatives are ineffective for a significant portion of patients (approximately 50%).

Surgical alternatives for treating chronic spine pain are challenging because the direct source of the pain is often nonspecific and cannot easily be localized. The spine is composed of a stacked collection of vertebral bones, linked together anteriorly with softer cartilaginous joints (intervertebral discs) and posteriorly with bilateral diarthrodial joints (facet joints). The spine is passively stabilized by a network of ligaments and fascia that surrounds and attaches adjacent vertebral bones. Active control and additional stabilization is provided through a series of muscle attachments. Most of these structures (bones, discs, facet joints, ligaments, and muscles) are innervated with pain-sensing nerves capable of triggering back pain (**Fig. 1**). Spinal motion causes mechanical deformation in all of these structures; thus, isolating a particular source of pain is challenging.

Notwithstanding the difficulty in localizing a particular pain source, up to 90% of spinal surgeries cite degenerated disc disease as a primary or secondary factor. These surgeries have an overall goal of reducing or eliminating pain induced by mechanical factors and often involve both a decompressive component intended to unload the nerves and spinal cord at the location of the affected segment, and a stabilization component intended to restore nonpainful motion to the patient. By far, the most common surgical approach to stabilizing the spine is spinal fusion. Spinal fusion consists of severely reducing or eliminating motion at a painful spinal segment and is typically achieved through the use of a combination of bone graft material and rigid fusion hardware (plates, rods, screws). Spinal fusion has relatively low patient satisfaction rates (approximately 50%), extended recovery time, reduction of spinal mobility, and has been linked with accelerated disc degeneration at adjacent spinal levels. Clinician and patient dissatisfaction with this surgery has prompted significant attention to development of new technologies for treating chronic spine pain.

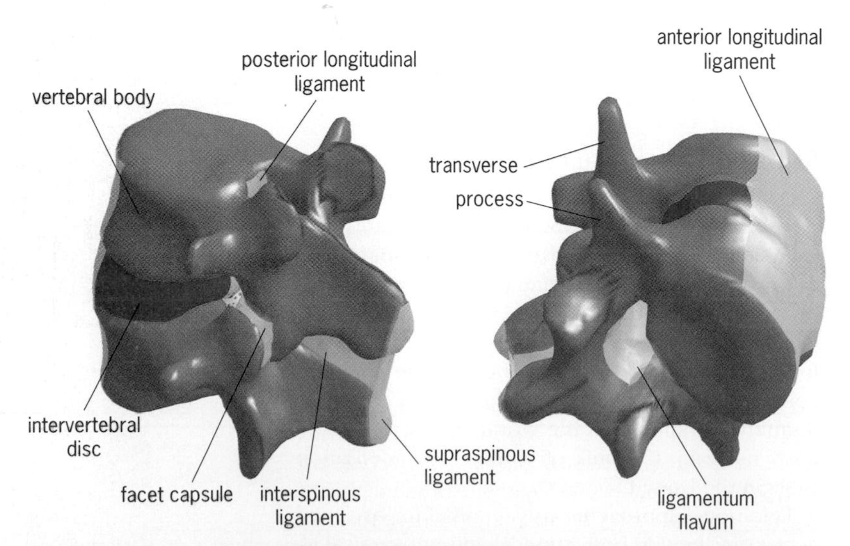

Fig. 1. Pain-sensing structures in the passive elements of the spine. In addition to these passive elements, the muscles contain additional pain sensors.

Motion preservation. Over the last 2 decades, there has been a growing paradigm shift in understanding regarding the treatment of chronic spine pain. This paradigm shift has been compared to the switch in treatment of hip osteoarthritis from hip fusion to hip replacement. A key observation is that chronic spine pain can often be reduced or eliminated while preserving or restoring spinal motion at the affected spinal segment. This can be accomplished using several techniques, including artificial disc replacement, nucleus replacement, facet replacement, and dynamic stabilization. The focus of this article is on artificial disc replacement (ADR).

ADR consists of replacing the damaged or degenerated intervertebral disc with a mechanical component. The concept of ADR is not new, and early efforts dating more than 50 years ago utilized a rigid metal ball that was implanted between adjacent vertebrae after surgical removal of the intervertebral disc. However, these early efforts were generally unsuccessful and often resulted in device subsidence (that is, the device sunk into the adjacent vertebral endplates and eventually resulted in a suboptimal spinal fusion). More recent efforts have incorporated an advanced understanding of the biomechanics of the spinal segment.

Biomechanics of artificial disc replacement. Unlike the more typical diarthrodial joints found in the hip and the knee, each spinal segment consists of a three-joint complex (the intervertebral disc and two facet joints). All three joints in the complex must work together synergistically to accomplish the mobility and protective functions of the spinal segment. Failure of any of the joints to function appropriately triggers degenerative changes in the other joints.

The intervertebral disc in particular is biomechanically complex in its function. Because it is a thick, cartilaginous joint, the intervertebral disc provides substantial shock dissipation and viscoelastic damping. Both the shape and the material components of the intervertebral disc are complex, and as a consequence the moment-rotation behavior (that is, the stiffness) of the intervertebral disc is nonlinear. For example, the intervertebral disc provides little resistance to small motions centered about a neutral posture and increasing resistance to extreme motions. Not only is the stiffness of the intervertebral disc nonlinear, it is also distinct in each mode of bending; flexion-extension stiffness is different from lateral bending stiffness and is significantly different from axial rotation stiffness. This anisotropic, nonlinear stiffness is important as it provides protection during activities where the sensitive spinal cord and associated nerve roots are more vulnerable to mechanical damage. As a further consequence of its material complexity, the center of rotation of the spinal segment changes throughout its range of motion. This shift in center of rotation preserves loading at the facets and equalizes the loading of adjacent spinal segments.

Ideally, an ADR will duplicate all of these functions, and by doing so will relieve pain by restoring natural loading to all of the surrounding tissues. From a practical standpoint, however, developing an artificial spinal disc is a difficult engineering challenge. Innovative design approaches are required to achieve the natural disc functions. Longevity is an important additional concern for an ADR. The intervertebral disc undergoes millions of bending cycles each year. Because of its sensitive location near the spinal cord and spinal nerve roots, revision surgery for an ADR can be especially challenging. Thus, designing an ADR with an extended functional life (preferably longer than that of the patient) is a paramount consideration. A sampling of the design considerations required for development of an ADR is given in the **table**.

Current status of artificial disc replacement. Most of the first generation of modern ADRs incorporated the metal-polyethylene bearing surface combination that is typically found in successful hip and knee replacements. Early results have been generally positive, with significantly reduced recovery times (days instead of weeks) and substantially higher patient satisfaction rates. The use of cervical ADR in particular has continued to grow and a recent clinical trial reported that cervical disc replacement showed superiority to anterior cervical decompression fusion surgery.

Nonetheless, first-generation ADR devices have demonstrated room for improvement. The biomechanics of the very low friction interfaces (for example, metal-polyethylene bearing surfaces) used in these devices does not duplicate the nonlinear stiffness exhibited by the natural disc. As a consequence, these devices have been linked with accelerated degeneration of both the facet joints at the operated level and the intervertebral discs at adjacent spinal segments. Some of these devices (for example, the DePuy Charité, **Fig. 2**) have incorporated a mobile core intended to duplicate the change in location of the segmental center of rotation exhibited by the natural disc, but many have not. Although one first-generation ADR (AcroMed Acroflex) incorporated viscoelastic shock absorption and damping through the use of a polyolefin rubber, most have not incorporated this biomechanical element into their design. Device longevity is a concern and first-generation devices have exhibited both component wear and fatigue damage.

Second-generation ADR devices are currently in various stages of testing and development and have been designed to address the limitations of first-generation ADRs. The design directions being pursued are varied and range from polymeric rubber components with graded material stiffness and integrated electronic circuitry to complex multi-spring components. A promising approach to ADR utilizes a compliant mechanism (**Fig. 3**). This device employs a shaped solid core that allows the center of rotation of the device to accurately mimic that of a healthy spinal disc. The solid core is connected to integrated flexures that reproduce the full, nonlinear

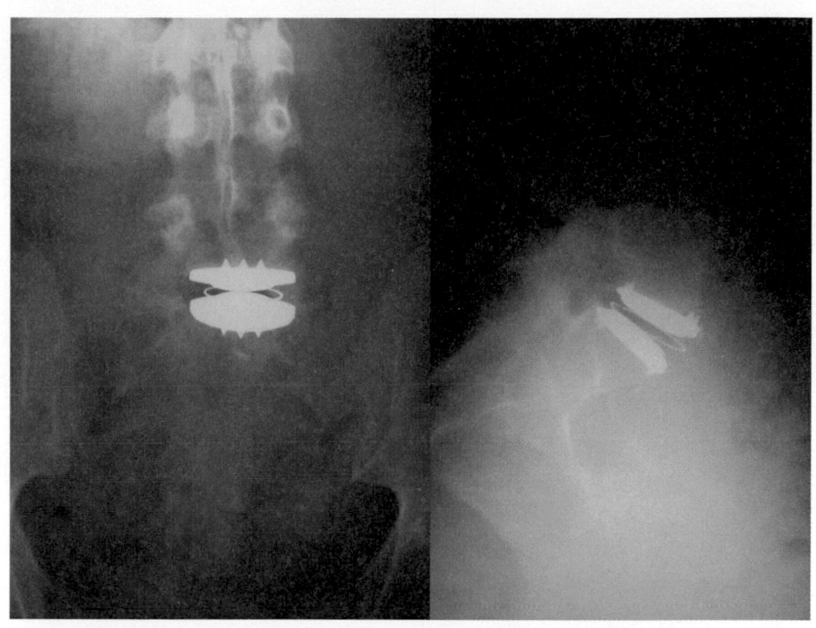

Fig. 2. X-ray images of an implanted first-generation artificial disc replacement. (*Charité, DePuy*)

Fig. 3. A second-generation artificial disc replacement. (*FlexBAC, Nexus TDR*)

Design considerations for artificial disc replacement

Viscous damping and shock dissipation	Surgical approach
Range of motion	Device placement sensitivity
Moment-rotational stiffness	Ease of revision
Anisotropic response	Indications for patient selection
Centrode location and path	Recovery time and protocol
Device fatigue	Manufacturing cost
Minimal wear debris generation	Regulatory considerations
Material biocompatibility	

moment-rotation stiffness and range of motion of the healthy spinal disc in flexion-extension, lateral bending, and axial rotation. Because it is a compliant mechanism, the ADR is designed such that there is no sliding friction (only rolling friction), which dramatically reduces component wear. As with virtually all ADR devices, it only replaces the degenerated disc, leaving the facet joints intact. This compliant mechanism ADR was designed to provide an infinite fatigue life under activities of normal daily living, and is scheduled to begin human trials in late 2014. Second-generation ADR devices are anticipated to provide improved long-term outcomes for patients suffering from chronic spine pain and to pave the way for biological and tissue-engineered solutions on the more distant horizon.

For background information *see* ARTHRITIS; BIOMECHANICS; BIOMEDICAL ENGINEERING; BONE; BONE DISORDERS; JOINT (ANATOMY); JOINT DISORDERS; PROSTHESIS; SKELETAL SYSTEM; SURGERY; VERTEBRA in the McGraw-Hill Encyclopedia of Science & Technology. Anton E. Bowden

Bibliography. S. Blumenthal et al., A prospective, randomized, multicenter Food and Drug Administration investigational device exemptions study of lumbar total disc replacement with the CHARITE™ artificial disc versus lumbar fusion: Part I: Evaluation of clinical outcomes, *Spine*, 30(14):1565–1575, 2005, DOI:10.1097/01.brs.0000170587.32676.0e; R. J. Davis et al., Cervical total disc replacement with the Mobi-C cervical artificial disc compared with anterior discectomy and fusion for treatment of 2-level symptomatic degenerative disc disease: A prospective, randomized, controlled multicenter clinical trial, *J. Neurosurg. Spine*, 19:532–545, 2013, DOI:10.3171/2013.6.SPINE12527; P. A. Halverson, A. E. Bowden, and L. L. Howell, A compliant mechanism approach to achieving specific quality of motion in lumbar total disc replacement, *Int. J. Spine Surg.*, 6:78–86, 2012, DOI:10.1016/j.ijsp.2012.02.002; C. Hirsch, B. E. Ingelmark, and M. Miller, The anatomical basis for low back pain: Studies on the presence of sensory nerve endings in ligamentous, capsular and intervertebral disc structures in the human lumbar spine, *Acta Orthop. Scand.*, 33:1–17, 1963, DOI:10.3109/17453676308999829; A. White and M. Panjabi, *Clinical Biomechanics of the Spine*, Lippincott Williams & Wilkins, Baltimore, 1990.

Autism and eye gaze abnormalities

Beginning in the first days of life, developing infants typically show a preference for looking at others' faces and eyes that persists into childhood and adulthood. Eye contact and preferential attention to social information are fundamental processes of social communication used to initiate and maintain interactions with others. Because attention to others is so integral to typical social experiences, atypical eye contact is, in contrast, one of the most striking diagnostic features of autism spectrum disorder (ASD), which is a developmental disorder defined by impairments in social communication. Indeed, atypical eye contact is often noted as a "red flag" (warning signal) for early developmental concerns.

The use of eye-tracking technology, which unobtrusively records eye movements to measure where an individual is looking, has facilitated the study of atypical gaze patterns in ASD in greater detail from infancy through adulthood. Studying how people spontaneously distribute their gaze in different contexts while engaged in unconstrained free-viewing tasks (that is, viewing videos according to their own interests and goals, without explicit instructions from the experimenter) indicates what individuals with ASD find salient within their environment. Those patterns help indicate how individuals with ASD are learning about the social world, and how they do so in ways that differ from their typically developing peers. Results have yielded insight into the development of social impairments in ASD and have implications for both the early diagnosis and treatment of the disorder.

Atypical eye gaze patterns in children and adults with ASD. Studies of eye gaze abnormalities in ASD have focused on understanding the differences in how individuals with ASD distribute their gaze relative to their more typically developing peers while viewing social stimuli, such as still images of faces or dynamic video clips of people interacting in naturalistic settings. To measure the distribution of visual attention, most eye-tracking studies calculate the percentage of total viewing time spent looking at different regions of interest, such as the eyes versus the mouth and other less socially pertinent background regions. Across all age groups, from toddlers to school-age children and adolescents to adults, results of these studies have indicated that individuals diagnosed with ASD look less at people's eyes than their typically developing peers do. Because the eyes often convey significant social cues about people's affects and intentions, individuals with ASD, by not attending to people's eyes, may miss important social information that can alter the meaning of social situations (**Fig. 1**). Accordingly, the amount of visual attention to people's eyes generally correlates with symptom severity: Decreased attention to people's eyes is associated with more severe social disability.

In contrast, results regarding the amount of time spent looking at people's mouths have varied more across studies. Some studies have found that individuals with ASD look more at people's mouths than typically developing peers do, but other studies have found either no differences between the groups or reduced mouth-looking in individuals with ASD. The inconsistency across studies has been attributed to variability in the clinical characteristics of each study sample and the type of social stimuli used (for example, still images or dynamic video clips). When

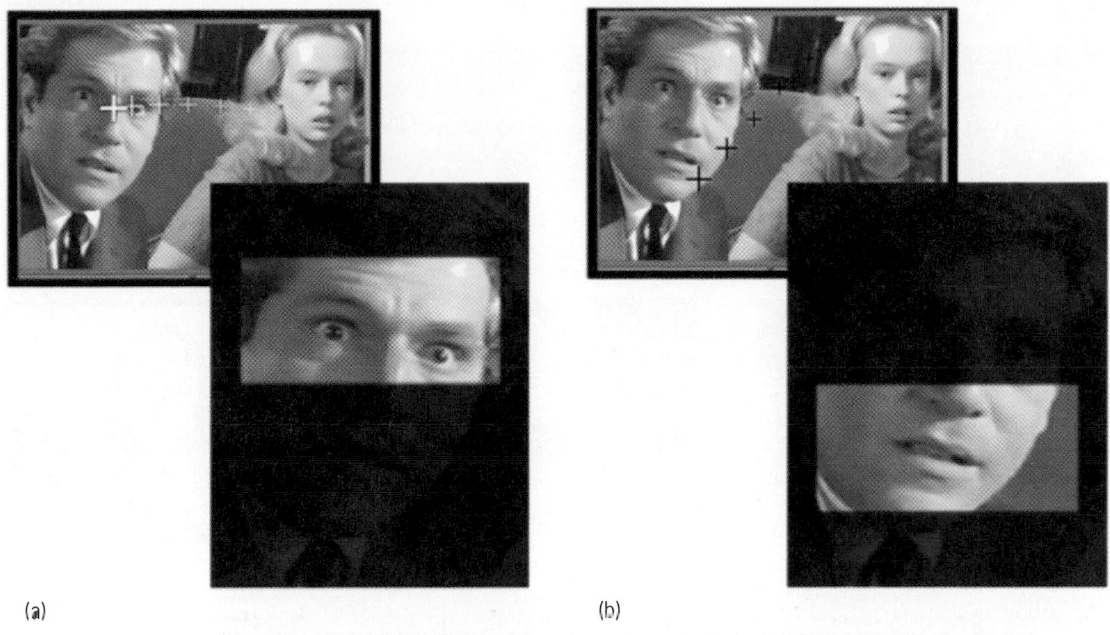

(a) (b)

Fig. 1. Examples of eye gaze patterns in adults. (*a*) Eye gaze pattern of a typically developing adult (*white crosshairs*). Focusing on the eyes reveals the character's surprise and horror. (*b*) Eye gaze pattern of an adult with autism spectrum disorder (ASD) [*black crosshairs*]. Focusing on the mouth suggests a neutral response. (*Reprinted with permission from A. Klin et al., The enactive mind, or actions to cognition: Lessons from autism, Phil. Trans. R. Soc. Lond. B, 358:345–360, 2003, DOI:10.1098/rstb.2002.1202*)

viewing dynamic video clips of social scenes, the amount of attention to the mouth tends to be related to the level of verbal functioning. For individuals with ASD and high verbal functioning, particularly older school-age children and adults, the increased attention to the mouth may be a compensatory strategy because increased attention to the mouth is associated with less severe social disability. By attending to the source of speech, that is, the mouth, these highly verbal individuals show evidence of an attempt to rely on others' words, rather than on the nonverbal social communication of the eyes, in order to navigate social interactions.

Development of atypical eye gaze patterns in infants. To examine the early development of atypical eye gaze patterns, even prior to the time when ASD can be reliably diagnosed at 18 to 24 months, researchers have focused on infants at high risk for developing ASD: specifically, on younger siblings of children already diagnosed with the disorder. Within the broader population, the rate of ASD diagnosis is approximately 1%. However, because of the high heritability of ASD, the diagnosis rate among younger siblings of children with ASD is closer to 20%.

In unexpected contrast to the consistent differences in eye gaze patterns identified between older individuals with ASD and their typically developing peers, which are present by the age of 18 to 24 months, studies of infants who later receive an ASD diagnosis have not identified cross-sectional differences from typically developing infants at any age before 12 months. Instead, infants who later receive an ASD diagnosis appear to look at people's eyes as much as typically developing infants do when viewing dynamic video clips of actresses portraying caregivers (**Fig. 2**). Critically, however, those cross-sectional similarities in the level of looking at eyes mask marked developmental differences between typically developing infants and those who are later diagnosed with ASD. Whereas attention to people's eyes remains fairly stable over time, from 2 to 24 months, in typically developing infants, infants who later receive an ASD diagnosis show a decline in attention to the eyes, leading to significantly reduced attention to people's eyes by 24 months. The degree of deviation from the normative developmental trajectory between 2 and 6 months was predictive of later ASD diagnosis, and it also correlated with the severity of social disability.

At present, the difference in the trajectory of attention to people's eyes represents the earliest identified marker for ASD. Future studies in larger samples will be necessary to confirm the utility of such a marker for early diagnosis. In addition, the lack of difference in eye gaze patterns at the earliest ages, in the first months after birth, suggests that basic neural mechanisms underlying the predisposition for social orienting may be initially intact in infants who are later diagnosed with ASD. Given this evidence, the social deficits in ASD may depend on differences in a postnatal phase of experience-dependent learning, relying upon a distinct set of neural mechanisms, which are now an important area of future research. These results hold promise that earlier detection and earlier treatment may

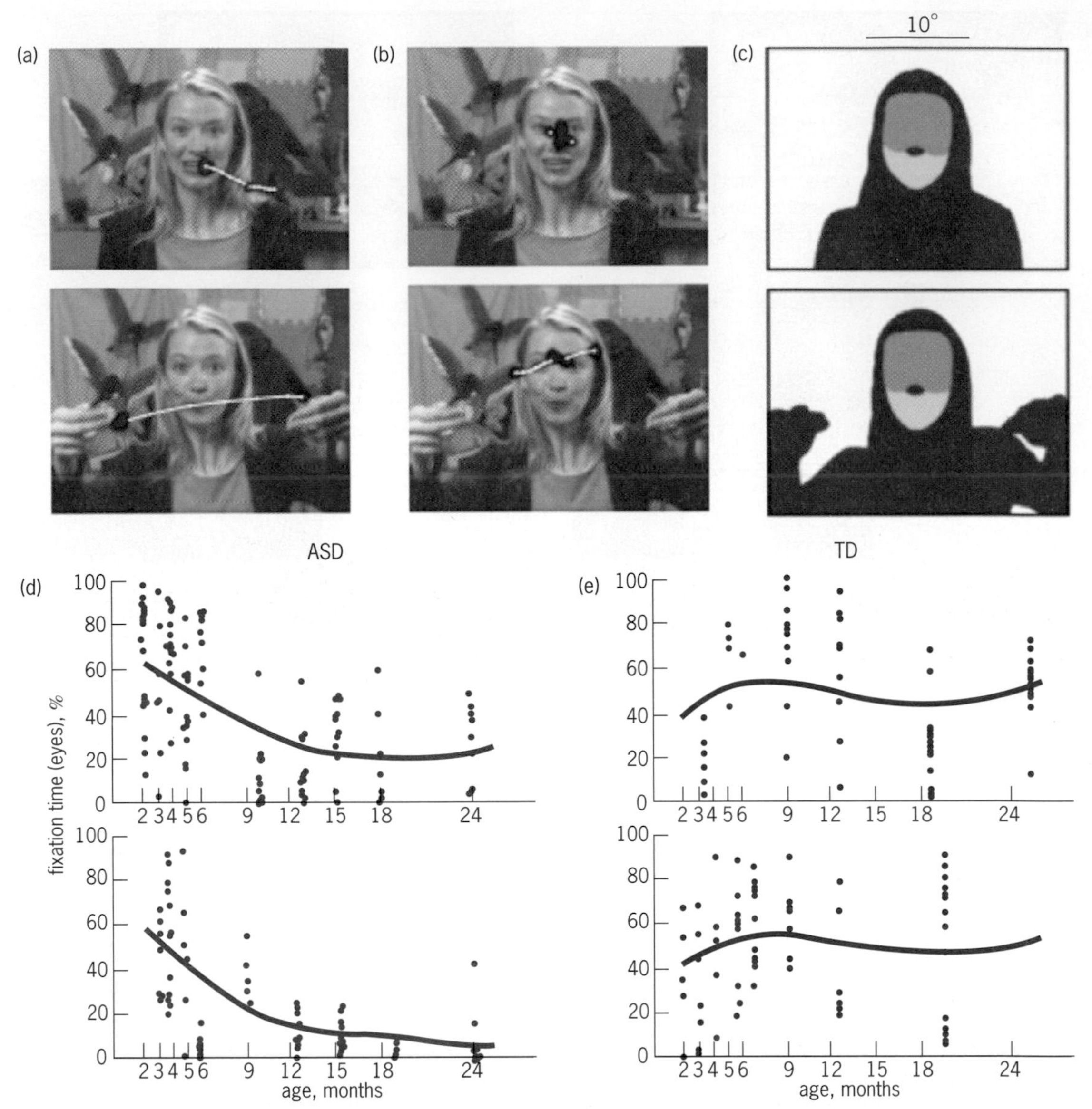

Fig. 2. Examples of eye gaze patterns in infants from 2 to 24 months of age. (*a*) Eye gaze patterns from a 6-month-old infant later diagnosed with autism spectrum disorder (ASD) (*red*). (*b*) Eye gaze patterns from a typically developing 6-month-old infant (*blue*). (*c*) Defined regions of interest, including the eyes, mouth, body, and background/object regions. (*d*) Percentage of fixation time on eyes for two children with ASD (*red*) followed longitudinally from 2 to 24 months of age. (*e*) Percentage of fixation time on eyes for two typically developing (TD) children (*blue*) followed longitudinally from 2 to 24 months of age. (*Reprinted with permission from W. Jones and A. Klin, Attention to eyes is present but in decline in 2–6-month-old infants later diagnosed with autism, Nature, 504:427–431, 2013, DOI:10.1038/nature12715*)

promote more positive outcomes in childhood and adulthood for individuals with ASD.

For background information *see* AUTISM; BRAIN; COGNITION; DEVELOPMENTAL GENETICS; DEVELOPMENTAL PSYCHOLOGY; EYE (VERTEBRATE); NERVOUS SYSTEM (VERTEBRATE); NEUROBIOLOGY; PERCEPTION; SOCIOBIOLOGY; VISION in the McGraw-Hill Encyclopedia of Science & Technology.

Jennifer Moriuchi; Warren Jones; Ami Klin

Bibliography. T. Falck-Ytter, S. Bölte, and G. Gredebäck, Eye tracking in early autism research, *J. Neurodev. Disord.*, 5:28, 2013, DOI:10.1186/1866-1955-5-28; W. Jones and A. Klin, Attention to eyes is present but in decline in 2–6-month-old infants later diagnosed with autism, *Nature*, 504:427–431, 2013, DOI:10.1038/nature12715; W. Jones, K. Carr, and A. Klin, Absence of preferential looking to the eyes of approaching adults predicts level of social disability in 2-year-old toddlers with autism spectrum disorder, *Arch. Gen. Psychiatr.*, 65:946–954, 2008, DOI:10.1001/archpsyc.65.8.946; A. Klin et al., The enactive mind, or actions to cognition: Lessons from autism, *Phil. Trans. R. Soc. Lond. B*, 358:345–360, 2003, DOI:10.1098/rstb.2002.1202; A. Senju and M. H. Johnson, Atypical eye contact in autism: Models, mechanisms, and development, *Neurosci. Biobehav. Rev.*, 33:1204–1214, 2009, DOI:10.1016/j.neubiorev.2009.06.001.

Barnacle reproduction

Barnacle reproduction involves the transfer of sperm from one individual barnacle to another to produce embryos that develop into planktonic larvae for recolonization and dispersal. Barnacles (infraclass Cirripedia) are crustaceans, which constitute a large group of arthropods, including familiar animals such as shrimps, lobsters, and crabs and less familiar (although no less common) ones such as ostracods (seed shrimp), branchiopods (fairy shrimp and water fleas), and copepods. Despite their crustacean heritage, barnacles differ from other crustaceans in two important ways. First, the most familiar barnacles (the stalked and acorn barnacles belonging to the superorder Thoracica) are typically hermaphrodites, whereas a separation of sexes is the norm for other higher taxa of the crustaceans and for primitive barnacle groups. Second, the Cirripedia (class Maxillopoda, subclass Thecostraca) is the only higher taxon of free-living (nonparasitic) crustaceans characterized by sessile (permanently attached) adults. Both of these differences greatly influence barnacle reproductive biology.

Reproductive modes and anatomy. Thoracican barnacles, which include the iconic stalked (Pedunculata) and acorn (Sessilia) types, are typically hermaphroditic, so each individual can produce both eggs and sperm. However, a mix of males and hermaphrodites (androdioecy) has evolved at least twice in both acorn and stalked barnacles. The separation of sexes (dioecy) has also evolved at least four times in stalked barnacles and was likely the ancestral state in the Thoracica. Interestingly, in all cases of androdioecy and dioecy, the male is much smaller than the female. Typically, paired testes that lie within the body (prosoma) produce the sperm, which is stored in large seminal vesicles before copulation. Eggs are produced by extensive paired ovaries that either ramify (branch) into the stalk of stalked barnacles or lie embedded in the mantle tissue lining the shell plates in acorn barnacles.

The sessile lifestyle of barnacles imposes a huge constraint on their reproduction. Because they must copulate somehow (an anachronism of their crustacean ancestry, where copulation is the rule), barnacles possess some of the largest penises, relative to their body size, of any animal (see **illustration**). Their long penises allow them to reach potential partners to mate. The tubular, muscular-walled penis can be extended up to eight body lengths by hydraulic pressure of the hemocoel (an expanded portion of an arthropod's blood system), which is similar to the way that their featherlike feeding legs are extended (see illustration). The penis tip bears an array of sensory hairs that are likely both tactile and chemosensory, so the penis can detect neighboring barnacles that might be willing to mate. The length of the penis can vary greatly throughout the year. It is maximal during the breeding season, but it may disappear almost entirely in the nonbreeding season. The penis length also varies with wave exposure in some intertidal species. Longer penises can reach more mates in quiet water, but they are more difficult to control in turbulent flow. Remarkably, these length differences are not genetic. Instead, they are actually induced by different flow conditions.

Mating behavior, fertilization, and development. Curiously, although thoracican barnacles are typically hermaphroditic, cross-fertilization appears to be obligatory in most species. A few species may self-fertilize when no neighbor is nearby, but the recent discovery that barnacles can capture sperm from the water, in addition to the normal mode of copulation, raises doubts about any reports of self-fertilization that have not been verified by genetic markers.

Although hermaphroditic, an individual barnacle appears to function as either a male or a female during mating. A functional male extends its long penis, sweeping the immediately surrounding substrata looking for a willing partner. The seminal fluid released into the mantle cavity of a functional female gels upon contact with the seawater and apparently induces egg release. Eggs are extruded into paired, thin-walled, porous, and membranous sacks (ovisacs). The activated sperm cells pass through the porous walls of the ovisacs to fertilize the eggs. The malleable ovisacs are then flattened into robust egg lamellae (thin, platelike structures) that lie within the mantle cavity until the larvae are released. The

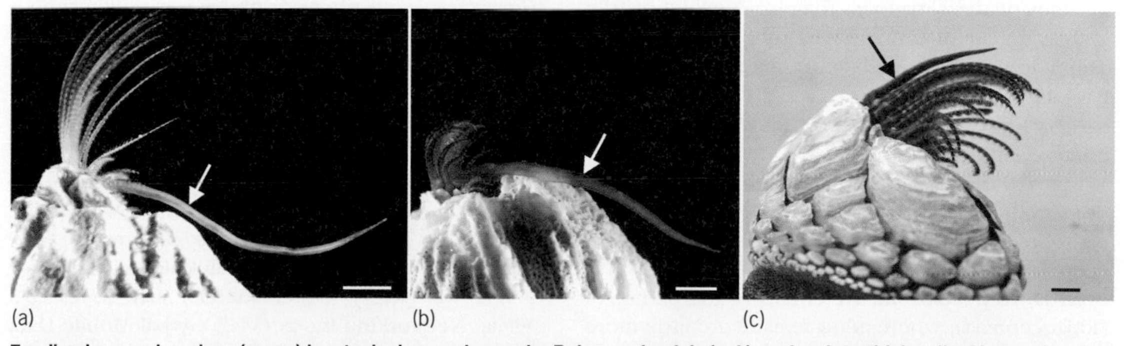

Feeding legs and penises (*arrow*) in a typical acorn barnacle, *Balanus glandula* (*a*, *b*), and an intertidal stalked barnacle, *Pollicipes polymerus* (*c*). Acorn barnacles of the same body size from protected bays (*a*) have longer and more slender legs and penises, whereas those from wave-swept, open-ocean shores (*b*) have shorter and stouter legs and penises. Scale bar: 2 mm (0.08 in.).

feeding larvae are released from the brooding parent when they reach the nauplius stage (which occurs after a few days to a few months, depending on the species and habitat). Next, they pass through six naupliar stages before transforming into the non-feeding cypris larva, which searches for a suitable settlement site. After settling, the cypris larva undergoes a remarkable metamorphosis and transforms into a minute barnacle within 24–36 h.

Egg size, fecundity, and energetics. Egg size varies greatly among the various species of barnacles. It also varies within species, depending on the barnacle's body size, habitat, and geographical latitude. Egg length is generally larger in stalked barnacles [commonly 180–500 μm (0.007–0.02 in.), but up to 1500 μm (0.06 in.) in some scalpellids] than in acorn barnacles [typically 120–300 μm (0.005–0.012 in.), but more than 500 μm (0.02 in.) in some Tetraclitidae].

Fecundity varies enormously, depending on the body size, food supply, season, and habitat. It is likely limited by the volume of the mantle cavity available for brooding. An individual acorn barnacle having a 10-mm (0.4-in.) basal shell diameter may produce 5000 embryos, whereas one of 28-mm (1.12-in.) diameter may produce approximately 75,000 embryos per brood. Brood sizes (embryos per mg dry body weight) also differ greatly among taxa: *Balanus* species, ~3500; *Chthamalus* species, ~2500; *Elminius* species, ~1800; *Tetraclita* species, ~1500; and *Pollicipes* species, ~800. In highly productive environments, an individual barnacle may produce more than one brood per year.

On an annual basis, barnacles devote considerable energy to their gametes (sex cells). In one acorn barnacle (*Balanus glandula*), the energy devoted to egg production (calories per individual per year) equaled the total energy devoted to body (prosoma) growth, shell growth, and molting. If a similar commitment is made to the sperm, then the total energy that a barnacle devotes to its gametes would exceed that devoted to its total body growth.

Reproduction in non-thoracican barnacles. Barnacles as a whole (that is, the Cirripedia) exhibit a stunning diversity of reproductive modes. This diversity is tied to different and highly unusual modes of life. The sexes are separate (dioecy) in the two lineages of the Cirripedia that arose evolutionarily before the great radiation of the hermaphroditic Thoracica (stalked and acorn barnacles). Boring barnacles (Acrothoracica), which were a puzzle to Charles Darwin, can bore into mollusk shells, coral skeletons, other barnacles, and even limestone. Minute dwarf males live around the apertural margin of much larger females and use long penises (up to eight times their body length) to release their sperm deep in the female's mantle cavity. Reproduction is even more peculiar in the endoparasitic barnacles (Rhizocephala), where adult females are little more than a diffuse tissue mass that ramifies throughout the host's body. When an endoparasitic female matures, she produces one or more body outgrowths (externae) that extend outside the host's body. A male cypris larva settles out of the plankton onto an externa and injects cells that transform into sperm-producing cells inside the externa. Curiously, the female produces two different sizes of eggs: small ones that become females and large ones that yield male cypris larvae.

For background information *see* ACROTHORACICA; ANIMAL REPRODUCTION; ASCOTHORACICA; CIRRIPEDIA; CRUSTACEA; FERTILIZATION (ANIMAL); HERMAPHRODITISM; PENIS; RHIZOCEPHALA; THORACICA in the McGraw-Hill Encyclopedia of Science & Technology. A. Richard Palmer

Bibliography. M. Barazandeh et al., Something Darwin didn't know about barnacles: Spermcast mating in a common stalked species, *Proc. R. Soc. B*, 280:20122919, 2013, DOI:10.1098/rspb.2012.2919; M. W. Kelly and E. Sanford, The evolution of mating systems in barnacles, *J. Exp. Mar. Biol. Ecol.*, 392: 37–45, 2010, DOI:10.1016/j.jembe.2010.04.009; C. J. Neufeld and A. R. Palmer, Precisely proportioned: Intertidal barnacles alter penis form to suit coastal wave action, *Proc. R. Soc. B*, 275:1081–1087, 2008, DOI:10.1098/rspb.2007.1760; M. Pérez-Losada, J. T. Høeg, and K. A. Crandall, Deep phylogeny and character evolution in Thecostraca (Crustacea: Maxillopoda), *Int. Comp. Biol.*, 52:430–442, 2012, DOI:10.1093/icb/ics051; Y. Yusa et al., Adaptive evolution of sexual systems in pedunculate barnacles, *Proc. R. Soc. B*, 279:959–966, 2012, DOI:10.1098/rspb.2011.1554.

Big data

The term "big data" has become almost ubiquitous in industrial and academic settings due in large part to the exponential growth of data moving through the Internet. New technological devices, such as smartphones and tablets as well as social networks, blogs, tweets, online archiving, and video streaming, are generating amounts of data that a few years ago were thought impossible to achieve. If, in addition to this, we add the large amount of data constantly generated by more traditional sources, such as financial transactions, point-of-sale information, research data, and computerized automated services (for example, weather sensors), we are talking about a digital universe, which by conservative estimates may reach 2.8 zettabytes [ZB] (2.8×10^{21}; that is, 2.8 multiplied by 10 followed by 20 zeroes) by 2020. As a rough estimate of the size of the data traffic, if we assume that a byte equals a grain of sand in a beach, 2.8×10^{21} ZB is about 57 times the amount of sand of all the beaches in the world. The McKinsey Global Institute in its 2011 report estimates that the data, on a global scale, will increase about 40% yearly. Other sources such as Cisco's Visual Networking Index (VNI) Global Mobile Data Traffic Forecast Update estimated that by 2012, the number of mobile-connected devices exceeded the number of people on Earth. According to Networked

Volume

- The most visible and important aspects of big data are its sheer size and the fact that it increases on a yearly basis, as exemplified by the February 2013 update of Visual Networking Index (VNI) forecast for 2012–2017, where it is expected that the annual global Internet protocol (IP) traffic will reach 1.3 zettabytes per year or 110.3 exabytes per month.
- By 2017, the global mobile data traffic will reach 11.2 exabytes per month (134 exabytes annually); growing 13-fold from 2012 to 2017, and will increase at an incredible pace every year thereafter.

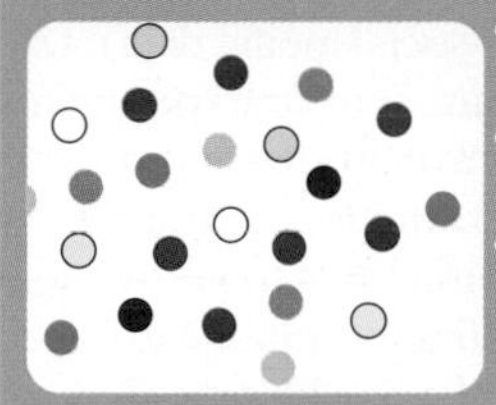

Variety

- Variety shows the mix of unstructured and multistructured data that comprises the volume of information.
- Unstructured data is unorganized data that cannot be used by traditional databases. This type of data is generally "text" as found in tweets, blogs, or Facebook messages.
- Multistructured data includes a wide range of data formats, generally the result of the interactions in social networks, and may include diverse combinations of images, text, and sounds of different types or formats. Data of this type can be generated by data streaming from social media or sensor data from satellites, GPS, or radio-frequency sensors.

Velocity

- Velocity refers to the increasing rate in which data flows into an organization. That is, this property captures data in "motion." When discussing velocity, it is necessary to consider both the incoming and outgoing data because the "tighter" the feedback loop, the greater the competitive advantage. In other words, the faster we can capture, analyze, and transform the data into useful information, the better. Although a vast amount of data is being collected in shorter time frames for later processing, real-time information is critical, but the rate of incoming or outgoing data is not always constant.

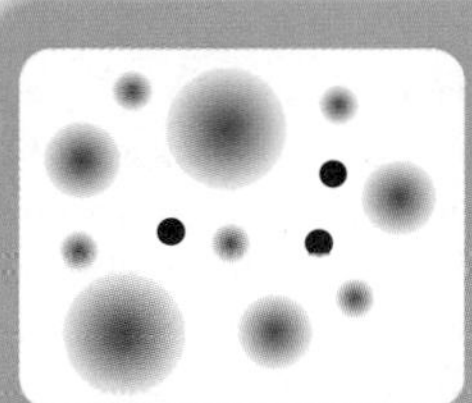

Veracity and Validity

- Big data veracity refers to the biases, noise, and abnormality in data. It refers to the uncertainty caused by data inconsistency and incompleteness, ambiguities, latency, deception, and model approximations. Until recently, the amount of information collected was not analyzed or archived properly. Because of increasing velocities and varieties of data, its flow can be highly inconsistent with periodic peaks.
- Another issue is validity in that the data being stored and mined should be meaningful to the problem or decisions that need to be made.

Value

Value is the usefulness and relevancy of the information extracted from the data for the decision-making process or the advancement of the research problem at hand, as opposed to the practice of collecting large volumes of data for regulatory or archival purposes where the data is not used to its full potential.

Volatility

Value and validity give rise to the issue of volatility. Volatility refers to how long the data is valid and for how long it should be stored. Analysis is required to find when in the future the data will no longer be relevant for a particular organization.

Fig. 1. Properties of big data.

European Software and Services Initiative (NESSI), the amount of mobile data traffic is expected to grow to 10.8 exabytes (10^{18} bytes) per month by 2016. The sheer amount of data generated is profoundly affecting businesses and the technology world; the former is trying to figure out how to make use of that tremendous amount of data for commercial purposes; the latter is trying to figure out the changes, at software and hardware levels, that are necessary to manipulate, transmit, process, protect, and store such large amounts of data. To address these issues it seems necessary to define what we mean by big

		Business Objective	
		Measurement	Experimentation
Data Type	Nontransactional data	Social analytics	Decision science
	Transactional data	Performance management	Data exploration

Fig. 2. Dimensions of big data. (*From Four strategies to capture and create value from big data, Ivey Bus. J., http://iveybusinessjournal.com/topics/strategy/four-strategies-to-capture-and-create-value-from-big-data#.U5IK53amVf6*)

data. However, the notion of "big" in big data is relative, because a data set that seems large today will certainly seem small in the near future. For example, as indicated by George Dyson in his book *Turing's Cathedral*, in 1953, there were only 53 kilobytes of storage in the entire world; it is estimated that globally during 2010 the new data stored by enterprises and individual consumers exceeded 13 exabytes (13×10^{18} bytes). Currently, there is no standardized definition of big data because of its complex and diverse nature. We also need to consider that what is understood as big data may vary depending on the sector under consideration and the software tools available as well as the "average" size of the data set common in a particular industry. However, for the purposes of this article, we will view big data in a dual role: first, as the ongoing collection of techniques to capture data from traditional and digital sources, and second, as the ongoing analysis of massive data sets that facilitate both long-term planning and real-time data-driven decision making. As pointed out by NESSI, big data presents a formidable challenge for current software and service providers whose aim is to ensure the availability and the security of service and data while providing, at the same time, quality of service (QoS). In other words, the goal of managing big data is to achieve full-time availability of services and resources based on data everywhere for anyone and at all times. For these reasons, big data, because of its complexity, diversity, and exponential growth, can be seen as a key enabler for providing resources anytime and anywhere as well as its adaptation to users' demand.

What makes big data big? The main characteristics related to the handling and processing of big data sets are known as the six V's of big data, as shown in **Fig. 1**.

Analysis of big data. The two dimensions used to create and analyze big data are business objectives and data types. In a normal course of operation, organizations capture and process data and store it in databases. This is known as operational or transactional data. In other instances, organizations deal with data that come from sources that are typically unstructured (for example, social-media data). Data can then be used for measurement or experimentation. Any organization, when analyzing and processing its data to generate information, may pose hypotheses and use scientific methods to verify them. Through the use of metrics, organizations can collect and analyze patterns and behavior of data. Combining these two dimensions results in four different strategies for analysis: performance management, data exploration, social analytics, and decision science, as shown in **Fig. 2**.

Big data framework. To manage the immense quantities and varieties of data and ever-faster expectations for analysis, businesses use data frameworks, which are software libraries and their associated algorithms, to enable distributed processing and analysis of big data. There are a growing number of technologies used to aggregate, manipulate, manage, and analyze big data. Among the most popular software are Apache™ Hadoop® and its different embodiments (MapR®, Google File Systems®, Hbase™). Hadoop is an open-source data-processing platform for the distributed systems, first used by Internet giants, including Yahoo®, Google®, and Facebook®.

Big data techniques. To make better use of big data it is necessary to aggregate, manipulate, analyze, and visualize it using several techniques and methods derived from artificial intelligence, applied mathematics, statistics, and economics. A summary of the techniques and methods most commonly used are shown in the **table**.

Big data techniques and technologies*

Techniques		Technologies	
A/B testing	Network analysis	BigTable	Mashup
Association rule learning	Optimization	Business intelligence	Metadata
Classification	Pattern recognition	Cassandra	Nonrelational database
Cluster analysis	Predictive modeling	Cloud computing	R
Crowdsourcing	Regression	Data mart	Relational database
Data fusion and data integration	Sentiment analysis	Data warehouse	Semistructured data
Data mining	Signal processing	Distributed system	SQL
Ensemble learning	Spatial analysis	Dynamo	Stream processing
Genetic algorithms	Statistics	Extract, transform, load (ETL)	Structural data
Machine learning	Supervised learning	Google File System	Unstructured data
Natural language processing	Simulation	Hadoop	Visualization
Neural networks	Time series analysis	HBase	
Unsupervised learning	Visualization	MapReduce	

*From Big data: The next frontier for innovation, competition and productivity, McKinsey Global Institute.

Economic and societal impact of big data. Big data analytics has started to impact all types of organizations, as it has the potential to extract knowledge from big amounts of data from diverse fields and domains, such as personal location data, health care, retail, and manufacturing. New technologies collect massive streams of data over time that makes it possible to extract patterns from that data. New web technologies are empowering citizens to rapidly create new content and easily report, blog, and send messages (such as tweets on Twitter), contributing to the huge amount of data to be stored for analysis.

SWOT (Strengths, Weaknesses, Opportunities, Threats) analysis. The strengths of big data include speed, scalability, mobility, security, flexibility, stability, as well as integration of both structured and unstructured data and real-time monitoring and forecasting of vast amounts of data collected from social media or remote sensors. The opportunities that big data present depend primarily on the increasing amount of new knowledge that can be obtained by different methods and techniques. New knowledge potentially affects every aspect of our lives; for example, the secondary uses of patient data could lead to the discovery of drugs or cures for a wide range of diseases and the prevention of others. Other uses of this new knowledge could lead to better ways of managing utilities (electric, water, and so forth) as well as managing both air and ground traffic. We need to take into account that from a company prospective, knowing more about its users' behavioral patterns or purchasing preferences may lead to a refinement of the marketing strategies used to increase revenues by altering sales pitches or by providing discounts for a particular set of products. However, gathering too much information about consumers may be perceived as a threat to their privacy. In this sense, users may feel that the data collected about them can be used either to disclose information that they may not want to be made public or to discriminate against them based on their preferences, economic, race, or religious profiles. Another aspect of big data that is troublesome for some is the lack of transparency of the data gathering and its potential use. Although we would like to assume fairness on the type of information that is collected, it is very difficult to anticipate how it could be misused. This seems to be a concern for which there is no clear solution. Some have proposed regulatory intervention by the government, while others strive for a self-regulated industry that seeks a balance between the privacy and utility of the data. The jury on this issue is still out.

For background information *see* ALGORITHM; ARTIFICIAL INTELLIGENCE; CLOUD COMPUTING; DATA MINING; DATA WAREHOUSE; DATABASE MANAGEMENT SYSTEM in the McGraw-Hill Encyclopedia of Science & Technology.

Pranshu Gupta; Ramon A. Mata-Toledo

Bibliography. T. H. Davenport, *Big Data at Work: Dispelling the Myths, Uncovering the Opportunities*, Harvard Business Review Press, Boston, 2014; G. Dyson, *Turing's Cathedral: The Origins of the Digital Universe*, Vintage, New York, 2012; V. Mayer-Schönberger and K. Cukier, *Big Data: A Revolution That Will Transform How We Live, Work, and Think*, Houghton Mifflin Harcourt, New York, 2013; J. Needham, *Disruptive Possibilities: How Big Data Changes Everything*, O'Reilly Media, Sebastopol, CA, 2013.

Big data and personalized medicine: the role of pathology

The current advances in molecular technology and data processing have shifted the understanding of cancer to the molecular level, challenging the existing tissue biopsy as the gold standard for cancer diagnosis. The sheer volume of data generated by these new techniques lends the name "big data" to this phenomenon. Conversely, the actual changes that are detected are at the nucleotide and single molecule level, and they are being assigned to individual patients, giving rise to the term "personalized medicine." The pathologist must in turn adopt a personalized approach—tacit recognition of the fact that current morphologic subtypes, for example, large or small cell carcinoma of the lung, are no longer sufficient for selection of therapy (see **illustration**).

Toward personalized pathology. This change of paradigm results from the convergence of DNA- and protein-detecting technologies with quantum leaps in computer power, bandwidth, and cloud computing. The former generates big data, whereas the latter enables the processing of big data. The Human Genome Project consumed $3 billion (U.S. dollars) and 13 years, and it was completed in 2003. It illustrated only the protein-coding portion of the DNA, just 1–2% of the human genome. The subsequent Encyclopedia of DNA Elements (ENCODE) project took 9 years, serving to identify many of the remaining sequences as regulatory factors. Personalized medicine encompasses the concept and the intent of finding the uniqueness in each patient's pathophysiology in order to tailor treatment to each individual.

Molecular technology is a powerful addition to histopathology. Microscopic examination of a tissue section by a skilled pathologist has survived the test of time, but that time is coming to an end. When the appearances of a group of cells have changed to an extent that can be visualized with a light microscope, disruption of the cellular pathways is extensive and irreversible. In situ hybridization and immunohistochemistry can sometimes detect cumulative changes of nucleotide sequences (mutations) or protein expressions that antedate typical morphologic changes. However, measurement of the real amount of such changes is semiquantitative at best, pending the arrival of true quantification. Molecular techniques offer advantages in this regard. Sequencing and array technologies can generate data in minute detail and large quantity, detecting

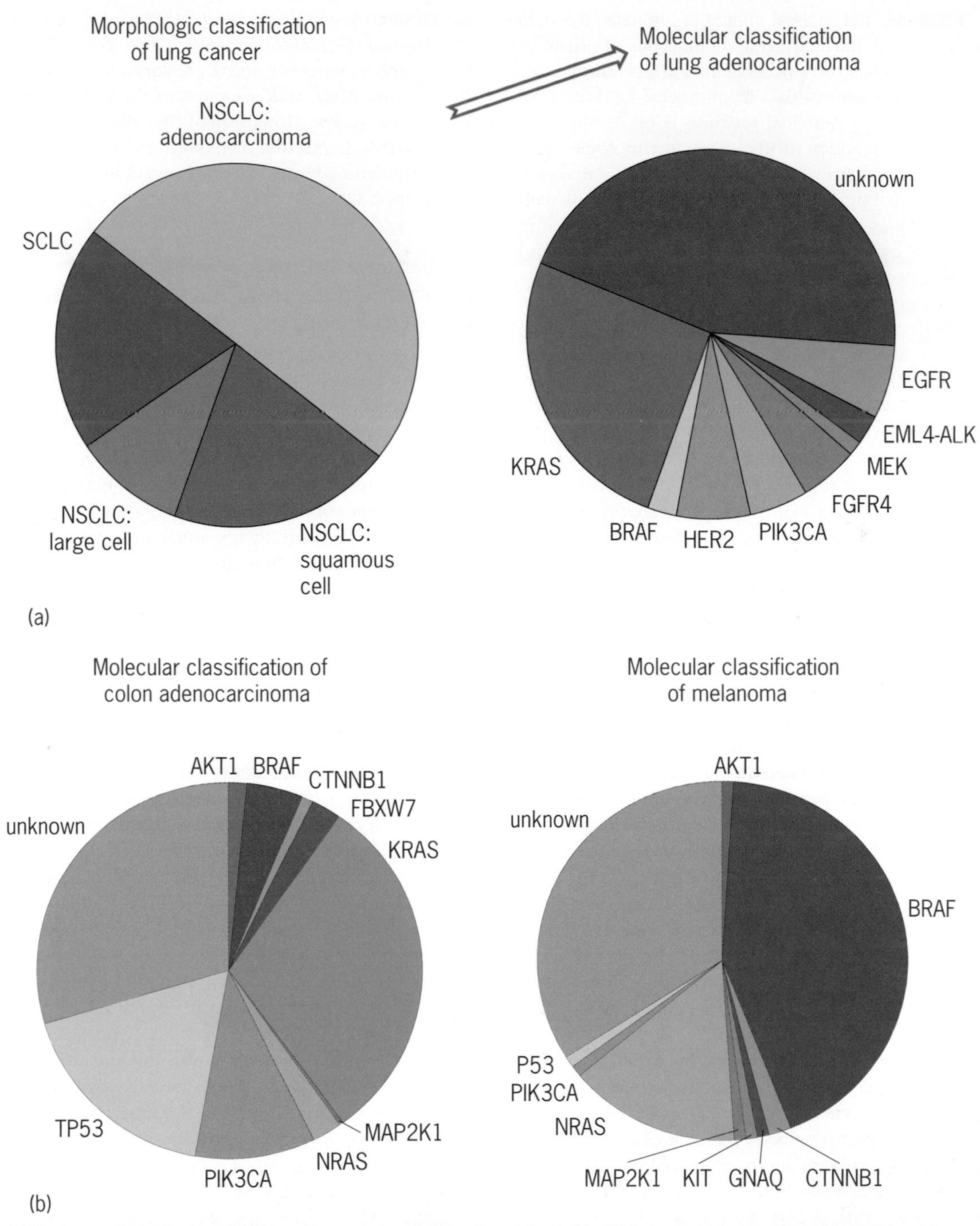

(*a*) Basic morphologic classification of lung cancer (*left*) is of no value in identifying responders to therapies that target defined molecular subtypes (*right*). SCLC, small cell lung cancer. NSCLC, non-small cell lung cancer. (*b*) Other major tumor types show comparable ranges of molecular alterations, many of which have led to the generation of targeted therapies (Table 2). (*Data from COSMIC: Catalogue of Somatic Mutations in Cancer; http://cancer.sanger.ac.uk/cancergenome/projects/cell_lines/*)

mutations, insertions, deletions, RNA alterations, noncoding DNA, and DNA methylation patterns that may predate morphologic changes. Such changes also may identify molecules that have the potential to serve as targets for new drugs, which currently is a powerful driving force in the pharmaceutical industry.

However, there are attendant disadvantages. Molecular methods are performed using tissue extracts, in which the exact nature and number of contributing (cancer) cells is only inferred. It is, therefore, difficult to separate the signatures of abnormal cells from admixed normal elements, limiting the practical detection sensitivity to samples containing at least 10% or 20% of cells having a particular mutation.

The study of cancer exemplifies the progress in molecular pathology. The diagnosis of cancer is

TABLE 1. Typical tumors and their driver gene mutations

Cancer type	Number of mutations
Colorectal	166
Lung (small cell)	163
Lung (non-small cell)	147
Melanoma	135
Esophageal (squamous)	79
Non-Hodgkin lymphoma	74
Head and neck	66
Esophageal (adeno)	57
Gastric	53
Endometrial	49
Pancreatic adenocarcinoma	45
Ovarian (serous)	42
Prostate	41
Hepatocellular	39
Glioblastoma	35
Breast	33
Chronic lymphocytic leukemia	12
Neuroblastoma	12
Acute lymphocytic leukemia	8
Medulloblastoma	8
Rhabdoid	4

central to the work of the pathologist. To date, molecular and genetic findings have extended our knowledge well beyond the current morphological classifications of cancer (see illustration). Literally thousands of alterations in the so-called cancer genome have been detected, revealing approximately 140 genes that, when mutated, can drive tumorigenesis (tumor formation). Of these, thus far, 71 are tumor suppressor genes and 54 are oncogenes. A typical tumor contains a few to more than a dozen of these "driver gene" mutations (**Table 1**), which can be assigned to a dozen signaling pathways that regulate cell fate, cell survival, or maintenance of the genome. Some cancer types carry relatively few mutations, for example, medulloblastomas, testicular germ cell tumors, and acute leukemias, whereas others, such as lung and colon cancers and melanomas, have occasionally as many as 100,000 mutations.

In a recent report of The Cancer Genome Atlas (TCGA) project, somatic variants across 3281 specimens of 12 tumor types were analyzed, identifying 127 significantly mutated genes (SMGs). Most SMGs were related to clonal sublineages of cancer cells, reflecting an extreme heterogeneity of expression. A subsequent study analyzed somatic point mutations from 4742 human cancers and their matched normal-tissue samples across 21 cancer types. Most known cancer genes were identified in these tumor types, as well as 33 genes that were not previously known to be significantly mutated in cancer. It was estimated that 600–5000 samples per tumor type must be analyzed in order to obtain a nearly complete profile of the genomic alterations for that tumor type.

Today, there is a clear consensus that it takes a very long time (decades) to develop metastatic cancer, and the cancer only becomes clinically apparent a few years or months before death. It is likely that successful detection at a much earlier phase in the evolution of the cancer may lead to a significant reduction in cancer deaths.

Pathological diagnosis for new therapy. In recent years, the proteins altered by driver mutations have become targets for successful anticancer drug development. These drugs include small-molecule inhibitor drugs that target kinases or monoclonal antibodies that target cell surface receptors. So far, at least 39 targeted therapy drugs for cancer have been approved by the U.S. Food and Drug Administration and another 45 are in clinical trials (**Table 2**). The demand for pathologists to employ suitable detection techniques, variously termed "predictive markers," "companion diagnostics," or "advanced personalized diagnostics," is inexorable.

A single example suffices, although new instances occur almost daily. In the early 2000s, human epidermal growth factor receptor 2 (HER2) emerged as the target for therapy with trastuzumab (Herceptin®; anti-HER2 human monoclonal antibody). With HER2-negative tumors, trastuzumab has no beneficial effect, but only causes toxicity. Thus, it is now more meaningful for pathologists to inform the oncologists of HER2 status than to subclassify the various recognized histologic subtypes. Lung cancer provides a dramatic example (see illustration). Only a small fraction of lung cancer patients have epidermal growth factor receptor (EGFR) gene mutations or anaplastic lymphoma kinase (ALK) gene translocations, and only these patients will respond to the appropriate drug. Morphology alone is unable to identify these responders.

The demands of personalized diagnostics are similar to those of high-quality immunohistochemistry. Consistent documented sample preparation is critical. Extraction of DNA and RNA from formalin-fixed paraffin-embedded (FFPE) tissue samples, although less ideal than from fresh samples, is increasingly common because FFPE tissue is often the only option. Changes that are initiated at the DNA and RNA levels are effected by the encoded proteins and often are within the power of detection by current technology [for example, immunohistochemistry or fluorescence in situ hybridization (FISH)]. There is another major, even essential, reason for the involvement of pathologists. Driver gene mutations are closely related to the ongoing evolution of subclones, with marked heterogeneity in many tumors. Understanding this heterogeneity will challenge pathologists to develop methods that integrate molecular and morphologic findings, whether by immunohistochemistry, in situ hybridization, or other molecular methods that retain a morphologic reference at the cellular level.

Conclusions. The traditional task of the pathologist is to assist physicians in making the correct diagnosis of diseases at the earliest possible stage, leading to the most effective treatment. In this respect, the traditional tissue diagnosis is just a tool, and one that increasingly appears inadequate. Meanwhile, molecular and genomic techniques have demonstrated clear value in the diagnosis of individual patients and in the treatment tailored specifically to these patients. Both active adaption and innovation are required.

TABLE 2. Targeted therapies and general modes of action*

Agent	Target	Application
(A) Block enzymes and growth factor receptors involved in cancer cell proliferation		
Imatinib mesylate (Gleevec®)	Tyrosine kinase enzymes	Gastrointestinal stromal tumor (GIST), leukemia, dermatofibrosarcoma protuberans, myelodysplastic disorders, systemic mastocytosis
Dasatinib (Sprycel®)	Tyrosine kinase	Chronic myeloid leukemia (CML) or acute lymphoblastic leukemia
Nilotinib (Tasigna®)	Tyrosine kinase	CML
Bosutinib (Bosulif®)	Tyrosine kinase	CML
Trastuzumab (Herceptin®)	HER2	Breast cancer, some gastric or gastroesophageal junction adenocarcinoma
Pertuzumab (Perjeta™)	HER2	Breast cancer
Lapatinib (Tykerb®)	HER2	Breast cancer
Gefitinib (Iressa®)	EGFR	Non-small cell lung cancer
Erlotinib (Tarceva®)	EGFR	Non-small cell lung cancer, pancreatic cancer
Cetuximab (Erbitux®)	EGFR	Head and neck or colorectal cancer
Panitumumab (Vectibix®)	EGFR	Colon cancer
Temsirolimus (Torisel®)	Serine/threonine kinase	Renal cell carcinoma
Everolimus (Afinitor®)	mTOR kinase	Kidney cancer
Vandetanib (Caprelsa®)	EGFR, VEGF, RET	Medullary thyroid cancer
Vemurafenib (Zelboraf®)	Serine/threonine kinase BRAF	Melanoma
Crizotinib (Xalkori®)	EML4-ALK	Non-small cell lung cancer
(B) Modify the function of regulatory proteins		
Vorinostat (Zolinza®)	Histone deacetylases (HDACs)	T-cell targeted therapies, block enzymes and growth factor receptors, cutaneous T-cell lymphoma (CTCL)
Romidepsin (Istodax®)	HDACs	CTCL
Bexarotene (Targretin®)	Retinoid X receptors	CTCL
Alitretinoin (Panretin®)	Retinoic acid receptors and retinoid X receptors	AIDS-related Kaposi sarcoma
Tretinoin (Vesanoid®)	Retinoic acid receptors	Acute promyelocytic leukemia
(C) Induce cancer cells to undergo apoptosis (programmed cell death)		
Bortezomib (Velcade®)	Proteasome	Myeloma and mantle cell lymphoma
Carfilzomib (Kyprolis™)	Proteasome	Multiple myeloma
Pralatrexate (Folotyn®)	RFC-1	T-cell lymphoma
Bevacizumab (Avastin®)	VEGF	Glioblastoma, non-small cell lung cancer, colorectal cancer, and metastatic kidney cancer
Ziv-aflibercept (Zaltrap®)	VEGF	Colorectal cancer
Sorafenib (Nexavar®)	VEGF	Renal cell carcinoma and some cases of hepatocellular carcinoma
Sunitinib (Sutent®)	VEGF	Renal cell carcinoma, GIST
Pazopanib (Votrient®)	VEGFR, c-KIT PDGFR	Renal cell carcinoma and soft tissue sarcoma
Regorafenib (Stivarga®)	VEGFR angiopoietin-1 R (TIE2), PDGFR, RET, c-KIT, RAF	Colorectal cancer
Cabozantinib (Cometriq™)	VEGFR, RET, MET, TRKB, TIE2	Medullary thyroid cancer
(D) Assist the immune system to destroy cancer cells		
Rituximab (Rituxan®)	CD20	B-cell non-Hodgkin lymphoma and chronic lymphocytic leukemia (CLL)
Alemtuzumab (Campath®)	CD52	B-cell CLL
Ofatumumab (Arzerra®)	CD20	CLL
Ipilimumab (Yervoy™)	Cytotoxic T-lymphocyte-associated antigen-4 (CTLA-4)	Melanoma
(E) Monoclonal antibodies that deliver toxic molecules to cancer cells		
Tositumomab and ^{131}I-Tositumomab (Bexxar®)	CD20	B-cell non-Hodgkin lymphoma
Ibritumomab tiuxetan (Zevalin®)	CD20	B-cell non-Hodgkin lymphoma
Denileukin diftitox (Ontak®)	IL-2 receptors	CTCL
Brentuximab vedotin (Adcetris®)	CD30	Systemic anaplastic large cell lymphoma and Hodgkin lymphoma

*Pathologists may expect to be responsible for identifying these target molecules with immunohistochemistry, in situ hybridization, polymerase chain reaction (PCR), DNA sequencing, or other molecular techniques that are not routinely available.

The pathology community has been alerted and several initiatives are being pursued.

Personalized medicine begets personalized pathology, or it should. The choice, and the onus of choosing, is in the hands of the pathology community.

For background information *see* CANCER (MEDICINE); CLINICAL PATHOLOGY; CLOUD COMPUTING; DISEASE; DRUG DELIVERY SYSTEMS; GENE; GENETICS; GENOMICS; MEDICINE; MOLECULAR PATHOLOGY; ONCOLOGY; PHARMACOLOGY; PUBLIC HEALTH in the McGraw-Hill Encyclopedia of Science & Technology.

Clive R. Taylor; Jiang Gu

Bibliography. J. Gu and C. R. Taylor, Practicing pathology in the era of big data and personalized medicine, *Appl. Immunohistochem. Mol. Morphol.*, 22:1-9, 2014, DOI:10.1097/PAI.0000000000000022; S. M. Hewitt et al., *Quality Assurance for Design Control and Implementation of Immunohistochemistry Assays: Approved Guidelines*, 2d ed., Clinical and Laboratory Standards Institute, Wayne, PA, 2011; F. Janku, D. J. Stewart, and R. Kurzrock, Targeted therapy in non-small-cell lung cancer—Is it becoming a reality?, *Nat. Rev. Clin. Oncol.*, 7:401-414, 2010, DOI:10.1038/nrclinonc.2010.64; S. Jones et al., Comparative lesion sequencing provides insights into tumor evolution, *Proc. Natl. Acad. Sci. USA*, 105:4283-4288, 2008, DOI:10.1073/pnas.0712345105; C. Kandoth et al., Mutational landscape and significance across 12 major cancer types, *Nature*, 502:333-339, 2013, DOI:10.1038/nature12634; M. S. Lawrence et al., Discovery and saturation analysis of cancer genes across 21 tumour types, *Nature*, 505:495-501, 2014, DOI:10.1038/nature12912; C. R. Taylor, From microscopy to whole slide digital images: A century and a half of images, *Appl. Immunohistochem. Mol. Morphol.*, 19:491-493, 2011, DOI:10.1097/PAI.0b013e318229ffd6; C. R. Taylor, Quantifiable internal reference standards for immunohistochemistry: The measurement of quantity by weight, *Appl. Immunohistochem. Mol. Morphol.*, 14:253-259, 2006; C. R. Taylor and R. J. Cote (eds.), *Immunomicroscopy: A Diagnostic Tool for the Surgical Pathologist*, 3d ed., Elsevier/Saunders, Philadelphia, 2006; C. R. Taylor and R. M. Levenson, Quantification of immunohistochemistry—issues concerning methods, utility and semiquantitative assessment II, *Histopathology*, 49:411-424, 2006, DOI:10.1111/j.1365-2559.2006.02513.x; The Cancer Genome Atlas Network, Comprehensive molecular characterization of human colon and rectal cancer, *Nature*, 487:330-337, 2012, DOI:10.1038/nature11252; P. J. Tonellato et al., A national agenda for the future of pathology in personalized medicine: Report of the proceedings of a meeting at the Banbury Conference Center on genome-era pathology, precision diagnostics, and preemptive care: A stakeholder summit, *Am. J. Clin. Pathol.*, 135:668-672, 2011, DOI:10.1309/AJCP9GDNLWB4GACI; B. Vogelstein et al., Cancer genome landscapes, *Science*, 339:1546-1558, 2013, DOI:10.1126/science.1235122.

Blue stain in wood

Millions of lumber dollars are lost every year because of the discoloration of freshly cut logs resulting from a fungal infection known as blue stain. The discoloration can be caused by a number of different wood-staining fungi belonging to several genera. Collectively, blue stain fungi (**Fig. 1**) are microscopic fungi that grow in the form of multicellular filaments, called hyphae, which infect the sapwood of freshly cut trees. A connected network of tubular branching hyphae, called a mycelium, will multiply with genetically identical nuclei and is considered a single organism. This is referred to as a colony.

There are thousands of known species of fungi. They derive energy not through photosynthesis but from the organic matter in which they live. Typically, fungi secrete hydrolytic enzymes, mainly from their hyphal tips. These enzymes degrade complex biopolymers, such as starch, cellulose, lignin, and other available organic compounds, into simpler substances, which can be absorbed by the hyphae. Although fungi grow on organic matter throughout nature, their presence is visible to the unaided eye only when fungal colonies grow.

The color of the stain typically varies from blue to gray to brown and black, depending on the infecting organism. The color can also depend on the tree species and moisture content, and it can include various shades of yellow, orange, purple, and red. The blue stain does not degrade the wood, so it has no effect on the strength properties of the wood. In general, the fungus is usually dead by the time that a log arrives at a sawmill. Any live fungi that may remain are killed during the kiln drying process. The fungus is not dangerous to humans.

In a cross section of wood infected with blue stain (**Fig. 2**), the discoloration is usually oriented radially, which corresponds to the direction of the wood ray cells. It can also appear as specks or streaks and infects both hardwoods and softwoods.

Conditions for attack. Temperature, oxygen, moisture, and nutrients are the main variables that determine whether blue stain fungi will attack. The optimal temperature range for attack is between 21°C (70°F) and 32°C (90°F), but blue stain can attack at any temperature between 4°C (39°F) and

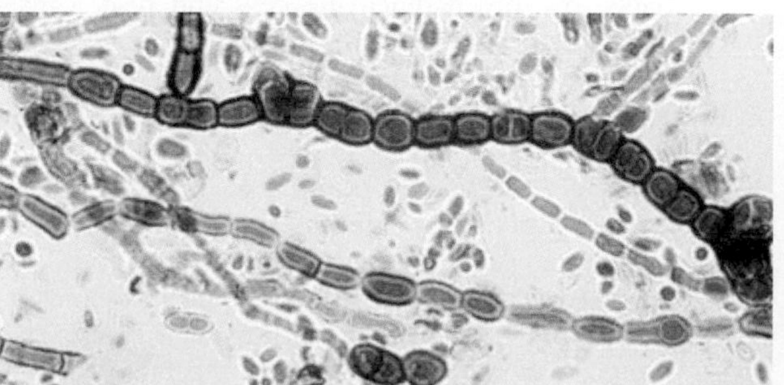
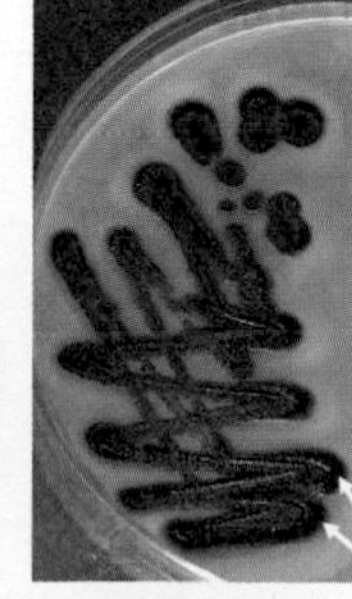

Fig. 1. A scanning electron image of a blue stain fungus (*left*) and a colony of blue stain fungi growing on a lab agar plate (*right*).

Fig. 2. Blue stain in the end grain of two conifer logs: winter (*left*); summer (*right*).

55°C (131°F). The moisture content of the wood is generally 20% or higher. In northern regions of the United States, blue stain is likely to be a problem between April and November. In southern regions of the United States, the problem can occur at any time of the year.

Some lumber standards limit the amount of blue stain that is permitted on structural lumber. Although the stain does not result in any reduction in structural strength, its appearance is not acceptable to many builders, even if the wood is going to be covered. Lumber that will be used to make furniture is usually rejected if it is infected with blue stain. This is especially important in very light woods, such as pine, fir, spruce, maple, basswood, aspen, ash, birch, and oak. This is also true for lumber that is selected for flooring.

Blue stain spores. Blue stain spores that are floating in the air rarely enter wood. Instead, the spores mainly infect conifers by tagging along with bark beetles (including mountain pine beetles). After the beetles penetrate the bark of the tree, they invade the sapwood, and the blue stain spores attain entry at the same time. For example, *Grosmannia clavigera*, which is a notable blue stain fungus, can spread to ponderosa pine, lodgepole pine, and Douglas fir from mountain pine beetles. The beetles are able to lay eggs while avoiding the defenses of the tree, and they infect the wood with the staining fungus.

Other blue stain fungi, primarily from genera such as *Ophiostoma* and *Ceratocystis*, rapidly invade freshly cut wood. Hyphae are found in all xylem cells, but ray parenchyma cells and resin canals are preferentially colonized, eventually blocking the water-conducting columns of the tree. The discoloration is a result of melanin-like pigments in the hyphae.

Aureobasidium pullulans, also called the "black yeast," produces a green melanin that turns black over time. Colonies are fast growing, smooth, and covered with slimy masses of conidia (asexual reproductive spores) that are brown or black. This fungus can grow on or through a painted surface. Fungi growing on the surface of the paint get their nutrients when passing through the paint film.

Prevention and control. The best means to prevent blue stain attack include the avoidance of any unfavorable conditions. For example, the wood should be kept dry, the wood should be protected from mountain pine beetles, and temperatures should be maintained above or below the ideal growing temperatures. Temperatures higher than 150°C (302°F) are lethal to blue stain fungi. In some cases, preventive chemicals may be employed.

It generally takes three days for the blue stain fungi to colonize. Therefore, any prevention methods need to be applied within this time frame. Beyond three days, the staining may have already occurred.

Successful control of the blue stain fungi can be accomplished by treating the ends of freshly cut timber with a nonpigmented strain of *Ophiostoma piliferum*. This is a very aggressive stain fungus that is colorless because it lacks the melanin-like compounds responsible for the discoloration. It competes with other blue stain fungi and does not allow them to colonize.

It is also possible to coat the ends of freshly cut logs to slow down the drying process, and this can prevent the blue stain fungi from colonizing. In addition, freshly cut logs can be kept wet by spraying with water. The high water content prevents the growth of the blue stain fungus by reducing the available oxygen supply.

Another preventive method is to put freshly cut logs in cold storage at a temperature below that needed for the fungus to survive. Although this is costly, it prevents the staining of very valuable logs that might be used for veneer production, high-end furniture, or musical instruments. In this regard, logging is often done in the cold winter months to prevent the growth of the stain fungi. However, if logging is done in warmer months, the logs should be transported to a sawmill to be cut into lumber and dried before the stain fungi are allowed to colonize.

It is also important to keep apart the logs of different species. Because conifers are more susceptible to attack by blue stain fungi, conifer logs should not be stored with other logs. This is especially true if the conifer logs are infected with the mountain pine beetle. Controlling the mountain pine beetle is the best way to control staining in softwoods.

There are several commercial fungicides that can be sprayed on the logs to prevent the colonization of the blue stain fungi. Although these fungicides are toxic to the fungi, they are also toxic to the environment and to animals, so appropriate safety measures must be taken. The chemical must be applied to the log ends and to any debarked portions of the logs within 24 h for maximum effectiveness.

Finally, colonizing white rot fungi are being employed as biological control agents to prevent blue staining in wood used for pulp and paper production. Fungi, including *Phlebiopsis gigantea* (previously referred to as *Peniophora gigantea*), have been used to treat pulpwood during shipping and storage. The treatment prevents stain fungi from attacking and causes beneficial changes in the wood that can help to facilitate the pulping process (for example, it reduces the required energy use during mechanical pulp production and it improves various paper qualities).

For background information *see* CELLULOSE; FUNGAL ECOLOGY; FUNGI; LIGNIN; LIGNIN-DEGRADING FUNGI; LUMBER; PLANT PATHOLOGY; TREE DISEASES; WOOD ANATOMY; WOOD DEGRADATION; WOOD ENGINEERING DESIGN; WOOD PRODUCTS; WOOD PROPERTIES; WOODWORKING; XYLEM in the McGraw-Hill Encyclopedia of Science & Technology.

Roger M. Rowell

Bibliography. C. J. Behrendt, R. A. Blanchette, and R. L. Farrell, Biological control of blue-stain fungi in wood, *Phytopathology*, 85(1):92–97, 1995, DOI:10.1094/Phyto-85-92; L. Safranyik and A. L. Carroll, The biology and epidemiology of the mountain pine beetle in lodgepole pine forests, pp. 3–66, in L. Safranyik and W. R. Wilson (eds.), *The Mountain Pine Beetle: A Synthesis of Biology, Management, and Impacts on Lodgepole Pine*, Natural Resources Canada, Canadian Forest Service, Pacific Forestry Centre, Victoria, British Columbia, 2006; O. Schmidt, *Wood and Tree Fungi: Biology, Damage, Protection, and Use*, Springer, Berlin, Germany, 2006; M. J. Wingfield, K. A. Seifert, and J. F. Webber (eds.), *Ceratocystis and Ophiostoma: Taxonomy, Ecology and Pathogenicity*, American Phytopathological Society, St. Paul, MN, 1993; R. A. Zabel and J. J. Morrello, *Wood Microbiology: Decay and Its Prevention*, Academic Press, San Diego, CA, 1992.

Capsule robots for endoscopy

Tens of thousands of people are affected daily by the hardships of gastrointestinal (GI) cancer. Doctors and biomedical engineers have been collaborating for years to improve the method of distribution, identification, and treatment of cancer in the GI tract. Recent advancements in miniaturization technology capabilities and increased efficiency in batteries, actuators, and electronics have brought researchers closer to clinical implementations of novel designs that address the complications of conventional colonoscopies. Limited access to the GI tract using flexible endoscopes and restricted maneuverability of the instrument within the colon pose the main problem, resulting in patient reluctance to undergo endoscopy.

The most threatening GI cancer is colorectal cancer (CRC), cancer of either the colon or rectum. With appropriate screening, CRC is preventable, treatable, and curable. Robotics in wireless capsule endoscopy (WCE) is raising the bar as the soon-to-be most effective method of CRC screening because of its patient tolerance and increased diagnostic and therapeutic capabilities. In the near future, a miniature pill-sized swimming, walking, and biopsying robot traversing the unconquered depths of the GI tract could be the solution leading to mass distribution of CRC screenings.

Need. Colorectal cancer is prevalent in men and women of all racial and ethnic groups and is life threatening. According to the National Cancer Institute, CRC is the third most common cancer in men and the second most common cancer in women worldwide. It is also the second deadliest cancer in the United States and fourth deadliest in the world, accounting for approximately 608,000 deaths of the 1.14 million documented cases annually. But with proper treatment, it is survivable. Most CRC is caused by an abnormal cell growth, known as a polyp, on the inner lining of the large intestine. If a screening, known as a colonoscopy, reveals polyps in their early stages of growth, CRC can be prevented, treated, and even cured.

CRC screening saves lives. If a polyp is identified early and appropriately treated, studies show that there is a 5-year survival rate of almost 90%. Yet if it is not identified early, the 5-year survival rate drops to 5%. As a result, routine screenings are recommended for people over 50 years of age. It has been documented that CRC screening rates are on the rise in the United States, but 22 million people are still bypassing recommended screenings and therefore are at risk. This disregard can be partially accredited to the discomfort and indignation that patients feel during a colonoscopy.

Flexible endoscopy. A conventional colonoscopy is performed using a colonoscope inserted through the anus of the patient. A colonoscope is a long flexible tube with a steerable endoscopic camera on the head. It is manually controlled from the proximal end by the endoscopist as it is navigated through the GI tract. In addition, a colonoscope has a working channel for inserting other medical instruments. During a colonoscopy if a polyp is identified, forceps are inserted through this channel to biopsy the tissue or remove the polyp.

A colonoscopy starts with bowel preparation and patient sedation. Bowel preparation is essential to clear the GI tract, maximizing visual inspection by the colonoscope. It requires the patient to sustain a clear liquid diet and to take a strong laxative the day before the procedure, which is unpleasant. Sedation is not necessary, but is used to increase patient tolerance and often results in increased technical success rates.

Overall, GI endoscopy has low complication rates. Because of patients' possible indignation, pain, and fear of sedation side effects, distribution remains the greatest clinical challenge for CRC screenings. In a colonoscopy, the greatest pain for the patient and difficulty for the doctor arises in the maneuvering of the endoscope. The colonoscope must traverse the curves of the colon, which applies pressure on its walls causing pain and frequently results in a looping of the scope, which inhibits further movement through the colon. Special maneuvers can be performed to minimize looping, making colonoscopy a

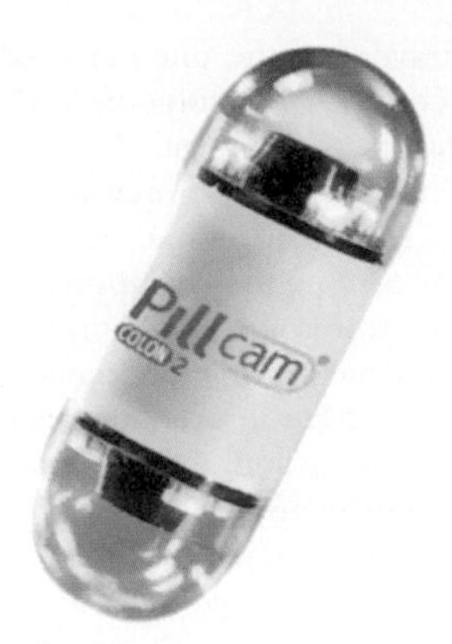

Fig. 1. PillCam® COLON 2 capsule. (*Courtesy of Given Imaging Ltd.*)

procedure that requires a great degree of technical skill and experience to safely perform. Another limitation is that biopsy samples can only be determined cancerous after being processed in a lab once the sample is removed from the patient's GI tract. Therefore if the sample is positive, the original biopsy location must be relocated in a second colonoscopy and then treated. Other possible complications include perforation, bleeding, and infection.

Since the introduction of the colonoscope in the late 1950s, the main technological improvements have been in image resolution and video technology, while the basic features of the instrument have progressed very little. Some alternative methods for CRC screening have been developed, but have ultimately failed to provide a better solution to conventional colonoscopy. The most notable of these is virtual endoscopy, an approach using an MRI scan to inspect the GI tract.

Wireless capsule endoscopy. In an effort to provide a more patient-friendly method of CRC screenings, advancements in technology have led to the invention and implementation of wireless capsule endoscopy (WCE). In this procedure, the patient swallows a miniature pill-sized camera that passively moves through the digestive system, providing images of the inside of the GI tract along the way. This gives endoscopists the ability to screen patients without the need for intubation or sedation, essentially eliminating most patient discomfort.

In 2001, Given Imaging Ltd. was the first company to receive approval for a capsule endoscope used in the small bowel (intestine). A wireless capsule endoscope consists of an external biocompatible shell, typically the size of a large antibiotic pill (11 mm in diameter and 26 mm in length), a camera, a control and communication unit, and an energy source. Antennas are placed on the patient to receive wireless data, and the camera images are stored on a device for the doctor to review on completion.

WCE has become the gold standard in endoscopy for the small intestine, where biopsies and active locomotion are of little importance for diagnosis. As for the large intestine, in February 2014, the FDA approved the Given Imaging PillCam® COLON (**Fig. 1**) for incomplete colonoscopies (that is, if the endoscopist fails to visualize the entire colon via standard colonoscopy, the patient can be referred for WCE examination).

During a colonoscopy, it is ideal for the endoscopist to be able to maneuver the camera view in order to further explore suspicious lesions. Current capsules lack this capability since they only move through peristalsis (the rhythmic wave caused by muscular contraction of the intestine). In addition, capsules also lack the ability to interact with the GI tissue.

In order to replace colonoscopy, future capsules must be able to move under the control of an external operator and be able to collect biopsy samples and clip polyps, all key components to identifying and treating CRC (**Fig. 2**).

Robotic WCE. To overcome these limitations, researchers are currently pursuing robotics in an effort to revolutionize CRC screenings.

To address the issue of passive locomotion, miniature onboard actuators and mechanisms are being explored to give the endoscopist control over the robot's movement in the bowels. Some innovative mechanisms include legged locomotion systems that move similar to an earthworm, and actuating systems using magnetic repulsion-attraction to propel the capsule forward. External locomotion is also a possibility using a large external magnetic field to apply a force to the capsule containing an onboard permanent magnet.

Robotic mechanisms are also necessary to address the current capsule incapability for tissue interaction. Some designs have been proposed and tested for biopsy sampling, while polyp clipping has yet to be tackled. For biopsy sampling, only 1 mm^3 of tissue must be collected, albeit this still presents a challenge for bioengineers because of the miniature size restriction of the capsule. The most notable capsule design thus far consists of a rotational tissue-cutting razor attached to a magnetic torsional spring and block system. Once activated, the razor is released

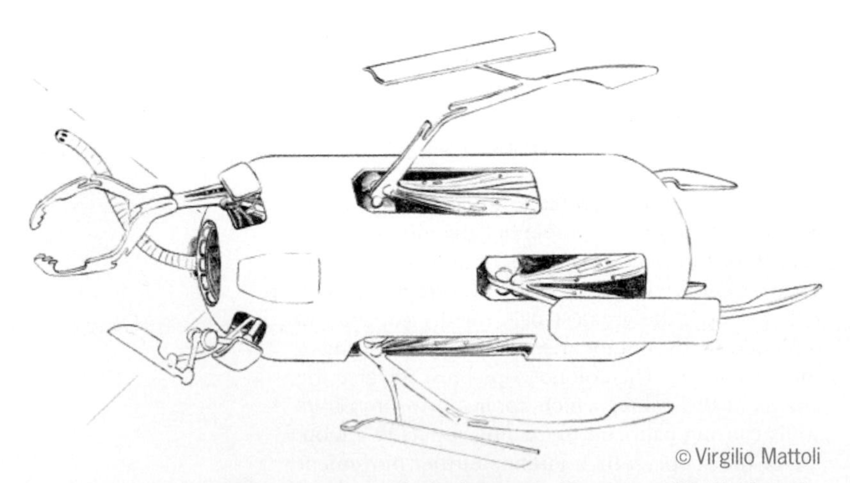

Fig. 2. Ideal representation of an advanced robotic capsule. (*Courtesy of Virgilio Mattoli*)

from the block and rotates to cut and collect a tissue sample. This system lacks a visual aid and was designed for random biopsy sampling (**Fig. 3**).

To be effective for CRC biopsy sampling and clipping, a robotic capsule providing real-time images is necessary to examine more specific polyp targets. The greatest challenge for tissue interaction mechanisms is system stabilization during the biopsy.

Wireless capsule endoscopes are increasing in capability with the addition of actuators, sensors, cameras, and other subsystems. These all require electrical power. Therefore, advancements in powering are also needed before WCE can be clinically applicable. Some future potential powering solutions include enhanced batteries, alternative solutions (for example, inductive coupling or fluid power), or possibly using the same kind of energy source that the human body uses.

Current efforts toward a new generation of capsule robots. Researchers at Vanderbilt University, supported by the National Science Foundation, are trying to systematize capsule robot design by creating a cyber-physical design environment that will lower the barriers to design space exploration, thus accelerating progress toward prototyping.

A systematic approach to the design of capsule robots is possible by outlining the crosscutting constraints that these systems must address. The main issues are (1) size, because ideally a capsule device should be small enough to swallow or to enter natural orifices without requiring a dedicated incision; (2) power consumption, as given the limited space available onboard, energy is limited; (3) communication bandwidth for wireless signals to be transmitted through the human body with a sufficient data rate; (4) fail safe operation, since the device is deep inside the human body, the user has no access to it; and (5) effective interaction with the target site, according to the specific functions the device is required to fulfill.

Given these common constraints, it is possible to identify a general system architecture for a capsule robot consisting of the following general modules: (1) a central processing unit (CPU) that can be programmed by the user to accomplish a specific task; (2) a communication submodule that links the device with user intent; (3) a source of energy that powers the system; and (4) sensors and (5) actuators, both of which interact with the surrounding environment to accomplish one or more specific tasks. It is also desirable for the designers to have a model of the environment, in order to predict the effectiveness of the specific design in accomplishing the desired task. Starting from this systematic approach, the researchers are creating a web-based cyber-physical design framework that will offer different options for the basic submodules of the capsule robot that can be integrated to obtain a simulation of the expected performance. The main goal of the design environment is to lower the barriers to design space exploration, accelerating progress to prototyping, and increasing the probability of success for each prototype of capsule robot.

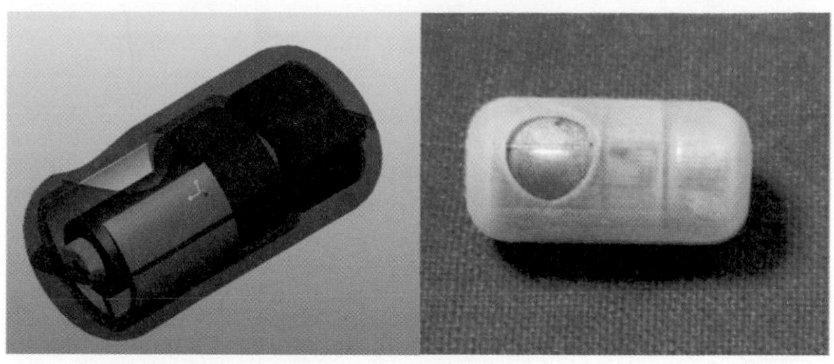

Fig. 3. A biopsying robotic capsule: 17 mm in length and 9.5 mm in diameter. (*Courtesy of M. Simi, G. Gerboni, A. Menciassi, and P. Valdastri, 2013*)

[Acknowledgment: This work was supported by the National Science Foundation under grant no. CNS-1239355.]

For background information *see* CAMERA; CANCER (MEDICINE); COLON; DIGESTIVE SYSTEM; GASTROINTESTINAL TRACT DISORDERS; INTESTINE; ONCOLOGY; ROBOTICS; SURGERY in the McGraw Hill Encyclopedia of Science & Technology.

Vanessa Nicole Valentine; Pietro Valdastri

Bibliography. S. R. Cai et al., Barriers to colorectal cancer screening: A case-control study, *World J. Gastroenterol.*, 15(20):2531–2536, 2009, DOI:10.3748/wjg.15.2531; F. A. Farraye et al., Update on CT colonography, *Gastrointest. Endosc.*, 69(3):393–398, 2009, DOI:10.1016/j.gie.2008.10.009; J. Ferlay et al., Estimates of worldwide burden of cancer in 2008: GLOBOCAN 2008, *Int. J. Cancer*, 127(12):2893–2917, 2010, DOI:10.1002/ijc.25516; S. Gorini et al., A novel SMA-based actuator for a legged endoscopic capsule, *Proc. IEEE/RAS-EMBS Int. Conf. Biomed. Robot. Biomechatron*, Feb. 20–22, Pisa, Italy, 443–449, 2006, DOI:10.1109/BIOROB.2006.1639128; J. L. Gorlewicz et al., Wireless insufflation of the gastrointestinal tract, *IEEE Trans. Biomed. Eng., IEEE Trans. on*, 60(5):1225–1233, 2013, DOI:10.1109/TBME.2012.2230631; H. H. Hopkins and N. S. Kapany, A flexible fiberscope, using static scanning, *Nature*, 173:39–41, 1954, DOI:10.1038/173039b0; D. A. Joseph et al., The colorectal cancer control program: Partnering to increase population level screening, *Gastrointest. Endosc.*, 73(3):429–434, 2011, DOI:10.1016/j.gie.2010.12.027; A. Loeve, P. Breedveld, and J. Dankelman, Scopes too flexible... and too stiff, *IEEE Pulse*, 1(3):26–41, 2010, DOI:10.1109/MPUL.2010.939176; M. Simi, G. Gerboni, A. Menciassi, and P. Valdastri, Magnetic torsion spring mechanism for a wireless biopsy capsule, *J. Med. Dev. Trans. ASME*, 7(4):041009-1–041009-9, 2013, DOI:10.1115/1.4025185; X. Wang and M. Q. H. Meng, An inchworm-like locomotion mechanism based on magnetic actuator for active capsule endoscope, *Proc. IEEE/RSJ IROS*, Oct. 9–15, Beijing, 1267–1272, 2006, DOI:10.1109/IROS.2006.281887.

Chromosome "fusions" in karyotype evolution

The karyotype is the complement of chromosomes characteristic of an individual, species, genus, or other grouping. Eukaryotic organisms vary in their number of chromosomes, from 1 to 600 per haploid genome (the genes that make up a haploid set of chromosomes). Haploid cells contain a single set of chromosomes. This is in contrast to diploid cells, which contain two complete sets of chromosomes. Although the number of chromosomes is usually species-specific, it can increase or decrease in populations or species. The number of chromosomes can increase through polyploidy (whole-genome duplication), that is, duplication of a whole chromosome set (or sets), or by centric fission, that is, centromere breakage, splitting the original chromosome into two chromosomes. The loss or gain of a single chromosome or a few chromosomes, called aneuploidy, can be caused by erroneous separation of chromatids or chromosomes during meiosis or mitosis. (Note that the term chromatid is used to describe either of the two daughter strands of a replicated chromosome; these strands are joined by a single centromere and will separate into individual chromosomes during cell division.)

The number of chromosomes can also be reduced as a consequence of DNA recombination between two different (nonhomologous) chromosomes. For the sake of simplicity, this process is inaccurately called chromosome "fusion" or "insertion," although telomere-capped chromosomes in standard cells cannot be fused without the presence of preceding double-strand breaks (DSBs) on both participating chromosomes. (Note that telomeres are DNA segments that occur at the ends of chromosomes and are used to regulate chromosome replication.) Thus, chromosome "fusions" and "insertions" are actually chromosome translocation events

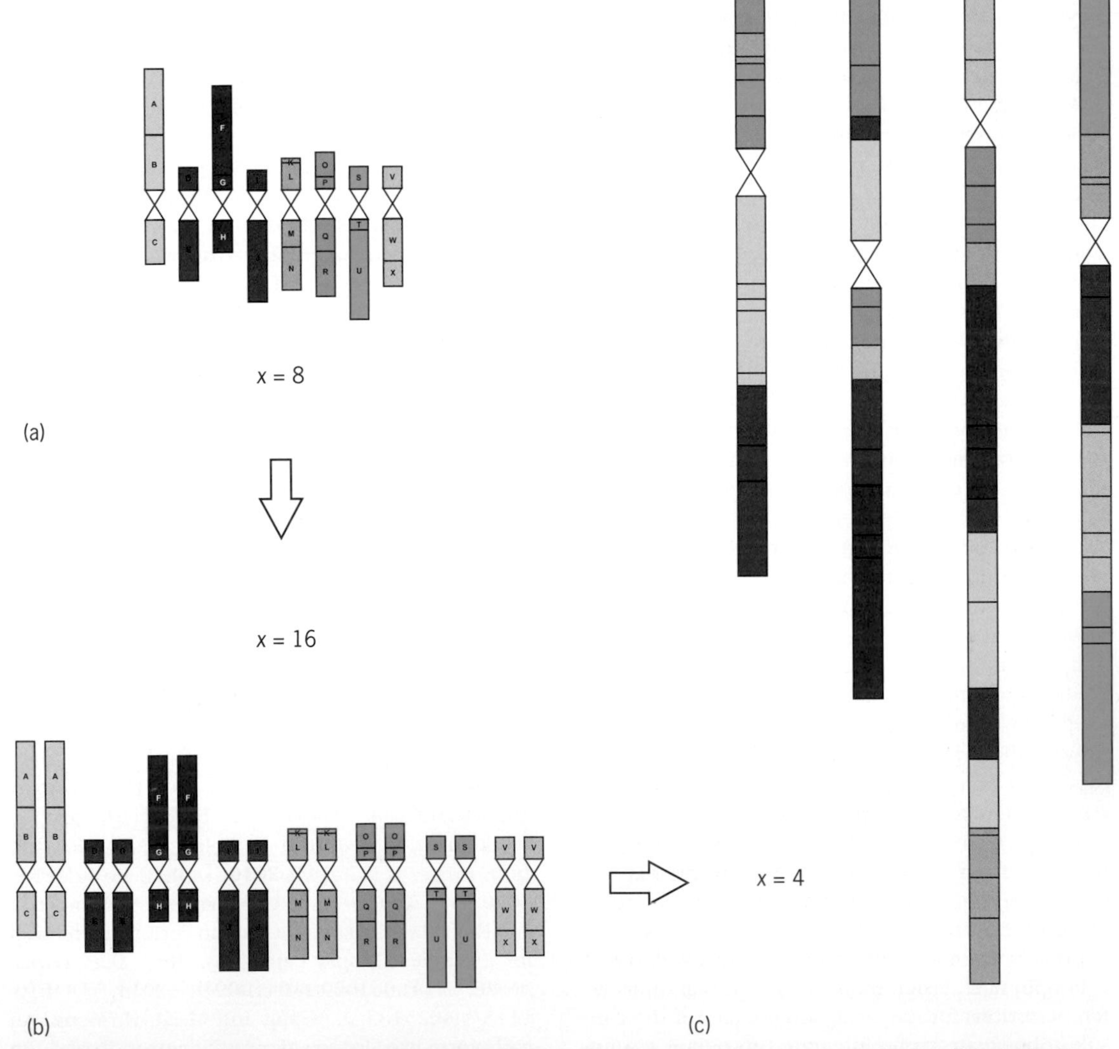

Fig. 1. Whole-genome duplication followed by genome diploidization and chromosome number reduction. Polyploidization of a diploid genome with 8 chromosomes and 24 genomic blocks (*a*) produced a tetraploid genome with 16 chromosomes and 48 blocks (*b*). (*c*) Diploidization of the polyploid genome was associated with a fourfold reduction of chromosome number. Chromosome translocations, inversions, and centromere inactivations and deletions mediated the reshuffling and/or split of the 48 genomic blocks and the removal of 12 centromeres. (*Modified from T. Mandáková et al., Fast diploidization in close mesopolyploid relatives of Arabidopsis, Plant Cell, 22:2277–2290, 2010, DOI:10.1105/tpc.110.074526*)

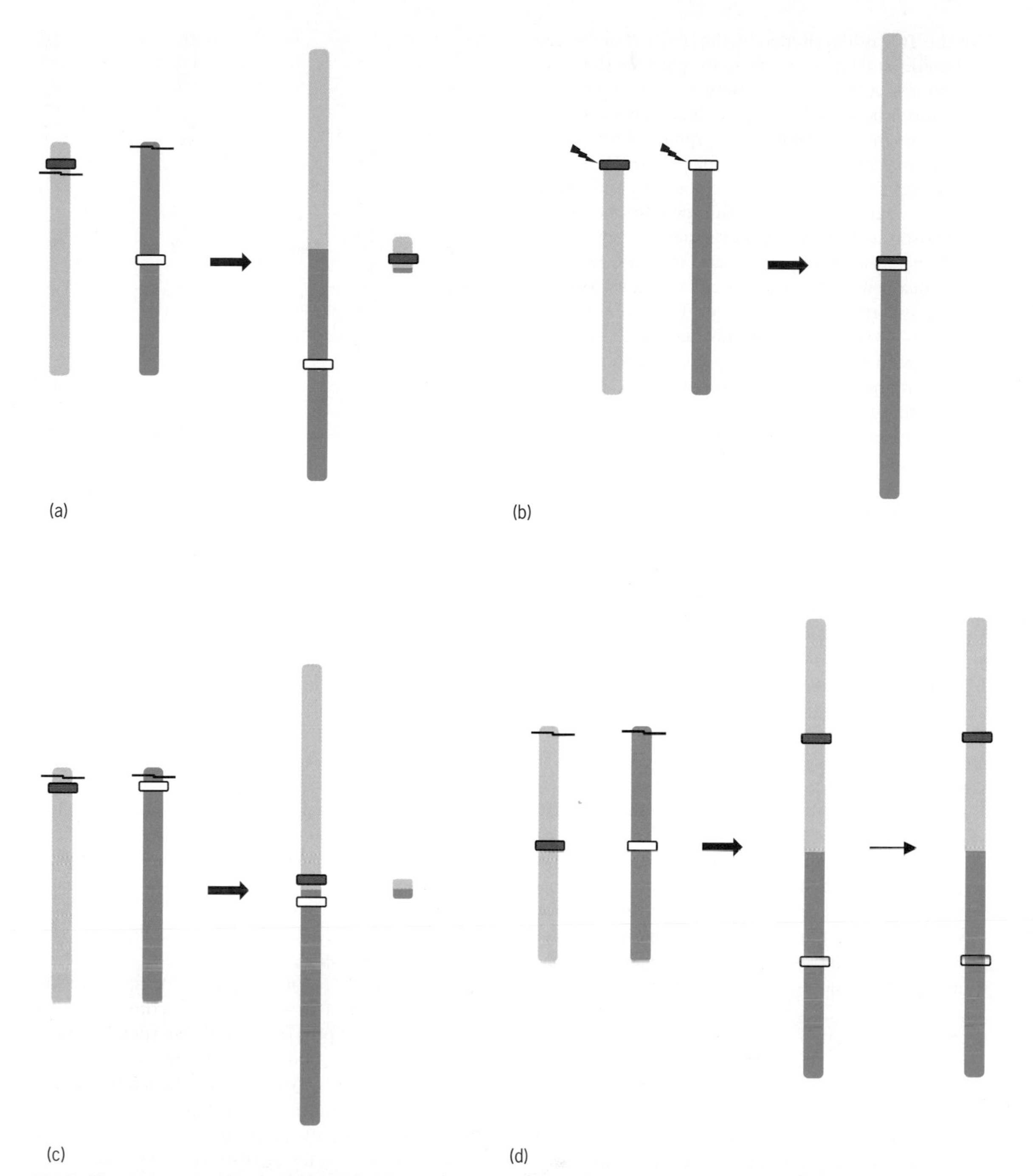

Fig. 2. Chromosome number reduction by terminal chromosome translocations (TCTs). (*a*) TCT resulting in a "fusion" chromosome and a minichromosome. (*b*) TCT with break points in the centromeres of two telocentrics produces a monocentric chromosome with a "hybrid" centromere. (*c*) The origin of a dicentric chromosome with two active centromeres resulting from an asymmetric TCT between two acrocentric chromosomes. (*d*) An asymmetric TCT between two metacentric chromosomes results in a dicentric chromosome, with one centromere becoming inactive.

resulting from a DSB misrepair by homologous recombination (HR) or nonhomologous end joining (NHEJ). The HR pathway is often based on interchromosomal sequence homology between (sub)telomeric, (peri)centromeric, ribosomal, or other repetitive sequences. Chromosome number reduction occurs most easily in polyploid organisms. In plants, which are the eukaryotes with the highest incidence of polyploidy, chromosome number reduction is particularly pronounced during genome diploidization following whole-genome duplication events. This gradual process is accompanied by extensive chromosome repatterning and reduction in chromosome number (for example, from $x = 8$ through $x = 16$ to $x = 4$; **Fig. 1**).

Terminal chromosome translocation. Terminal chromosome translocations (TCTs), often called end-to-end fusions, include recombination events between the terminal regions (ends) of two chromosomes. A TCT results in a monocentric ("fusion") chromosome and a centric fragment or in a dicentric chromosome and an acentric fragment. The outcome of the TCT is determined by the position of the centromeres and break points on the participating chromosomes. A translocation between the centric end or short arm of a telocentric or an acrocentric chromosome and one end of any type of chromosome results in a large ("fusion") chromosome comprising most parts of the original chromosomes and a small centric fragment (**Fig. 2***a*). This minichromosome is supposed to contain a centromere, two telomeres, and dispensable sequences, and is supposed to be meiotically unstable because of its inability to segregate regularly during the first or the second meiotic division. It is conceivable that such a centric fragment, with telomeres at both ends, becomes fixed in a genome as an extra chromosome (known as a B chromosome) by accumulation of sequences from other chromosomes and from mitochondrial and/or plastid genomes. If TCT break points are located within the centromeres of two telocentric chromosomes, parts of both centromeres can stay active in the resulting metacentric chromosome (Fig. 2*b*). An asymmetric TCT between the centric ends or short arms of two telocentric or acrocentric chromosomes, respectively, will result in a dicentric ("fusion") chromosome. If the centromeres are in close proximity, both can remain active (Fig. 2*c*). However, the larger the distance between the centromeres, the higher the probability that, during nuclear divisions, the dicentric chromosome will break in the region between the centromeres if the sister centromeres are pulled to opposite poles. An asymmetric reciprocal translocation between two (sub)metacentric chromosomes with terminal break points will result in a large ("fusion") dicentric chromosome and a small acentric fragment. In this case, the risk that the dicentric chromosome will break between distantly positioned centromeres is high. If such a chromosome should segregate regularly, one of its two centromeres has to become inactive (Fig. 2*d*; also see the text below). TCTs between the centric ends of two telocentric or acrocentric chromosomes form a (sub)metacentric chromosome and are called Robertsonian translocations or centric fusions.

Nested chromosome insertion. A translocation event involving two nonhomologous chromosomes, whereby one whole chromosome appears to be "inserted" into or near the centromere of another (recipient) chromosome, is called nested chromosome insertion (NCI) [**Fig. 3**]. This type of insertion requires at least three DSBs: two at the telomeres of the inserted chromosome and one at the (peri)centromeric region of the recipient chromosome. The insertion may target the core centromere or regions close to the centromere. If the recipient centromere remains intact, the resulting "fusion" chromosome becomes dicentric. Although the fate of the centromere of the recipient chromosome is typically unknown, it is assumed that it becomes inactive through epigenetic modifications (that is, alterations in gene expression that are not the result of changes in DNA sequences). NCIs have been shown to mediate chromosome number reduction in different subfamilies and genera of the grass family (Poaceae) and have largely contributed to the present-day genome structures of maize, sorghum, wheat, and other grass species. In addition to their activity in the grasses, NCIs have been reported to play a role in the origin of "fusion" chromosomes in the gourd family (Cucurbitaceae) and in crucifers (Brassicaceae). NCIs as a mechanism of chromosome number reduction in plants might be more common than suggested by the limited amount of comparative cytogenomic data.

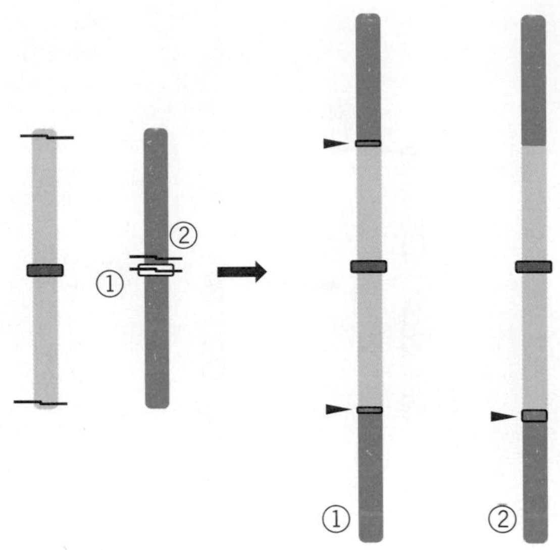

Fig. 3. Nested chromosome insertion (NCI). NCI with a break point in the core centromere (①) or pericentromere (②) of the recipient chromosome. The arrowheads indicate sequence remnants of the recipient chromosome centromere.

Centromere inactivation and loss in dicentric "fusion" chromosomes. Both TCT- and NCI-based chromosome number reductions may generate dicentric "fusion" chromosomes. If the two centromeres are in close physical proximity, both can remain active, and the dicentric chromosome becomes functionally monocentric. However, this generally occurs with low frequencies. The long-term survival and heritability of dicentrics are usually ensured only if one of the centromeres (usually the smaller one) becomes inactive. Although the molecular mechanisms of centromere suppression remain elusive, they are most likely based on epigenetic modifications. Alternatively, centromere loss is thought to be associated with the removal of centromere-specific sequences (tandem repeats and retrotransposons). Centromere inactivation is usually manifested by the depletion of centromere-specific histones (positively charged proteins that package DNA) and kinetochore proteins, by the loss of certain histone phosphorylations, and by increased cytosine hypermethylation of centromeric sequences. Also, elevated levels of specific histone dimethylations and trimethylations, as well as the loss of other histone phosphorylations, have been observed in plants. It is assumed that the epigenetic inactivation is coinciding with or followed by the removal of centromere-specific sequences and pericentromeric heterochromatin (specialized chromosome material), collectively called centromere decay. It remains unknown exactly how and when the centromeric sequences are excised. Unequal homologous intrachromosomal and interchromosomal recombinations [the former combined with the formation of extrachromosomal circular DNA (eccDNA) molecules] are plausible options. Centromere inactivation and loss have been documented to occur in yeast, mammals, and plants (maize and crucifer species).

For background information *see* CELL BIOLOGY; CELL DIVISION; CHROMOSOME; CHROMOSOME ABERRATION; DEOXYRIBONUCLEIC ACID (DNA); DNA METHYLATION; GENE; GENETIC MAPPING; HISTONE; HUMAN GENETICS; MEIOSIS; MITOSIS; POLYPLOIDY; RECOMBINATION (GENETICS) in the McGraw-Hill Encyclopedia of Science & Technology. Martin A. Lysak

Bibliography. J. A. Birchler and F. Han, Meiotic behavior of small chromosomes in maize, *Front. Plant Sci.*, 4:505, 2013, DOI:10.3389/fpls.2013.00505; S. Fu et al., Dicentric chromosome formation and epigenetics of centromere formation in plants, *J. Genet. Genomics*, 39:125–130, 2012, DOI:10.1016/j.jgg.2012.01.006; J. Jiang and J. A. Birchler (eds.), *Plant Centromere Biology*, Wiley-Blackwell, Ames, IA, 2013; P. Kalitsis and K. H. Choo, The evolutionary life cycle of the resilient centromere, *Chromosoma*, 121:327–334, 2012, DOI:10.1007/s00412-012-0369-6; M. C. Luo et al., Genome comparisons reveal a dominant mechanism of chromosome number reduction in grasses and accelerated genome evolution in Triticeae, *Proc. Natl. Acad. Sci. USA*, 106:15780–15785, 2009, DOI:10.1073/pnas.0908195106; M. A. Lysak, Live and let die: Centromere loss during evolution of plant chromosomes, *New Phytologist*, 203:1082–1089, 2014; M. A. Lysak and I. Schubert, Mechanisms of chromosome rearrangements, in I. J. Leitch et al. (eds.), *Plant Genome Diversity*, vol. 2, Springer, Vienna, 2013; T. Mandáková et al., Fast diploidization in close mesopolyploid relatives of *Arabidopsis*, *Plant Cell*, 22:2277–2290, 2010, DOI:10.1105/tpc.110.074526; I. Schubert and M. A. Lysak, Interpretation of karyotype evolution should consider chromosome structural constraints, *Trends Genet.*, 27:207–216, 2011, DOI:10.1016/j.tig.2011.03.004; K. M. Stimpson, J. E. Matheny, and B. A. Sullivan, Dicentric chromosomes: Unique models to study centromere function and inactivation, *Chromosome Res.*, 20:595–605, 2012, DOI:10.1007/s10577-012-9302-3.

Color of ancient cephalopod ink

Coleoids are members of the cephalopod subclass Coleoidea. They include octopods, squids, cuttlefish, and the extinct belemnites. Chemical techniques have revealed that the ink sacs found in some Jurassic-age coleoids contained pigment that was dark brown to black in color. This 160 million-year-old ink is chemically similar to that found in the ink sacs of modern cuttlefish and is the oldest melanin known in the fossil record.

Melanin as a coloring agent. Melanin is a colored biological pigment that is found in a diverse array of organisms, including bacteria, plants, fungi, and animals. Melanin protects against radiation and ultraviolet light, provides camouflage, plays a role in sexual and social displays, quenches reactive oxygen species, and binds otherwise toxic metal ions. The effectiveness of melanin in these roles and the spectrum of colors available to a given organism depend on the pigment's precise chemical composition.

In animals, melanins can be classified into two main groups: eumelanin and pheomelanin. Eumelanin, an insoluble black-to-brown pigment composed of indolic units (indole is an organic heterocyclic compound with repeating units of a six-membered benzene ring fused to a five-membered pyrrole ring), is derived from the oxidation of tyrosine (a phenolic amino acid). Pheomelanin, an alkali-soluble yellow-to-reddish-brown pigment composed of benzothiazines (which are heterocyclic compounds), is derived from the oxidation of cysteinyldopa (a sulfur-containing melanin precursor). Both groups of melanin are produced in a process known as melanogenesis. The initial steps of melanogenesis are detailed by the Raper-Mason scheme in **Fig. 1**. Melanosomes (melanin-bearing organelles) are filled with these pigments, but only the structures of the colorless monomers (the repeating units within a polymer) are unambiguously known.

Biomarker preservation. In the past, melanin was studied primarily in the feathers, melanoma lesions, and pigmented regions of modern organisms because it was not known to persist in the fossil record, but was presumed to be lost along with other primary organic components of ancient life (for example, proteins, lipids, and DNA). In recent years, though, new evidence has emerged. The canonical view that organic materials, such as melanin, break down in the fossil record has been challenged by the discovery of intact organic constituents in certain fossils. Rather than succumbing to the normal processes of bacterial degradation or mineral replacement, the skin, hair, feathers, melanin, or other soft-tissue components have been preserved, and the processes normally involved in degradation have been arrested.

Even in the most exceptionally well preserved soft-tissue fossils, many organic constituents are lost. Those soft tissues that are preserved have distinctive structural attributes: highly cross-linked structures, large hydrophobic (water-hating) regions, and polymeric backbones (main chains) made of many smaller monomeric units. Melanins, keratin (from within skin, nails, and hair), and collagen (primarily from under the skin) are among the few organic constituents that can persist in the fossil record. As

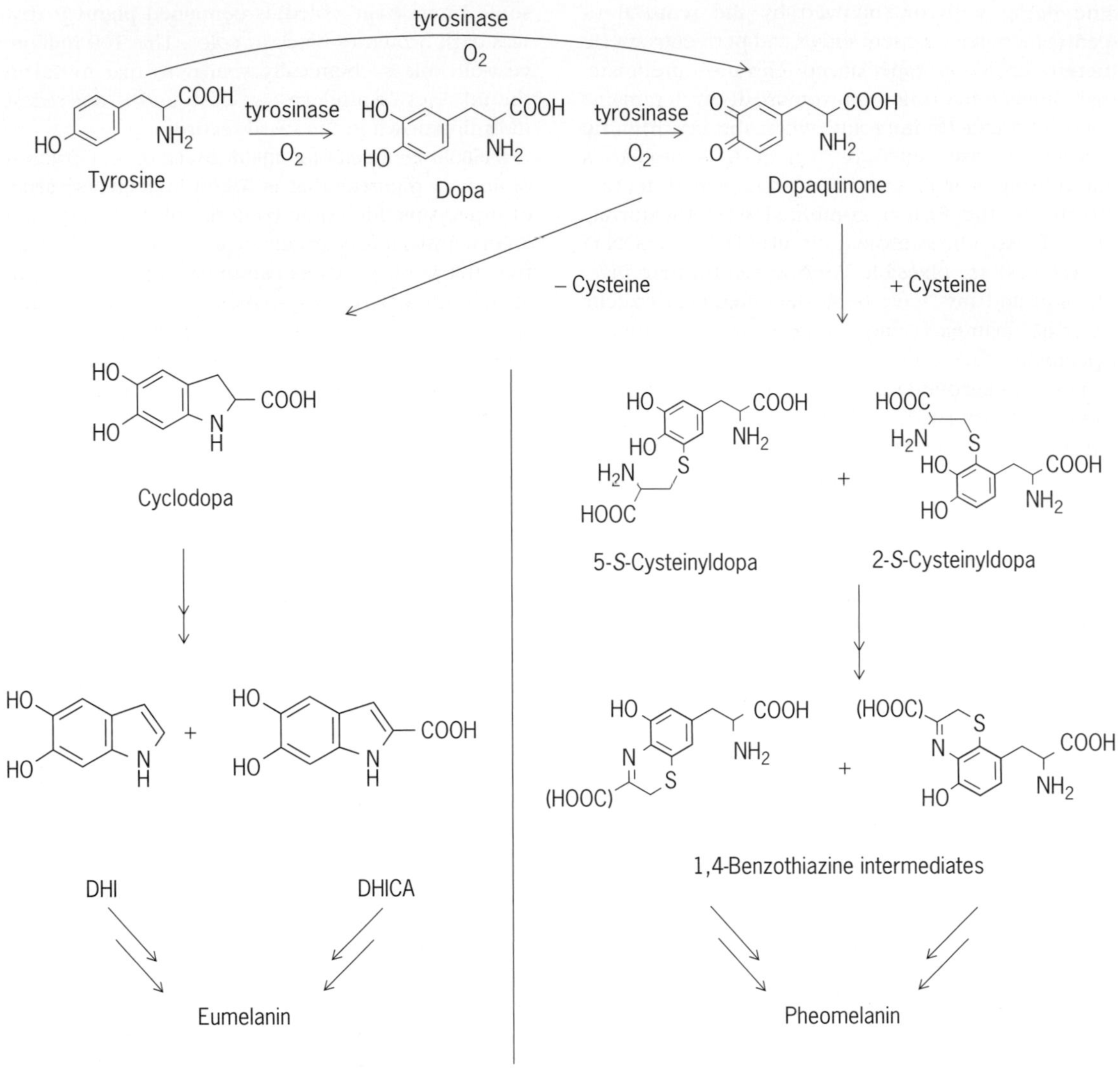

Fig. 1. Abbreviated depiction of the early steps of melanogenesis: the Raper-Mason scheme. Dopaquinone is the branch point between the formation of eumelanin and pheomelanin. The final pigment formed depends on the concentration of cysteine that is available.

scientists have become aware of the intact organic constituents residing in the fossil record, new analytical techniques have been applied that have allowed the study of ancient life in unprecedented detail. Part of the challenge in this research is to differentiate an organism's original molecules from bacterial degradation products and modern contaminants.

Morphological evidence for melanin in fossils. Pigmented fossils offer insights into the coloration of ancient organisms and the role (or roles) that melanin played in those organisms. Scanning electron microscopy (SEM), an imaging technique used for morphological analysis of submicron structures, has enabled the detection of structures that are morphologically similar to melanin. In 2008, the SEM imaging of fossil feathers revealed rod-shaped and spherical microstructures that were interpreted as fossilized melanosomes. Since then, structures of approximately the same size and shape as melanosomes and melanin granules have been found in an array of fossils. Using the dimensions of these structures to extrapolate color through discriminant analysis (for example, rod-shaped structures indicate black-brown eumelanosomes, whereas spheroidal structures indicate reddish-brown pheomelanosomes), the studies appear to offer powerful insights into ancient coloration. However, none of these early studies provided definitive chemical evidence to support the claims, leaving the possibility that some of the structures may actually represent the remains of bacteria, the remnants of keratin from feathers, or other similarly sized microstructures.

Chemical evidence for melanin in fossils. In order to meet the demand for chemical evidence to prove the presence of melanosomes and melanin granules in the fossil record, analytical techniques that capitalize on chemical signals associated with melanin have been developed. There are several particularly promising techniques, including trace-metal analysis of bound ions, time-of-flight secondary ion mass spectrometry (ToF-SIMS), and alkaline hydrogen peroxide oxidation.

Trace-metal analysis involves the spatial detection of metal ions such as calcium, copper, iron, and zinc that are known to bind to melanin. The analysis of modern melanins has shown that melanin binds these metal ions tightly and prevents their circulation through the body. Trace-metal analysis may involve a variety of nondestructive chemical-imaging techniques: synchrotron rapid-scanning x-ray fluorescence (SRS-XRF) gives full-sample metal identification, whereas x-ray absorption near-edge structure (XANES) and extended x-ray absorption fine structure (EXAFS) provide localized metal mapping. The techniques have been successfully applied to the study of melanin in fossil feathers. By taking advantage of long-lived metal ions that may continue to bind melanin postmortem, trace-metal analysis provides a new tool for verifying the melanic nature of microstructures identified by SEM. The disadvantage of trace-metal analysis is that microorganisms of a size and shape similar to those of melanosomes are also able to absorb trace-metal ions and may interfere with melanin identification.

In ToF-SIMS, a beam of ions is pulsed over the surface of a sample. Ions from the sample surface (secondary ions) are released and funneled into a mass spectrometer, where their masses are analyzed. This technique provides a nondestructive chemical fingerprint of the sample surface. By comparing the chemical fingerprint of a specimen of interest to a related melanin standard, it is possible to determine whether melanin is present or not. ToF-SIMS has been successfully applied to the study of melanin in fossilized reptile skin and fish eyes. The only detriment to this technique is that the chemical signatures of melanin in these systems may not differ substantially from those of melanin derived from other sources, including bacteria. Melanin-containing bacteria introduced during decay could replace the original melanin components, resulting in false positives.

Unlike trace-metal analysis and ToF-SIMS, alkaline hydrogen peroxide oxidation is a destructive assay commonly used to determine the presence and concentration of melanins in modern organisms. This technique can be confidently used to distinguish melanin from nonmelanin components because the chemical markers of eumelanin and pheomelanin have never been produced by the alkaline hydrogen peroxide oxidation of any other biological material. It is also capable of differentiating between melanins from animal sources and melanins from bacteria, fungi, and plant sources, which often do not contain nitrogen.

Recently, alkaline hydrogen peroxide oxidation was used to establish the presence of melanin in the preserved ink sacs of Jurassic cephalopods (**Fig. 2**). As shown in the **table**, only the chemical markers for

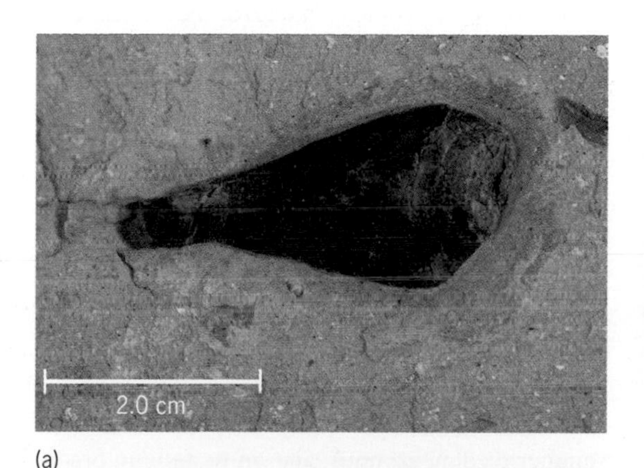

(a)

(b)

(c)

(d)

Fig. 2. Photographs of Jurassic coleoid ink sacs. Specimens from the United Kingdom: (*a*) GSM 122841 (Oxford Clay Formation, Wiltshire) and (*b*) GSM 120386 (Blue Lias Formation, Dorset). Specimens from Germany: (*c*) YPM 221212 (Posidonia Shale Formation, Baden-Württemberg) and (*d*) YPM 221210 (Posidonia Shale Formation, Baden-Württemberg).

Alkaline hydrogen peroxide oxidation products*

Specimen	PDCA	PTCA	isoPTCA	PTeCA
Sepia officinalis melanin	704	15,710	390	2090
GSM 122841	58	342	47	438
GSM 120386	173	1710	245	2220
YPM 221212	0.73	0.9	2.7	1.3
YPM 221210	2.06	0.5	5.2	1.2
GSM 122841 sediment	1.7	2.2	<2.2	<1.3
GSM 120386 sediment	2.4	3.2	<3.9	<1.5
YPM 221212 sediment	<0.1	0.3	<0.5	<0.4
YPM 221210 sediment	<0.1	0.4	<0.5	<0.5

*Chemical markers for eumelanin: pyrrole-2,3-dicarboxylic acid (PDCA); pyrrole-2,3,5-tricarboxylic acid (PTCA); pyrrole-2,3,4-tricarboxylic acid (isoPTCA); and pyrrole-2,3,4,5-tetracarboxylic acid (PTeCA). The quantity of each oxidation product is listed in ng/mg of specimen.

eumelanin [pyrrole-2,3-dicarboxylic acid (PDCA), pyrrole-2,3,5-tricarboxylic acid (PTCA), pyrrole-2,3,4-tricarboxylic acid (isoPTCA), and pyrrole-2,3,4,5-tetracarboxylic acid (PTeCA)] were detected in the ancient ink. Modern cephalopod ink from the cuttlefish *Sepia officinalis* also contains only the pigment eumelanin. The analyses showed that modern cephalopod ink is chemically and morphologically identical to the fossil ink from the UK specimens (Fig. 2, parts *a* and *b*), but morphologically different from the fossil ink from the German specimens (Fig. 2, parts *c* and *d*). After comparison of the burial histories of the fossil ink sacs, these morphological differences were attributed to thermal sediment maturation.

All these approaches provide accurate detection and characterization of pigment in fossils. Accompanied by the analysis of entombing sediments and corroborating analytical techniques, each approach has the potential to elucidate the roles of melanin in fossilized organisms, including the melanin found in Jurassic coleoid ink sacs.

For background information *see* CEPHALOPODA; COLEOIDEA; COLOR; FOSSIL; JURASSIC; ORGANIC GEOCHEMISTRY; PALEONTOLOGY; PIGMENTATION; SCANNING ELECTRON MICROSCOPE; SEPIOIDEA; TAPHONOMY in the McGraw-Hill Encyclopedia of Science & Technology.

John D. Simon; Keely Glass; Shosuke Ito; Philip Wilby

Bibliography. M. d'Ischia et al., Melanins and melanogenesis: Methods, standards, protocols, *Pigm. Cell Melanoma Res.*, 26(5):616–633, 2013, DOI:10.1111/pcmr.12121; K. Glass et al., Direct chemical evidence for eumelanin from the Jurassic period, *Proc. Natl. Acad. Sci. USA*, 109(26):10218–10223, 2012, DOI:10.1073/pnas.1118448109; K. Glass et al., Impact of diagenesis and maturation on the survival of eumelanin in the fossil record, *Org. Geochem.*, 64:29–37, 2013, DOI:10.1016/j.orggeochem.2013.09.002; S. Ito and K. Wakamatsu, Chemistry of mixed melanogenesis—pivotal roles of dopaquinone, *Photochem. Photobiol.*, 84:582–592, 2008, DOI:10.1111/j.1751-1097.2007.00238.x; M. E. McNamara, The taphonomy of colour in fossil insects and feathers, *Palaeontology*, 56(3):557–575, 2013, DOI:10.1111/pala.12044; M. H. Schweitzer, Soft tissue preservation in terrestrial Mesozoic vertebrates, *Annu. Rev. Earth Planet. Sci.*, 39:187–216, 2011, DOI:10.1146/annurev-earth-040610-133502.

Connected wildlands: corridors for survival

Long before people arrived in the wild expanses of North America, its continent-spanning mountain ranges, vast prairies, serpentine river systems, and lush coastlines were key components in a heavily traveled system of ecological "highways." Of course, those ancient routes had no traffic signs and were not traversed by any vehicles. Instead, they were used intuitively by wide-ranging native wildlife species as a multilane throughway, connecting them with distant relatives, seasonal food sources, new mates, and preferred weather.

The landscape: then and now. Prior to the settlement of North America by Europeans, grizzly bears, mountain lions, wolves, and jaguars, as well as massive herds of bison, antelope, and elk, regularly traveled in safety through uninterrupted wild landscapes. They traversed stream-filled canyon bottoms, vast grasslands, fir-covered mountain tops, and marsh-fringed shores on their journeys from protective winter ranges, to spring birthing habitats, to summer feeding grounds, and on to autumn breeding areas. Overhead, crowded migration flyways served as another type of pathway for hundreds of bird species, with many traveling thousands of miles, nourished by the friendly landscapes below.

Today, however, some of the native species populations (including wolves, grizzlies, jaguars, and bison) that formerly roamed across those North American wildlife pathways have been reduced dramatically because of excessive hunting or eradication efforts. Such remnant populations are often corralled into strictly bounded and isolated parks and reserves. Nevertheless, these magnificent species' instincts to roam widely still remain, and the vast, historic wildlife corridors through which they once maneuvered also still exist. However, millions of people now also use those same traditional wildlife pathways for recreation, travel, and development. Those trails continue to be shared, of course, with other wildlife species that remain relatively abundant, such as black bear, cougar, coyote, deer, elk, and many bird species. Yet even these hardy survivors regularly face life-threatening obstacles as they negotiate long-distance travel along their ancestral, but now human-affected, pathways.

Challenges for survival. Biologists have determined that the disruption of age-old wildlife habitat corridors by highways, housing developments, canals, mining operations, agricultural fencing, hunting, off-road vehicle travel, and other infrastructure is a substantive precursor of future extinction or extirpation for many species. Inability to access new food sources, inability to find safe havens, and inability to boost genetic strength through breeding with other unrelated members of their species now threaten many animals' ability to survive into the distant future.

In addition, challenges to such survival posed by climate change will be among the greatest problems that many animals will face over time. Climate change is indeed forcing wildlife of all types to travel greater migration distances and to make significant changes in their durations of stay in normal habitats. These impacts have resulted in the need for expansion of existing connecting corridors to accommodate the uncertainties of future climatic disruption. However, if the blockage of existing corridors continues to increase, then movement adaptation to changing climates, or the simple need for normal wildlife movement patterns, will become even more difficult for affected species.

This complex dilemma is already playing out across North America. A prominent example is in the United States–Mexico borderlands region. Here, many impenetrable security barriers block much of the United States–Mexico border to wildlife migration, severely limiting movement adaptation to climate change or normal migratory movements. Furthermore, many species in Mexico, including jaguars and ocelots, are now greatly deterred from crossing the border in their natural efforts to recolonize their former ranges in the United States. Without an open pathway for these wide-ranging species to travel, border security fencing could ultimately cause reduced species population numbers or even future extinctions.

Recent catastrophic and habitat-altering forest fires, the frequency and severity of which can also be associated with climate change, are another reason for wildlife corridor protection, particularly in the western United States. In this case, large-scale blockage of surrounding movement corridors can essentially trap species as they attempt to escape from fire or other ongoing climatologically extreme events. As unnaturally severe fire drives some animals to make habitat changes, or to simply avoid immediate fire threats, they must be able to move quickly and safely along protected pathways to surrounding landscapes.

Conservation and ecological health of wildlands. Continuing to protect new, yet isolated, parks and reserves that are disconnected from other safe wildlife habitats will not be enough to allow the return of the now-missing (and ecologically necessary) native species that once roamed North America's vibrant wildlife corridors. Conservation science has indisputably shown that, if populations of large, native carnivores are reduced or eliminated, the natural benefits that they provide to keep ecosystems (and their dependent human communities) healthy will be lost. For example, keeping harmful rodent populations in check, maintaining forest health, and reducing the overbrowsing (eating of vegetation) done by elk and deer that damages water retention in key watersheds are not possible without the presence of carnivores.

Overcoming these paramount threats to wildlife survival and to overall ecological health has now become a focus of extensive conservation efforts in Canada, the United States, and Mexico, not to mention around the globe. Key developments include the following efforts:

1. Federal, state, and local governments have begun incorporating the identification and protection of safe passage corridors for wildlife into planning and zoning rules;
2. Nonprofit conservation organizations, led by the Wildlands Network, are producing detailed science-based maps showing specific, regional wildlands complexes, including six North American "Wildlands Network Designs," which detail how core wildlands protected areas can be linked to one another via wildlife corridors;
3. Individuals are volunteering to act as "citizen scientists" in tracking where wildlife moves across local landscapes and contributing their data to online repositories;
4. "Wildlife bridges" across corridor-severing highways are being promoted, planned, and constructed, often with federal and state funding;
5. National petition efforts are in progress to urge passage of a federal "Wildlife Corridor Protection Act";
6. Land trusts are forming for the specific purpose of protecting privately owned corridor lands by acquiring or placing conservation easements on such properties;
7. Ranchers are collaborating to link huge, neighboring open spaces for wildlife movement;
8. An international network of wildlands protection organizations is implementing a unique project (using a combination of the aforementioned actions) to protect a 7000-mi (11,265-km) wildlife corridor called the "Western Wildway," stretching from Alaska to northern Mexico (see **illustration**).

Future outlook. Despite these concerted, ongoing efforts to create safe landscape passage for wide-ranging wildlife, many conservation scientists continue to worry that infrastructure development within these ancient wildlife connections is fast outpacing the needed habitat protections. With prospects for near-term federal or state protection designations of large-landscape wildlife corridors unreliable, many researchers believe that the

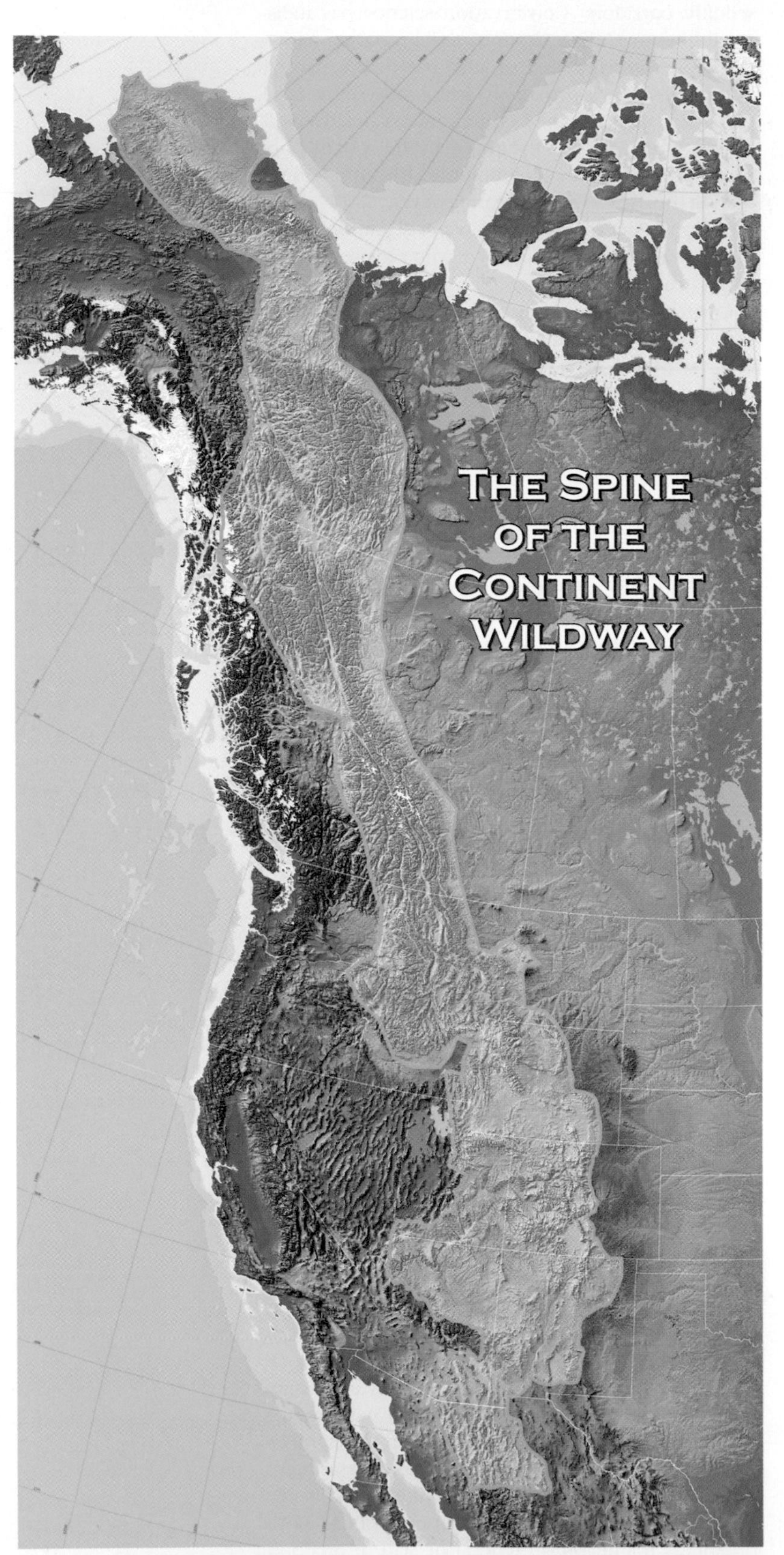

The Spine of the Continent Wildway (also known as the "Western Wildway") is a proposed 7000-mi (11,265-km) wildlife habitat corridor, stretching along the Rocky Mountains and associated ranges from Mexico to Alaska.

responsibility for wildlife pathway safekeeping lies squarely with the private sector. According to acclaimed conservation biologist Michael Soulé, people must recognize the importance of allowing for the safety and survival of North America's most magnificent native species, not just for their own survival, but for the health of communities and people as well. "What we need," Soulé reminds us, "are networks of people protecting networks of connected wildlands."

For background information *see* BIODIVERSITY; CONSERVATION OF RESOURCES; ECOLOGICAL MODELING; ECOLOGICAL SUCCESSION; ECOLOGY, APPLIED; ECOSYSTEM; ENDANGERED SPECIES; EXTINCTION (BIOLOGY); LAND RECLAMATION; NORTH AMERICA; POPULATION ECOLOGY; POPULATION VIABILITY; PREDATOR-PREY INTERACTIONS; RESTORATION ECOLOGY in the McGraw-Hill Encyclopedia of Science & Technology. Kim Vacariu

Bibliography. D. Foreman, *Rewilding North America: A Vision for Conservation in the 21st Century*, Island Press, Washington, DC, 2004; J. Hilty, W. Lidicker, and A. Merenlender, *Corridor Ecology: The Science and Practice of Linking Landscapes for Biodiversity Conservation*, Island Press, Washington, DC, 2006; M. Soulé and J. Terborgh, *Continental Conservation: Scientific Foundations of Regional Reserve Networks*, Island Press, Washington, DC, 1999; J. Terborgh and J. Estes, *Trophic Cascades: Predators, Prey, and the Changing Dynamics of Nature*, Island Press, Washington, DC, 2010.

Coordination chemistry of the actinide elements

The actinides (An) comprise elements 89 to 103 in the periodic table. All of the actinides are radioactive; however, several possess isotopes with long half-lives (for example, ^{232}Th, $t_{1/2} = 1.4 \times 10^{10}$ y; ^{238}U, $t_{1/2} = 4.5 \times 10^9$ y; ^{237}Np, $t_{1/2} = 2.1 \times 10^6$ y; ^{239}Pu, $t_{1/2} = 2.4 \times 10^4$ y), allowing their chemistry to be explored using standard synthetic techniques. That said, the higher radioactivity of neptunium (Np) and plutonium (Pu) requires the use of specialized containment procedures for these elements to be handled safely. As a result, most research on the actinides has been performed with uranium (U).

The coordination chemistry of the actinides is unique in comparison to other blocks in the periodic table, and the actinide elements are often described as a hybrid of the transition-metal and lanthanide elements. For instance, like the lanthanide elements, the actinides feature large ionic radii and large coordination numbers (coordination numbers greater than 8 are common; **Fig. 1**). However, like the transition metals, several actinide elements (particularly U, Np, and Pu) feature a range of oxidation states and complicated redox (reduction-oxidation) chemistry (Fig. 1). Perhaps more interestingly, the actinides possess both valence 5*f* and 6*d* orbitals, and both can be involved in metal–ligand bonding. In addition, their $6p$ "semi-core" orbitals can also be involved in bonding; $6p$ orbital involvement is commonly invoked to explain the electronic structure and geometry of the actinyl ions, $[AnO_2]^{2+}$ (discussed below). To our knowledge, the invocation of core orbitals in bonding is unique to the actinide elements. To further complicate matters, the ligand field splitting of the $5f$ orbitals is typically quite small. As a result, the f orbitals are nearly degenerate and spin-orbit coupling can be significant. This contrasts with the transition metals, where spin-orbit coupling can usually be ignored because of the large splitting of the d orbitals. Overall, these combined effects make for a rich and exciting bonding picture within actinide systems.

Covalency and the nature of the actinide–ligand bond. For several decades, it was accepted that actinide-ligand (An–L) bonding was dominated by ionic interactions, with only a few exceptions, such as actinyl ions. However, recent research has reignited a long simmering debate about whether the actinide elements can participate more broadly in covalent bonding (defined as the sharing of valence electrons between the metal and ligand). One reason this question has taken so many years to resolve is that it is actually quite challenging to measure covalency experimentally; however, a number of techniques have been developed in the last few years to address this issue, including, most successfully, x-ray absorption near-edge spectroscopy (XANES). In particular, studies on the model system $[UCl_6]^{2-}$ reveal measurable amounts of $6d$ and $5f$ orbital participation in the U–Cl bonds, consistent with a moderate level of covalency. Interestingly, these measurements also show that there is three times more $6d$ orbital character in the U–Cl bonds than $5f$ orbital character. Optical spectroscopy (that is, UV/Vis/NIR spectroscopy) confirms that only small amounts of f orbital participation are present in uranium-ligand bonding. For instance, in the U(V) complexes $[UX_6]^-$ X = F, Cl, Br, O^tBu, $N{=}C(^tBu)Ph$), f-orbital participation contributes only 12–17 kcal/mol to the strength of each U–X bond, which is a small fraction of the total U–X bond strength. Overall, these studies suggest that the d orbitals play the largest role in An–L bonding, while f-orbital participation is small. However, it should be noted that this picture is at odds with other studies, including electronic structure calculations on the U(VI) alkyl complex $U(CH_2SiMe_3)_6$, and the U(VI) terminal nitride, $[U(N)\{N(CH_2CH_2NSi^iPr_3)_3\}]$, which suggest the opposite picture, namely, that the f orbitals play a greater role in bonding than the d orbitals. No doubt, future research will allow us to understand this apparent discrepancy better.

The actinyl ions, $[AnO_2]^{n+}$ (An = U, Np, Pu, Am; n = 1, 2). The cationic trans di(oxo) fragment, $[AnO_2]^{n+}$ (An = U, Np, Pu, Am; n = 1, 2), is known as the actinyl ion. It is an incredibly common fragment; in fact, more than 50% of all reported uranium coordination complexes contain this moiety. The An–O bonds in the actinyl fragment are short (about

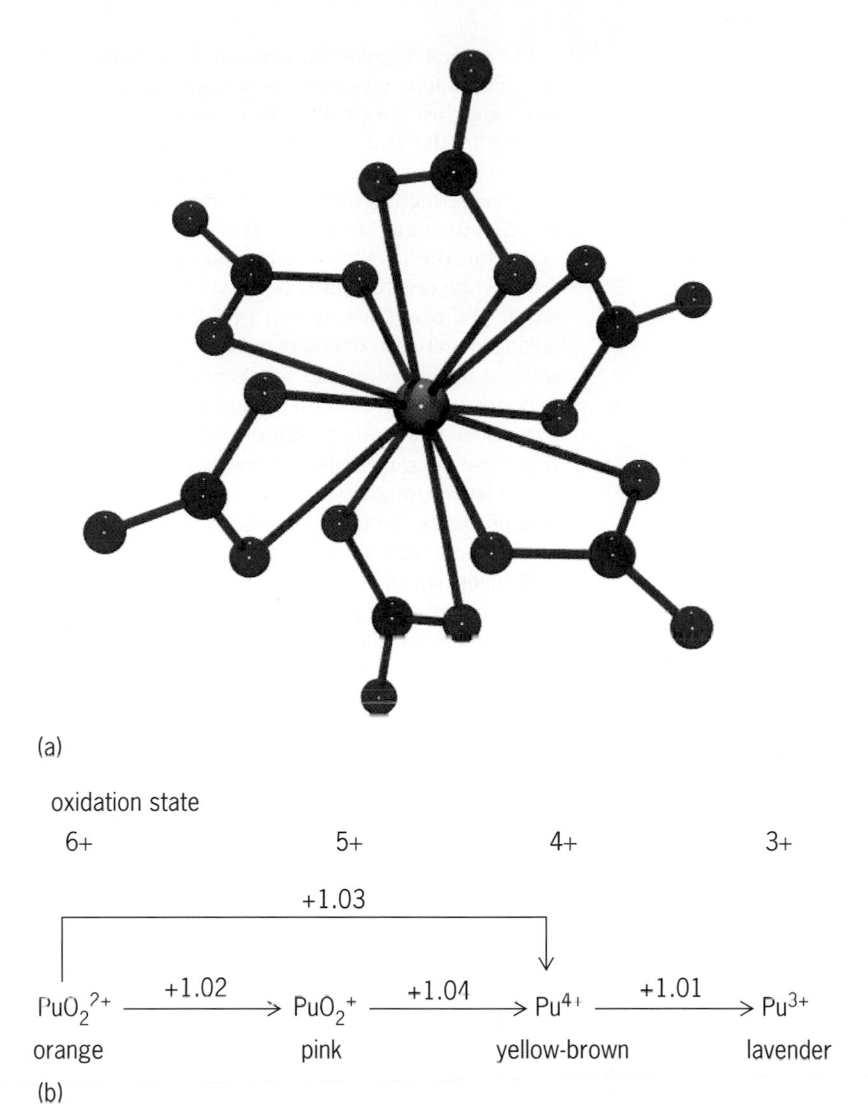

Fig. 1. (*a*) Structure of the 12-coordinate $[Th(NO_3)_6]^{2-}$ anion. This complex features a distorted icosahedral geometry. Atom color code: gray = Th; red = O; blue = N. (*b*) Partial Latimer diagram for plutonium.

1.80 Å) and exceptionally strong. Numerous experimental and computational studies support the presence of a covalent triple bond between An and O, in which both the actinide $5f$ and $6d$ orbitals interact strongly with the oxygen $2s$ and $2p$ orbitals. The trans arrangement of two strong-field oxo ligands is also unusual and has its origins in the participation of the "semi-core" $6p$ orbitals in the U–O σ-bond, which enforces the 180° angle between the two oxo ligands. In addition, the strong An–O (oxo) bonds stabilize the high formal oxidation state (either 6+ or 5+) of the metal ion, and as a result, the actinyl ions are surprisingly weak oxidants. Ligands can coordinate readily to the actinyl fragment, but they are restricted to the plane that is orthogonal to the O–An–O vector. Equatorial coordination numbers of 4 and 5 are the most common (**Fig. 2**), and generally speaking, actinide-ligand bonding within the equatorial plane is quite weak. As a result, these ligands have a tendency to be quite labile.

For several decades, the coordination chemistry of the actinyl ions was essentially restricted to ligand-exchange reactions within the equatorial plane, due mostly to the fact that the An-O (oxo) bonds in actinyl are so strong. That changed in 2008, when the first reductive silylation reaction was reported for the uranium derivative (Fig. 2). In reductive silylation, the $[UO_2]^{2+}$ ion is reduced by one electron, while one (or both) of its oxo ligands are silylated. The reaction is driven by the formation of a strong Si-O bond, which provides the driving force needed to overcome the breaking of the two U-O π bonds that occurs upon silylation. Since the original report, many other examples of reductive silylation have been described. Because of its generality and relative simplicity, this reaction may eventually find use in the processing of spent nuclear fuel and in environmental remediation.

Actinide–ligand multiple bonds. Similar to the transition metals, the actinide ions are able to form multiple bonds (including triple bonds) to a single ligand. For example, the terminal nitride complex $[U(N)\{N(CH_2CH_2NSi^iPr_3)_3\}]$ (**Fig. 3**) features a formal triple bond between the uranium center and the nitrogen atom. This description is supported by the U-N bond length in the solid state [1.799(7) Å], which is among the shortest U-N bonds reported. In addition, electronic structure calculations reveal a substantial amount of 5*f*-orbital participation in the U-N triple bond, and suggest a bond order of nearly 3. Several other elements, including sulfur, selenium, and phosphorus, are also able to form multiple bonds to uranium atoms. Included in this class of compounds are a number of actinyl analogs, such as $[U(NR)_2]^{2+}$ (R = alkyl, aryl) and $[OUE]^{2+}$ (E = S, Se) [Fig. 3]. Like uranyl, these complexes feature short U-E axial bonds and a trans arrangement of the multiply bonded groups, which is indicative of a uranyl-like electronic structure. As with uranyl, computational studies suggest that the axial bonding is strongly covalent and involves both the uranium 5*f* and 6*d* orbitals. These studies confirm the importance of covalency in actinide-ligand bonding, and further solidify our understanding of the technologically important uranyl ion, $[UO_2]^{2+}$.

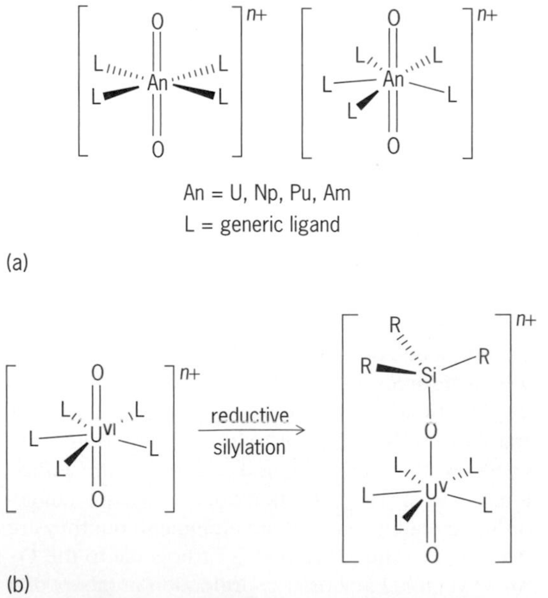

Fig. 2. (*a*) Structure of the 6- and 7-coordinate actinyl ions. (*b*) Schematic of the reductive silylation reaction. The "=" notation between An and O is not meant to imply a specific bond order. Instead, it is used to indicate the oxidation or valence state of the oxygen atom.

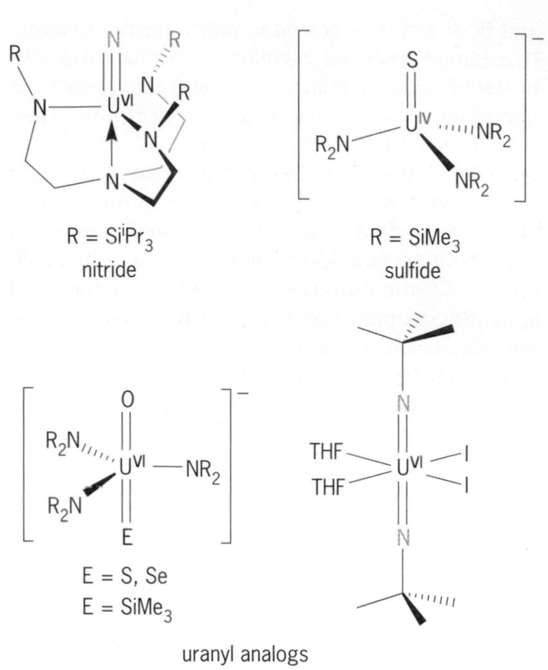

Fig. 3. Examples of molecules possessing actinide–ligand multiple bonds.

Outlook. Our understanding of actinide coordination chemistry has improved greatly in the last 20 years, and the unusual chemical and physical properties of the actinides elements are now being harnessed for a variety of applications, including single-molecule magnetism and catalysis. However, there is still much we do not understand, especially when it comes to actinides other than uranium. In this regard, it is clear that we will need to study the other actinides in greater detail if we want to truly understand the unique nature of this corner of the periodic table.

For background information *see* ACTINIDE ELEMENTS; CHEMICAL BONDING; COORDINATION CHEMISTRY; COORDINATION COMPLEXES; LIGAND; LIGAND FIELD THEORY; MOLECULAR ORBITAL THEORY; OXIDATION-REDUCTION; PERIODIC TABLE in the McGraw-Hill Encyclopedia of Science & Technology.

Danil E. Smiles; Trevor W. Hayton

Bibliography. P. L. Arnold et al., Reduction and selective oxo-group silylation of the uranyl dication, *Nature*, 451:315–317, 2008, DOI:10.1038/nature06467; T. W. Hayton, Recent developments in actinide-ligand multiple bonding, *Chem. Commun.*, 49:2956–2973, 2013, DOI:10.1039/C3CC39053E; T. W. Hayton et al., Synthesis of the imido analogs of the uranyl ion, *Science*, 310:1941–1943, 2005, DOI:10.1126/science.1120069; S. T. Liddle et al., Isolation and characterization of a uranium(VI)-nitride triple bond, *Nat. Chem.*, 5:482–488, 2013, DOI:10.1038/nchem.1642; W. W. Lukens et al., Quantifying the π and σ interactions between U(V) f orbitals and halide, alkyl, alkoxide, amide and ketimide ligands, *J. Am. Chem. Soc.*, 135:10742–10754, 2013, DOI:10.1021/ja403815h; S. G. Minasian et al., Determining relative f and d orbital contributions to M-Cl covalency in MCl_6^{2-} (M = Ti, Zr, Hf, U) and $UOCl_5^{-}$ using Cl K-edge x-ray absorption spectroscopy and time-dependent density functional theory, *J. Am. Chem. Soc.*, 135:5586–5597, 2012, DOI:10.1021/ja2105015; L. A. Seaman et al., A rare uranyl(VI)–alkylate complex $[Li(DME)_{1.5}]_2[UO_2(CH_2SiMe_3)_4]$ and its comparison with a homoleptic uranium(VI)-hexaalkyl, *Angew. Chem. Int. Ed.*, 52:3259–3263, 2013, DOI:10.1002/anie.201209611.

CRISPR/Cas9 gene editing

The CRISPR/Cas9 system has been used as a general tool to engineer the genomes of many different organisms; the system is derived from naturally occurring components of a common prokaryotic immune system. Clustered regularly interspaced short palindromic repeats (CRISPRs) and cas9 are genes found in Bacteria and Archaea, and they mediate immunity in these species. cas9 is a gene encoding a DNA-cutting enzyme, whereas the CRISPR gene encodes many RNA cofactors on the Cas9 enzyme. These RNA cofactors provide specificity to the enzyme. Just 20 nucleotides of these RNA cofactors mediate DNA-cleavage specificity, and these nucleotides can be modified by the user to create new specificities. If DNA cleavage occurs in eukaryotic cells, the cleaved DNA is repaired through an error-prone mechanism called nonhomologous end-joining. Nonhomologous end-joining introduces mutations in the genomic DNA, altering the target gene's function in ways that are useful to the geneticist or genetic engineer.

History. The function of the CRISPR/Cas9 system was first identified by food microbiologists. They found that rare bacteria that survived infection by a bacteriophage (bacterial virus) had acquired immunity to future infection by the infecting bacteriophage, but not to other bacteriophage types. This demonstrated that the bacteria had developed an acquired immunity to infection (**Figs. 1** and **2**). These researchers further discovered that 20 base-pair pieces of DNA that were identical in sequence to the bacteriophage DNA had become incorporated into the bacterial genome at the CRISPR gene in the bacteriophage-resistant bacterial cells. They were able to demonstrate that these short pieces of DNA within the CRISPR gene were both necessary and sufficient to confer immune resistance to infection. This result suggested that these short pieces within the CRISPR gene were used to target and disable the bacteriophage. Although three types (I, II, and III) of CRISPR/Cas systems have now been identified in diverse bacterial and archaeal species, these systems share functional and structural homologies.

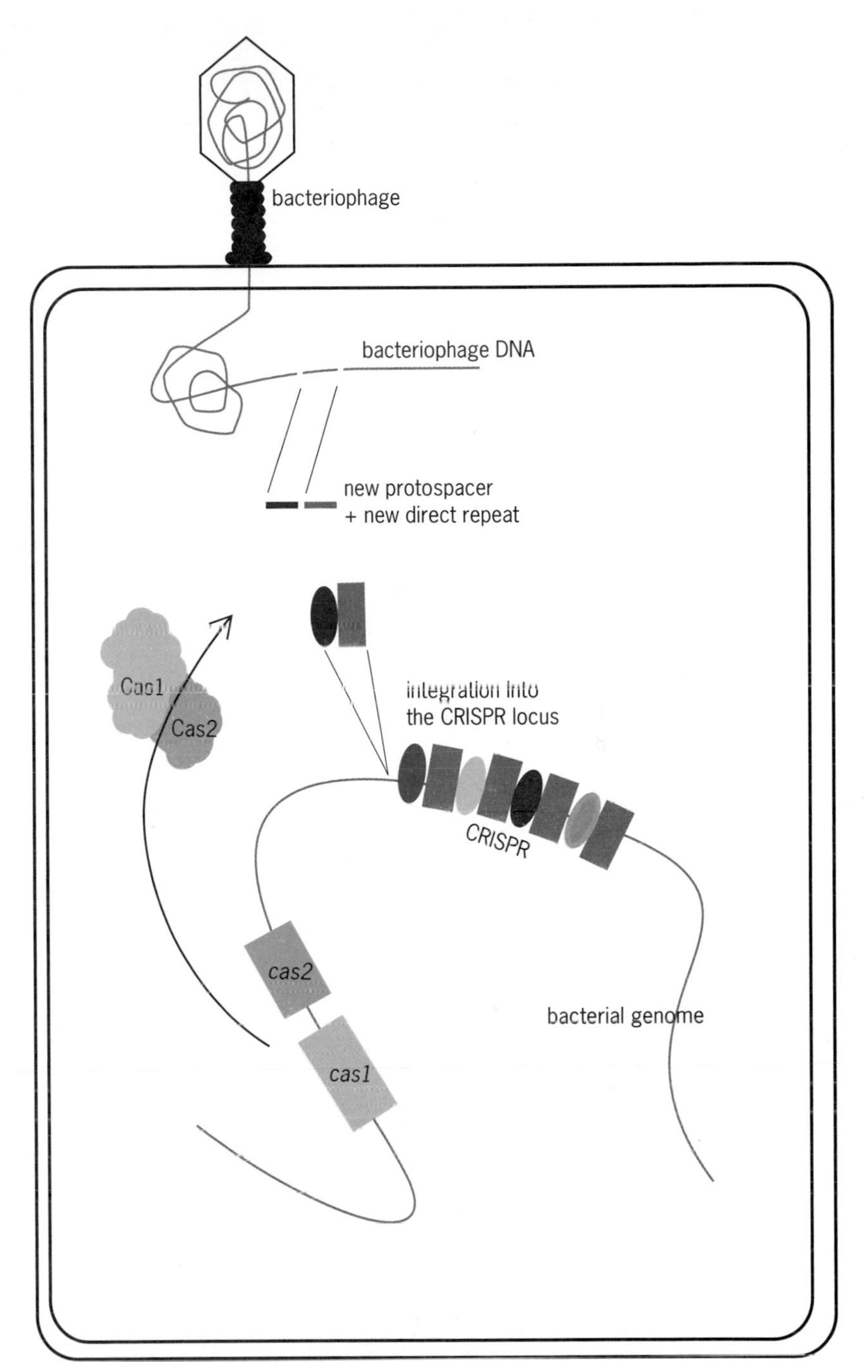

Fig. 1. Acquisition of immunity in the CRISPR/Cas9 system. Bacterial cells that are not immune to a particular bacteriophage and that survive infection acquire immunity through the function of the *cas1* and *cas2* genes. Through unknown mechanisms, Cas1 and Cas2 proteins mediate the fusion of short fragments of the bacteriophage genome, which become protospacers (*ovals*), to a direct repeat (*blue box*). This protospacer–direct repeat product is integrated into the 5′ region of the CRISPR locus. Thus, the CRISPR locus contains a record of each survived bacteriophage infection because each protospacer sequence is derived from an infection. The sequence of the protospacer base-pairs specifically to the bacteriophage genome, thereby recognizing future infections by that bacteriophage species and facilitating the cleavage and degradation of the infectious bacteriophage genome.

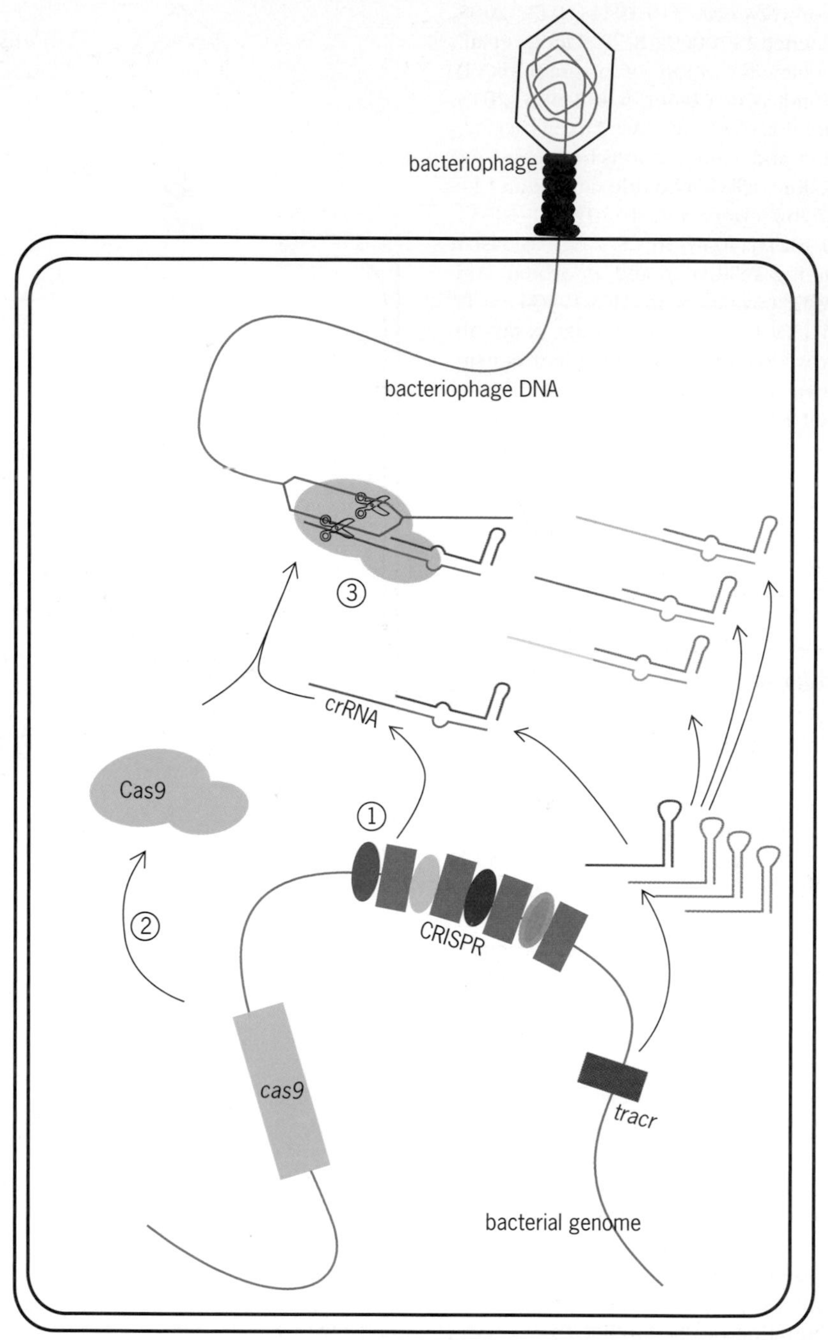

Fig. 2. The function of the CRISPR/Cas9 system in bacterial immunity. Bacterial cells that are immune to particular bacteriophage species contain protospacer sequences that match or closely match the bacteriophage genome. (1) The bacterial genome contains CRISPR repeats, which contain many copies of the direct repeat sequence (*blue boxes*). Each direct repeat is preceded by sequence diverse protospacers (*ovals*), which are derived from previous failed bacteriophage infections. The CRISPR repeats are transcribed into RNA and then cleaved into individual crRNAs containing the protospacer sequence at the 5′ end and a 3′ region derived from the direct repeat that base-pairs to the tracrRNA. The tracrRNA is expressed from an adjacent gene. (2) The *cas9* gene is transcribed and translated into the Cas9 protein, which is an enzyme that cleaves DNA. (3) The complex between the crRNA–tracrRNA complex and the Cas9 protein base-pairs to the bacteriophage DNA from an infecting phage particle and cleaves the DNA on both strands, destroying the infectious genome.

The CRISPR locus consists of a series of direct repeats (the same sequence repeated in the same orientation) interleaved with sequences of uniform size, but diverse sequence, called protospacers (Figs. 1 and 2). Protospacers are acquired from pathogens (bacteriophages or plasmids) by the action of the products of CRISPR-associated genes 1 and 2 (*cas1* and *cas2*). The protospacers confer sequence specificity to this immune system. The CRISPR locus is transcribed into CRISPR RNA (crRNA). In the type-II system, two adjacent genes are essential for pathogen interference: One produces transactivating CRISPR RNA (tracrRNA) and the other encodes the protein endonuclease Cas9 (Fig. 2). The crRNA–tracrRNA–Cas9 complex targets specific DNA sites through the recognition of a motif adjacent to the 3′ end of the protospacer [protospacer adjacent motif (PAM)] and through the base-pairing of the crRNA to the target DNA. The two endonuclease domains of the Cas9 then cleave each strand of the targeted genomic DNA (Fig. 2).

Genome engineering with CRISPR/Cas9. The discovery of the CRISPR/Cas9 function led to its very rapid development as a genome-engineering tool (**Fig. 3**). This tool is predicted to have as great an impact as restriction enzymes did in enabling genetic engineering in vitro. Restriction enzymes are not suitable for genome engineering because their 4- to 8-base-pair specificity cuts the genome in many pieces. Genome engineering, however, requires a specificity of 16 base pairs or longer. The CRISPR/Cas9 system has targeting specificities of 20 base pairs, allowing single cleavage events to occur in a genome.

Essentially all functional sequences in a genome can be targeted. The only constraint on the CRISPR/Cas9 system is the PAM at the 3′ end of the target sequence. The PAM is required for Cas9 binding. Depending on the particular bacterial species, this sequence can vary. For example, the three-nucleotide sequence for the commonly used Cas9 of *Streptococcus pyogenes* is NGG (where "N" stands for any base). This sequence occurs on average every 8 base pairs. Therefore, a crRNA can be generated to cleave DNA roughly every 8 base pairs.

Execution of the CRISPR/Cas9 genome editing is straightforward. The system can be simplified by linking the crRNA and tracrRNA together as a single guide RNA (sgRNA) [Fig. 3]. The target component of the sgRNA is designed by the scientist and synthesized chemically as a DNA oligomer (a molecule made up of a relatively small number of monomer units). Genome editing occurs when a cell has both the sgRNA and Cas9 protein. The two components are delivered as in vitro–transcribed RNA or through expression vectors. The system has been used extensively for in vivo genome engineering. The enzyme often rapidly cleaves both copies of the diploid genome. Because of the cleavage and the subsequent DNA repair, a process is initiated that results in small deletions or insertions at the cleavage site. These changes to the genome are often mutagenic and disrupt the function of the gene.

Although the system is relatively new and is undergoing numerous engineering changes, it has found practical uses in several contexts. The first use has been to confer bacteriophage resistance to commercially important bacteria used in yogurt and cheese manufacturing. This has improved the robustness of the bacterial strain to bacteriophage destruction. Another important use has come from the generation of designed sgRNA molecules, which have been used to generate mutations in cell culture systems. These site-specific mutations can be used to study gene functions in cell culture systems. Moreover,

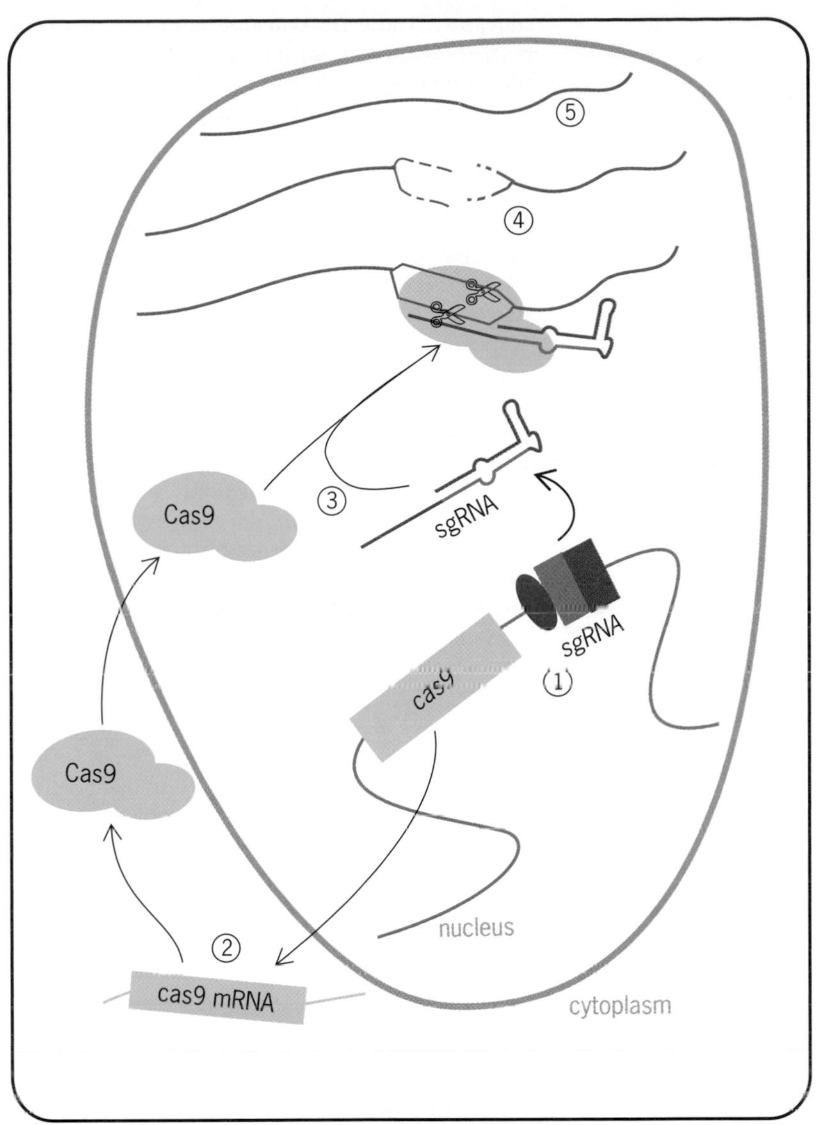

Fig. 3 The function of the CRISPR/Cas9 system in genome engineering. Multiple methods are used to engineer genomes using the CRISPR/Cas9 system. One common method is illustrated here. (1) An engineered *cas9* gene and an engineered *crRNA–tracrRNA* fusion gene, called the single guide RNA or sgRNA, are expressed in the targeted cell. The targeting sequence (*red*) is chosen by the scientist and matches the site for mutagenesis in the genome. The *cas9* gene and the sgRNA can be expressed by using a lentiviral expression system in cultured cells or by direct injection of the cas9 mRNA and sgRNA in zygotes (not shown). (2) The Cas9 protein is translated from the cas9 mRNA and (3) associates with the sgRNA. (4) The complex of cas9 and the sgRNA cleaves the targeted site in the genome of the host. Through unknown mechanisms, the genomic DNA ends are "chewed back" or extended (not shown) by a few base pairs and (5) ligated together. Steps 4 and 5 occur through the process known as nonhomologous end-joining (NHEJ).

this approach can be performed on a gene-by-gene basis, or large libraries of sgRNAs can be created to target essentially all genes, allowing whole genome phenotypic screening experiments. The approach has also been used at the level of whole organisms, notably in the genetic model organisms favored by scientists, that is, the fruit fly, the nematode, the zebrafish, and the mouse. It has also been successfully used in other models, including livestock. In addition, through engineering of the Cas9 gene, it is possible to generate modified Cas9 proteins that encode DNA-nicking activities (and not DNA-breaking activities). These DNA-nicking Cas9 proteins have been shown to further increase the specificity and allow homologous recombination at the targeted locus. Further engineering of the Cas9 gene can generate enzymatically dead Cas9 proteins, which have been shown to site-specifically bind DNA in vivo and prevent transcription of the bound locus. By adding other functionalities to the dead Cas9 proteins, scientists have been able to repurpose the system to suppress or activate transcription, and they have been able to visualize the spatial and temporal dynamics of specific DNA elements in live cells.

The CRISPR/Cas9 system is also being developed to correct genetic errors. If successful, this tool will be of huge clinical significance, facilitating gene therapies for patients with genetic disease.

For background information *see* BACTERIAL GENETICS; BACTERIOPHAGE; DEOXYRIBONUCLEIC ACID (DNA); DNA REPAIR; ENZYME; GENE; GENETIC ENGINEERING; GENETICS; GENOMICS; MUTATION; PROTEIN; RECOMBINATION (GENETICS); RESTRICTION ENZYME; RIBONUCLEIC ACID (RNA) in the McGraw-Hill Encyclopedia of Science & Technology.

Patrick S. Page-McCaw; Wenbiao Chen

Bibliography. A. Du Toit, Activating and guiding Cas9, *Nat. Rev. Microbiol.*, 12:236–237, 2014, DOI: 10.1038/nrmicro3237; J. Marx, New bacterial defense against phage invaders identified, *Science*, 315:1650–1651, 2007, DOI:10.1126/science.315.5819.1650a; J. van der Oost, New tool for genome surgery, *Science*, 339:768–770, 2013, DOI:10.1126/science.1234726.

Cruise ships

A cruise ship is a passenger ship designed exclusively for transportation of passengers in overnight accommodations on a sea voyage to at least one port of call other than the port of embarkation or disembarkation. This article mainly addresses traditional ocean-going cruise ships (**Figs. 1***a* and 1*b*), but there are also many other types of cruise ships, including river cruise ships (Fig. 1*c*), expedition cruise ships (Fig. 1*d*), sailing cruise ships (Fig. 1*e*), and coastal cruise ships (Fig. 1*f*). These ships come in a variety of shapes and sizes and are designed to accomplish specific missions while providing the hotel services expected of a large ocean-going cruise ship.

Cruise ship design continues to evolve rapidly as additional features are demanded and developed. These features include continuous innovations in the areas of passenger needs and attractions, safety enhancements, and measures to reduce the environmental footprint of these ships and their operations.

Design and unique features. The public areas for entertainment and the provision of onboard services are important considerations to the design. In essence, the payload of a cruise ship is analogous to that of a resort. Therefore, once the number of passengers has been specified, the deck areas and the internal volumes necessary to accommodate those passengers are predominant factors of the design. Cabins, dining facilities, and common spaces must also be provided for the crew. **Figure 2** is an illustration of the number of passengers and the relative number of crew carried to serve them as a function of ship size.

A cruise ship has both hotel and ship functions. The hotel functions include passenger cabins, restaurants, shops, outdoor deck spaces, and other common areas for passengers. For passengers to get to and from these areas, adequate "traffic lanes" must be provided; these include corridors, stairways, and elevators. To operate successfully as a resort at sea, service functions also need to be considered. Galleys, pantries, store rooms, and laundry facilities must be incorporated in the design. It is common policy for many crew members to remain separated from passengers when they are not in uniform, so separate crew corridors and stairways are also provided.

The mission of the ship-related functions is to safely transport the passengers from port to port. The propulsion machinery and navigational equipment are examples of systems that are vital to accomplishing this task. The ship must also produce electricity, produce hot and cold freshwater, condition air, treat sewage, and handle garbage all on its own. Safety and redundancy are crucial in the layout of all important ship systems.

Hull form. Because passenger capacity is a base criterion for cruise ship design, the hull of a cruise ship has to meet the demand of large deck areas and internal volumes to accommodate a large number of passengers. A cruise ship's gross tonnage is an indication of the internal volume of the ship, and the average gross tonnage for the world's cruise fleet is on the rise as larger ships are delivered every year than those in the fleet they typically replace or augment (**Fig. 3**). Another important consideration is that most destination ports have relatively shallow waters; therefore the hull must be designed to have a minimal draft. Additional hull dimensions with respect to gross tonnage are shown in **Fig. 4**.

Watertight subdivision. To provide the ship with the highest probability of staying afloat in the event of hull damage, the hull below the ship's bulkhead deck is separated into watertight compartments. While the ship is maneuvering or at sea, all watertight doors that provide entry into these compartments must remain closed, except in special limited circumstances.

Propulsion. Cruise ships require a substantial amount of electrical power to meet the demand of the hotel and auxiliary loads, which includes lighting, air conditioning, sewage and water treatment systems, and so forth. As a result, modern cruise ships typically utilize a diesel-electric power plant to provide electrical power for both propulsion and the ship's hotel and auxiliary loads. Modern cruise ships generally have either twin propellers and twin rudders or azimuthing propulsion units [a configuration of propellers that can be rotated to any

Fig. 1. Cruise ships. (*a*) Ocean-going cruise ship *Oasis of the Seas* (*Royal Caribbean International*). (*b*) Ocean-going cruise ship *Norwegian Pearl* (*Norwegian Cruise Line*). (c) River cruise ship *AmaDagio* (*AmaWaterways*). (*d*) Expedition cruise ship *Ocean Nova* (*Adventure Shipping Ltd.*). (e) Sailing cruise ship *Wind Surf* (*Windstar Cruises*). (*f*) Coastal cruise ship *American Glory* (*American Cruise Lines*).

horizontal angle (azimuth)], along with thrusters for either type of propulsion system, to propel and maneuver the ship. An attractive feature of using an electric drive is the convenience of controlling the propeller speed and maneuvering the ship from a number of remote locations, which allows for the system to be controlled directly by the hands of the bridge team.

Traditional diesel-electric propulsion used on cruise ships generally consists of two propellers and two rudders for propulsion and steering (**Fig. 5*a***). The electrical power produced by the ship's diesel-driven generators is distributed from the main switchboard to the frequency converters, which control the speed and torque of the propulsion motors and thus the propellers.

Azimuthing propulsion units, also known as azipods or azimuth thrusters, are podded propulsion units designed to rotate 360° around a vertical shaft and that have an electric motor built into the hub (Fig. 5*b*). This design incorporates a much shorter propeller shaft within the unit, thus eliminating the need for long shaft lines. The rotational characteristic of these units also eliminates the need for

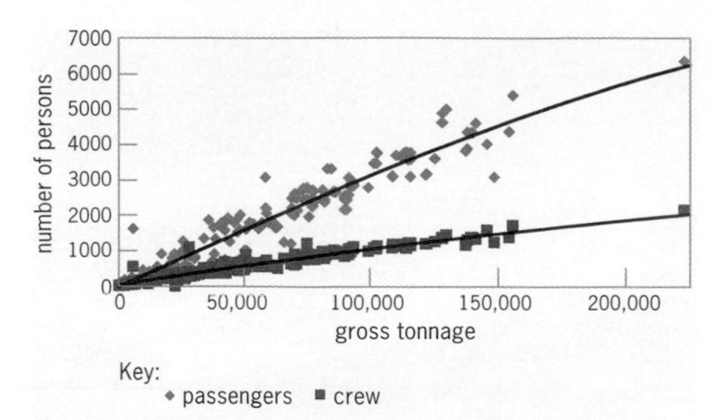

Fig. 2. Passenger and crew capacity of cruise ships as a function of gross tonnage. Gross tonnage is the entire internal volume of a ship, less certain exempted spaces, expressed in units (tons) of 100 ft^3 (2.83 m^3).

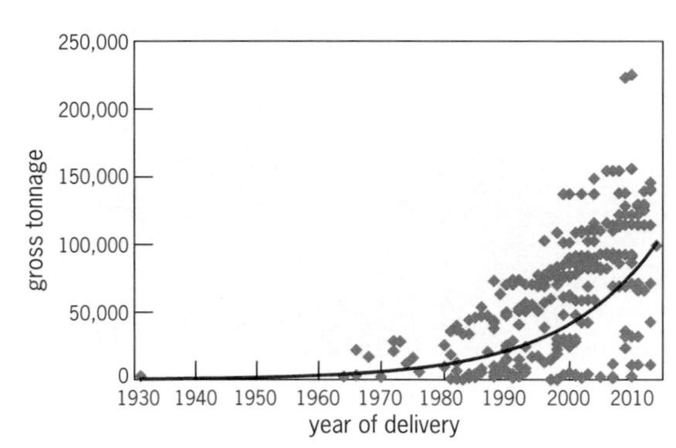

Fig. 3. Size of the current world cruise fleet.

rudders. The podded propulsion units tend to be of the "pulling" type, which has the added benefit of an undisturbed wake field. Other advantages include few vibrations and little noise, excellent maneuvering capabilities, and typically lower fuel costs due to a reduction in power demand than for similarly sized conventional propulsion systems.

Life-saving survival craft. Chapter III of the International Convention for the Safety of Life at Sea (SOLAS) prescribes the types and capacities of survival craft required to be carried aboard ships. For cruise ships, the capacity of the lifeboats on each side of the ship (**Fig. 6**) must be at least 37.5% of all persons onboard, thus totaling 75% of all persons onboard as a minimum. Inflatable or rigid life rafts may be installed to provide the rest of the survival-craft capacity onboard so that the total capacity covers a minimum of 125% of the ship's allowed passengers and crew. A cruise ship may also be outfitted with a marine evacuation system (MES), by which persons can slide into the system's survival craft by way of slides or chutes.

Cabins. The size and number of cabins in which to accommodate passengers is another important design consideration of cruise ships. Generally, standard cabin sizes can range from 10 m^2 (108 ft^2) in the budget category to 35 m^2 (377 ft^2) in the luxury category. Cabins can be located on the outside, where they can have windows and frequently balconies, or on the interior, with no windows. Cabins are individually fabricated in specialized workshops. During hull assembly, the finished cabin unit, including its furniture, is then lifted by crane into the ship through temporary openings cut into the side shell. When installing cabins in this fashion, the spacing needed for insulation, wiring, and structural components between the cabin wall and the side shell must be considered.

Public passenger spaces. Public passenger spaces aboard cruise ships include dining rooms, theaters, outdoor deck areas, spas, pools, indoor and outdoor cafes, bars and lounges, nightclubs, casinos, and fitness clubs. Means of access for passengers from their cabins to public spaces and all areas in between is another important consideration for a cruise ship's arrangement. The stairways and elevators must be easy to find, and all pathways should be well guided.

Crew and service spaces. On cruise ships, sufficient space for food preparation, laundry, and store rooms is vital to providing the necessary services to the guests. These spaces, along with the crew corridors, should not be used by passengers.

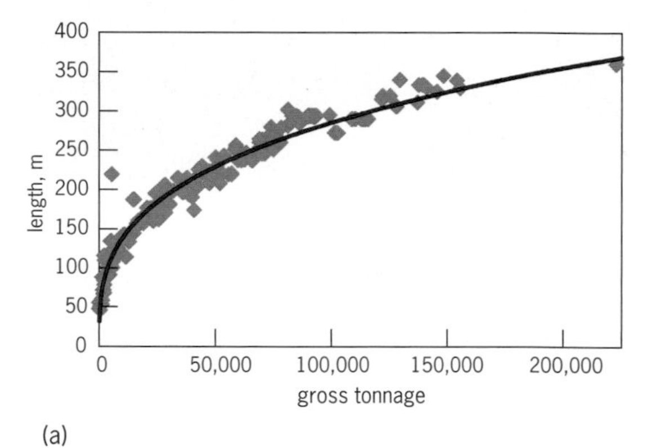

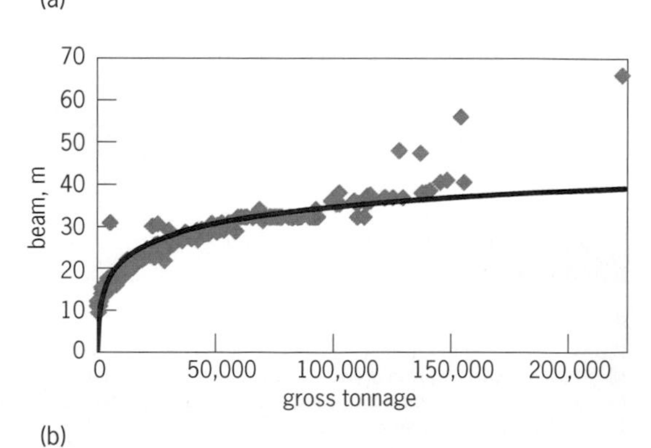

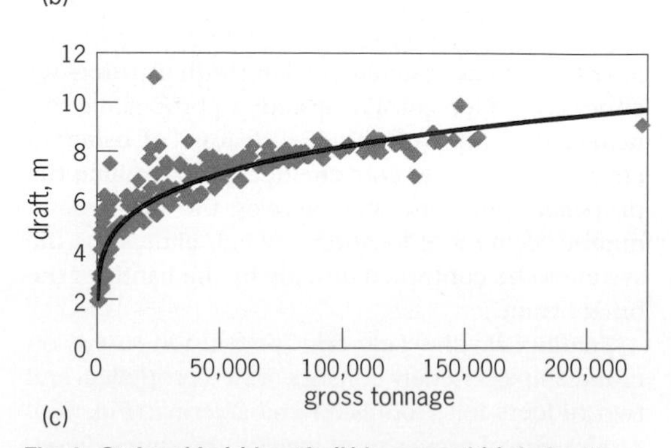

Fig. 4. Cruise ship (*a*) length, (*b*) beam, and (*c*) draft with respect to ship size. 1 m = 3.28 ft.

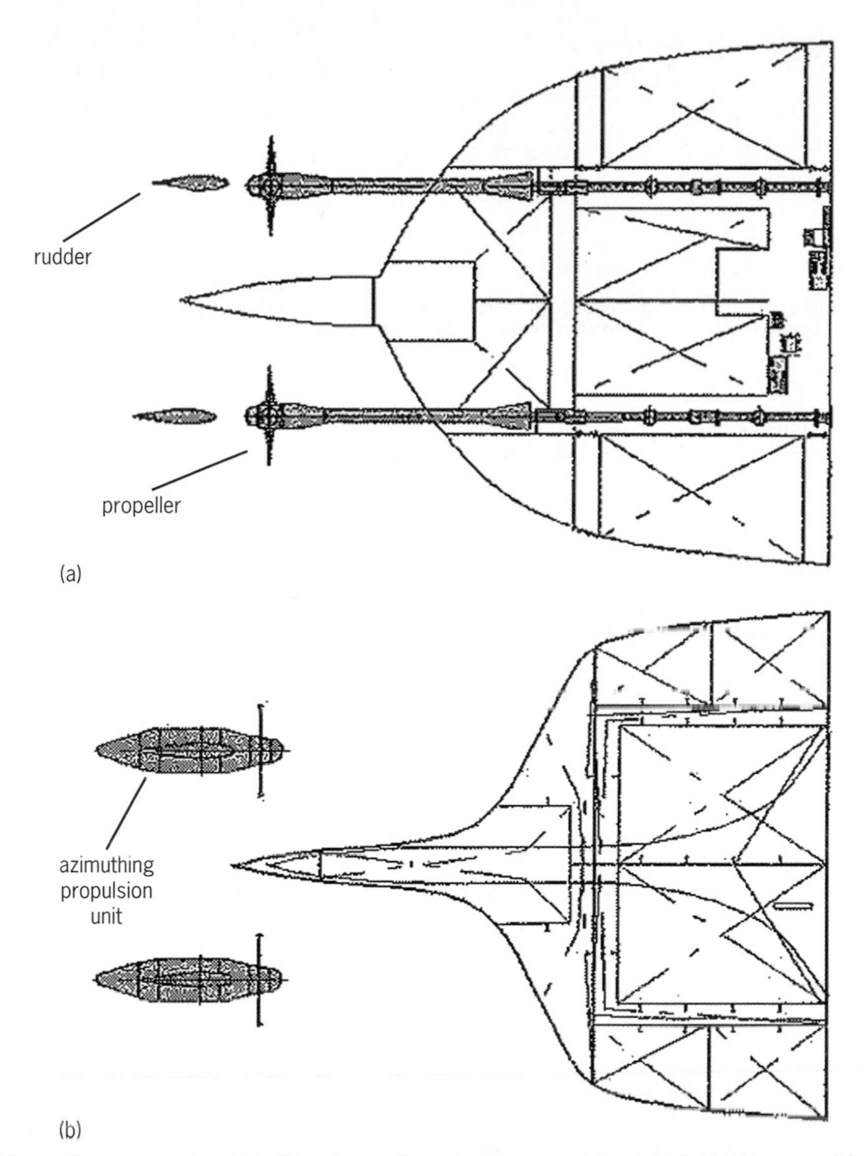

Fig. 5. Propulsion unit arrangements. (*a*) Traditional propeller and rudder arrangement. (*b*) Azimuthing propulsion unit arrangement. These views are horizontal cutaway sections of the ship as if you were looking down at the propulsion unit from above.

Fire protection. All cruise ships are required to be subdivided longitudinally into main vertical zones, also known as fire zones. These boundaries are classified as Class A-60 bulkheads, meaning that they are insulated with approved noncombustible materials such that the average temperature of the unexposed side will not rise about 180°C (324°F) above the original temperature with 60 min of exposure. The length of a main vertical zone can be extended to 48 m (157 ft) if the fire zone then ends in line with a watertight bulkhead, or if the space being enclosed contains a large public space (**Fig.** 7), provided that the total deck area between bulkheads does not exceed 1600 m^2 (17,222 ft^2).

All spaces within main vertical zones are separated from each other with different class divisions, depending on the designated usage of the space. Bulkheads and overheads are lined with noncombustible materials. All accommodation spaces and service spaces, including corridors and stairways, must be equipped with automatic fire detection and

Fig. 6. A row of lifeboats aboard a cruise ship.

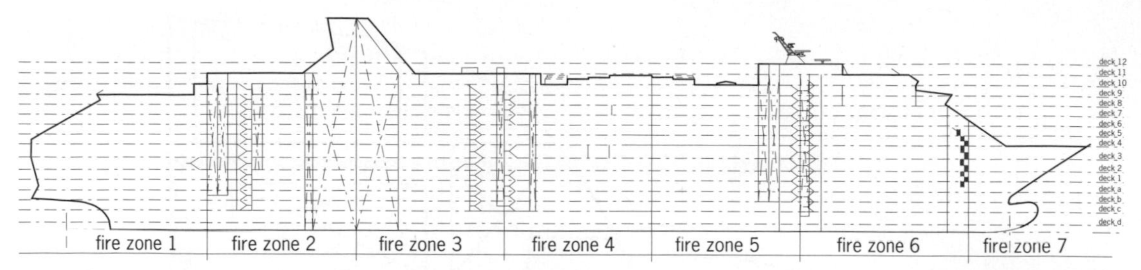

Fig. 7. General main vertical zone arrangement on a cruise ship showing boundaries. This view is a vertical cutaway section of the ship as if you were looking at the fire zones from the side of the ship.

extinguishing systems. In addition, a fixed-pressure, water-spraying fire extinguishing system must be installed on balconies.

For background information *see* MARINE MACHINERY; PROPELLER (MARINE CRAFT); SHIP DESIGN; SHIP POWERING, MANEUVERING, AND SEAKEEPING in the McGraw-Hill Encyclopedia of Science & Technology.

Kierstin M. Del Valle; Charles V. Darr

Bibliography. J. A. Beverley, Electric propulsion drives, pp. 304–309, in R. L. Harrington (ed.), *Marine Engineering*, Society of Naval Architects and Marine Engineers, Jersey City, NJ, 1992; K. Levander, Passenger ships, pp. 37-1–37-39, in T. Lamb (ed.), *Ship Design and Construction*, vol. 2, 3d ed., Society of Naval Architects and Marine Engineers, Jersey City, NJ, 2004; K. Van Dokkum, *Ship Knowledge: Ship Design, Construction and Operation*, 8th ed., Dokmar Maritime Publishers, Enkhuizen, the Netherlands, 2013; D. Ward, *Berlitz Cruising & Cruise Ships 2015*, 23d ed., Ingram Publisher Services, La Vergne, TN, 2014.

Dementia

Dementia is a syndrome characterized by a generalized decline in cognition that is severe enough to cause functional impairment in daily activities. Dementia can be caused by a variety of conditions common in late life, including vascular diseases or strokes, Parkinson's disease, Lewy body disease (dementia associated with abnormal brain cell structures called Lewy bodies), alcoholism, and Alzheimer's disease (which is, by far, the most common cause of dementia).

Alzheimer's disease. Alzheimer's disease was first described by Alois Alzheimer in a 1907 case report of presenile dementia in a 51-year-old woman. Until the 1970s, it was generally believed that the condition was a relatively rare phenomenon that occurred in middle age. Today, it is known that Alzheimer's disease and related dementias are very common, especially in advanced industrial countries that have rapidly growing geriatric populations. The rates of Alzheimer's disease and other dementias increase dramatically with age. By age 95 and older, at least two-thirds of people will have enough cognitive impairment to significantly diminish their ability to function independently. The rising number of older persons with dementia who are being cared for by family members or who have been placed in institutional settings has captured the attention of the public and policy makers around the globe. The high cost of long-term care alone is a compelling reason for medical researchers and society to be interested in the "epidemic" of Alzheimer's disease.

Pathophysiology and diagnosis. Alzheimer's disease affects the structure of the brain. Two main types of neuropathologic changes, or lesions, are seen in the brains of affected persons, particularly areas of the brain that are used for memory and other cognitive functions. The disease is characterized by the formation of amyloid plaques (dense, largely insoluble deposits of protein and cellular material) outside and around the brain's neurons, and neurofibrillary tangles (insoluble twisted fibers) that build up inside neurons. A definitive diagnosis of Alzheimer's disease by autopsy involves counting the number of plaques and tangles found in specific brain structures. Increased brain volume atrophy that occurs secondary to neuronal death is another diagnostic characteristic of the disease. Although the postmortem analysis of brain tissue is currently the "gold standard" for diagnosing Alzheimer's disease, it has two significant limitations. First, Alzheimer's disease plaques and tangles are sometimes found in persons who did not exhibit symptoms of the disease before death. Second, both brain atrophy and the development of plaques and tangles are associated with general aging, making it difficult to discriminate between the pattern of neuropathologic lesions that occurs with Alzheimer's disease and the pattern of atrophy and lesions that may occur in normal older adults.

The most common early clinical symptoms of Alzheimer's disease include a loss in recent (short-term) memory that is accompanied by problems with language and judgment. The affected individual may exhibit personality changes and may begin to have difficulty with familiar activities, such as operating household appliances, preparing meals, or managing finances. As the disease progresses, disturbances in behavior, including emotional outbursts, wandering, or agitation, may occur. Eventually, as more areas of the brain become affected, the patient may become bedridden, incontinent, and totally dependent on others for all aspects of daily care.

Clinicians rely on a variety of tools to diagnose "probable" or "possible" Alzheimer's disease in persons experiencing progressive changes in their memory and cognition. Once dementia is suspected, physicians obtain a comprehensive medical history and conduct a thorough physical examination, including laboratory tests, to rule out other causes of cognitive decline. Today, there is still no accepted single test for diagnosing Alzheimer's disease. In many cases, neuropsychological testing is conducted to measure memory, language skills, and other abilities related to brain functioning. Brain imaging tests, such as computed tomography scanning and magnetic resonance imaging, are often conducted to look for evidence of stroke or other structural brain diseases.

Causes. The most important risk factor for Alzheimer's disease is age. However, a growing body of research into the etiology and risk factors for development indicates that Alzheimer's disease is most likely caused by the interaction of a number of genetic and environmental factors. Gene mutations on three different chromosomes (1, 14, and 21) have been identified as the cause of many familial (inherited), early-onset cases. This form of Alzheimer's disease, which generally affects persons between the ages of 30 and 60, is very rare, occurring in less than 1% of all cases. The condition typically occurs in families because the inheritance of any one of these mutations means that an individual will develop the associated form of early-onset Alzheimer's disease.

There is currently no evidence that the gene mutations associated with early-onset Alzheimer's disease play a major role in the more common, nonfamilial (or sporadic) form of late-onset Alzheimer's disease. However, genetics do contribute to the risk of development of sporadic Alzheimer's disease. The apolipoprotein E (ApoE) gene, which is found on chromosome 19, codes for a protein that helps carry blood cholesterol through the body. It is also found in excess amounts in the plaques that accumulate in the brains of persons with sporadic Alzheimer's disease. There are three common forms (alleles) of the ApoE gene: $\varepsilon 2$, $\varepsilon 3$, and $\varepsilon 4$. Scientists have found that persons who have one or two copies of the ApoE $\varepsilon 4$ allele have an increased risk of developing sporadic Alzheimer's disease. In contrast, the relatively rare ApoE $\varepsilon 2$ allele may be associated with a decreased risk of developing sporadic Alzheimer's disease and with a later onset when the disease does develop. Unlike the inheritance of gene mutations associated with early-onset, familial Alzheimer's disease, the mere inheritance of one or two ApoE $\varepsilon 4$ alleles does not guarantee that an individual will develop late-onset, sporadic Alzheimer's disease. Likewise, the presence of ApoE $\varepsilon 2$ alleles does not ensure that sporadic Alzheimer's disease will not occur. In addition to the ApoE gene on chromosome 19, recent advances in genetic research have uncovered susceptibility areas located on many other chromosome genes. These advances may help to provide a better understanding of how the disease develops, but they do not explain very much of the risk that a person has in developing Alzheimer's disease. Overall, the finding that risk of disease development is to some degree linked with inheritance of various alleles may help explain some of the variations pertaining to the age of disease onset and the prevalence of disease among various racial and ethnic groups. It also provides for research into its diagnosis, treatment, and prevention.

In addition to research into the genetics of Alzheimer's disease, there is considerable interest in possible nongenetic or environmental factors that may alter disease risk. Risk factors that have been identified in the scientific literature include a history of serious head trauma, depression, and occupational exposure to neurotoxins. Lower levels of education are associated with an earlier onset of the disease, and advanced levels of education seem to delay onset. Recent studies have suggested that lower levels of brain maturation related to social and economic deprivation during early childhood development (when brain development, particularly in the regions responsible for memory and higher-order function, occurs) are associated with an increased risk of Alzheimer's disease. There is also growing evidence that the presence of other neurological, cerebrovascular, or inflammatory diseases may increase the clinical severity of Alzheimer's disease or perhaps its development. In combination, these findings suggest that the onset and progression of Alzheimer's disease are the result of a combination of genetic and nongenetic factors that affect total brain reserve and age-related decline (for example, cell loss and interneuronal activity). Thus, factors that reduce brain reserve (resistance to damage) and speed cognitive decline, such as exposure to neurotoxins or development of multiple small brain strokes, may increase the risk of development of Alzheimer's disease, particularly in genetically predisposed individuals. Conversely, other factors that promote general health and well-being show great promise for delaying disease onset and possibly slowing the rate of decline. In addition, a reduction in vascular risk is emerging as a so-called modifiable risk factor.

Other dementia subtypes. Vascular dementia, related to cerebrovascular disease, is the second most common form of age-related dementia. This type of dementia is not caused by a single condition. Instead, it may result from one or more large-vessel infarcts (localized areas of dead tissue caused by an obstructed inflow of arterial blood) or strokes, or from the presence of widespread small-vessel disease or infarcts, particularly in the frontal and subcortical portions of the brain. Vascular dementia also frequently coexists with other causes of dementia, especially Alzheimer's disease. The type, extent, and severity of cognitive impairment found in cases of vascular dementia are associated with the location and size of the strokes and vascular changes in the brain. For example, patients with left-brain lesions are more likely to show problems with verbal naming and fluency tasks, whereas patients with right-brain lesions exhibit more problems with visuospatial skills. Frontal or subcortical lesions also may produce changes in personality and judgment. A diagnosis is generally made in cases in which there are external changes detected during a neurologic examination, there is evidence of vascular disease or infarcts on neuroimaging tests, and there is a history of vascular risk factors accompanied by clinical symptoms of progressive dementia. A more common feature observed among older persons is that patients display dementia resulting from various causes. That is, the brain at autopsy will show signs of Alzheimer's disease, frontotemporal dementias, Parkinson's disease changes, and vascular degeneration in various combinations.

Frontotemporal dementias (including Pick's disease) are associated with a significant atrophy of the frontal and temporal lobes of the brain. This atrophy is usually evident on neuroimaging. The frontotemporal dementias are the second most common degenerative dementias (after Alzheimer's disease) in persons who develop dementia before the age of 65. Subcortical dementia refers to the progressive cognitive changes associated with a variety of conditions, such as Parkinson's disease, Lewy body disease, supranuclear palsy, combined-system disease, and Huntington's disease. These diseases are characterized by changes in subcortical or deep-brain structures, including the basal ganglia, caudate, substantia nigra, and putamen. A diagnosis is made primarily based on the results of a neurologic examination and the clinical course. Huntington's disease, which is caused by an abnormal gene on chromosome 4, can also now be confirmed by genetic testing. In comparison with patients who have Alzheimer's disease, individuals with frontotemporal or subcortical dementias tend to have more prominent impairments in executive functioning (for example, problem solving or planning), visual perceptions, mental and motor speed, and language fluency. Mood and personality changes, particularly depression, are common. Memory difficulties, although present, may be less severe than for the patient with Alzheimer's disease and can often be compensated for with recognition cues and repetition. Patterns of cognitive decline can vary widely among individuals with frontotemporal or subcortical diseases.

In older persons who are at risk for developing multiple dementia-related illnesses, a differential diagnosis can be particularly difficult. In fact, "mixed" dementia (usually referring to a common form of dementia caused by a combination of Alzheimer's disease and cerebrovascular disease) frequently coexists with Parkinson's disease and Lewy body disease in varying combinations. There is a general need for research to improve diagnostic methods for differentiating among vascular, frontotemporal, and subcortical dementias. More accurate diagnoses are particularly important if more effective treatments become available.

Treatment. It is important to provide ongoing care for individuals with dementia. Virtually every dementia patient needs to have safety and security issues addressed. Treatment has to be tailored to the individual patient, their family, and their surroundings. Medications to slow decline have been developed and promoted with great enthusiasm, but most patients and their families are disappointed because benefits are not evident and the medications have side effects. In general, these drugs are designed to prevent the breakdown of acetylcholine (a neurotransmitter) and to improve its action in the central nervous system. Newer drugs are continuously in development. Epidemiologic studies have focused on a host of other drugs that might help prevent or delay the disease. Thus far, the evidence is disappointing and the prospect of a single drug, or magic pill, as a cure for this common ailment does not appear evident at this time. Lifestyle factors and the treatment of other conditions common in old age seem to offer greater promise based on current knowledge. Exercise, including walking for exercise, is perhaps the most agreed-upon treatment to delay onset at this time. Various medications are also used to treat some of the behavioral symptoms that are common in patients with Alzheimer's disease, including agitation, depression, and sleep disorders.

Nonmedication treatments for dementia are typically focused on helping the patient and caregiver maintain an optimal quality of life by reducing patient behavior problems and caregiver burden. Behavioral approaches to understanding, preventing, and intervening with psychiatric symptoms and behavior problems have been successfully used in both community and institutional settings. Good medical care to treat or prevent other common age-related illnesses can also help patients with dementia remain functional for as long as possible. For example, patients who maintain an appropriate daily level of physical activity may reduce the risk for fall-related injuries and have fewer problems with sleep disturbance, depression, and motor restlessness. Other age-related conditions (including daytime urinary incontinence, vision and hearing loss, and muscular or arthritic pain) that interfere with a patient's quality of life are exacerbated by dementia and thus may be improved with proper medical evaluation and treatment.

For background information *see* ACETYLCHOLINE; AGING; ALLELE; ALZHEIMER'S DISEASE; BRAIN; COGNITION; GENE; HUMAN GENETICS; HUNTINGTON'S DISEASE; MENTAL DISORDERS; MUTATION; NERVOUS SYSTEM DISORDERS; PARKINSON'S DISEASE in the McGraw-Hill Encyclopedia of Science & Technology.

Eric B. Larson

Bibliography. G. J. Kennedy, Dementia, pp. 1079–1093, in C. K. Cassel et al. (eds.), *Geriatric Medicine*, 4th ed., Springer, New York, 2003; E. B. Larson, K. Yaffe, and K. M. Langa, New insights into the dementia epidemic, *New Engl. J. Med.*, 369(24):2275–2277, 2013, DOI:10.1056/NEJMp1311405; V. M. Moceri et al., Early-life risk factors and the development of Alzheimer's disease, *Neurology*, 54:415–420, 2000, DOI:10.1212/WNL.54.2.415;

H. W. Querfurth and F. M. LaFerla, Alzheimer's disease, *New Engl. J. Med.*, 362:329–344, 2010, DOI:10.1056/NEJMra0909142; S. E. Tom et al., Characterization of dementia and Alzheimer disease in an older population: Updated incidence and life expectancy with and without dementia, *Am. J. Publ. Health*, in press, 2014; F. Zou et al., Brain expression genome-wide association study (eGWAS) identifies human disease-associated variants, *PLoS Genet.*, 8(6):e1002707, 2012, DOI:10.1371/journal.pgen.1002707.

Dialectical behavior therapy

Dialectical behavior therapy (DBT) is a psychotherapy developed by psychologist Marsha M. Linehan in the 1990s and designed originally to treat borderline personality disorder (BPD). DBT derives from a cognitive-behavioral theoretical foundation and uses four main behavior-change strategies: skills training, exposure therapy, cognitive therapy, and contingency management. Through behavioral theory, DBT focuses on increasing skills to process emotions, solve problems, understand the functions of behaviors, and change contingency patterns. Balancing acceptance and change in DBT is thought to create a therapeutic dynamic in which patients feel understood and accepted and can work toward changing their behavior.

DBT is guided by three overarching theories: the biosocial theory of development and maintenance of BPD behaviors, behavioral theory, and dialectical philosophy. In 1993, Linehan presented the biosocial theory of BPD to understand the intense emotions and behaviors frequently displayed by individuals with BPD. This theory posits that a predisposition to biological dysfunction in the form of heightened emotional sensitivity, greater emotional intensity, and a slower return to emotional baseline, exacerbated by an invalidating environment, creates a pattern over time that leads ultimately to extreme attempts to regulate emotion in these individuals. Recently, Linehan has updated this theory with more research to suggest that a vulnerability that is expressed in childhood as impulsivity is also an important biological component of BPD.

Treatment. Patients in standard DBT treatment receive three modes of treatment: individual therapy, skills groups, and phone coaching sessions with a DBT therapist. A typical length of treatment in a DBT program can last up to a year, although research suggests that behavioral control can be achieved within 4 to 8 months of comprehensive DBT. DBT is broken down into four stages of treatment, with targeted goals in each stage. In stage 1, the goal is to move from behavioral dyscontrol to behavioral control. The first treatment target in this stage is to reduce and eventually eliminate life-threatening behaviors or risky behaviors that put the patient at serious physical risk. These behaviors include suicide attempts, nonsuicidal self-injury, and suicidal ideation. Once these behaviors are significantly reduced or eliminated, the treatment moves to the second target, which is to reduce therapy-interfering behaviors, such as missing appointments, not completing homework assignments, not collaborating with the therapist, and avoiding the discussion of treatment goals at sessions. Additionally, eliminating or reducing the use of hospitalization for crisis management is another target during this stage. The third target during stage 1 of treatment is to decrease behaviors that reduce the patient's quality of life. This target includes phobias, avoidant behaviors, isolation, eating disorders, neglecting physical or medical problems, not attending work or school, destroying relationships, and lack of social support. Patients are also encouraged to increase behaviors that improve their quality of life.

Stage 2 treatment focuses on moving the patient from avoiding emotions or dissociating to being able to experience emotions fully as they occur. This is the stage where a DBT therapist may use other empirically supported treatments to treat posttraumatic stress disorder (PTSD) or other co-occurring disorders. In stage 3, patients start to address life problems, such as employment, marital or social conflict, career goals, and health goals. Stage 4, the final stage of therapy, is directed toward solving existential problems that may leave individuals feeling incomplete, increasing meaning and meaningful connections in their lives.

DBT uses four sets of skills that are arranged around the acceptance/change dialectic. Distress-tolerance skills help patients tolerate unpleasant emotions when their full participation in the emotional experience is not possible or effective for the patient. Mindfulness skills assist patients to remain in the present moment and to tolerate crises without engaging in behaviors that may worsen a situation. Emotion-regulation skills teach patients how to identify, tolerate, and regulate their emotional experience. Interpersonal effectiveness skills help patients manage relationships, increase assertion skills, and gain confidence and practice in self-respect by decreasing "game-playing" or behaviors that undermine their self-respect and increase feelings of guilt and shame.

Current research findings. DBT is considered an empirically supported treatment for BPD and nonsuicidal self-injury and suicidal behaviors. To date, 12 randomized controlled trials have investigated the efficacy of DBT, which has been shown to be effective in comparison to treatment-as-usual (standard practice interventions) at reducing self-injurious behaviors, suicidal ideation, treatment dropout, length and frequency of hospitalization, and medical severity of self-injurious behaviors.

Altered or adapted versions of DBT have also been tested. A 16-session DBT skills class for treatment-resistant depression in adults demonstrated efficacy versus a wait-list control group (that is, a group assigned to a waiting list that will receive an intervention after the active treatment group does).

Efficacy was also seen in a 24-week randomized controlled trial examining DBT plus medication versus medication alone. In addition, a values-focused DBT skills group has been presented and tested. Finally, DBT has been adapted for individuals with co-occurring substance abuse and BPD, eating disorders, treatment-resistant depression, and attention-deficit hyperactivity disorder, as well as for adolescents, elderly people, and prison populations.

Limitations and opportunities for development. Compared to other treatments for BPD, DBT has the most extensive evidence base in support of its efficacy. However, there are criticisms and requirements for future research in DBT. First, DBT encompasses a wide range of strategies and skills. Some of these are theoretically contradictory, such as exposure therapy and distraction techniques. It is possible that only portions of these strategies are necessary for successful outcomes in therapy. Therefore, dismantling studies are needed to determine which components are most useful in producing the desired changes. Second, DBT is a complex, multimodal, multistage treatment that requires a significant amount of therapist training, commitment from the patient, and organizational commitment to be implemented in the manner intended by Linehan. Currently, there is insufficient evidence to determine whether all modes and stages of treatment are necessary for all DBT patients.

For background information *see* ANXIETY DISORDERS; BRAIN; COGNITION; EMOTION; MENTAL DISORDERS; PSYCHOLOGY; PSYCHOTHERAPY; PUBLIC HEALTH; SUICIDE in the McGraw-Hill Encyclopedia of Science & Technology. Amy Yule Cameron

Bibliography. S. Kliem, C. Kröger, and J. Kosfelder, Dialectical behavior therapy for borderline personality disorder: A meta-analysis using mixed-effects modeling, *J. Consult. Clin. Psychol.*, 78:936–951, 2010, DOI:10.1037/a0021015; M. M. Linehan, *Cognitive-Behavioral Treatment of Borderline Personality Disorder*, Guilford Press, New York, 1993; M. M. Linehan and H. Schmidt III, The dialectics of effective treatment of borderline personality disorder, pp. 553–584, in W. T. O'Donohue and L. Krasner (eds.), *Theories in Behavior Therapy: Exploring Behavioral Change*, American Psychological Association, Washington, DC, 1995; M. M. Linehan et al., Cognitive behavioral treatment of chronically parasuicidal borderline patients, *Arch. Gen. Psychiatry*, 48:1060–1064, 1991, DOI:10.1001/archpsyc.1991.01810360024003; M. M. Linehan et al., Dialectical behavior therapy for patients with borderline personality disorder and drug-dependence, *Am. J. Addict.*, 8:279–292, 1999, DOI:10.1080/105504999305686; T. R. Lynch et al., Dialectical behavior therapy for depressed older adults: A randomized pilot study, *Am. J. Geriatr. Psychiatry*, 11:33–45, 2003, DOI:10.1097/00019442-200301000-00006; T. R. Lynch et al., Treatment of older adults with co-morbid personality disorder and depression: A dialectical behaviour therapy approach, *Int. J. Geriatr. Psychiatry*, 22:131–143, 2007, DOI:10.1002/gps.1703; S. L. Rizvi, L. M. Steffel, and A. Carson-Wong, An overview of dialectical behavior therapy for professional psychologists, *Prof. Psychol. Res. Pract.*, 44:73–80, 2013, DOI:10.1037/10029808.

DNA vaccination

Vaccination has proven to be one of the cheapest and most effective defenses against infectious diseases. However, most vaccines are composed of proteins, and proteins are very sensitive to changes in temperature. In the 1990s, investigators made a surprising observation that led to a revolution in the way that vaccines are thought about and designed. Specifically, when mice were injected intramuscularly with plasmid DNA, genes on the plasmid could be expressed in the host's muscle tissues. Plasmids are small autonomously replicating circular pieces of bacterial DNA that are easy to manipulate and purify. Scientists have been genetically engineering plasmids for decades to develop tools to understand how genes and their RNA and protein products function in living cells. What was surprising was the ability of mammalian muscle cells to take up and express these foreign bits of DNA. Subsequent studies have shown that the expression of foreign genes following injection with plasmid DNA can provoke the body's immune system to mount both antibody and T-cell (T-lymphocyte) responses to the foreign proteins (antigens) expressed by the plasmid. This led to the birth of a completely new field of vaccine research that has been variously termed DNA, nucleic acid, or genetic vaccination.

Conventional vaccines. Most conventional vaccines consist of either a weakened (attenuated) or killed version of the disease-causing microbe. Attenuated pathogens are rarely associated with disease. However, their ability to provoke effective immune responses requires that they actively replicate in and cause damage to host cells. Because live attenuated vaccines reproduce in the body, they can pose a risk to individuals with immature or weakened immune systems. In contrast, killed vaccines, which can be composed of either the whole pathogen or an immunogenic protein or proteins derived from the pathogen, do not replicate in the body. Therefore, killed vaccines are generally safer than live attenuated vaccines. Although they are safer, killed vaccines elicit very weak immune responses and often need to be mixed with adjuvants (materials that enhance the action of a vaccine) to stimulate responses from the body's immune system that will protect the host from the disease-causing organism. Even when mixed with adjuvants, most nonreplicating vaccines often fail to stimulate the development of CD8+ cytotoxic T cells, which are critical for protection against many viruses. Thus, many nonreplicating vaccines are ineffective against viral pathogens.

Nucleic acid vaccines. Nucleic acid vaccines, which can be composed of either DNA or RNA, offer a number of theoretical advantages over more conventional live attenuated and killed vaccines. Large-scale manufacturing of nucleic acids is inexpensive, and vaccines against multiple antigens can be produced using the same isolation and purification methodologies. Nucleic acid sequences also can be easily manipulated to correspond to genetic sequences in rapidly mutating pathogens, such as HIV and influenza. This could facilitate rapid responses to global pandemics and could lead to the development of individually tailored designer vaccines. The most important perceived advantage of nucleic acid vaccines is their thermostability (stability at high temperatures). Almost all of the protein-based vaccines used today need to be kept cold to maintain their potency. This requirement for refrigeration adds significantly to the cost of the vaccine and makes it difficult to get vaccines to the people who need them most. Nucleic acid vaccines can maintain their effectiveness even when they are not refrigerated. Hence, they should be less expensive to distribute than conventional vaccines.

Nucleic acid vaccines have been shown in numerous animal species to induce potent and protective T-cell and B-cell (B-lymphocyte) responses against a variety of pathogen- and cancer-derived antigens. This has led to considerable interest in the commercialization of plasmid-based vaccines for use in both animals and humans. Although a few plasmid vaccines have been licensed for veterinary use (for example, an infectious hematopoietic necrosis virus vaccine for fish, a melanoma cancer vaccine for dogs, and a vaccine against West Nile virus for horses), clinical trials of DNA vaccines in humans have been largely unsuccessful. For reasons that are somewhat unclear, the immune responses elicited by DNA vaccines in humans are much weaker than those that can be achieved by vaccination with conventional protein-based vaccines. Consequently, considerable effort has gone into the development of techniques to improve the immunogenicity of DNA vaccines.

Improving the physical delivery of DNA. In order for a plasmid-encoded gene to be transcribed into messenger RNA (mRNA) and translated into protein, the plasmid first needs to be taken into the host cell nucleus. This process requires transport of the plasmid across two separate lipid bilayer barriers, namely, the outer cell membrane and an inner nuclear membrane. Improved vaccination of animals has been achieved by complexing (packaging) DNA within artificial lipid bilayer membranes and cationic polymers and by encapsulation of DNA in biodegradable microparticles and nanoparticles. Moreover, in a process known as electroporation, an externally applied electric field has been used successfully to increase the permeability of cells near the injection site, thereby increasing the uptake and subsequent expression of the injected DNA.

High-pressure delivery systems, in which DNA is either suspended in solution or dried onto gold or tungsten nanoparticles and shot directly into skin cells, have also been effective for delivering plasmid vaccines to the interior of the cell. An additional benefit of this approach is the elimination of the need for needles and syringes, thereby reducing the cost and minimizing the trauma associated with vaccination.

Increasing the expression and processing of DNA-encoded antigens. In attempts to improve the overall efficiency of DNA vaccination, experimental plasmid vaccines have been designed to study how various genetic elements that control gene expression, mRNA stability, and translation of mRNA into protein affect the resulting immune response. In addition, genetic modification of the expressed protein has been observed to be an effective way to enhance both B- and T-cell responses to plasmid-encoded antigens. For example, the addition of a sequence encoding ubiquitin, a small protein that enhances protein degradation via the proteasome and the processing of the resulting peptide fragments for presentation to CD8+ T cells, has been proven to enhance the ability of plasmid vaccines to elicit CD8+ responses to the expressed antigen.

Modulating the immune response. For all vaccines, robust activation of the immune system is essential to achieve strong, long-lasting immunity. As investigators have begun to understand the molecular mechanisms that control inflammation and how these mechanisms affect B- and T-cell responses to foreign antigens, this information has been exploited to improve the clinical effectiveness of DNA vaccines. Plasmids encoding a wide array of cytokines and chemokines (small hormone-like proteins that control the activation, migration, and proliferation of immune cells) have been studied for their ability to amplify specific B- and T-cell responses to DNA vaccines. The coadministration of these genetic adjuvants along with an antigen-expressing plasmid has, in many instances, resulted in both quantitative and qualitative improvements in the immune response against the plasmid-encoded antigen. Thus, the ability to use the knowledge of how the immune system works in order to amplify and modify the body's response to specific antigens has led to a concept known as rational vaccine design. This concept, which was extensively tested using plasmid-based vaccines, is now being applied to the development of more effective protein-based vaccines against a variety of diseases, including AIDS, tuberculosis, malaria, cancer, and autoimmune diseases.

Safety concerns associated with DNA vaccination. DNA vaccines carry the risk of plasmids randomly integrating into the genome of the transfected cell. If integration disrupts a gene involved in controlling cellular replication or survival, then it could lead to the development of cancer. Although the probability of genomic integration is extremely small, it could become a potentially serious safety issue if hundreds of thousands of doses of vaccine were administered, as might be expected for a vaccine against a typical infectious disease. Other potential safety issues include the length of time that the plasmid persists in the body and the length of time that the encoded gene is expressed. Plasmids have been documented to persist in the tissues of vaccinated animals for several months, and prolonged expression of the plasmid-encoded antigen has been hypothesized to be detrimental to the immune response by chronically stimulating antigen-specific T cells. Although the phenomenon has never been observed following vaccination with plasmid DNA, overstimulation of T cells during chronic viral infections can lead to the development of dysfunctional T cells in a process known as T-cell exhaustion. Although T-cell exhaustion is a theoretical possibility, the demonstration of long-lived functional T-cell responses in mice following multiple administrations of an experimental antiviral DNA vaccine argues against the possibility of T-cell exhaustion occurring in humans.

Future of DNA vaccines. In general, there has been tremendous excitement about the possibility that genetically engineered plasmids could lead to a new generation of inexpensive, stable, and effective anticancer and antimicrobial vaccines. However, recent clinical trials have reported varied results, and there are no vaccines of this type approved for human use. For example, the Allovectin-7® vaccine consists of a plasmid expressing the human leukocyte antigen B-7 and beta-2-microglobulin formulated with an immunogenic lipid. It was designed for use in late-stage melanoma (skin cancer) patients. The proteins expressed by the plasmid are normally present on the surface of most cells and function to present cell-derived peptides (protein fragments) to cytotoxic CD8+ T cells. The plasmid-lipid complexes were injected directly into melanoma tumors, where the plasmid was expected to be expressed by tumor cells, thereby increasing the ability of cytotoxic T cells to recognize and kill cancer cells expressing the vaccine antigens. The failure of Allovectin-7 to achieve clinically relevant responses in late-stage melanoma patients was a significant disappointment. Although large-scale phase III trials of Allovectin-7 were unsuccessful, results with a DNA vaccine called TransVax™ (ASP0113), which is intended to be given to bone-marrow transplant recipients at risk of life-threatening cytomegalovirus infections, have been somewhat more encouraging. The TransVax vaccine consists of a plasmid expressing two antigens from the cytomegalovirus formulated with a cationic polymer. Although the initial study reported no significant differences in bone-marrow transplant patients who received the vaccine and those who received a placebo, a follow-up study showed both significant reductions in the frequency of viral reactivation in the bloodstream and lengthening of the average time to reactivation in the vaccinated group. A more extensive phase III trial is currently in progress. These two examples typify the mixed results seen with DNA vaccines in humans. Despite tremendous progress in the field, it remains to be seen whether this technology will become a significant contributor to global efforts to eliminate cancer and infectious diseases.

For background information *see* ANTIBODY; ANTIGEN; ANTIGEN-ANTIBODY REACTION; BIOLOGICALS; BIOTECHNOLOGY; CELLULAR IMMUNOLOGY; DEOXYRIBONUCLEIC ACID (DNA); GENE; GENETIC ENGINEERING; GENETICS; IMMUNOLOGY; NUCLEIC ACID; PLASMID; PROTEIN; VACCINATION in the McGraw-Hill Encyclopedia of Science & Technology.

Daniel E. Hassett

Bibliography. M. L. Bagarazzi et al., Immunotherapy against HPV16/18 generates potent TH1 and cytotoxic cellular immune responses, *Sci. Transl. Med.*, 4(155):155ra138, 2012, DOI:10.1126/scitranslmed.3004414; M. A. Kharfan-Dabaja et al., A novel therapeutic cytomegalovirus DNA vaccine in allogeneic haemopoietic stem-cell transplantation: A randomised, double-blind, placebo-controlled, phase 2 trial, *Lancet Infect. Dis.*, 12(4):290–299, 2012, DOI:10.1016/S1473-3099(11)70344-9; F. Saade and N. Petrovsky, Technologies for enhanced efficacy of DNA vaccines, *Expert Rev. Vaccines*, 11(2):189–209, 2012, DOI:10.1586/erv.11.188; N. Y. Sardesai and D. B. Weiner, Electroporation delivery of DNA vaccines: Prospects for success, *Curr. Opin. Immunol.*, 23(3):421–429, 2011, DOI:10.1016/j.coi.2011.03.008; D. Tang, M. DeVit, and S. A. Johnston, Genetic immunization is a simple method for eliciting an immune response, *Nature*, 356:152–154, 1992, DOI:10.1038/356152a0; J. B. Ulmer et al., RNA-based vaccines, *Vaccine*, 30(30):4414–4418, 2012, DOI:10.1016/j.vaccine.2012.04.060.

Dutch elm disease

The Dutch elm disease (DED) pathogen infects the vascular tissues of elms and disrupts their water transport, causing wilt symptoms and tree death (**Fig. 1**). DED devastated American elms across the eastern United States in the mid-1900s. Despite the huge numbers killed by DED, elms remain important in our forests and cities, and the use of disease-tolerant elms in our landscapes is becoming more feasible.

History. The disease first appeared in northwest Europe in the early 1900s, causing the death of European elms, and was given the name Dutch elm disease. In the 1920s, the causal fungal pathogen (*Ophiostoma ulmi*) and the smaller European elm bark beetle were introduced to North America on imported logs. The disease spread across Europe, southwest Asia, and North America in a pandemic of elm death. By the 1940s, the rate of elm mortality declined in Europe. However, in the 1970s, there was a sharp increase in the incidence of DED. Simultaneously, in North America, elms began dying in greater numbers, and it was recognized by the late 1970s that there was a second, more aggressive fungal species, *O. novo-ulmi*, that was causing the increased mortality. Molecular studies later revealed that there were separate North American and European races of *O. novo-ulmi*, and that hybridization

Fig. 1. An American elm street tree (in Minnesota) dying from Dutch elm disease. The leaf symptoms advance from wilting to yellowing to browning of the leaves. Eventually, the entire crown is affected. (*Photo courtesy of Joseph O'Brien*)

between *O. ulmi* and *O. novo-ulmi* had resulted in changes in the fungal population. A related fungus, *O. himal-ulmi*, is present in the Himalayas, where it causes little damage to native elms, but appears to be highly virulent on European and American elms. Both *O. ulmi* and the North American strain of *O. novo-ulmi* were recently discovered in Japan, but without record of wilting trees or visible disease. It is quite likely that *O. ulmi* and *O. novo-ulmi* are native to Asia, particularly because Asian elms have a much higher tolerance to the pathogens, but this has not been confirmed. However, it is known that there are multiple species and subspecies of *Ophiostoma* that cause DED, and these fungi hybridize and exchange genetic material.

Overall, DED is a complex of multiple pathogens acting on many hosts that grow in different habitats. This article will focus primarily on the biology of DED and opportunities for recovery of the American elm in North America.

Disease cycle. The American elm (*Ulmus americana*) is very highly susceptible to DED, and no native elms within North America are immune. Other North American elms (winged elm, September elm, slippery or red elm, rock elm, and cedar elm) range from susceptible to somewhat tolerant. European and Asiatic elms are less susceptible than the American elm. DED is considered to now be present throughout the entire natural range of the American elm.

The disease cycle of DED is linked to the life cycles of elm bark beetles, including the smaller European elm bark beetle (*Scolytus multistriatus*) and the native elm bark beetle (*Hylurgopinus rufipes*) [**Fig. 2**]. The more recently introduced banded elm bark beetle (*S. schevyrewi*) can also carry spores of *Ophiostoma*. Adult elm bark beetles are attracted to stressed, dying, or dead elms to complete the breeding stage of their life cycle. The adult beetles tunnel in and lay their eggs in the inner bark; then, the eggs hatch and the larvae feed in the inner bark and cambium (the layer of cells between the phloem and xylem of most vascular plants), leaving diagnostic patterns of galleries. If a beetle-attacked tree has DED, the fungus produces sticky spores in the beetle galleries. When the larvae mature and emerge from the trees as adults, they often carry spores in and on their bodies. Adult beetles then visit healthy trees, feed in twig crotches or branch inner bark, and introduce the fungus to severed vessels as they feed, thereby infecting the tree and perpetuating the cycle of disease. Once an elm tree is diseased, the

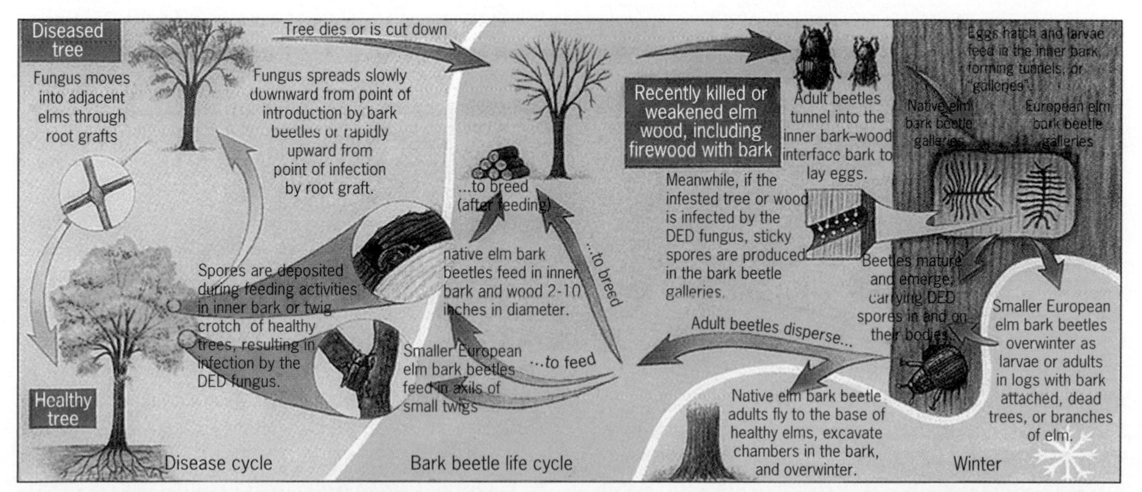

Fig. 2. The disease cycle of Dutch elm disease is closely linked to the life cycle of elm bark beetles. Elm bark beetles initiate new infections in healthy trees. Dead and dying elms provide the habitat for elm bark beetles to reproduce. (*Illustration courtesy of the USDA Forest Service*)

pathogen moves through the vascular system of the tree and spreads to adjacent elms through connected roots.

The American elm is adapted to grow in a wide variety of soils and environments, including floodplains and uplands. It grows quickly and produces prolific seed at a young age. Despite the effects of DED, small-diameter elm trees are still common in fence rows and forests. The persistence of elms in the environment ensures the survival of the species, but it also enables the persistence of disease and vectors (pathogen-carrying agents) on the landscape. Outbreaks of DED occur when the elm population increases in the landscape and then comes under stress from drought or other factors, allowing elm bark beetle populations to build up and spread disease. The pathogen, however, does not need stressed trees to infect. In fact, it actually spreads faster within a tree when there is adequate moisture and the vessel elements have a large diameter and are open.

Potential for management of the disease. Management of DED should focus on breaking the disease cycle of the pathogen by management of the vectors, fungi, or host. Because of the strong relationship between the vectors and the spread of DED, sanitation to remove woody debris containing elm bark beetles can be quite successful in limiting the severity of outbreaks. Chemical control of the vectors with insecticides is used to a limited degree.

Spread of the fungus to adjacent trees through connected roots can be stopped by breaking the grafts or connections between trees, often with a vibratory plow. Certain fungicides injected properly into valuable elms can protect them from fungal spores that are carried by elm bark beetles. This protection can last up to 3 years. However, the treatment itself poses some risk to the ongoing health of the tree.

Managing the host to reduce the risk of disease is primarily accomplished by planting individual trees with tolerance to the disease. Injection of individual elms with a nonpathogenic strain of another fungus appears to provide temporary acquired resistance, but the utility of this treatment is limited.

Opportunities for recovery. For American elms, the ability to survive infection by DED is usually described as tolerance, rather than resistance, because surviving trees may be infected by the fungus, but are able to tolerate the infection and survive with minimal symptoms. The mechanism or mechanisms that control this tolerance are not yet fully understood, but may be related to the ability of the tree to isolate the pathogen by compartmentalizing it. Only a few putatively DED-tolerant pure American elm cultivars are commercially available, including Valley Forge, New Harmony, Princeton, Prairie Expedition, American Liberty, and Jefferson.

Large surviving American elms are occasionally found in the landscape. Some of these trees may have escaped infection by DED through luck or seclusion, but some may have innate characteristics that enable them to survive. Several research groups, including the Elm Recovery Project at the University of Guelph (Ontario, Canada), the Elm Selection Program at the University of Minnesota, and the U.S. Forest Service Northern Research Station, have collected scion wood (that is, the section of a woody plant, usually a stem or bud, which is attached to the stock in grafting) from promising "survivor elms" to test for DED tolerance. The current testing procedure takes many years, requiring the test elms to be grown past the stage of juvenile resistance before being challenged by the DED pathogen. Through this screening process, more varieties of DED-tolerant American elms should become available.

The American elm has a different number of chromosome copies than other elm species, which has complicated breeding programs that would cross the American elm with more tolerant elms of other species. However, recent research at the National Arboretum in Washington, DC, has revealed that the number of copies of chromosomes varies within the American elm population. Further research in this rapidly emerging field may yield new opportunities for elm breeding and DED tolerance.

Conclusions. Despite the ravages of DED, there is hope that the American elm may one day be restored to its former ecological niches. The use of traditional resistance breeding and new techniques for genetic manipulation may provide us with American elms that can survive, despite the disease, in cities and forests. Even now, cultivars of American elm are available for urban utilization and wild land restoration. Caution is advised, however, because there are very few available cultivars and the pathogen is highly adaptable. It is prudent to use the available tolerant elms in mixtures with other tree species.

For background information *see* BREEDING (PLANT); COLEOPTERA; ELM; FOREST AND FORESTRY; FOREST MANAGEMENT; FORESTRY, URBAN; FUNGAL VIRUS; FUNGI; FUNGISTAT AND FUNGICIDE; GENETICALLY ENGINEERED PLANTS; INVASION ECOLOGY; PLANT GROWTH; PLANT PATHOLOGY; POPULATION ECOLOGY; PRIMARY VASCULAR SYSTEM (PLANT); REFORESTATION; TREE; TREE DISEASES in the McGraw-Hill Encyclopedia of Science & Technology.

Linda Haugen

Bibliography. C. M. Brasier and K. W. Buck, Rapid evolutionary changes in a globally invading fungal pathogen (Dutch elm disease), *Biol. Invasions*, 3:223–233, 2001, DOI:10.1023/A:1015248819864; L. M. Haugen, How to identify and manage Dutch elm disease, USDA Forest Service publication NA-PR-07-98, 1998; S. A. Merkle et al., Restoration of threatened species: A noble cause for transgenic trees, *Tree Genet. Genomes*, 3(2):111–118, 2007, DOI:10.1007/s11295-006-0050-4; R. J. Scheffer, J. G. W. F. Voeten, and R. P. Guries, Biological control of Dutch elm disease, *Plant Dis.*, 92(2):192–200, 2008, DOI:10.1094/PDIS-92-2-0192; A. T. Whittemore and R. T. Olsen, *Ulmus americana* (Ulmaceae) is a polyploid complex, *Am. J. Bot.*, 98(4):754–760, 2011, DOI:10.3732/ajb.1000372.

Earliest hominins

A major issue in paleoanthropology concerns the ability to identify the first members of our evolutionary lineage, the earliest hominins (see **illustration**). Based on both genetic and morphological studies, it is well established that hominins are closely related to the African apes, especially to the panins (chimpanzees and bonobos). Coupled with the fragmentary nature of the fossil record, the morphological similarities among fossil specimens, extant African great apes, and extinct Miocene hominoids make it exceedingly difficult to distinguish the earliest hominins from other hominoid lineages.

Australopithecus/Praeanthropus. Brain size was once considered to be a hominin Rubicon (a dividing line), wherein the initial divergence between hominins and their primate relatives involved a marked expansion of brain size. The South African fossil hominin, *Australopithecus africanus*, named by Raymond Dart in 1925, was initially sidelined by many of the established anthropologists of the time as a juvenile gorilla because of its well-preserved, albeit small, apelike braincase. It was not until Robert Broom and others worked tirelessly to recover additional early hominins that *Au. africanus* was accepted by the paleoanthropological community as the then-earliest known hominin. Unlike humans and other members of the genus *Homo*, early hominins possessed relatively small brains that were more similar in size to those of chimpanzees. Based on the anterior position of the foramen magnum (the large opening) at the base of the skull and a suite of distinctive postcranial (below-the-head) traits, Dart and others inferred upright posture and bipedal locomotion (bipedalism) in the South African hominins. An older hominin, *Au. afarensis* (or *Praeanthropus afarensis*), was discovered in Tanzania and Ethiopia and described by Donald Johanson and colleagues in 1978. A number of specimens, including the partial skeleton known as "Lucy," were shown to demonstrate strong evidence of

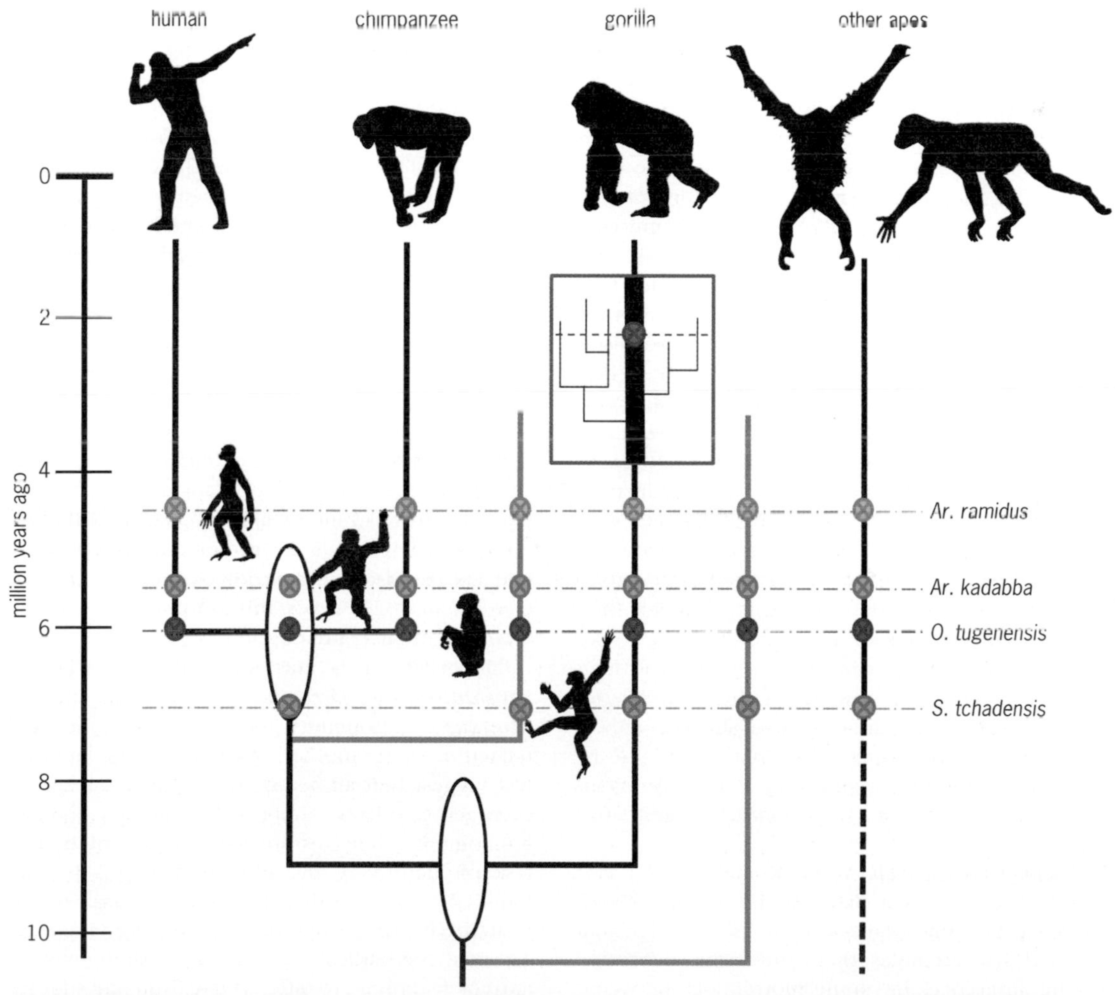

Possible positions of the earliest putative hominins [shown horizontally along their estimated dates (million years ago)] superimposed on evolutionary lineages of living African apes and humans (left to right: *Homo sapiens*, *Pan*, and *Gorilla*). Approximate divergence times (with conservative ranges) are shown. Note that the gorilla lineage is zoomed in upon to show that each lineage contains additional branches. These branches represent hypothetical sublineages of now-extinct species, any one of which a fossil can fall upon. The line on the right is intended to represent an "outgroup" lineage that might include orangutans, gibbons (lesser apes), or various fossil Miocene hominoids. (*Image courtesy of Milena R. Shattuck; silhouettes from www.phylopic.org*)

bipedalism, but they retained chimpanzee-like skull features: an even smaller brain and a more projecting jaw than *Au. africanus*. Dating to 4.1–4.2 million years ago (MYA), the oldest known definitive hominin is *Au. anamensis*, which is known from craniodental and postcranial remains, including a tibia evincing adaptations for a bipedal gait.

Ardipithecus ramidus. Older fossils have been recovered, but their placement on the hominin lineage is less well established. Associated teeth of a putative hominin were discovered in Ethiopia and assigned to *Au. ramidus* in 1994. Following the recovery of new specimens, including a relatively complete partial skeleton, the species was designated in a new genus, *Ardipithecus*. The initial discoveries have been supplemented with hundreds of specimens, representing several dozen individuals from the Middle Awash and Gona fossil areas in Ethiopia. *Ardipithecus ramidus* is well dated at Aramis in the Middle Awash to 4.4 MYA, and between 4.3 and 4.6 MYA at Gona. The female partial skeleton, nicknamed "Ardi," is badly crushed and took more than a decade to extract carefully from the surrounding matrix and reconstruct. The cranium, in particular, is highly fragmented and required meticulous virtual reconstruction. In a series of papers describing Ardi and other specimens published in 2009, Tim White and colleagues interpreted the combination of a short iliac blade (upper part of the pelvis) with an anterior inferior iliac spine (the origin of the rectus femoris muscle, which is an important hip flexor) and a grasping, nonpropulsive hallux (big toe), among other features, to suggest a strong retention of an arboreal lifestyle (living in trees) and facultative bipedalism in this species. Additionally, a recent study of the basicranial morphology in *Ar. ramidus* may support the presence of a short, broad cranial base with anterior foramen magnum placement that had been suggested previously as a complex of traits linking *Ar. ramidus* with later hominins. Whereas brain size is chimpanzee-like at 300 cm^3 (cubic centimeters), *Ar. ramidus* males apparently lacked large, tusklike canines with the premolar honing complex that characterizes chimpanzees, gorillas, and other sexually dimorphic (occurring in two distinct forms) primates. Instead, all canines in the assemblages are relatively small and diamond-shaped, rather than dagger-shaped, suggesting that canine size is more monomorphic (having only a single form) in this species, as it is in *Australopithecus* and later hominins.

Ardipithecus kadabba. At Asa Koma, an older site in the Middle Awash dated to between 5.6 and 5.8 MYA, Yohannes Haile-Selassie discovered dental material that resembles *Ar. ramidus*. Subsequently, noting differences in canine morphology between the two assemblages, Haile-Selassie elevated this material from the subspecies to species level, designating it *Ar. kadabba*. In addition to the dental material, upper-limb material from the same site and a 5.2-million-year-old proximal pedal phalanx (a toe bone) from another site are known. This latter bone has been interpreted as evidence for bipedalism based on dorsal canting (sloping) of the articular surface, a feature suggesting dorsiflexion (backward bending) of the joint, presumably during bipedalism. However, this trait may be present in a nonbipedal Miocene hominoid, *Sivapithecus*.

Orrorin tugenensis. Slightly older material (dating to 5.7–6.0 MYA) from the Lukeino Formation in Kenya was discovered in 2000 by Brigitte Senut and Martin Pickford. It was named *Orrorin tugenensis* (and nicknamed "Millennium Man") and is argued to be a hominin based on the morphology of three partial femora (thigh bones). Specifically, the proximal femur morphology resembles *Australopithecus* in demonstrating a long femoral neck and small femoral head. However, more recent work indicates that the overall shape is in some ways similar to certain fossil Miocene hominoids, suggesting that its femoral morphology is perhaps more generalized than initially thought. The *Orrorin* assemblage includes a distal humerus (upper arm bone) and two finger bones: a curved, apelike proximal phalanx and a homininlike distal phalanx of the pollex (thumb).

Sahelanthropus tchadensis. Dating to 7 MYA, the oldest putative hominin is *Sahelanthropus tchadensis*. It was discovered in 2001 by Michel Brunet and colleagues far west of the East African Rift Valley at Toros-Menalla in northern Chad. "Toumaí," as the cranium and best-known specimen is nicknamed, is relatively complete but crushed, and it thus required virtual reshaping to correct for significant deformation. It possesses a mixed suite of African apelike features, including a small (380-cm^3) braincase; widely spaced eye orbits with a large, barlike ridge running above them; and a relatively large crest along the back of the cranium for neck (nuchal) musculature attachment. However, its face is relatively short, its nuchal crest is lower on the cranium than in gorillas and chimpanzees, and it has relatively small canines that seem to lack functional honing facets. Although no postcranial bones are preserved, bipedalism is inferred based on an anterior position of the foramen magnum. However, this morphological configuration is not accepted by all researchers.

Conclusions. It is currently unknown whether *Sahelanthropus*, *Orrorin*, *Ar. kadabba*, and *Ar. ramidus* are hominins and, if so, how they are related to one another. Although some authorities suggest that all belong to the same lineage or even genus, others caution that multiple—perhaps nonhominin—lineages are likely represented. The task of identifying the earliest hominins is confounded by the fact that shared resemblance is correlated with time since the last common ancestry. As such, the earliest hominins and panins will be extremely difficult to differentiate from one another and from members of related, now-extinct lineages. Bipedalism is considered the current Rubicon of the hominin lineage, but it is not yet known when it evolved or in what form. Continuing morphological, developmental, and genetic work on living hominoids and studies of existing and newly discovered fossils will shed a brighter light on the earliest hominins and our origins.

For background information *see* ANTHROPOLOGY; APES; AUSTRALOPITH; DATING METHODS; EARLY MODERN HUMANS; FOSSIL; FOSSIL APES; FOSSIL HUMANS; FOSSIL PRIMATES; PHYSICAL ANTHROPOLOGY; PRIMATES in the McGraw-Hill Encyclopedia of Science & Technology. Scott A. Williams

Bibliography. S. Almécija et al., The femur of *Orrorin tugenensis* exhibits morphometric affinities with both Miocene apes and later hominins, *Nature Commun.*, 4:2888, 2013, DOI:10.1038/ncomms3888; D. R. Begun (ed.), *A Companion to Paleoanthropology*, Wiley-Blackwell, Chichester, UK, 2013; D. R. Begun, The earliest hominins—Is less more?, *Science*, 303:1478–1480, 2004, DOI:10.1126/science.1095516; M. Brunet, Two new Mio-Pliocene Chadian hominids enlighten Charles Darwin's 1871 prediction, *Phil. Trans. R. Soc. B*, 356:3315–3321, 2010, DOI:10.1098/rstb.2010.0069; J. G. Fleagle, *Primate Adaptation and Evolution*, 3d ed., Elsevier/Academic Press, Amsterdam, 2013, W. H. Kimbel et al., *Ardipithecus ramidus* and the evolution of the human cranial base, *Proc. Natl. Acad. Sci. USA*, 111:948–953, 2014, DOI:10.1073/pnas.1322639111; J. Reader, *Missing Links: In Search of Human Origins*, Oxford University Press, New York, 2011; K. E. Reed, J. G. Fleagle, and R. E. Leakey (eds.), *The Paleobiology of Australopithecus*, Springer, Dordrecht, the Netherlands, 2013; S. C. Reynolds and A. Gallagher (eds.), *African Genesis: Perspectives on Hominin Evolution*, Cambridge University Press, Cambridge, UK, 2012; T. D. White et al., *Ardipithecus ramidus* and the paleobiology of early hominids, *Science*, 326:75–86, 2009, DOI:10.1126/science.1175802; B. Wood and T. Harrison, The evolutionary context of the first hominins, *Nature*, 470:347–352, 2011, DOI:10.1038/nature09709.

Early Cretaceous insect camouflage

The ability to move about undetected is a remarkable asset. To the cloaked individual, it immediately confers an advantage over those individuals who are not so endowed. Hiding oneself comes in a variety of forms, ranging from camouflage to outright mimicry, and is a considerably complicated evolutionary transition to achieve. In any manifestation, the evolution of disguise involves concerted changes in behavior, morphology, and sometimes physiology and biochemistry. Although spectacular examples abound of cryptic morphologies, ranging from mimetic stick insects and thorn bugs to various beetles and flies mimicking stinging wasps, the use of exogenous (external) materials for camouflage is much less widespread. Camouflage involves the use of materials that conceal the body and resemble the surroundings of the local environment. In some instances, the construction of a nest can serve similar purposes, with the materials of the structure blending into the surrounding environment and thereby acting as an indirect form of camouflage. However, direct and purposeful camouflage of the body, independent of a refuge or roost, is more uncommon and embodies a very different series of behavioral and anatomical attributes, as well as a fundamentally divergent suite of underlying genetic and physiological components.

Camouflage and the fossil record. Unlike animal architecture, which by its very nature produces a corporeal (bodily) structure that may be entombed and preserved under specific conditions, many behaviors do not involve physical constructs nor require specialized anatomies that can be detected in fossils. Specialized behaviors and physiologies, such as the catalepsis (rigidity and fixity of posture) and swaying motions of stick insects, are ephemeral and do not leave a trace on the environment, making them virtually impossible to discover in the fossil record. Mimesis (resemblance) and mimicry are more readily observed in fossils because the specialized anatomies of the individuals involved are modified to take on the appearance of a model organism or object in the coeval (contemporaneous) environment. Unlike mimesis, the intricate behavioral suite associated with camouflaging is more difficult to discern when it is not tied to specific morphologies. Concrete evidence of camouflage typically requires the unique discovery of organisms with anatomical features permitting the attachment of exogenous materials to the body, as well as evidence of such environmental debris harvested specifically for such purposes and not associated merely because of taphonomic (fossil preservation) factors. Therefore, definitive evidence of camouflage in the fossil record requires very specific conditions, which are not often met by most modes of preservation.

Camouflage among insects. Camouflaging behavior among invertebrates is known from groups as disparate as sea urchins, gastropods, crabs, and immature insects. In most of these cases, the camouflage is specifically used as a defensive mechanism in order to avoid predators. Among immature insects, however, several species also use camouflage as a means of stealthily approaching their prey, permitting them more ease in capturing their victims while simultaneously eluding their own predators. The immature stages of green lacewings (order Neuroptera, family Chrysopidae) display this type of camouflaging behavior. Because of their direct crypsis (the ability of an organism to avoid observation) and their sometimes "wolf-in-sheep's-clothing" behavior, the immature green lacewings have become the focus of much attention for understanding the ecological and evolutionary advantages of camouflage. Although adult green lacewings are delicate creatures that largely feed on nectar and pollen, their larval stages are rapacious predators of aphids, insect eggs, mites, and other minute arthropods. They are sufficiently potent predators of many plant pests that they have been employed as effective biological control

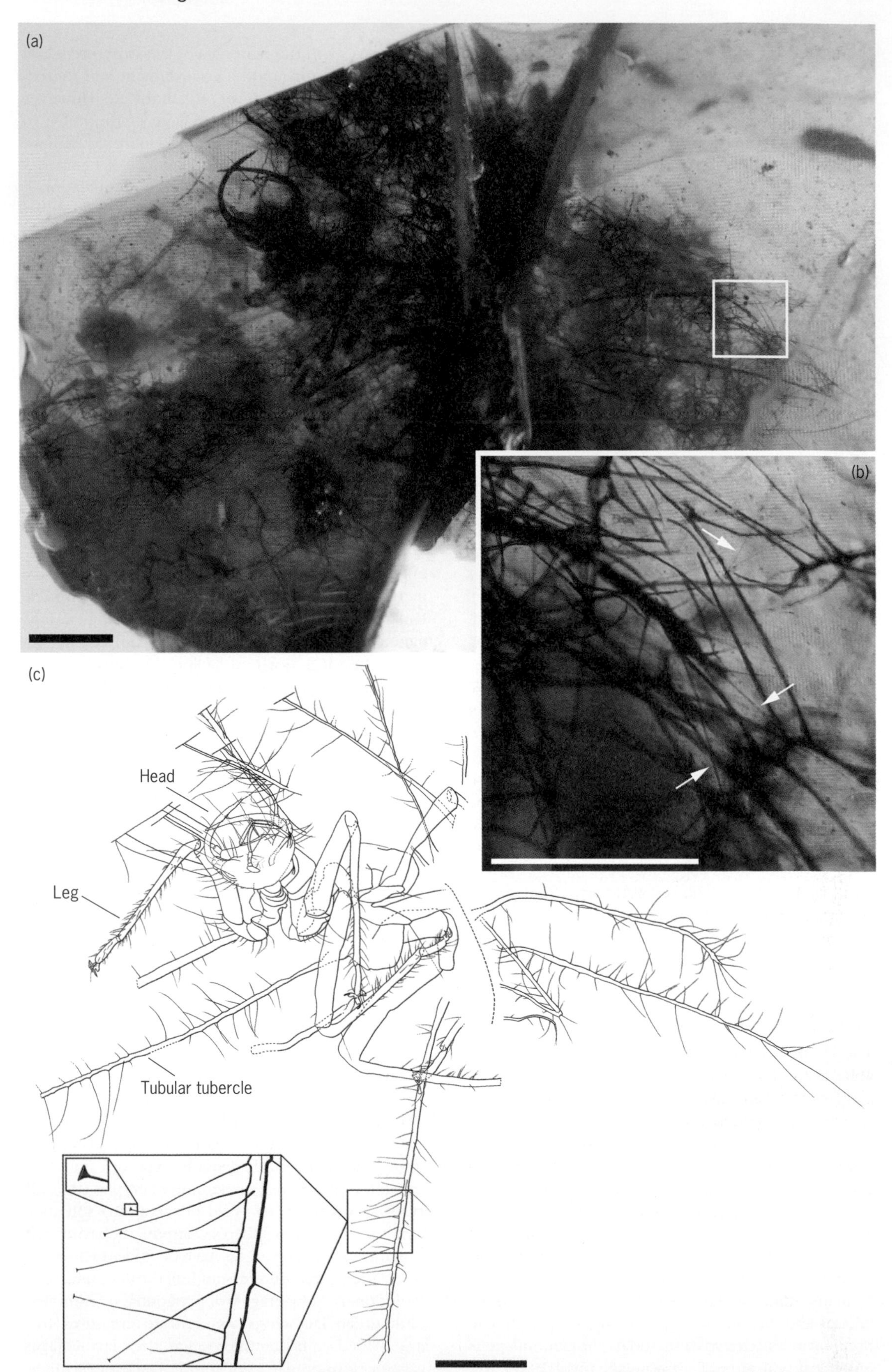

An Early Cretaceous green lacewing larva, *Hallucinochrysa diogenesi*. (*a*) Fossil as preserved in amber from northern Spain. (*b*) Detail of fern trichomes composing the camouflage packet. The arrows indicate some of the elongate hairs of the larva that are entangled among the trichomes. (*c*) Drawing of the larva without the camouflage packet and detailing the specialized structures for entangling debris. Scale bars: (*a*, c) 1 mm (0.04 in.); (*b*) 0.5 mm (0.02 in.). (*Illustration by R. Pérez-de la Fuente and courtesy of R. Pérez-de la Fuente et al., Early evolution and ecology of camouflage in insects, Proc. Natl. Acad. Sci. USA, 109:21414–21419, 2012, DOI:10.1073/pnas.1213775110*)

agents. The larvae of green lacewings selectively collect plant material, detritus, and even the remains of their prey from their environment, and these are carefully manipulated onto the larva's back, where specialized extensions of the insect's cuticle (hard outer skeleton) are used to entangle and hold the debris in place. A completed packet effectively covers most of the larva, forming a defensive shield and camouflaging the insect from its predators and sometimes its own prey. Camouflaging behavior has certainly been one of the many factors contributing to the evolutionary success of green lacewings, which collectively comprise more than 1200 species and have ancestors extending back into the Jurassic Period (approximately 200 to 145 million years ago).

Early Cretaceous camouflage of green lacewings. It has been determined that camouflage evolved early among green lacewings, with the complete suite of behaviors and associated anatomies for entangling and transporting exogenous materials appearing in the Early Cretaceous Period (approximately 145 to 65 million years ago). A preserved green lacewing larva of *Hallucinochrysa diogenesi* (see **illustration**) was discovered in 110-million-year-old amber (fossil resin derived from a coniferous tree) from northern Spain. Given the fidelity of preservation in amber, the finer aspects of the larva's morphology are discernible and reveal a complex assortment of specialized structures specific to the attachment of exogenous debris to the dorsal (hind) surface of the body. Such structures include greatly elongate, serial, and paired cuticular tubercles (small knoblike prominences) covered by scattered fine hairs. The hairs of the tubercles are trumpet shaped at their apices (tips), which, along with their flexible construction, permit them to become more easily entangled with the materials comprising the camouflaging packet. More significant yet, the fossil was preserved with its covering of plant trichomes (filamentous appendages) partially intact, demonstrating conclusively the larva's harvesting of such materials to form a dorsal shield. As in some living species, *H. diogenesi* also selectively harvested a single material from its environment, which it used to build its camouflage, demonstrating an early specialization within this behavior.

In its entirety, the fossil establishes the early evolution of camouflaging behavior and an intricate ecology of insect-plant interactions. Given that *H. diogenesi* exhibits such highly specialized anatomies for the transport of debris, it is assured that camouflage must have first arisen even earlier. The pervasiveness of the trait across the diversity of Chrysopidae suggests that this behavior may extend back into its earliest representatives among the Jurassic faunas of Eurasia. A limited number of Jurassic lacewing larvae are known, but none have yet been found that exhibit modifications for camouflage or preserve an associated packet of debris. Indeed, such fossils are preserved solely as compressions in stone and lack the degree of fidelity known for *H. diogenesi*. *Hallucinochrysa diogenesi* remains a singular discovery, but one that reveals much about the early appearance and long-term stasis of a complex behavioral repertoire. It also is a significant evolutionary novelty, promoting the diversification of a prominent family of insects.

For background information *see* AMBER; BEHAVIORAL ECOLOGY; CRETACEOUS; ECOLOGICAL COMPETITION; FOSSIL; INSECT PHYSIOLOGY; INSECTA; NEUROPTERA; PALEOECOLOGY; PREDATOR-PREY INTERACTIONS; PROTECTIVE COLORATION; TAPHONOMY in the McGraw-Hill Encyclopedia of Science & Technology.

Michael S. Engel

Bibliography. S. J. Brooks and P. C. Barnard, The green lacewings of the world: A generic review (Neuroptera: Chrysopidae), *Bull. Brit. Mus. Nat. Hist. Entomol.*, 59:117–286, 1990; M. Canard, Y. Séméria, and T. R. New, *Biology of Chrysopidae*, Dr. W. Junk Publishers, The Hague, the Netherlands, 1984; T. Eisner et al., "Wolf-in-sheep's-clothing" strategy of a predaceous insect larva, *Science*, 199:790–794, 1978, DOI:10.1126/science.199.4330.790; D. Grimaldi and M. S. Engel, *Evolution of the Insects*, Cambridge University Press, Cambridge, UK, 2005; R. Pérez-de la Fuente et al., Early evolution and ecology of camouflage in insects, *Proc. Natl. Acad. Sci. USA*, 109:21414–21419, 2012, DOI:10.1073/pnas.1213775110.

Electronic communication and identity development

Electronic communication via social media applications [for example, Facebook, Instagram, Twitter, and blogs (personal Web pages, wherein new entries are posted before older ones)] has become immensely popular among adolescents and emerging adults. Given the pervasiveness of such communication and the finding that online interactions among youth occur with offline peers and acquaintances, researchers have proposed that digital venues are an important social context in the lives of young persons. Research over the last several years has begun to show that youth use electronic communication in the service of key developmental tasks, including those related to sexuality, identity, and intimacy. The pioneering developmental psychologist Erik H. Erikson first proposed that formulating a coherent and stable identity was a key task that began during adolescence. According to Erikson, it was important for youth to explore alternative roles and identities, and such identity exploration occurred in a social context. As the transitional period between adolescence and adulthood has elongated, it has become clear that identity development continues beyond adolescence and into what is now called emerging adulthood. Thus, it is important for educators, policy makers, and others interacting with youth to have access to information on the role of electronic communication in identity development during adolescence and emerging adulthood.

Online identity and self-presentation. When online communication contexts first emerged, interactions occurred via text, and users were afforded the opportunity to be disembodied and anonymous. Thus, investigators speculated that individuals would take advantage of these characteristics and would "leave their physical bodies behind" to create alternative selves in an online setting. However, research on the early online applications, such as chat rooms and blogs, suggested that youth did not use electronic communication to experiment actively with alternative or false identities. Instead, they seemed to be capitalizing on the affordances of online contexts for identity presentation. Within text-based anonymous teen chat rooms, identity presentation was nonetheless very frequent, and users employed creative means to present key details of their offline identity, such as their age, gender, and location. A content analysis of adolescent blogs suggested that youth bloggers used their Web entries to construct self-narratives about the people and events in their lives. Finally, avatars or virtual characters within online games provide another avenue for identity exploration and self-presentation. Interestingly, online avatars tend to be similar to offline identity and personality characteristics. The exception to this is the practice of "gender bending," which occurs when an online player assumes the gender opposite to the offline self for his or her avatar.

As the nature of online contexts has changed, so too has the electronic communication within them. Social networking media, including Facebook, Twitter, and Instagram, are multimodal and allow users to upload and share text, pictures, videos, and other kinds of user-generated content with other members in one's network of "friends." They also make it easy to provide and receive feedback about content that is presented online. In one survey, 91% of adolescents reported posting a picture of themselves along with other identifying information (for example, their school name or email address). Research also suggests that youth use photographs, status updates, and "wall" posts (that is, online posts in the form of a community notice board) on social media sites (for example, Myspace and Facebook) to present aspects of themselves, including their physical attractiveness, interests, gender, ethnicity, and social identity. Such online presentation may also vary with age, and there is some indication that younger youth may present more self-related content online.

Connection between online self-presentation and self-development. Research suggests that the online self-presentation by young persons may be relevant to self-development. In the context of offline behavior, different kinds of self-presentation before an audience may help individuals to modify and consolidate their selves. Indeed, it appears that youth engage in impression management online just as they do offline, and they often do so in an intentional and strategic manner. Often, their online self-presentation is consistent with who they are offline. On the other hand, it also varies. For instance, investigations on the development of the self have found that youth engage in presenting different selves at different times. Similarly, youth report that they present different aspects of their self, such as their real self, ideal self, and aspects of their false self, on Facebook. In line with research on offline impression management, the online self-presentation by youths is often motivated by the desire to present authentic, positive, and professional selves in socially acceptable ways.

Youth online self-presentation is also related to their offline well-being and may be motivated by the desire for feedback from their network of friends. One study of publicly available profiles on Myspace found that college students disclosed symptoms of depression and did this more frequently when they received reinforcement from their audience. Self-esteem, identity, and personality traits (for example, neuroticism) appear to be related to online self-presentation on Facebook. In general, young persons stating that they had a more coherent identity reported presenting their real self to a greater extent, and those with a less coherent sense of the self, lower self-esteem, and higher levels of neuroticism reported presenting their false self to a greater extent.

Conclusions. In sum, the research to date suggests that youth use electronic communication within digital contexts in the service of identity construction and development. Contrary to early speculation, youth are not creating alternative identities within the virtual space. Instead, they capitalize on the particular features of the online context as they consolidate their individual selves. In particular, they engage in different kinds of self-presentation, and this self-presentation is related to aspects of their personality, identity, and well-being. Questions remain about the impact of such online self-presentation on identity development, and longitudinal studies are needed to determine whether early exposure to electronic communication facilitates or delays the formation of a coherent identity.

For background information *see* BRAIN; COGNITION; DEVELOPMENTAL PSYCHOLOGY; ELECTRICAL COMMUNICATIONS; HUMAN-COMPUTER INTERACTION; INFORMATION TECHNOLOGY; LEARNING MECHANISMS; MOTIVATION; PSYCHOLOGY in the McGraw-Hill Encyclopedia of Science & Technology.

Kaveri Subrahmanyam

Bibliography. E. H. Erikson, *Identity and the Life Cycle*, Norton, New York, 1959; M. Madden et al., *Teens, Social Media, and Privacy*, Pew Internet and American Life Project, Washington, DC, 2013; A. M. Manago et al., Self-presentation and gender on MySpace, *J. Appl. Dev. Psychol.*, 29(6): 446–458, 2008, DOI:10.1016/j.appdev.2008.07.001; K. Subrahmanyam and D. Smahel, *Digital Youth: The Role of Media in Development*, Springer, New York, 2011, DOI:10.1177/110330881202000304; K. Subrahmanyam, D. Smahel, and P. Greenfield, Connecting developmental constructions to the Internet: Identity presentation and sexual exploration in online teen chat rooms, *Dev. Psychol.*, 42(3):395–406, 2006, DOI:10.1037/0012-1649.42.3.000.

Elephant phylogeny

The order Proboscidea was first described in 1811 by Johann Karl Wilhelm Illiger, who named members of this group for their long proboscis, or trunk, which is so characteristic of many species (Greek *proboskis* = *pro*, which means forward or before; and *bosko*, which means to feed or nourish). Proboscideans are a taxonomic order of mammals containing the living Asian and African elephants. It is one of the most derived groups of mammals and also one of the earliest modern orders to appear in the fossil record. The classification of proboscideans is frequently revised, with new fossils and new investigations, and some relationships within the order remain unclear. At least 177 species and subspecies of proboscideans, classified in 43 genera, are now recognized, although the majority of these are extinct species.

Fossil record and evolutionary history. Proboscideans have a long and diverse fossil record, and the evolutionary history (phylogeny) of the order has been proposed as a series of adaptive shifts in response to changing environments. Evolutionary trends in the skull and dentition, hallmarks of the order, have been linked to increasing diversifications and specializations in diet. The earliest known proboscidean is *Eritherium azzouzorum*, from Moroccan fossil sites dating to approximately 60 million years ago (MYA). *Phosphatherium escuillei* and *Daouitherium rebouli* are Late Paleocene proboscideans also from Moroccan fossil deposits. These early animals are very small, about the size of a fox, and have the general dentition pattern ancestral to proboscideans, which includes enlarged second incisors (tusks in later species) and loss of the first premolar. *Eritherium* also represents a major leap in proboscidean evolution, with the development of true lophodonty (transverse ridges on the molars, instead of cusps) [**Fig. 1**] and a body form with graviportal adaptations (that is, adaptations for supporting great body weights).

Fig. 1. Molars of *Loxodonta africana* (left) and *Elephas maximus* (right) showing multiple ridges, or lamellae, on the surface of the molar teeth, instead of cusps as in other mammals.

Proboscideans diversified during the Eocene and Oligocene, with several new species from the Fayum Depression in Egypt, including *Numidotherium*, *Barytherium*, *Moeritherium*, and *Paleomastodon*. Recent analyses of some fossils attributed to *Numidotherium* have placed them into a new genus, *Arcanotherium savagei*. Additionally, a recent stable-isotope analysis of these ancient proboscideans has revived the hypothesis that these may have been semiaquatic in nature, further linking them to the sirenians (manatees and dugongs) and extinct desmostylians (extinct marine herbivore mammals) and supporting an aquatic origin for the African Paenungulata [Proboscidea, Sirenia, and Hyracoidea (hyraxes)].

It was during the Miocene, however, that proboscideans underwent a significant adaptive radiation, spreading out from Africa to southern Asia and ultimately becoming global in distribution. Miocene elephantiforms include the families Mammutidae, Deinotheriidae, Stegodontidae, and the gomphotheres (the largest group). This diverse group evolved rapidly during the Miocene and includes the shovel-tuskers (such as *Ambelodon* and *Platybelodon*), the straight-tusked *Anancus*, the four-tusked *Gomphotherium*, and many others. The taxonomy of this group is unclear and remains unresolved, with many different hypotheses for relationships within the group as well as with regard to other proboscideans. The gomphotheres spread from Africa into Eurasia and then to North America and South America throughout the Miocene. During the Pliocene and Pleistocene, these families were replaced by the Elephantidae, with the exception of the deinotheres and stegodonts, which persisted into the Pleistocene in Africa and Asia (**Fig. 2**).

Since their origin in Africa, early human species have undoubtedly shared ecosystems with proboscideans, eating their meat, which was obtained initially by scavenging carcasses and then later in time by intentional hunting. Cave paintings in Europe depict images of mammoths being hunted. Overhunting by humans is one of the theories for the extinction of the megafauna at the end of the Pleistocene, but a new hypothesis of second-order predation suggests an ecological basis for these extinctions, in which climate change, increased mosaic habitats (multiple habitat types), and predation operate together to cause the extinctions. By the end of

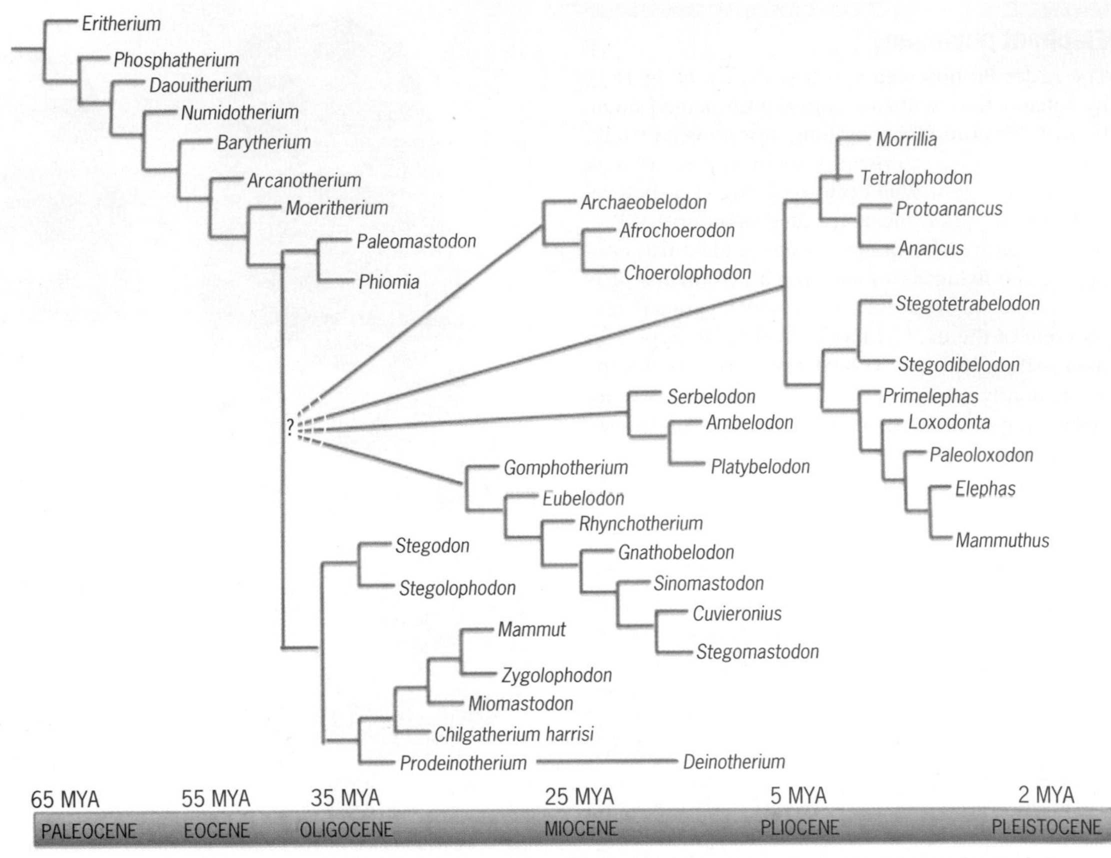

Fig. 2. A composite phylogeny of the Proboscidea based on published phylogenetic trees. An approximate timescale is shown, although the phylogeny is not drawn to scale. However, the majority of the recognized genera are included. MYA = million years ago.

the last glacial period, all species of proboscideans were extinct, except for *Loxodonta africana*, *Loxodonta cyclotis*, and the various subspecies of *Elephas maximus*.

Closest relatives. A close affinity among the Proboscidea, Sirenia, Desmostylia, and Hyracoidea is evident from unique morphological characters (synapomorphies) that they all share. For example, the carpal (wrist) bones are arranged one directly on top of another; this pattern is unlike that seen in other mammals, where the bones are staggered. In addition, the position and structure of the internal ear are unique, and the anterior border of the eye orbit is positioned over or in front of the first upper molar. These animals also share similarities in molar replacement and enlargement of incisors as tusks.

Evolutionary trends. Early proboscideans were small in body size. However, over the course of 60 million years of evolution, they experienced a gradual increase in size and the development of graviportal features in their limbs, including fixed pronation (downward and inward motion) of the limbs, solid trunklike limbs, and specialization in the feet for supporting heavy weight. Some of the largest species include *Mammuthus columbi*, reaching 4 m (13.1 ft) in height, and *Deinotherium giganteum*, which had a height as tall as 5 m (16.4 ft). Some lineages of proboscideans isolated on islands reversed this trend and underwent dwarfing in size. These include a number of elephantid species throughout the Pleistocene, such as those found on Mediterranean islands and the Channel Island mammoths off the coast of California. There is also evidence for a dwarfed *Stegodon* in Indonesia.

The hallmark features of the Proboscidea are the long, muscular proboscis, or trunk, and the elongated second incisors (tusks). There is little evidence for a long trunk in early proboscideans. However, aspects of the rostrum (snout) suggest a prehensile upper lip, and a combination of cranial features of the premaxilla and frontal bones suggests the presence of a proboscis, which may have evolved independently in different lineages. The extant elephants have very long and complex trunks, which consist of thousands of individual muscle fibers. The loss of the central incisors and the enlargement of the second incisors occur quite early, and proboscideans evolved a wide variety of tusk sizes and shapes, from elongated and straight, to shovel-shaped, to downward curving scythe-like tusks, to having both upper and lower tusks, to the magnificent mammoth tusks that spiral over in front. These tusks are used for defense, sparring, sexual display, rooting through soil for minerals, stripping bark off of trees, and other food-gathering functions.

In addition to the enlarged tusks, all proboscideans are characterized by horizontal displacement and replacement of the molar teeth. The teeth are positioned as if they are moving on a conveyor belt, with new teeth erupting behind the teeth in wear, which are eventually shed at the front of the mouth. All 6 cheek teeth increase in size, and the number of cusps (which become lamellae, or plates) increases through time, with as many as 30 plates in mammoths. The large sixth molar in Pleistocene mammoths may weigh more than 5 kg (11 lb).

The skull grows to an extraordinarily large size in later proboscideans and is accompanied by a shortening of the neck to accommodate the greater weight. Elongation of the lower jaw occurs initially in proboscidean evolution, but reverses as the skull increases in size. This shortening of the jaw is accompanied by a posterior shift in the center of gravity to balance the large and heavy skull on the neck, as well as to support the long tusks.

Living elephants. Three living elephant species are present today: *Loxodonta cyclotis* (forest African elephant), *Loxodonta africana* (bush African elephant), and *Elephas maximus* (Asian elephant). Three subspecies of *Elephas maximus* have been recognized: *E. m. maximus* (Sri Lankan elephant), *E. m. sumatranus* (Sumatran elephant), and *E. m. indicus* (mainland elephant). A group of Asian elephants in Borneo has been identified recently as being genetically distinct from other *E. maximus* elephants. They are smaller in size and may comprise another subspecies. In addition, a possible third species of African elephant in West Africa is suggested by analysis of DNA extracted from dung. This analysis also suggests that West African elephants have been isolated for as long as 2.4 million years.

Elephants communicate via infrasonic sound, which is sound below the frequency of human hearing and which can travel many kilometers across ecosystems. Many individual calls have been identified by researchers. The social behavior of elephants is highly complex, with small close-knit family groups led by matriarchs. Elephant infants are cared for by all members of the family unit, and very close lifetime bonds are formed. They have a highly complex brain, with large temporal lobes compared to other brain areas, and these lobes are known to function as memory centers. Elephants exhibit many higher behaviors often not associated with animals: they grieve for the dead, they form long-lasting and strong bonds, they are known to use and make tools, they have excellent memories, and some even paint. Workers at elephant sanctuaries in Southeast Asia have trained elephants to paint to raise money for the continued conservation efforts of these sanctuaries.

Field observations on living elephants reveal that they play a pivotal role in their ecosystems as keystone species (that is, species critical for the structural maintenance and survival of an ecological community). They knock down trees, providing access for other smaller animals. This speeds up ecological succession and maintains biodiversity in these habitats. They also make clearings in rainforests at waterholes, which are then utilized by other species. There are also species of plants that will only germinate in elephant dung, or after passing through an elephant's digestive tract. Thus, elephants contribute to the survival of other species living in the same ecosystem.

All three species of living elephants are critically endangered and are listed on the *Convention of the International Union for the Conservation of Nature* (IUCN) Red List and Appendix I of the *Convention on International Trade in Endangered Species of Wild Fauna and Flora* (CITES). Loss of critical habitat and poaching for the ivory tusks of these creatures are the main reasons for their endangered status. It is estimated that there are less than 35,000 Asian wild elephants and approximately 600,000 African elephants left in the wild, and habitat loss, human-elephant conflict, and illegal hunting will continue to contribute to their loss unless significant global conservation efforts are undertaken.

For background information *see* ADAPTATION (BIOLOGY); ANIMAL EVOLUTION; DENTITION; DESMOSTYLIA; ELEPHANT; ENDANGERED SPECIES; FOSSIL; HYRACOIDEA; MACROEVOLUTION; MAMMALIA; PHYLOGENY; PROBOSCIDEA; SIRENIA; TOOTH in the McGraw-Hill Encyclopedia of Science & Technology. Nancy E. Todd

Bibliography. P. Grubb et al., Living African elephants belong to two species: *Loxodonta africana* (Blumenbach, 1797) and *Loxodonta cyclotis* (Matschie, 1900), *Elephant*, 2:1-4, 2000; V. J. Maglio, Origin and evolution of the Elephantidae, *Trans. Am. Phil. Soc.*, 63(part 3):1-149, 1973; S. Mahboubi et al., Was the Early Eocene proboscidean *Numidotherium koholense* semi-aquatic or terrestrial?: Evidence from stable isotopes and bone histology, *C. R. Palevol*, 13:501-509, 2014, DOI:10.1016/j.crpv.2014.01.002; H. F. Osborn, The evolution, phylogeny and classification of the Proboscidea, *Am. Mus. Novit.*, 1:1-15, 1921; J. Shoshani, Understanding proboscidean evolution: A formidable task, *Trends Ecol. Evol.*, 13(12):480-487, 1998, DOI:10.1016/S0169-5347(98)01491-8; J. Shoshani and P. Tassy, Advances in proboscidean taxonomy & classification, anatomy & physiology, and ecology & behavior, *Quaternary Int.*, 126-128:5-20, 2005, DOI:10.1016/j.quaint.2004.04.011; J. Shoshani and P. Tassy, *The Proboscidea: Evolution and Palaeoecology of Elephants and Their Relatives*, Oxford University Press, Oxford, UK, 1996; N. E. Todd, Trends in proboscidean diversity in the African Cenozoic, *J. Mamm. Evol.*, 13(1):1-10, 2006, DOI:10.1007/s10914-005-9000-4; N. E. Todd and V. L. Roth, Origin and radiation of the Elephantidae, pp. 193-202, in J. Shoshani and P. Tassy (eds.), *The Proboscidea: Evolution and Palaeoecology of Elephants and Their Relatives*, Oxford University Press, Oxford, UK, 1996.

Emerging energy technologies

Electricity generation is the lifeblood of the modern world. It has been predicted that over the next 50 years the world will have to find 50% more energy to satisfy growing needs from emerging nations and the power-demanding developed world. This, together with the looming crisis of climate change and the fact that oil will soon reach peak production, has sparked a new generation of emerging energy technologies, many of them geared around renewable forms of energy. There is now no doubt among climate scientists that anthropogenic carbon dioxide and particulates, generated largely from the burning of coal and other fossil fuels, are responsible for our present changing climate. As a result, new ways must be found to generate electricity and to power our transportation system. Climate change is without doubt the most severe problem that our civilization has ever had to face, and with it comes the devastating scourges of drought, flooding, famine, lack of clean water, disease, overpopulation, and human migration, all on an unprecedented scale.

Emerging energy technologies include the exploitation of shale gas, coalbed methane, methane hydrates, wind energy, as well as solar photovoltaics, solar concentrated power, solar hot-water systems, geothermal wells, tidal currents, wave energy, improved nuclear fission processes, and potentially important nuclear fusion. Apart from these energy sources there are exciting new developments in battery technology, fuel cells, hydrogen as a carrier, carbon capture, and "smart" grids. Added to these technologies are the various types of energy-storage systems that must be developed if we are to commit seriously to investment in wind, solar, tidal, or wave renewable energies.

Shale gas. Shale gas (mostly methane) is found trapped in very small pores in some fine-grained sedimentary rocks of very low porosity, all over the world. The gas is linked to organic matter laid down many millions of years ago. The low permeability of the shale has acted as a cap, trapping the gas. The gas is often found in narrow seams that are not always horizontal. Over the past two decades, technological developments in drilling techniques have made it possible to exploit these shales for their methane content. After the vertical wellbore has been drilled to the level of the gas-rich seam, holes, which may be horizontal or at an angle, are drilled into the seam. Next, mud and chemicals are injected to induce a pressure pulse that fractures the seam and increases its surface area, thereby releasing the methane gas. The United States is today the leader in the field of shale gas exploitation, also known as hydraulic fracturing or fracking. Shale gas exploitation in the United States began in the 2000s, and it has been estimated that by 2040 it will be responsible for over 50% of the U.S. natural gas production. It is not only changing the fossil-fuel import and export profile of the United States but also its domestic energy use. Other countries are following in the wake of the U.S. bonanza.

Coalbed methane. Methane invariably occurs in coal deposits, and over the past 20 years coalbed methane (CBM) has become an important energy resource, especially in the United States. Currently, over 40 countries with coal reserves are initiating some CBM program. The extraction process involves drilling vertical and horizontal wells into coal seams, followed by fracture stimulation techniques, including fluid, pressure, and CO_2 displacement. At present, about 5% of U.S. natural gas is obtained from CBM developments.

Methane hydrates. Methane hydrate is a term for a solid material formed from water and methane and involves methane molecules enclosed in a cagelike structure of hydrogen-bonded water. The methane molecules are not chemically bonded to the water. These cagelike (clathrate) compounds form under high pressure and can be found under seas and deep underground in natural gas fields. They were discovered in the 1970s, as the result of the clogging up of gas wells. The total volume of natural gas tied up in hydrate formations is enormous, and these formations occur all over the world in onshore and offshore sites. However, there are major difficulties in exploiting these hydrates; and at present, there is no commercial exploitation of methane hydrates. Current research involves chemical treatment and depressurization and is being pursued by many countries, including Japan, the United States, and Canada, because the potential rewards are great, especially for countries without oil reserves.

Wind energy. Wind energy is a relatively new energy source that has increased by leaps and bounds over the past 15 years. In 1997, the annual installed worldwide capacity was less than 2 gigawatts (GW). In 2012, this had risen to over 100 GW, which is equivalent to 100 reasonably large conventional coal or oil power stations. The size of the turbine defines the potential energy output; a turbine with 30-m blades produces 300 kilowatts (kW), while a 12-m turbine is rated at 5000 kW. At the moment, the cranes necessary to erect these turbines appear to be the limiting factor in their installation. However, 150-m [10-megawatt (MW)] and 250-m (20-MW) turbines are being considered. Wind turbines are proving to be successful, and offshore and onshore wind energy farms are most likely to be major contributors to future renewable energy resources.

Tidal-current energy. Tidal currents are created by water flowing from one body of water to another as a result of the need to equalize levels. The difference in levels is a result of tidal activity. If the speed of water flow is high enough, then the tidal flow can be harnessed for electricity generation. Such situations are found in many parts of the world, and it is surprising that only one successful tidal barrage exits—the La Rance Barrage in Brittany, which has been operating for nearly 50 years. The barrage contains 24 underwater turbines and produces on average about 60 MW of electricity. The United Kingdom is perhaps the leader in tidal current research, and it is very likely that the next tidal power station will be built in the United Kingdom, although the United States, Canada, South Africa, and South Korea all have ongoing project proposals.

Solar energy (photovoltaics). Photovoltaic (PV) energy is based on the photoelectric effect, whereby the Sun's energy excites electrons in metals (and semimetals or metalloids), thus producing electricity directly. Most PV devices are made from silicon, and modern cells are capable of energy-conversion efficiencies of 25%. The cells are relatively expensive to produce, so other PV materials have been developed. They are made from a combination of cadmium, indium, gallium, tellurium, copper, and silicon. Such cells achieve an efficiency of 20%. Special triple-junction cells (involving three different layers of PV material, aimed at different sections of the solar spectrum) can now achieve efficiencies of over 40%. These are, as expected, expensive to produce.

In the 1970s and 1980s, the cost of producing PV cells was a major problem, but this is no longer the case and it is predicted that within the next few decades the energy derived from PVs will cost less than that from fossil fuels. PV cells are expected to make a significant contribution to world energy in the years to come.

Solar energy (concentrated solar power). Concentrated solar power (CSP) is obtained by focusing the Sun's energy to create high temperatures in a material (called an absorber) from which the heat energy is transferred to a fluid (known as a heat-transfer fluid) that is used to generate electricity in a conventional steam turbine power plant. The focusing is usually done by mirrors that track the Sun, and only direct radiation can be concentrated in these optical systems. The CSP industry began in the 1970s, and by the end of 2012, worldwide operating CSP installations amounted to 2.7 GW, with 2.3 GW under construction and 31.7 GW planned. At the moment, the United States and Spain are the leaders of CSP. There are many designs of CSP systems, such as parabolic troughs and absorber tubes; curved mirrors and absorber troughs; heliostats with solar towers; parabolic reflector dishes; linear Fresnel mirrors; direct steam generation; and molten-salt energy storage facilities. The potential for solving the world's global warming crisis is great, and it has been estimated that if only 1% of the Sahara desert were used for CSP, there would be sufficient energy to satisfy the world's electricity needs.

Solar energy (water heating). Solar water heaters (SWHs) collect solar radiation from the Sun and this energy is then transferred to water and stored for future use. It has been estimated that global installed capacity is greater than 200 gigawatt thermal (GW_{th}), with China (60%), Europe (18.5%), and the United States and Canada (8%) having the largest installations. In terms of the capacity per 1000 inhabitants, the top three contenders were Cyprus (577 GW_{th} per 1000 inhabitants), Israel (397 GW_{th} per 1000 inhabitants), and Austria (388 GW_{th} per 1000 inhabitants). Many designs are available, including flat-plate collectors; evacuated-tube collectors; concentrated collectors; active and passive systems; combined PV/thermal systems; solar-assisted heat pumps; combined water heating and space heating; and even district heating systems. Most countries in the world would benefit from these water heaters and, where practical it should be mandatory to install such SWH systems in all new industrial, public, or private buildings.

Energy storage technologies. With the advent of intermittent energy sources, such as wind, tidal, and solar power, it is necessary to be able to store energy to eliminate peaks in power supplies to national grids. The importance of this area has not been recognized by governments, resulting in, as an example, the closure of wind turbines at certain times to avoid these peaks. Energy storage can take many forms, with pumped-storage hydroelectric power being the most common. This involves pumping water into storage dams in times of high energy supply and recouping the energy by releasing the water through electricity-generating turbines. Other techniques include compressed-air storage; flywheel energy storage; battery energy storage; making hydrogen; and phase-changing compounds (melting or vaporizing). Far too little has been done in the area of energy storage, and this is beginning to stifle the development of wind and solar renewable energy.

Carbon dioxide capture. One way to mitigate the carbon dioxide (CO_2) produced from burning fossil fuel and from cement manufacture is to capture and store the CO_2 (known as carbon capture and storage; CCS) in deep saline formations or in depleted hydrocarbon fields or even under oceans. Estimates for worldwide CCS capacity is orders of magnitude larger than the annual global CO_2 emission levels of 34 gigatons, so the potential does exist for storing CO_2 for centuries to come. It is unfortunately a relatively expensive process. There are at present large-scale CCS projects in Norway (Sleipner and Snohvit projects), Algeria (In-Salah project), and Canada (Weyburn project), but this is just a drop in the ocean compared to what is required.

Battery and fuel-cell technologies. If we are to reduce the amount of CO_2 currently being produced in the world, we must find alternative energy sources for trains, planes, and vehicular transport, as these are producing 13% of all CO_2 emissions. Although this is half the amount produced by electric power stations, it is a significant amount. The only contenders as replacements for hydrocarbon fuels are batteries and fuel cells. Much work is being done in these areas, but no definitive alternative has emerged. Lithium-ion batteries appear to be the most successful for cars and buses and are slowly being introduced in hybrid cars. They do have the disadvantage that unless non-fossil-fuel energy sources are used in making and charging the batteries, there is no reduction in CO_2 emissions.

Other technologies. Other emerging technologies include geothermal energy; biomass- or coal- or methane- or waste-to-transport fuel; liquefied natural gas as a transport fuel; new coal processing technologies; new developments in deepwater and arctic oil and gas exploration; oil sands; nuclear fusion; smart grids; the beaming of energy from outer space to Earth via laser technology; new thorium nuclear fission reactors, and nuclear fusion (predicted to come on-stream in the second half of the twenty-first century), to mention some that have either reached maturity or could one day play an important role in our future energy options.

For background information *see* BATTERY; CLATHRATE COMPOUNDS; COALBED METHANE; ELECTRIC POWER GENERATION; ENERGY STORAGE; FLYWHEEL; FUEL CELL; GEOTHERMAL POWER; GLOBAL CLIMATE CHANGE; HYDRATE; HYDROGEN; METALLOID; METHANE; NATURAL GAS; NUCLEAR FUSION; NUCLEAR POWER; PHOTOVOLTAIC CELL; PHOTOVOLTAIC EFFECT; PUMPED STORAGE; SHALE; SOLAR ENERGY; SOLAR HEATING AND COOLING; TIDAL POWER; TIDE; WATERPOWER; WIND POWER in the McGraw-Hill Encyclopedia of Science & Technology.

Trevor M. Letcher

Bibliography. T. M. Letcher (ed.), *Climate Change: Observed Impacts on Planet Earth*, Elsevier, Oxford, UK, 2009; T. M. Letcher (ed.), *Future Energy: Improved, Sustainable and Clean Options for our Planet*, 2d ed., Elsevier, Waltham, MA, 2013.

Encephalitis (arboviral)

The arboviral encephalitides comprise several different families of arthropod-borne viruses (arboviruses) that cause encephalitis, which is an inflammation of the brain tissue in humans and vertebrate animals. The arboviruses are transmitted from one vertebrate host to another by insects, ticks, or other arthropods. The biology and behavior of the arthropods and vertebrate hosts greatly influence the ecology and epidemiology of these viruses. Monitoring for arboviral encephalitis involves the organized tracking of levels of virus activity, vector (pathogen-transferring agent) populations, and infections in vertebrate hosts, along with human cases, weather conditions, and other factors, to detect or predict changes in the transmission dynamics of the infection.

Infectious agents. There are approximately 500 species of arboviruses in the world. More than 100 arbovirus species have been isolated from humans, although not all cause serious disease. At least 44 arboviruses are known to infect livestock or other domestic animals. Historically, there are four common encephalitis viruses in the United States: La Crosse encephalitis, St. Louis encephalitis, and Western and Eastern equine encephalomyelitis. Venezuelan equine encephalomyelitis (VEE) occasionally moves northward into the southwestern United States. Several other viruses (Jamestown Canyon, Cache Valley, and Powassan viruses) sometimes cause disease in humans or domestic animals in the United States. In 1999, an Old-World arbovirus, West Nile virus (WNV), appeared in New York. It spread rapidly throughout North America, and presently West Nile virus is the most common arboviral disease in humans in the United States. In 2009, another arbovirus, Heartland virus, apparently transmitted by ticks, was detected in Missouri.

Transmission cycles. Arbovirus transmission cycles are usually complex (see **illustration**). The normal (natural) cycle may involve several bird or mammal species and several mosquito species [or, in a few cases, tick species (not described in this article)]. Weather, food resources, predators and parasites, and space resources influence the arbovirus transmission cycle. Enzootic virus transmission (that is, transmission within the natural habitat) may occur at a low intensity among the normal vertebrate hosts and vectors within specific habitats in rural or suburban environments. Thus, transmission may remain undetected by most monitoring programs. However, when favorable weather or other environmental conditions lead to an abundance of both vertebrate hosts and mosquitoes, and when few vertebrate hosts are immune to the virus, transmission may increase in intensity and expand in distribution, producing an epizootic (transmission to animals outside the natural habitat and host range). If epizootics begin early in the transmission season, and if localized centers of epizootic activity (referred to as foci) expand into urban centers with adequate host and vector populations, the risk of human involvement increases. (When humans are involved, the term epidemic is used.)

Zoonotic diseases. Zoonotic diseases (that is, diseases that typically occur in vertebrate hosts, but can be transmitted to humans), and particularly the vector-borne zoonoses, are usually separated in time and space from the human population. This means the virus has usually already spread through the wild vertebrate population by the time that it begins to infect human or domestic animal hosts. To be effective, however, prevention activities generally must target the zoonotic portion of the cycle, often at the level of the vectors. Thus, monitoring programs must track the zoonotic portion of the cycle. By the time cases are diagnosed in humans, the epidemic will be well underway and no longer preventable. Even the smallest delays can diminish the impact of prevention or control measures on the course of a potential epidemic.

Encephalitis epidemics tend to be cyclic, recurring every 10–20 years. The time lag makes it difficult to prepare for the next occurrence. Without adequate monitoring, health agencies can be lulled into the assumption that the viruses have disappeared and are no longer a problem.

Arbovirus surveillance. In order to design an arbovirus monitoring program, planners must decide what kinds of data to collect, what kinds of methods and equipment to use, and how often and in how many locations to collect the data. They also must decide how they will integrate the resulting

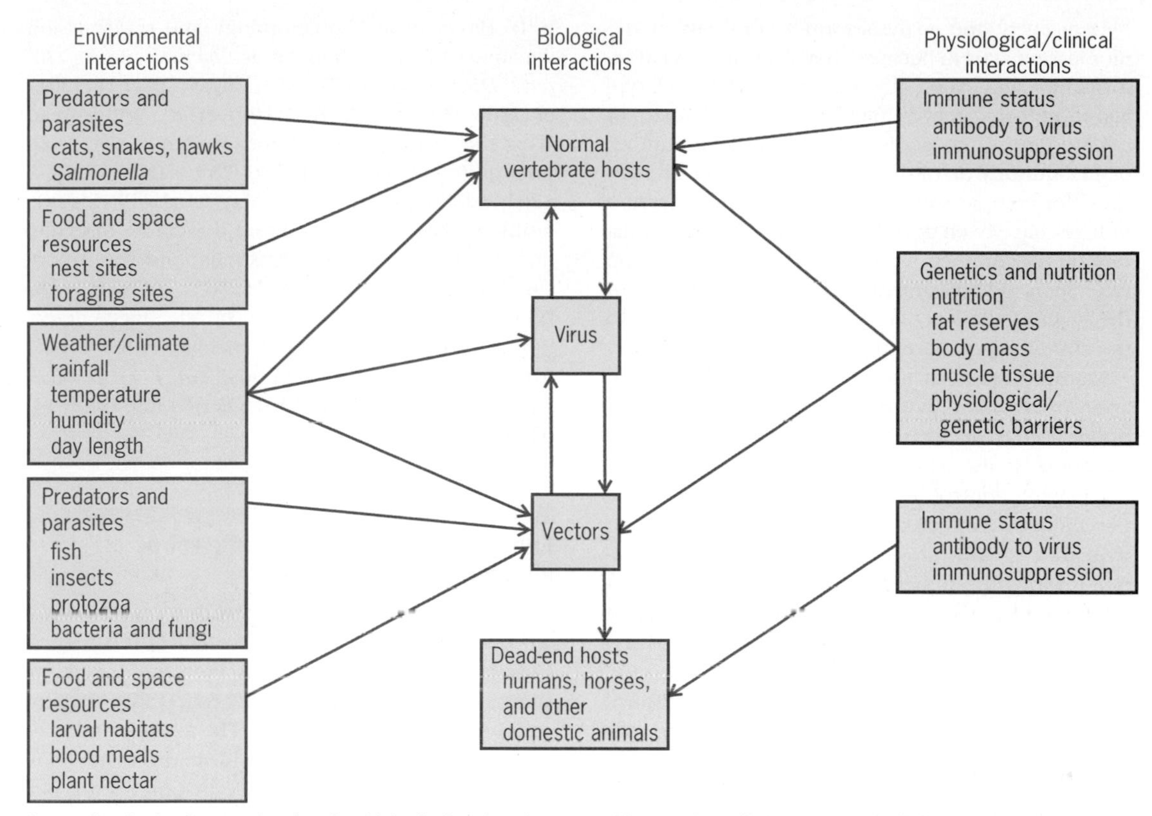

Generalized arbovirus cycle, showing biological and environmental interactions. (*Adapted from C. G. Moore et al., Guidelines for Arbovirus Surveillance Programs in the United States, Division of Vector-Borne Infectious Diseases, National Center for Infectious Diseases, Centers for Disease Control and Prevention, Public Health Service, U.S. Department of Health and Human Services, Fort Collins, CO, 1993*)

data into the larger mission of health protection. To get the most accurate estimate of the potential for epidemics, it is preferable to have multiple measures [for example, weather conditions, mosquito collections, and sentinel flocks (groups of birds that are set up to be bitten by mosquitoes and develop antibodies to the virus if it is present, but without contracting the full-blown disease, thereby acting as "sentinels" of virus transmission)]. A single measure may not reveal the full extent of virus activity. Two common monitoring methods are mosquito trapping and the use of sentinel chicken flocks. Specimen collection must be rapid, economical, and within the capabilities of the agency doing the work. Monitoring programs also must fit within the budgetary constraints of the local agency. Information useful in predicting arbovirus activity includes environmental factors (rainfall, mountain snowpack, temperature, and humidity), vector populations (abundance, age, and infection rates), wild (normal) vertebrate hosts (abundance, immune status, infection rates, and age), incidental hosts (infection rates), and viral strain differences.

Often, biased sampling methods can be used to collect specimens in a particular physiological state. For example, some mosquito traps primarily collect gravid (egg-bearing) females. Gravid females have taken at least one blood meal (because they need blood to produce eggs), so they are more likely to have become infected with a virus. This can reduce the effort needed to detect virus activity in the community, but it may give an imperfect picture of the status of the entire vector population.

Mosquitoes collected in the monitoring program are tested for evidence of viruses. Collections are separated into groups (or "pools") based on location, date, species, and sex. They are ground up in a solution containing buffers and other reagents, and then processed by one of several different systems. "Dipstick" antigen detection systems are economical and sensitive, and they can be used at the local program level. Other methods, including DNA amplification techniques such as the polymerase chain reaction (PCR), are more sensitive, but may be too cost-prohibitive for use in local health or vector control programs.

In the vertebrate hosts, changes in the immune system (production of antibodies) can be measured to determine whether the animal has been infected recently (or at some time in the past) with a particular virus. Enzyme-linked immunosorbent assay (ELISA) tests can be used to detect immunoglobulins M and G (IgM and IgG). The presence of IgM can be a good indicator of recent infection, whereas the presence of IgG gives a measure of long-term virus activity in the host population. In the case of West Nile virus, the dipstick methods described previously have been used to detect viral antigens in oral and cloacal swabs from dead birds, including crows and magpies.

Data entry and analysis form a core part of the monitoring system because they determine what information will eventually reach the end user. A typical monitoring report might show the numbers of infected mosquito pools or birds, the total number of mosquitoes or birds tested, and historical averages (for comparison purposes). Recently, infection indexes have been developed that combine information on the number of mosquitoes and the infection rate. Such indexes appear to be good predictors of the level of virus activity and, therefore, of risk to the community where the monitoring takes place.

Monitoring information is useless unless it is used appropriately and in a timely fashion. The information is used to decide when to begin specific control measures. It also can be used to justify withholding or delaying control. Ideally, the data (including the percentages of infected hosts and vectors, and which mosquito species are involved) will provide accurate information to health planners, who can determine whether they are dealing with either isolated hot spots or widespread activity. This information then guides the prevention and control response. For example, the U.S. Centers for Disease Control and Prevention maintains the ArboNET reporting system, which provides rapid data exchange at the national level.

Arboviral encephalitis prevention. Information from monitoring programs allows health planners and agencies to respond in a timely, organized, and effective fashion using all available tools. Integrated mosquito management (IMM) programs generally focus on the entire range of prevention technologies. At the first sign of virus activity, the typical control response should begin with public service messages designed to reduce human-mosquito contact. These messages should encourage people to use repellents, wear long-sleeved shirts and long pants, repair door and window screens, and avoid certain localities. Other important community activities include the removal, draining, or covering of mosquito larval habitats, and the reporting of neighborhood mosquito problems. Veterinarians should remind owners to vaccinate horses if equine-infecting viruses are active. As indicators become more severe, local agencies should begin adult mosquito control or other prevention activities as appropriate. The local community should be aware of the results of the monitoring program on a daily basis, especially as virus activity increases and the likelihood of activities such as adult mosquito control becomes necessary.

For background information *see* ANIMAL VIRUS; ARBOVIRAL ENCEPHALITIDES; ARTHROPODA; BRAIN; DISEASE ECOLOGY; EPIDEMIC; EPIDEMIOLOGY; VIRUS; VIRUS CLASSIFICATION; WEST NILE VIRUS; ZOONOSES in the McGraw-Hill Encyclopedia of Science & Technology. Chester G. Moore

Bibliography. E. R. Deardorff et al., Powassan virus in mammals, Alaska and New Mexico, USA, and Russia, 2004-2007, *Emerg. Infect. Dis.*, 19(12): 2012-2016, 2013, DOI:10.3201/eid1912.130319; E. B. Hayes et al., Epidemiology and transmission dynamics of West Nile virus disease, *Emerg. Infect. Dis.*, 11(8):1167-1173, 2005, DOI:10.3201/eid1108.050289a; A. A. Marfin et al., Widespread West Nile virus activity, eastern United States, 2000, *Emerg. Infect. Dis.*, 7(4):730-735, 2001, DOI:10.3201/eid0704.017423; F. J. May et al., Phylogeography of West Nile virus: From the cradle of evolution in Africa to Eurasia, Australia, and the Americas, *J. Virol.*, 85(6):2964-2974, 2011, DOI:10.1128/JVI.01963-10; H. M. Savage et al., First detection of Heartland virus (Bunyaviridae: *Phlebovirus*) from field collected arthropods, *Am. J. Trop. Med. Hyg.*, 89(3):445-452, 2013, DOI:10.4269/ajtmh.13-0209.

Environmental life-cycle impact of alternative aviation fuels

As world fuel prices increase and people become more aware of the role of fossil fuels in creating carbon emissions, there is an increased need for alternative aviation fuels that are renewable, sustainable, and economically viable. The aviation industry currently creates 2% of worldwide carbon emissions and 12% of emissions from the transport sector, but alternative aviation fuels produced from renewable feedstocks have the potential to reduce fuel life cycle emissions by up to 80%. While it has been demonstrated that commercial flights can be flown successfully using a range of different alternative fuels and blends, feasibility assessments of such fuels must include an evaluation of compliance with regulatory bodies' requirements and quantification of reductions in environmental emissions. One method that identifies the potential environmental impacts of producing aviation fuel from alternative sources is life-cycle assessment (LCA). LCA accounts for emissions from the entire life cycle of the fuel, from growing and harvesting the feedstock in the field through refining the fuel to its combustion in an aircraft. LCA studies are becoming increasingly popular as more alternative fuels are investigated; one example is the study conducted in 2010 by Russell W. Stratton, Hsin Min Wong, and James I. Hileman, which analyzed greenhouse gas emissions for a range of potential aviation fuel feedstocks in the United States, including soybean, palm, jatropha, and algae oil. The study also applied different LCA methods and feedstock yield rates to each pathway in order to show the sensitivity of the results to various model assumptions. There are numerous possible feedstock-to-fuel pathways that can be modeled, many of which are still on a lab scale and not yet ready for commercialization. *See* GREEN AVIATION.

Feedstock overview. The alternative fuel production and combustion cycle is a closed-loop system in which fuel combustion emissions are assumed to be absorbed by the feedstock crops during the natural growth cycle, providing the potential for lower carbon emissions over the fuel life cycle

(**Fig. 1**). Guidelines for selecting future feedstocks include a lack of impact on food and water security, minimal indirect land use change (ILUC), and a lower greenhouse gas impact than equivalent fossil fuels. Australian candidate feedstocks that can be selected as examples to illustrate the application of the LCA process to various fuel production pathways include molasses from sugarcane, autotrophic algae, and the oilseed tree *Pongamia pinnata*.

Sugarcane is mechanically harvested and milled into raw sugar and molasses, and the molasses is then fermented and passed through a hydrocracking process to produce aviation fuel plus secondary fuel sources, diesel and naphtha. The milling waste product, bagasse, is combusted to produce electricity and steam (**Fig. 2***a*).

Autotrophic algae can be grown in open-air ponds on nonarable land with a supply of groundwater, seawater, and carbon dioxide. In one possible formulation, algal oil and phospholipids are extracted from the algae paste via a centrifuge and hexane extraction process before the algal oil is refined into aviation fuel, diesel, and naphtha via a hydrocracking and distillation process. The algal waste products from the oil extraction process can be digested and combusted to produce electricity and steam (Fig. 2*b*).

Pongamia pinnata produces seeds that are dried and crushed to extract pongamia oil and a phospholipid coproduct; then the pongamia oil is further refined into aviation fuel, diesel, and naphtha (Fig. 2*c*).

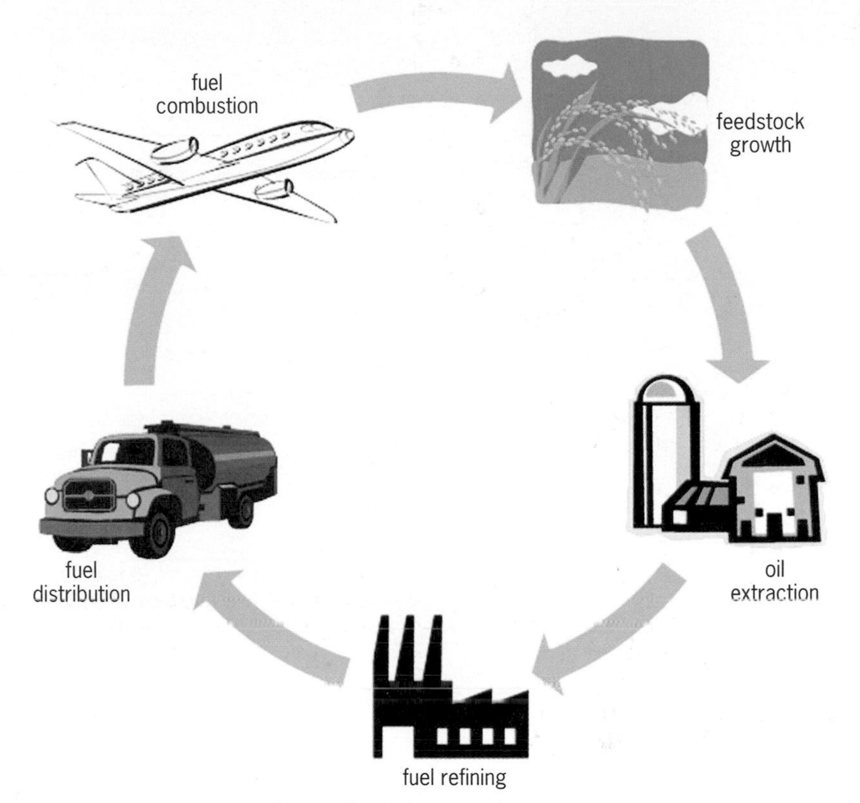

Fig. 1. Life cycle of alternative aviation fuels.

Assigning impacts to coproducts. One important step in an LCA is to assign environmental impacts to products and coproducts throughout the system. There are two methods of doing this, namely, allocation and system expansion. The allocation method seeks to allocate environmental impacts to the products and coproducts based on a common factor such as mass, energy, or economic value, whereas system expansion "expands" the system boundary to include a marginal product that has been offset by a coproduct, resulting in a global warming potential (GWP) credit to the system.

There is ongoing debate over which method is optimal. While the International Organization for Standardization (ISO) recommends the use of system expansion whenever possible, followed by allocation, the choice of method can depend on the purpose of the model, the feedstock selected, and any coproducts generated during the life cycle. System expansion is considered to provide a prospective or forward-looking result for LCA, as it takes into account how a change in production volume can have a flow-on effect to other products and markets. However, selection of the marginal products that have been affected can be highly subjective and uncertain, and the approach can be difficult to apply to some feedstock pathways. On the other hand, the allocation method is considered a more retrospective or status quo approach that merely allocates environmental impacts across each of the coproducts of the system and does not factor in any external impacts on marginal products. The results can be more straightforward to interpret, and the method is easily applied across many feedstock processes.

For example, one potential way of applying the system expansion method to the sugarcane pathway is to assign 100% of the impacts of raw sugar production to raw sugar; since molasses has value as an animal feed, the system boundary is expanded to include the offset production of sorghum when the molasses is used as animal feed. However, in the aviation fuel process, molasses is actually the product of interest, and as the molasses is now diverted away from use as an animal feed, extra sorghum must be produced to make up the deficit in supply to the animal feed industry. Therefore, the impact of producing aviation fuel from sugarcane is actually the impact of additional sorghum production plus the impact of the aviation fuel refining process. This gives a unique result that is not seen with many other feedstock processes, since the sugarcane growing process does not contribute at all to the GWP result for aviation fuel.

In the case of algae and pongamia, 100% of the impact of algae or pongamia oil production could be assigned to the oil product, which then flows through to the fuel process. The system boundary is expanded to include the offset production of lupins when they are used as a phospholipid substitute, resulting in a small credit to the system that then flows through to the aviation fuel process.

Results. Normalized well-to-wake (WTW) results for the three feedstock pathways using the system expansion method show that sugarcane and pongamia have the potential to produce higher

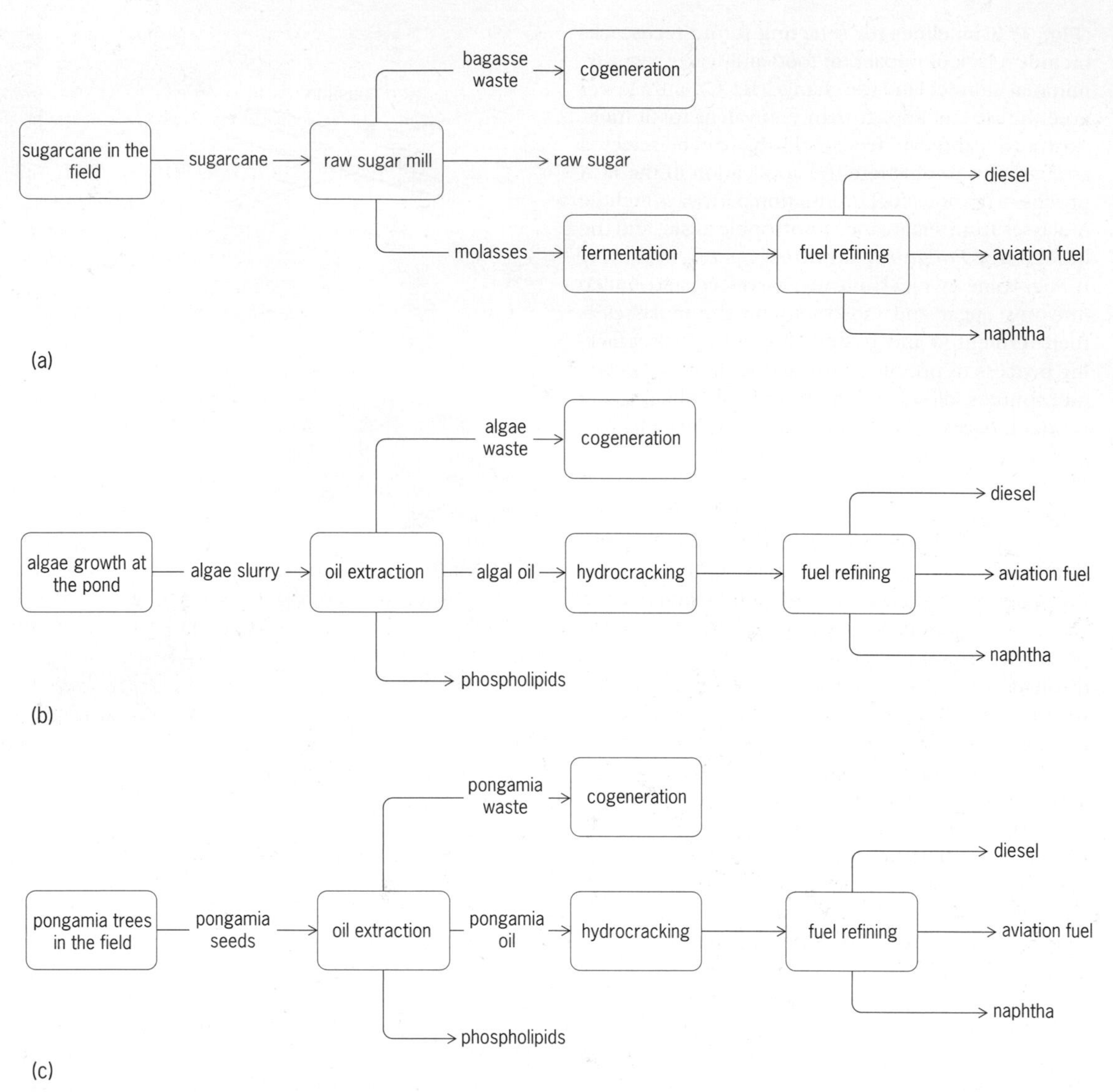

Fig. 2. Process flow diagrams for alternative aviation fuels. (a) Sugarcane. (b) Autotrophic algae. (c) The oilseed tree, *Pongamia pinnata*.

carbon emissions over the fuel life cycle than conventional aviation fuel (**Fig. 3**). Selection of the marginal offset product is significant in swaying system expansion results, particularly in the case of sugarcane, as it is the sorghum production that replaces molasses that makes the most significant contribution to the results for this feedstock pathway. The algae feedstock produces a 20% reduction in carbon emissions compared to conventional aviation fuel, and the process modeled assumes a relatively energy-intensive centrifuge system for extracting algae from the pond, so that other processes that are under development could improve the results even further.

Electricity supply is a key driver of carbon emissions for the different feedstocks. However, these emissions also depend on the type of electricity available at the feedstock and fuel production sites. For example, approximately 77% of electricity in Australia is generated from coal, which increases the carbon emissions for each feedstock

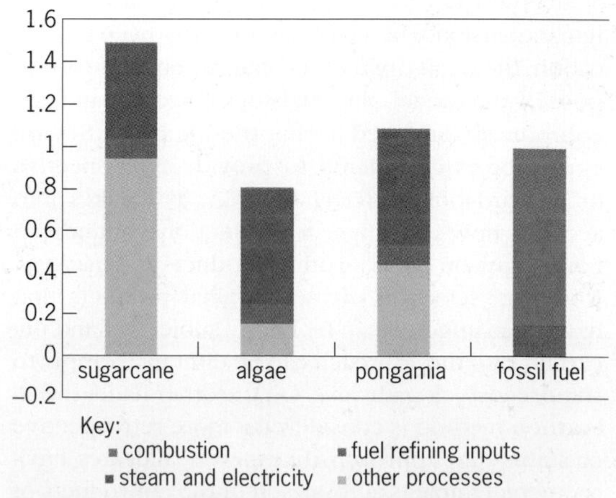

Fig. 3. Normalized global warming potential (GWP) contributions of feedstock pathways and conventional aviation fuel for the system expansion method.

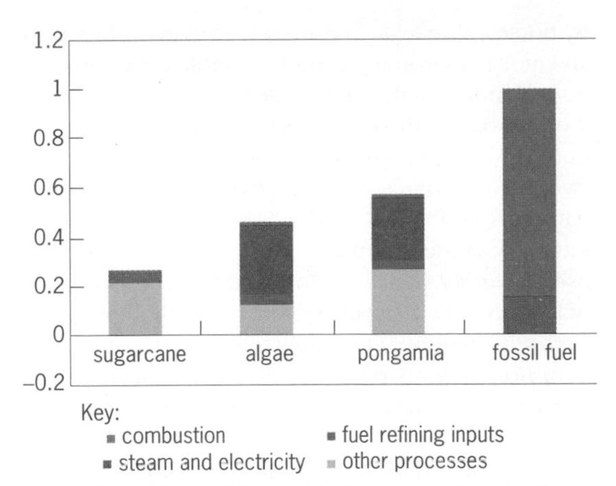

Fig. 4. Normalized GWP contribution comparison for the allocation method.

pathway, and undertaking the same feedstock analysis for another geographic location with a higher proportion of renewable or alternative energy sources may provide a lower GWP result.

An alternative to system expansion is to apply economic allocation across the coproducts of each system, which can result in a significantly different GWP impact from system expansion (**Fig. 4**). In this case, each feedstock results in lower carbon emissions over the fuel life cycle than conventional aviation fuel and therefore may be environmentally viable in terms of GWP. Sugarcane in particular has significantly reduced emissions, since this result excludes any impact from sorghum production. The first problem with economic allocation is that economic value can change drastically as new products flood the market; hence, this model would be subject to change in the future. The second drawback of economic allocation is that it simply takes a snapshot of a current process and does not take into account the impact that the change in the production of aviation fuel and its coproducts may have on similar products in the market.

It is also worth noting that the production of alternative fuels generally produces higher carbon emissions than the production of fuel from fossil sources. This is because the production of alternative fuels includes the agricultural growing and oil refining processes, which can require significant resources; in contrast, the fossil fuel source is extracted from the earth and refined, so that it requires fewer resources to produce and can appear "cleaner." However, once the fuel combustion process is included in the cycle, we see the potential for alternative fuels to reduce carbon emissions over the life cycle of the fuel, since the carbon dioxide released into the atmosphere through combustion is taken up by the plant crop through the natural growth process.

Conclusion. Studies of alternative fuels are becoming increasingly popular and will continue to highlight the potential benefits and drawbacks of future alternative fuel options. LCA studies can produce varied results, since they are highly sensitive to many factors, including the LCA method used; the marginal offset product selected; farming, oil extraction, and refining processes; and the proposed geographic location and local energy sources. Thus, the results presented in this article simply give a snapshot of the possible pathways. The LCA process is a powerful tool in defining these environmental impacts, and the results should be carefully interpreted and utilized by industry leaders in their decision-making processes on the future of the alternative fuel industry throughout the world.

For background information *see* AIRCRAFT FUEL; ALGAE; BIOMASS; CLIMATE MODIFICATION; COGENERATION; DIESEL FUEL; DISTILLATION; FERMENTATION; HYDROCRACKING; NAPHTHA; PHOSPHOLIPID; RENEWABLE RESOURCES; SUGAR; SUGARCANE; SORGHUM in the McGraw-Hill Encyclopedia of Science & Technology. Kelly Cox

Bibliography. *Beginner's Guide to Aviation Biofuels*, Air Transport Action Group, Geneva, Switzerland, 2009; *Environmental Management—Life Cycle Assessment—Principles and Framework*, ISO 14040, International Organization for Standardization (ISO), Geneva, Switzerland, 2006; P. Graham et al., *Flight Path to Sustainable Aviation: Towards Establishing a Sustainable Aviation Fuels Industry in Australia and New Zealand*, CSIRO Energy Transformed Flagship, Mayfield West, New South Wales, Australia, 2011; R. W. Stratton, H. M. Wong, and J. I. Hileman, *Life Cycle Greenhouse Gas Emissions from Alternative Jet Fuels*, Partnership for Air Transportation Noise and Emissions Reduction Project 28, Massachusetts Institute of Technology, Cambridge, MA, 2010.

Equine glanders

Glanders is a highly contagious and fatal bacterial infection of equines that results in chronic suppurative (pus-producing) lesions of the skin and mucous membranes, pneumonia, and septicemia (a clinical syndrome in which infection is disseminated through the body in the bloodstream). The disease was first described in 330 BC by Aristotle, who named it "malleus." In general, three clinical forms of glanders are observed: the nasal form (**Fig. 1**), characterized by a yellowish green mucopurulent nasal discharge commonly associated with the formation of nodules and ulcers in the nasal septum; the pulmonary form, associated with a persistent dry cough accompanied by labored breathing; and the cutaneous or skin form, also known as farcy (**Fig. 2**), which presents as ulcerative nodules, especially on the inner aspect of the hind limbs (**Fig. 3**). In the field, the three forms of the disease usually do not manifest distinctly and can occur together. Donkeys and mules are more susceptible to infection than horses and may succumb to the disease within a few days. In contrast, the disease generally takes a more chronic course in horses, which may survive for several years. Infection may also be transmitted to goats, camels, and carnivores. Cattle and

Fig. 1. The nasal form of glanders showing blood-tinged nasal secretions from a mule in Uttar Pradesh, Northern India.

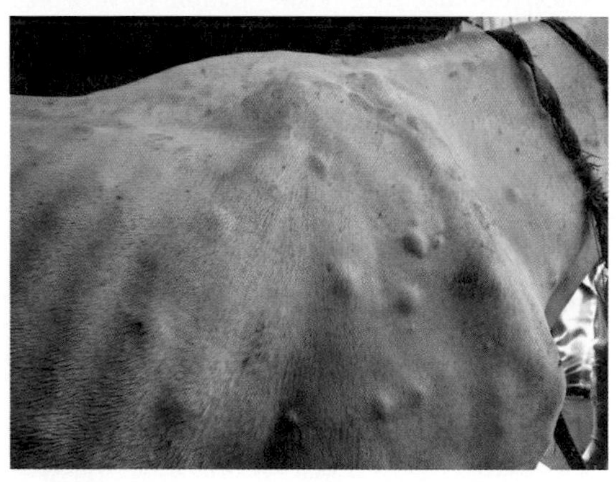

Fig. 2. The cutaneous or skin form of glanders (farcy) showing swollen nodules on the shoulder and neck region of a mule.

pigs are resistant to glanders. The disease is of great zoonotic significance because humans may become infected through direct contact with the causative organism and prolonged contact with infected animals. Veterinarians, horse caretakers, abattoir workers, and researchers are the main risk groups. An extremely high rate of mortality can occur in untreated humans.

Causative agent. The causative agent of glanders, *Burkholderia mallei*, was isolated first by the German microbiologists Friedrich Loeffler and Wilhelm Schultz in 1882. It is a nonmotile, gram-negative, facultative intracellular bacterium and is susceptible to drying, heat, sunlight, and many common disinfectants, including sodium hypochlorite, ethanol, and glutaraldehyde. Typically, the organism does not survive long outside of its natural hosts, that is, horses, donkeys, and mules. However, in warm and moist environments such as stable beddings, manure, water troughs, and fodder, it may survive up to 2 or 3 months. *Burkholderia mallei* is a genetically uniform species, and its genome is distributed across two chromosomes. Comparative genomics studies have revealed that *B. mallei* is a pathoadaptive evolutionary clone of a much more diverse species, *B. pseudomallei*, which is an opportunistic pathogen of humans and the primary cause of melioidosis. The *B. mallei* genome continues to evolve through random homologous recombinations, which may contribute to the differences in virulence among geographically distinct strains.

Mode of transmission. Transmission of the pathogen among equines occurs primarily by ingestion of skin exudates and respiratory discharges and by oronasal mucous membrane exposure. In rare cases, it may occur by inhalation. Sharing of feed and water troughs and snorting may facilitate transmission. The disease can also be transmitted sexually from an infected partner and vertically from an infected mare to a foal. A long-term exposure to asymptomatic chronically infected equines and the handling of contaminated fomites (inanimate objects that may be contaminated with infectious organisms), such as grooming tools, hoof-trimming equipment, husbandry equipment, bedding, and veterinary equipment, without precautions could become a rare but perilous risk to humans.

Diagnosis and control. The identification of cases in the field includes the observation of typical clinical symptoms and mallein testing of equines by the intradermopalpebral route, that is, an intradermal injection of a test reagent (mallein) into the skin of the eyelid. Malleinization was extensively used during earlier eradication programs. However, the test has been discontinued in many places because of limited availability of the mallein purified protein derivative (PPD) and possible interference with serological tests. Nevertheless, this method can be used in remote field conditions with sufficient reliability by a trained practitioner. In general, the gold standard consists of the isolation and identification of

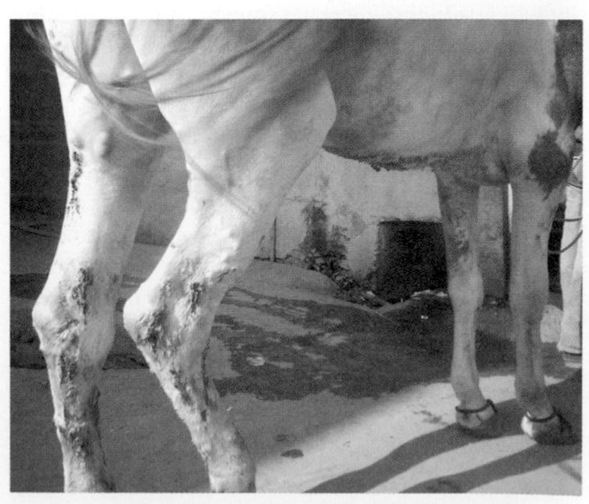

Fig. 3. Ulcerative lesions on the hind limbs of a horse.

the bacterial organism from fresh cutaneous lesions, nasal discharges, and various lesions at necropsy. It should be noted, however, that the identification is a laborious and time-consuming process and is not always attainable during the early stages of the infection or in clinically inapparent cases. Although polymerase chain reaction (PCR)–based assays and DNA sequencing offer a quick and specific identification of the organism, their application is limited in the surveillance of equine populations by the need for containment laboratory facilities, trained personnel, and financial resources.

Presently, the complement-fixation test (CFT) is the only officially recognized serological laboratory test for the international trade of equines, and it remains the preferred diagnostic tool for surveillance in eradication programs. However, the CFT is associated with a notable number of nonspecific results, especially with donkey and mule serums. Recently, recombinant protein–based enzyme-linked immunosorbent assays (ELISAs) have been developed for the safe, accurate, and quick serological diagnosis of glanders in equines. Other tests include a lipopolysaccharide-based Western blot and a monoclonal antibody–based competitive ELISA. These assays have yet to be validated and accepted officially.

Because of the unavailability of a prophylactic vaccine or effective treatment regime, *B. mallei* is considered a potential biological weapon and has been included in the B list of bioterrorism agents by the U.S. Centers for Disease Control and Prevention (CDC). Recognizing the high fatality associated with the disease and its public health importance, many countries have enacted special legislation to control and eradicate glanders. This includes mandatory testing, quarantine of suspicious cases, elimination of infected animals, and restriction on movement of animals to and from infected zones.

Prevalence of disease and eradication strategies. Great Britain became the first country to eradicate glanders in 1928. Subsequently, the disease was eradicated from the United States, Canada, and Western Europe, and it is now rare in the developed world. However, glanders is still endemic in parts of Asia, Africa, South America, and Eastern Europe. Between 2000 and 2013, glanders was reported in 12 countries in Asia, including Afghanistan, the United Arab Emirates, Bahrain, India, Iran, Iraq, Kuwait, Lebanon, Mongolia, Myanmar (Burma), Pakistan, and the Philippines. It also was reported in South America (Brazil and Bolivia), Africa (Eritrea and Ethiopia), and Eastern Europe (Lithuania and Russia). In addition, outbreaks have been reported in tigers and lions in zoological parks in Iran and in domestic camels in Bahrain. Considering its persistence in so many areas of the world, and occasional reports of its transboundary spread, there is continuing potential for its entry into glanders-free countries through importation.

Chronically and subclinically infected solipeds (firm-hoofed animals) are the only natural reservoir of *B. mallei* and are a source of infection to others. Identification of these animals is of great importance for eradication and containment, especially given that treatment is forbidden in many countries and vaccines for specific prevention have not been developed.

Although national campaigns of "test and slaughter" of infected solipeds have been used in affected countries for many decades, persistence of glanders in these countries poses a serious concern and warrants more intense and better funded control programs. Lack of awareness among horse and donkey owners and the limited availability of veterinary services are key factors responsible for underreporting of the disease. Also, the disease can be difficult to differentiate from many other treatable respiratory ailments, and thus asymptomatic, undiagnosed carriers are sources of infection for other susceptible animals. Hence, eradication of glanders needs strict implementation of a control policy, comprising education and awareness of owners and veterinarians, continuous veterinary education programs, rigorous disease surveillances, inclusion of suitable compensation for culling of infected equines, and traceability of infected equines to facilitate control of movement and further spread of the disease. Among these factors, compensation is most complicated because of vague directives on this issue, historically outdated policies, and inordinate delays in the disbursement of funds. As a consequence, many owners show no interest in culling infected animals, which may be the basis of their livelihoods. Eradication of glanders must be a priority program. Simple and effective compensation policies should be evolved and declared explicitly, along with organization of an awareness campaign to recruit active participation of owners. The application of universal hygienic procedures during handling of infected equines is also advisable. Finally, the development and availability of sensitive and specific diagnostic tests are crucial to the success of control programs.

For background information *see* DISEASE; EPIDEMIOLOGY; GLANDERS; HORSE PRODUCTION; INFECTIOUS DISEASE; MEDICAL BACTERIOLOGY; MEDICAL PARASITOLOGY; PERISSODACTYLA; ZOONOSES in the McGraw-Hill Encyclopedia of Science & Technology.

Harisankar Singha; Praveen Malik; Raj Kumar Singh

Bibliography. M. C. Elschner et al., Use of a Western blot technique for the serodiagnosis of glanders, *BMC Vet. Res.*, 7:4, 2011, DOI:10.1186/1746-6148-7-4; P. Khaki et al., Glanders outbreak at Tehran Zoo, Iran, *Iran J. Microbiol.*, 4(1):3–7, 2012; P. Malik et al., Emergence and re-emergence of glanders in India: A description of outbreaks from 2006 to 2011, *Vet. Ital.*, 48(2):167–178, 2012; R. A. Mota et al., Glanders in donkeys (*Equus asinus*) in the state of Pernambuco, Brazil: A case report, *Braz. J. Microbiol.*, 41(1):146–149, 2010, DOI:10.1590/S1517-83822010000100021; H. Singha et al., Optimization and validation of indirect ELISA using truncated TssB protein for the serodiagnosis of glanders amongst equines, *Sci. World J.*, 2014, DOI:10.1155/2014/469407.

Evolution of aplacophoran mollusks

The phylum Mollusca is a highly species-rich and morphologically diverse animal grouping that includes wormlike groups (the aplacophorans), eight-shelled polyplacophorans (chitons), and the primarily single-shelled conchiferans [including monoplacophorans, scaphopods (tusk shells), bivalves (mussels and clams), gastropods (snails and slugs), and cephalopods (squids and octopuses)]. According to one prevailing traditional hypothesis, the shell-less aplacophorans (see **illustration**) constitute the earliest molluscan offshoot. As a consequence, the last common ancestor to all mollusks was likewise imagined as a simple "worm." However, recent phylogenomic studies have argued against the "aplacophoran-first" hypothesis and instead suggest a basal split of Mollusca into two distinct groups: the univalved Conchifera and the Aculifera, which includes the aplacophorans and the polyplacophorans. Accordingly, three scenarios concerning the morphology of the last common molluscan ancestor appear principally possible: The "urmollusk" (last common ancestor to all mollusks) was (i) a mono-shelled conchiferan-like animal, (ii) a wormy shell-less aplacophoran-like mollusk, or (iii) an eight-shelled polyplacophoran-like creature.

Analysis of shell plates. To illuminate this longstanding issue concerning the dawn of molluscan origins, important new data have become available from two different sources, specifically developmental biology and paleontology. Detailed analyses of *Kulindroplax*, a wormlike fossil with seven shell plates, have determined that this animal is a member of the otherwise shell-less aplacophorans and not the polyplacophorans (as one might have assumed). This demonstrates that shell plates are not confined to polyplacophorans alone, but were also present in at least some extinct aplacophoran lineages. However, the determination of whether the seven aplacophoran and the eight polyplacophoran shell plates originated from a common ancestor, or evolved independently, remained elusive. Interestingly, polyplacophorans develop seven shells simultaneously in the larval stage, whereas the eighth plate forms considerably later (often weeks after metamorphosis). This may be interpreted as an ontogenetic reflection of polyplacophoran evolution from a seven-shelled ancestor. If true, this would strengthen the view that seven shell plates are the basal condition for polyplacophorans and aplacophorans, and thus the origin of this sevenfold armor preceded the evolutionary split of the two taxa from each other. Seen in the light of the evolution of the entire phylum Mollusca, this would reduce the number of possible evolutionary scenarios to two: The urmollusk had either seven plates or one single shell, but it was not naked (shell-less).

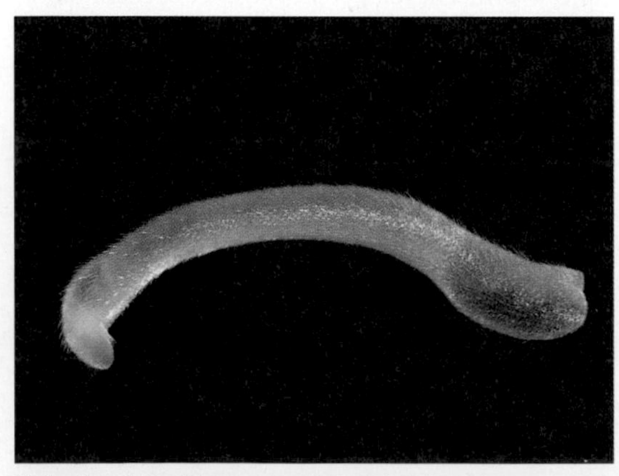

Helluoherpia aegiri, a neomeniomorph aplacophoran. Anterior is to the right. Note the spicules and the lack of shell plates. (*Image courtesy of Maik Scherholz, University of Vienna, Austria*)

Analysis of muscular architecture. One might argue that it is premature to base such far-reaching conclusions on data from one single fossil. Accordingly, to provide an independent set of data, a detailed analysis into the early life history of an aplacophoran representative was carried out on the neomeniomorph *Wirenia argentea*. This animal is rather small (having a length of a few millimeters) and is not easy to recover from its natural habitat [muddy seafloor, approximately 300 m (984 ft) below sea level], but the real challenge is to make these creatures reproduce under controlled laboratory conditions and to rear them from birth (hatching) through larval life and beyond metamorphosis. With respect to the larvae, these are free-swimming, ciliated, so-called trochophore-like life-cycle stages having a total size between 100 and 200 μm. Investigations of their ontogeny have revealed great surprises concerning the development of their musculature. Whereas adult aplacophorans have a relatively simple-structured, slender, cylindrical body with outer ring, inner longitudinal, and interspersed oblique muscles, supplemented by numerous repetitive dorsoventral muscles along their longitudinal axis, their larvae show a much more complex muscular architecture. This consists of specific muscular subunits that had so far been known from larval or adult polyplacophorans only, and include a predominant dorsal longitudinal system ("rectus muscle"), paired lateral enrolling muscles, a pair of ventrolateral muscles, and a single ventromedian muscle. Interestingly, the latter two are transitory in both polyplacophoran and *Wirenia* larvae, and they disappear at metamorphosis. Both muscles are only known from these two animal groups. The rectus muscle is integrated into the adult polyplacophoran muscular body plan, but becomes remodeled during *Wirenia* metamorphosis and eventually forms part of the longitudinal muscles of the body wall of the adult. In *Wirenia*, the same is true for the paired ventrolateral muscles and the enrolling muscles, with the latter becoming a prominent lateral system in adult polyplacophorans. These findings clearly demonstrate that the wormlike body musculature of *Wirenia* is a derived condition that arose by modification from an ancestor that had a much more elaborated muscular anatomy, resembling that of modern-day polyplacophorans.

However, this was not the only message conveyed by the *Wirenia* muscular development. Focusing on the question as to whether the aculiferan ancestor had shell plates or not, the chronology of formation of the dorsoventral musculature provided significant enlightenment. In all shell-bearing mollusks, the shell is connected to the main body by prominent muscles that penetrate the animal in a dorsoventral direction. Accordingly, these dorsoventral muscles, which are often paired, are called "shell muscles." The number of such muscle pairs may range from eight in adult polyplacophorans and some bivalves to one in many gastropods. Because they are intimately linked to the shell (or shells), it is no surprise that seven such pairs form synchronously following metamorphosis in juvenile polyplacophorans after the first seven shell plates have been established. The eighth muscle pair develops later in conjunction with the respective plate. Interestingly, in *Wirenia*, which does not develop shell plates, seven pairs of dorsoventral muscles have been likewise found to emerge simultaneously in the early larva, whereas the multiple seriality (sequence) of these muscles is only established during later development. Being in line with the morphology of the seven-shelled fossil aplacophoran *Kulindroplax*, as well as with the phenomenon of the first-forming seven shell plates and the corresponding muscles in polyplacophorans, a clear scenario emerges in which a seven-shelled polyplacophoran-like ur-aculiferan (the last common ancestor to all aculiferans) ancestor gave rise to two subclades: the aplacophorans by secondary simplification and shell loss in most cases (but not in all cases; see *Kulindroplax*); and the polyplacophorans, which evolved a secondary eighth shell.

While the origin of the univalved conchiferans remains unclear, the eight seriality of the shell musculature in the monoplacophorans (a putatively basal conchiferan group) may indicate that a similar kind of seriality of shell muscles (and possibly shells) was also part of the conchiferan plan. If true, this could push back the chitonesque (chiton-like) morphology to the earliest mollusk on Earth, which emerged from a deep protostomian ancestor.

For background information *see* ANIMAL EVOLUTION; APLACOPHORA; INVERTEBRATE PHYLOGENY; MACROEVOLUTION; MOLLUSCA; MONOPLACOPHORA; MUSCLE; NEOMENIOMORPHA; ORGANIC EVOLUTION; POLYPLACOPHORA in the McGraw-Hill Encyclopedia of Science & Technology. Andreas Wanninger

Bibliography. G. Haszprunar and A. Wanninger, Molluscs, *Curr. Biol.*, 22:R510–R514, 2012, DOI:10.1016/j.cub.2012.05.039; W. F. Ponder and D. R. Lindberg (eds.), *Phylogeny and Evolution of the Mollusca*, University of California Press, Berkeley, 2008; M. Scherholz et al., Aplacophoran mollusks evolved from ancestors with polyplacophoran-like features, *Curr. Biol.*, 23:2130–2134, 2013, DOI:10.1016/j.cub.2013.08.056; S. A. Smith et al., Resolving the evolutionary relationships of molluscs with phylogenomic tools, *Nature*, 480:364–367, 2011, DOI:10.1038/nature10526; M. D. Sutton et al., A Silurian armoured aplacophoran and implications for molluscan phylogeny, *Nature*, 490:94–97, 2012, DOI:10.1038/nature11328.

Evolutionary transition from C_3 to C_4 photosynthesis in plants

Photosynthesis in terrestrial plants is undergoing a dramatic transformation. Photosynthesis originated in bacterial species in a prehistoric environment under high carbon dioxide (CO_2) and low oxygen (O_2) conditions, which was an environment very different from the recent past and current conditions on Earth. Most plant species today still utilize the ancestral C_3 photosynthetic mechanism acquired through symbiosis with chloroplast (cell plastid)–transformed bacteria. In the C_3 system, Rubisco (ribulose-1,5-bisphosphate carboxylase/oxygenase, which is the primary enzyme that feeds carbon into the metabolic Calvin cycle) directly fixes CO_2 from the atmosphere to produce phosphoenolpyruvate, a three-carbon compound. However, Rubisco is an inefficient catalyst for CO_2 fixation, only turning over about twice per second when at maximum capacity. Additionally, Rubisco has oxygenase activity leading to a competing reaction, thereby producing a toxic by-product phosphoglycolate when the enzyme processes O_2 instead of CO_2; this process is referred to as photorespiration. Photorespiration undermines the efficiency of photosynthesis in the current high O_2 atmosphere and becomes exacerbated as temperatures increase, dramatically limiting the efficiency of photosynthetic production. Thus, some plants have evolved new methods to reduce this competing oxygenase reaction by excluding O_2 and concentrating CO_2 around Rubisco in order to run the enzyme at full capacity and full carbon fixation efficiency.

C_4 photosynthesis. Plants have evolved C_4 photosynthesis as a mechanism to increase the efficiency and productivity of carbon fixation. This mechanism separates the carbon fixation process into two cells: the mesophyll cell, which is exposed to the atmosphere and can capture carbon in a four-carbon acid; and the bundle sheath cell, which excludes atmospheric oxygen and utilizes Rubisco for carbon entry into the Calvin cycle. The bundle sheath and the mesophyll cells form concentric rings around the vasculature. This specific cellular arrangement is referred to as the Kranz anatomy (*kranz* is the German word for wreath). C_4 mesophyll cells upregulate native enzymes (that is, increase the response to native enzymes), such as carbonic anhydrase (converting CO_2 and water to bicarbonate), phosphoenolpyruvate carboxylase, and malate dehydrogenase, which are used to directly fix CO_2 into malate, a stable four-carbon compound, using phosphoenolpyruvate as the carbon skeleton to which the CO_2 is added by phosphoenolpyruvate carboxylase. Malate then moves from the mesophyll cells into

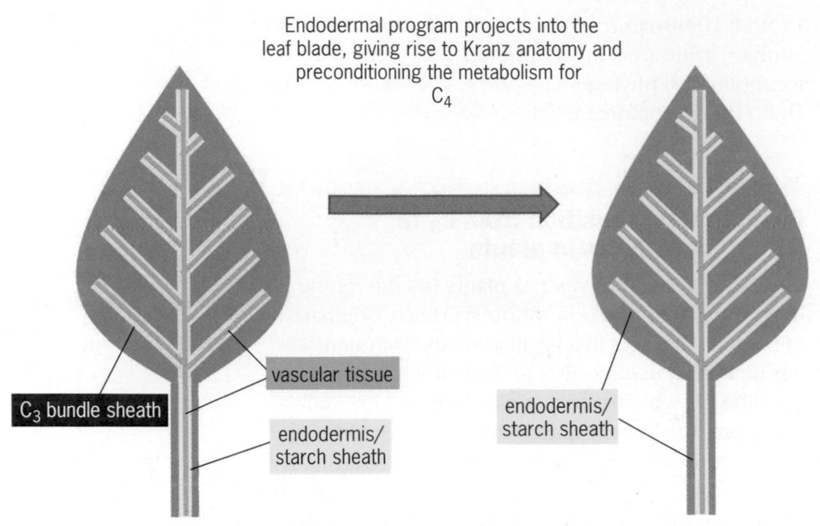

Fig. 1. Model for the evolution of Kranz-type C_4 photosynthesis in eudicots. A normal C_3 eudicot leaf is represented on the left. The endodermis/starch sheath (*yellow*) is present on the vasculature (blue) of the petiole and lower leaf zone, but is absent in the leaf blade/upper leaf zone. The hypothesized shift from C_3 to Kranz-type C_4 photosynthesis may arise when the endodermal/starch sheath program extends out from the petiole/lower leaf zone into the leaf blade. (*From T. L. Slewinski, Using evolution as a guide to engineer Kranz-type C_4 photosynthesis, Front. Plant Sci., 4:212, 2013, DOI:10.3389/fpls.2013.00212*)

the bundle sheath cells through plasmodesmata (narrow channels across cell walls), allowing cytosolic exchange between the cells. Within the chloroplast of the bundle sheath, malate is decarboxylated by either nicotinamide adenine dinucleotide (NAD)– or nicotinamide adenine dinucleotide phosphate (NADP)–malic enzyme to release CO_2 and concentrate it around the active site of Rubisco. These enzymes are also present in C_3 plants, usually performing "housekeeping" functions in other tissues of the plant. In addition, they may have a more ubiquitous role in the CO_2 recycling system that reclaims the CO_2 generated from the respiring tissues such as roots and stems.

In C_4 plants, the mesophyll chloroplasts contain the normal complement of the photosystem I and II reaction centers and have stacked thylakoid grana (multilayered membrane units that occupy the main body of chloroplasts). Thus, they are almost identical to the C_3 mesophyll chloroplasts. Bundle sheath cells

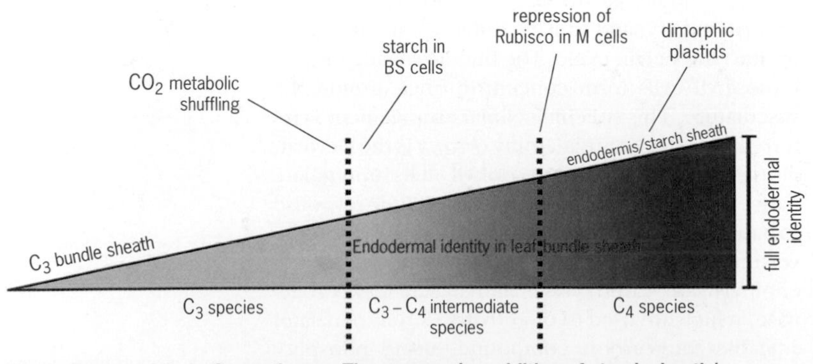

Fig. 2. Proposed C_3 to C_4 continuum. The progressive addition of starch sheath/endodermis identity to the leaf bundle sheath also confers many of the classical features commonly associated with Kranz-type C_4 photosynthesis. Terms: M cells, mesophyll cells; BS cells, bundle sheath cells.

form a single layer around the vein, preferentially accumulate starch compared to mesophyll cells, usually have a suberin (waxy hydrocarbon) layer in their cell walls that limits water and gas exchange, and have modified chloroplasts that have limited photosystem II activity and unstacked thylakoid grana. Unlike C_3 plants, many (but not all) C_4 plants display this characteristic dimorphic plastid arrangement.

Evolutionary aspect. Interestingly, Kranz-type C_4 photosynthesis is one of the most remarkable examples of convergent evolution in the history of life. This mechanism has independently arisen more than 70 times in the angiosperms, in both eudicots and monocots, over the past 35 million years. Additionally, C_4 photosynthesis appears rapidly in many plant species, suggesting that only a few changes are necessary for the C_4 mechanism to arise from a C_3 background. The repeated and rapid evolution of C_4 photosynthesis is most likely a modification of an inherent program that is common to all angiosperms because all of the requisite enzymes involved in C_4 photosynthesis are already inherent in C_3 plants.

In her classic textbook, *Plant Anatomy*, Katherine Esau described all bundle sheaths as having some properties of endodermal tissues, and this is most pronounced in roots and stems. She also pointed out that some species have a "starch sheath" extending from the petiole into the leaf blade. Interestingly, these plants that have the starch sheath in the leaf blades as described turn out to be C_4 photosynthetic plants, specifically corn and sorghum. In all plant tissues, radial patterning centered on veins appears to have a conserved underlying developmental mechanism. The primary genes involved in endodermis formation in the roots are called *Scarecrow* and *Short-root*, and they are also central to the development of the starch sheath in the stems and petioles. These genes have an identical expression pattern in C_3 leaves when compared to roots and stems, which suggests that the underlying developmental mechanism for the endodermis is already present in C_3 leaves, including the associated genetic framework necessary for starch sheath development. Thus, it appears that only small changes in the regulatory framework are necessary or sufficient to establish Kranz-type C_4 plants from underlying components that are already in place in the leaf. In addition, it has been shown that mutations in the orthologs of the *Scarecrow* and *Short-root* genes of maize, a C_4 species, affect the development of the Kranz anatomy in the leaves. Hence, these studies support the hypothesis that the Kranz anatomy and C_4 physiology are derived from the endodermis/starch sheath that is common to all angiosperms. In this hypothesis, there is a continuum of C_4-related traits, both anatomical and physiological, which are transferred into C_3 leaves along with the endodermis/starch sheath program originating from the stem and petiole. Therefore, the more endodermal tissue identity that a leaf bundle sheath has, the more C_4-like the leaf becomes.

A new hypothesis for the repeated and rapid evolution of C_4 photosynthesis was developed from these observations. Specifically, the Kranz-type C_4 photosynthetic mechanism arises when the endodermal/starch sheath program extends into photosynthetic structures, such as leaves, where it is normally repressed or underdeveloped. This leads to a synergistic interaction that can produce the novel C_4 pathway from underlying components of both the C_3 photosynthetic program and anatomical and metabolic features of the endodermis/starch sheath. A visual representation of this hypothesis is depicted in **Fig. 1**. This theory suggests that the anatomical and metabolic features of the starch sheath/endodermis program are compatible with the C_4 photosynthetic mechanism, but only a subset of these features are actually required for function. For example, some C_4 plants have dimorphic chloroplasts, resulting in a stronger starch sheath/amyloplast (starchy cell plastid) identity within the bundle sheath. Other C_4 bundle sheath cells are suberized (undergo suberin deposition) in a similar manner to the endodermis in the stems and roots.

C_3 to C_4 transition. This new model for C_4 evolution proposes thresholds of endodermal/starch sheath identity on the C_3 to C_4 continuum. One threshold moves a C_3 plant into a C_3–C_4 intermediate state, and another threshold moves C_3–C_4 intermediates into the full C_4 state (**Fig. 2**). The more starch sheath/endodermal identity that is transitioned into the bundle sheath, the more C_4 photosynthetic-associated characteristics are installed into these cells. The variations in C_4-associated anatomy and biochemistry may just be a reflection of the degree of the starch sheath/endodermis program projecting into the leaf blade. Overall, the transition from C_3 to Kranz-type C_4 photosynthesis may be relatively simple, only requiring small changes in the developmental programming of cells, which is a feature inherent to all angiosperms. In turn, this information may be used by plant genetic engineers in the future to breed the C_4 pathway into C_3 crops (for example, rice) to improve yield, especially in the face of global warming.

For background information *see* BOTANY; ENDODERMIS; GENE; GENETICALLY ENGINEERED PLANTS; GENETICS; GENOMICS; PHOTORESPIRATION; PHOTOSYNTHESIS; PLANT; PLANT ANATOMY; PLANT CELL; PLANT METABOLISM; PLANT PHYSIOLOGY; PLANT RESPIRATION in the McGraw-Hill Encyclopedia of Science & Technology. Thomas L. Slewinski; Miya D. Howell

Bibliography. N. J. Brown et al., C_4 acid decarboxylases required for C_4 photosynthesis are active in the mid-vein of the C_3 species *Arabidopsis thaliana*, and are important in sugar and amino acid metabolism, *Plant J.*, 61:122–133, 2010, DOI:10.1111/j.1365-313X.2009.04040.x; H. Cui et al., An evolutionary conserved mechanism delimiting SHR movement defines a single cell layer of endodermis in plants, *Science*, 318:801–806, 2007, DOI:10.1126/science.1139531; K. Esau, *Plant Anatomy*, Wiley & Sons, New York, 1953; J. M. Hibberd and W. P. Quick, Characteristics of C_4 photosynthesis in stems and petioles of C_3 flowering plants, *Nature*, 415:451–454, 2002, DOI:10.1038/415451a; J. Lim et al., Conservation and diversification of SCARECROW in maize, *Plant Mol. Biol.*, 59:619–630, 2005, DOI:10.1007/s11103-005-0578-y; M. T. Morita et al., Endodermal-amyloplast less 1 is a novel allele of SHORT-ROOT, *Adv. Space Res.*, 39:1127–1133, 2007, DOI:10.1016/j.asr.2006.12.020; R. F. Sage, P.-A. Christin, and E. J. Edwards, The C_4 plant lineages of planet Earth, *J. Exp. Bot.*, 62:3155–3169, 2011, DOI:10.1093/jxb/err048; R. F. Sage, T. L. Sage, and F. Kocacinar, Photorespiration and the evolution of C_4 photosynthesis, *Annu. Rev. Plant Biol.*, 63:19–47, 2012, DOI:10.1146/annurev-arplant-042811-105511; T. L. Slewinski et al., *Scarecrow* plays a role in establishing Kranz anatomy in maize leaves, *Plant Cell Physiol.*, 53:2030–2037, 2012, DOI:10.1093/pcp/pcs147; T. L. Slewinski et al., *Short-root1* plays a role in the development of vascular tissue and Kranz anatomy in maize leaves, *Mol. Plant*, 7:1388–1392, 2014, DOI:10.1093/mp/ssu036; T. L. Slewinski, Using evolution as a guide to engineer Kranz-type C_4 photosynthesis, *Front. Plant Sci.*, 4:212, 2013, DOI:10.3389/fpls.2013.00212.

Exoplanet research

An exoplanet is a planet external to our solar system. Exoplanets are found orbiting other stars as well as free-floating between those stars. Most exoplanets have names like beta Pictoris b, or HR 8799 e; the name usually starts with the name of the star followed by the lowercase letter b for the first one to be discovered around that star, c for the second, and so on.

When an exoplanet is first discovered, it is considered to be a "candidate" because several follow-up observations and calculations need to be performed to ensure that it is not a "false positive" (a signal that looks like a true exoplanet but is not, for example, a noise glitch or an eclipsing binary). After sufficient follow-up testing, the candidate can be promoted to "confirmed" status, whereby it is classified as a real exoplanet. As of mid-2014, there were a total of about 4600 exoplanets, with 37% confirmed and 63% candidates.

Beginning with the first exoplanet discovery in 1989, the total number of exoplanets has been growing at a rate of about 40% per year (**Fig. 1**). Looking to the future, we expect to find that there are more exoplanets than there are stars in the sky, which is roughly 400 billion in our Milky Way Galaxy alone.

Discovery methods. There are four leading methods for discovering an exoplanet. The transit method has found the most to date, about 87% of the total. The radial-velocity method accounts for about 11%. Direct imaging and gravitational microlensing each contribute less than 1%. Other methods are astrometry, transit timing variations, eclipse timing

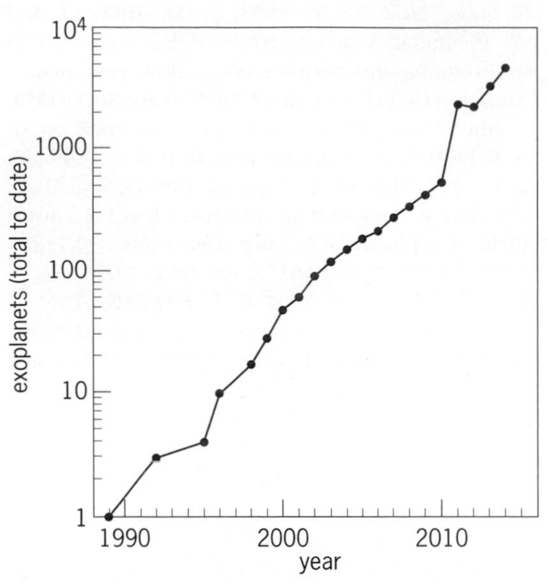

Fig. 1. The total number of confirmed and candidate exoplanets, from 1989 through 2014, by all methods of discovery. The average rate of increase is about 40% per year over this 26-year period.

variations, pulsar timing variations, pulsation timing variations, and orbital brightness modulations. **Figure 2*a*** shows the relative number of exoplanets broken down by the method of discovery. For planets with measured radii, Fig. 2*b* shows the breakdown by radial size, ranging from smaller than the Earth to larger than Jupiter.

With the transit method, astronomers observe the brightness of a star continuously, searching for a decrease in brightness caused by a planet crossing the disk of the star. For the few orbits that do produce a transit (roughly 0.1 to 10%) along our line of sight, the relative decrease in brightness tells us the area of the planet as a fraction of the star's area, and the time between transits tells us the period of the orbit. For example, an Earth transiting a star like the Sun would cause the star's brightness to decrease by a factor of about 0.00008 or 80 parts per million, and the transit would last for about 13 h. For a Jupiter transiting a Sun-like star, the decrease would be about 0.010 and would last for 30 h.

With the radial-velocity method, astronomers measure the spectrum of a star, searching for a periodic variation in the wavelength of spectral lines as a planet orbits the star, giving the spectrum a small but repeatable positive and negative Doppler shift. If the mass of the star is known, then the mass and period of the planet can be found. The Doppler shift for the Earth–Sun system would be about 0.1 m/s with a 1-year period, and for the Jupiter–Sun system it would be 12 m/s with a 12-year period.

The transit and radial-velocity methods are most successful when a planet is close to its star, so most of the exoplanets known today, about 98%, are in orbits that are closer to their star than the Earth is to the Sun, that is, in orbits with a semimajor axis less than 1 au (astronomical unit) and periods less than 1 year. In future observations, it will be a challenge to find planets with larger separations and longer periods.

The direct-imaging method uses a telescope with a coronagraph or starshade to block out starlight and its diffraction pattern, allowing the planet to be imaged as a point of light. The challenge here is that the planet is faint and close to its star. For example, the contrast (planet brightness divided by star brightness) of the Earth–Sun system is about 2×10^{-10} and the separation (for a star at a distance of 10 parsecs) is 0.1 arcsec. For Jupiter, the values are 1×10^{-9} and 0.5 arcsec. Such faint, mature exoplanets have not yet been directly imaged. Seeing these would require a coronagraph on a stable telescope in space, above the Earth's atmosphere. Younger planets are still warm from the energy released from their initial formation; they have been directly imaged with large ground-based telescopes equipped with technology to correct for image distortion caused by the Earth's atmosphere.

The gravitational-microlensing method uses a telescope on the ground or in space to stare at many stars for a long period of time, watching for the occasional brightening that occurs when a foreground

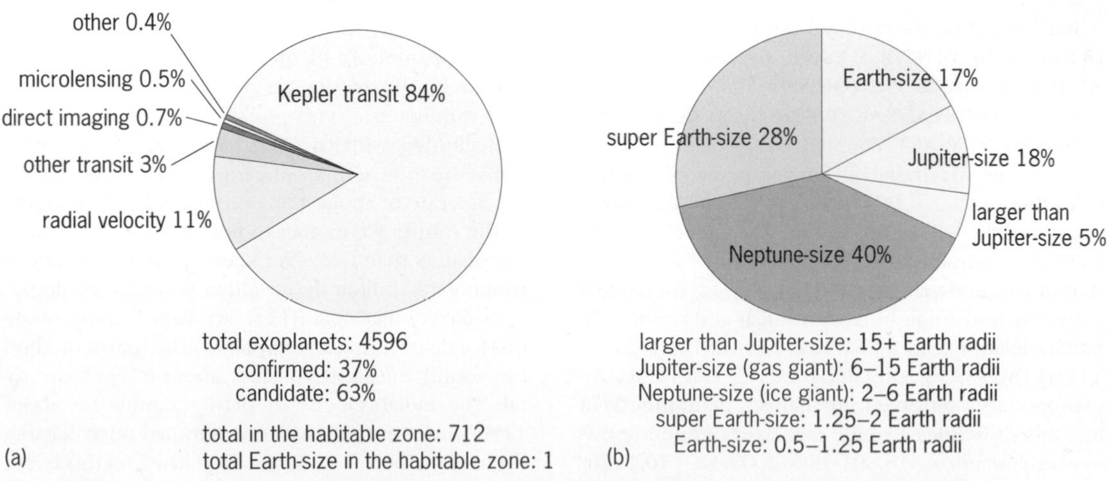

Fig. 2. Exoplanet scorecard. Total exoplanets are shown (*a*) by discovery method, and (*b*) by size, as of mid-2014.

star passes almost directly in front of a background star. The gravity field of the nearby star bends the light of the more distant one, focusing it slightly, and making it several times brighter for a few weeks. If there is a planet around the nearby star, it will cause a similar brightening, but for about a day. From these signals, the mass of the planet and its separation from the nearby star can be estimated.

Most transit detections have been made with the Kepler mission, and most radial-velocity detections have been made with ground-based telescopes. The WFIRST-AFTA space mission, with a 2.4-m (94-in.) telescope and a launch planned for 2024, will carry a direct-imaging coronagraph as well as a gravitational microlensing instrument.

Planetary orbits. The shape of a planet's orbit is described by its eccentricity. The orbit can range from perfectly circular, with an eccentricity of 0, to a highly elongated ellipse, with an eccentricity approaching 1. In the solar system, the eccentricities range from about 0.01 to 0.09, except for Mercury, whose orbital eccentricity is about 0.21. The case seems to be similar for exoplanets, where single-planet systems have a median eccentricity of about 0.2, which is an ellipse about 2% longer than it is wide. However, planets in systems where there are many planets, for example, 8, as in the solar system, tend to have an eccentricity of about 0.05, very similar to the solar system. The reason for these smaller eccentricities may be that the gravitational attraction of neighboring planets tends to smooth out the extremes of highly elliptical orbits. This is an interesting example of how we might learn about the history of a planetary system simply by looking at its appearance.

Another interesting aspect of a planet's orbit is the angle between the plane of that orbit and the plane of the equator of its rotating star, or the average plane of other planets in the system. This angle is called the inclination. In the solar system, where we compare the orbital planes of planets to the plane of the Earth's orbit, the inclination of most planets is very small, about 0.5–3° (with Mercury again the exception at 7°), meaning that the orbital planes of these planets align pretty closely to the Earth's. However, in some exoplanet systems, the inclination is found to be as large as 90°, and sometimes even larger. What could cause this apparent departure from what we might consider to be normal? Here again it is believed that a gravitational interaction with another planet or companion star is responsible for changing the orbit's inclination. This could be another clue to the system's history.

Planetary masses. The mass of a planet is an important value. We expect that low-mass planets, like Earth, will tend to have relatively thin atmospheres. Massive planets, like Jupiter, which is 311 times the mass of Earth, will tend to have massive atmospheres, as Jupiter does. Likewise, low-mass planets are presumed to be "rocky" like the Earth, whereas massive planets are presumed to be "ice giants" like Uranus and Neptune, or "gas giants" like Jupiter and Saturn. For the approximately 1000 planets for which we know the mass, about 85% are in the mass range greater than Uranus or Neptune (~0.05 Jupiter mass), so we believe that these are all ice or gas giants. For planets more massive than about 13 Jupiters, the deuterium in their highly compressed and hot interiors will cause nuclear fusion to occur, further heating up the planet like a small, very faint star, at which stage the planet is called a brown dwarf, and is usually not considered to be a true planet.

Planetary radii. The radius of a planet is also important, because along with mass it tells us the average density of the planet, with rocky planets having high average densities, around 5 g/cm^3, that is, between the densities of rock and iron, and gas giants having low average densities, around 1 g/cm^3, about the density of water. A cold sphere of any material can grow only to a certain radial size, beyond which point adding more mass simply helps crush the planet instead of making it larger. For a rocky planet, this size limit is about 3 Earth radii, and for a gas-giant planet, it is about 12 Earth radii, a bit larger than Jupiter. In spite of this theoretical limit, some objects are found that appear to have radii as large as 20 or more Earth radii (**Fig. 3**). A possible explanation for these huge objects is that they are heated, either by conduction from the inside with leftover heat from their formation, or by radiation from the outside with the heat of a nearby star; another possible explanation is that they are brown dwarfs; and a third explanation is that they might be false positives (for example, background eclipsing binary stars).

Planetary atmospheres. The atmosphere of a rocky planet is the layer of gas and cloud above the solid surface, with, for example, a pressure at the bottom of that layer of 0.006 bar for Mars, 1 bar for Earth, and 90 bars for Venus. (A bar is approximately 1 atmosphere of pressure; 1 bar = 10^5 Pa.) The atmosphere of a gas giant extends down to the small core of the planet, which could be rocky, with a pressure in the millions of bars. For the portion of an atmosphere that can be seen with a telescope, by thermal radiation in the infrared or by reflected or transmitted light in the visible part of the spectrum, we typically are looking at only the top 1 bar.

Planetary spectra. The spectrum of a planet comprises the planet's own thermal light, emitted in the infrared, plus its reflected light from the planet's star, mainly in the visible. The planet's spectrum contains absorption or emission lines that originate in the gases of the atmosphere. So from the spectrum we can inventory these gases. For example, a spectrum that contains mainly strong absorption lines of methane (CH_4) will tell us that the planet is likely to be an ice or gas giant, but a spectrum that contains absorption lines of water and oxygen will tell us that we may have found a planet with an atmosphere similar to the Earth's.

Habitable zones. The habitable zone is that region around a star where a planet could have liquid water on its surface, given sufficient atmospheric pressure.

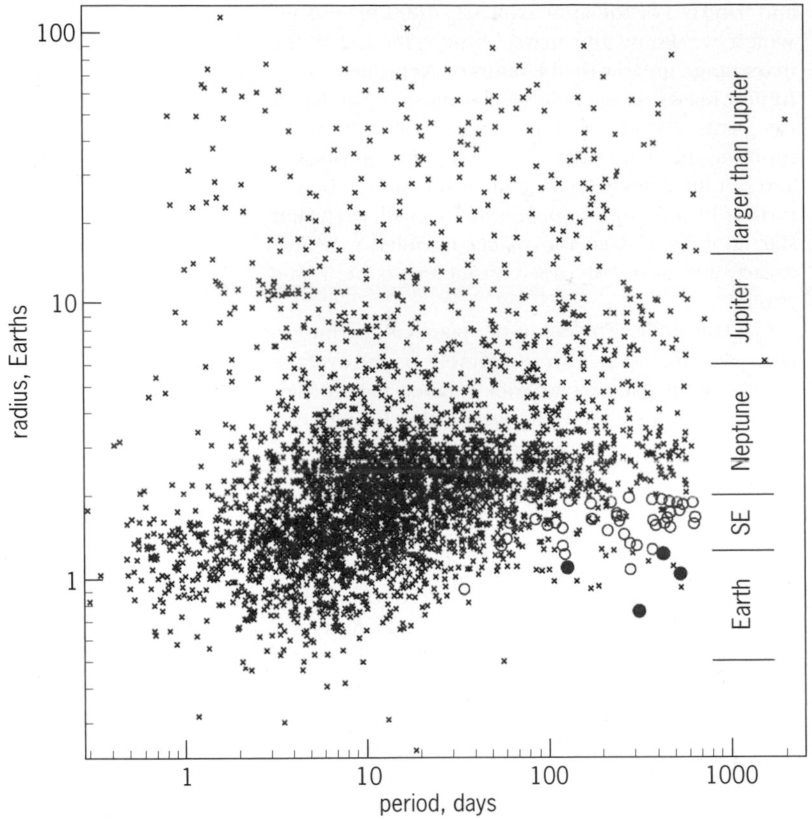

Fig. 3. Exoplanet radius versus orbital period. A total of about 3700 planets with known radii are shown. The conventional classification of planets by size is given along the right side of the chart (SE = super Earth). There is a large concentration of planets in the range 1–3 Earth radii, and relatively few Jupiters. The Kepler mission is sensitive to the full range of planets larger than about 0.5 Earth radius, except for the area in the lower right of this figure, where the 4-year lifetime of the mission, and the measurement noise of the instrument, preclude finding more than a handful of small planets at long periods, even though there may be many such planets around Kepler's target stars. The solid red circles are Earth-size planets in their star's narrow habitable zone (only one of these is confirmed), and the open red circles are Earths and super Earths in their star's wide habitable zone (all of these are still candidate planets).

The habitable zone is defined in this way because we believe that liquid water is essential for life, and hence habitability. On Earth, water vapor and carbon dioxide (CO_2) provide a greenhouse warming that gives us an average surface temperature of about 15°C (59°F), allowing liquid water; today's Venus surface is about 462°C (864°F, too hot for liquid water), and Mars is about −53°C (−63°F, too cold). However, the exact definition of the habitable zone is controversial, owing to the twin issues of unknown planetary albedo and unknown greenhouse effect, in addition to the energy incident at the top of the planet from the star. We recognize this uncertainty by defining two types, a "narrow" and a "wide" habitable zone.

The narrow habitable zone is defined as the orbital range between a runaway greenhouse (with hot water vapor) at the inner edge and a barely sustainable greenhouse (with cold water vapor and CO_2) at the outer edge. In the present solar system, the narrow habitable zone runs from 0.95 to 1.68 au, that is, from the Venus–Earth range to just outside the orbit of Mars at 1.52 au.

The wide habitable zone is defined as an adjustment of the narrow range, based on our understanding of the histories of Venus and Mars. The inner edge is set by the observation that Venus may have lost its water only about 1×10^9 years ago, when the Sun was only 8% fainter than it is now. The outer edge is set by the observation that Mars probably had liquid water on its surface some 3.8×10^9 years ago, when the Sun was 25% fainter than it is now. In the present solar system, this wide habitable zone runs from 0.75 to 1.77 au, that is, from near Venus at 0.72 au to a bit beyond Mars.

As of mid-2014, only one confirmed planet has been found that is Earth-like in size (by convention, a radius between 0.5 and 1.25 times the Earth's radius) and in the narrow habitable zone around a star. This planet is Kepler 186 f, so named for the mission that discovered it in 2014, and with the "f" indicating that it is the fifth planet discovered in that system. This is shown as the 129-day solid red circle in Fig. 3; the other solid red circles are candidates that are Earth-like in size and fall in their star's narrow habitable zone. The open red circles are candidates that are Earths or super Earths (1.25–2 times the Earth's radius), and fall in their star's wide habitable zone.

In time, doubtless other such planets will be found, and when they are, we will want to use a large telescope to collect the light, a coronagraph or starshade to isolate the planet's light from that of its star, and a spectrometer to analyze the spectrum of the planet. Only then will we have clues telling us whether that planet could be habitable.

Is there life on other planets? To answer this age-old question, we must obtain the spectra of exoplanets. From water vapor absorptions we will know whether an exoplanet is habitable. Oxygen absorption features in its spectrum will be a possible sign of life.

For background information *see* ALBEDO; ASTROBIOLOGY; ASTRONOMICAL SPECTROSCOPY; CORONAGRAPH; DOPPLER EFFECT; EARTH; ECLIPSING VARIABLE STARS; EXTRASOLAR PLANETS; GRAVITATIONAL LENS; GREENHOUSE EFFECT; MARS; ORBITAL MOTION; PLANET; RETROGRADE MOTION (ASTRONOMY); SOLAR SYSTEM; VENUS in the McGraw-Hill Encyclopedia of Science & Technology. Wesley A. Traub

Bibliography. L. Billings, *Five Billion Years of Solitude: The Search for Life Among the Stars*, Penguin Books, New York, 2013; S. Seager (ed.), *Exoplanets*, University of Arizona Press, Tucson, AZ, 2010.

Eye development (vertebrate)

The vertebrate eye is an extraordinary organ in terms of its structure, function, and development. Vision is acquired during embryonic development as a result of the coordinated formation and growth of several different eye tissues. The mature eye consists of anterior and posterior sectors (**Fig. 1**). The major parts of the anterior sector are the cornea, anterior chamber, iris, and crystalline lens. The posterior sector

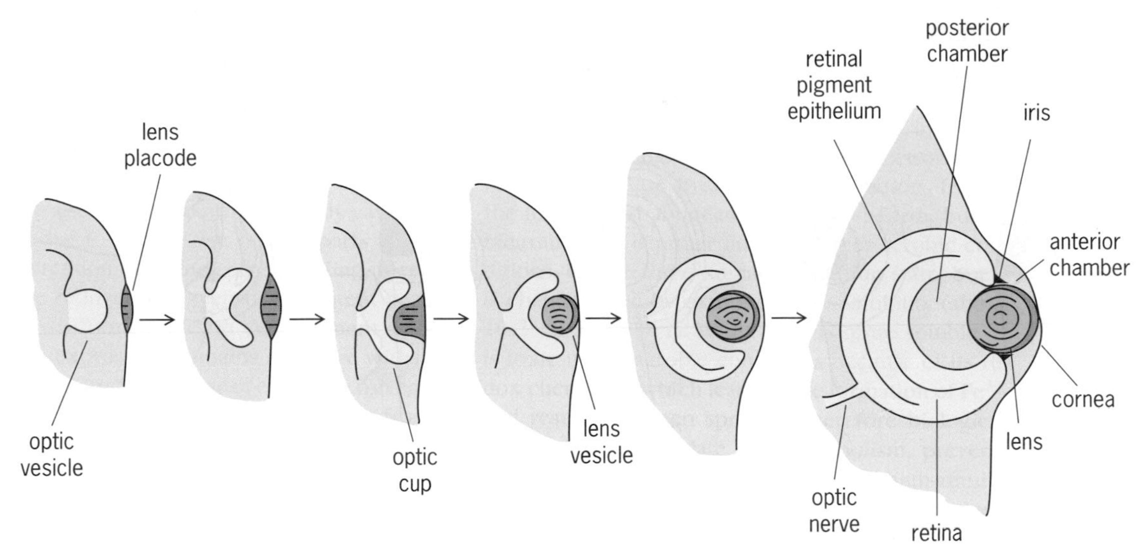

Fig. 1. Eye development in the surface-dwelling form of *Astyanax mexicanus*.

consists of the posterior chamber, retina, and retinal pigment epithelium (RPE). Eye tissues are derived from three different embryonic sources. The lens and the external part of the cornea originate from the surface ectoderm. The interior part of the cornea and a part of the iris are derived from the neural crest. The retina, RPE, and the other part of the iris are formed from the optic vesicle, which is derived from the anterior neural ectoderm.

The optic vesicle, a lateral protrusion of the forebrain, invaginates to form the optic cup. The retina develops on the layer of the optic cup that will later face the lens, and the RPE develops on the opposite side of the optic cup. The connection to the forebrain is the optic stalk, which becomes the optic nerve. Distinct cell layers differentiate in the retina, consisting of neural ganglion cells, glial cells, bipolar cells, horizontal cells, and the rod and cone photoreceptor cells. The lens develops from the lens placode, which is a thickening in the surface ectoderm that forms the lens vesicle. The lens is induced by a complex series of tissue interactions beginning during gastrulation and continuing until the time of its first contact with the optic vesicle. The lens vesicle consists of lens epithelial cells, which divide during lens growth, and lens fiber cells, which undergo differentiation and produce the lens crystallin proteins. Later, the cornea forms by differentiation of the surface ectoderm above the lens, and the iris develops at the margin of the optic cup. Once the lens and optic cup make contact, further eye development is mediated by reciprocal signaling between the lens, retina, and RPE. These tissues transmit and receive signals from each other to orchestrate eye growth and differentiation. In some vertebrates, such as fishes, the eye grows continuously throughout life. In other vertebrates, such as mammals, the eye reaches its maximum size shortly after birth.

Genetic mechanisms of eye development. Many genes are involved in eye development. The genes encoding the proteins Chordin, Noggin, Follistatin, and Cerberus mediate neural induction by antagonizing the bone morphogenetic protein (BMP) signaling pathway. Following neural induction, early eye specification involves the expression of Otx2, Pax6, Rx, and Six3 transcription factors in the anterior neural plate. Cyc signaling emanating from the prechordal mesoderm beneath the neural plate induces Shh expression along the ventral midline, which inhibits Pax6 expression and splits the single optic field into the right and left presumptive eye domains. Antagonistic BMP and fibroblast growth factor (FGF) signaling, along with the Mitf, Chx10, and Pax6 transcription factors, are involved in retina specification. Under the influence of retinoic acid, reciprocal repression between Pax2 and Pax6 specifies the boundaries between the retina, RPE, and optic stalk within the optic vesicle.

The lens placode is specified by the concerted action of several transcription factors, including Pax6, Otx2, and Sox2, and the BMP, FGF, and retinoic acid signaling pathways. FGF signaling is also required for the early phases of lens development, including lens induction. The Sox1, Sox2, and Sox3 transcription factors are intricately involved in lens development. Sox2 and Sox3 are expressed prior to lens placode formation in the ventral surface ectoderm bordering the presumptive lens region. Sox2 and Pax6 form a functional complex that regulates lens crystallin gene expression. Other transcription factors involved in various aspects of lens induction are Six3, Prox1, and Maf, which also regulate lens crystallin formation. There are many different lens crystallin proteins, which are classified as alpha-, beta-, or gamma-crystallins. The alpha-crystallins are one of the most abundant proteins in the lens. The alpha A-crystallin gene encodes a chaperone protein, which is required for lens survival.

Eye degeneration. Many optic diseases involve the degeneration of the eye or one of its components.

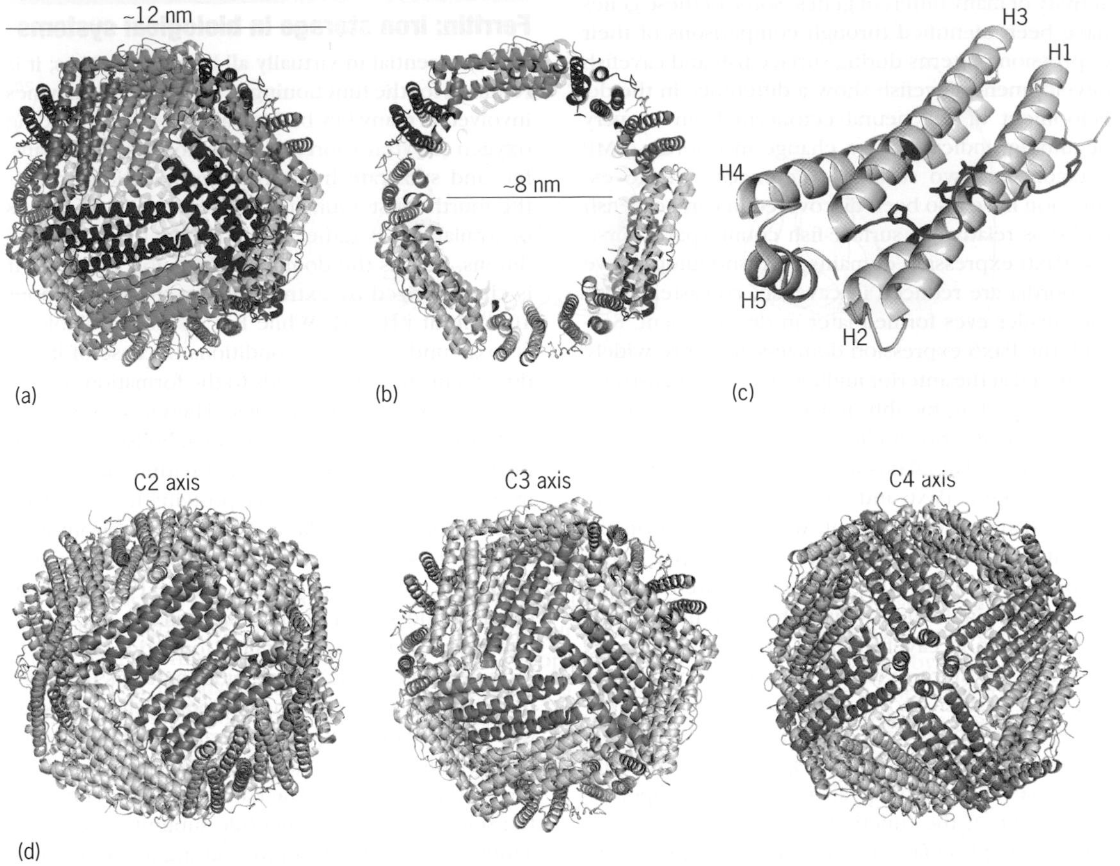

Fig. 1. Different views of maxiferritins. (*a*) External view of the nanocage. (*b*) The large inner cavity. (*c*) The ferritin subunit, with the four-helix bundle helices (H1–H4) and the short helix H5 shown in gray; the iron ligands at the ferroxidase site in red; and the external loop in blue. (*d*) In green, a dimer at the C2 axis (left panel); in cyan, three symmetry-related subunits form the C3 ion channel (middle panel); and in magenta, four symmetry-related subunits generate C4 pores through their H5 helices (right panel). Drawn using PyMOL and the Protein Data Bank file 4DAS.

are composed of bacterial and plant maxiferritins are catalytically active. The catalytic activity of H and M subunits of maxiferritins derives from a ferroxidase (or oxidoreductase) site hosted in the central part of the four-helix bundle.

In addition to the maxiferritins, there are also the heme-containing bacterioferritins, found only in eubacteria, and the smaller miniferritins [12-subunit Dps (DNA protection during stress/starvation) proteins], present only in prokaryotes. Bacterioferritins are iron-storage proteins that share with maxiferritins the function and presence of an intrasubunit catalytic ferroxidase center. Normally, they contain heme groups located at each of the 12 twofold interfaces between subunits, possibly playing an electron-transfer role to facilitate reduction of the iron core. At variance, the primary proposed role of Dps proteins is in protecting DNA against the combined action of Fe^{2+} and hydrogen peroxide (H_2O_2). Dps proteins share with maxiferritins a similar four-helix bundle structure, with the exception of an extra short helix located between H2 and H3 in the monomer bundle, but their active site is located at the interface between two subunits. The central cavity has a diameter of 4–5 nm, while the outer diameter is about 9 nm, with a maximum storage capacity of only 500 iron atoms.

Ferritin function can be divided into two main phases: (1) iron uptake, which requires Fe^{2+} entry, catalytic reaction with O_2, and $Fe^{3+}O$ formation and biomineralization; and (2) iron (Fe^{2+}) release via mineral reduction/dissolution.

Each event occurs in different structural domains of the protein-mineral complex and on different time scales, with the catalytic reaction between Fe^{2+} and O_2 being very fast (<0.5 s) and the nucleation-mineralization process being significantly slower (minutes to hours).

Iron biomineralization. Nanocage properties common to all organisms govern the various processes, which are strongly dependent on electrostatic gradients along the iron pathways and on the presence of several weakly coordinating sites that might assist iron trafficking through the cage. The external surface of ferritin is characterized by an overall negative electrostatic potential that is attractive to cations. The uptake of Fe^{2+} by ferritin occurs spontaneously in solution, but in vivo it is probably mediated by chaperone proteins. In solution, Fe^{2+} transfer to the ferroxidase site is mediated by the hydrophilic C3 channels, with a negative electrostatic potential at the outer region lining their entrance. Positive areas surround the negative opening of the C3 channel, giving rise to an electrostatic field gradient that

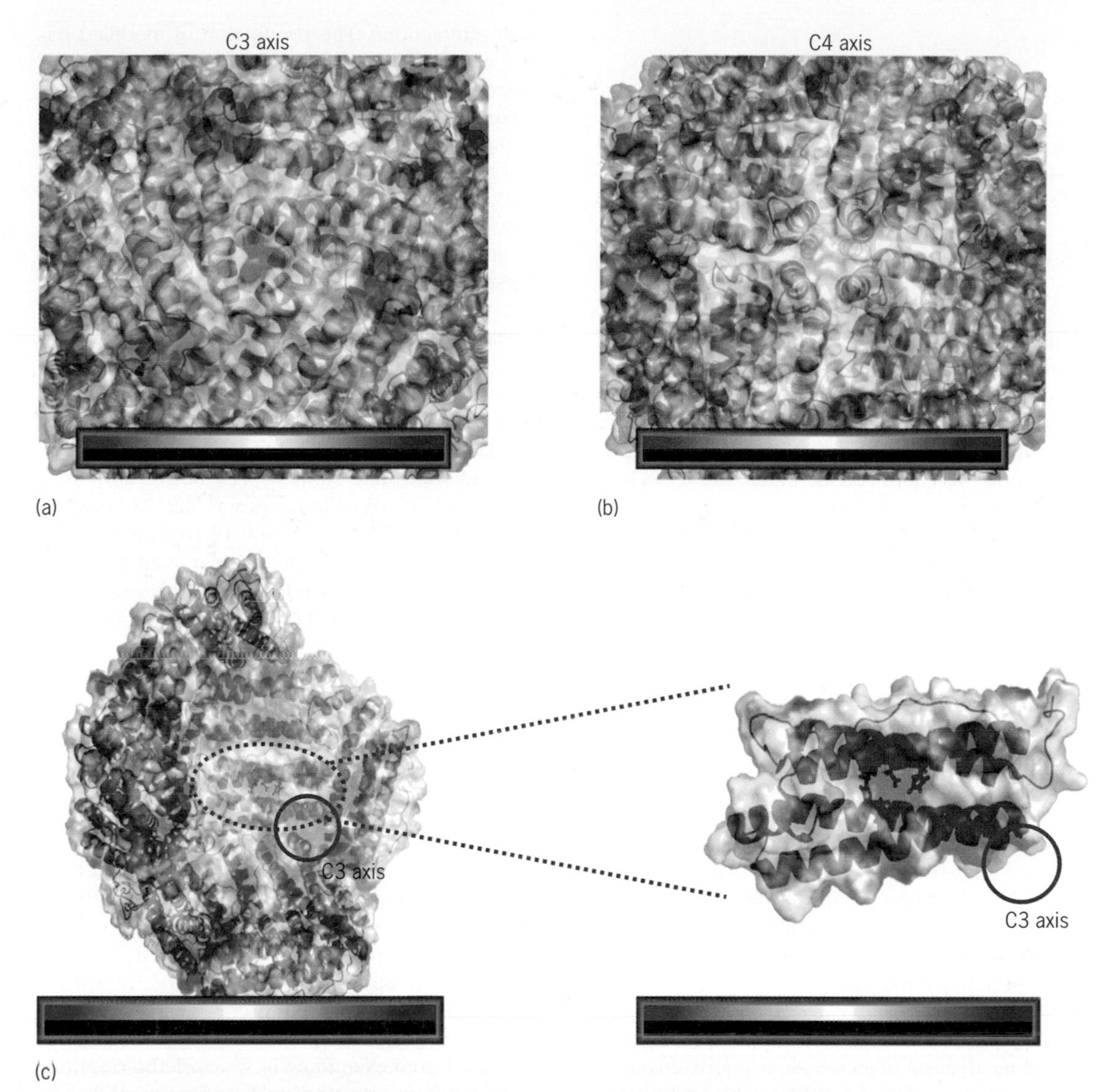

Fig. 2. Electrostatic potential surface in frog M ferritin calculated using PyMOL and the Protein Data Bank file 4DAS. View of the external surface potential (*a*) at the C3 channel and (*b*) at the C4 channel. (*c*) The internal surface potential is shown in a ferritin cross section (left panel), one of the monomers forming the C3 channel is enlarged (right panel) to help in visualizing the location of the ferroxidase site (iron ligands are shown as sticks).

strongly directs Fe^{2+} toward the channel entrance (**Fig. 2*a***). The negative electrostatic potential extends along the wide channel, running from the outside to the inside. The inner cavity surface is also characterized by areas with different electrostatic properties. A cluster of negatively charged amino acids, exposed to the internal surface of the cavity and located in the central part of each subunit in correspondence to the active site, may attract Fe^{2+} toward the reaction center (Fig. 2*c*).

The C4 channels are characterized by a largely hydrophobic inner surface (Fig. 2*b*), ruling out their possible involvement in Fe^{2+} entry. An exception is plant ferritins with hydrophilic C4 channels. In these proteins, both the C4 and the C3 hydrophilic channels are necessary for iron diffusion and oxidation.

The ferroxidase site is located in the interior of the four-helix bundle channel, about halfway between the N- and C-terminals. It is a di-iron center formed by a bridging glutamate and a few other protein ligands [a monodentate glutamate and a histidine in one site and a bidentate glutamate in the other as observed for the M type subunits (PDB 3RBC)], whose side chains possess a certain degree of flexibility to accommodate both the prereactive diferrous and postreaction diferric species, which are characterized by different intermetal distances (Figs. 1*c* and 2*c*). Overall, the site is weakly coordinating, as is functionally required for a reaction in which the metal acts as the substrate. Additional amino acids proximal to the ferroxidase site have been suggested to play a role in delivering Fe^{2+} toward its dinuclear binding site via electrostatic or weak coordination guidance.

The ferroxidase site binds two Fe^{2+} ions, which readily react with O_2 to form transient peroxo complexes (observable in eukaryotic ferritins) that decay in seconds to form diferric-oxo/hydroxo species

$$2Fe^{2+} + O_2 \longrightarrow [Fe^{3+} - O\text{–}O\text{–}Fe^{3+}]$$

$$[Fe^{3+} - O\text{–}O\text{–}Fe^{3+}] + H_2O \longrightarrow H_2O_2 + [Fe^{3+} - O\text{–}Fe^{3+}]$$

$$[Fe^{3+} - O\text{–}Fe^{3+}] + (H_2O)_{x+2} \longrightarrow Fe_2O_3(H_2O)_x + 4H^+$$

$$2Fe^{2+} + O_2 + (H_2O)_{x+3} \longrightarrow Fe_2O_3(H_2O)_x + 4H^+ + H_2O_2$$

Fig. 3. Iron biomineralization in ferritin is a multistep process.

(**Fig. 3**). Under the effect of new incoming Fe^{2+} in subsequent catalytic turnovers, the diferric-oxo species are released from the active site and proceed toward the cavity to begin mineral nucleation and growth. The path from the catalytic site to the nucleation center has not been identified, but it has been proposed that it extends from the ferroxidase center toward H5. The nucleation sites might correspond to negatively charged areas in the cavity. In vertebrates, the number and position of such negatively charged areas depend on the number of catalytically active H subunits of maxiferritin, with respect to the L subunits.

Ferritins can use all the catalytic centers at the same time under in vitro conditions when fast iron increments are given, but the increments inside the cells are probably much slower, so that the presence of 24 catalytic sites in one cage is not necessary. This hypothesis is supported by the evolution in vertebrates of the L subunit, which is catalytically inactive but characterized by negative groups on the cavity surface that accelerate mineralization and enhance the transfer of iron from the ferroxidase center of the H subunits to the iron core.

Iron release. Although it is well known that stored iron is readily available for cellular needs, the physiological mechanism of its release is poorly characterized. In solution, the iron core is stable and requires reducing agents to trigger mineral dissolution. In-vitro studies suggest that the C3 channels, following a localized unfolding of the external part, may represent the main iron exit pathway. Recent findings have identified a role for the C4 channels in the control of iron release. In cells, ferritin iron seems to be mainly released after proteolytic degradation of the cage via the proteasome as well as lysosomal autophagy.

Outlook. The large ferritin cavity has been exploited for the production of biologically compatible magnetic resonance imaging probes (via encapsulation of paramagnetic Gd(III) complexes or Mn(II) oxide/hydroxide) or theranostic carriers, which deliver therapeutic and imaging agents simultaneously. Selective delivery of ferritin-based carriers exploits two main strategies: (1) the different expression of receptors of the ferritin transporting route in different cell types, and (2) surface modification that makes the ferritin nanocage able to specifically interact with characteristic molecules exposed on the surface of pathogenic cells and to release their cargo as a consequence of the conformational changes induced by such interactions. The production of modified ferritin cages is a promising delivery nanoplatform for cancer theranostics.

For background information *see* AMINO ACIDS; AUTOPHAGY; BIOINORGANIC CHEMISTRY; BIOLOGICAL OXIDATION; ELECTRON-TRANSFER REACTION; ENZYME; FREE RADICAL; IRON; IRON METABOLISM; LIGAND; LYSOSOME; MOLECULAR CHAPERONE; PEPTIDE; PROTEASOME; PROTEIN; TRANSCRIPTION in the McGraw-Hill Encyclopedia of Science & Technology.

Caterina Bernacchioni, Paola Turano

Bibliography. S. C. Andrews, The ferritin-like superfamily: Evolution of the biological iron storeman from a rubrerythrin-like ancestor, *Biochim. Biophys. Acta*, 1800(8):691–705, 2010, DOI:10.1016/j.bbagen.2010.05.010; D. Lalli and P. Turano, Solution and solid state NMR approaches to draw iron pathways in the ferritin nanocage, *Acc. Chem. Res.*, 46(11):2676–2685, 2013, DOI:10.1021/ar4000983; E. C. Theil, Ferritin: The protein nanocage and iron biomineral in health and in disease, *Inorg. Chem.*, 52(21):12223–12233, 2013, DOI:10.1021/ic400484n.

Finite element analysis in paleontology

Biologists have long commented on how the vertebrate skeleton appears to be designed in accordance with engineering principles. Bones and their associated soft tissues are adapted to transmit, resist, and take advantage of the many forces that the skeleton experiences during normal function. How bones respond to forces, that is, how they are stressed and strained, is therefore linked to their function. A great deal of functional information can be gained from examining how a skeleton responds to stresses and strains, yet paleontologists are limited because fossil bones are petrified (and thus have different properties than biomaterials in living animals), and skulls in particular are complex shapes not applicable to straightforward mathematical analysis. Recently, however, biologists and paleontologists have begun to borrow from the sophisticated design tool kit of engineers in order to examine skeletal construction.

Finite element analysis (FEA) is one such tool to examine skeletal construction. FEA is a means by which stress, strain, and displacement in a two- or three-dimensional structure may be deduced using mathematical principles. FEA can be performed using predesigned computer software or specific user-written code. The exponential increase in computing power over the past 10–20 years has increased the accessibility of FEA to researchers who traditionally lie outside the engineering sphere. By creating digital models of skeletons, FEA can potentially offer a new avenue of inquiry to vertebrate paleontologists who are interested in the function of extinct animal skeletons and the mechanical principles and evolutionary pressures underlying their construction.

Finite element analysis (FEA) model of an *Allosaurus fragilis* skull. Elements are represented by each individual triangle. Geometric equations can be used to calculate the strain and stress in simple structures. FEA permits the application of these geometric equations to complex, nongeometric shapes by calculating the stress and strain in each individual element. (*Image reprinted with permission; copyright © Emily J. Rayfield*)

How FEA works. In the first stage of FEA, a digital two- or three-dimensional model of the structure to be tested is divided into a finite number of simple geometric shapes, which are called elements. Elements are joined to each other at discrete points, called nodes, and the element and node construct is known as a mesh (see **illustration**). Specific material properties that approximate those of the actual structure, such as stiffness and density, are applied to the element mesh. Boundary conditions are then applied to the mesh: Constraints are applied to prevent the structure from moving in a particular direction; and loadings are applied as forces, pressures, or accelerations, which mimic loads that the structure would experience during life or use. This process of model creation is known as preprocessing.

The next stage is analysis, which involves the calculation of force vectors and displacements at each individual node, taking into account the material properties of the structure. Stress and strain at each nodal point are subsequently calculated to provide a composite picture of the mechanical behavior of the structure. Mathematical solvers integrated into the FEA software perform this stage of the analysis.

Finally, during postprocessing, results are visualized and interpreted. Emphasis is placed upon the checking of errors and refinement of the original mesh and boundary conditions to ensure that the model represents the original structure as accurately as possible.

FEA in zoology and paleontology. The basic principles of FEA were originally derived by engineers in the late 1950s and early 1960s. Since the early 1970s, this method has been used widely in orthopedic medicine and bioengineering. An offshoot of FEA, known as finite element scaling analysis (FESA), which is concerned with the calculation of displacements and not stress and strain, has been used since the mid-1970s to quantitatively examine shape change. A 200-element FEA model of the bill of a shoebill (a type of stork), published in the mid-1980s, appears to represent the first application of FEA to zoology or paleontology. Stress plots and displacements were displayed for two bill-loading regimes; however, actual loads and material properties were not specified. Not until the late 1990s did further zoological studies appear, mainly focused on primate lower jaws and teeth and horses' hooves. The first published application of FEA in paleontology was an investigation into the shell strength of ammonites (an extinct group of mollusks) in the late 1990s. FEA showed that increasingly complex septal (internal shell wall) construction weakened ammonite shells (with weakening indicated by higher stress magnitudes). Later, more complex models that were better able to represent the complexity of ammonite suture morphology and used more accurate material properties demonstrated that complex septal morphology was in fact stronger than simple septal morphology. These results highlight accuracy problems in model geometry and element choice, which are situations that all users of FEA must face.

Current status of FEA in paleontology. Since the early 2000s, the use of FEA in paleontology has increased at some pace. From studies that looked at two- and three-dimensional models of stress and strain on dinosaur skulls, a whole range of fossil forms have now been subject to FEA. These extend from carnivorous theropod dinosaurs to herbivorous theropods, ornithopods, ceratopsians, and sauropods. A wealth of studies on primates and hominins now exist, highlighting questions that relate to feeding mechanics and dietary adaptation in our earliest ancestors (for example, what were the functions of expanded brow ridges or robust molar teeth?). Innovative studies of substrate deformation are also building a picture of the gait and locomotion of dinosaurs and hominins. The functions of unusual features, such as the tail clubs of armored ankylosaurs and the bone-headed skulls of pachycephalosaurs, have been subject to analysis with FEA. A range of other extinct animals have also been subject to FEA. For example, carnivores (such as extinct bears, dogs, hyaenids, and lions) have been investigated, in some cases to compare their performance to their marsupial counterparts (such as the extinct marsupial lion and thylacine); niche selectivity in fossil marine crocodiles has been established based on differences in stress, strain, and strength in their snouts; and new information on the feeding and locomotory performance in giant extinct birds (including the terror birds of the Americas and the moas of New Zealand) has been revealed. Furthermore, the skulls of synapsids (early mammalian relatives) have been analyzed with regard to predatory and herbivorous forms, and studies on the elaborate head crests of pterosaurs have been carried out. Numerous studies on living animals have also enlightened our understanding of the function and evolution of their fossil relatives. These examples and others reveal how FEA is now firmly established as a tool in the paleobiologist's armory to test hypotheses of anatomical and functional evolution.

These studies have revealed much about the behavior of individual fossil organisms. A further notable advantage of the technique is that it can also be used to inform on comparative trends in functional and anatomical evolution. For example, multiple FEA models of different animals from the same clade, or natural group, can be studied to test how biomechanical traits (such as peak or average stresses in the skull) are distributed across the phylogeny (family tree) of the group. Comparative phylogenetic methods can allow researchers to test how quickly certain traits evolved across the phylogenetic tree. Such large comparative studies are just being realized (for example, in early tetrapods, bats, and cats), although this use of FEA model data is in its infancy. However, these analyses offer great potential to explore biomechanical evolution in organisms. FEA models are also now subject to integration with other types of software. More commonly, this involves the creation of three-dimensional digital models from computed tomography (CT) scan data sets using industry-standard or open-source software. A further computational tool, multibody dynamics analysis (MDA), has been employed to deduce muscle activation patterns and bite and joint forces in living animals (including lizards and *Sphenodon*). It has great potential to simulate muscle action and joint loading in extinct animals. Finally, geometric morphometrics has been used to inform on the extremes of shape variation expressed in a sample of organisms. It also can be used as a tool to quantify the relative differences in stress, strain, and deformation in similar models subject to variable loading conditions.

Potential problems. The aforementioned examples provide a review of the current status of FEA in vertebrate paleontology. Use of the technique will surely increase because FEA has the potential to address both specific and wide-ranging questions concerning the functional morphology and evolution of fossil animals. However, a number of technical and theoretical problems face future FEA users. First, experimental evidence from living animals has shown that not all structures are adapted to the functions that they undertake, and functional signatures are often muddled in bones that undertake numerous functional tasks. With structure decoupled from function, the elucidation of skeletal stress patterns may not yield satisfactory hypotheses of function. Nevertheless, FEA still bears the potential to test predetermined hypotheses of function and adaptation, and the strengths of the technique lie in this particular area.

Second, the creation of model geometry is difficult, and problems are faced in deciding how abstract a model should be when created. Material properties of bone and other tissues must be estimated from analogs in living animals. Boundary conditions (constraints and loading forces) also must be estimated (ideally in absolute terms; if not, then in relative ones). Moreover, there are technical issues involving element choice, mesh size, and position of constraints that could potentially influence the FEA output. Ground-truthing (validating) the model provides an insight into what degree the output of the model reflects reality, and a sensitivity analysis informs on how modifications of input parameters can change the model results. A number of studies have measured in vivo or ex vivo (from cadavers) the strains generated when a load is applied to an animal skeleton. FEA modeling is then used to best represent the experimental loading conditions, and the computationally derived stresses and strains are compared to those measured during the experiment. Studies so far have shown that most FEA models can replicate the overall mechanical behavior of a bone or other hard parts, but the computation of exact stresses and strains is very difficult. Typically, an increase in the accuracy of model material properties creates a better fit between experimental and computational results; however, it is impossible to know for certain the mechanical behavior of fossil materials. Moreover, it is not clear how FEA model results are influenced by the presence of sutures, joints, and soft tissues, and this remains an area of active study. It is therefore most appropriate to frame scientific questions asked of paleontological FEA with the results of validation studies in mind.

Future prospects. Taking these cautionary notes into account, the importance of FEA as a tool to investigate the mechanical behavior and function of extinct animal skeletons is evident. Paleobiologists are able to use FEA, a technique relatively new to paleontology, to address old, fundamental questions, such as why the skeletons of extinct animals were shaped the way that they were and how, and why did the shape of skeletons and other tissues evolve in a particular manner in deep time.

For background information *see* ANIMAL EVOLUTION; COMPUTER-AIDED ENGINEERING; DIGITAL COMPUTER; DINOSAURIA; FINITE ELEMENT METHOD; FOSSIL; PALEONTOLOGY; PHYLOGENY; SKELETAL SYSTEM; STRESS AND STRAIN in the McGraw-Hill Encyclopedia of Science & Technology.

Emily J. Rayfield

Bibliography. R. M. Alexander, *Bones: The Unity of Form and Function*, Macmillan, New York, 1994; C. McGowan, *A Practical Guide to Vertebrate Mechanics*, Cambridge University Press, Cambridge, UK, 1999; E. J. Rayfield, Finite element analysis and understanding the biomechanics and evolution of living and fossil organisms, *Annu. Rev. Earth Planet. Sci.*, 35:541–576, 2007, DOI:10.1146/annurev.earth.35.031306.140104; E. J. Rayfield et al., Cranial design and function in a large theropod dinosaur, *Nature*, 409:1033–1037, 2001, DOI:10.1038/35059070; E. F. Weibel, C. R. Taylor, and L. Bolis (eds.), *Principles of Animal Design: The Optimization and Symmorphosis Debate*, Cambridge University Press, Cambridge, UK, 1998.

Foodborne disease

The U.S. Centers for Disease Control and Prevention's current estimate for foodborne disease cases acquired annually in the United States is 47.8 million episodes. This includes 38.4 million episodes caused by unspecified agents and 9.4 million episodes caused by 31 major pathogens. In total, there are an estimated 127,839 hospitalizations and 3037 deaths associated with foodborne illnesses annually.

Of the major foodborne pathogens causing illness, most episodes are caused by noroviruses (58%), nontyphoidal *Salmonella* species (11%), *Clostridium perfringens* (10%), and *Campylobacter* species (9%). Most food-associated hospitalizations and deaths are the result of *Salmonella* species. Other major bacterial causes of foodborne illnesses include *Staphylococcus aureus*, *Shigella* species, Shiga toxin–producing *Escherichia coli* (including *E. coli* O157:H7), *Yersinia enterocolitica*, *Bacillus cereus*, and *Listeria monocytogenes*. Parasites largely responsible for foodborne illnesses include *Toxoplasma gondii*, *Giardia intestinalis*, *Cryptosporidium* species, and *Cyclospora cayetanensis*.

Noroviruses. Noroviruses are the leading cause of foodborne illnesses in the United States, with more than 350 foodborne norovirus outbreaks on average reported annually. (Note that a foodborne disease outbreak occurs when two or more cases of a similar illness result from the consumption of a common food.) Leafy vegetables (33%), fruits and nuts (16%), and mollusks (13%) have been implicated as the most common vehicles in which a single commodity was identified. Infected food handlers were the source of contamination in approximately half of the outbreaks, with the majority of foods likely contaminated during preparation and service.

Noroviruses have a low infectious dose (<50 viral particles) and are shed in feces or vomitus in large numbers (10^5–10^{11} viral particles per gram of feces) by ill persons. The viruses have relatively long survival rates, remaining infectious for up to 2 weeks on surfaces and for more than 2 months in water. They also are resistant to many common disinfectants, such as low doses of chlorine. Key locations for the intervention of contaminated foods are restaurants, delicatessens, and other commercial settings (including cruise ships) where norovirus-associated foodborne outbreaks most often occur.

***Salmonella* species.** The leading known bacterial cause of foodborne illness is *Salmonella*, which is responsible for an estimated 1 million cases of food-associated gastroenteritis per year. The most notable *Salmonella* serotypes (microorganism variations distinguished by a characteristic set of antigens) include Enteritidis, Heidelberg, Typhimurium, Javiana, and Newport. Eggs, fresh produce, and poultry are among the vehicles most frequently associated with cases of salmonellosis. Cooking foods and proper food handling practices are the best approaches to preventing foodborne salmonellosis. Cooking poultry to 74°C (165.2°F) and ground beef to 71.5°C (160.7°F) will kill salmonellae.

Nontyphoid *Salmonella* infections most commonly result in gastroenteritis, which generally occurs within 8 to 72 h after ingestion of the pathogen. However, chronic conditions, such as Reiter's syndrome, ankylosing spondylitis, and reactive arthritis, can develop in a small number of cases. Development of antibiotic resistance in *Salmonella* species has been a growing public health concern because of the potential development of untreatable illnesses caused by antibiotic-resistant *Salmonella*. This problem is particularly serious when multiple-antibiotic-resistant salmonellae are implicated in outbreaks of enteric fever, in which there are unusually high fatality rates.

***Campylobacter* species.** Although *Campylobacter* is responsible for an estimated 850,000 cases of foodborne illness annually, the pathogen is associated with very few foodborne outbreaks. Rather, sporadic (isolated) infections account for more than 99% of the *Campylobacter* infections in the United States. Poultry is the most common vehicle of these sporadic illnesses, whereas raw milk has been the most frequently identified vehicle of outbreaks of *Campylobacter* infections. Fresh produce, seafood, and water are also documented vehicles of outbreaks. *Campylobacter jejuni* is the most common species associated with foodborne illnesses, although *C. coli* and *C. fetus* have been implicated in foodborne outbreaks as well. Diarrhea, abdominal cramps, and fever are the typical symptoms of illness, although bloody diarrhea and vomiting have also been reported. The medium incubation time for onset of illness ranges from 3 to 168 h, and the median duration of illness is 2 to 336 h.

Campylobacter jejuni has also been associated with Guillain-Barré syndrome, which is the leading cause of acute neuromuscular paralysis in the United States. Guillain-Barré syndrome is an autoimmune disease that is triggered by several factors, including an acute infection of the gastrointestinal tract. It is estimated that 25–50% of the cases of the syndrome are precipitated by *C. jejuni* infections, accounting for more than 2000 cases annually.

In general, *Campylobacter* is not a hardy bacterium when outside the intestinal tract of its host. The bacterium is sensitive to desiccation, freezing, and prolonged exposure to atmospheric concentrations of oxygen. Most strains grow at 37°C (95°F), with optimal growth at 42–45°C (107.5–113°F), and they do not grow at temperatures below 30°C (86°F). *Campylobacter* species are microaerobic organisms that grow best in an atmosphere of 3–5% oxygen and 5–10% carbon dioxide. Hence, they do not normally grow on food during processing and storage.

Attribution of foodborne illnesses. The results of recent attribution studies of foodborne illnesses have been revealing. Of more than 120,000 outbreak-associated illnesses that occurred between 1998 and

PLANT ORGANS; POLLEN; POLYPODIALES in the McGraw-Hill Encyclopedia of Science & Technology.

Vaughn M. Bryant

Bibliography. V. M. Bryant and G. D. Jones, Forensic palynology: Current status of a rarely used technique in the United States of America, *Forensic Sci. Int.*, 163(3):183–197, 2006, DOI:10.1016/j.forsciint.2005.11.021; D. C. Mildenhall, P. E. J. Wiltshire, and V. M. Bryant, Forensic palynology: Why do it and how it works, *Forensic Sci. Int.*, 163(3):163–172, 2006, DOI:10.1016/j.forsciint.2006.07.012; L. Milne et al., Forensic palynology, in *Forensic Botany: Principles and Applications to Criminal Case*, Boca Raton, FL, 2005.

Functional electrical stimulation therapy

Functional electrical stimulation (FES) is defined as the application of an electrical stimulus to paralyzed nerves or muscles following stroke or any other neuromuscular disorder. Its main purpose is to restore and achieve function, and it has long been used in neurological rehabilitation. The efficacy and therapeutic application of FES are well documented and supported by a significant amount of literature. It can be used to stimulate nerves in the arms, legs, trunk, and buttocks in order to achieve a range of functional movements. FES is most often paired with task-specific practice. A common example is the use of FES for motor recovery of upper extremity function in poststroke subjects. FES also can be used to aid in walking because foot drop (the inability to lift the front part of the foot) is one of the most common mobility problems from stroke and other neuromuscular disorders.

Description of FES. FES uses a minute amount of electrical current to stimulate nerves or muscles that connect to the paralyzed muscles and cause the muscles to contract. The application of electric current to tissues creates an action potential that propagates along cellular membranes. It is well known that some tissues are more sensitive to electricity than others. For example, membranes of the nerves require 100 times less current for activation than membranes of the muscle. It is for this reason that motor point activation is desired, and not direct muscle fiber stimulation, in order to achieve muscle movement.

The most common way of delivering current is to use electrodes placed on the skin over the nerve. The electrical impulse is delivered in a direct relationship with performance of function, and the user may be required to wear an electrical impulse device as a neuroprosthesis or hybrid orthosis (an artificial or mechanical aid to assist movement of an injured body part). It also can be used within the movement of a piece of equipment such as a cycle ergometer (a stationary cycling instrument).

In the application of any form of electrical stimulation, it is important to recognize that physiological contraction differs from an electrically driven contraction. The action potential generated from an electrically driven contraction travels both toward the neuromuscular junction and the anterior horn cell (a motor neuron in the spinal cord) in an antegrade (forward) and retrograde (backward) fashion, respectively. The recruitment of motor units also differs in both number and type. Voluntary muscle contractions will preferentially recruit type I fibers. Type I fibers are slow-contracting, fatigue-resistant motor units compared to type II fibers, which are fast-contracting, fatigable, and forceful motor units. One has to keep in mind that fatigue occurs more rapidly in an electrically generated contraction because a greater portion of fatigable motor units is recruited. Traditionally, atrophic muscles are characterized by type II B fibers. Strengthening through electrically activated contraction allows conversion to type I fibers.

Benefits and applications of FES. One of the more encouraging benefits of FES is the appearance of incremental cellular changes in the nervous system. For example, rodent studies using electrical stimulation have produced the birth of endogenous neural progenitors in the white matter of the spinal cord. Furthermore, large numbers of these cells have developed into myelin-producing oligodendrocytes [glial cells of the central nervous system that produce myelin (an insulating sheath)], which provide great hope for the restoration of axonal remyelination of damaged neurons.

There are many other benefits of FES, including neuromuscular reeducation, edema reduction, stimulation of sensory fibers, improvement of circulation, and increased range of motion. Moreover, FES is used to prevent and reverse disuse muscle atrophy, improving strength and muscle mass during or following periods of inactivity. FES is also known to reduce the effects of spasticity.

There are many therapeutic applications of FES. In one randomized controlled trial, the effectiveness of myoelectrically controlled functional electrical stimulation (MECFES) was clearly demonstrated for the rehabilitation of upper limbs in poststroke subjects. Eleven poststroke hemiparetic patients (that is, patients with weakness on one side of the body) who received three to five treatment sessions per week showed improvements in their arm action test scores, which was a condition that was maintained at follow-up.

Another study reported on the effect of body weight support treadmill training (BWSTT) with power-assisted FES on functional movement and gait in stroke patients. Thirty subjects were randomly assigned to either a standard rehabilitation program or BWSTT with power-assisted FES. Those receiving FES as part of their therapy showed significant improvements in their functional movement and gait pattern.

Lower extremity FES cycling also has been shown to promote physical and functional recovery in patients with neurogenic paralysis from spinal cord injury (SCI). In this study, a retrospective cohort with 25 SCI subjects received FES during cycling. FES was found to be associated with an increase in muscle size and force generation potential, reduced spasticity, and improved quality of life.

Patient evaluation for FES. Of course, it is critically important to evaluate a patient's medical history when determining whether a patient should be a candidate for FES treatment. Active metastases and pregnancy may exclude a patient for a limited time. Contraindications include thrombosis, hemorrhage, epilepsy, cancer, osteomyelitis (bone infection), and a history of an implanted device. It is highly important to evaluate the risks and benefits prior to starting FES treatment. Once it has been decided that a patient is a suitable candidate, an electrical stimulation trial is carried out as part of the evaluation. The standard trial includes assessment of a patient's response and tolerance. Documentation should include the parameters used and which muscles were tested. The muscle response is documented as follows: no contraction, twitch contraction, tetanic contraction, or tetanic contraction with full joint movements. Basic parameters for neuromuscular electrical stimulation include frequency, waveform, pulse width, amplitude, ramp, and duration. The goal is to generate the lowest possible amount of current while maintaining the desired response.

Future outlook. Overall, new roles and applications for FES in the context of advancing technologies are being discovered and implemented. As is the case with any intervention, FES intervention is intended to complement all functional treatment goals.

For background information *see* BIOMEDICAL ENGINEERING; BRAIN; ELECTRODIAGNOSIS; ELECTROMYOGRAPHY; MEDICAL CONTROL SYSTEMS; MUSCLE; MUSCULAR SYSTEM; MUSCULAR SYSTEM DISORDERS; NERVE; NERVOUS SYSTEM (VERTEBRATE); NEURON; SPINAL CORD; SPINAL CORD DISORDERS in the McGraw-Hill Encyclopedia of Science & Technology.

John W. McDonald; Albert Recio

Bibliography. D. Becker et al., Functional electrical stimulation helps replenish progenitor cells in the injured spinal cord of adult rats, *Exp. Neurol.*, 222(2):211–218, 2010, DOI:10.1016/j.expneurol.2009.12.029; H. J. Lee, K. H. Cho, and W. H. Lee, The effects of body weight support treadmill training with power-assisted functional electrical stimulation on functional movement and gait in stroke patients, *Am. J. Phys. Med. Rehabil.*, 92(12):1051–1059, 2013, DOI:10.1097/PHM.0000000000000040; R. Martin et al., Functional electrical stimulation in spinal cord injury: From theory to practice, *Top. Spinal Cord Inj. Rehabil.*, 18(1):28–33, 2012, DOI:10.1310/sci1801-28; C. L. Sadowsky et al., Lower extremity functional electrical stimulation cycling promotes physical and functional recovery in chronic spinal cord injury, *J. Spinal Cord Med.*, 36(6):623–631, 2013, DOI:10.1179/2045772313Y.0000000101; R. Thorsen et al., Myoelectrically driven functional electrical stimulation may increase motor recovery of upper limb in poststroke subjects: A randomized controlled pilot study, *J. Rehabil. Res. Dev.*, 50(6):785–794, 2013, DOI:10.1682/JRRD.2012.07.0123.

Functional inorganic nanomaterials

Over the last few decades, considerable attention has been devoted to the fabrication of inorganic nanomaterials and their application in a wide variety of technologies. As a special class of materials, inorganic nanomaterials consist of particles of metals, metal oxides, or metal chalcogenides (metal sulfides, selenides, or tellurides) with at least one dimension in the 1- to 100-nanometer (10^{-9} m) range. These nanomaterials are highly interesting because they exhibit properties that bridge the gap between bulk and molecular structures. Unlike their bulk counterparts, nanomaterials exhibit tunable size- and shape-dependent optical, electronic, and magnetic properties. Because of these unique properties, nanomaterials are being explored to address key global challenges in areas such as energy conversion, catalysis, medicine, sensing, and environmental remediation.

In this article, we present a general overview of the four main types of functional inorganic nanomaterials: plasmonic nanomaterials, quantum dots, magnetic nanoparticles, and electrocatalytic nanometals.

Plasmonic nanomaterials. When the size of a metallic material (typically, but not exclusively, silver or gold) is reduced to the nanometer range, interesting plasmonic properties emerge as a result of the collective oscillation of the conduction electrons, which is stimulated by incident light. This oscillation leads to a strong plasmon absorption band, known as localized surface plasmon resonance (LSPR). The spectral position and intensity of the LSPR band can be tuned from the ultraviolet (UV) to near-infrared (IR) wavelengths by changing the composition, size, and shape of the material (**Fig. 1**).

The theoretical basis for plasmonic particles (gold colloids) was first described by Gustav Mie in the first decade of the twentieth century. However, the application of plasmonic materials dates far back in time, as exemplified by the use of colloidal metallic nanoparticles in the fabrication of the Gothic stained-glass rose window of the Notre-Dame de Paris cathedral and the Lycurgus cup (fourth century), which exhibits different colorations depending on the illumination direction. There is also evidence of the ancient use of plasmonic nanomaterials from the Egyptian dynasties and Mayan civilizations. These early applications were realized without an in-depth understanding of the origin of the optical effects, which was established only later with the development of nanoscience at the end of the twentieth century. Since then, the application of plasmonic nanomaterials has been expanded to areas of sensing, biomedical research, telecommunication, and energy conversion.

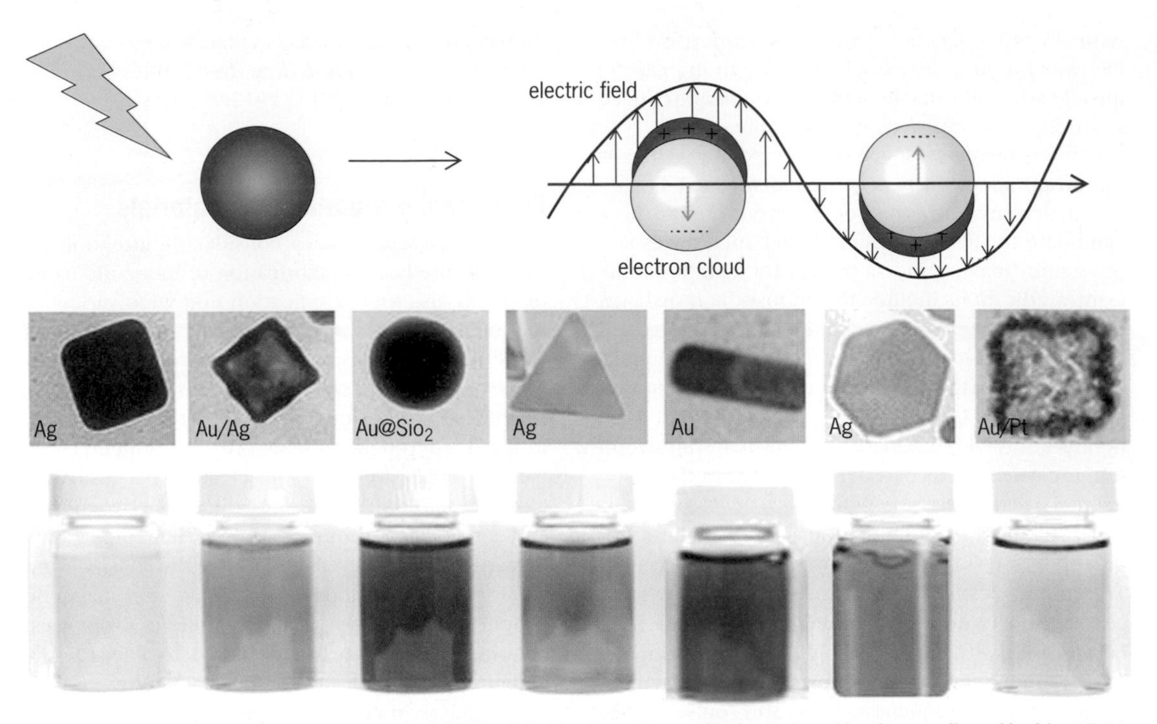

Fig. 1. Localized surface plasmon resonance (LSPR) originating from the oscillation induced by the coupling of incident light with the free conduction electrons in the metal nanoparticle. The LSPR absorption band can be tuned in the UV to near-IR spectral range by varying the size, composition, and shape of the metal nanoparticle.

Quantum dots. When a semiconductor material is irradiated with light of an appropriate wavelength, excited electron-hole pairs (excitons) are generated, which are separated by a material-dependent distance known as the Bohr exciton radius. As the size of the semiconductor material is reduced to dimensions smaller than its Bohr exciton radius, quantum confinement occurs, leading to a transition from continuous to discrete energy levels.

Quantum dots are zero-dimensional semiconductor materials, typically metal chalcogenides (MX, where M = Cd, Pb and X = S, Se, Te), with diameters in the 2- to 10-nm size range. These nanomaterials exhibit size-quantized optoelectronic properties. For example, decreasing the size of a quantum dot leads to an increase in the effective band gap of the material, which determines the spectral range of emitted light. More specifically, as the size of the quantum dot decreases, the emission wavelength shifts from red to blue (**Fig. 2**).

Other unique properties of quantum dots include (1) narrow, Gaussian-shaped emission spectra; (2) a broad excitation band, which allows different-sized quantum dots to be simultaneously excited with a single light source; (3) a high extinction coefficient (absorptivity), which leads to a brighter emission signal for increased assay sensitivity; and (4) excellent photostability with continuous high-energy excitation. The highly tunable and unique optical properties of quantum dots have brought a lot of attention to the study of these functional nanomaterials for their use in light-emitting diodes (LEDs), solar cells, and biomedical imaging applications.

Magnetic nanoparticles. Bulk magnetic materials exhibit a multidomain structure, where regions of uniform magnetization are separated by domain

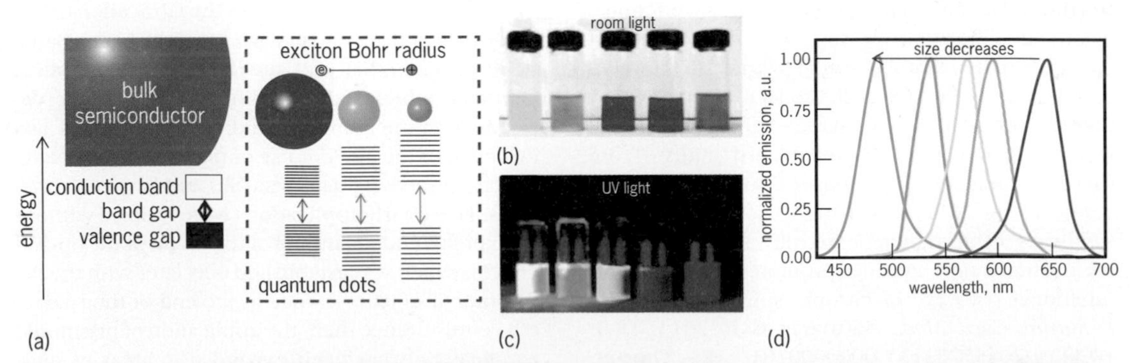

Fig. 2. Quantum confinement effect in semiconductor materials, resulting in (*a*) the transition from continuous to discrete energy levels. Quantum-dot solutions of increasing particle sizes (left to right) when exposed to (*b*) room light and (*c*) UV light. (*d*) Emission spectra of the respective quantum-dot solutions.

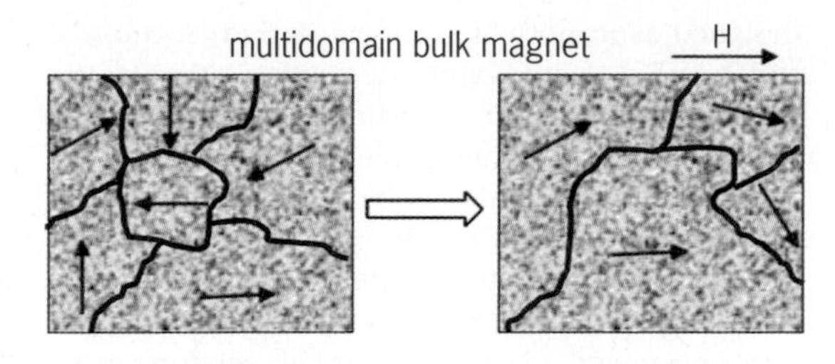

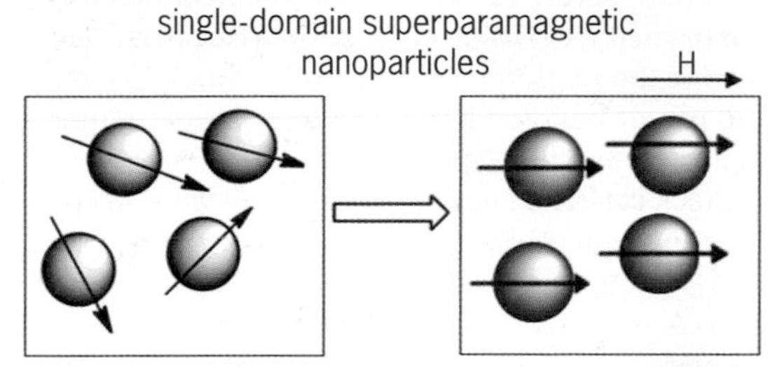

Fig. 3. Comparison of the magnetic properties of bulk multidomain magnets and single-domain superparamagnetic nanoparticles. H represents the applied magnetic field direction. As illustrated by the behavior of the nanoparticles in solution, when an external field is applied to the magnetic fluid, the nanoparticles respond and are attracted to the magnet; when the field is removed, the nanoparticles resume their random orientations and re-suspend freely in solution.

walls. As such, when an external magnetic field is applied, the change in magnetization is influenced not only by the realignment of the magnetic dipoles, but also by the movement of the domain walls. However, when the size of the magnetic material is reduced to a critical size (below 100 nm), the formation of domain walls becomes thermodynamically unfavorable, with the result that the material can support only a single magnetic domain structure and the magnetic moments of all the atoms of the particle are aligned in one direction. Consequently, remagnetization is dependent on coherent spin rotation only. As the size of the magnetic nanomaterial is further reduced (to tens of nanometers), the magnetic moment reversal can be readily activated thermally, and the system becomes superparamagnetic; that is, when an external magnetic field is applied, the nanoparticle behaves like a paramagnetic material with a giant (super) magnetic moment (**Fig. 3**), and without an external magnetic field, the magnetic moment is zero.

Superparamagnetic nanomaterials (for example, superparamagnetic iron-oxide nanoparticles) are of great interest in biomedical research because of their sensitive response to external magnetic fields, which is currently explored in magnetic resonance imaging (MRI) and magnetic hyperthermia (for example, cancer treatment) applications. In addition, their lack of residual magnetization after magnetic field removal, which prevents aggregation and makes them ideal materials for drug delivery and environmental remediation applications.

Electrocatalytic nanometals. The design and fabrication of efficient catalytic materials is of high importance in the area of alternative energy technologies. In particular, electrocatalytic metal nanomaterials (platinum and palladium based) are being explored in fuel-cell research. For these types of materials, the specific surface area plays a crucial role, as it determines the overall adsorption and electron-transport properties, which have a direct impact on the materials' electrocatalytic performance. Moreover, the catalytic activity can be further optimized by precisely controlling the morphology of the nanomaterials, as the electrocatalytic process has been shown to be highly structure sensitive. Additionally, to prevent poisoning of the highly active metal surface, the local electronic structure of the electrocatalyst is usually modified by the incorporation of other elements and the formation of alloys. Not only does the alloying of electrocatalytic metal nanomaterials provide a means to prepare a rich diversity of compositions, but different atomic-ordering patterns can also be adapted (**Fig. 4**) to create a large library of electrocatalysts that can be tailored for specific applications. This leads to the discovery of electrocatalysts with greatly enhanced sensitivity and selectivity, which facilitates the expansion of this field to applications in biomedical research and environmental monitoring technologies.

Outlook. The development of functional inorganic nanomaterials with unique and tunable properties is an expanding area of research. Current trends involve the fabrication of multicomponent nanostructures comprised of different types of inorganic nanomaterials to obtain hybrids that not only combine

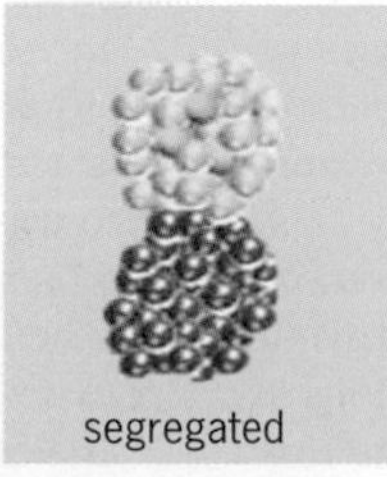

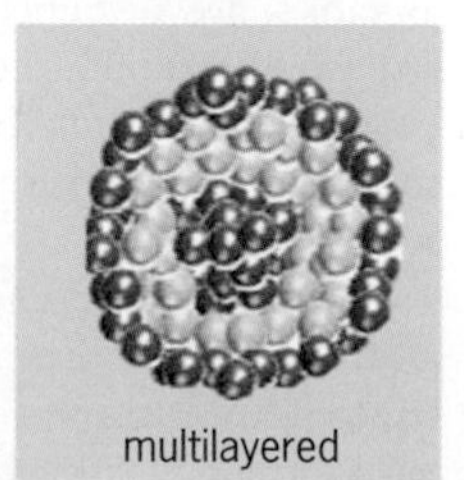

Fig. 4. Different types of atomic ordering observed in metal nanoalloys.

the characteristic behavior of their components, but often exhibit entirely new properties. In addition, considerable research focuses on the fabrication of hybrid organic-inorganic nanomaterials that provide better processability and easier handling.

For background information *see* ALLOY; BAND THEORY OF SOLIDS; CATALYSIS; ELECTRON-HOLE RECOMBINATION; ELECTRON SPIN; EXCITON; FUEL CELL; LIGHT-EMITTING DIODE; MAGNETIC RESONANCE; MAGNETIC SUSCEPTIBILITY; NANOPARTICLES; NANOTECHNOLOGY; PLASMON; PARAMAGNETISM; SEMICONDUCTOR; SOLAR CELL in the McGraw-Hill Encyclopedia of Science & Technology.

Adriana Popa; Anna Cristina S. Samia

Bibliography. C. Burda et al., Chemistry and properties of nanocrystals of different shapes, *Chem. Rev.*, 105(4):1025–1102, 2005, DOI:10.1021/cr030063a; R. Ferrando, J. Jellinek, and R. L. Johnston, Nanoalloys: From theory to applications of alloy clusters and nanoparticles, *Chem. Rev.*, 108(3):845–910, 2008, DOI:10.1021/cr040090g; X. Gao et al., In vivo cancer targeting and imaging with semiconductor quantum dots, *Nat. Biotechnol.*, 22:969–976, 2004, DOI:10.1038/nbt994; X. M. Lin and A. C. S. Samia, Synthesis, assembly and physical properties of magnetic nanoparticles, *J. Magn. Magn. Mater.*, 305(1):100–109, 2006, DOI:10.1016/j.jmmm.2005.11.042; D. Yoo et al., Theranostic magnetic nanoparticles, *Accounts Chem. Res.*, 44(10):863–874, 2011, DOI:10.1021/ar200085c.

Fundamental design of reinforced concrete water structures

Reinforced concrete structures are used in numerous water and wastewater applications, including solar-heating reservoirs, cisterns, catch basins, water tanks, water towers, small reservoirs, sewage treatment and collection systems, and swimming pools, to name a few. The design of water structures is no different from other structures except in the care about water-side sections and the quality of concrete in order to avoid reinforcement corrosion and water leakage by controlling cracking. The control of cracks can be achieved by using adequate concrete cover, limiting the maximum stresses in tension steel, and using well-distributed smaller deformed reinforcement bars over the zone of maximum concrete tension.

Design concept of water structures. Reinforced concrete structures containing water or liquids, such as tanks, are designed to withstand internal forces and environmental conditions, including causes of corrosion of reinforcement (steel). For simplicity, this type of structure is addressed here as a water structure.

The basic difference between water structures and other structures is illustrated by the following example of an open channel in **Fig. 1**. Sections 1 and 4 in Fig. 1 are air-side sections; that is, the tensile stresses are on the air side and hence, they are designed as in normal sectional design. Sections 2 and 3 in Fig. 1 are water-side sections; that is, the tensile stresses are on the water side. The design of these sections should avoid reinforcement corrosion and fluid seepage. This can be achieved by controlling the surface crack width.

Factors affecting surface crack width. The crack width, **Fig. 2**, can be reduced by (1) increasing the concrete cover, (2) decreasing the reinforcement bar diameter, (3) using deformed surface bars, (4) reducing the reinforcement stresses, and (5) avoiding the use of bundled bars and satisfying the requirements of minimum spacing between bars.

Crack control for environmental conditions. According to the American Concrete Institute (ACI) code, ACI 350-06, water-side sections are classified into two categories according to their degree of environmental exposure: normal and severe exposure. For liquid retention, normal environmental exposure is defined as exposure to liquids with a pH greater than 5, or exposure to sulfate solutions of 1000 ppm or less. Severe environmental exposures are conditions in which the limits defining normal environmental exposure are not fulfilled.

For protection of reinforcement against corrosion, and for aesthetic reasons, many fine hairline cracks are preferable to a few wide cracks. Cracks have to be controlled particularly when the yield stress of reinforcement exceeds 280 megapascals (MPa). Current good detailing practices will usually lead to adequate crack control even when reinforcement of 420 MPa is used. It has been demonstrated experimentally that the crack width at service loads is proportional to the steel stress. The significant variables reflecting steel detailing were found to be the thickness of the concrete cover and the area of concrete in the zone of maximum tension surrounding individual reinforcing bars.

The calculated stress, f_s, in reinforcement closest to a surface in tension at service loads shall not exceed that given by Eqs. (1) and (2), as defined

$$f_{s,\max} = \frac{56{,}000}{\beta\sqrt{s^2 + 4(50 + d_b/2)^2}} \qquad (1)$$

$$f_{s,\max} = \frac{45{,}000}{\beta\sqrt{s^2 + 4(50 + d_b/2)^2}} \qquad (2)$$

by ACI 350-06, and shall not exceed a maximum of 250 MPa. The calculated flexural stress in reinforcement at service load, f_s, shall be computed as the unfactored moment divided by the product of steel area and internal moment arm.

In normal environmental exposure areas as defined earlier [Eq. (1)], walls and slabs need not be less than 140 MPa for one-way and 165 MPa for two-way walls and slabs. In severe environmental exposure areas as defined earlier [Eq. (2)], they need not be less than 115 MPa for one-way and 140 MPa for two-way walls and slabs.

In Eqs. (1) and (2) the strain gradient amplification factor, β, which is the ratio of distances to the neutral

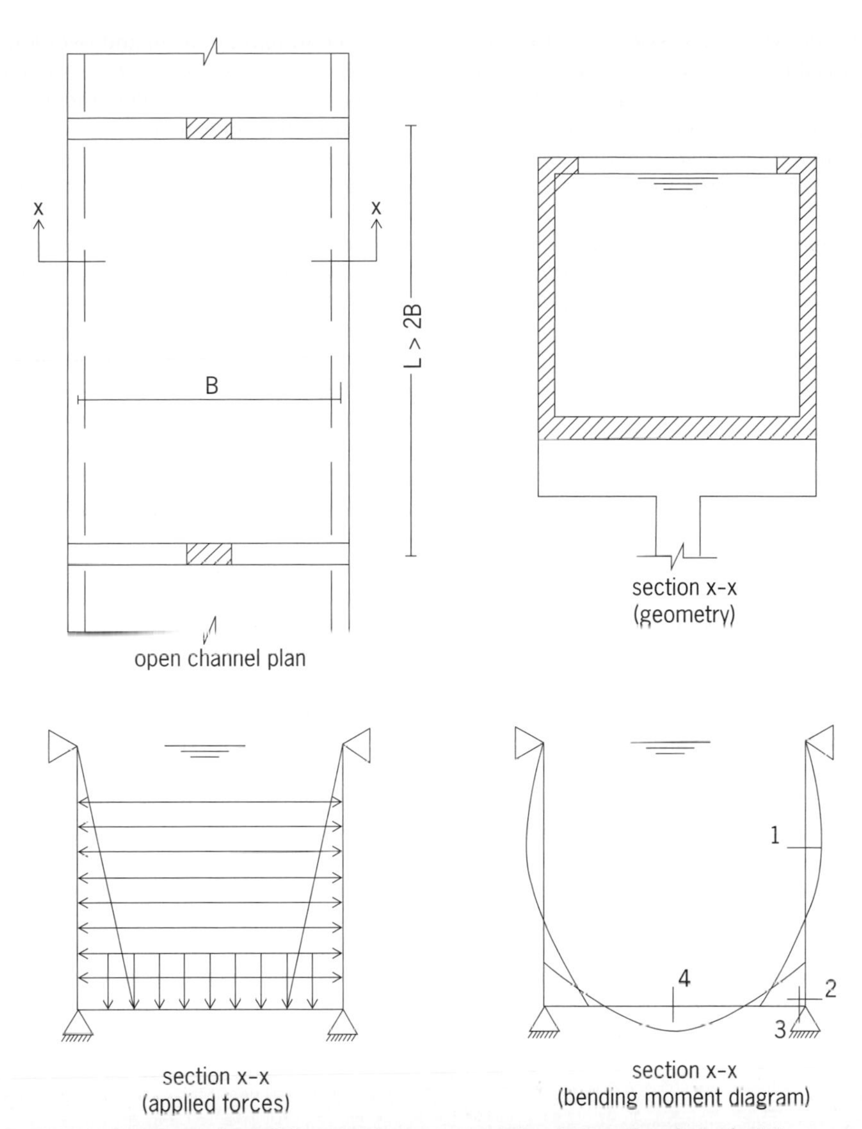

Fig. 1. Example of an elevated open channel. L = span between supporting frames. B = channel width.

axis from the extreme tension fiber and from the centroid of the main reinforcement, shall be given by Eq. (3), where c is the distance from extreme

$$\beta = \frac{h - c}{d - c} \qquad (3)$$

compression fiber to the neutral axis, calculated at service loads, h is the overall thickness of member, d is the distance from the extreme compression fiber to the centroid of tension reinforcement, and s is the spacing of tension reinforcement closest to the surface. It shall be permitted to use the value 16,000 for the term $4(50 + d_b/2)^2$ as a simplification. In addition, it shall be permitted to use β equal to 1.2 for $h \geq 400$ mm and 1.35 for $h < 400$ mm in Eqs. (1) and (2).

Where appearance of the concrete surface is of concern and concrete cover exceeds 75 mm, the service load flexural tension stress must not exceed the values stated earlier, and the spacing of tension reinforcement closest to the surface shall not exceed

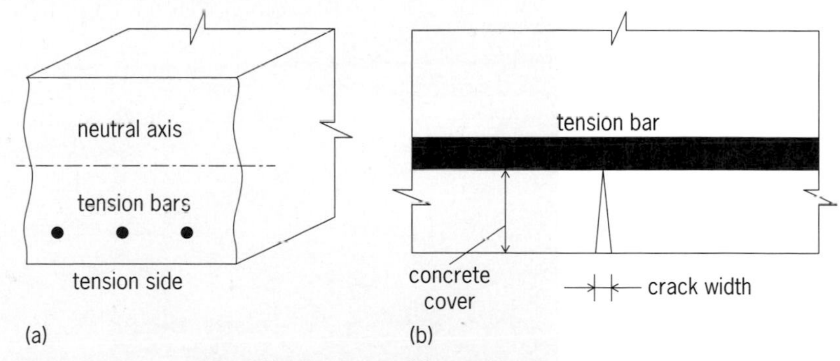

Fig. 2. Surface cracking. (*a*) Wall segment. (*b*) Part side view.

that given by Eq. (4) but not greater than 300 mm,

$$s = \frac{94{,}600}{f_s} - 2.5c_c \qquad (4)$$

where c_c is the clear cover from the nearest surface in tension to the surface of the flexural tension steel.

For background information *see* CONCRETE; CORROSION; PH; PRESTRESSED CONCRETE; REINFORCED

Consortium (DPAC), which coordinates the efforts of several European institutes and data processing centers. This work involves some 400 people, organized in 9 coordination units and handling data simulations, processing of the raw data, preliminary analysis of the processed data, and publication. The satellite itself was built by an industrial consortium led by Airbus DS, formerly known as Astrium Toulouse, with contributions from specialized industries in Europe. The European Space Operations Centre in Darmstadt, Germany, handles the daily operations. The initial stages of the data processing take place at the European Space Astronomy Centre near Madrid, Spain, from which the data are distributed to other data processing centers in France, Spain, the United Kingdom, and Italy.

For background information *see* ALPHA CENTAURI; ASTROMETRY; BINARY STAR; CEPHEID VARIABLES; CHARGE-COUPLED DEVICES; EXTRASOLAR PLANETS; MAGNITUDE (ASTRONOMY); MILKY WAY GALAXY; PARALLAX (ASTRONOMY); PARSEC; SATELLITE (SPACECRAFT) in the McGraw-Hill Encyclopedia of Science & Technology. Floor van Leeuwen

Bibliography. L. Lindegrin et al., The Gaia mission: Science, organization and present status, in W. J. Jin, I. Platais, and M. A. C. Perryman (eds.), *A Giant Step: From Milli- to Micro-arcsecond Astrometry*, pp. 217–223, Proceedings of the International Astronomical Union, IAU Symposium 248, Shanghai, China, October 15–19, 2007, Cambridge University Press, Cambridge, UK, 2008; F. Mignard et al., Gaia: Organisation and challenges for the data processing, in W. J. Jin, I. Platais, and M. A. C. Perryman (eds.), *A Giant Step: From Milli- to Micro-arcsecond Astrometry*, pp. 224–230, Proceedings of the International Astronomical Union, IAU Symposium 248, Shanghai, China, October 15–19, 2007, Cambridge University Press, Cambridge, UK, 2008.

Gene therapy (cancer)

Since 1970, the median survival of patients having advanced cancer with optimal standard systemic treatment has remained less than 1 year. To increase survival rates, gene therapy has emerged as a new cancer treatment plan. Gene therapy is defined as the introduction of a functional copy of a curative gene into a patient in order to achieve a therapeutic objective. In 1973, a technique to introduce DNA into mammalian cells was developed. The first preclinical attempt to use gene insertion for therapy in animals was done in 1994. In this case, a growth hormone gene was inserted into cells and injected into mice with deficient growth hormone production. The growth hormone gene was introduced successfully, and a functional protein product was expressed as a result. Since then, gene therapy has progressed, providing opportunities for many new directions in cancer management. One opportunity can be characterized as a process involving "personalized molecular therapeutics." There are six key alternatives of the neoplastic physiolome (a set of physiologic features) that dictate malignant growth: (1) self-sufficiency; (2) insensitivity to growth inhibition, including immune escape (the ability to avoid immune-mediated rejection); (3) independence from programmed cell death (apoptosis); (4) unlimited replicative potential; (5) sustained angiogenesis (the development of blood vessels); and (6) local/distal tissue invasion capacity.

The process of cancer management involving personalized molecular therapies involves the identification (within an individual's cancer) of signature genome/expressed protein patterns necessary for cancer survival, growth, and spread that are distinguishable from nonmalignant tissue. Such a characterization of the cancer-specific molecular profile will then enable the patient to be therapeutically managed with gene replacement or gene blockade of the significant cancer signals in order to critically cripple the cancer pathway and induce tumor response.

Signal targeting. Tumorigenesis is a multistep process, reflecting genetic and epigenetic alterations that drive the transformation of normal cells into malignant cells. Each cancer is a complex adaptive network of dynamically evolving spatial-temporal biomolecular interactions. Although it appears intuitive that disruption of any one of these global physiologic capabilities would provide a therapeutic opportunity, each cancer is a robust system capable of maintaining its functional characteristics following internal or external perturbation. Cancer cells can buffer genetic modifications by providing redundant functional pathways in which different structural elements have overlapping functions. This process is termed degeneracy. Positive and negative feedback controls allow for further robustness. Multileveled functional complementation results from modules at each organizational level [gene level (genome) → RNA transcript level (transcriptome) → protein level (proteome) → metabolite level (metabolome)], which interrelate in a functional organizational hierarchy. Therapeutic targeting based on unique individual dynamic genomic–proteomic patterns (as opposed to traditional chemotherapy principles) are projected to be employed in clinical investigations.

The correlation of gene expression patterns with disease outcome (survival) has been demonstrated in a variety of cancers. Unfortunately, in order to truly assess the oncogenic function of a specific gene, a better understanding of nucleic acid–protein and protein–protein interactions is needed. Gene transcript levels often show poor correlation with protein levels, and they cannot predict posttranslational modifications. Proteins assemble themselves into networks through a variety of protein–protein interactions. Disruptions of these coupling events may be the "functional" target of drug therapy. Pathogenic gene mutations and gene deletions, duplications, and amplifications can result in defective, absent, and overexpressed proteins. These proteins realign within the normal cellular protein network in an alternative or degenerative redundant pattern, producing an oncopathologic hostile takeover. Proteomic technologies are currently successfully employed for drug discovery and biomarker identification. However, despite their sophistication, these technologies have substantial limitations when they are applied to tissue and blood samples.

Gene therapeutics. Somatic gene-based therapy offers the promise of revolutionizing modern medicine and, in particular, oncology by targeting and modifying the underlying relevant function-dependent pathway (or pathways) of cancer cells, rather than ubiquitously targeting proliferative cells. For gene-based therapeutic approaches directed at the cancer cell, effectiveness is dependent on three criteria: (1) identification of a genetic or epigenetic defect that provides a pivotal survival advantage to the cancer population; (2) introduction of a potent gene-based agent whose expression can reverse the oncogenic phenotype; and (3) utilization of an effective global delivery mechanism. So far, clinical investigations of both turn-on and knockdown expressive gene therapeutics have demonstrated activity and safety involving both targeted immune activation and critical procancer signal disruption.

RNA interference. RNA interference (RNAi) is defined as a mechanism of gene silencing (regulation of gene expression via mechanisms that include suppression at transcriptional and posttranscriptional levels) produced by small RNAs, which include endogenous microRNA (miRNA) and exogenous small interfering RNA (siRNA) or short hairpin RNA (shRNA). Gene silencing is an evolutionarily conserved process and is highly dependent on gene sequence.

The mechanism of RNAi has been thoroughly investigated. Briefly, a double-stranded small RNA is incorporated into the pre-RISC (RNA-induced silencing complex). This is followed by the cleavage-dependent (matched guide and passenger strands) or cleavage-independent (unmatched guide and passenger strands) release of the passenger strand, forming the guide strand that contains the RISC. The guide strand (antisense strand) then guides the RISC to the complementary or near-complementary region of the target messenger RNA (mRNA). In general, siRNA (from the cleavage-dependent RISC), with a perfect match to its target, cleaves the target mRNA via the endonuclease Argonaute 2 (Ago2) [note that an endonuclease is an enzyme that degrades DNA or RNA by breaking linkages within the polynucleotide chains]. On the other hand, miRNA, with an imperfect match to its target, induces mRNA degradation (or sequestration) in the P-body (a specific type of processing body) and translational inhibition.

As an alternative strategy to siRNA, shRNA has been developed to allow for long-term gene silencing. In this method, shRNA is transcribed in the nucleus from an expression vector bearing a short double-stranded DNA sequence with a hairpin loop. The shRNA transcript is processed and loaded into the RISC in the cytoplasm, and then the same cytoplasmic RNAi process is followed as described for siRNA.

Recently, a third novel approach called bifunctional shRNA (bi-shRNA) has been developed to exploit both the cleavage and translational inhibition mechanisms of RNAi. It consists of two stem-loop shRNA structures: one cleavage-dependent unit with a passenger strand and a guide strand that are perfectly matched, and one cleavage-independent unit composed of a mismatched double strand. The two shRNA units are embedded in a microRNA (miR-30) scaffold and are encoded by a plasmid vector. The mature transcript of the cleavage-dependent unit is loaded onto the RISC, incorporating Ago2. In contrast, the processed transcript of the cleavage-independent unit functions as a microRNA (miRNA) by binding to RISC without cleavage of the guide strand because of the strategic placement of the mismatch site, incorporating Ago1–4 without endonuclease function, inducing mRNA degradation and P-body sequestration, or translational inhibition. In principle, bi-shRNA is able to induce cleavage- and noncleavage-mediated degradation of the target mRNA (decapping and deadenylation) and inhibit translation concurrently, leading to a more rapid onset of gene silencing and higher efficacy and greater durability when compared with either siRNA or shRNA.

So far, clinical trials utilizing both siRNA and bi-shRNA in cancer have demonstrated evidence of safety and efficacy in phase I and phase II testing.

For background information *see* APOPTOSIS; CANCER (MEDICINE); DEOXYRIBONUCLEIC ACID (DNA); GENE; GENE SILENCING; GENETIC ENGINEERING; GENETICS; MICRORNA; ONCOGENES; ONCOLOGY; PROTEIN; RIBONUCLEIC ACID (RNA); RNA INTERFERENCE (RNAI); SMALL INTERFERING RNA (SIRNA) in the McGraw-Hill Encyclopedia of Science & Technology.

John Nemunaitis

Bibliography. A. Cervantes et al., Phase I dose-escalation study of ALN-VSP02, a novel RNAi therapeutic for solid tumors with liver involvement, *J. Clin. Oncol.*, 29(15):3025, 2011; A. Fire et al., Potent and specific genetic interference by double-stranded RNA in *Caenorhabditis elegans*, *Nature*, 391(6669):806–811, 1998, DOI:10.1038/35888; F. L. Graham and A. J. van der Eb, A new technique for the assay of infectivity of human adenovirus 5 DNA, *Virology*, 52(2):456–467, 1973, DOI:10.1016/0042-6822(73)90341-3; D. Hanahan and R. A. Weinberg, Hallmarks of cancer: The next generation, *Cell*, 144(5):646–674, 2011, DOI:10.1016/j.cell.2011.02.013; N. S. Lee et al., Expression of small interfering RNAs targeted against

HIV-1 *rev* transcripts in human cells, *Nat. Biotechnol.*, 20(5):500–505, 2002, DOI:10.1038/nbt0502-500; P. B. Maples et al., FANG vaccine: Autologous tumor vaccine genetically modified to express GM-CSF and block production of furin, *BioProcessing J.*, 8(4):4–14, 2010; M. Miyagishi and K. Taira, U6 promoter-driven siRNAs with four uridine $3'$ overhangs efficiently suppress targeted gene expression in mammalian cells, *Nat. Biotechnol.*, 20(5):497–500, 2002; F. L. Moolten, Drug sensitivity ("suicide") genes for selective cancer chemotherapy, *Cancer Gene Ther.*, 1(4):279–287, 1994; D. D. Rao et al., siRNA vs. shRNA: Similarities and differences, *Adv. Drug Delivery Rev.*, 61(9):746–759, 2009, DOI:10.1016/j.addr.2009.04.004; Z. Wang et al., RNA interference and cancer therapy, *Pharm. Res.*, 28(12):2983–2995, 2011, DOI:10.1007/s11095-011-0604-5; J. Y. Yu, S. L. DeRuiter, and D. L. Turner, RNA interference by expression of short-interfering RNAs and hairpin RNAs in mammalian cells, *Proc. Natl. Acad. Sci. USA*, 99(9):6047–6052, 2002, DOI:10.1073/pnas.092143499.

Geologic history of reefs

Reefs in the geologic sense are ancient marine ecosystems that have experienced profound changes during 500 million years of prosperity and collapse. Reefs have experienced high levels of carbon dioxide (CO_2) during their geologic history. The buildup of this greenhouse gas leads to global warming and ocean acidification. Ancient reefs responded to these changes, which can be seen by their frequent collapse followed by long periods in which reefs were absent (reef gaps). Reefs today and in the past were controlled by nutrient levels, sedimentation, sunlight, and temperature. They develop best in tropical settings between latitudes of 30°N and 30°S of the equator. This occurs as a result of an important relationship with symbiotic algae living in the tissues of corals and other reef organisms. These algae exploit photosymbiosis (a symbiotic relationship between two organisms, with one being able to perform photosynthesis), providing substantial nutritional benefit to their reef hosts while also greatly accelerating calcification rates. Photosymbiosis explains why reefs exist today and may well explain their successes and failures through time. Understanding the dynamics of ancient reef systems is central to understanding the current reef crisis.

The earliest reefs? During the Precambrian Era, several billion years before the first complex organisms appeared, the closest approximations to reefs were calcimicrobes (calcifying microbes) or cyanobacteria (blue-green algae). They produced deposits of carbonate rock during the Middle and Late Proterozoic eons. At the advent of the Phanerozoic Eon (approximately 540 million years ago), the first calcifying metazoans (primitive multicellular animal organisms) appeared. Early Cambrian reeflike mounds constructed by archaeocyathids (extinct marine sponges) and calcimicrobes existed for more than 10 million years. Debates have centered on whether these Cambrian mounds were reefs like those of today and whether they were capable of photosymbiosis. They disappeared after the Early Cambrian and then mostly microbes remained, making a reef gap for the rest of the Cambrian and Early Ordovician periods.

After the shift from microbes to metazoans, the Ordovician Period witnessed a biodiversification event and the start of stony sponge–coral–algal reef associations. During the Late Ordovician, reef complexes were dominated by corals, stony sponges, and a variety of filter-feeding organisms, such as bryozoans and brachiopods, which rimmed shallow reef platforms. The end of the Ordovician experienced pulses of mass extinction and reef collapse brought on by climate change and glacial cooling.

Mid-Paleozoic reefs. Reefs resumed after the Ordovician extinctions, establishing a new, long-lived reef ecosystem of the Silurian and Devonian periods (see **illustration**). This new ecosystem consisted of coral–stromatoporoid (a type of extinct calcareous sponge)–red algae associations that were characterized globally by huge reef tracts far greater in size than present-day reefs. They existed for the longest of any reef association, lasting almost 100 million years. Reefs proliferated during a "supergreenhouse" interval of global warming, and spectacular reef systems developed. Large Paleozoic coral colonies and stony sponges created an ecosystem similar to that of modern coral reefs. In addition, the evidence strongly supports the existence of photosymbiosis. During this time, reefs expanded to latitudes of 60°N and 60°S of the equator because high CO_2 levels resulted in warm temperatures. A mass extinction and cooling near the end of the Devonian Period led to ecosystem collapse. Afterwards, the reef belt shrank to scattered sponge–microbial biotic associations, and major framework builders were absent for more than 20 million years.

Late Paleozoic reefs. The Carboniferous to Early Permian period was a lengthy icehouse (cold and glacial) interval of low CO_2 levels (see illustration). Large-scale reefs were absent in the Carboniferous, but there were patches of calcified algae, sponges, bryozoans, noncolonial invertebrates, and problematic organisms (that is, organisms that are difficult for investigators to classify), as well as deep-water mounds. The Early Permian displayed some corals, but the period was dominated by calcified algae, foraminifers, brachiopods, chaetetid sponges, bryozoans, and many problematic organisms. Many patches and reef mounds lacked much organic framework, perhaps as a result of the previous mass extinction that had decimated most of the larger reef-builders. This extinction also may have extinguished many photosymbiotic associations responsible for reef growth.

The Middle to Late Permian reefs contained a wide variety of organisms in many different associations.

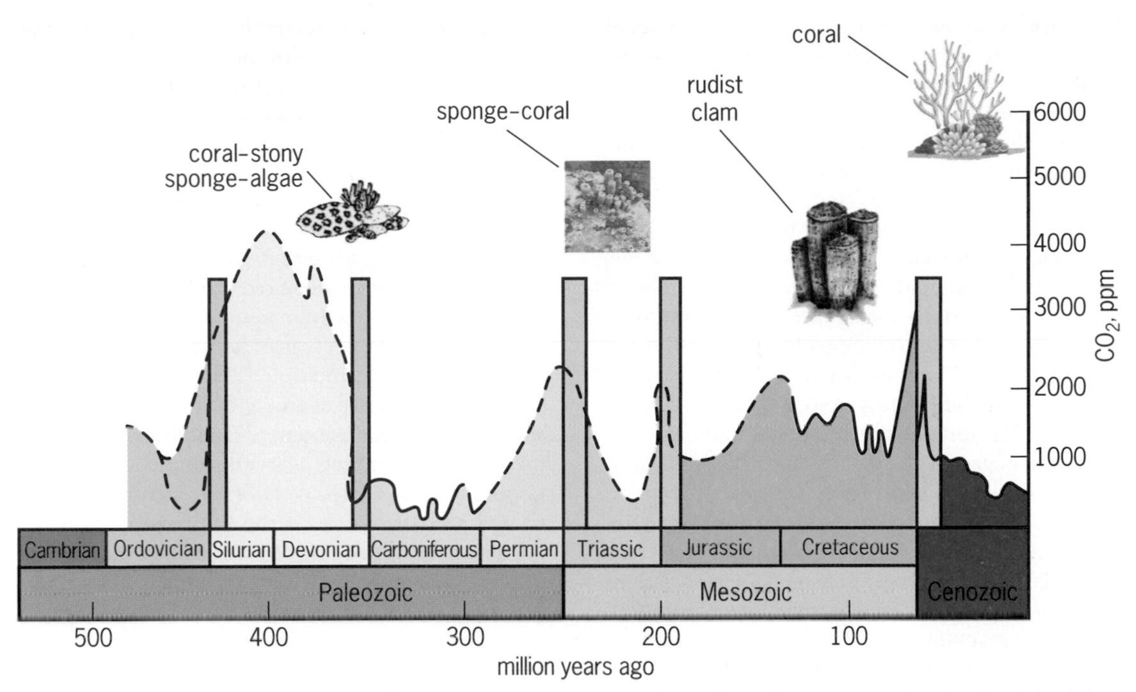

Reef ecosystems through geologic time. The top sketches show four of the major reef ecosystems after the Cambrian. The gray vertical bars are reef gaps immediately following the five largest mass extinctions. The lines show postulated levels of CO_2 in parts per million (ppm). The solid lines indicate better certainty, whereas the dashed lines indicate less certainty. Note that the Cenozoic Era can be divided into smaller epochs, including the Paleocene, Eocene, Oligocene, Miocene, Pliocene, Pleistocene, and Holocene (Anthropocene). (*Modified from J. E. N. Veron, A Reef in Time: The Great Barrier Reef from Beginning to End, Belknap Press, Cambridge, MA, 2008*)

However, large calcified fossils, such as those of the Devonian Period, were absent. Some Late Permian reefs included thickets of branching rugose extinct Paleozoic corals. More typically, though, there were chambered sponges, brachiopods, bryozoans, and calcified algae. The end-Permian mass extinction was devastating, extinguishing all corals and reefs. The Early Triassic postextinction time reveals vast oceans devoid of both reefs and much carbonate sediment.

Mesozoic reefs. The reef gap of the Early Triassic lasted 7–8 million years, and scientists have been intrigued why reefs took so long to recover. Discoveries in Nevada and other regions of the world have revealed small mounds of potentially reef-building organisms, such as sponges and algae, occurring just 1.5 million years after the extinction. However, protracted low oxygen conditions and a sea chemistry adverse to reefs and carbonate production may have held back potential reef-builders until these conditions ended in the Middle Triassic.

The first Mesozoic reefs in the Middle Triassic contain sponges, algae, and problematic taxa similar to those of the Permian. They also contain a new group of modern corals called scleractinians. These ancestors of all living reef-building species mysteriously appeared in different parts of the world. The sudden appearance of modern corals in the Middle Triassic Period has been explained by the "naked coral" hypothesis. This theory proposes that lineages of soft-bodied "anemones," with polyps similar to those of hard corals, inhabited Paleozoic seas, but were not preserved as fossils. Paleozoic corals and more than 90% of all life died out in the end-Permian mass extinction, but these soft anemones survived. In the Middle Triassic, the coral debut was likely a calcification event among soft-bodied forms prompted by responses to geochemical changes in the oceans.

During the Middle to Late Triassic, reefs and mounds were constructed by calcimicrobes, encrusting foraminifers, calcareous algae, chambered sponges, corals, bryozoans, bivalves, and diverse assemblages of encrusting taxa that defy classification into groups. A biotic changeover or minor extinction in the early part of the Late Triassic led to larger scale reefs (as seen in the later part of the Late Triassic) and the emergence of scleractinian corals as the primary builders of those reefs. These were the first major coral reefs since the Devonian Period. The Late Triassic reef expansion led to large reef complexes built by colonial corals, chambered sponges, calcareous algae, and other encrusting taxa. The latest Triassic "reef bloom" took place as corals emerged as the major builders of reefs, and reef belts expanded into higher latitudes. High levels of CO_2 are suggested (see illustration), indicating warm temperatures, but other ideas posit that these reefs prospered in a cold climate.

At the end of the Triassic Period, this coral-dominated reef ecosystem collapsed suddenly in a mass extinction, which was followed by another reef gap. Only a few coral and reef organisms survived into the Early Jurassic Period, and very few reefs have been discovered within this interval. Later, a recovery led to extraordinary diversity among corals, and impressive Middle to Late Jurassic reefs appeared. Jurassic reefs were paleogeographically and paleoecologically complex and varied in composition, being controlled by sea-level fluctuations, climate, sedimentation, nutrients, and water depth. Some important reef components were supplied by sponges, microbialites (calcium carbonate-forming microorganisms), and corals. Two major radiations and reef expansions took place at this time, leading to the maximum reef development in the Late Jurassic Period. The coral-dominated reefs also show good evidence of photosymbiosis.

During the Jurassic to Cretaceous interval, a reef transition began, with corals being replaced by large bivalves called rudistids. These reef-adapted bivalves increased in warm tropical settings during the Cretaceous Period and eventually replaced corals. Most rudistids lived in gregarious reeflike associations. Similar to corals, they grew upright (see illustration), with some reaching more than a meter in size. Their shells show adaptations to light, suggesting that they employed photosymbiosis. Corals remained diverse during the period of rudistid dominance, but they rarely produced reefs. CO_2 levels spiked at the end of the Cretaceous, leading to wider tropical zones under a super-greenhouse climate. Here, rudistids formed large reeflike bioconstructions. Although debates center on whether they produced anything comparable to coral reefs, these bivalves dominated until near the end of the Cretaceous, when a mass extinction extinguished all of them. Corals, conversely, survived the extinction, although their diversity was substantially reduced.

Cenozoic reefs. The CO_2 levels dropped and temperatures cooled in the Early Cenozoic Era. However, recovery ensued soon after, and photosymbiotic corals began forming new reeflike associations. Photosynthetic coralline (red) algae helped bind the reef framework together, and vast amounts of carbonate sediment resulted.

The Early Cenozoic was marked by drastic and sudden climatic fluctuations of warming–cooling cycles and changes in ocean circulation. Coral reefs experienced species turnover (a change in species composition) during the Early Cenozoic (Eocene–Oligocene) and then in the Oligocene-to-Miocene transition. This transitional period can be regarded as the time of modernization of coral reef organisms, and both reef expansions and extinctions took place. A major adaptive radiation of photosymbiotic corals and the reefs that they built took place in the Early to Middle Miocene, when the Mediterranean, Caribbean, and Indo-Pacific provinces (reef regions) were forming. The Neogene Period (Miocene and Pliocene) saw further expansions of modern reefs and corals, which appear to have been resistant to climatic changes and tectonic events (including the rising of the Isthmus of Panama). Corals and reefs were amazingly resolute even during the severe fluctuations of climate and sea level of the Pleistocene (the time of the most recent Ice Age). Their success may lie in the evolution of different genetic strains of photosymbionts with an adaptive tolerance to climate change. Some researchers are of the opinion that the masters of the reef must be the symbionts, and not the corals.

Future outlook. The resiliency of reefs is being put to the test by forces already underway on our planet. In our present-day epoch, appropriately named the Anthropocene (so-called because of the collective impact of human activities; also known as the Holocene), reefs may be pushed toward the brink of oblivion by multiple effects, including ocean acidification, the rise of greenhouse gases, and global warming. Bleaching and new coral diseases are also increasingly emerging as overfishing and pollution go unchecked. Collectively, these factors have led to dire predictions about the future of coral reefs. For example, the World Resources Institute has indicated that more than 90% of coral reefs will be adversely affected by 2030. In other scenarios, coral reefs are predicted to disappear by 2050. Unable to ever recover, reef species of the world may join other taxa of the "walking dead," all moving inextricably toward extinction. Therefore, ancient reefs may provide lessons from the past to help us understand the speed and biotic dynamics of the impending reef crisis. Perhaps, if we are not too late in acting, the Anthropocene can become the interval of recovery rather than death.

For background information *see* BIODIVERSITY; CARBON DIOXIDE; CLIMATE HISTORY; DEPOSITIONAL SYSTEMS AND ENVIRONMENTS; ECOLOGICAL COMMUNITIES; ECOLOGICAL MODELING; EXTINCTION (BIOLOGY); FOSSIL; GEOLOGIC TIME SCALE; MARINE ECOLOGY; PALEOCLIMATOLOGY; PALEOECOLOGY; REEF; SCLERACTINIA; SCLEROSPONGES; SEDIMENTOLOGY in the McGraw-Hill Encyclopedia of Science & Technology.

George D. Stanley, Jr.

Bibliography. R. Bradbury, World without coral reefs, p. A17, *New York Times*, July 14, 2012; P. Copper, 100 million years of reef prosperity and collapse: Ordovician to Devonian interval, *Paleontol. Soc. Pap.*, 17:15–32, 2011; N. Knowlton and J. B. C. Jackson, Shifting baselines, local impacts, and global change on coral reefs, *PLoS Biol.*, 6:215–220, 2008, DOI:10.1371/journal.pbio.0060054; C. Perrin, Tertiary: The emergence of modern reef ecosystems, pp. 587–621, in W. Kiessling, E. Flügel, and J. Golonka (eds.), *Phanerozoic Reef Patterns*, SEPM, Tulsa, OK, 2002; G. D. Stanley, Jr., Photosymbiosis and the evolution of modern coral reefs, *Science*, 312:857–860, 2006, DOI:10.1126/science.1123701; G. D. Stanley, Jr. (ed.), *The History and Sedimentology of Ancient Reef Systems*, Kluwer/Plenum, New York, 2001; G. D. Stanley, Jr. and J. H. Lipps, Photosymbiosis: The driving force for reef success and failure, *Paleontol. Soc. Pap.*, 17:33–60, 2011; J. E. N. Veron, *A Reef in Time: The Great Barrier Reef from Beginning to End*, Belknap Press, Cambridge, MA, 2008.

Graphene devices for quantum metrology

Graphene is a one-atom-thick layer of carbon atoms arranged in a hexagonal honeycomb crystal lattice. It was first isolated in 2004 by two scientists at Manchester University, Konstantin Novoselov and Andre Geim, by peeling it from a piece of graphite using a piece of sticky tape. This simple material has led to an unprecedented revolution in condensed-matter physics, owing to a myriad of unique properties and potentially exciting applications. The Manchester pair was awarded the Nobel Prize for Physics in 2010, only 6 years after their initial discovery.

Graphene is the first truly two-dimensional crystal to be discovered, and not too dissimilar from the previously discovered carbon allotropes, nanotubes (a rolled up sheet of graphene) and C_{60} fullerenes (60 carbon atoms arranged in sphere). Its conductivity is higher than that of silver and its strength is greater than that of steel. Yet the material is very flexible and transparent, absorbing only about 2% of light over a very broad spectral range. All these extreme properties imply that graphene could be used in many applications such as touchscreen displays, printable ink, ultrafast transistors, gas sensors, reinforcement of composites, and supercapacitors. The list is almost endless.

These unique properties find their origin in the unusual band structure of graphene, which is characterized by a linear relation between momentum and energy. For most materials, this relationship is quadratic, which results in a finite effective mass for the charge carriers. In stark contrast, the effective mass is zero in graphene, and the charge carriers behave more like massless photons. The charge carriers in graphene move not at the speed of light but about 300 times more slowly.

Quantum electrical metrology. One of the first phenomena discovered in graphene was the quantum Hall effect. The quantum Hall effect is one of the most fundamental phenomena in condensed-matter physics and occurs only in systems in which the motion of charge carriers is confined to two dimensions. The effect occurs at extremely low temperatures (close to absolute zero) and under the application of a strong magnetic field (typically 100,000 times stronger than the Earth's magnetic field). Under these extreme conditions, the resistance R_H of the material becomes independent of the material parameters and depends only on two fundamental constants of nature: the elementary charge, e, and the Planck constant, h (the precise relationship is $R_H = h/ie^2$, where i is an integer). This last fact is of particular interest to fundamental metrology, which is concerned with the realization of measurement units. Fundamental constants are ideal as a basis for defining a measurement unit because we believe that they are the same everywhere in the universe and invariant in time, which results in unprecedented precision and repeatability. For this reason, the SI unit for electrical resistance, the ohm (symbol Ω), has been based on the quantum Hall effect since 1990. At present, a large international effort is ongoing to base all seven base units of our measurement system (the SI) on fundamental constants of nature.

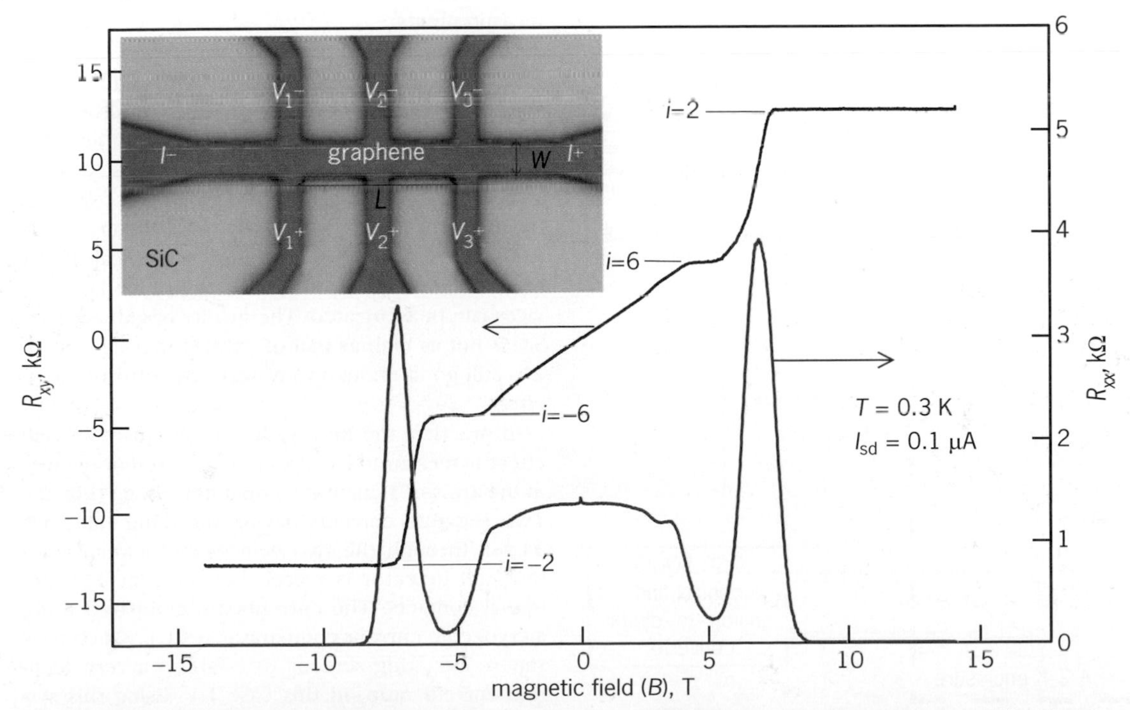

Fig. 1. Quantum Hall effect measurement for a graphene-on-silicon carbide device (inset with $L = 180\ \mu m$ and $W = 35\ \mu m$). A magnetic field B is applied perpendicular to the plane of the graphene device and a source-drain current, I_{sd}, is passed from I^+ to I^-. The voltage is measured between contacts V_2^+ and V_2^- for the Hall voltage, V_H, and V_1^+ and V_3^+ for the longitudinal voltage. $R_H = R_{xy} = (V_2^+ - V_2^-)/I_{sd}$ and $R_{xx} = (V_1^+ - V_3^+/I_{sd}$. The temperature is 0.3 K. The integers, i, indicate the expected quantized values for $R_H = h/ie^2$.

Until the discovery of graphene, the quantum Hall effect had been observed only in a few semiconductor systems, in particular, gallium arsenide (GaAs) heterostructures and silicon metal-oxide semiconductor field-effect transistors (MOSFETs). Graphene is of particular interest to metrology for two reasons. First, compared to semiconductor systems, the energy quantization in graphene is very strong (about five times stronger for the same magnetic field). This implies that in graphene the quantum Hall effect can be observed at much lower magnetic fields and significantly higher temperatures than in semiconductor systems, which makes the experimental apparatus much simpler and cheaper. Second, the band structure of graphene is fundamentally different from that of semiconductors (that is, the linear versus quadratic energy momentum dispersion), and this provides the opportunity to perform a stringent experimental universality test on the quantum Hall effect. In a universality test, the quantum Hall resistance in one material is compared to that measured in another material to the highest possible accuracy. This demonstrates the material independence of $R_H = h/ie^2$, something that is very difficult to prove theoretically. Demonstrating universality of the quantum Hall effect is important because the relationship $R_H = h/ie^2$ also enters in the new proposed realization of another important SI base unit, the kilogram.

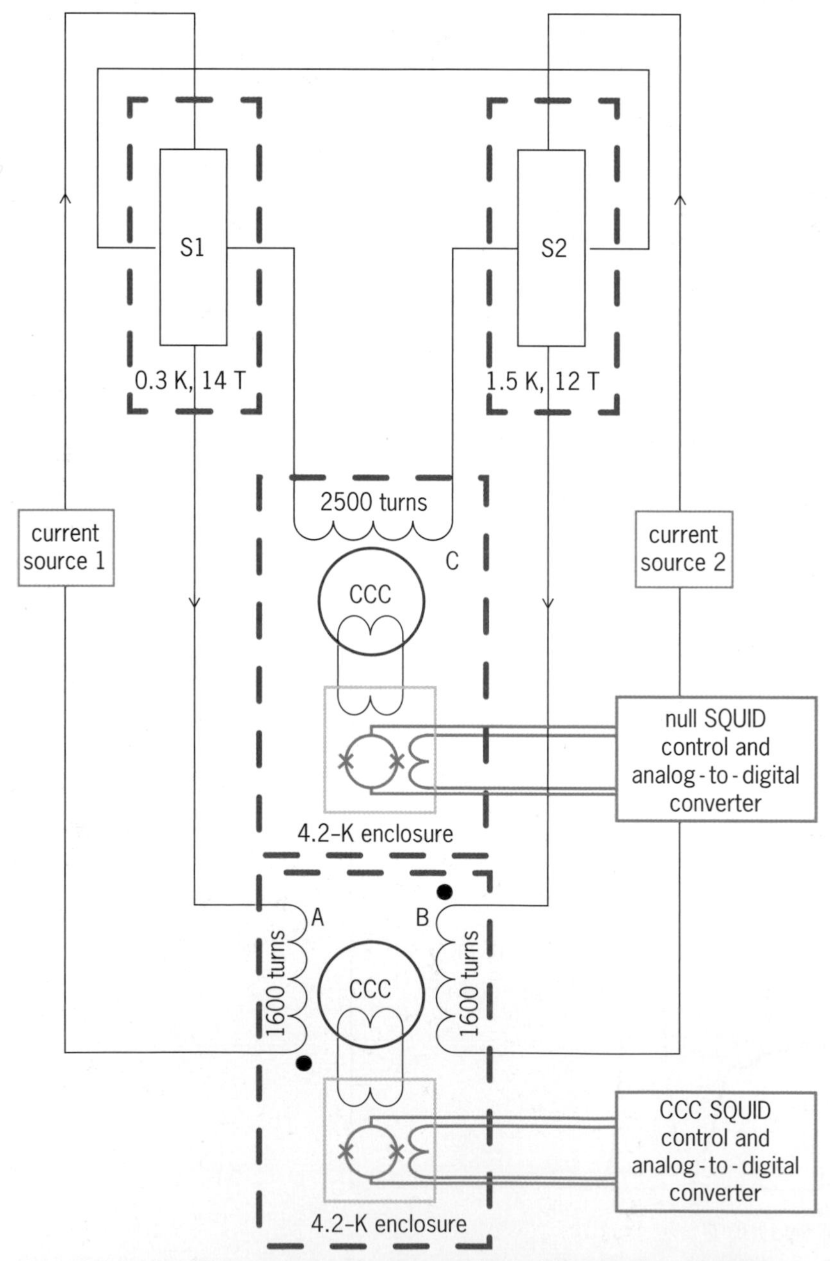

Fig. 2. Schematic of the measurement system used for the high-precision universality test of the quantum Hall effect. S1 is the graphene-on-silicon carbide device, S2 is a GaAs-GaAlAs heterostructure. The lower cryogenic current comparator (CCC) established the accurate current ratio between the two sides of the bridge. The top CCC is configured as a sensitive null detector.

Application of graphene in metrology. To make a high-precision measurement, a large signal-to-noise ratio in the experimental system is required. In a quantum Hall effect experiment, one measures the transverse voltage, V_H, generated when a bias current, I_{sd}, is passed from source to drain (**Fig. 1**), and R_H is defined as V_H/I_{sd}. Consequently, a larger bias current results in a larger signal-to-noise ratio. At some critical source-drain current, the quantum Hall effect disappears; for graphene, this current turns out to be extremely large, about 0.5 mA, which corresponds to a current density of 50 A/m in a material that is only one atom thick. (For normal conductors, the current density is in A/m^2; in a 2D material, it is in A/m.) For semiconductor systems, the equivalent maximum current density is almost an order of magnitude smaller. Other properties, such as charge carrier density, mobility, and electrical contact resistance, are also important for high-precision measurements.

Graphene can be produced via various methods. The original mechanical exfoliation technique results in very pure crystals with ideal properties. Unfortunately, these crystals are very small, and this severely limits the maximum currents these devices can carry. Graphene can also be grown, for example, by high-temperature sublimation of a silicon carbide (SiC) crystal. Using this technique, wafer-scale layers of graphene can be produced from which large devices can be fabricated. The quality of graphene on SiC is not as high as that of exfoliated material, but it is still good enough to observe the quantum Hall effect.

In practice, the universality of the quantum Hall effect is measured by placing two different devices in the arms of a current comparator bridge (**Fig. 2**). Two separate currents having the same value are passed through the two devices and a very sensitive null detector is placed between the Hall potential contacts. The currents are compared using a cryogenic current comparator (CCC), which uses superconducting screens to establish a very accurate current ratio, in this case 1:1. Using this system, the relative difference in resistance between the graphene and the GaAs device was shown to be smaller than 8.7×10^{-11}. This is a significant improvement on previous estimates and is actually limited by the maximum source-drain current that the semiconductor device can sustain. The result has demonstrated that the quantum Hall effect is a truly universal property of 2D systems.

Single-electron transport. A new definition is being proposed for the standard of electrical current, the ampere, based on the elementary charge, *e*. This definition can be realized by passing an exactly known number of electrons through a circuit such that the current $I = nef$, where n is the number of electrons transported per cycle of the drive frequency f. The electron charge is very small, and therefore the challenge is to pass or pump as many electrons as possible to generate a current that can be measured accurately (a 1-GHz frequency results in only ~160 pA of current). A number of methods are being developed to achieve this standard. One of these is a so-called double-dot pump in which two small islands are connected via tunnel barriers (**Fig. 3**). The device relies on the fact that, if one electron is present on an island, one has to overcome a significant amount of energy to add a second electron to the same island. This charging energy depends inversely on the capacitance of the island, getting larger for smaller islands. The tunnel barriers separate the islands, preventing the electrons from freely flowing away. The tunnel times of these barriers set the maximum speed at which the pump can be operated and depend on the resistance and capacitance of the barriers. A smaller resistance makes the barriers more transparent and the pump can operate at higher frequencies, but this also results in more pump errors. So far, most electron pumps have been made from metals with thin oxide layers as tunnel barriers with fairly high resistance and capacitance, resulting in low operating frequencies. Graphene offers some interesting advantages because of its very small dimensions and tunable barrier resistance. The first graphene single-electron pumps have been demonstrated, showing accurate operation up to gigahertz frequencies.

Recently, first reports have been published on the observation of the Josephson effect in graphene. Here, a voltage is generated that is proportional to a microwave drive frequency and the ratio of h/e, such that $V = n(h/2e)f$. This could realize the third electrical quantum standard in graphene and at the same time give the tantalizing prospect of directly comparing the three quantum standards in one device against each other via Ohm's law, $R = V/I$. This is known as the quantum electrical metrology triangle and allows us to test the consistency of the fundamental constants. Such an experiment is not yet in reach because the best accuracies for single-electron transport and the Josephson effect in graphene are worse than the target uncertainty of 1 part in 10^8.

It is remarkable that this relatively new material, only a decade after its discovery, has had such an immediate impact in electrical metrology. This will almost certainly not be the end of the application of graphene to metrology, because many more properties have yet to be fully explored.

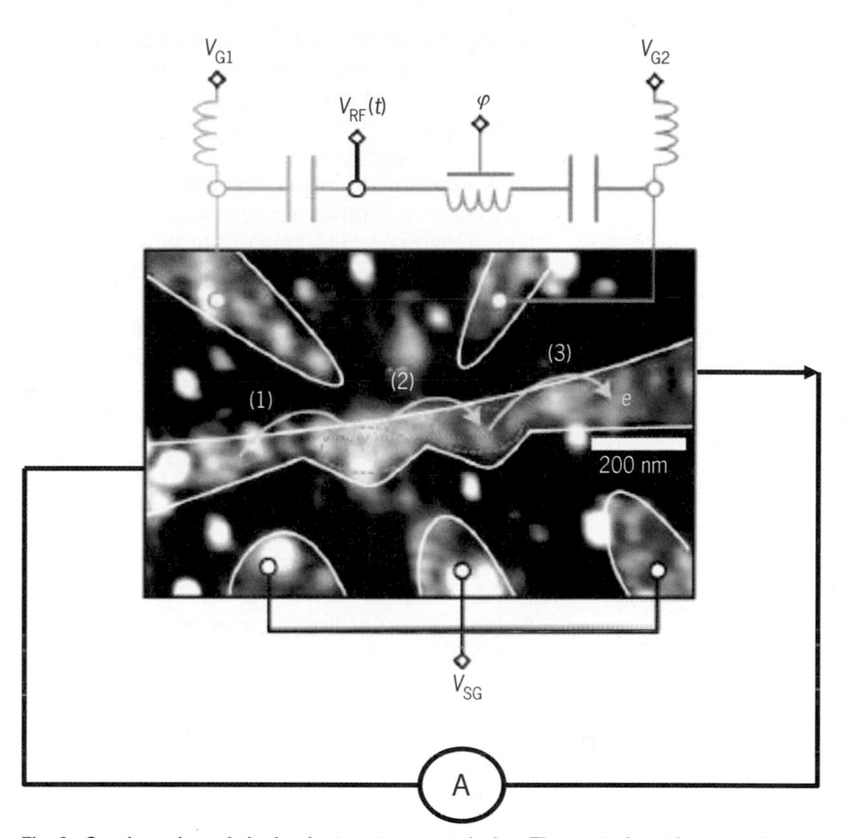

Fig. 3. Graphene-based single-electron transport device. The central part is a scanning electron micrograph of the device showing two quantum dots in series, marked out by green and red (white contour lines are added for clarity). Electrons can be transported from left to right via a series of jumps (1–3) across tunnel barriers by applying two phase-shifted periodic signals to finger gates G1 and G2. V_{SG} is used as a side gate to control the charge carrier density in the graphene.

For background information *see* CARBON NANOTUBES; ELECTRICAL UNITS AND STANDARDS; FULLERENE; FUNDAMENTAL CONSTANTS; GRAPHENE; HALL EFFECT; JOSEPHSON EFFECT; PHYSICAL MEASUREMENT; PLANCK'S CONSTANT; SEMICONDUCTOR HETEROSTRUCTURES; TRANSISTOR; TUNNELING IN SOLIDS in the McGraw-Hill Encyclopedia of Science & Technology.

Jan-Theodoor Janssen

Bibliography. A. H. Castro Neto et al., The electronic properties of graphene, *Rev. Mod. Phys.*, 81:109–162, 2009, DOI:10.1103/RevModPhys.81.109; A. K. Geim and K. S. Novoselov, The rise of graphene, *Nat. Mater.*, 6:183–191, 2007, DOI:10.1038/nmat1849; T. J. B. M. Janssen et al., Quantum resistance metrology using graphene, *Rep. Progr. Phys.*, 76:104501 (48 pp.), 2013, DOI:10.1088/0034-4885/76/10/104501; I. M. Mills et al., Redefinition of the kilogram, ampere, kelvin and mole: A proposed approach to implementing CIPM recommendation 1 (CI-2005), *Metrologia*, 43:227–246, 2006, DOI:10.1088/0026-1394/43/3/006.

Gravity Probe B: final results

Gravity Probe B (*GP-B*), launched on April 20, 2004, was a space-based, cryogenic physics experiment testing two predictions of Albert Einstein's geometric theory of gravity, general relativity. In 1960,

Leonard I. Schiff showed that an ideal gyroscope in orbit about the Earth would undergo two very slow general relativistic spin-axis precessions: (1) a geodetic drift in the orbit plane, caused by the massive Earth distorting space-time, and (2) a frame-dragging drift, caused by the rotation of the Earth dragging space-time. By contrast, the spin-axis orientation of an ideal gyroscope in a Newtonian universe would not drift, but rather, would remain fixed. The GP-B experiment was designed to measure gyroscope drift and to isolate it from the effect of imperfections in the implementation of the sensors.

Experimental design. **Figure 1** shows the experiment configuration. The spin-axis orientation was measured relative to the direction to a distant "guide star," which provided a nearly fixed reference. For the experiment's polar orbit, the two relativistic effects are perpendicular to each other, and a single gyroscope measures both effects. The predicted geodetic effect in *GP-B*'s 642-km-high (399-mi) orbit was −6606.1 milliarcseconds (mas)/yr, and the predicted frame-dragging effect was −39.2 mas/yr. [Here, 1 mas = 4.848×10^{-9} rad, or approximately the angle subtended by a human hair when viewed from a distance of 16 km (10 mi).] In practice, four gyroscopes provided redundancy and cross-checks of the science result.

Each gyroscope incorporated a spinning rotor that was electrostatically suspended. The rotor was a sphere of fused quartz, 38 mm (1.5 in.) in diameter, coated with a superconducting niobium thin film. It was placed within two housing halves containing three sets of electrodes for rotor levitation, a channel for gas to spin up each rotor, and a pickup loop for spin-axis measurement (**Fig. 2**). To approach ideal performance, the effect of disturbing torques had to be nearly eliminated by (1) reducing the acceleration level on the vehicle from the typical 10^{-8} to 10^{-11} g (where g is the acceleration of gravity at the Earth's surface), (2) reducing magnetic fields to 10^{-11} tesla (T), (3) reducing gas pressure to 10^{-12} pascal (Pa), (4) limiting electric charge to 10^{-8} coulomb (C), and (5) rolling the spacecraft about the line of sight to the star.

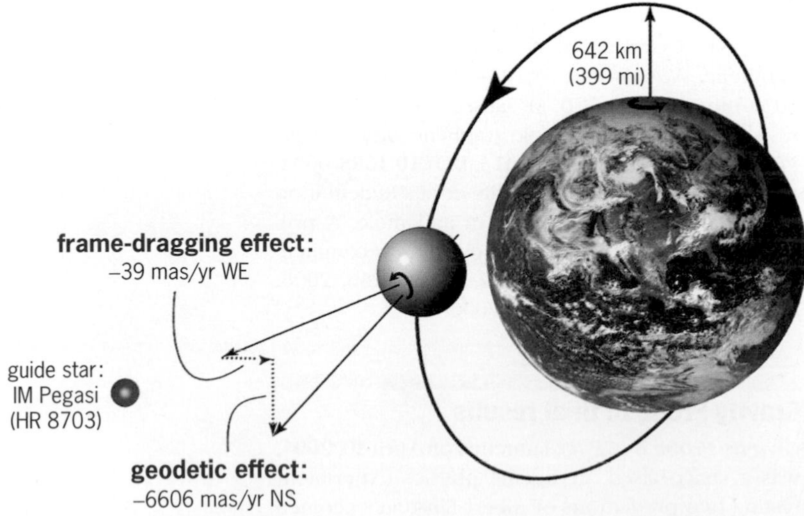

Fig. 1. Precessions of *Gravity Probe B* gyroscopes predicted by general relativity.

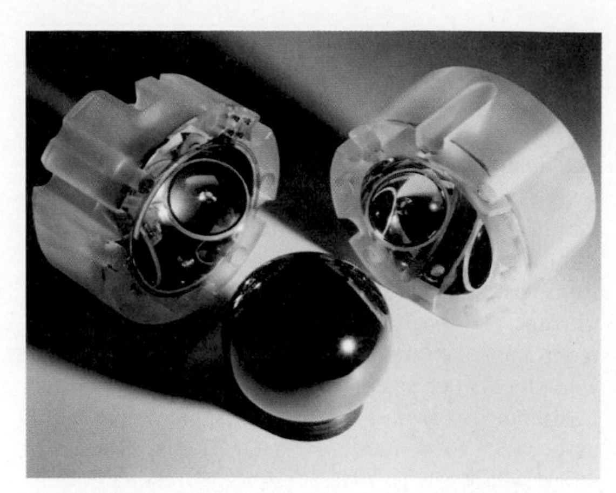

Fig. 2. *GP-B* gyroscope with rotor and housing halves.

The rotor's superconducting coating provided a magnetic pointer for the measurement of spin-axis orientation. This magnetic marker, the London moment, arises in rotating superconductors. The orientation of the London moment is intrinsically aligned with the spin axis so that changes in spin-axis direction were reflected in the London moment signal, allowing measurement by a superconducting quantum interference device (SQUID).

The gyroscope-angle-to-SQUID-output-voltage ratio is called the scale factor and was determined to high precision using the known pointing variation of the *GP-B* satellite caused by starlight aberration. As the satellite orbited around the Earth and the Earth orbited around the Sun, starlight aberration caused the apparent direction to the guide star to change in a known way. The scale factor was determined by comparing the resulting effect upon the SQUID signal to the change in the guide star's apparent position as inferred by onboard GPS and astronomical measurement.

To provide superconductive gyroscope and SQUID operation, the system was passively cooled to a temperature of ~2 K using superfluid helium. A 2400-L (634-gal) liquid helium storage dewar cooled an evacuated inner structure containing the gyroscopes and SQUIDs.

In orbit, the hold time of the helium defined the experiment's duration. Operations were divided into three phases: 4.2 months for experiment setup, 11.5 months for science data collection, and 1.6 months for calibrations designed to place limits on systematic effects. The experiment setup included spinning up the four gyroscopes to rates ranging from 61.82 Hz (gyro 2) to 82.09 Hz (gyro 3), readout initialization, and locking on to the guide star. During science data collection, spacecraft activity was minimized to limit gyroscope disturbances.

Figure 3 shows the spin-axis histories prior to calibration. The geodetic drift is visible in the slope of orientation versus time. Without relativity, there would be zero slope.

The raw orientation histories deviate from the straight lines predicted by general relativity. To understand the source of these deviations, in-orbit calibration tests were performed, leading to a mathematical description of the underlying effects and, with it, an improved relativity assessment.

Three perturbing effects. Three important perturbing effects, differing in detail for each gyroscope, were identified in the raw histories: an evolving scale factor and two electrostatic torques. The evolving scale factor affected the spin-axis measurement; the torques affected the spin-axis orientation.

Evolving scale factor. Before launch, it was understood that a small magnetic field would remain frozen on the gyroscope rotor as it cooled below its superconducting transition temperature, and that this "trapped flux" would contribute to the readout scale factor. When a gyroscope spins, its rotation axis moves slowly along a closed-loop path through the gyroscope body, even though the spin-axis orientation remains fixed when viewed from the outside. Without energy dissipation, this "polhode" motion would be defined by a repeating path in the body, giving rise to periodic scale-factor variation. Instead, in orbit, a very small amount of dissipation (less than 1 pW) caused the polhode to slowly dampen, resulting in an evolving contribution of the trapped flux to the scale factor.

The evolving scale factor is determined along with the trapped flux distribution and polhode for each rotor by considering a general representation for the flux distribution. The spinning gyroscope produces a signal in the pickup loop at the rotation frequency. This signal evolves over time as the rotor's polhode continuously modifies the orientation of the trapped flux relative to the pickup loop. A loss-free description of the dynamics of the rotor can be extended to include damped motion by use of an additional term. Fitting the evolving spin signal to this extended description gives the evolving scale factor, and, with it, accurate spin-axis orientation profiles.

Spurious torques. The two torques resulted from the interaction of electrically charged patches on the rotor and housing. Prelaunch studies indicated that interactions between the rotor, with the intrinsic patch size governed by near-microscopic surface morphology, and the housing support voltages would not cause significant torques. Surprisingly, after launch, large-scale intrinsic patches were found on the rotor and the housing.

A general representation of the torques is found by considering the electrostatic energy stored in the gap between the rotor and the housing arising from arbitrary patch distributions. To determine the specific influence of each torque on the changing gyroscope orientation, detailed knowledge of the gyroscope's changing polhode is required. Fortunately, with trapped flux and charged patches both physically fixed in the gyroscope body, this information is known from the scale-factor analysis described previously.

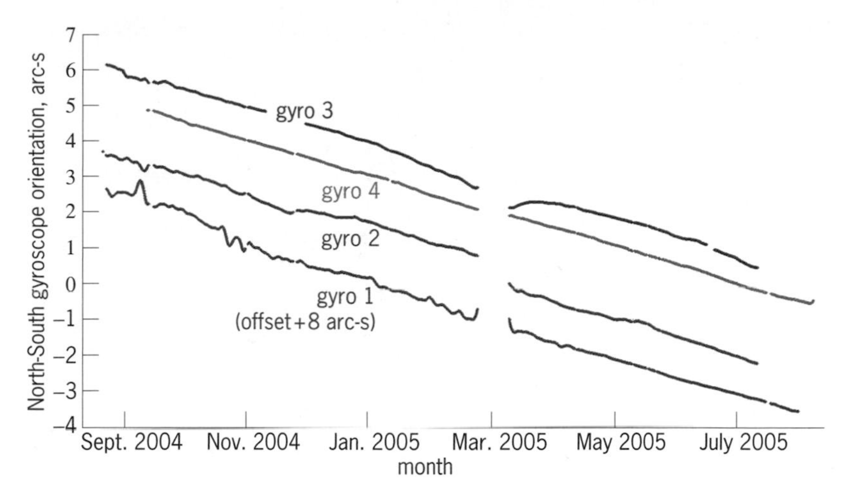

Fig. 3. Raw north-south gyro orientation histories. Data gaps, most apparent in March 2005, were caused by computer reboots and other spacecraft anomalies.

Final results. To determine the relativistic drift rates, the orientation time histories for each of the four gyroscopes are set equal to the sum of three terms: relativity, modeled as a straight line, and the two torque terms. A least-squares fitting analysis gives independent relativistic drift rates and torque estimates for each gyroscope. The **table** lists the four relativistic drift rate estimates and the final joint result with 1σ (1 standard deviation) uncertainties. These results are also plotted as 95% confidence ellipses in **Fig. 4**. The values are corrected for the small proper motion of the guide star relative to even more distant extragalactic objects.

The quoted drift rates for each gyroscope are the weighted means of 10 analyses, each based upon evaluation of the full yearlong data record. In the analyses, the number of parameters used to represent the scale factor and the two torques is varied, leading to small drift-rate differences, which are then averaged to give the final results.

The 1σ uncertainties in the table and the 95% confidence ellipses in Fig. 4 are defined by the root-mean-square sums of statistical uncertainty, parameter sensitivity, and unmodeled uncertainty. Statistical uncertainty is provided by the fitting analyses. Parameter sensitivity is given by the scatter in drift rates, as found in the previous paragraph. Uncertainty due to unmodeled systematic effects was limited by separate considerations to less than 1 mas/yr. For example, drift due to the mass unbalance of the

Final experiment results

Source	r_{NS}, mas/yr	r_{WE}, mas/yr
Gyroscope 1	-6588.6 ± 31.7	-41.3 ± 24.6
Gyroscope 2	-6707.0 ± 64.1	-16.1 ± 29.7
Gyroscope 3	-6610.5 ± 43.2	-25.0 ± 12.1
Gyroscope 4	-6588.7 ± 33.2	-49.3 ± 11.4
Joint	-6601.8 ± 18.3	-37.2 ± 7.2
General relativity prediction	*−6606.1*	*−39.2*

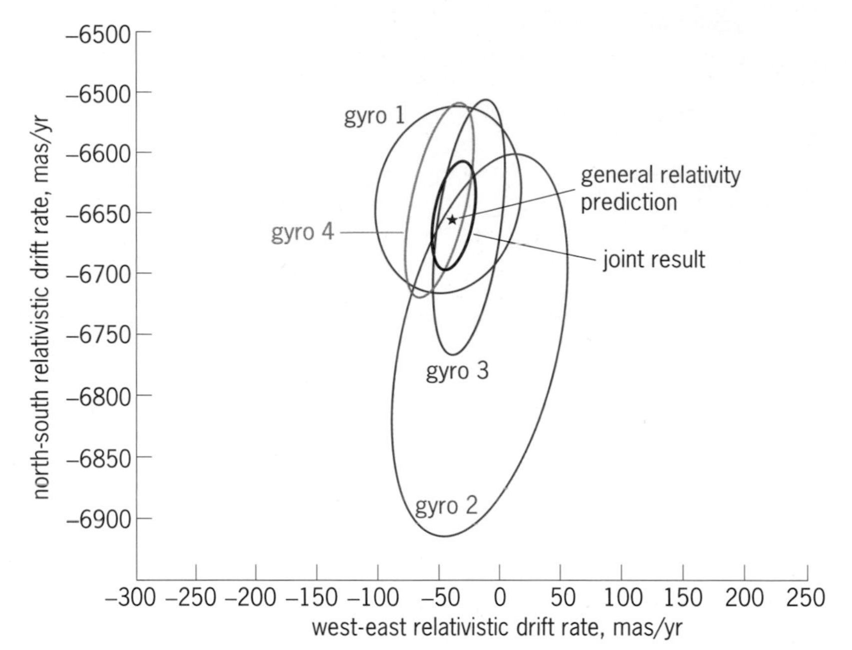

Fig. 4. Final experiment results: North-south and west-east relativistic drift rate estimates (95% confidence) for the four individual gyroscopes (colored ellipses) and the joint result (black ellipse).

rotor was limited to less than 0.1 mas/yr by in-orbit measurement of the mass unbalance and the support force.

The relativity results are cross-checked in various ways. Data subsections give consistent results, orbit-by-orbit gyroscope orientations give the exotic Cornu spirals predicted for one of the torques, and a 21 ± 7-mas estimate of the gravitational deflection of light by the Sun is consistent with the 21.7-mas value predicted by general relativity. Two separate data analysis methods give a final cross-check.

Significance. General relativity gives a remarkable geometric view of gravity. Tests of general relativity about the Earth are challenging because Newtonian physics provides a very good approximation for this "weak field" environment. Einstein, in his book *The Meaning of Relativity*, published in 1953, wrote, "these effects, which are to be expected in accordance with Mach's ideas, are actually present according to our theory, although their magnitude is so small that confirmation of them by laboratory experiments is not to be thought of." Seven years after he wrote these words, an experiment was proposed. Two years later, work began on *Gravity Probe B*, ultimately providing four independent measurements of the geodetic and frame-dragging effects, with net accuracies of 0.28 and 19%, respectively.

For background information *see* ABERRATION OF LIGHT; GYROSCOPE; LEAST-SQUARES METHOD; MAGNETISM; RELATIVITY; RIGID-BODY DYNAMICS; ROOT-MEAN-SQUARE; SQUID; STATISTICS; SUPERCONDUCTING DEVICES; SUPERCONDUCTIVITY; TORQUE in the McGraw-Hill Encyclopedia of Science & Technology.

Barry Muhlfelder

Bibliography. C. W. F. Everitt et al., Gravity Probe B: Final results of a space experiment to test general relativity, *Phys. Rev. Lett.*, 106:221101 (5 pp.), 2011, DOI:10.1103/PhysRevLett.106.221101; G. M. Keiser et al., Gravity Probe B, *Riv. Nuovo Cimento*, 32(11):555–589, 2009, DOI:10.1393/ncr/2009-10049-y; L. I. Schiff, Possible new experimental test of general relativity theory, *Phys. Rev. Lett.*, 4:215–217, 1960, DOI:10.1103/PhysRevLett.4.215.

Gravity Recovery and Interior Laboratory (GRAIL) mission

Measuring the gravitational fields of the planets is a very powerful tool for remotely investigating the structure of their interiors. Planetary gravity is typically measured via radio tracking of spacecraft near their targets, either orbiting or performing a fly-by, and sensing the pull from the mass of the planet. When a spacecraft transmits a radio signal to Earth, the Doppler effect causes a shift in the frequency of the signal received at Earth that is proportional to the motion of the spacecraft. Perturbation to the trajectory by the gravitational forces causes the detectable motion. This "line-of-sight tracking" method requires that the spacecraft be in view of the Earth, so this method cannot be employed at the Moon. The Moon is synchronously locked in its orbit around Earth and shows us only its near side, with no possible Doppler tracking from the far side. Having two spacecraft, instead of one, solves this challenge. They fly in formation and track each other as first demonstrated in Earth orbit by the Gravity Recovery and Climate Experiment (GRACE) mission using twin orbiters. NASA's Jet Propulsion Laboratory provided a version of the GRACE payload to fly on the Gravity Recovery and Interior Laboratory (GRAIL) lunar mission. The gravity instrument carried on each GRAIL spacecraft conducts dual one-way ranging measurements, which are used to develop the lunar gravity field map. This required precisely measuring how much and how fast the separation distance between the two GRAIL spacecraft changes. The measurable distance is about the size of a blood cell, and the velocity is clocked to about a millionth of a meter per second.

History of the mission. The GRAIL project is part of NASA's Discovery Program, in which NASA competitively selects Principal-Investigator-managed projects for development and operations within a defined project cost cap. The selection of GRAIL was announced on December 10, 2007, after an earlier selection of three proposed missions from 25 submitted concepts. As proposed, the GRAIL flight system consisted of twin spacecraft manufactured by Lockheed Martin. The spacecraft were derived from a classified program, the Experimental Small Satellite 11 (XSS-11); as implemented, the avionics were derived from the *Mars Reconnaissance Orbiter*. GRAIL was proposed with six science investigations; the first four, the science floor that is mandatory for mission success, were successfully completed in May 2012, and investigations

5 and 6, desired for full mission success, were completed as planned in June 2013. In addition, GRAIL executed an unprecedented education and public outreach program, in which participating middle-school students across the nation were able to target the onboard cameras to image features of the lunar surface. This program, managed by Sally Ride Science, has involved 3116 student operations centers at 2853 schools. GRAIL took more than 100,000 images for this purpose. The twin GRAIL spacecraft were named *Ebb* and *Flow* by a competition among students, with the winning entry coming from fourth-graders at the Emily Dickinson Elementary School in Bozeman, Montana.

After launch together on September 10, 2011, from Cape Canaveral in Florida, the twin orbiters embarked on independent trajectories on a 14-week-long cruise to the Moon via the EL-1 Lagrange point in order to save on the needed energy. *Ebb* was inserted into polar orbit on December 31, 2011, and *Flow* on January 1, 2012. After a succession of a record high 19 maneuvers, they lined up to collect the science data on March 1, 2012, at an average height above the lunar surface of 55 km (34 mi). Gravity field data were acquired globally for approximately 3 months; then the primary mission was completed with 99.99% of available science data acquired. Another 3-month period followed where the spacecraft did not collect science data due to the Sun-spacecraft-Earth geometry; then an extended science mission started for 3 more months, this time at an average altitude of 23 km (14 mi). In the final mapping or "endgame" phase, the spacecraft were lowered to an average altitude of 11 km (7 mi). This was quite close to the surface, as the spacecraft orbited only a couple of kilometers above the highest topography. Such low-altitude mapping resulted in greatly increased sensitivity to measuring the gravity due to the "one over *r* squared" effect. During the extended mission, over 99.99% of available science data were also collected. Subsequent to the endgame, with the fuel running out, the two spacecraft were deliberately impacted in a controlled fashion into the lunar surface on December 17, 2012. The previously unnamed GRAIL impact site was named after Sally K. Ride, America's first woman in space and GRAIL's Education and Public Outreach lead.

Results. Due to the success of the GRAIL mission, scientists have been able to construct the gravitational field of the Moon to unprecedented accuracy for any known body in the universe (**Fig. 1**). This field reveals large mass concentrations, called mascons, associated with major impact basins. Discovered during the early stage of lunar exploration, mascons perturbed early satellites from their orbits as they mapped the lunar surface to identify landing sites for the Apollo astronauts. But the new map shows much more than those early observations. Scientists use a mathematical formulation called spherical harmonic coefficients and expect the GRAIL results to exceed a record 1000 degree and order field globally, a much finer resolution than for other terrestrial planets, including Earth. The resolution of the Moon's current gravity field is approximately 5 km (3 mi), which reveals many features not previously resolved, including tectonic structures, volcanic landforms, basin rings, crater central peaks, and numerous simple bowl-shaped craters, the most abundant landform on the Moon's surface. It was discovered that, for the kilometer to hundred kilometer spatial scales, the vast majority (98%) of

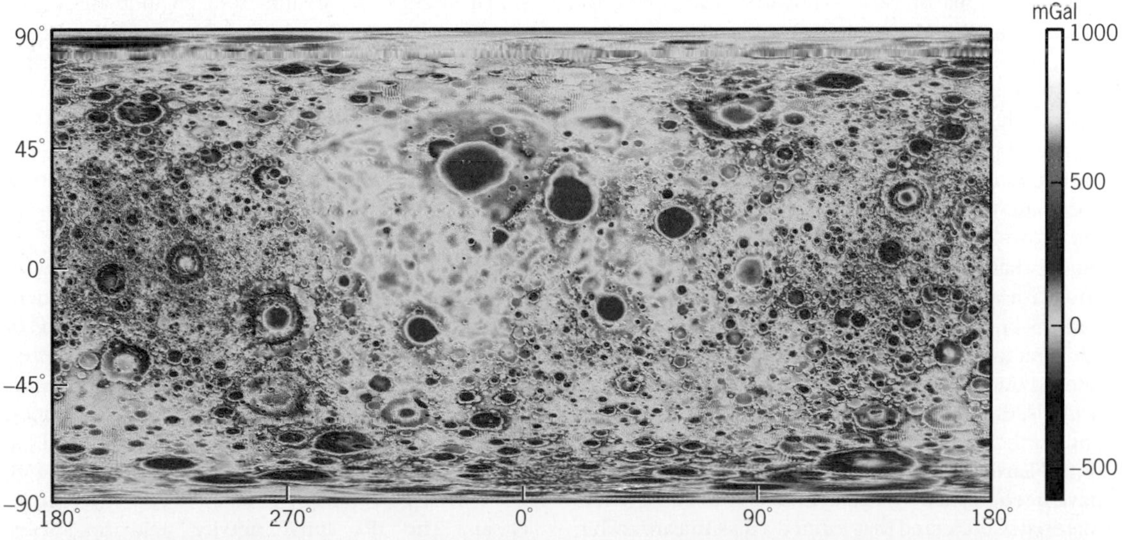

Fig. 1. Gravity field of the Moon from the Gravity Recovery and Interior Laboratory (GRAIL) mission. This represents how the gravity at the surface of the Moon varies from one location to another due to the surface and subsurface properties. The map is in cylindrical projection, with the near side of the Moon in the center and the far side on either side. The map is to spherical harmonic degree and order 900, and resolves blocks of 6 km (4 mi) on the lunar surface. Gravity is in units of milligalileos (mGal), where 1 galileo = 1 cm/s^2. Red corresponds to mass excesses relative to an average, and blue corresponds to mass deficiencies. The large red circular features on the near side are lunar mascons, large positive gravitational anomalies associated with impact basins.

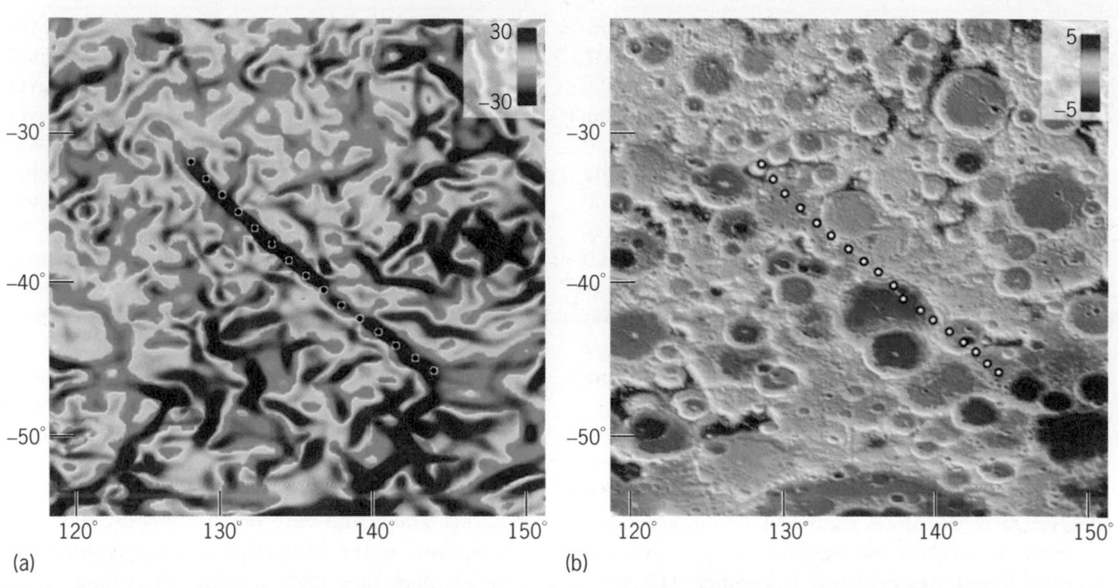

Fig. 2. Comparison of (*a*) horizontal gradient of Bouguer gravity (gravity with attraction of surface topography removed) and (*b*) surface topography from the *Lunar Orbiter Laser Altimeter*. Dots show position of a buried dike, visible in gravity but not topography. Horizontal gradient of Bouguer gravity is in Eötvös units, where 1 Eötvös unit = 10^{-9} galileo/cm. Surface topography is in kilometers. (*From J. C. Andrews-Hanna et al., Ancient igneous intrusions and the early expansion of the Moon revealed by GRAIL gravity gradiometry, Science, 339:675–678, 2013, DOI:10.1126/science.1231753*)

the small-scale gravitational signature is associated with topography, a result that reflects the preservation of crater relief in a highly fractured crust. The remaining 2% represents fine details of structure beneath the surface not previously known.

Results to date also show that the bulk density of the Moon's highlands crust is 2550 kg/m^3 (2.550 g/cm^3), substantially lower than previously assumed. When combined with remote sensing and sample data, this density implies an average crustal porosity of 12% to depths of at least a few kilometers. Knowing the porosity, or amount of empty space in the material, is important for planets where liquids—such as magma or water—flowed during geologic history. On the Moon, there is now evidence that some major faults penetrated as deep as this, providing conduits for magma formed at depth to reach the surface. It was also observed that lateral variations in crustal porosity correlate with the largest impact basins, whereas lateral variations in crustal density correlate with crustal composition. The density finding allows construction of a global crustal thickness model that satisfies the Apollo seismic constraints, and with an average crustal thickness between 34 and 43 km (21 and 27 mi), the bulk composition of refractory elements (that is, elements that condensed out of the solar nebula at high temperatures, such as aluminum) is about the same for the Moon and Earth. This result has implications for the origin of the Earth-Moon system. The Moon is believed to have been formed by a giant impact into Earth when material was ejected that formed a disk in Earth orbit that accumulated into the Moon. From the similarity of refractory elements we now know that the impactor mixed very well with the Earth.

The earliest history of the Moon is poorly preserved in the surface geologic record because of numerous impacts from other objects, but aspects of that history may be preserved in subsurface structures. GRAIL gravity revealed a population of long linear gravity anomalies, with lengths of hundreds of kilometers (**Fig. 2**). These are interpreted to be ancient vertical tabular intrusions, called dikes, formed by magma flow in combination with extension of the lithosphere (the Moon's rigid outer shell). Cross-cutting relationships support a pre-Nectarian to Nectarianage (greater than about 4×10^9 years), preceding the end of the heavy bombardment of the Moon. The distribution, orientation, and dimensions of the intrusions indicate a globally uniform extensional stress state arising from an increase in the Moon's radius by 0.6–4.9 km (0.4–3.0 mi) early in lunar history, consistent with predictions of thermal models.

For background information, *see* COSMOCHEMISTRY; DOPPLER EFFECT; EARTH, GRAVITY FIELD OF; GRAVITATION; MOON; PLANET; SOLAR SYSTEM; SPACE PROBE; SPHERICAL HARMONICS in the McGraw-Hill Encyclopedia of Science & Technology.

Sami W. Asmar; Maria T. Zuber

Bibliography. J. C. Andrews-Hanna et al., Ancient igneous intrusions and the early expansion of the Moon revealed by GRAIL gravity gradiometry, *Science*, 339:675–678, 2013, DOI:10.1126/science.1231753; A. S. Konopliv et al., JPL high-resolution lunar gravity fields from the GRAIL Primary and Extended Missions, *Geophys. Res. Lett.*, 41:1452–1458, 2014, DOI:10.1002/2013GL059066; A. S. Konopliv et al., The JPL lunar gravity field to spherical harmonic degree 660 from the GRAIL primary mission, *J. Geophys. Res.*, 118:1415–1434, 2013, DOI:10.1002/jgre.20097; F. G. Lemoine et al., GRGM900C: A degree-900 lunar gravity model from GRAIL primary and extended mission data, *Geophys.*

Res. Lett., 41:3382–3389, 2014, DOI:10.1002/2014GL060027; F. G. Lemoine et al., High-degree gravity models from GRAIL primary mission data, *J. Geophys. Res.*, 118:1676–1698, 2013, DOI:10.1002/jgre.20118; M. A. Wieczorek et al., The crust of the Moon as seen by GRAIL, *Science*, 339:671–675, 2013, DOI:10.1126/science.1231530; M. T. Zuber et al., Gravity field of the Moon from the Gravity Recovery and Interior Laboratory (GRAIL) mission, *Science*, 339:668–671,2013, DOI:10.1126/science.1231507; M. T. Zuber et al., Gravity Recovery and Interior Laboratory (GRAIL): Mapping the lunar interior from crust to core, *Space Sci. Rev.*, 178:3–24, 2013, DOI:10.1007/s11214-012-9952-7.

Green aviation

Green aviation is the technology that addresses environmental issues in the design and operation of aircraft, particularly the problems of aircraft emissions of gases, such as carbon dioxide (CO_2) and nitrogen oxides (NO_x), and of noise abatement. In the next few decades, air travel is forecast to experience the fastest relative growth of any mode of transportation (automobiles, buses, rail, ships, airplanes, and so forth) because of the manyfold increase in demand in the major developing nations of Asia and Africa. Air travel accounted for less than 3% of the world passenger traffic of 5.5×10^{12} passenger kilometers traveled (PKT) by all modes of transportation in 1960, increased to 9% of the PKT of 23.4×10^{12} in 1990, and is expected to be 25% of the estimated PKT of 53×10^{12} in 2020 and nearly 40% of the estimated PKT of 103×10^{12} in 2050. (1 km = 0.62 mi.) These projections are based on 3% annual growth in world gross domestic product (GDP), 5.2% annual growth in passenger traffic, and 6.2% annual increase in cargo movement.

As a result of the threefold increase in air travel by 2025, it is estimated that the total carbon dioxide (CO_2) emissions due to commercial aviation may reach $1.2–1.5 \times 10^9$ metric tons annually by 2025, up from its current (2013) level of 670 million metric tons (1 metric ton = 1000 kg = 2205 lb). The amount of nitrogen oxides around airports, generated by aircraft engines, may rise from 2.5 million metric tons in 2000 to 6.1 million metric tons by 2025. The number of people who may be seriously affected by aircraft noise may rise from 24 million in 2000 to 30.5 million by 2025. Therefore, there is urgency in addressing the problems of emissions and noise abatement through technological innovations in the design and operation of commercial aircraft. The ground infrastructure for sustainable aviation, including the concept of "sustainable green airport design," is also important in this context.

To meet the environmental challenges of the twenty-first century that result from the growth in aviation, the Advisory Council for Aviation Research and Innovation in Europe (ACARE) has set the following three goals for reducing noise and emissions by 2020: (1) reduce the perceived noise to one-half of current average levels, (2) reduce the CO_2 emissions per passenger kilometer (PKM) by 50%, and (3) reduce the NO_x emissions by 80% relative to 2000. NASA has more aggressive goals for 2020, as shown in the **table** for (N + 2)-generation aircraft.

The achievement of these goals will not be easy; it will require the cooperation and involvement of airplane manufacturers, the airline industry, regulatory agencies such as the ICAO and the FAA, and research and development organizations, as well as political will by many governments and public support.

Recently, 25 new technologies, initiatives, and operational improvements were identified that may make air travel one of the greenest industries by 2050. The most important among them, with

NASA subsonic transport system level metrics for improving noise, emissions, and performance using technology and operational improvements

Technology benefits	Technology generations (technology readiness level = 4–6)[1]		
	N + 1 (2015)	N + 2 (2020)[2]	N + 3 (2025)
Noise (cumulative margin relative to stage 4)[3]	−32 dB	−42 dB	−52 dB
LTO NO_x emissions (relative to CAEP 6)[4]	−60%	−75%	−80%
Cruise NO_x emissions (relative to 2005 best in class[5])	−55%	−70%	−80%
Aircraft fuel/energy consumption[6] (relative to 2005 best in class[5])	−33%	−50%	−60%

SOURCE: NASA Aeronautics Research.

[1]Technology readiness level (TRL) = 4–6 implies that the technology should be ready to allow the transition to the prototype aircraft, and typically needs another 5–10 years to be included as part of a new commercial aircraft.

[2]For (N + 2)-generation aircraft, the accomplishment of the stated goals by 2020 will require a TRL of 4–6 for "long-pole" technologies by 2015; long-pole technologies are those that are complex and risky, and may require the longest time to develop.

[3]The noise reduction goals are referenced to stage 4 noise goals established by the Federal Aviation Administration (FAA) and the International Civil Aviation Organization (ICAO).

[4]The ICAO Committee on Aviation Environmental Protection (CAEP) has established a series of standards for aircraft emissions around airports; NO_x emissions at landing and takeoff (LTO) are the most critical near the airport. The CAEP 6 report provides the latest standards in the area of emissions.

[5]For the N + 1 and N + 3 goals, the reference "best in class" airplane is the B737-800 with CFM56-7B engines. For the N + 2 goals, the reference "best in class" airplane is the B777-200 with GE90 engines.

[6]CO_2 emissions are tracked by the CAEP as part of the aircraft fuel/energy metric, which includes the fuel life cycle as well as other emissions such as methane and N_2O by converting them to a CO_2 equivalent (CO_2e). CO_2 emission benefits are dependent on the life-cycle CO_2e emission per megajoule for the fuel or energy source used, as well as on aircraft fuel/energy consumption.

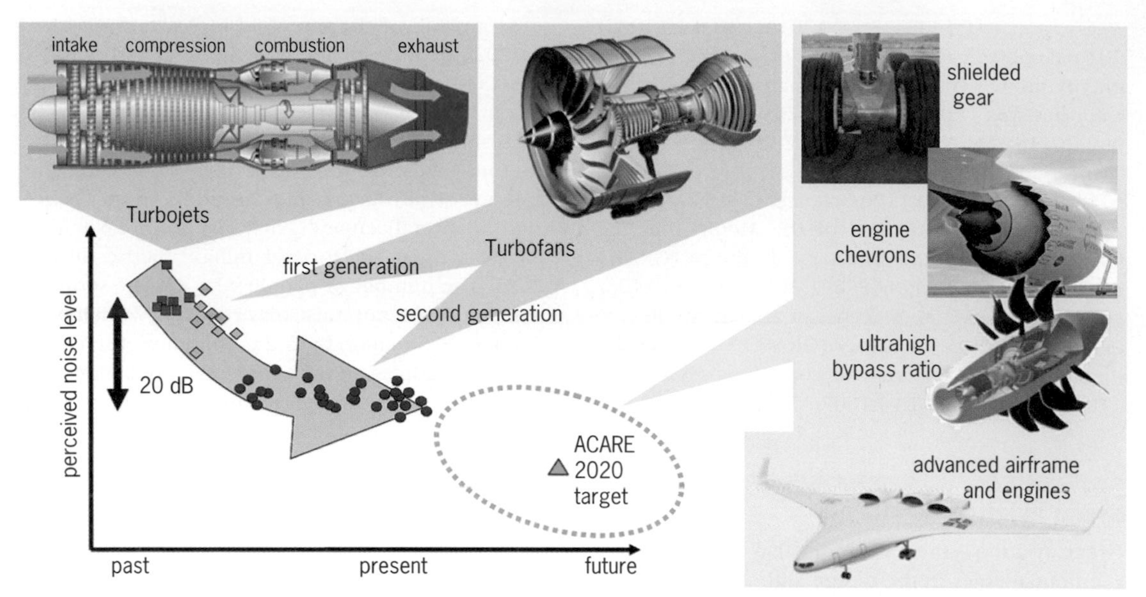

Fig. 1. Evolution of engine noise reduction technologies. (*Illustration courtesy of T. G. Reynolds, Environmental challenges for aviation: An overview, in Low Cost Air Transport Summit, London, June 11–12, 2008, http://bullfinch.arct.cam.ac.uk/Documents/LCATS_Reynolds.pdf*)

maximum impact, require technological innovation in aircraft design and engines, use of alternative fuels and materials, and improvements in operations by deploying an advanced air-traffic management system.

Noise and its abatement. Historically, the reduction in airplane noise has been a major focus of airplane manufacturers because of its health effects and its impact on the quality of life of communities, especially those in the vicinity of major metropolitan airports. As a result, there has been significant progress in achieving major reductions in the noise levels of airplanes in the past five decades. The noise level of jet airplanes built in the 1950s (for example, the Boeing B707, the B720, and the DC8-20) was in the range of 110–120 EPNdB (effective perceived noise in decibels); these airplanes employed turbojet engines. The airplanes built during 1970–1980 (for example, the B737-200, the B747-100, and the DC9-20) employed the first generation of turbofan engines, which reduced the noise levels by nearly 10 EPNdB. The second generation of improved turbofan engines, employed in the B777-200 and the Airbus A310-300, resulted in an additional reduction of nearly 10 EPNdB to approximately 90–95 EPNdB. This trend is continuing with the use of ultrahigh bypass (UHB) ratio engines using turbofans in the latest aircraft, the B787. Although the engine is a major contributor to aircraft noise, significant gains in noise reduction have also been achieved by reducing the noise from the airframe and undercarriage, as well as by making changes in operations (especially in the vicinity of airports).

In order to meet the ACARE and NASA goals of reducing the perceived noise by almost 50% of the current level by 2020, several new technology ideas are being investigated by airplane and engine manufacturers to both reduce and shield the noise sources (**Fig. 1**). The most promising for the near future are chevron nozzles, shielded landing gears, and UHB engines with improved fans (the geared fan and the contra fan) and fan exhaust-duct-liner technology. Advanced airframe designs such as the blended wing body (BWB) for (N + 2)-generation aircraft can also aid in the further reduction of aircraft noise. In addition, new flight path designs in ascending and descending flight can reduce perceived noise levels in the vicinity of airports.

Emissions and fuel burn. Aviation worldwide now consumes nearly 238 million metric tons of jet-kerosene per year. Jet-kerosene is only a very small part of the total world consumption of fossil fuel or crude oil. The world consumes 85 million barrels per day in total [1 barrel (bbl) = 42 gallons (gal) = 159 liters (L)], of which aviation consumes only 5 million. At present, aviation contributes only 2–3% to the total CO_2 emissions worldwide; however, it contributes 10% of emissions by the entire transportation sector (**Fig. 2**). With air travel forecast to become 40% of the total PKT in 2050, it will become a major contributor to greenhouse gas emissions.

It is clear that there are three key drivers of emissions reductions (**Fig. 3**): (1) innovative engine technologies and aircraft designs, (2) improvement in air-traffic management and operations, and (3) alternative fuels, for example, biofuels.

Innovative engine technologies. In the cruise condition, the amount of fuel burn varies in inverse proportion to propulsion efficiency and lift/drag ratio. The greatest gains in fuel burn reduction in the past 60 years (since the appearance of the jet engine) have come from improvements in engine designs. According to the International Airport Transport Association (IATA), new aircraft such as the B777 and B787, using second-generation turbofan engines, are 70%

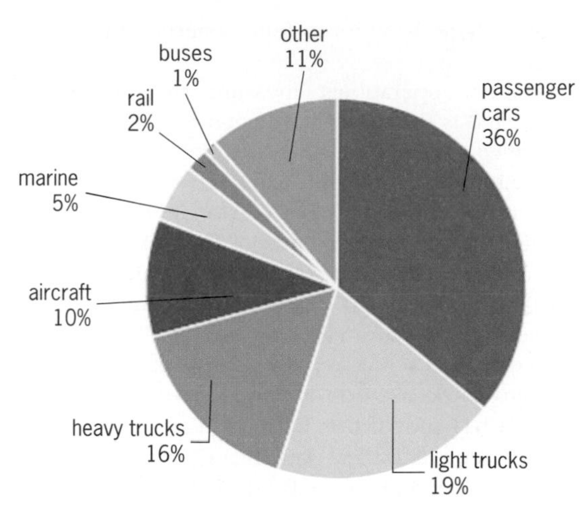

Fig. 2. CO_2 emissions worldwide contributed by various transportation sectors in 2000. (*From Table 1-14 in Inventory of U.S. Greenhouse Gas Emissions and Sinks: 1990–2000, U.S. Environmental Protection Agency, Office of Atmospheric Programs, Washington, DC, 2002, http://www.epa.gov/climatechange/Downloads/ghgemissions/02CR.pdf*)

more fuel efficient than the aircraft of 40 years ago, such as the B707 and DC8, which employed turbojet engines. In 1998, passenger aircraft such as the B757 averaged 4.8 L of fuel per 100 km per passenger; the newest aircraft—the A380 and B787—use only 3 L. The current focus is on making turbofans even more efficient by leaving the fan in the open. Such a ductless "open rotor" design (essentially a high-technology propeller) would make larger fans possible; however, the noise problem may need to be addressed, as well as the problem of how to fit such engines on the airframe. It is worth noting that, in the mid-1980s, General Electric invested significant effort into advanced turboprop technology. The unducted fan (UDF) on a GE36 UHB engine on the MD-81 aircraft at the Farnborough Air Show in 1988 (**Fig. 4**) generated enormous interest in the air transportation industry. However, in spite of its potential for 30% savings in fuel consumption over existing turbofan engines with comparable performance at speeds up to Mach 0.8 and altitudes up to 9000 m (30,000 ft), for a variety of technical and business reasons, the advanced turboprop concept never quite got off the ground.

Innovative aircraft designs. It is well established that increasing the lift (*L*)/drag (*D*) ratio is one of the most powerful means of reducing fuel burn. The three ways of increasing *L*/*D* are to (1) increase the wingspan, (2) reduce the vortex drag factor, and (3) reduce the profile drag area. Current swept-wing aircraft are highly developed, and there is little scope for further improvement. The third option for increasing *L*/*D*, reducing the profile drag of the aircraft, is currently seen as the option with the greatest mid-term and long-term potential. For large aircraft, the adoption of a BWB layout reduces profile drag by about 30%, providing an increase of around 15% in *L*/*D* (estimates of 15–20% have been published). The work on such configurations, both by Boeing [the X-48B, wind tunnel and flight tested at model scale by NASA (**Fig. 5**)] and by Airbus within the NACRE (New Aircraft Concepts Research) project, is proceeding. The first applications of the Boeing BWB are envisaged to be in military roles or as a freighter, with 2030 suggested as the earliest entry-to-service date for a civil passenger aircraft.

The other well-known approach to reducing the profile drag is through the use of laminar flow control in one of its three forms: natural, hybrid, or full. The aerodynamic principles are well understood, but the engineering of efficient, reliable, lightweight suction systems for laminar flow control requires further work. It has been estimated that for an (N + 1)-generation conventional small twin aircraft [162 passengers and a range of 2940 nautical miles (1 nm = 1.852 km)], a 21% reduction in fuel burn can be achieved by using advanced propulsion

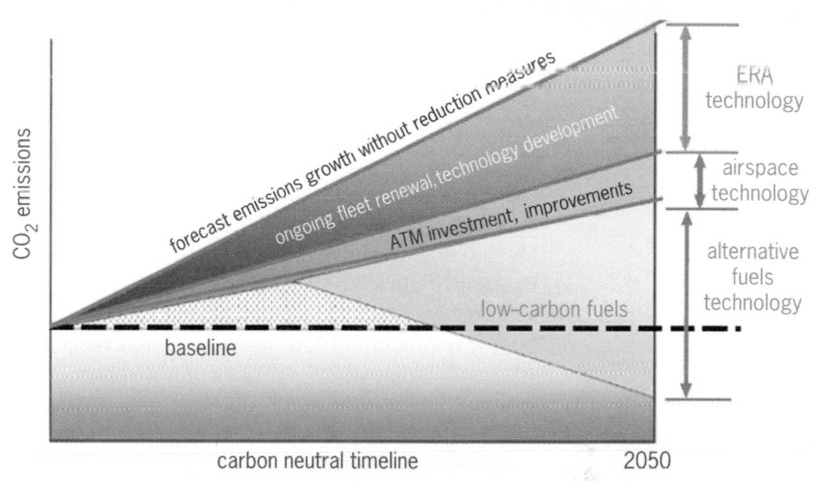

Fig. 3. Schematic forecasts of aircraft CO_2 emissions, with and without various reduction measures, showing key drivers for emissions reductions. ERA technology = environmentally responsible aviation technology; ATM = air-traffic management. (*From A. Strazisar, Overview of goals for the Integrated System Research Program (ISRP), Environmentally Responsible Aviation (ERA) Project, NRC Meeting of Experts on NASA's Plans for System-Level Research in Environmental Impact Mitigation, National Harbor, MD, May 14–15, 2009, NASA, http://www.aeronautics.nasa.gov/pdf/2009_05_14_nrc_tony_strazisar3.pdf*)

Fig. 4. MD-81 aircraft with unducted fan (UDF) on a GE36 turboprop demonstrator engine. Inset shows detail of the UDF. (*Copyright © 2009 by Burkhard Domke, reprinted with permission*)

Fig. 5. Boeing/NASA X-48B BWB technology demonstrator aircraft. (*NASA photo by Carla Thomas*)

technologies and advanced materials and structures, and by improvements in aerodynamics and subsystems. For an (N + 2)-generation aircraft (300 passengers and 7500-nm range) flying at a cruise speed of Mach 0.85, a 40% savings in fuel burn relative to the baseline of a B777-200ER airplane with GE90 engines can be achieved by a combination of a hybrid wing-body configuration (with an all-composite fuselage), advanced engine and airframe technologies, embedded engines with boundary-layer-ingested (BLI) inlets, and laminar flow.

Operational improvements and changes. Many improvements in operations that can reduce CO_2 emissions significantly are being introduced or will be introduced in the relatively near future.

Foremost among operational improvements is the reduction of inefficiencies in air-traffic management that give rise to routes with dog legs, stacking at busy airports, queuing for a departure slot with engines running, and so forth. The U.S. Next Generation Air Transportation System (NextGen) architecture and the European air-traffic control infrastructure modernization program, SESAR (Single European Sky Air-Traffic Management Research Program), are ambitious and comprehensive attacks on this problem. NextGen's computerized air transportation network stresses adaptability by enabling aircraft to immediately adjust to ever-changing factors such as weather, traffic congestion, aircraft position via GPS, flight trajectory patterns, and security issues. By 2025, all aircraft and airports in U.S. airspace will be connected to the NextGen network and will continually share information in real time to improve efficiency and safety, and to absorb the predicted increase in air transportation. Operational measures, which can apply to almost the entire world fleet, can have a greater and more rapid impact than the introduction of new aircraft and engine technologies, which can take perhaps 30 years to fully penetrate the world fleet.

Another operational measure that has been suggested is the use of medium-range aircraft with intermediate stops for long-haul travel. It has been estimated, using a simple parametric analysis, that undertaking a journey of 15,000 km in three hops in an aircraft with a design range of 5000 km would use 29% less fuel than doing the trip in a single flight in a 15,000-km design. In order to avoid the intermediate refueling stops, air-to-air refueling (AAR) has been suggested as a means of enabling medium-range designs to be used on long-haul operations. R. K. Nangia has shown that an aircraft with $L/D = 20$ would require 20,932 kg (46,147 lb), 73,150 kg (161,269 lb), and 119,328 kg (263,073 lb) of fuel to cover ranges of 3000, 6000, and 9000 nm, respectively. With AAR, it would require 41,864 kg (92,294 lb) and 62,796 kg (138,441 lb) of fuel for ranges of 6000 and 9000 nm, respectively, indicating savings of 43% and 47% in fuel burn relative to that required without AAR. However, with AAR, it is paramount that the absolute safety of the aircraft is assured. Therefore, although AAR has been employed in military aircraft operations for some time, it is not practical for a transport aircraft and is unlikely to be implemented.

The possibility of using close formation flying (CFF), akin to the formation flight of a flock of birds, to reduce fuel burn or to extend aircraft range is well known in aerodynamics. The obvious benefit of flying in formation is a more uniform downwash velocity field, which minimizes the energy transferred into it from propulsive energy consumption. Another benefit is the cancellation of vortices shed from the wingtips of individual airplanes, except the two outermost ones. How effective this cancellation will be depends upon the practicality of achievable spacing among the aircraft. There would also be a substantial benefit in elimination of vortex contrails and cirrus clouds. Recently, NASA conducted tests on two F/A-18 aircraft formations. For a two-aircraft formation, the maximum fuel savings were 4% with a tip-to-tip gap between the aircraft equal to 10% of the span, and 10% with a tip overlap equal to 10% of the span. For the three-aircraft inverted-V formation, the maximum fuel savings were about 7% with tip-to-tip gaps equal to 10% of the span, and about 16% with tip overlaps equal to 10% of the span. However, flying in formation will require extreme safety measures based on sensors coupled automatically to control systems of individual aircraft.

Another operational measure for saving fuel is the concept of "tailored arrivals," which can reduce fuel burn, lower the controller workload, and allow for better scheduling and passenger connections. To optimize tailored arrivals, additional controller automation tools are needed. In 2009, Boeing completed the trial of Speed and Route Advisor (SARA) with the Dutch air-traffic control agency, LVNL (Luchtverkeersleiding Nederland), and Eurocontrol. SARA delivered traffic within 30 s of planned time on 80%

of approaches at Schiphol Airport in the Netherlands compared to within 2 min on a baseline of 67%. At San Francisco airport, more than 1700 complete and partial tailored arrivals were completed between December 2007 and June 2009 using B777 and B747 aircraft. Tailored arrivals saved an average of 950 kg (2094 lb) of fuel and approximately $950 per approach. Complete tailored arrivals saved approximately 40% of the fuel used in arrivals. For a 1-year period, four participating airlines saved more than 524,000 kg (1,155,000 lb) of fuel and reduced the carbon emissions by 1.6×10^6 kg (3.5×10^6 lb).

Fuel savings through use of composites. Reducing the weight of an aircraft is one of the most powerful means of reducing the fuel burn. The replacement of structural aluminum alloy with carbon fiber composite is the most powerful weight-reducing option currently available to the aircraft designer working toward a given payload-range requirement. The B787 and A350 both have wings and fuselage made with carbon fiber composites, and most new designs are likely to take this path. Moreover, aircraft manufacturers are investing in advanced composites, which have the prospect of being even lighter and stronger than carbon fiber composites.

Alternative fuels. In recent years, there has been emphasis on improving energy efficiency by replacing jet-kerosene fuel with biofuels. However, until recently, biofuels have not been considered cost competitive with jet-kerosene. An important, much-desired characteristic of an alternative fuel is its ability to be used without any change to the aircraft or its engines. The attractions of such a drop-in fuel are clear: It does not require the delivery of new aircraft, but the environmental impact of all aircraft flying today can be significantly reduced. However, the alternative fuels need to meet specific aviation requirements and essentially should have the key chemical characteristics of kerosene; that is, they should not freeze at flying altitude, and they should have a high enough energy content to power an aircraft's jet engine. In addition, the alternative fuel should have good high-temperature thermal stability characteristics in the engine and good storage stability over time. Airplane and engine manufacturers believe that the present engine technology can operate on biofuels (tests are very promising) and that the aviation industry can convert to such fuels within 5–15 years. Recently, key fuel properties of currently used Jet A/Jet A-1 fuel were compared with those of bio-SPK fuel, which is based on synthesized paraffinic kerosene. The bio-SPK fuel was derived from three different feedstocks (Jatropha, Jatropha/Algae, and Jatropha/Algae/Camelina) for neat fuel, and from 50% blends (50% Jet A/Jet A-1 fuel and 50% second-generation biofuel). It was demonstrated that all bio-SPK blends met or exceeded the aviation jet fuel requirements. Since 2008, there have been many experimental flights using both neat and blended biofuels in one or both engines of a twin-engine aircraft that have clearly demonstrated that the energy efficiency and compatibility of biofuels is comparable to that of jet-kerosene. Methane and liquefied natural gas (LNG) are also being considered as alternatives to jet fuel. Although methane and LNG are not carbon neutral, they are less greenhouse-gas-emission-intensive than jet fuel and are considerably cheaper than jet fuel at present. *See* ENVIRONMENTAL LIFE-CYCLE IMPACT OF ALTERNATIVE AVIATION FUELS; OPTIONS FOR FUTURE METHANE- AND LNG-FUELED AIRCRAFT.

Electric-, solar-, and hydrogen-powered aircraft. For many years, there have been several exploratory studies to build and fly aircraft using sources of energy other than jet-kerosene or synthetic kerosene (biofuels). There have been several notable successes in recent years. In 2008, a successful test flight was conducted of a crewed all-electric aircraft powered by proton exchange membrane (PEM) hydrogen fuel cells. In 2009, the prototype of a solar-powered crewed aircraft was unveiled in Switzerland. This airplane is designed to fly both day and night without the need for fuel. It is powered by 12,000 solar cells mounted on the wing to supply renewable solar energy to the four 7.5-kW (10-hp) electric motors. During the day, the solar panels charge the plane's lithium polymer batteries, allowing it to fly at night. An around-the-world flight of the aircraft is planned for 2015. *See* HYBRID ELECTRIC AND UNIVERSALLY ELECTRIC AIRCRAFT CONCEPTS; TECHNOLOGIES FOR A NEW GENERATION OF SMALL ELECTRIC AIRCRAFT.

The idea of using liquid hydrogen (H_2) as a propellant has been around for many decades, but it is unlikely to become feasible for commercial aircraft because of the many technical and operational issues that would have to be overcome. Cryogenic hydrogen is the only possibility for a hydrogen-powered airplane, since the high-pressure tanks for gaseous hydrogen would be too heavy. Liquid hydrogen occupies 4.2 times the volume of jet fuel for the same amount of energy; therefore, the liquid hydrogen tanks would have to be huge. However, jet fuel weighs 2.9 times more than liquid hydrogen for the same amount of energy; therefore, the reduced weight would partly compensate for the increased aerodynamic drag of the tanks. Nevertheless, the hydrogen-powered plane would have less range and speed than a comparable-weight A310. It would have higher empty weight. Furthermore, whatever energy source was used, 30% would be lost in hydrogen liquefaction. In addition, the cost, infrastructure, and passenger acceptance issues would have to be addressed. The main advantage of using a hydrogen-powered airplane is the reduced emissions, since the use of hydrogen does not produce any CO_2.

Sustainable airports. The airports and associated ground infrastructure constitute an integral part of green aviation. To address the issues of energy and environmental sustainability, the Clean Airport Partnership (CAP) was established in the United States in 1998 and is the only not-for-profit corporation in the United States devoted exclusively to improving environmental quality and energy efficiency at airports.

Airport expansion and the development of new airports should include consideration of both the environmental costs and the life-cycle costs. Sustainable growth of airports requires that they be developed as intermodal transport hubs as part of an integrated public transport network. The ground infrastructure development should include low-emission service vehicles, LEED (Leadership in Energy and Environmental Design)-certified green buildings with low energy requirements, and recyclable water usage. There should be effective land-use planning of the area around the airports, with active investments in the surrounding communities. Airport expansion must also consider the issue of noise and its impact on the surrounding communities, and should be involved in its mitigation by engaging in flight path design. The air quality near the airports should be monitored, and measures for its continuous improvement should be put in place.

Opportunities and future prospects. It is clear that the expected threefold increase in air travel in the next 20 years offers an enormous challenge to all the stakeholders—airplane manufacturers, airlines, airport ground infrastructure planners and developers, policy makers, and consumers—to address the urgent issues of energy and environmental sustainability. The emission and noise mitigation goals enunciated by ACARE and NASA can be met by technological innovations in aircraft and engine designs, by the use of advanced composites and biofuels, and by improvements in aircraft operations. Some of the changes in operations can be easily and immediately put into effect, such as tailored arrivals. Some innovations in aircraft and engine design, the use of advanced composites, the use of biofuels, and the overhauling of the air-traffic management system may take time but are achievable through the concerted and coordinated efforts of government, industry, and academia. They may require significant investment in research and development.

For background information *see* ACOUSTIC NOISE; AERODYNAMIC FORCE; AERODYNAMIC SOUND; AIR POLLUTION; AIR TRANSPORTATION; AIRCRAFT DESIGN; AIRCRAFT ENGINE PERFORMANCE; AIRCRAFT FUEL; AIRCRAFT PROPULSION; AIRPORT NOISE; BIOMASS; BOUNDARY-LAYER FLOW; CLIMATE MODIFICATION; COMPOSITE MATERIALS; FUEL CELL; GENERAL AVIATION; GREENHOUSE EFFECT; LIQUEFIED NATURAL GAS (LNG); METHANE; TURBINE ENGINE SUBSYSTEMS; TURBINE PROPULSION; TURBOFAN; TURBOJET; WING STRUCTURE in the McGraw-Hill Encyclopedia of Science & Technology. Ramesh K. Agarwal

Bibliography. R. K. Agarwal, Sustainable (green) aviation: Challenges and opportunities, William Littlewood Lecture, *SAE Int. J. Aero.*, 2:1-20, March 2010, DOI:10.4271/2009-01-3085; F. Collier et al., Environmentally responsible aviation—Real solutions for environmental challenges facing aviation, Paper ICAS 2010-1.6.1, in *Proceedings of the 27th Congress of the International Council of the Aeronautical Sciences*, September 19-24, 2010, Nice, France, Curran Associates, Inc., 2011; Green Issue, *Aero. Int.*, Royal Aeronautical Society, UK, March 2009; D. L. Greene and A. Schafer, *Reducing Greenhouse Gas Emissions from U.S. Transportation*, Pew Center on Global Climate Change, Arlington, VA, 2003; R. K. Nangia, *Way Forward to a Step Jump for Highly Efficient and Greener Civil Aviation—An Opportunity for the Present and a Vision for the Future*, RKN-SP-2009-130, November 2009; T. G. Reynolds, *Environmental Challenges for Aviation: An Overview*, Low Cost Air Transport Summit, London, June 11-12, 2008; A. Schafer et al., *Transportation in a Climate-Constrained World*, MIT Press, Cambridge, MA, 2009.

Heterogeneous computing

In the context of scientific computing, heterogeneous computing refers to the use of different types of processors, such as the central processing unit (CPU) and the graphics processing unit (GPU), in a given scientific computing task. The objective is to improve performance by assigning different parts of the task to specialized processors. Heterogeneous computing is an emerging technology with great potential in many scientific and engineering fields where large-scale computation plays an important role. Examples include, but are not limited to, fluid dynamics, molecular dynamics, and quantum mechanics. Today, a large percentage of computers used for scientific computing, including most workstations and many clusters, are equipped with both CPUs and GPUs, which makes heterogeneous computing possible. On the other hand, the increase of computing power in CPUs has significantly slowed down in recent years due to both architecture limitations and physical constraints (for example, the "power wall"), which motivates the use of other available resources. Originally designed for image rendering, GPUs specialize in parallel vector operations on very large data sets. In the past decade, a number of languages, compilers, and libraries have been developed for programming GPUs for scientific computing, and significant speedup (up to 100-fold) has been achieved in various applications by assigning part of the computing task to GPUs.

Processor architectures. The key idea in heterogeneous computing is to exploit the advantages of different processor architectures. So far, the most widely used combination is CPU and GPU, although other architectures exist. The present article focuses on this combination. Modern CPUs typically include a small number (2-16) of high-frequency (usually between 2 and 4 GHz) processor cores, which can execute different instructions independently for different data input. This architecture of parallelization is referred to as multiple instruction, multiple data (MIMD) in the well-known Flynn's taxonomy. In order to reduce the time spent on memory access, CPUs are also equipped with a large cache of up to 6 megabytes (MB). CPUs are generalists that perform well for a wide range of applications including latency-sensitive sequential workloads and coarse-grained task-parallel or data-parallel workloads. On the other hand, GPUs are composed of a large number (hundreds) of low-frequency (usually around 1 GHz) processing units that execute the same instruction in parallel for different data input. This architecture is referred to as single instruction, multiple data (SIMD). Because of this specific architecture, GPUs are particularly attractive for large-scale vector operations (for example, $y = ax + b$), in which the same operation is applied to each and every component of a vector. This type of operation is indeed the most intensive computing task in many computer graphics applications, such as 3D rendering, for which GPUs were originally designed.

Programming tools and issues. Currently, a GPU is always managed by a host CPU, and a specialized programming language is required to access this powerful computing device. Several programming platforms have been developed in this regard. They usually include specialized languages, compilers, libraries, and user interfaces. The most widely used platforms are the Compute Unified Device Architecture (CUDA), developed by NVIDIA, and the Open Computing Language (OpenCL), developed by the Khronos Group. CUDA is developed for GPUs produced by NVIDIA. It provides a number of mathematics libraries including cuFFT (a fast Fourier transform library), cuBLAS (the complete BLAS library implemented for GPUs), cuSPARSE (a sparse matrix library), and cuRAND (a random number generator library). OpenCL can be used to program all major GPUs and other heterogeneous devices such as digital signal processors (DSPs). Subroutines to be executed on the heterogeneous device, defined as kernel functions, are automatically optimized for the specific device. Several mathematical libraries have also been developed for OpenCL, including APPML (a general math library) and ViennaCL (a dense and sparse linear algebra library).

Even with these available tools, in most cases, parallelizing a sequential scientific program for combined CPU-GPU computing is still not trivial. Major issues include task assignment, vectorization, and data management. GPUs are fast in large-scale vector operations, but have lower frequency compared to CPUs. Therefore, to harness the power of the GPU, a decision needs to be made as to which parts of the computing task should be moved onto the GPU, and which parts should remain on the CPU. Moreover, existing implementations in a sequential code may need to be vectorized, that is, rewritten in the form of vector operations. Furthermore, a GPU is typically attached to the host by a PCI Express (Peripheral Component Internet Express, abbreviated PCIe) bus (**Fig. 1**); as a result, data transfer between its onboard memory and the host is much slower than local data transfer within onboard memory. In practice, data transfer through a PCI-Express bus should be minimized.

Applications. In general, high speedup can be expected for computing tasks that satisfy three

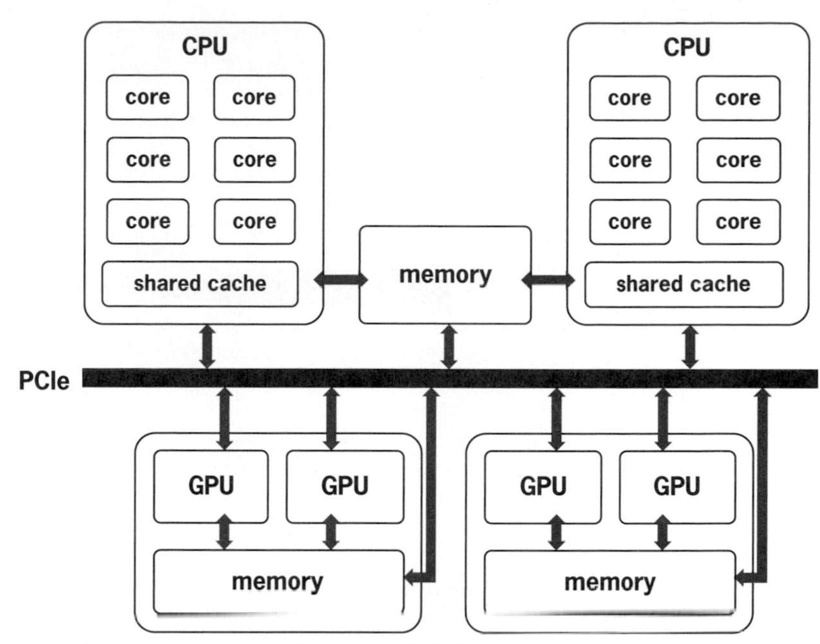

Fig. 1. Hardware architecture of combined CPU-GPU computing platform.

conditions. First, the task needs to be highly parallelizable such that all the processing units on a GPU can be effectively used. In the context of scientific computing, this usually means the computational grid needs to have a large number of grid points. Second, the task needs to be predictable in order to exploit the SIMD architecture of a GPU. In practice, this typically means that the algorithm should not be dominated by "if", "else" routines. Third, memory usage should be small enough to avoid frequent data transfer between the GPU and the host.

Considerable efforts have been made in various disciplines to modify or redesign existing algorithms for GPU computing. At the point of this writing (2014), for most scientific computing problems, the optimal task assignment between CPUs and GPUs is still a research topic that has yet to be explored; however, significant speedup has been achieved in many applications by moving most, if not all, the computational tasks to GPUs.

For example, in computational fluid dynamics (CFD), 2D and 3D compressible inviscid flows were solved in 2008 using primarily an NVIDIA 8800GTX GPU (128 processing units, 1.35 GHz). Compared to CPU-only simulations on an Intel Core 2 Duo E6600 (2.4 GHz, 4 MB L2 Cache), 40-fold speedup was obtained for subsonic flow around a 2D airfoil, while 20-fold speedup was obtained for hypersonic flow around a 3D model aerial vehicle (**Fig. 2**). It was also shown that higher speedup can be achieved on larger computational meshes, as expected. In this work, three types of kernel functions involved in the simulation were identified and optimized for GPUs.

In the same application area, it was shown in 2010 that substantial speedup can also be obtained by

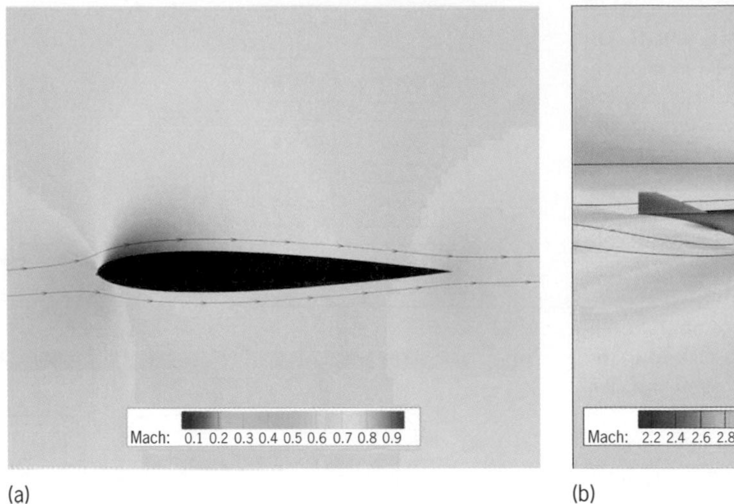

(a)

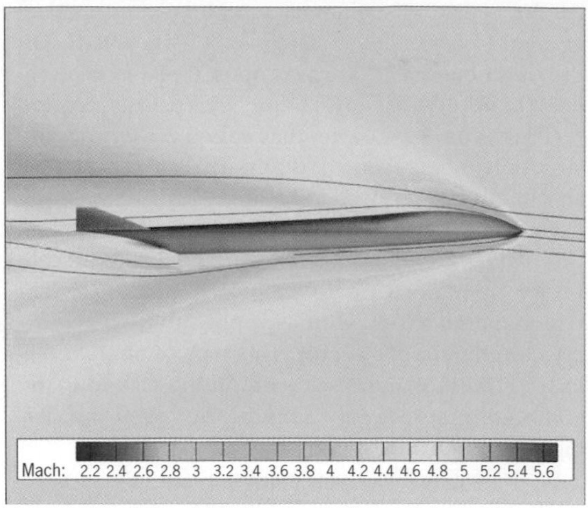

(b)

Fig. 2. Steady flows computed on an NVIDIA GPU. (*a*) Subsonic flow (Mach 0.6) around an airfoil. (*b*) Supersonic flow (Mach 5.0) around a 3D model. (*Reprinted from E. Elsen, P. LeGresley, and E. Darve, Large calculation of the flow over a hypersonic vehicle using a GPU, J. Comput. Phys., 227:10148–10161, 2008, with permission from Elsevier*)

using multiple GPUs connected using the Message Passing Interface (MPI). More specifically, in this work, 128 GPUs with a total of 30,720 processing units were used to solve 2D and 3D incompressible Newtonian flows. It was shown that, compared to CPU simulations on 8 cores, MPI-GPU simulations on 8 GPUs produce a 68-fold speedup.

Most recently, a number of commercial scientific and engineering software companies have started supporting combined CPU-GPU computing in their products. For example, users of the latest version of ANSYS Fluent, version 15.0, can speed up their fluid dynamics simulations using NVIDIA GPUs, enabling designs of higher quality and greater efficiency for aircraft, automotive vehicles, electronic devices, and many other systems.

For background information *see* COMPUTATIONAL FLUID DYNAMICS; COMPUTER ARCHITECTURE; COMPUTER GRAPHICS; CONCURRENT PROCESSING; DIGITAL COMPUTER; MULTIPROCESSING; SUPERCOMPUTER in the McGraw-Hill Encyclopedia of Science & Technology.

Kevin G. Wang

Bibliography. E. Elsen, P. LeGresley, and E. Darve, Large calculation of the flow over a hypersonic vehicle using a GPU, *J. Comput. Phys.*, 227:10148–10161, 2008, DOI:10.1016/j.jcp.2008.08.023; D. A. Jacobsen, J. C. Thibault, and I. Senocak, An MPI-CUDA implementation for massively parallel incompressible flow computations on multi-GPU clusters, American Institute for Aeronautics and Astronautics, AIAA 2010-522, pp. 6151–6167, *48th AIAA Aerospace Sciences Meeting Including the New Horizons Forum and Aerospace Exposition*, January 4–7, 2010, Orlando, Florida, Curran Associates, Red Hook, NY, 2010, DOI:10.2514/6.2010-522; J. E. Stone, D. Gohara, and G. Shi, OpenCL: A parallel programming standard for heterogeneous computing systems, *Comput. Sci. Eng.*, 12(3):66–72, 2010, DOI:10.1109/MCSE.2010.69.

High-energy astrophysical neutrinos at IceCube

Since their discovery in 1912 by Viktor Hess, the origin of the highest-energy cosmic rays has remained one of the most enduring mysteries in physics. These subatomic particles, predominantly nuclei, can reach extraordinary energies, with those of observed cosmic rays exceeding 10^{20} electron-volts (eV), 10 million times the energies achieved at the Large Hadron Collider (LHC). These particles must originate in the universe's most energetic objects, such as gamma-ray bursts or the supermassive black holes in the center of active galaxies. However, which of these types of objects are the actual sources, and how these cosmic rays are produced, remains unknown.

Cosmic-ray astronomy. Direct measurements of the sources of these particles are challenging because of the electric charges of the atomic nuclei that come to us as cosmic rays. Our galaxy is not known to contain any objects capable of accelerating cosmic rays to the highest energies, and so it is likely that they come from great distances. As the cosmic rays travel, their paths are scrambled by magnetic fields in the Milky Way Galaxy as well as the intergalactic medium. Cosmic rays arriving at Earth thus do not in general point back to their sources and, after a century of observations, the cosmic-ray sky has remained enigmatic.

Astrophysical neutrinos. Neutrinos were recognized as a potential solution to the problem of identifying cosmic-ray sources almost immediately after their discovery in 1956. Unlike cosmic rays, they are electrically neutral and travel without deflection to Earth. Their low interaction rate with matter means that they can travel out of even the densest and most violent environments undisturbed. And, because they are produced only by interactions of

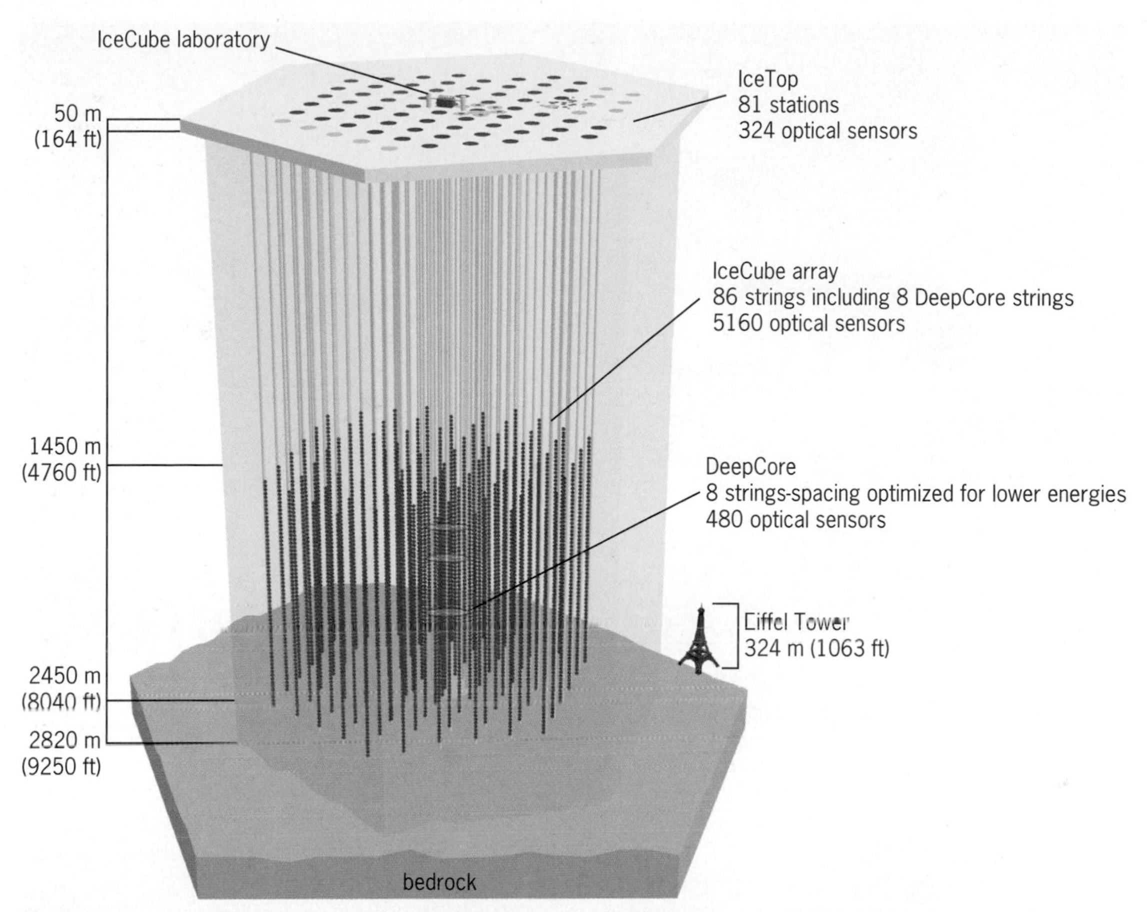

Fig. 1. Image of the IceCube array. The detector is located at depths of between 1.5 and 2.5 km (0.9 and 1.6 mi) in the south polar ice cap and uses 5160 photomultiplier tubes (dots) attached to cables in 86 vertical bore holes.

high-energy hadrons, that is, of cosmic rays, observation of high-energy neutrinos provides a "smoking gun" signature of cosmic-ray acceleration.

Detection of these astrophysical neutrinos requires extremely large-volume detectors because of the rarity of neutrino interactions with matter. Currently operating high-energy (teraelectronvolt) neutrino detectors are the NT200, ANTARES, and IceCube neutrino observatories. (1 TeV = 10^{12} eV = 10^6 MeV. By comparison, the highest energy of solar neutrinos is about 14 MeV, and the energy of protons at the LHC is expected to reach 6.5 TeV in 2015.) The largest of these detectors, IceCube, is a cubic kilometer (1 km = 0.6 mi) in size and continuously observes a gigaton of material for the rare collisions of neutrinos passing through. Structures of this scale cannot be built; each of these instruments uses sensitive photodetectors to observe neutrino interactions in a naturally occurring body. ANTARES and NT200 use the sea and a deep lake, respectively. IceCube, completed at the end of 2010, instruments a cubic kilometer of the south polar ice cap, 2 km (1.2 mi) from the United States's Amundsen-Scott research station at the geographic South Pole (**Fig. 1**).

Detection principle. When neutrinos interact, they produce high-energy charged particles that, in a process analogous to a sonic boom called Cerenkov radiation, produce visible light in transparent material. These charged particles acquire a large fraction of the high energies and momenta of the interacting neutrinos. As a result, their energies and directions follow those of the original neutrino and can be used to discover the neutrino's source and properties.

A key aspect of the design of astrophysical neutrino detectors is to avoid background particles produced in interactions of cosmic rays in the atmosphere, which, when entering the detector, can resemble neutrinos. When cosmic rays strike the Earth's atmosphere, they produce a shower of nuclear debris containing a wide variety of particles. Observation of these so-called air showers provides the principal means of detection of cosmic rays; it also poses the principal challenge to detection of the astrophysical neutrinos that may reveal their origin.

Cosmic-ray air showers attenuate rapidly once they hit the ground. Placement of the detector deep below the surface, at the bottom of the sea or ice cap, thus eliminates most of the particles in the shower. Some shower particles, however, particularly muons, can penetrate even to depths of multiple kilometers. A common approach to avoid confusion from these is to look for events coming through the Earth, which is nearly transparent to neutrinos. While IceCube may be under only 2 km (1.2 mi) of ice when viewed from one side of the Earth, 12,700 km (7900 mi) of rock lie on top of it from

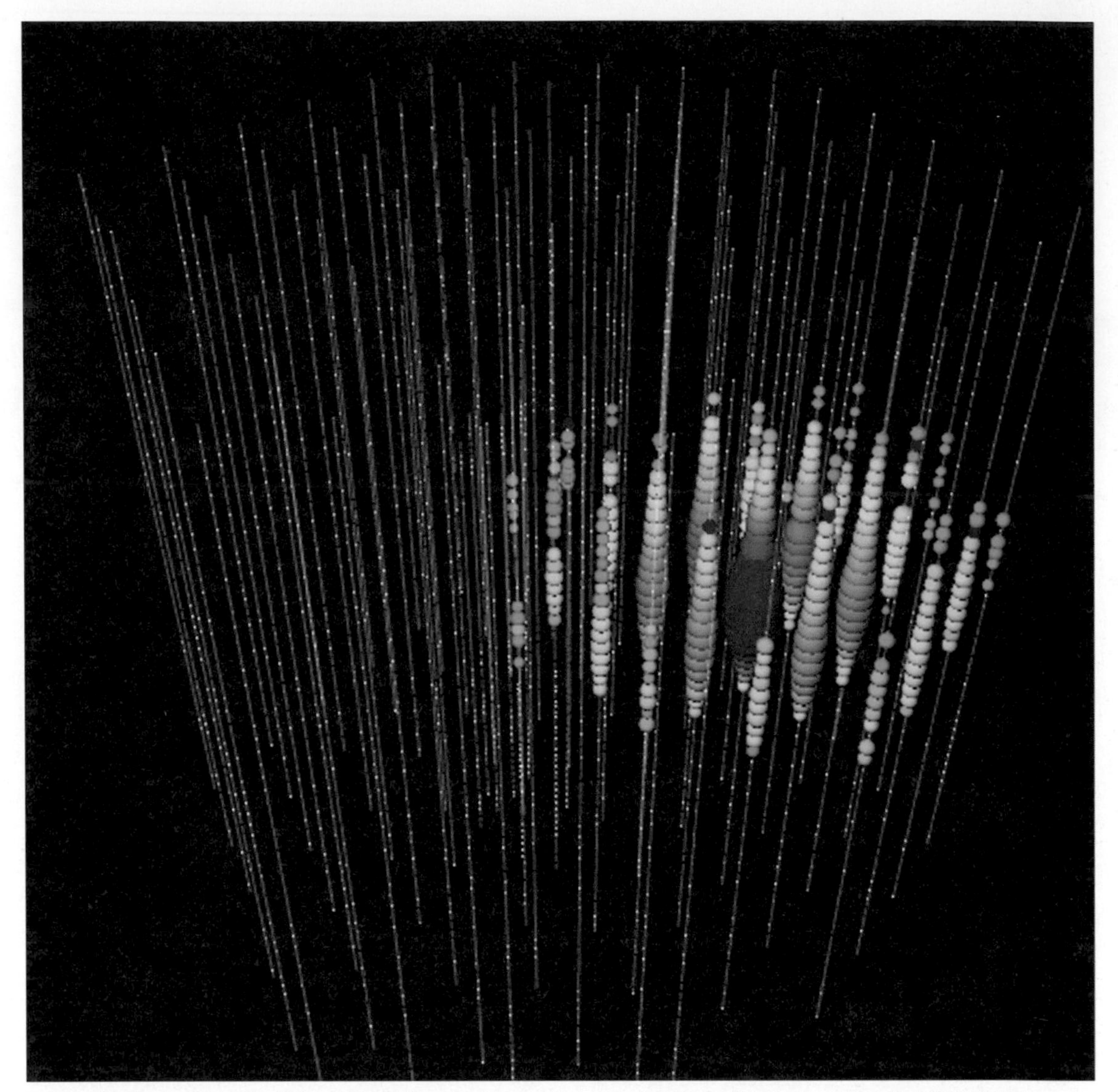

Fig. 2. The observation at IceCube of the highest-energy neutrino ever detected. This neutrino deposited 1100 TeV of energy into the IceCube detector on January 3, 2012.

the other. However, even a planetary mass cannot eliminate the air shower completely: In the same types of interactions believed to make neutrinos in the sources (proton-proton collisions), the cosmic rays also produce neutrinos when they collide with the Earth's atmosphere.

Rejection of atmospheric neutrinos. Thus, once they have a neutrino sample in hand, the principal goal of neutrino astronomers is to identify which neutrinos come from cosmic-ray interactions at their sources and which come from much more common interactions in the Earth's atmosphere.

Atmospheric neutrinos are made primarily in the decay of π and K mesons produced in cosmic-ray interactions. For these neutrinos to be produced, their parent mesons must avoid losing their energy through interactions with the atmosphere or ground before decaying. This becomes more and more difficult with increasing energy: As a result of relativistic time dilation, the mean decay lifetime of π and K mesons increases with energy. Therefore, fewer and fewer mesons decay before interacting at high energies, and atmospheric neutrinos become progressively rarer.

An astrophysical neutrino flux, however, would be produced in space, where interactions with the atmosphere will cause no such suppression. As a result, at sufficiently high energies, astrophysical neutrinos should become more common than atmospheric ones. Above approximately 100 TeV, only one background atmospheric neutrino interaction is expected per gigaton of detector mass per year. Any more than that would be a telltale sign of the observation of extraterrestrial neutrinos.

Atmospheric neutrinos are also a part of cosmic-ray air showers and are produced along with many other particles. While the muons from an atmospheric neutrino from below the detector have long since been absorbed by the Earth, any atmospheric neutrino from above will likely be accompanied by muons from the same cosmic-ray air shower that produced it. A neutrino coming from above without any such accompaniment is thus very likely to originate from beyond the atmosphere.

IceCube detection of cosmic neutrinos. In 2013, IceCube released the first two years of data from a search for astrophysical neutrinos with the completed detector. The goal of this search was to detect an excess at energies above 100 TeV and, in particular, to observe the Southern Hemisphere, above IceCube, for neutrinos unaccompanied by cosmic-ray air showers. Observation of a large number of neutrinos at such high energy with no connection to production by cosmic rays would then provide definitive evidence for the observation of neutrinos of extraterrestrial origin.

In two years of data, 28 interactions were detected by IceCube relative to only 11 expected from cosmic-ray air showers. These 28 interactions had energies far higher than expected for the atmospheric background. The two most energetic neutrinos had energies above 1000 TeV (**Fig. 2**). These are the most energetic neutrinos ever observed—over 100 times what any human-made accelerator has produced and an order of magnitude beyond what was expected to be the end of the atmospheric background. Most important, 24 neutrinos arrived from the Southern Hemisphere, above IceCube, and all but one of these had no observed air-shower counterpart.

Cosmic neutrino sources. These data are just the beginning of a new field and at present offer the means to draw only two conclusions with certainty: that we are beginning to see high-energy neutrinos associated with cosmic-ray acceleration, and that, after many years, we have reached the level of sensitivity to see many more in the future.

The main question of interest going forward is what the natures of the neutrino and cosmic-ray sources are. The neutrino energy spectrum and spatial distribution will be the main tools to answer this. The energy spectrum, for example, will tell us whether the neutrinos are coming from cosmic-ray interactions with gas or radiation and thereby the sources' environments. Even with the current limited statistics, the data are already in tension with certain radiation-dominated models, and more data will give us greater understanding of the processes by which cosmic rays are made.

The directions from which the neutrinos arrive will also provide critical information. Correlation with known high-energy astrophysical objects would allow the identification of cosmic-ray accelerators. Even with the current data, we already have some hints. No statistically significant correlations were observed with the plane of the Milky Way Galaxy, which suggests that at least some of these neutrinos come not only from beyond the atmosphere, but from beyond the Milky Way. Determining whether these neutrinos are from beyond the Galaxy would resolve a number of long-standing questions. The cosmic-ray energy range associated with these neutrinos is approximately 1000–10,000 TeV, a range in which it is not clear whether the cosmic rays are of galactic or extragalactic origin.

The next question is what the nature of those sources may be. New, larger detectors, now on the drawing board after the first detection of cosmic neutrinos, will give increased statistics and usher in an era of precision neutrino astronomy. Follow-up observations with other tools, in particular, high-energy gamma rays, are also ongoing: With a neutrino map in hand, we will now know for the first time where at least some cosmic-ray accelerators are. Soon, perhaps, nature's most enigmatic particle will bring us the answer to one of her greatest puzzles.

For background information *see* ASTROPHYSICS, HIGH-ENERGY; BLACK HOLE; CERENKOV RADIATION; COSMIC RAYS; ELECTRONVOLT; GALAXY, EXTERNAL; GAMMA-RAY ASTRONOMY; GAMMA-RAY BURSTS; MILKY WAY GALAXY; NEUTRINO; NEUTRINO ASTRONOMY; PARTICLE ACCELERATOR; SOLAR NEUTRINOS in the McGraw-Hill Encyclopedia of Science & Technology. Nathan Whitehorn

Bibliography. F. Halzen and D. Hooper, High-energy neutrino astronomy: The cosmic ray connection, *Rep. Progr. Phys.*, 65:1025–1078, 2002, DOI:10.1088/0034-4885/65/7/201; IceCube Collaboration, Evidence for high-energy extraterrestrial neutrinos at the IceCube detector, *Science*, 342:1242856 (7 pp.), 2013, DOI:10.1126/science.1242856; K. H. Kampert and A. A. Watson, Extensive air showers and ultra high-energy cosmic rays: A historical review, *Eur. Phys. J. H*, 37:359–412, 2012, DOI:10.1140/epjh/e2012-30013-x; C. Spiering, Towards high-energy neutrino astronomy, *Eur. Phys. J. H*, 37:515–565, 2012, DOI:10.1140/epjh/e2012-30014-2.

History of insect body size

Insects are the most successful animal group, accounting for nearly three-quarters of all described species and occupying pivotal roles in many terrestrial ecosystems. The body sizes of living insects also span an enormous range, from 0.15-mm-long (0.006-in.-long) mymarid wasps (known as fairyflies) to the Atlas moth with its wingspan of nearly 30 cm (12 in.). Even larger extinct insects such as *Meganeura* and other extinct relatives of dragonflies (sometimes called "griffinflies") have been discovered. The factors that permitted them to reach such enormous wingspans [up to 70 cm (28 in.), as large as a modern crow] have been the subject of great interest.

Factors affecting insect body size. The largest extinct insects lived from the end of the Carboniferous Period to the Early Permian Period, between 310 and 270 million years ago (MYA), which was a time when atmospheric oxygen concentrations (as reconstructed from geochemical models) are estimated to have reached 30% or more (the present value is 21%). The presence of giant insects during a time of peak oxygen levels led to the hypothesis that gigantism

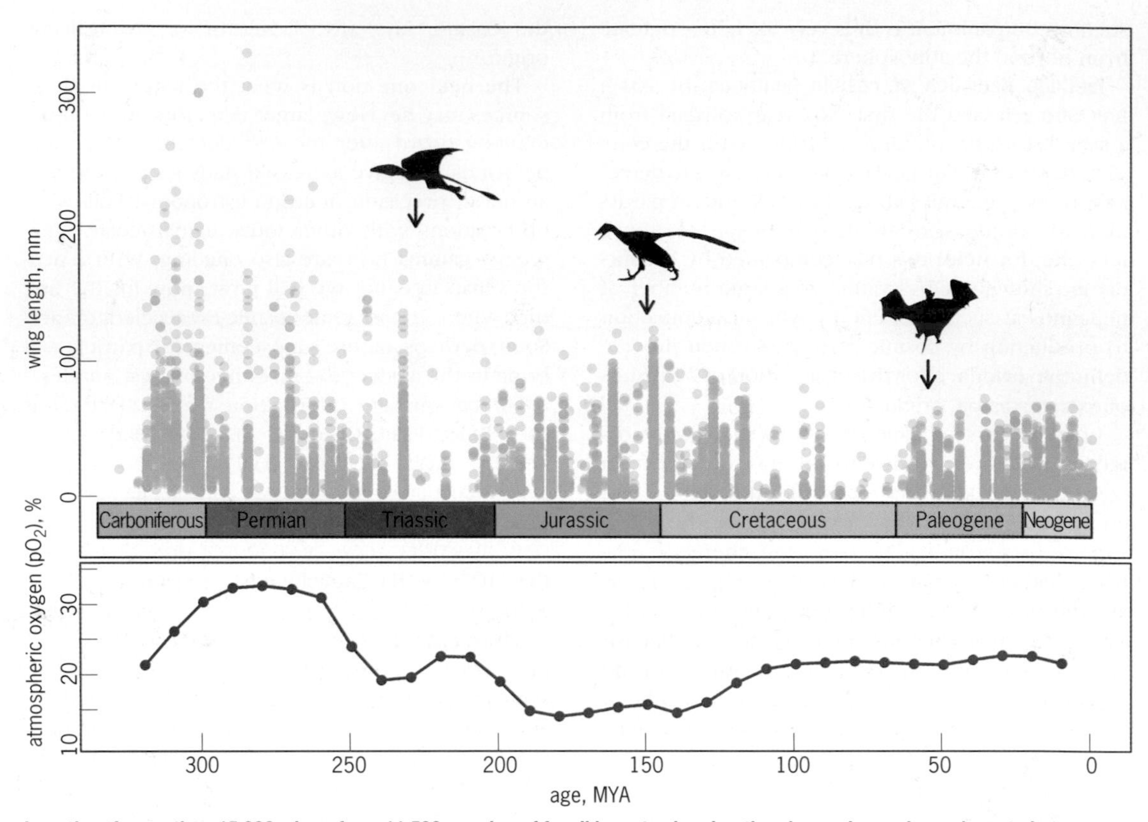

Lengths of more than 15,000 wings from 11,500 species of fossil insects showing the change in maximum insect size over their evolutionary history. The atmospheric oxygen (pO_2) curve is an estimate based on geochemical modeling. Animal silhouettes indicate the first appearance of pterosaurs (flying reptiles, 230 MYA in the Triassic Period), birds (150 MYA in the Jurassic Period), and bats (50 MYA in the Early Paleogene Period).

was enabled or caused by the high oxygen concentration. In addition, this hypothesis is supported by extensive experimental studies on the physiology of living insects. The relationship between oxygen levels and fossil insect size has also been investigated quantitatively using a large database of measured fossil insect wings (see **illustration**). Prior to the Cretaceous Period (from 320 to 150 MYA), the size of the largest insects closely tracked the increases and decreases in atmospheric oxygen, supporting the hypothesized importance of oxygen levels.

However, the insect size and atmospheric oxygen trends diverged in the Early Cretaceous Period (from 140 to 100 MYA), when the largest insects became smaller, although the geochemical model predicts that oxygen concentrations increased (see illustration). This divergence suggests that other factors may also have been important. Its timing, occurring during the evolution and diversification of early birds in the Cretaceous Period, raises the possibility of interactions (such as predation or competition) with other flying animals. Making a simple assumption that predation or competition would impose a consistent constraint on size allows numerical models to be fitted and compared using statistical techniques. The most accurate model identified oxygen as the best explanation for the size changes prior to 140 MYA and concluded that the maximum wing length was constant from 140 to 90 MYA and from 90 MYA to the present. Models that included shifts in size around 230 MYA, coinciding with the evolution of pterosaurs (flying reptiles), or 50 MYA, coinciding with the evolution of bats, were not as well supported by the statistical testing. The model fittings and comparisons suggest that oxygen was the most important control on the size of the largest insects for the first 150 million years of their evolutionary history, but predatory or competitive interactions with birds became increasingly important as avian flight became specialized between 140 and 90 MYA. Predatory interactions likely are a more powerful driver of size evolution, but the current data cannot disentangle the effects of different types of biological interactions. Larger insects are less maneuverable, either in flight or when taking off from a surface, so they would have been more vulnerable to predation by flying birds. The largest insects since the Jurassic Period have often been dragonflies. Because they are predators of smaller species, there may have been competition with early birds for food resources.

Other flying groups did not appear to have as large of an impact on insect size. Although insect size did decrease substantially between the Middle Triassic and the Early Jurassic periods, it is difficult to assess the relative importance of pterosaur evolution (approximately 230 MYA) and a Late Triassic fall in atmospheric oxygen (approximately 210–190 MYA) because insect fossils are very sparse in the interval from 230 to 210 MYA (see illustration). The evolution of flight in bats around 50 MYA did not coincide with any change in the size of the largest insects. However, the largest insects are almost exclusively dragonflies and grasshoppers, which are mostly active during the day. It seems plausible that the appearance of bats had an effect on evolution in nocturnal insects, but it is difficult to identify nocturnal insects based only on fossils. In addition, groups like moths (often the target of bat predation) do not have a rich fossil record.

Further role of oxygen. The fossil record of insects also provides opportunities to test how variations in oxygen concentration may have influenced body size evolution. Oxygen concentration most likely acted as a ceiling to limit the size of the largest insects, rather than promoting evolution of larger or smaller size in all insects. Most fossil insects were small, even when oxygen levels were high, suggesting that their body size was adapted in response to other constraints or to fill other ecological roles. Body size in insects is affected by a multitude of factors. In addition to oxygen, biotic interactions (including food supply), temperature, and aerodynamic limitations are factors that likely played important roles. The maximum size in small-bodied groups, such as flies, beetles, or wasps, also did not track the fluctuations in oxygen concentrations, further suggesting that the size in those groups responded to different pressures. Biological interactions also would have disproportionately affected the largest insects, acting more as an upper limit on size than as a pressure on all groups.

Oxygen can also influence insects at different stages in their life cycle (most insects have a larval stage that metamorphoses into an adult with a different morphology). The traditional view holds that oxygen levels limit the energy that is available for activity in adult insects, especially in active flying groups with comparatively inefficient respiratory systems. As an insect grows larger, it can no longer extract sufficient oxygen from the air to support energy-intensive activities, such as flight, because its mass increases more quickly than the respiratory surface for oxygen extraction. Alternatively, oxygen levels may more strongly influence the larval stage in groups with aquatic larvae (such as dragonflies, mayflies, or stoneflies) because the larvae cannot regulate oxygen uptake as well as the adults. Fossil size data are more consistent with the traditional hypothesis because parallel trends in maximum size have occurred in dragonflies (aquatic larvae) and grasshoppers (land-dwelling juvenile stages) since the Triassic Period and because stonefly size (aquatic larvae) does not track oxygen levels closely.

For background information *see* ANIMAL EVOLUTION; AVES; CARBONIFEROUS; CRETACEOUS; ECOLOGICAL COMPETITION; FOSSIL; INSECT PHYSIOLOGY; INSECTA; JURASSIC; MACROEVOLUTION; OXYGEN; PALEOECOLOGY; PERMIAN; POPULATION ECOLOGY; PREDATOR-PREY INTERACTIONS; PTERYGOTA; TRIASSIC in the McGraw-Hill Encyclopedia of Science & Technology.

Matthew E. Clapham

Bibliography. S. L. Chown and K. J. Gaston, Body size variation in insects: A macroecological perspective, *Biol. Rev.*, 85:139–169, 2010, DOI:10.1111/j.1469-185X.2009.00097.x; M. E. Clapham and J. A. Karr, Environmental and biotic controls on the evolutionary history of insect body size, *Proc. Natl. Acad. Sci. USA*, 109:10927–10930, 2012, DOI:10.1073/pnas.1204026109; J. F. Harrison, A. Kaiser, and J. M. VandenBrooks, Atmospheric oxygen level and the evolution of insect body size, *Proc. Royal Soc. B*, 277:1937–1946, 2010, DOI:10.1098/rspb.2010.0001.

Hybrid electric and universally electric aircraft concepts

Hybrid electric and universally electric concepts, whose adoption is supported by the progress made in electrical component technology, will serve to enable solutions in meeting challenges related to reducing the environmental impact of transport aircraft. The prospects for these aircraft for future market segments need to be evaluated in terms of the available scope for fuel-burn reduction, energy consumption, efficiency, and operating economics. However, when assessing these concepts, many of the paradigms developed for aircraft powered by conventional fuel must be reconsidered, since the integration of such innovative electric propulsion systems perturbs the classical design, sizing, and performance of the aircraft. *See* GREEN AVIATION.

Propulsion system architecture. Several advanced electric propulsion system architectures can be considered for the design of hybrid electric and universally electric aircraft. Their selection will depend not only on the benefits they provide to the aircraft, but also on the availability, scalability, and potential evolution of the electrical components. Indeed, a decisive factor in the commercial success of hybrid and universally electric airliners will be the ability to upgrade the system when new, enhanced electrical components become commercially available. Currently, experimental electric aircraft exist, but they are generally retrofitted versions of general aviation aircraft. *See* TECHNOLOGIES FOR A NEW GENERATION OF SMALL ELECTRIC AIRCRAFT.

A wide variety of propulsion system architectures can be considered (**Fig. 1**), distinguished by both the component (or component chain) enabling the generation of shaft power and the devices utilizing it. The type of electric power system integrated at the aircraft level determines the hybrid electric or universally electric nature of the concept. If electric power is the only power source considered, the aircraft will be categorized as a universally electric aircraft (or simply an electric aircraft). If another source of power (typically fuel power) is used in combination with electric power, the aircraft will be classified as a hybrid electric aircraft.

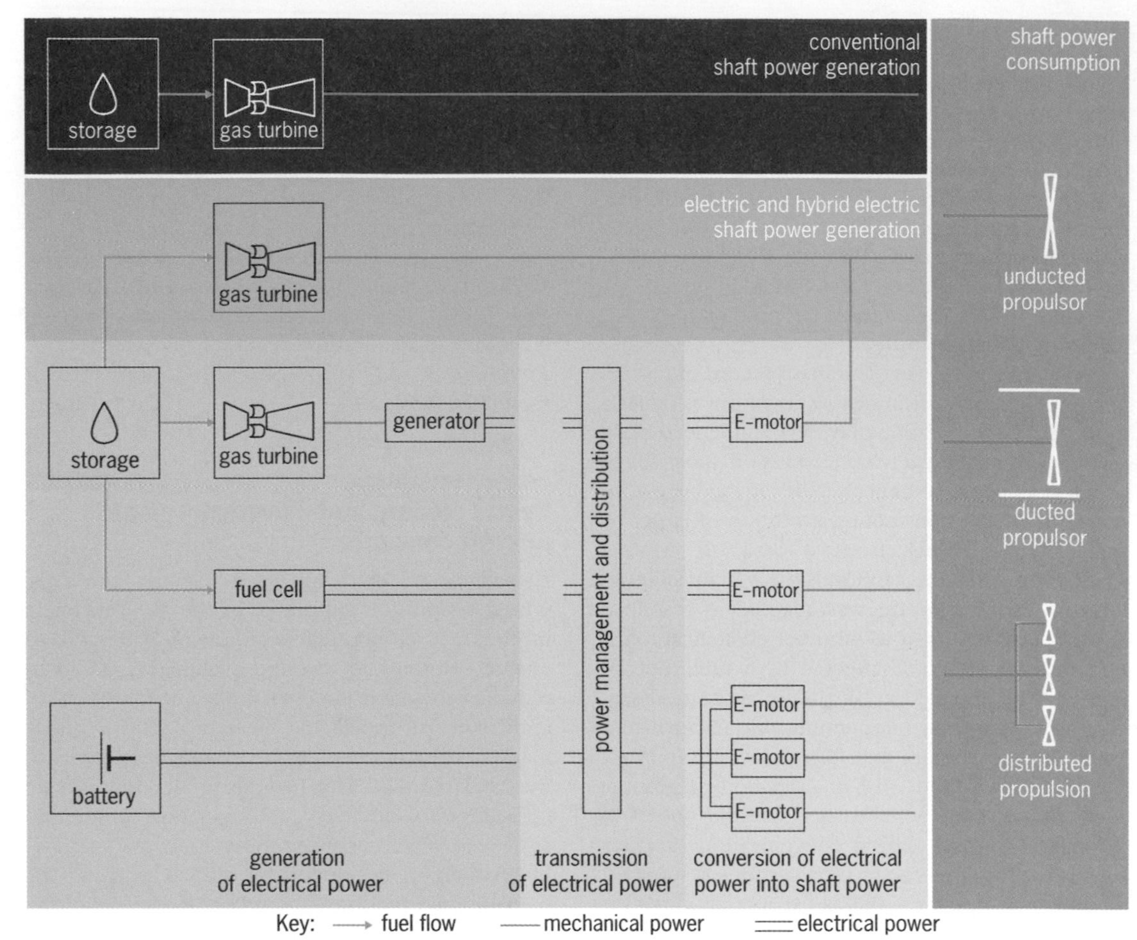

Fig. 1. Conventional, hybrid electric, and universally electric propulsion systems.

In the propulsion system, the consumer of the power is the propulsor device, typically either unducted or ducted, which provides the required thrust. The propulsor device consumes power in the form of mechanical power transmitted at its shaft, hence the terminology of shaft power. Conventionally, shaft power is created by burning fuel in a combustion engine, typically a reciprocating engine or a gas turbine.

An electrically useful power source is required for hybrid electric and universally electric propulsion. Most commonly, the electric power is provided by one of the following devices or possible combinations of them:

1. A fuel cell, which generates electric power through a chemical reaction facilitated by a fuel (such as natural gas or hydrogen) and oxygen.

2. An electrochemical battery, which converts stored chemical energy into electric energy.

3. A generator driven by a conventional fuel-powered combustion engine. This system architecture is categorized as turboelectric (also called a serial system).

The power management and distribution (PMAD) system conveys and monitors the electric power. It is generally composed of cables, connectors, controllers, converters, and electrical buses. Its architecture also takes care of redundancy in the system through appropriate switching in abnormal modes. Because of the utilization of electric power and devices, the risks to circuitry arising from potential exposure to electromagnetic fields and radiation need to be investigated.

In order to drive the propulsor device, the useful electric power needs to be converted into shaft power. In electrical system architectures, this role is taken by an electric motor. It can drive the shaft of the propulsor device directly, be installed on the shaft of a conventional engine to support its operation (an arrangement called parallel system integration), or even drive the shaft itself during some segments of the complete flight profile.

Because of the nature of electric power, which can be easily distributed, new degrees of freedom in the integration are possible, resulting in enhanced synergies between the airframe and the propulsion system. Consequently, the field of distributed propulsion, which concerns the distribution of useful electric power to many integrated propulsor devices, has drawn some attention for advanced studies.

The complexity of the system, the gain in overall propulsion system efficiency, the aerodynamic improvement, and the change in weight will determine the optimal integration at the aircraft level.

Technology requirements and development. The development of the electrical component characteristics will determine the feasibility of electric and hybrid electric systems for commercial air transportation. The specific power (amount of power delivered per unit of mass), efficiency, and volume are critical. In addition, the evolution of the specific energy (the amount of energy contained per unit of mass of the storage device) characteristic of the energy source will be decisive for the application of electric and hybrid electric aircraft in different market ranges. Using the example of a battery, the potential will depend greatly upon the specific energy that the batteries achieve. A specific energy of around 250 Wh/kg is reached by state-of-the-art batteries. However, specific energies of at least 1.0 kWh/kg will be required to ensure the competitiveness of hybrid electric and universally electric transport aircraft in the regional and short-range market segments. Leading innovation in electrical component technology will be one of the main challenges for industries, universities, and research institutions.

Hybrid electric aircraft concept. One of the first advanced hybrid electric transport concepts was the SUGAR Volt, studied by Boeing Research and Technology (**Fig. 2**). (SUGAR stands for subsonic ultragreen aircraft research.) The aircraft was designed for 154 passengers, with entry into service (EIS) in 2030–2050. A variety of hybrid electric propulsion system architectures were considered, notably, fuel cell–gas turbine and battery–gas turbine concepts. For both systems, the electric motor is mounted with a gearbox on the gas turbine shaft (parallel integration), giving the system the capability to operate in all-gas turbine, all-electric, or combined modes. Detailed analysis, sizing, and engineering trade-off studies of the battery–gas turbine concept were undertaken. The increase in maximum takeoff weight (MTOW) caused by the high weight sensitivity of the electrical system and particularly of the batteries was highlighted. For a mission with a range of 900 nautical miles (1667 km), a 70% fuel-burn reduction compared to a state-of-the-art aircraft was achieved.

However, it is not sufficient to focus only on fuel-burn reduction (which is equivalent to CO_2 emission reduction). Indeed, the achievable reduction results mainly because a different source of energy, namely, electric energy, is used to power the aircraft, not because the efficiency of the aircraft system has been significantly improved. This important aspect is captured by the vehicular efficiency metric, which relates the distance traveled to the total energy consumption (including fuel and electric energy) and the number of passenger seats. Mainly because of the large increase in aircraft weight, the total energy consumption might be much larger than that of a conventional aircraft even if the fuel consumption is reduced. This consideration is essential, since the electric energy may not be produced solely by renewable energy sources in the future and therefore should be accounted for in any life-cycle assessment of energy and emissions. *See* ENVIRONMENTAL LIFE-CYCLE IMPACT OF ALTERNATIVE AVIATION FUELS.

Fig. 2. SUGAR Volt hybrid electric concept from Boeing Research and Technology. (*Boeing image*)

An investigation of an advanced hybrid electric concept, as presented by the author on behalf of Bauhaus Luftfahrt, with batteries powering the propulsors electrically during cruise only, demonstrated that the regional market segment is the most suitable for fuel-battery hybrid electric aircraft, as a significant reduction in the block fuel (total fuel required for a flight) was found to be achievable concurrently with an in-flight neutral energy consumption compared to an advanced conventional aircraft with EIS in 2035.

Within the regional market segment, a hybrid electric commuter concept for 50 passengers, the NXG-50 with EIS in 2025, has been investigated by Georgia Tech. It utilizes a ducted fan powered by an electric motor via a gearbox coupled to a turboelectric generator and batteries. A reduction in life-cycle energy consumption by 15% and a life-cycle cost 4.2% lower than the reference Bombardier CRJ200 was projected.

Universally electric aircraft concept. The Ce-Liner, a universally electric aircraft concept with EIS in 2035, carrying 189 passengers up to 900 nautical miles (1667 km), has been studied by Bauhaus Luftfahrt (**Fig. 3**). A pair of high-temperature superconducting electric motors driving ducted fans powered by modularly stored advanced batteries within the fuselage was proposed.

A characteristic of universally electric aircraft concepts is mass invariance. As the aircraft is driven by electric power only, its mass remains constant during its overall mission. This aspect has implications not only for the structural design and flight technique optimization, but also for the energy consumption. The advantages of the electric aircraft are its zero in-flight emissions and its highly efficient electrical propulsion system. However, the MTOW

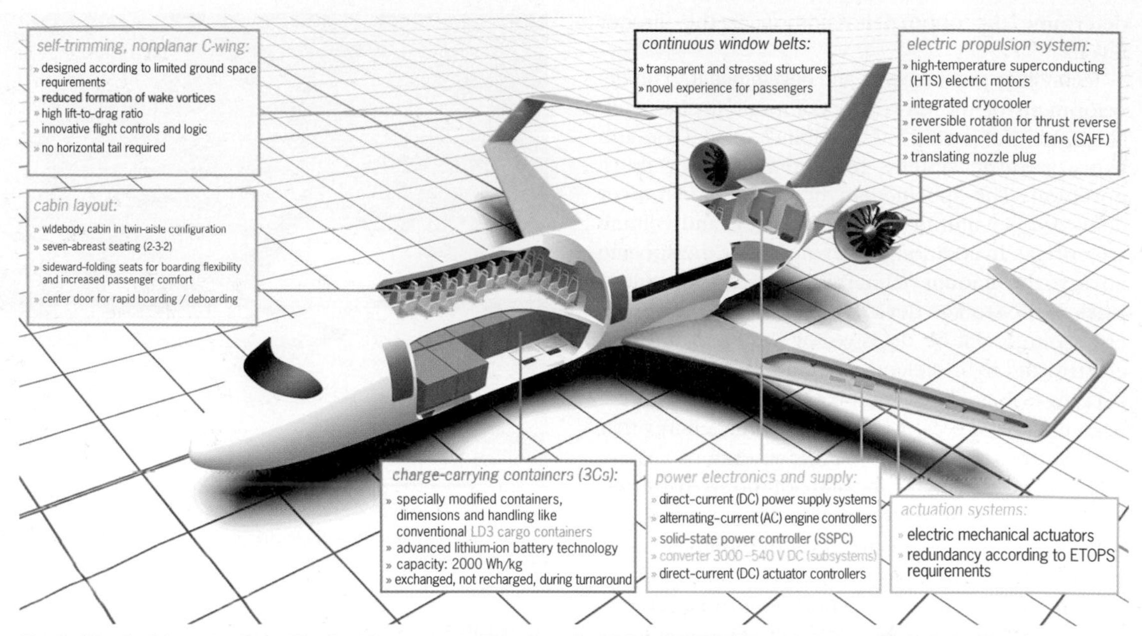

Fig. 3. The Ce-Liner, a universally electric concept of Bauhaus Luftfahrt. ETOPS (Extended range Twin Operations) refers to rules that allow twin-engine airliners to fly long-distance routes. (*Bauhaus Luftfahrt Neue Wege*)

is dramatically increased because of the use of batteries only, as is the size of the electrical system, since it is driven by takeoff power requirements (considered an important finding). As a result, the energy consumption may be degraded even if the propulsion system efficiency is greater than in a conventional approach. Consequently, the vehicular efficiency needs to be considered to fairly compare the performance of a universally electric aircraft with that of a conventional aircraft targeting the same EIS. According to the projected battery technology evolution, it seems that battery-electric transport aircraft will be limited to the short-haul and regional market segments, up to 900 nautical miles (1667 km).

Conclusion. With respect to availability, development, and testing, the integration of the electric components will evolve from general aviation toward servicing regional aircraft and short- to medium-range segments. According to the forecasted technology evolution, the application of fuel-battery hybrid electric and universally electric propulsion to the long-range segment seems unlikely because of the detrimental weight impact of such systems.

Greater attention to the application of electric and hybrid electric concepts to commercial transport aircraft is relatively recent. Consequently, advanced electric propulsion systems and innovative integration still need to be investigated further in order to define the potentials. In view of the change in aircraft mass, energy use, emissions, and systems architecture, sufficiently rigorous economic analysis will be essential in order to establish "best and balanced" hybrid and universally electric concepts. Finally, the investigation has to go beyond the aircraft system to include life-cycle analysis in order to determine the influence of the hybrid electric and universally electric philosophy on the complete energy chain.

For background information *see* AIRCRAFT DESIGN; AIRCRAFT ENGINE PERFORMANCE; AIRCRAFT PROPULSION; BATTERY; FUEL CELL; INTERNAL COMBUSTION ENGINE; MOTOR in the McGraw-Hill Encyclopedia of Science & Technology. Clément Pornet

Bibliography. T. Banning et al., *2012–2013 FAA Design Competition for Universities Electric/Hybrid-Electric Aircraft Technology Design Category, NXG-50*, GT Aircraft, Inc., Georgia Institute of Technology, Atlanta, 2013; M. K. Bradley and C. K. Droney, *Subsonic Ultra Green Aircraft Research: Phase I Final Report*, NASA/CR-2011-216847, Boeing Research and Technology, Huntington Beach, CA, 2011; M. K. Bradley and C. K. Droney, *Subsonic Ultra Green Aircraft Research Phase II: N + 4 Advanced Concept Development*, NASA/CR-20120217556, Boeing Research and Technology, Huntington Beach, CA, 2012; A. R. Gibson et al., The potential and challenge of turboelectric propulsion for subsonic transport aircraft, American Institute of Aeronautics and Astronautics, AIAA 2010-276, pp. 3208–3229, in *48th AIAA Aerospace Sciences Meeting Including the New Horizons Forum and Aerospace Exposition*, Orlando, FL, January 4–7, 2010, Curran Associates, Red Hook, NY, 2010, DOI:10.2514/6.2010-276; A. S. Gohardani, G. Doulgeris, and R. Singh, Challenges of future aircraft propulsion: A review of distributed propulsion technology and its potential application for the all electric commercial aircraft, *Progr. Aero. Sci.*, 47(5):369–391, 2011, DOI:20.1016/j.paerosci.2010.09.001; C. Pornet et al., Integrated fuel-battery hybrid for a narrow-body sized transport aircraft, *Aircraft Eng. Aero. Tech.*, vol. 86, iss. 6, 2014, DOI:10.1108/AEAT-05-2014-0062.

Imaging of developmental biological processes

Organismal development typically begins with a single cell and finishes with a highly organized three-dimensional structure. Through coordinated cell division, guided cell migration, cell segregation, and differentiation, cells are organized into distinct tissues. These tissues then create complex, integrated organ systems that function in a cooperative manner to promote life. Microscopy has proven an indispensable tool to visualize and analyze individual cell and tissue behaviors within the living embryo, thus elucidating complex developmental biological processes.

Fluorescence-based imaging. Critical to the understanding of early development is the need to first examine a broad, expanded view of tissue architecture before focusing on specific, individual cell behaviors. This allows the observer to better understand the individual cell behaviors in relation to their anatomical surroundings. This is especially relevant to the study of embryonic development, which comprises myriad distinct cellular events separated in both time and space. Whereas advances in light microscopy have been incremental, the advent of fluorescence-based imaging has greatly accelerated the ability to visualize and analyze complex developmental processes by permitting a broad view of tissue architecture, yet providing single cell resolution.

Fluorescence-based imaging has revolutionized the use of microscopy as an important enabling tool. However, there remain certain roadblocks that impede its full potential in the in vivo study of the developing embryo. First, phototoxicity associated with exposing living tissue to high amounts of laser light is detrimental to living cells and can interfere with normal developmental processes. Second, poor tissue penetration [limited to 150–200 micrometers (μm) for most standard confocal microscopes] has limited the ability to visualize events deep within the growing embryo. Poor depth resolution makes it difficult to resolve single cells, limiting the tracking of cell trajectories. This article discusses recent advances in imaging technologies and techniques that

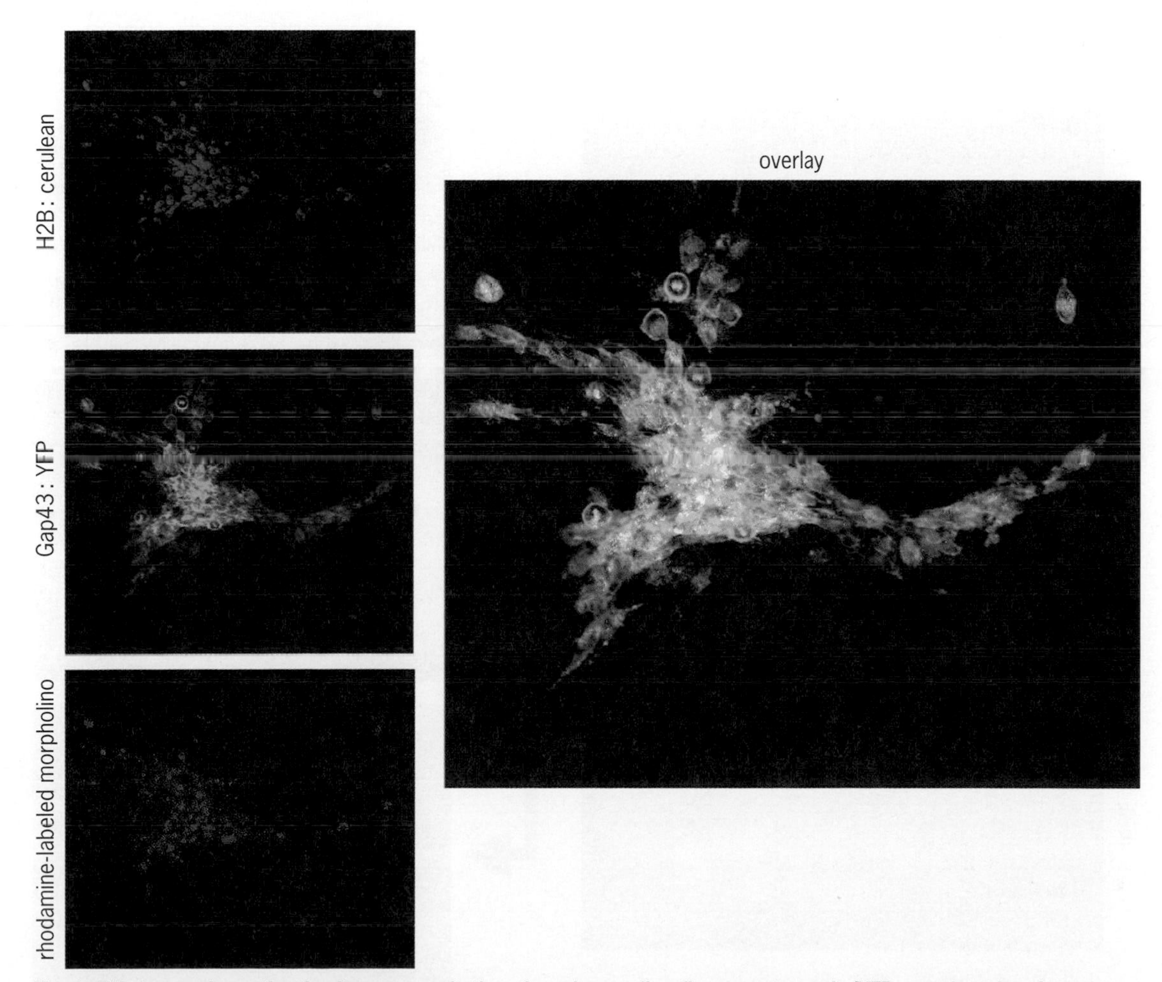

Fig. 1. This image shows the simultaneous excitation of cerulean, yellow fluorescent protein (YFP, pseudo-colored green in image), and rhodamine (red) fluorophores using a tunable Ti:S laser at 850 nm and coupled to an optical parametric oscillator. Two-photon imaging utilizes a pulsed femtosecond laser to excite fluorophores using roughly double the required excitation wavelength. Thus, a fluorophore (such as GFP) typically excited with a 488-nm-wavelength laser can be excited by a wavelength around 950 nm. Because this is a nonlinear process, several different fluorophores can be excited by a single wavelength.

help overcome these limitations and thus enhance the role of microscopy in the study of embryonic development in real time.

Imaging innovations and strategies. Overcoming phototoxicity requires shorter exposure to damaging laser light. This can be achieved in several ways. First, the development of improved, brighter fluorescent proteins can reduce the amount of energy required for excitation. Second, nonlinear two-photon excitation, which is distinct from standard one-photon excitation confocal microscopy, can be achieved for many fluorescent proteins using near-infrared excitation wavelengths. Near-infrared light is less damaging and penetrates deeper into tissues than shorter visible and ultraviolet (UV) wavelengths, simultaneously reducing phototoxicity and enhancing depth resolution. Two-photon imaging can also be used to simultaneously excite multiple fluorophores (fluorescent molecules) [**Fig. 1**], thus reducing the overall light necessary to excite multiple colors in the same biological sample. Third, the use of techniques that employ planar excitation (in contrast to point excitation) can significantly reduce the exposure time of harmful light to live tissue. Recently, this technique, called selective-plane illumination microscopy (SPIM), has been combined with two-photon excitation. This combination offers reduced phototoxicity, increased depth penetration for improved axial resolution, and lower background fluorescence. Two-photon SPIM can quickly image large structures (such as a developing embryo) with high resolution and has been used to image various developmental model systems, including the fly and the zebrafish.

Another key to successful in vivo imaging is the ability to label and track a single cell or group of cells in three dimensions over time (4D imaging). However, distinguishing one cell from another during cell–cell contact or as two cells cross trajectories can be challenging. One simple, yet highly effective method to separate cells is the use of multiple fluorescent colors. Multicolor approaches can be effective in separating single cells among a population, tracking cell movement and lineage, and resolving cellular subdomains (**Fig. 2**). A beautiful example of this technique is the introduction of multiple colors of fluorescent proteins into the neurons of mice. Termed "Brainbow," these

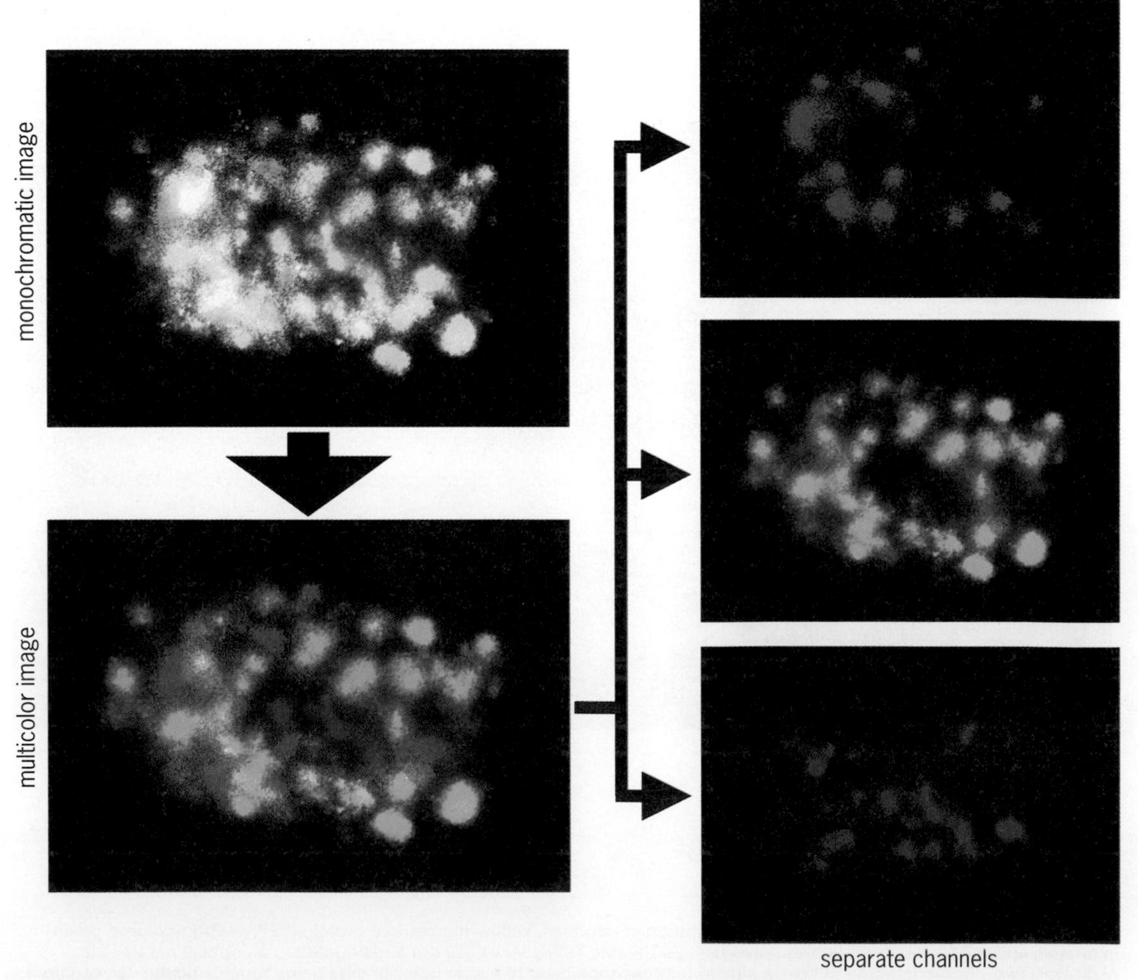

Fig. 2. This image shows how multiple colors can visually separate cells in a tightly grouped cluster. Equal amounts of cells were soaked in vital dyes of blue, green, and red, and then recombined to form a cluster. The monochromatic image is provided as a comparison to the multicolor image.

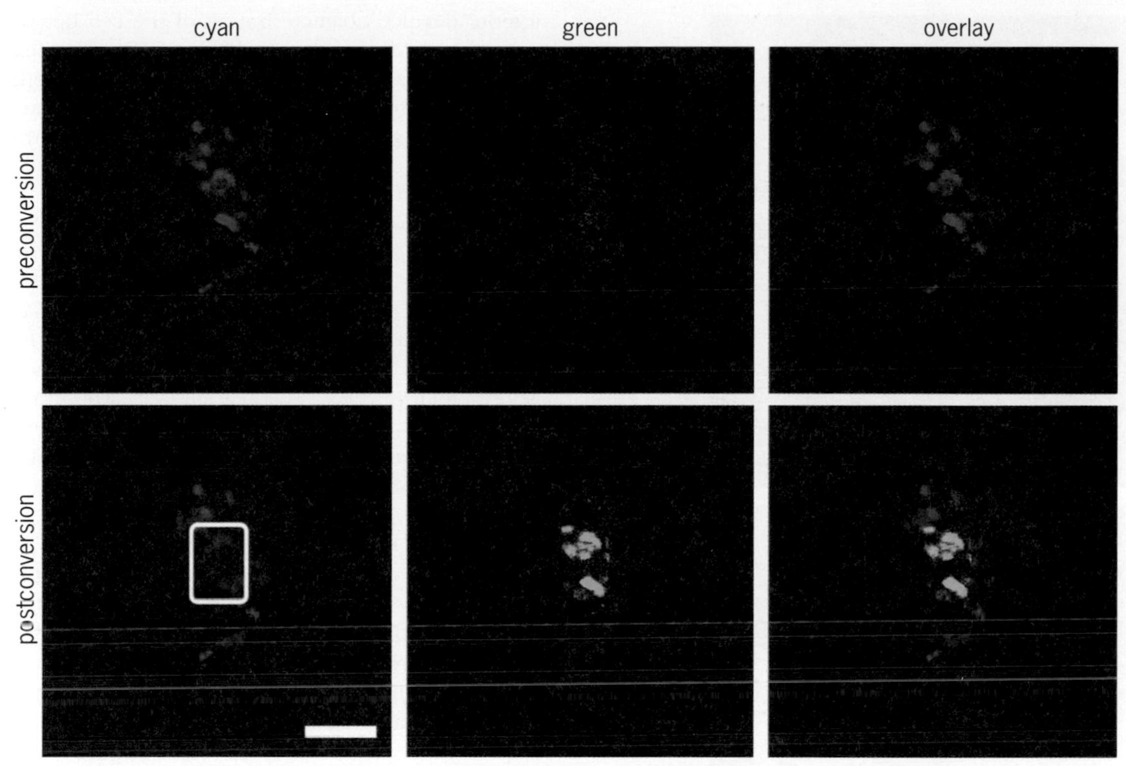

Fig. 3. This image shows the photoconversion of PSCFP2-labeled cells by two-photon excitation at 760 nm. Photoconversion is possible in vivo, and converted cells can be tracked for up to 48 h. The white box represents the region that was photoconverted. The scale bar represents 50 μm.

transgenes (genes or genetic materials that have been experimentally transferred from one organism to another) stochastically label neurons with different colors, allowing biologists to observe complex neuronal circuitry in live tissue with unprecedented resolution.

Another powerful method for distinguishing single cells or cell populations over time is the use of photoactivatable or photoconvertible fluorescent proteins. Although other cell-marking strategies, such as fluorescent dye injection or targeted electroporation of labeling constructs, can be effective, the spatial and temporal accessibility of target cells in live embryos presents limitations on these techniques. The use of photoactivatable or photoconvertible proteins can overcome these obstacles. Photoactivatable proteins, such as photoactivatable green fluorescent protein (GFP), display a significant increase in fluorescence intensity following exposure to near-UV light, whereas the emission spectrum of photoconvertible proteins, such as photoswitchable cyan fluorescent protein (PSCFP), shifts toward the red light spectrum following exposure to the proper activating wavelength of light. Importantly, many of these proteins can also be "activated" by two-photon laser excitation using near-infrared light. The use of two-photon excitation facilitates subcellular accuracy and deeper tissue penetration, with reduced phototoxicity owing to the fact that exposure to UV light is not necessary. Because photoconverted proteins can retain their red-shifted color for greater than 48 h, this technique allows for specific cell tracking and lineage studies in vivo over time (**Fig. 3**).

Integration of gene expression analysis. In addition to enhanced and novel approaches to visualizing cell behaviors during development, significant advances have been made recently in the ability to integrate gene expression information with observed cell behaviors. Rapid progress over the last several years has been made in gene expression analysis of single cells. By combining whole transcriptome (RNA transcript) analysis of single cells with in situ single cell isolation techniques, such as laser capture microdissection (LCM), researchers can now identify and label specific cells based on behaviors of interest, and then can remove those cells for gene expression analysis. By comparing the gene expression signatures of cells displaying unique behaviors, candidate genes that may regulate a specific behavior can be identified. As an example, the chick embryonic developmental model system has been utilized to study invasive behaviors of human cancer cells. As the chicken egg provides easy access to the developing embryo, both in vivo imaging and embryo manipulation are less challenging with the chick model than mammalian systems. Thus, this model system provides an in vivo microenvironment into which investigators can transplant human cancer cells and watch their behavior in real time. Using LCM, it is possible to then harvest cells of interest and perform comparative gene profiling experiments that may provide clues into the process of cancer metastasis (**Fig. 4**). The combined approaches of

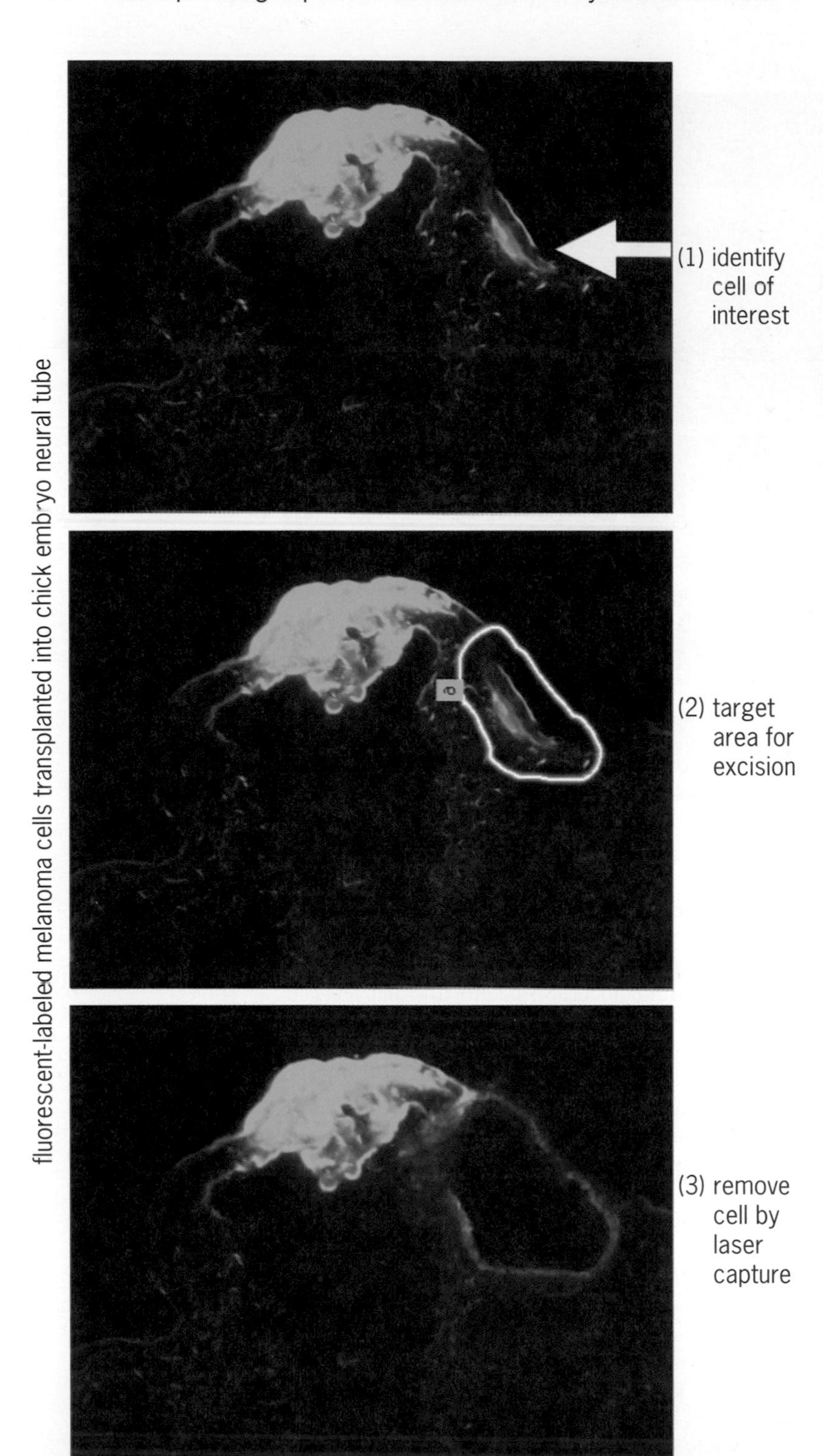

Fig. 4. This image shows the extraction of a single fluorescently labeled cell by laser capture microdissection (LCM). Following transplantation into the chick embryo, fluorescent-labeled cancer cells were allowed to invade the host tissue. The embryo was then cryosectioned (that is, subjected to frozen sectioning procedures), and cells of interest were extracted by LCM and profiled by comparative gene expression analysis.

microscopic observation, manipulation, and gene expression analysis will continue to advance the understanding of the intricate processes associated with embryonic development.

Future outlook. In summary, recent advances in imaging technology have enabled biologists to build a more detailed characterization of the cell behaviors and molecular signals that underlie complex developmental processes. Continued advancement that will permit the linking of these multiscale data should yield greater understanding into how intricacies at the molecular level dictate cell and tissue behaviors that produce the complete embryo.

For background information *see* CELL (BIOLOGY); DEVELOPMENTAL BIOLOGY; EMBRYOLOGY; FLUORESCENCE; FLUORESCENCE MICROSCOPE; GENE; GREEN FLUORESCENT PROTEIN; LASER PHOTOBIOLOGY; LASER SPECTROSCOPY; MICROSCOPE; MOLECULAR BIOLOGY in the McGraw-Hill Encyclopedia of Science & Technology. Caleb M. Bailey; Paul M. Kulesa

Bibliography. C. M. Bailey, J. A. Morrison, and P. M. Kulesa, Melanoma revives an embryonic migration program to promote plasticity and invasion, *Pigment Cell Melanoma Res.*, 25:573–583, 2012, DOI:10.1111/j.1755-148X.2012.01025.x; P. M. Kulesa et al., Multispectral fingerprinting for improved in vivo cell dynamics analysis, *BMC Dev. Biol.*, 10:101, 2010, DOI:10.1186/1471-213X-10-101; J. Livet et al., Transgenic strategies for combinatorial expression of fluorescent proteins in the nervous system, *Nature*, 450:56–62, 2007, DOI:10.1038/nature06293; I. C. Macaulay and T. Voet, Single cell genomics: Advances and future perspectives, *PLoS Genet.*, 10(1):e1004126, 2014, DOI:10.1371/journal.pgen.1004126; P. Mahou et al., Multicolor two-photon tissue imaging by wavelength mixing, *Nat. Methods*, 9:815–818, 2012, DOI:10.1038/nmeth.2098; J. A. Morrison, C. M. Bailey, and P. M. Kulesa, Gene profiling in the avian embryo using laser capture microdissection and RT-qPCR, *Cold Spring Harb. Protoc.*, 2012(12), 2012, DOI:10.1101/pdb.prot072140; G. H. Patterson and J. Lippincott-Schwartz, A photoactivatable GFP for selective photolabeling of proteins and cells, *Science*, 297:1873–1877, 2002, DOI:10.1126/science.1074952; D. A. Stark and P. M. Kulesa, An in vivo comparison of photoactivatable fluorescent proteins in an avian embryo model, *Dev. Dynam.*, 236:1583–1594, 2007, DOI:10.1002/dvdy.21174; T. V. Truong et al., Deep and fast live imaging with two-photon scanned light-sheet microscopy, *Nat. Methods*, 8:757–760, 2011, DOI:10.1038/nmeth.1652.

Incorporating exposure science into life-cycle assessment

Life-cycle assessment (LCA) is used to estimate the potential for environmental damage that may be caused by a product or process, ideally before the product or process begins. LCA includes all of the steps from extracting natural resources through manufacturing through product use, disposal, and recycling. LCA is increasingly used in environmental and sustainability decision making. In the past, some LCAs have employed human and ecological toxicity as a surrogate for risk; however, risk is a function of both toxicity and exposure to the toxic agent. For more complete assessments, LCAs estimate the potential for environmental damage by incorporating environmental fate, transport, and an agent's potential contact with humans and other species with toxicity. LCA calculations of human toxicity and ecotoxicity can be significant, especially considering all of the substances and exposure pathways which can be involved at each time and place where the toxin is released. *See* ENVIRONMENTAL LIFE-CYCLE IMPACT OF ALTERNATIVE AVIATION FUELS.

Adding exposure information to inventories. Part of the challenge in making LCAs more useful as predictive tools has been incomplete and unreliable life-cycle inventories (LCIs). An LCI describes and quantifies the flows of matter and energy into and out of the systems being assessed. This includes matter entering by water and air flows, as well as releases to air, land, water, and biota. LCIs describe and quantify the flows of matter and energy into and out of the systems being assessed. This includes natural resources used or consumed throughout the life cycle, as well as releases to air, land, water, and biota. LCIs exist in many different formats, often distributed through commercial software (such as GaBi and SimaPro). Recently, open-source software is making data more accessible to a larger audience and improving transferability of data.

There are a number of ways that exposure and toxicity can be combined in LCAs, and this integration is continuing to develop. In the past, the U.S. EPA's *Exposure Factors Handbook* has been incorporated into risk assessment spreadsheets such as CALTOX and developed into LCA characterization factors such as TRACI's original human health characterization factors.

Currently, other tools are available with several being evaluated as part of the U.S. EPA's Systematic Empirical Evaluation of Models (SEEM) framework. SEEM applies models to extrapolate exposure and internal dose for multiple-exposure scenarios, routes, and pathways. These high throughput predictions can be made by using data and models in a forward manner, or be inferred in a reverse manner, such as from biomarkers. However, these tools have not yet been utilized within LCA.

Recently it has been recognized that microenvironments (μE) may have a significant impact on human health, but have not been included in LCAs. The initial focus to estimate exposure to chemicals in products used in μE encountered in their daily habits and routines necessitates a "systems" model to delineate data needs arising from numerous knowledge bases to integrate product formulations, purchasing and use activities, and human activities. This will indicate products likely to be in different μEs and the potential for human contact with chemicals in these products. Modeling consumer product exposures requires a wide array of data, including (1) product ingredients, (2) pharmacokinetic factors, (3) consumer product category-specific "exposure factor surrogates," and (4) time/activity estimates (human factors). These different data streams must then be integrated within an interface such that different exposure scenarios for individual, population, or occupational time-use profiles can be interchanged and quantitatively explored to screen tailored chemicals for potential exposure, and ultimately dose. This allows estimates of multichemical signatures of exposure, internalized dose (uptake), remaining dose or body burden, and elimination.

The resulting exposure data can then be mapped onto toxicity data. For example, EPA's ToxCast system predicts and prioritizes chemicals based on their potential toxicity by using advanced science tools to help to understand how normal human body processes are impacted by exposures to chemicals and to help determine which exposures are most likely to lead to adverse health effects. Exposure information can be combined with toxicity information in various ways ranging from qualitative to quantitative (*see* **illustration**). Although the approaches described in the illustration were developed for human exposure, similar approaches may be used for integrating exposure and ecotoxicity. For example, potential exposures to sentinel or endangered species may not differ substantially from human exposures since the target is a single species. Thus, the LCA may account for a chemical agent in a product that is intentionally applied (for example, a pesticide) or one that migrates to a habitat (for example, released from a stack or outfall structure and then transported to the habitat). Thereafter, the activities of the species (such as predator-prey, migration, and bioaccumulation factors) and pharmacokinetic factors within the organism can be modeled and estimated.

More comprehensive exposure assessment within LCA. Throughout the history of LCA, there has been a search for the appropriate level of sophistication and comprehensiveness concerning how to incorporate human health. The evolution from the exclusive use of toxicity information to the current practice acknowledges that most LCAs continue to lack incorporation of important environmental and human health risks, since they do not include all of the exposure pathways of agents, especially the thousands of chemical compounds in the marketplace. In addition to the exposure pathways found in the *Exposure Factors Handbook* and the pathways traditionally included in LCA, current and future research is focusing on consumer product exposures (for example, cosmetics), indoor air releases (such as off-gassing of building materials), and occupational exposures. This greater comprehensiveness in human and environmental exposure information should improve the utility and predictive ability of LCAs and make them more valuable risk management tools.

Disclaimer. The United States Environmental Protection Agency through its Office of Research and Development funded and managed the research described here. It has been subjected to Agency review and approved for publication. Mention of trade names or commercial products does not constitute endorsement or recommendation for use.

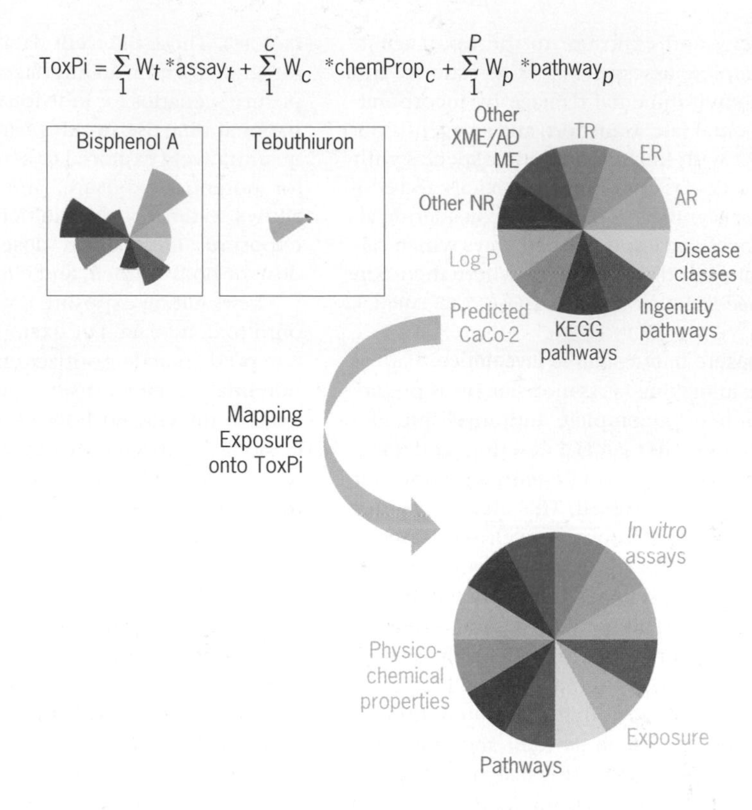

$$E = \Sigma\left[\left(\int_{t=t_1}^{t=t_1} C(t)dt\right)_{\text{Activity 1}} + \left(\int_{t=t_1}^{t=t_1} C(t)dt\right)_{\text{Activity 2}} + \ldots \left(\int_{t=t_1}^{t=t_1} C(t)dt\right)_{\text{Activity 2}}\right]$$

where

E = personal exposure during time period from t_1 to t_2
$C(t)$ = concentration at interface, at t.

(a)

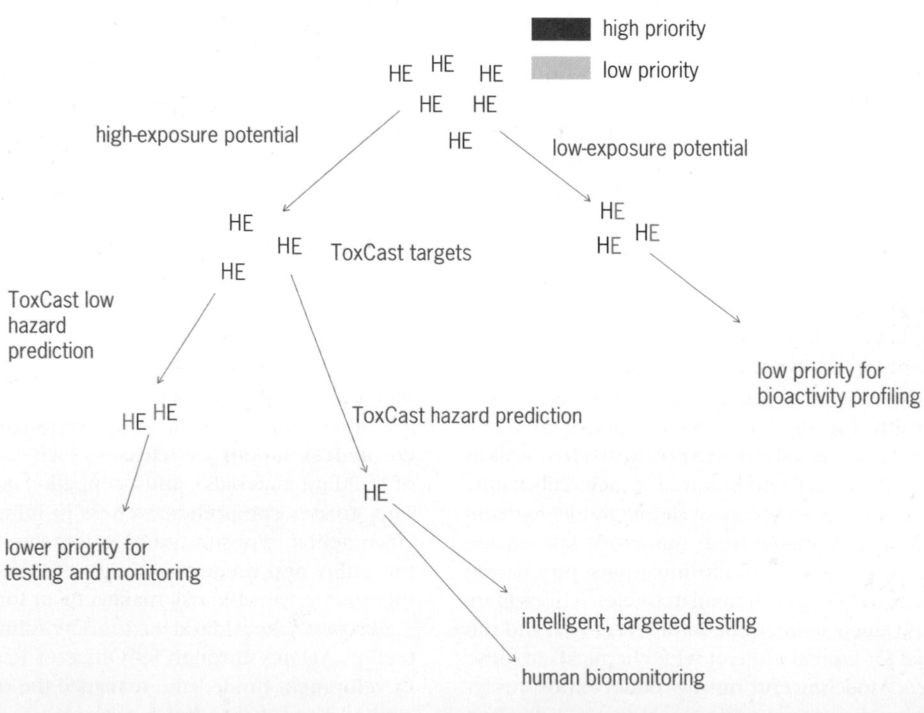

(b)

A variety of approaches are being considered ranging from a qualitative approach to semiquantitative toxicity estimation and prioritization (known as ToxPi) scoring approach (*S. Gangwal et al., 2012*) to more sophisticated quantitative approaches (*R. Judson, Personal communications, EPA, 2011*), whereby overlapping distributions of hazard and exposure distributions are evaluated. (*a*) Mapping exposure to ToxPi (toxicological priority index). (*b*) Qualitative approach for combining ExpoCast exposure (E) and ToxCast hazard (H) information.

For background information *see* ENVIRONMENTAL ENGINEERING; ENVIRONMENTAL TOXICOLOGY; MODEL THEORY; MUTAGENS AND CARCINOGENS; RISK ASSESSMENT AND MANAGEMENT; SIMULATION; TOXICOLOGY in the McGraw-Hill Encyclopedia of Science & Technology. Jane C. Bare; Daniel A. Vallero

Bibliography. J. Bare, Life cycle impact assessment research developments and needs, *Clean Technol. Environ. Policy*, 12(4):341–351, 2010, DOI:10.1007/s10098-009-0265-9; J. C. Bare and T. P. Gloria, Environmental impact assessment taxonomy providing comprehensive coverage of midpoints, endpoints, damages, and areas of protection, *J. Cleaner Product.*, 16(10):1021–1035, 2008; J. C. Bare et al., TRACI—The tool for the reduction and assessment of chemical and other environmental impacts, *J. Ind. Ecol.*, 6(3):49–78, 2008, DOI:10.1162/108819802766269539; European Commission, European reference life-cycle database (ELCD), 2008; R. Frischknecht et al., The ecoinvent database: Overview and methodological framework, *Int. J. LCA*, 10(1):3–9, 2005, DOI:10.1065/lca2004.10.181.1; S. Gangwal et al., Incorporating exposure information into the toxicological prioritization index decision support framework, *Sci. Total Environ.*, 435-436(Oct):316–325, 2012, DOI:10.1016/j.scitotenv.2012.07.003; M. Goedkoop et al., ReCiPe 2008, A life cycle impact assessment method which comprises harmonised category indicators at the midpoint and the endpoint level, First edition Report 1: Characterisation, January 6, 2009, http://www.lcia-recipe.net; J. Guinée et al., *Handbook on Life Cycle Assessment: Operational Guide to the ISO Standards*, Dordrecht, Netherlands, Kluwer Academic Publishers, 2002; M. Z. Hauschild et al., Identifying best existing practice for characterization modeling in life cycle impact assessment. *Int. J. LCA*, 18:683–697, 2012, DOI:10.1007/s11367-012-0489-5; S. Hellweg et al., Integrating human indoor air pollutant exposure within life cycle impact assessment, *Environ. Sci. Tech.*, 43(6):1670–1679, 2009, DOI:10.1021/es8018176; M. A. J. Huijbregts et al., Human population intake fractions and environmental fate factors of toxic pollutants in life cycle impact assessment, *Chemosphere*, 61(10):1495–1504, 2005, DOI:10.1016/j.chemosphere.2005.04.046; W. Ingwersen et al., Integrated metrics for improving the life cycle approach to assessing product system sustainability, *Sustainability*, 6(3):1386–1413, 2014, DOI:10.3390/su6031386; T. E. McKone, *CalTOX, A Multimedia Total Exposure Model for Hazardous-Waste Sites*, Lawrence Livermore National Laboratory, Livermore, CA, 1993; J. Mitchell et al., Comparison of modeling approaches to prioritze chemicals based on estimates of exposure and exposure potential, *Sci. Total Environ.*, 458-460:555–567, 2013, DOI:10.1016/j.scitotenv.2013.04.051; National Academy of Sciences, *Science in the EPA—The Road Ahead*, National Academies Press, 2012; National Research Council, *Sustainability and the US EPA*, Washington DC, The National Academies Press, 2011; R. R. Tice et al., Improving the human hazard characterization of chemicals: A Tox21 update, *Environ. Health Perspect.*, 121:756–765, 2013, DOI:10.1289/ehp.1205784; H. A. Udo de Haes, A. Wegener Sleeswijk, and R. Heijngs, Similarities, differences and synergisms between HERA and LCA—An analysis at three levels, *Hum. Ecol. Risk Assessment*, 12(3):431–449, 2006; United Nations, A 10-year framework of programmes on sustainable consumption and production patterns: Outcome of the Rio +20 UN Conference on Sustainable Development, 2012; U.S. Department of Energy, US Life-Cycle Inventory Database, 2008; U.S. Environmental Protection Agency, *Exposure Factors Handbook 2011 Edition (Final)*, Washington, DC, EPA/600/R-09/052F, 2011; J. Wambaugh, et al., High-throughput models for exposure-based chemical prioritization in the ExpoCast project, *Environ. Sci. Tech.*, 47:8479–8488, 2013, DOI:10.1021/es400482g.

Influence of nanostructural features on the properties of magnetic nanoparticles

Magnetic nanostructures may exhibit novel magnetic phenomena that do not occur in the corresponding bulk materials. These phenomena arise from the interplay of quantum, finite-size interactions with surface and interface effects. In particular, magnetic nanoparticles serve as powerful building blocks that have led to many nanotechnology applications in fields such as ultrahigh-density magnetic recording, biomedicine (for example, guided drug delivery, and cancer treatment through hyperthermia or cell separation and purification), and magnetic resonance imaging (MRI). The key challenges are to understand how nanostructural features (such as topography, composition, crystallinity, and surface chemistry) influence the physical properties of the nanoparticles (magnetic, electronic, and so forth), and how this insight can be used to induce or improve their functionality in different applications. For example, a detailed knowledge of specific surface modifications of nanoparticles is relevant in their functionalization using biomolecular interactions, or in the manipulation of interfacial cross-coupling between magnetic and electric properties in nanocomposites that exhibit simultaneously two or more ferroic cooperative phenomena, such as ferromagnetism and ferroelectricity. Meeting such challenges calls for advanced particle growth methods and an adequate combination of characterization techniques with increased spatial and chemical resolution, as well as high sensitivity to magnetic and electronic properties.

Nanoparticle requirements. As the size of magnetic particles is reduced below about 100 nm, strong deviations from bulk properties are widely observed (hereinafter referred to as particlelike behavior), often resulting in an increase of the magnetic disorder with respect to their bulk counterparts, and that are usually attributed to finite-size effects and surface spin disorder due to symmetry breaking and charge rearrangement. For transition-metal oxide nanoparticles, some typical magnetic anomalies include: (1) saturation magnetization of a factor of 2 or more smaller than in the bulk; (2) spin-glass-like freezing resulting from site disorder and frustration of magnetic interactions; and (3) high-field irreversibility in the hysteresis loops due to modified effective anisotropy energy barriers. Such particlelike behavior may induce undesirable effects in many applications. In particular, for biomedical purposes, the nanoparticles must meet the following requirements: (1) superparamagnetic behavior (meaning that the spatial orientation of the nanoparticle magnetization can rapidly fluctuate under thermal activation) near room temperature, to avoid particle aggregation due to interparticle interactions; (2) high saturation and low magnetic anisotropy, to achieve a high magnetic response under the application of moderate magnetic fields; (3) a limited size range for in vivo applications, 10–70 nm; and (4) biocompatibility and functionality to reach specific targets inside the body. Research in this field is mainly focused on new synthesis methods enabling the preparation of nanoparticles with high crystal quality and reduced magnetic disorder.

Iron oxide nanoparticles. Magnetite (Fe_3O_4) and maghemite (γ-Fe_2O_3) nanoparticles are of great interest because of their relevant magnetic properties, their potential for spin-filter devices and catalysis, and, crucially, their biological applications, which are made possible by their low toxicity and their ease of functionalization. Therefore, intensive efforts are being devoted to obtain size-controlled iron oxide nanoparticles with bulklike properties. Nevertheless, many questions remain open, such as the half-metallic character of these nanoparticles, the orbital contribution to their magnetic moments, and identification of the limiting factors to actually achieve the theoretically predicted large-spin polarization of Fe_3O_4, which is of relevance for spintronic devices.

Chemical synthesis based on the high-temperature decomposition of iron organometallic precursors in organic media has proved successful for the preparation of Fe_3O_4 nanoparticles. The presence of molecules of surfactants (fatty acids) in the solution allows control over the size and shape of the particles during their growth, hinders unwanted chemical reactions for specific molecular geometries, and avoids particle agglomeration. Depending on the temperature profile, final synthesis temperature (about 250–320°C or 482–608°F), reaction time, and chemical concentrations of precursors, solvents, and surfactants, $Fe_{3-x}O_4$ nanoparticles can be obtained with well-controlled shape and size, the latter ranging from a few nanometers to about 50 nm. A key factor is to use surfactants and organic solvents with higher boiling points than those of the organic precursors. These nanoparticles are typically hydrophobic due to the nonpolar organic acids used as surfactants. In contrast, iron oxide nanoparticles synthesized by coprecipitation of two iron salts in water enables the formation of hydrophilic nanoparticles with an even wider size range, but at the expense of much degraded magnetic properties on account of poorer crystal quality and a marked tendency to nanoparticle agglomeration. Therefore, one of the main challenges is to improve synthesis routes in order both to extend the range of achievable sizes of the nanoparticles without significant deterioration of their magnetic properties and to ligand-exchange the nanoparticles to a hydrophilic coating to make them biocompatible.

Magnetic nanoparticles with bulklike properties. The effects of the synthesis method and surface coating of nanoparticles on their structural quality and magnetic and electronic properties have been studied by comparing two sets of $Fe_{3-x}O_4$ nanoparticles with diameters of 5–50 nm. The first set was synthesized by thermal decomposition of an organic iron precursor (Fe[III] acetylacetonate) in an organic medium (dibenzyl ether, octyl ether, 1-octadecene, or trioctylamine) with a fatty acid (oleic or decanoic acid) used as a surfactant covalently bonded to the particle surface. The second set was synthesized by coprecipitation of Fe[II] and Fe[III] salts in an alkaline aqueous medium, with poly(vinyl alcohol) [PVA] used as a protective coating against oxidation adsorbed at the nanoparticle surface but without any chemical bond. High-resolution transmission electron microscopy (TEM) [**Figs. 1***a*, 1*b*, and **2**], magnetic measurements (**Fig. 3**), and synchrotron-based x-ray absorption spectroscopy (XAS) and x-ray magnetic circular dichroism (XMCD) showed that covalently bonded nanoparticles displayed very high crystal quality up to the particle surface and bulklike magnetic and electronic properties (for example, hysteresis loops resembling those of bulk Fe_3O_4 and barely dependent on the particle size), while nanoparticles with adsorbed coatings (PVA, dextran, or tetramethylammonium hydroxide) synthesized at lower temperature showed poor crystallinity (Fig. 1*c* and *d*) and particlelike properties (low magnetization, high anisotropy, and glassy behavior) [inset of Fig. 3]. The types of surface bonds were confirmed by the charge transfer analysis obtained from XAS measured at the Fe absorption $2p_{3/2} \rightarrow 3d$ (L_3) and $2p_{1/2} \rightarrow 3d$ (L_2) edges. Oleic acid-coated, nearly spherical, 5-nm $Fe_{3-x}O_4$ nanoparticles showed a saturation magnetization at low temperature of 80–87 emu/g (1 emu/g $= 1\ A \cdot m^2/kg$), close to that expected for bulk Fe_3O_4 (98 emu/g), and in contrast to the very much reduced value found for PVA-protected nanoparticles (about 50 emu/g). The XMCD results confirmed the dependence of the magnetic moment on the surface bonds and suggested that the orbital moment was effectively quenched in covalently bonded nanoparticles. In contrast, PVA-coated nanoparticles showed about a threefold increase in the orbital contribution, as often observed in low-dimensional structures, surfaces, and atomic aggregates.

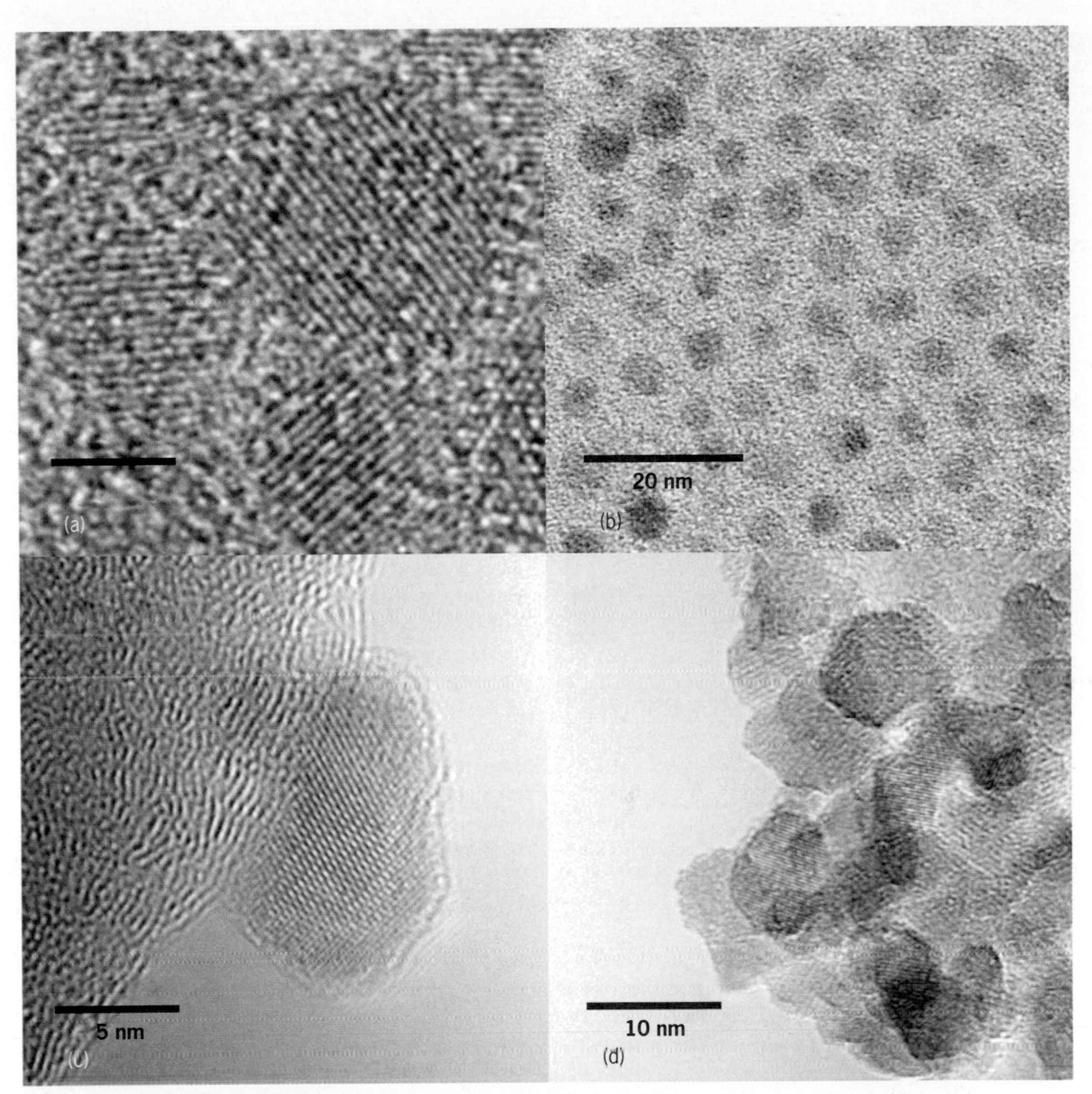

Fig. 1. High-resolution transmission electron microscopy (TEM) images of nanoparticles. (*a*, *b*) Oleic acid-coated $Fe_{3-x}O_4$ nanoparticles prepared by thermal decomposition, showing high crystalline quality and individual coating. (*c*, *d*) PVA-protected $Fe_{3-x}O_4$ nanoparticles produced by coprecipitation, showing poor crystallinity and particle aggregation. (*From X. Batlle et al., Magnetic nanoparticles with bulklike properties (invited), J. Appl. Phys., 109:07B524, 2011, DOI:10.1063/1.3559504; reprinted with permission from American Institute of Physics*)

It should be emphasized that none of the previously described particlelike behavior is observed in single-phase, highly crystalline nanoparticles down to a few nanometers. Therefore, magnetic disorder phenomena in nanoparticles should not generally be regarded as intrinsic surface-induced or finite-size effects, but rather as resulting from a lack of crystallinity and homogeneity of the chemical phases. This conclusion is further supported by Monte Carlo simulations of a single-particle model showing that intrinsic finite-size and surface effects are relevant only for sizes below about 5 nm.

Direct imaging and characterization of single particles. The apparently strong relationship between the structural parameters and compositional homogeneity of the individual nanoparticles and their improved magnetic response emphasizes the importance of single-particle experiments in order to get a deeper and consistent insight into the individual and collective properties. The magnetic and electronic properties of single nanoparticles can be addressed by combining photoemission electron microscopy (PEEM) with synchrotron-based, polarization-dependent XAS. This highly sensitive, element-specific technique provides quantitative information about the chemical composition, structural parameters, electronic structure, and magnetism (spin and orbital moments separately), both static and time-resolved, of single nanoparticles down to about 5 nm, provided that the average interparticle distance is larger than about 3 times the typical PEEM lateral resolution (about 50 nm). The main goals are to investigate the evolution of the electronic structure, chemical bonding, and surface modification of single nanoparticles as a function of size, shape, and crystalline structure; to study

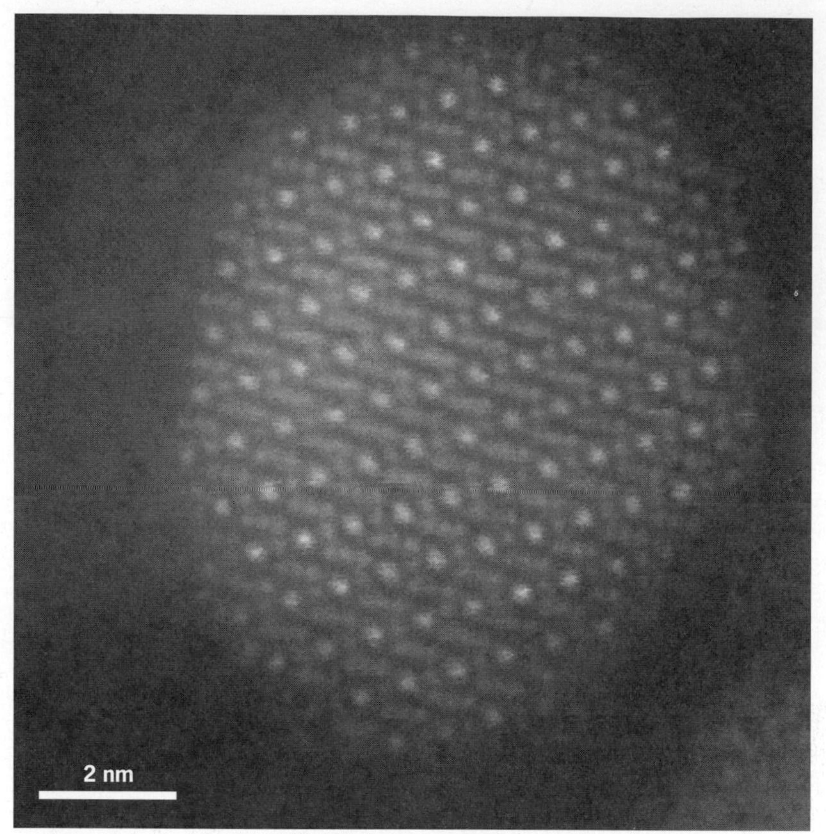

Fig. 2. Aberration-corrected annular dark-field scanning transmission electron microscopy (STEM) image of an Fe_3O_4 nanoparticle with high crystalline quality up to the particle surface. (*Courtesy of Maria Varela, Universidad Complutense de Madrid and Oak Ridge National Laboratory*)

the influence of different surfactants on the particle surface; and, crucially, to correlate the existence of structural defects (for example, oxygen vacancies and cationic distribution) to the magnetic properties. For example, the dependence of the latter on surface modifications such as the type and strength of the surface bonds is crucial to the nanoparticle functionalization using biomolecular interactions.

A quantitative analysis of the absorption spectra from single, covalently bonded $Fe_{3-x}O_4$ nanoparticles with mean diameters of 15 and 24 nm, deposited on bare and carbon-coated silicon substrates, respectively, can be obtained from fitting the line shapes and relative heights of the L_{3A}/L_2 and L_{3A}/L_{3B} peaks (the subscripts A and B correspond to the double-peak structure of the L_3 edge in **Fig. 4*a***) to a weighted sum of reference bulk spectra of different iron oxide species. Interestingly, the data show that even for particles of the highest crystal quality, the variation of some synthesis parameters may significantly alter the cationic distribution and the uniformity of magnetic phases (Fig. 4). In particular, it is found that 15-nm nanoparticles, synthesized by thermal decomposition of Fe[III] acetylacetonate in 1-octadecene, using oleic acid as a surfactant and 1,2-hexadecanediol as a stabilizer, are very homogeneous in stoichiometry, compatible with an Fe_3O_4 core surrounded by a thin γ-Fe_2O_3 shell (Fig. 4*a*). In contrast, 24-nm nanoparticles (Fig. 4*d*), synthesized using a different solvent (benzyl ether) and no stabilizer, are inhomogeneous, composed of either a thick FeO or metallic iron core surrounded by a thin Fe_3O_4 shell (50% and 10% of the nanoparticles, respectively), or an Fe_3O_4 core surrounded by a thin γ-Fe_2O_3 shell (40% of the nanoparticles). X-ray PEEM also enables measurement of the magnetization orientation of single nanoparticles through XMCD. Black and white contrast in Fig. 4*c* indicates opposite magnetization orientations, whereas any other gray level reflects either intermediate orientations or a nonstable magnetic contrast due to thermal fluctuations during measurement time, but a quantitative, accurate determination of their spin and orbital moments is still lacking due to the hitherto poor XMCD spectral quality.

A deeper insight regarding the enhanced magnetization at the particle surface in highly crystalline particles was obtained through a direct characterization of the structural, chemical, and magnetic properties of single $Fe_{3-x}O_4$ nanoparticles, combining aberration-corrected scanning transmission electron microscopy (STEM) and electron energy loss spectroscopy (EELS) with subnanometer resolution (**Figs. 5*a*** and **5*b***). Through these measurements, electron magnetic chiral dichroism (EMCD) analysis could be carried out (Figs. 5*c* and 5*d*), yielding a profile of the local magnetization. (The magnetization along the nanoparticle is proportional to the curve in the lower part of Fig. 5*d*.) In pioneering work, EMCD magnetization maps of single, covalently bonded $Fe_{3-x}O_4$ nanoparticles show that the surface of the nanoparticle is highly ferrimagnetic, with a magnetization of about 70% of that of the core, suggesting that surfactant organic molecules restore magnetism in metal-oxide nanoparticle surfaces (Fig. 5*d*). The comparison of the experimental

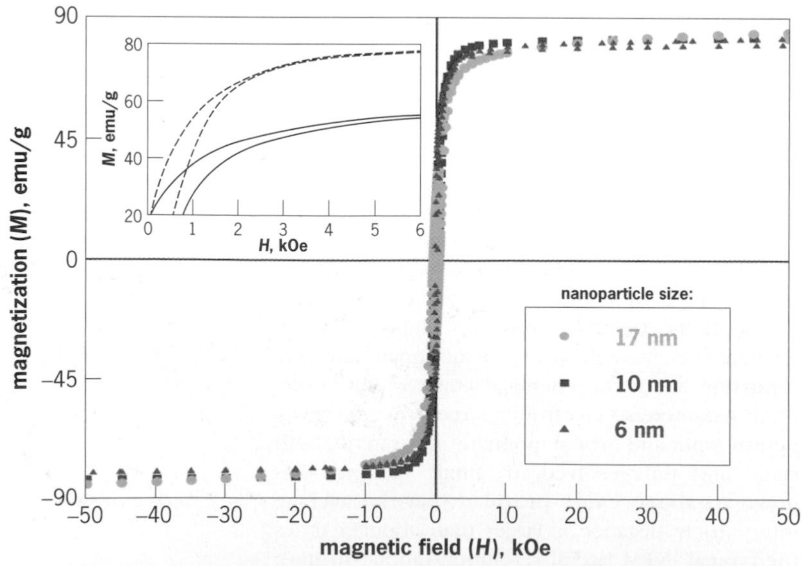

Fig. 3. Hysteresis loops at temperature $T = 5$ K of $Fe_{3-x}O_4$ nanoparticles of different particle sizes obtained by thermal decomposition. The inset compares magnified regions of the hysteresis loops corresponding to 5-nm nanoparticles prepared by coprecipitation (solid line) and thermal decomposition (broken line), showing that reduced magnetization and much higher values of the field at which the onset of irreversibility takes place are observed in the coprecipitation case. 1 kOe (kilooersted) = 7.96×10^4 A/m; 1 emu/g = 1 $A \cdot m^2/kg$.

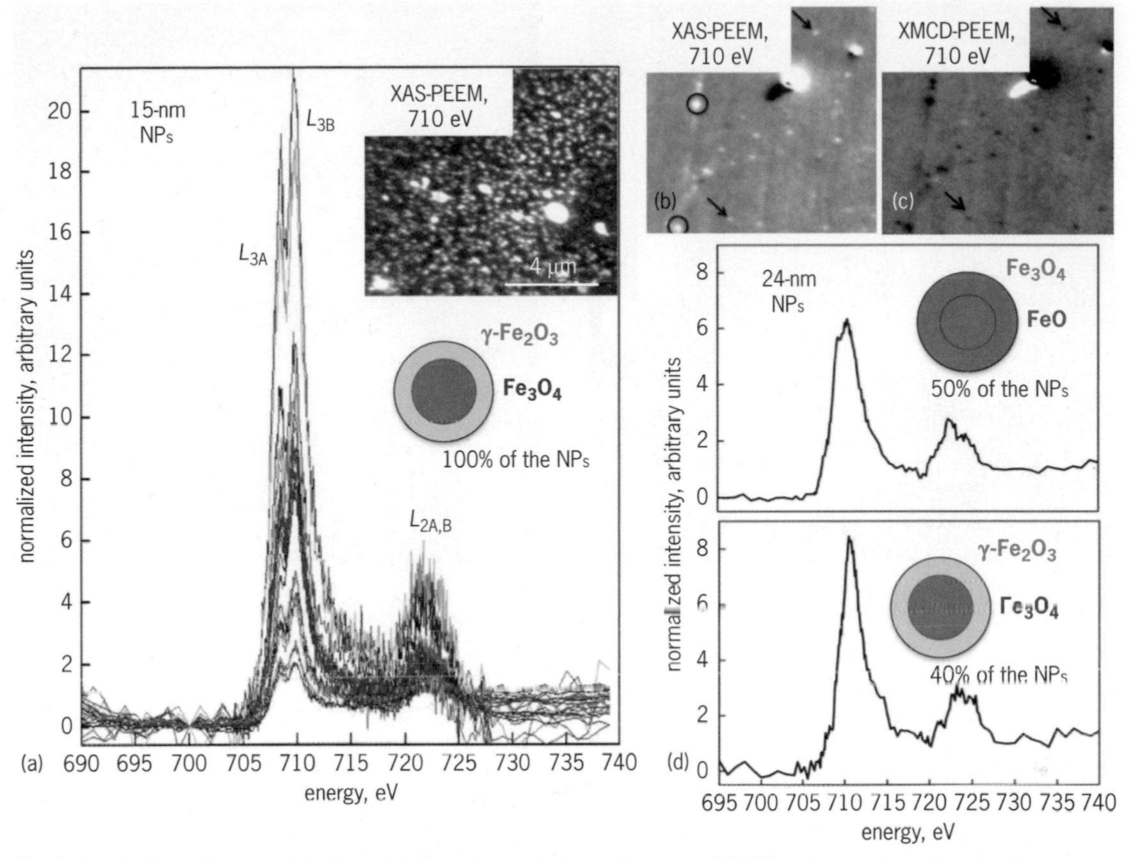

Fig. 4. Results based on a combination of photoemission electron microscopy (PEEM) and x-ray absorption spectroscopy (XAS). (*a*) Normalized local XAS of a selection of single 15-nm Fe_3O_4-based nanoparticles (NPs) obtained from a series of PEEM images recorded around the Fe L_3 and L_2 edges. The spectral shape of all 100% of the analyzed nanoparticles is compatible with an Fe_3O_4 core surrounded by a thin γ-Fe_2O_3 shell. The inset shows the elemental PEEM contrast (XAS-PEEM) of a single nanoparticle obtained by pixelwise division of two images recorded at the Fe L_3 absorption edge (710 eV) and at the pre-edge region (703 eV). (*b*) Elemental contrast obtained as in the inset of part a. (c) XMCD-PEEM contrast obtained by pixelwise division of two elemental contrast images (part *b*) with right and left circularly polarized x-rays. (*d*) Normalized local XAS of two single 24-nm nanoparticles, indicated by arrows in parts *b* and c: a single FeO/Fe_3O_4 nanoparticle and a single γ-Fe_2O_3/Fe_3O_4 nanoparticle, representative of 50% and 40% of the nanoparticles in the ensemble, respectively. While a double-peak structure is evident in the L_3 edge (L_{3A} and L_{3B} peaks), the resolution is not sufficient in the L_2 edge, and that single peak is referred to as $L_{2A,B}$ ($L_2 = L_{2A,B}$).

results with density functional theory calculations demonstrates the stabilization of the surface magnetization by the strong surface bond between the Fe and O ions in the carboxylic group of the oleic acid, yielding an O-Fe bond configuration and atomic distances close to those of bulk Fe_3O_4.

All of the foregoing indicates the key role of surface chemistry in determining the physical properties of ferrimagnetic nanoparticles and, in turn, in assisting in the design of novel nanoparticles with optimized magnetic properties.

Biological applications. It is possible to make hydrophobic nanoparticles that are water dispersible by replacing the oleic acid molecules on the surface of the nanoparticles by hydrophilic dimercaptosuccinic acid (DMSA). Although such a ligand exchange process reduces the saturation magnetization by about 10%, probably because of surface oxidation, and promotes the formation of a small amount of aggregates because of the shorter chain of DMSA as compared to oleic acid, the magnetic response is still much higher than that of the nanoparticles prepared by coprecipitation. These DMSA-coated nanoparticles display the improved relaxation rates of the surrounding water proton spins (the so-called MRI relaxivity) as compared to those of commercially available contrast agents. Moreover, it is also possible to obtain high-contrast MRI images of the livers and brains of mice by injecting the particles intravenously, or making them overcome the brain-blood barrier by osmotic disruption. The high magnetic signal of these particles enables biodistribution studies by means of magnetization measurements, providing an excellent alternative to the usual inductively coupled plasma mass spectroscopy (ICPMS) analysis, which does not discriminate among different iron species.

To illustrate this, the magnetization curves at 5 K of samples of lyophilized liver, kidney, and spleen of mice are shown in **Fig. 6**, where distinct contributions from the mouse tissue, which is a diamagnetic material (and therefore gives rise to curves with negative slope), paramagnetic natural ferritin (rounded curves), and ferromagnetic particles (squared curves) are clearly observed. Further, a comparison of the signal between injected and control animals allows accurate determination of the percentage of internalization of the nanoparticles

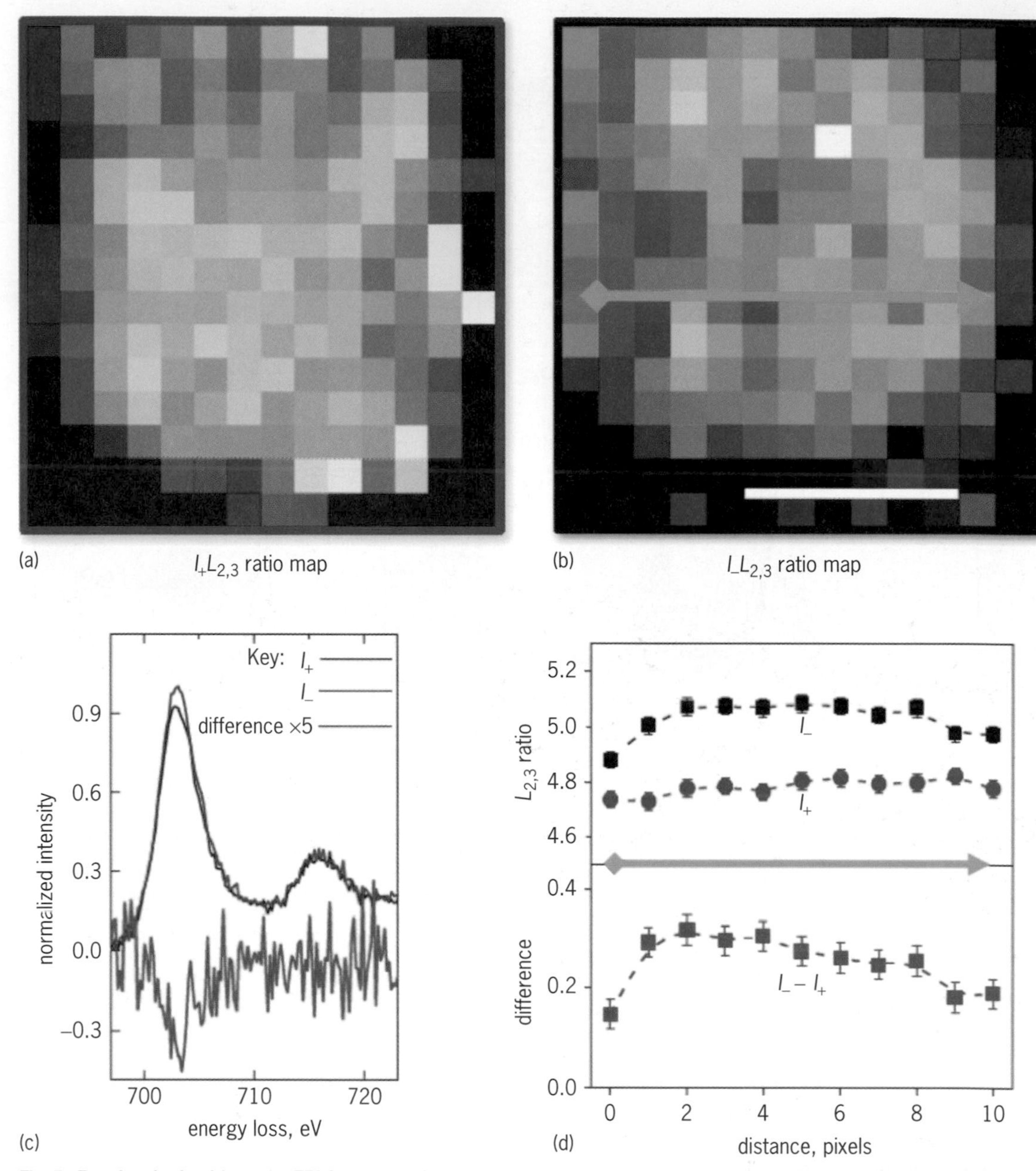

Fig. 5. Results obtained from the EELS data acquired at symmetric positions, I_+ and I_-, in the electron diffraction image of a nanoparticle with a cubic shape (not shown). (*a, b*) Color-coded $L_{2,3}$ ratio maps obtained at (part *a*) I_+ and (part *b*) I_-. ($L_{2,3}$ ratio increases from dark up to light colors; $L_{2,3}$ ratio stands for the L_3 to L_2 ratio of intensities.) Pixelwise division in the maps corresponds to minimum resolved areas. (*c*) Averaged I_+ and I_- EELS spectra around the Fe $L_{2,3}$ edges and the resultant dichroic signal, $(I_- - I_+) \times 5$. (*d*) Profiles along the direction of the arrow in part *b* of the I_- and I_+ $L_{2,3}$ ratios, and of the difference between them, $I_- - I_+$. *(Reprinted with permission from J. Salafranca et al., Surfactant organic molecules restore magnetism in metal-oxide nanoparticle surfaces, Nano Letters, 12:2499–2503, 2012, DOI:10.1021/nl300665z; copyright © 2012, American Chemical Society)*

with respect to the total dose down to an excellent mass fraction resolution of $10^{-4} - 10^{-5}$ (Fig. 6*d*). In the case of the kidney, the amount of particles is so small that it is the diamagnetic component that determines the magnetic behavior at high magnetic fields. A large amount of ferritin is naturally found in the spleen, so even the control samples show a predominantly paramagnetic behavior for all field values, with a superimposed ferromagnetic contribution for injected mice. The signal of the kidney and liver controls starts as ferromagnetic but becomes predominantly diamagnetic at moderate fields because of the lower amounts of ferritin found in these organs. The intravenous route, displaying the greatest uptake (a strong ferromagnetic signal), appears to be the most efficient form of administration to the liver, while the uptake was lower in the spleen and much lower in the kidney. By contrast, subcutaneous injection of the same nanoparticle dose yields no appreciable uptake in these three organs as compared to controls.

In conclusion, bulklike structural, magnetic, and electronic behavior in magnetic nanoparticles is strongly correlated with the crystal quality,

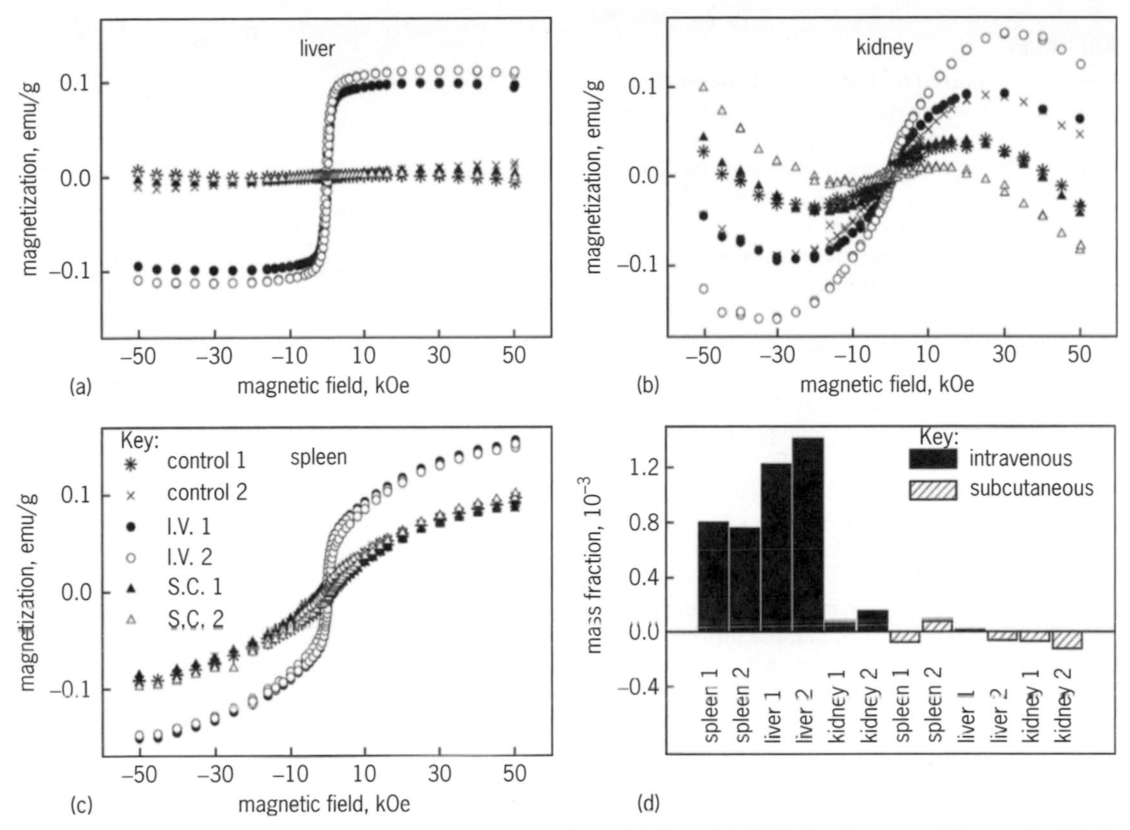

Fig. 6. Magnetization curves at 5 K of lyophilized samples of (*a*) liver, (*b*) kidney, and (*c*) spleen from two different mice (labels 1 and 2) after intravenous (I.V.) and subcutaneous (S.C.) nanoparticle treatments, compared with those of control mice (no nanoparticles injected). 1 kOe (kilooersted) = 7.96 × 10^4 A/m; 1 emu/g = 1 A · m^2/kg. (*d*) Mass fraction of accumulated iron oxide nanoparticles relative to the total organ mass. (*From R. Mejías et al., Liver and brain imaging through dimercaptosuccinic acid-coated iron oxide nanoparticles, Nanomedicine, 5:397–408, 2010, DOI:10.2217/nnm.10.15; reprinted with permission from Future Medicine Ltd.*)

composition homogeneity, and surface chemistry. In particular, for sizes larger than about 5 nm, particlelike behavior, such as strongly reduced magnetization, high magnetic anisotropy, and glassy behavior, is observed only in nanoparticles with poor crystallinity and structural defects. It has also been shown that thermal decomposition is a very versatile method for producing high-quality nanoparticles with bulklike properties and sizes of 5–50 nm that are suitable for biomedical applications. Finally, the results suggest that the magnetic disorder phenomena often observed in single-phase nanoparticles larger than about 5 nm should not be regarded as intrinsic finite-size or surface effects.

For background information, *see* CRYSTAL GROWTH; CRYSTAL STRUCTURE; DIAMAGNETISM; ELECTRON MICROSCOPE; EXTENDED X-RAY ABSORPTION FINE STRUCTURE (EXAFS); FERRIMAGNETISM; FERROMAGNETISM; MAGNETIC RESONANCE; MAGNETISM; MAGNETITE; MAGNETIZATION; NANOCHEMISTRY; NANOPARTICLES; NANOSTRUCTURE; PARAMAGNETISM; SPIN GLASS; SPINTRONICS; SURFACE AND INTERFACIAL CHEMISTRY; SURFACTANT; SYNCHROTRON RADIATION; X-RAY MICROSCOPE in the McGraw-Hill Encyclopedia of Science & Technology.

[Acknowledgments: This work was supported by the Spanish MICINN and MINECO (MAT2009-08667 and MAT2012-33037), Catalan DIUE (2009SGR856) and European Union FEDER funds (Una manera de hacer Europa). A. Fraile Rodríguez acknowledges support from the Spanish MICINN "Ramón y Cajal" Programme.]

Xavier Batlle; Amílcar Labarta; Arantxa Fraile Rodríguez

Bibliography. X. Batlle et al., Magnetic nanoparticles with bulklike properties (invited), *J. Appl. Phys.*, 109:07B524 (6 pp.), 2011, DOI:10.1063/1.3559504; X. Batlle and A. Labarta, Finite-size effects in fine particles: Magnetic and transport properties, *J. Phys. D: Appl. Phys.*, 35:R15–R42, 2002, DOI:10.1088/0022-3727/35/6/201; A. Fraile Rodríguez et al., Size-dependent spin structures in iron nanoparticles, *Phys. Rev. Lett.*, 104:127201 (4 pp.), DOI:10.1103/PhysRevLett.104.127201, 2010; R. Mejías et al., Liver and brain imaging through dimercaptosuccinic acid-coated iron oxide nanoparticles, *Nanomedicine*, 5:397–408, 2010, DOI:10.2217/nnm.10.15; N. Pérez et al., Nanostructural origin of the spin and orbital contribution to the magnetic moment in $Fe_{3-x}O_4$ magnetite nanoparticles, *Appl. Phys. Lett.*, 94:093108 (3 pp.), 2009, DOI:10.1063/1.3095484; J. Salafranca et al., Surfactant organic molecules restore magnetism in metal-oxide nanoparticle surfaces, *Nano Lett.*, 12:2499–2503, 2012, DOI:10.1021/nl300665z.

Karrikins

Fires have occurred on Earth for at least 400 million years, during which time the vegetation has been dominated by seed plants, initially conifers and more recently flowering plants. Fires were particularly prevalent during the Cretaceous Period, when major angiosperm groups were evolving, and it is believed that many fire-responsive traits in plants arose during that period. Although fires occur throughout the world, they are prevalent in Mediterranean-type environments, including Australia, California, the Mediterranean Basin, and South Africa.

Plants have developed many different strategies in their response to fire, including resprouting from shoot meristems that survive the fire, release of seeds from woody cones or fruits (serotiny or bradyspory), breakage of physical seed dormancy by disrupting tough seed coats, stimulation of flowering in plants (geophytes) with underground storage organs, and chemically induced germination of dormant seeds in the soil. Some plants depend on fire for seed germination, and many also respond to soil disturbance. The seeds of fire-stimulated species can remain in the soil for many years or decades, until they are stimulated to germinate by the next fire (**Fig. 1**).

Fig. 1. Postfire emergence and reproduction of *Anthocercis littorea*, the yellow tailflower of Western Australia. In the spring, following a bushfire, the yellow tailflower germinates from seed, grows, and produces new seed that falls to the ground. The plants die after one or two years, but the seeds remain in the soil until the next fire. This plant is a member of the Solanaceae, with flowers (*inset*) typical of members of that family.

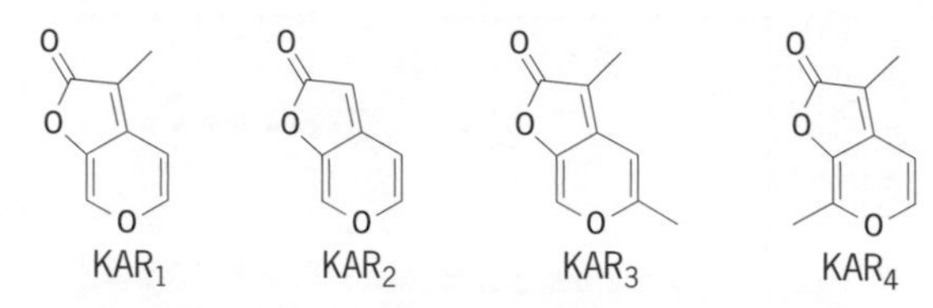

Fig. 2. Structures of karrikins. The upper ring structure is a butenolide, and the lower ring is a pyran. Karrikins are abbreviated as KAR, with KAR$_1$ being the original parent compound, also known as karrikinolide.

Discovery of karrikins. Smoke application is thought to have been used to promote seed germination by native peoples for hundreds of years, but it was studies in South Africa that resolved the action of smoke from fire. Further studies in the United States and Germany laid the foundation for the isolation of the active compounds in smoke. The first compound was isolated and confirmed through chemical synthesis by Australian scientists in 2004. It was found to contain both a pyran and a butenolide ring (**Fig. 2**). Later, it was found that smoke contains several similar active butenolides, and these were collectively named karrikins, from an Australian aboriginal word, ***karrik***, meaning "smoke." The original parent compound is often referred to as karrikinolide, reflecting the butenolide structure.

Fire and smoke also generate other known germination stimulants, including nutrients such as nitrate and gaseous compounds such as ethylene and nitrogen oxides. However, an additional component was discovered in 2011 as a result of its ability to stimulate germination in some species that do not respond to karrikins. This compound is glyceronitrile, a cyanohydrin, which releases small amounts of cyanide when dissolved in water. It is believed that the cyanide stimulates seed germination. Research suggests that karrikins are the most important of the chemical cues responsible for stimulating seed germination after fires.

Production and stability of karrikins. Karrikins are produced by heating or burning plant material. As little as 175°C (347°F) for 30 min is sufficient to produce karrikins. The burning of cellulose and sugars also generates karrikins in the laboratory, so it is believed that such carbohydrates are the source of the karrikins in nature. Pure karrikinolide is a solid at room temperature. However, in the heat of a fire, it will exist in the vapor phase in smoke. In general, most of it remains on the soil surface bound to soil particles and carbonized residues. Therefore, the prominent effects of fire on seed germination occur within centimeters of the site of the fire rather than at a distance in unburnt patches. Pure karrikins are unstable in ultraviolet light, but they are more stable in smoke and organic materials and may be resident in dry soil for several years. Rainfall will wash karrikins into the soil profile to commence the process of stimulating seed germination of the soil seed bank. Although most species that respond to karrikins have seed in the soil seed bank, some species with canopy-stored seed can also be promoted by application of karrikins.

Effects of karrikins on seeds and seedlings. Karrikins will trigger seed germination when other environmental factors, including water and temperature, are suitable. They do not necessarily break seed dormancy. Instead, karrikins stimulate germination in nondormant but germination-quiescent seed. In some cases, cycles of wetting and drying of seeds will cue seeds for karrikin stimulation of germination.

One surprising discovery is that the ability of karrikins to stimulate germination is not restricted to seeds of fire-stimulated species. Seeds of plant species that respond to physical disturbance events, including agricultural weeds, also respond to karrikins. Even species with nondormant seeds, such as carrots and tomatoes, are stimulated by karrikins with earlier, more uniform, and often robust germination. Therefore, the karrikin response may reflect a general effect of such butenolides on seed germination.

Several studies report that karrikins promote seedling growth and vigor. Stimulation of rapid germination will clearly give a seedling an advantage over seeds that germinate later, but the karrikin effect also seems to act directly on the seedling to promote root and shoot growth. Studies with the cress *Arabidopsis thaliana* showed that karrikins change the sensitivity of seeds and seedlings to light, so the growth of the seedling is more suited to the light regime in a postfire landscape.

Molecular mode of action of karrikins. The molecular mode of action of karrikins was determined by the discovery that the cress *A. thaliana* responds to karrikins, even though it is not known to be a fire-stimulated species (**Fig. 3**). *Arabidopsis* provides a powerful research system for gene discovery that involves the selection of mutants that are insensitive to karrikin. This approach led to the identification of the genes responsible for karrikin recognition and to the discovery that karrikins can be detected by a similar mechanism to that which detects strigolactone hormones. Strigolactones are chemically similar to karrikins, but they control shoot and root development. In addition, by exudation from the roots, they stimulate the formation of symbioses with mycorrhizal fungi and nitrogen-fixing bacteria. They also stimulate germination of parasitic weeds such as *Striga* and *Orobanche* species.

The karrikin and strigolactone receptors in *Arabidopsis* are very closely related enzymes with hydrolase (hydrolysis-catalyzing) activity. It is believed that the hydrolysis of strigolactones and potentially also karrikins leads to changes in the conformation of these enzymes, triggering their interaction with a protein complex that targets other proteins for destruction. In the case of strigolactones, one target is a protein that is thought to control the transcription of genes that regulate the growth of lateral

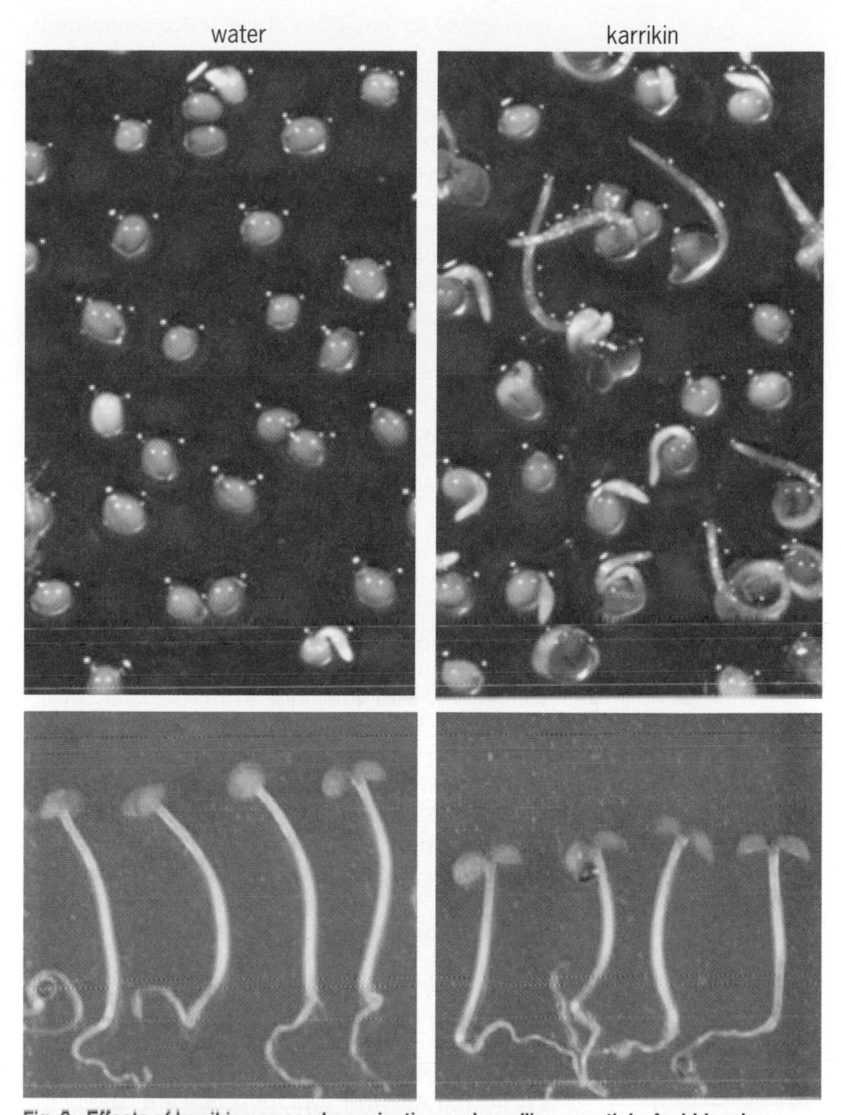

Fig. 3. Effects of karrikins on seed germination and seedling growth in *Arabidopsis thaliana*. Karrikin stimulates the germination of seeds that have some primary dormancy (*top row*). Karrikin treatment of seedlings growing in low light results in shorter hypocotyls and expanded cotyledons (*bottom row*).

meristems. In the case of karrikins, similar proteins are implicated as targets for destruction.

The interaction of karrikin signaling with other factors controlling seed germination is thought to be indirect. For example, both light and gibberellins (plant growth regulators) are required for karrikin stimulation of *Arabidopsis* seed germination, but these components act in parallel with one another so that the seed can integrate different signals depending on the environmental conditions. Abscisic acid is an inhibitor of seed germination, but any antagonistic effects between karrikins and abscisic acid are thought to be indirect. The precise mechanism by which karrikins stimulate seed germination is unknown, but it appears to be novel.

Applications. Aerosol smoke and smoke water are used on a small scale to stimulate seed germination in horticulture, nursery production, and sometimes landscape revegetation. However, the application of smoke or large volumes of smoke water is impractical for broadacre (large-scale) applications, and smoke water can be inhibitory unless diluted appropriately. Therefore, the discovery of karrikins and cyanohydrins opens up the possibility of using the pure compounds on a large scale for regeneration projects. Furthermore, because karrikins can stimulate the germination of weed seeds, they can potentially be used in agriculture to trigger the germination of the soil weed seed bank, allowing the weeds to be destroyed before crop planting.

For background information *see* ABSCISIC ACID; AGRICULTURAL SCIENCE (PLANT); DORMANCY; FIRE; FOREST ECOSYSTEM; GIBBERELLINS; PHYSIOLOGICAL ECOLOGY (PLANT); PLANT GROWTH; PLANT HORMONES; PLANT METABOLISM; PLANT PHYSIOLOGY; SEED GERMINATION; SMOKE in the McGraw-Hill Encyclopedia of Science & Technology.

Kingsley W. Dixon; Steven M. Smith; Gavin R. Flematti

Bibliography. S. D. S. Chiwocha et al., Karrikins: A new family of plant growth regulators in smoke, *Plant Sci.*, 177:252–256, 2009, DOI:10.1016/j.plantsci.2009.06.007; M. G. Kulkarni, M. E. Light, and J. Van Staden, Plant-derived smoke: Old technology with possibilities for economic applications in agriculture and horticulture, *S. Afr. J. Bot.*, 77(4):972–979, 2011, DOI:10.1016/j.sajb.2011.08.006; D. C. Nelson et al., Regulation of seed germination and seedling growth by chemical signals from burning vegetation, *Annu. Rev. Plant Biol.*, 63:107–130, 2012, DOI:10.1146/annurev-arplant-042811-10; M. T. Waters et al., The origins and mechanisms of karrikin signalling, *Curr. Opin. Plant Biol.*, 16(5):667–673, 2013, DOI:10.1016/j.pbi.2013.07.005.

Liquefied natural gas as a marine fuel

Natural gas is an important component of the global energy economy. For international transport, purified gas is chilled to approximately −164°C (−263°F), which condenses it to a liquid. Specialized ships can then carry it between ports with handling facilities to receive it. In recent years, interest has grown in using liquefied natural gas (LNG) as a vessel fuel, both for LNG carriers, to eliminate separate cargo and fuel tanks, and as a clean, efficient fuel for smaller vessels.

There are more than 100 LNG-fueled ships in operation or on order worldwide. Many are smaller vessels such as passenger ferries, patrol vessels, and platform supply vessels, but that is beginning to change, initially with orders for container ships and chemical tankers. This move away from the traditional use of heavy fuel oil offers significant advantages to shipowners in terms of costs and environmental performance.

The global merchant fleet currently consumes around 330 million metric tons of fuel annually, 80–85% of which is heavy fuel oil with high sulfur content. Fuel is a major expense for the shipping industry. A ship burns up to 250 tons of heavy fuel oil per day for the largest container vessels, and with oil prices expected to keep increasing in the years ahead, this fuel consumption will continue to be the most expensive part of operating a ship. By 2015, it is expected that as much as 80% of the daily costs of owning and operating a ship will be fuel.

Based on economic studies, the classification society DNV GL (Det Norske Veritas-Germanischer Lloyd) predicts that at least 1 in 10 new ships built in the next eight years will be delivered with gas-fueled engines. This equates to about 1000 ships. Furthermore, another 600–700 ships are predicted to be retrofitted with gas engines. Current growth in the LNG-fueled ship market is in line with these projections.

Environmental advantages. Using LNG as marine fuel eliminates sulfur oxide (SOx) and particulate matter emissions, nets a 20% reduction in greenhouse gas emissions, and diminishes nitrogen oxide (NOx) emissions by 85–90%. The certainty of increasingly strict environmental legislation, connected to regulations of the International Maritime Organization and to political bodies such as the European Union, and aimed at reducing air pollution and greenhouse gas emissions, is making LNG particularly desirable. Many ships entering Emission Control Areas (ECAs) in European waters must already switch from heavy fuel oil to cleaner fuels, and from 2015 the sulfur content in marine fuel must be lower than 0.1% in ECA areas around the world. From 2020, the sulfur limit is expected to drop to 0.5% for all international waters outside ECAs. The use of LNG as a marine fuel therefore addresses both local and global pollution issues.

Orders for LNG-fueled engines. The main maritime engine manufacturers are already introducing more new LNG-fueled engine models than heavy fuel oil engine models. The current order book for new ships, consisting of more than 50 vessels, includes propulsion systems with pure gas engines from Rolls Royce and Mitsubishi Heavy Industries and dual-fuel engines from Wärtsilä and MAN Diesel and Turbo. Some of the platform supply vessels on order have both dual-fuel and diesel engines, but the dual-fuel engines typically produce more than 90% of the energy on board with the diesels acting as backup most of the time.

Cost of LNG-fueled ships. Additional capital expenditure is required by shipowners wanting LNG-fueled vessels, since extra equipment is required, including advanced fuel tanks, a gas conversion and distribution system, and double-walled piping (**Fig. 1**). Nonetheless, in addition to its environmental benefits, LNG is proving to be economically favorable for shipowners based on cost savings of LNG relative to conventional fuels. Based on international energy analysts' expectations regarding oil and gas prices, LNG is expected to remain competitive for the lifetime of new vessels entering the market. So, even though an LNG-fueled ship will be some 10–15% more expensive to build than a ship with conventional fuel, the investment will be

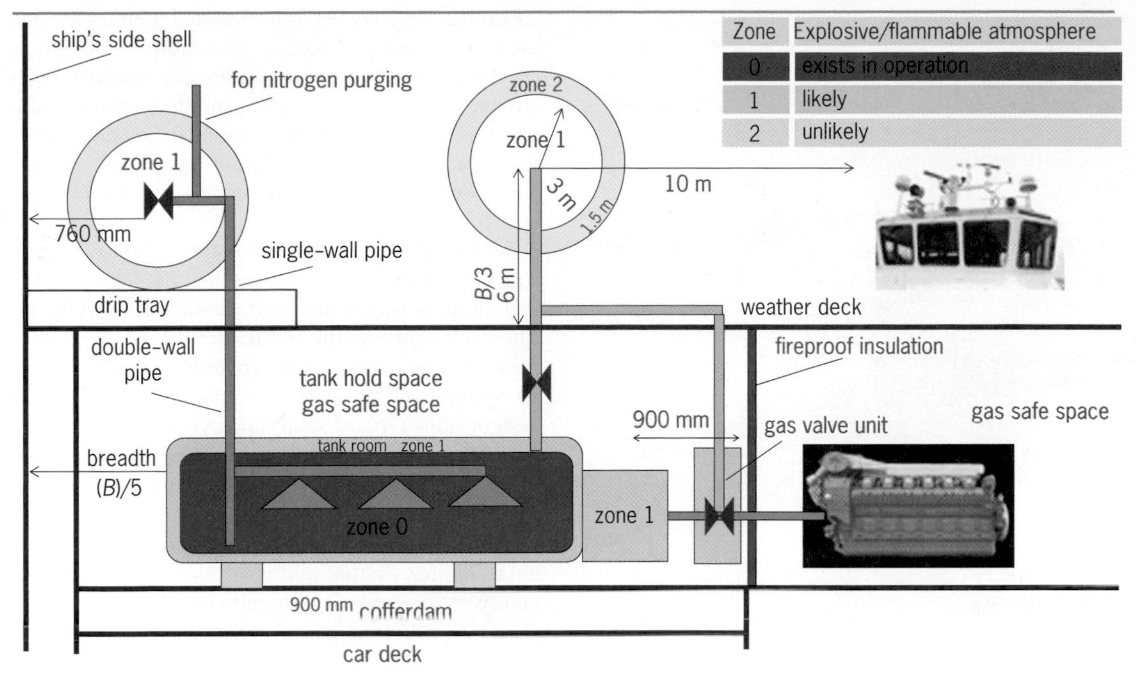

Fig. 1. Schematic profile (not to scale) showing arrangement of a typical liquefied natural gas (LNG) fuel system. (*Courtesy of Tony Teo*)

returned through a cleaner propulsion system, less maintenance, and most probably a cheaper fuel for the life of the vessel.

Risk management. The risks associated with the use of LNG are manageable. On land, there is commonplace acceptance of technologies such as liquefied petroleum gas (LPG)-fueled cars and LNG fueling stations for trucks located close to populated areas. At sea, the risks can take a different form, but boil-off gas has been used as a fuel for LNG carriers for over 40 years. There is over 10 years of experience in Norway with LNG-fueled ferries, offshore supply vessels, and patrol vessels. With appropriate design and operational safety measures in place, LNG is a clean solution without the potential for oil spills.

Standards and concept studies. The major ship classification societies have begun to promulgate and maintain standards for the construction, operation, and maintenance of fully or partially LNG-fueled vessels, as well as conducting research and concept studies. For example, DNV GL demonstrated the feasibility of large LNG-fueled ships through concept studies: 6200- and 9000-TEU (20-ft equivalent unit) container ships through the Quantum study (**Fig. 2**); *Triality*, a very large crude-oil tanker, which is also ballast-free; *Ecore*, a very large iron ore–carrying ship (**Fig. 3**); and *Oshima 2020*, a bulk carrier.

Shipbuilding. A number of shipowners in the United States have made the commitment to switching to LNG fuel in anticipation of the strictly limited emissions to air allowed under both the North American ECA requirements and phase II of California's Ocean-Going Vessel Clean Fuel Regulation. For example, Crowley Maritime is building two new LNG-powered container and roll-on cargo ships for the Puerto Rico trade. Matson has also decided to move forward with the construction of two new container ships at Aker Philadelphia Shipyard for trade to Hawaii.

Infrastructure. Internationally, infrastructure to support LNG-fueled vessels is increasing. The world's largest bunkering (fuel storage) hub, Singapore, is preparing to supply LNG and almost all of the large North European ports have communicated plans for LNG availability by the end of 2014. This includes Rotterdam, the world's third largest bunkering hub.

Within the United States, the Department of Transportation's Maritime Administration (MARAD) has made grants for experimental vessel conversions, and to analyze the issues and challenges associated with LNG bunkering, and the landside infrastructure

Fig. 2. *Quantum*, a 9000-TEU (20-ft equivalent unit) conceptual container ship, powered by liquefied natural gas (LNG) fuel. (*Courtesy of DNV GL*)

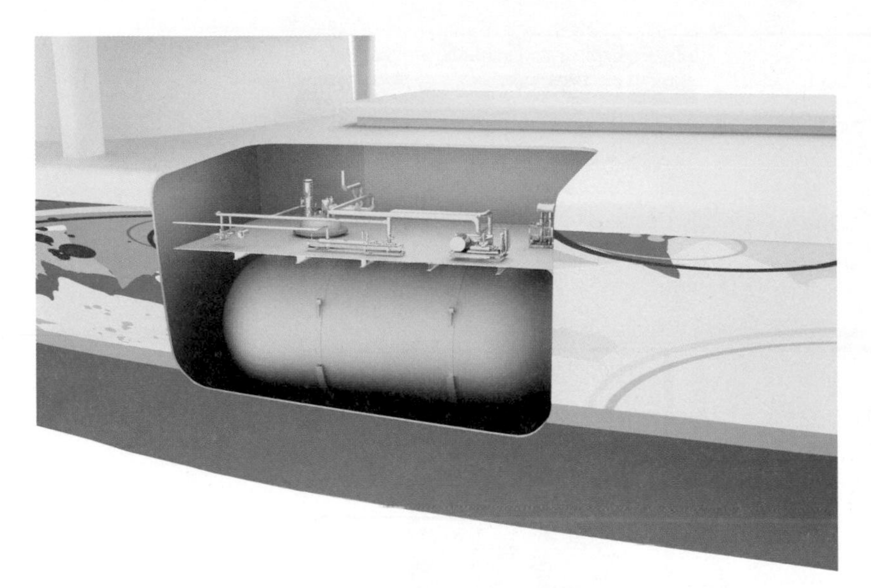

Fig. 3. Liquefied natural gas (LNG) tank (type C) with associated piping aboard *Ecore*, a conceptual ore carrier. (*Courtesy of DNV GL*)

needed to store and distribute LNG. *See* SHIP MODERNIZATION AND CONVERSION.

Current growth in the LNG-fueled ship market is exponential. While it is acknowledged that fuel choices will diversify over time, LNG is expected to be the early winner in the new alternative fuels race.

For background information *see* AIR POLLUTION; LIQUEFIED NATURAL GAS (LNG); MARINE CONTAINERS; MARINE ENGINE; MERCHANT SHIP; SHIPBUILDING; SHIP PROPULSION, MANEUVERING AND SEAKEEPING in the McGraw-Hill Encyclopedia of Science & Technology.

Wendy Laursen; Tony Teo

Bibliography. F. Adamchik, LNG as marine fuel, in *17th International Conference & Exhibition on Liquefied Natural Gas*, Houston, April 16–19, 2013; J. Herdzik, LNG as a marine fuel—possibilities and problems, *J. KONES Powertr. Transp.*, 18(2):169–176, 2011; P. Semolinos, G. Olsen, and A. Giacosa, LNG as marine fuel: Challenges to be overcome, in *17th International Conference & Exhibition on Liquefied Natural Gas*, Houston, April 16–19, 2013.

Long-span hybrid suspension and cable-stayed bridges

Bridges are important elements of modern civilizations. They enhance a nation's economic growth and help raise the standard of living of its inhabitants by facilitating the efficient transport of people, goods, and services. In times of natural or human-made calamities, bridges become essential lifelines for disaster relief and emergency evacuation. Depending on the locations, terrain conditions, material availability, required span lengths, and budgetary considerations, a variety of bridge forms can be considered. For spans that exceed 150 meters, the bridge form used most often is the cable-supported bridge system.

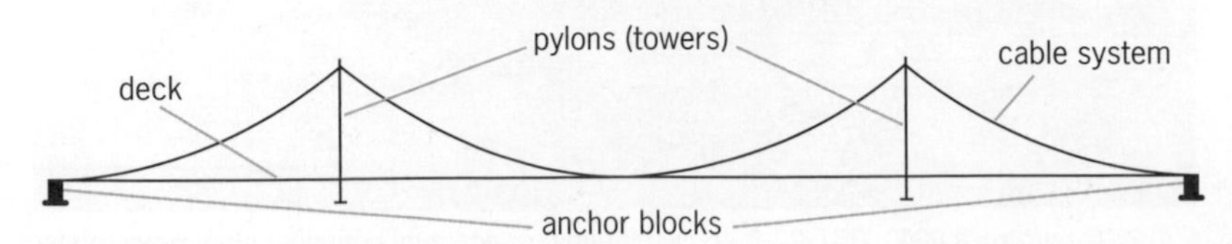

Fig. 1. Cable-supported bridge components.

Cable-supported bridge systems. A typical cable-supported bridge system consists of a deck, a cable system, towers (pylons), and anchor blocks (**Fig. 1**). The deck supports the traffic, the cable system supports the deck, the towers (pylons) support the cable system, and the anchor blocks stabilize the cable system. This type of structural system has been the natural choice for bridges with large spans because of its efficient use of structural materials and aesthetic appeal. Advances in design technology, quality of materials, and construction techniques have led to longer, lighter, and more slender bridges. Recently, bridges of magnificent sizes have been proposed in different parts of the world to improve human interactions and keep pace with social, political, and economic developments. The ambition is to further increase span lengths and to use shallower and more slender decks in future bridges.

There are different types of cable-supported bridges. The principal classification is based on the configuration of the cable system. Accordingly, they are categorized as suspension, cable-stayed, and hybrid suspension and cable-stayed (that is, hybrid cable-supported) bridge systems.

The suspension system (**Fig. 2**) consists of a main cable and vertical hangers suspended from the main cable to support the deck. In the cable-stayed system (**Fig. 3**), the cables or stays connect the deck directly to the pylons. These two cable systems dominate the realm of the cable-supported bridges currently in use, with the suspension system leading in the maximum span range (see **table**).

The third type, the hybrid suspension and cable-stayed system (**Fig. 4**), consists of a combination of the traditional suspension and cable-stayed systems.

A notable example of a hybrid suspension and cable-stayed bridge is the Brooklyn Bridge (**Fig. 5**). Opened to traffic in 1883, this 1825-m-long bridge is arguably regarded as the ancestor of modern suspension bridges. It was designed by the renowned bridge engineer John A. Roebling (1806–1869). He adopted this system based on his investigations on mitigating the aerodynamic problem faced by conventional suspension bridges. In this system, both the stayed and the suspension systems share the deck load throughout the span (Fig. 4*a*).

Another hybrid cable-supported bridge system, in which the stayed and suspension portions are completely separated, was proposed in 1938 by Franz Dischinger (1887–1953) for a cable-supported bridge with a 750-m main span to be built across the Elbe River in Hamburg. In this system, the central part of the bridge is carried by a suspension system, while the rest is supported by stays (Fig. 4*b*). As reported by Y. Meng and coworkers, the first modern hybrid cable-supported bridge of this type was built in 1997 in China with a main span of 288 m.

A number of hybrid cable-supported bridges were proposed in the past, including the Great Belt East (Storebæltsbroen) Bridge in Denmark, the Messina Strait Bridge in Italy, the Izmit Bay Bridge in Turkey, the Tagus River Bridge in Portugal, the Willamette River Bridge in the United States (**Fig. 6**), the Gibraltar Bridge between Europe and Africa, and some strait-crossing bridges in Japan. In the twenty-first century, numerous long- and super-long-span bridges have been planned for sea-crossing projects around the world. Most of these bridges are expected to endure extreme natural conditions, such as violent typhoons, soft-soil foundations, and deep-water foundations, which are unfavorable for conventional cable-stayed and suspension bridges.

Advantages of hybrid cable-supported bridges. Because it is a combination of two different but synergistic cable systems, the hybrid suspension and cable-stayed system benefits from some of the main advantages of both suspension and cable-stayed bridges. Proper design of this hybrid system can also mitigate certain deficiencies in the structural behavior, construction, economy, and wind stability of the traditional suspension and cable-stayed bridges.

Hybrid versus suspension system. When compared with a pure suspension system, the hybrid system has the following advantages.

The hybrid system leads to material savings, as less materials will be needed to carry the same load by stay cables than by suspension cables and hangers. The hybrid system also allows the use of a more optimum pylon height, as the pylons need not be so high for stiffness and sag-span ratio requirements as for the suspension system. And as the load in the main cable is reduced because of the stays, the load from the deck can be dispersed down the towers leading to further reduction in the weight of the towers. For the Dischinger's system, in particular, different structural materials can be used in the respective regions of stayed and suspension systems, leading to further material savings.

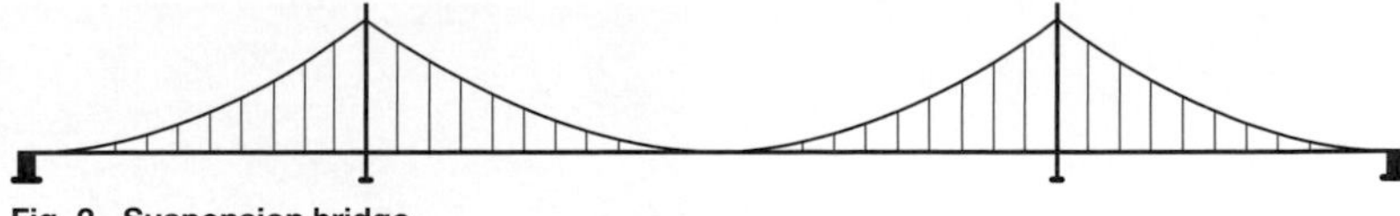

Fig. 2. Suspension bridge.

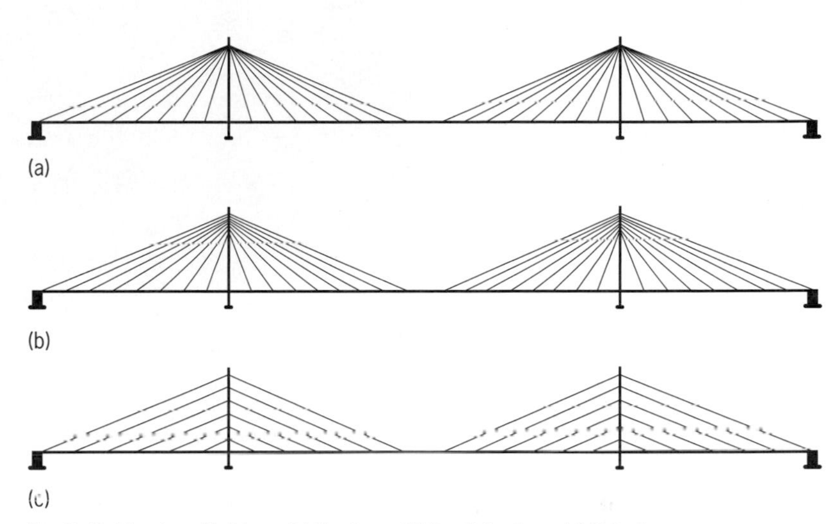

Fig. 3. Cable-stayed bridges. (*a*) Fan type. (*b*) Semi-fan type. (*c*) Harp type.

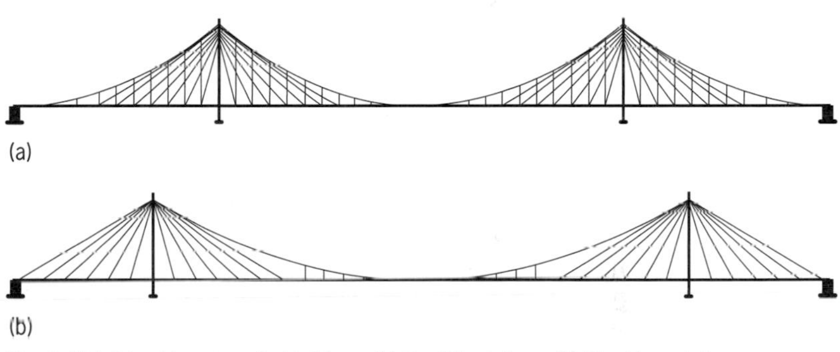

Fig. 4. Hybrid cable-supported bridges. (*a*) Roebling's type. (*b*) Dischinger's type.

World's longest cable-supported bridges by central span

No.	Name of bridge	Location	Central span, m	Year of completion
Suspension Bridges				
1	Akashi Kaikyo Bridge	Kobe-Awaji Route, Japan	1991	1998
2	Xihoumen Bridge	Zhoushan Archipelago, China	1650	2009
3	Storebælt Bridge	Halsskov-Sprogø, Denmark	1624	1998
Cable-Stayed Bridges				
1	Russky Bridge	Vladivostok, Russia	1104	2012
2	Sutong Bridge	Nantong, China	1088	2008
3	Stonecutters Bridge	Rambler Channel, Hong Kong	1018	2009

Fig. 5. View of the Brooklyn Bridge showing its main cables and vertical hangers and its stayed cables, which connect the deck to the tower. (*Photo courtesy of Niels Blumel*)

Although the unit costs, which include erection and materials, might be higher for stay cables, an overall saving can usually be obtained, as the reduction in cable steel quantity is often quite large for long-span bridges.

The load shared by the stays greatly reduces the tensile forces in the main cables, which in turn decreases the construction cost of the main cables, the massive anchors, and the difficulty of constructing them under water. This makes it possible to construct hybrid cable-supported bridges on relatively softer soil foundations.

Hybrid cable-supported bridges have improved stiffness and aerodynamic stability. Studies have shown that a considerable increase in critical wind speed and a reduction in the influence of higher modes (particularly the torsional mode) on the dynamic response can be realized.

Fig. 6. Hybrid cable-supported bridge design proposed for the Willamette River Transit Bridge. (*Photo courtesy of Rosales + Partners*)

In terms of maintenance, replacing individual cables is not as complex and prohibitive as in suspension bridges.

Hybrid versus cable-stayed system. When compared to a pure cable-stayed system, the hybrid system offers the following advantages.

The axial compression forces in the deck are reduced and hence lighter decks are possible. Even though introducing compression in the deck adjacent to the pylons generally proves advantageous, a considerable magnitude of compression elsewhere could lead to stability (buckling) problems.

The cantilever lengths needed for erection are reduced, thus leading to a reduction in crane capacity and an improvement in bridge stability during erection.

For the Roebling's system, in particular, the stay cables can be clamped to the vertical hangers at their points of intersections and this can help reduce sag variations in the stays caused by live loads. In addition, wind-excited oscillations of individual cables will be repressed and can possibly be eliminated.

Height of the towers, particularly for long spans, can be significantly reduced. This is because the stays need not be provided along the whole length of the bridge and thus a more favorable inclination angle for the stays can be maintained even when the towers are lowered.

It can be inferred from these advantages that the hybrid system can be an attractive design alternative for long- and particularly super-long-span bridges.

Optimum design of hybrid cable-supported bridges. In spite of the many advantages mentioned earlier, the hybrid system has not gained popularity as a result of some shortcomings associated with the two hybrid systems shown in Fig. 4.

The most favorable Roebling's type hybrid system (Fig. 4*a*) for a three-span bridge (discussed by N. J. Gimsing and C. T. Georgakis) has relatively short side spans (1/4 to 1/3 of the main span), continuous deck and main cables from one end to the other, centering devices placed at anchor blocks, main cables clamped to the deck at mid-span and fixed at the pylon tops along with all the stays, and no stays extending to the anchor blocks (**Fig. 7**). The erection of such a system is quite complex, and the amount of savings is not appreciable.

Because the fan system often uses less materials than those of the suspension system, supporting the applied loads by stay cables where they are the most effective (near the pylons where the inclination angle is the highest) could reduce construction complexity and result in more savings. Moreover, since the vertical-to-horizontal force components are the highest for stay cables close to the pylons, smaller cable sizes can be specified. Omitting the hangers from regions where the stay cables are the most effective results in a Dischinger's system (Fig. 4*b*).

The Dischinger's system, although economical from the above perspective, poses some structural and aesthetic problems. Because of the sudden change in stiffness at the points where the cable support changes from a stayed to a suspension system, a large unfavorable discontinuity in the bending moment on the deck can occur. In addition, the sudden change in the structural systems can create aerodynamic complexity and is not particularly eye pleasing.

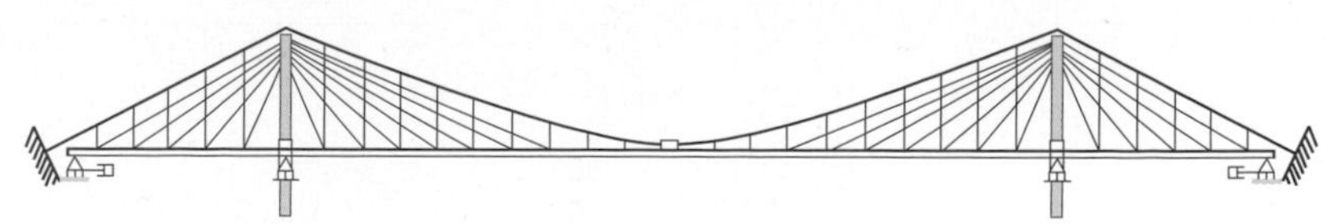

Fig. 7. Roebling's type hybrid cable system discussed by N. J. Gimsing and C. T. Georgakis.

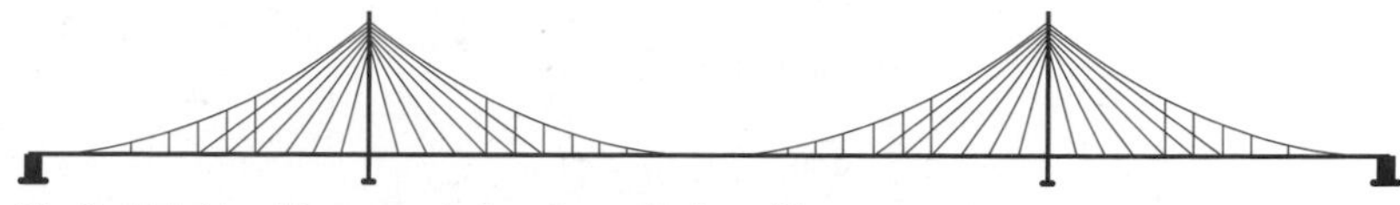

Fig. 8. Hybrid cable-supported system with transition zones.

A hybrid system with a transition zone from the stayed region to the suspension region will be more efficient and economical, especially for long- and super-long-span bridges. Such a system will have transition zones that blend the suspension and stayed systems to alleviate the sudden discontinuity in the bending moment, avoid the sudden change in system stiffness, and give the bridge a better aesthetic appeal (**Fig. 8**).

As an example, consider three cases of a hybrid cable-supported bridge with a main span length of 500 m and side-to-main span ratio of 0.45 (**Fig. 9**). The suspension region supports 40% of the deck span. Hybrid transition zones of 0%, 50%, and 100% are considered for the rest of the span and the resulting moment distributions are shown in **Fig. 10**. As can be seen, the Dischinger's system (0% transition zone) results in an unfavorable moment distribution at the points where the type of cable system

Fig. 9. Finite element model of a bridge with (*a*) 0%, (*b*) 50%, and (*c*) 100% hybrid transition zones.

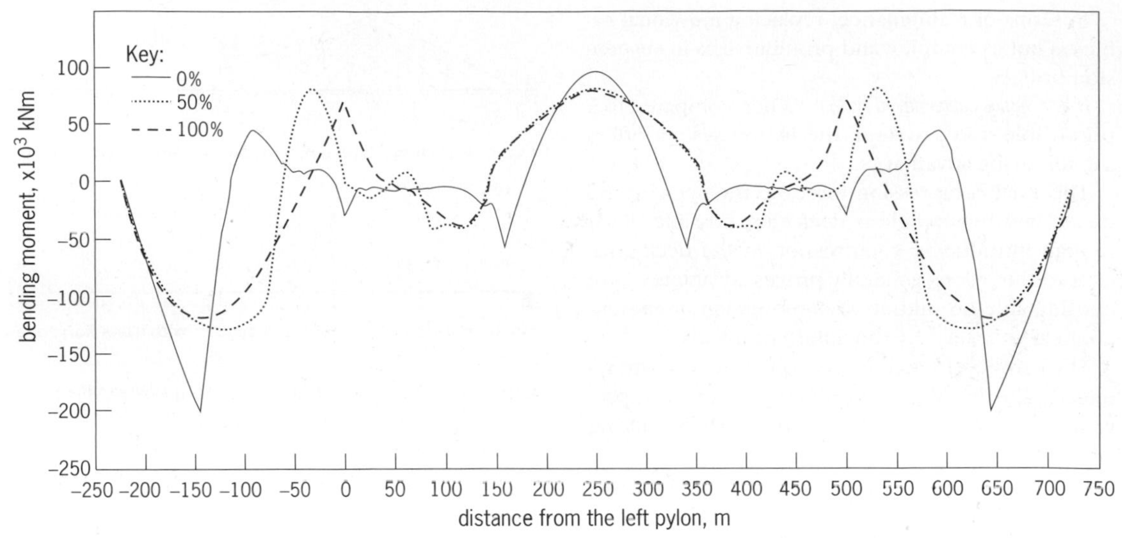

Fig. 10. Bending moment distribution along the deck with 0%, 50%, and 100% hybrid transition zones.

changes. On the other hand, a 50% transition zone yields almost a similar moment distribution as the Roebling's type (100% transition zone), but with a considerable reduction in the number of steel cables used. Hence, a careful selection of the extent of the transition zones could result in an economical and efficient design.

Because of the complex nature of this new hybrid system, the use of the conventional trial-and-error design approach is not advisable. Instead, a design technique that involves sensitivity analysis and optimization needs to be done to arrive at an optimal design.

For background information *see* BRIDGE; CANTILEVER; CONSTRUCTION ENGINEERING; FINITE ELEMENT METHOD; FOUNDATION; STRUCTURAL ANALYSIS; STRUCTURAL DESIGN in the McGraw-Hill Encyclopedia of Science & Technology.

Zekarias Tadesse Woldegebriel; Eric M. Lui

Bibliography. W. F. Chen and L. Duan (eds.), *Bridge Engineering Handbook*, 2d ed., CRC Press, Boca Raton, FL, 2014; N. J. Gimsing and C. T. Georgakis, *Cable Supported Bridges: Concept and Design*, 3d ed., John Wiley & Sons, West Sussex, UK, 2012; M. Hansford, Cable stay revolution: Experts debate viability of long span cable-stayed bridges, *New Civ. Eng.*, July 5, 2012, http://www.nce.co.uk/news/structures/cable-stay-revolution-experts-debate-viability-of-long-span-cable-stayed-bridges/8632515.article; H. S. Kim et al., Development of optimal structural system for hybrid cable-stayed bridges using ultra high performance concrete, *Engineering*, 5(9):720–728, 2013, DOI:10.4236/eng.2013.59086; P. Lonetti and A. Pascuzzo, Optimum design analysis of hybrid cable-stayed suspension bridges, *Adv. Eng. Softw.*, 73:53–66, 2014, DOI:10.1016/j.advengsoft.2014.03.004; Y. Meng, D. Liu, and S. H. Sun, Study on the design of long-span cable-stayed-suspension hybrid bridges, *J. Chongqing Jiaotong Instit.*, 18(4):8–12, 1999; F. Nieto, S. Hernandez, and J. A. Jurado, Optimum design of long-span suspension bridges considering aeroelastic and kinematic constraints, *Struct. Multidisciplin. Optimiz.*, 39:133–151, 2009, DOI:10.1007/s00158-008-0314-8.

Luminescent liquid crystals in OLED display technology

Organic light-emitting diodes (OLEDs) are the basis for a new display technology that was developed over the last 3 decades from initial prototypes to commercial displays found on some mobile phones today. The chief competing technology is the ubiquitous liquid-crystal display (LCD), but among the advantages offered by OLED technology is the possibility of flexible displays in novel formats. OLED displays are robust, energy-efficient, and offer excellent contrast because, being active displays, light is emitted only at those pixels that are "on." OLED technology also attracts significant interest for lighting applications. Around 20% of electrical consumption is used in developed countries for lighting, owing to the inefficiency of conventional methods, but OLED-based lighting offers large gains in energy efficiency and allows environmentally undesirable materials, such as mercury used in fluorescent lighting, to be avoided. For lighting purposes, emission of white light (that is, across the whole of the visible spectrum) is required.

How OLED displays work. The basis of the process is the emission of light from the OLED material, which is induced using an applied voltage. Although the OLED material is only one of several layers in the device (the others aid the efficient operation of the device), it is possible to understand the light emission by considering only the emissive layer.

As with all of the layers, the emissive layer is very thin (in the range of 1–50 nanometers), yet even at this thickness it will contain billions of molecules. Nonetheless, we can illustrate its behavior using an

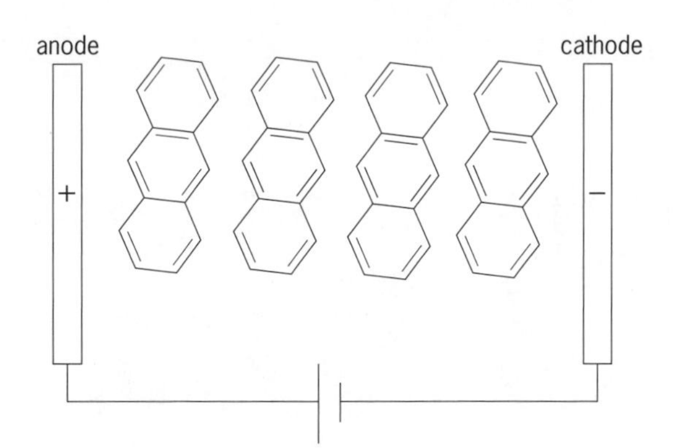

Fig. 1. Schematic representation of molecules within an OLED display.

imaginary layer in which only four molecules are sandwiched between the two electrodes (**Fig. 1**).

Each molecule contains many molecular orbitals, each of which has the possibility to be occupied by a pair of electrons. Some of these orbitals are filled and others are vacant, with the lowest-energy orbitals being filled preferentially. There is then an energy gap between the highest occupied molecular orbital (the HOMO) and the lowest unoccupied molecular orbital (the LUMO). Energy can be provided to allow an electron to be promoted from the HOMO to the LUMO, creating an exciton (excited state). This exciton is not stable, and the electron quickly drops back to the HOMO together with emission of light (luminescence) for which the wavelength (λ) is defined by the energy gap according to $\lambda = hc/E$, where h is Planck's constant, c is the speed of light, and E is the energy. In an OLED device, an electric field is used to create the excitons, hence the term electroluminescence. This is explained using **Fig. 2**, which shows the HOMO and LUMO for the four molecules of Fig. 1.

Thus, upon application of an electric field (Fig. 2*a*), an electron (e^-) will be injected into the LUMO of the molecule closest to the cathode, while at the anode, an electron will be removed from the HOMO [in the language of physics, this is also considered as the injection of a positive "hole" (h^+)]. In so doing, an excited state is created.

Because all electrons are driven from cathode to anode (and holes from anode to cathode) by the electric potential, they will hop from one molecule to another as indicated in Fig. 2*b*. At some point in this migration, the opposite charges (hole and electron) will encounter one another (Fig. 2*c*) and then combine to form an exciton (Fig. 2*d*). As the LUMO electron falls back to the HOMO, the material emits light, completing the cycle. The other layers of the OLED device are there to facilitate the efficient injection of charges from the electrodes and to aid their migration and balance in the device. In reality, the process is a little more involved than this, but

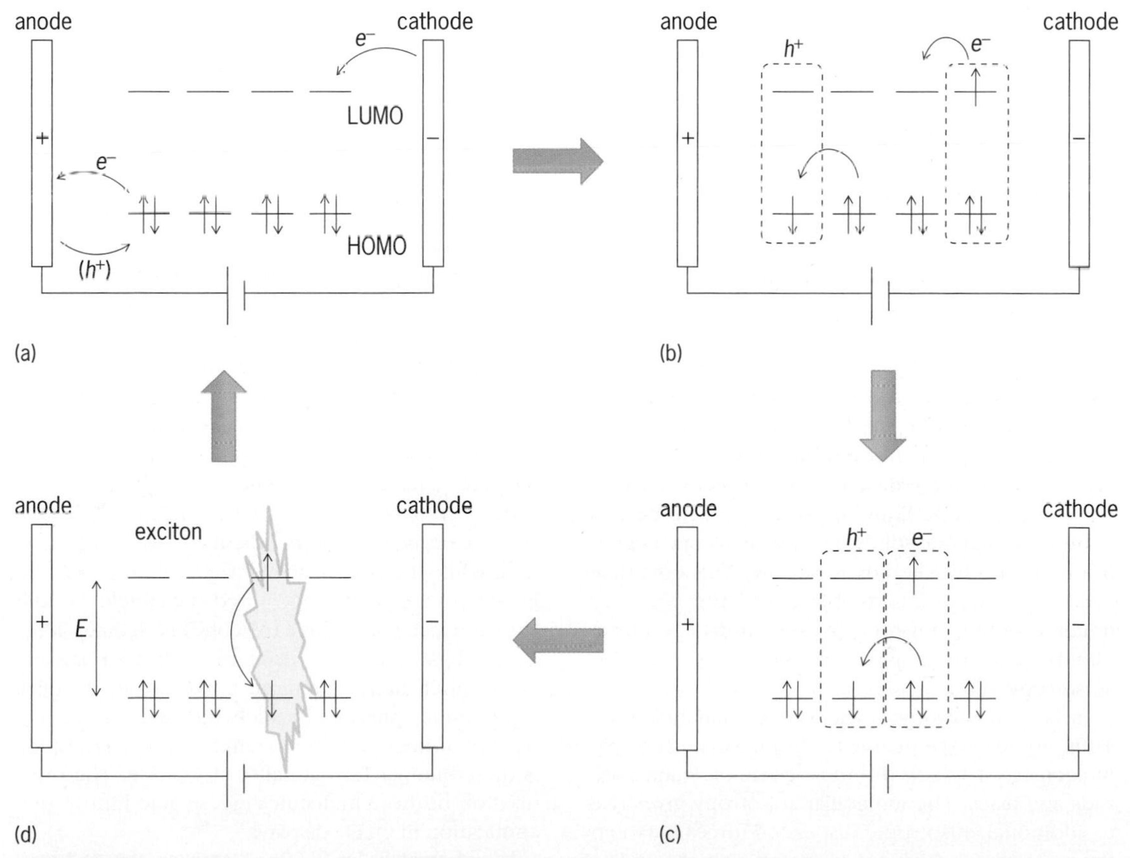

Fig. 2. Schematic representations of the charge transport in the process of electroluminescence, where e^- are electrons and h^+ are holes.

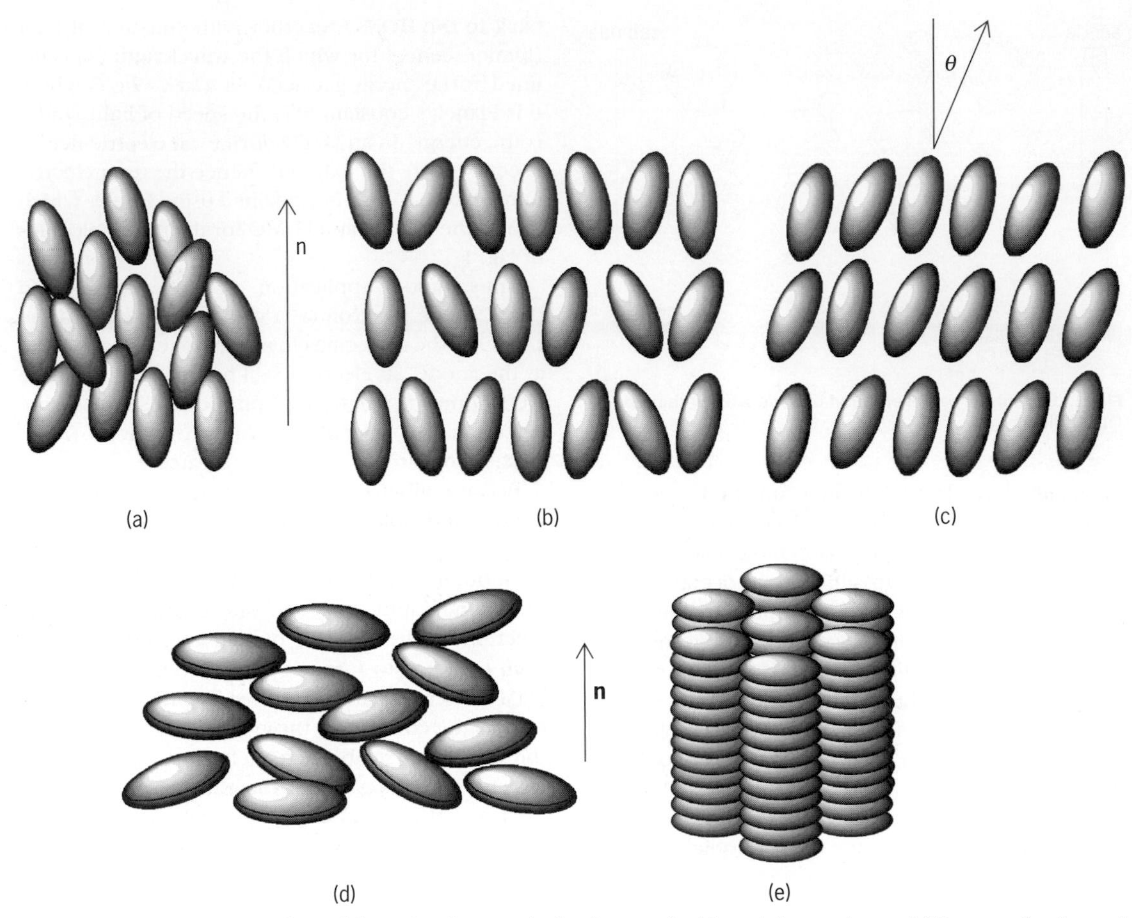

Fig. 3. **Schematic representations of the molecular organization in some liquid-crystal mesophases: (*a*) the nematic phase of rods; (*b*) the SmA phase; (*c*) the SmC phase; (*d*) the nematic phase of discs; and (*e*) the Col_h phase.**

the extra detail is not required to appreciate how liquid crystals can become involved.

Figure 2 implies that there is a great deal of charge transport during the operation of the OLED display. The hopping process depends strongly on orbital overlap, which in turn depends on how neighboring molecules are organized with respect to each other, and it is here that liquid-crystalline properties can be exploited.

Liquid crystals. Formally, liquid crystals represent the fourth state of matter, existing between the solid and liquid states for something like 80,000 materials. Intermediate between the solid and liquid states, at lower and higher temperatures, respectively, it is not surprising that liquid crystals retain properties of both, so that like solids they possess some degree of order, and like liquids they flow. This combination of order and fluidity means that they are very much ordered, anisotropic (direction-dependent) liquids, and their physical properties are also anisotropic.

In fact, anisotropy is the key to their behavior. Thus, liquid-crystal phases are formed of anisotropic molecules, of which the most common shapes are rods and discs. The molecular anisotropy gives rise to additional anisotropic dispersion forces between the molecules, which are strong enough to stabilize these intermediate states of organization.

To a first approximation, the organization of the phases depends to a large degree on the molecular shape. Rodlike (calamitic) molecules form two main types of phase—the nematic phase and a whole family of smectic phases (**Fig. 3**). In the nematic phase, the molecules possess a single dimension of orientational order but are free from positional order, with the average direction of orientation being defined by the director, ⓝ. It is the nematic phase that is used in the vast majority of liquid-crystal displays. There are many smectic phases, which are characterized by the occurrence of partial positional ordering in addition to the orientational order. Two examples are given as Fig. 3*b* and *c*. The former is the smectic A (SmA) phase, in which the molecules are aligned, on average, perpendicular to the layer direction, while the latter is the SmC phase, in which the molecules are, on average, tilted at an angle θ to the layer normal. For disclike (discotic) molecules, a nematic phase (Fig. 3*d*) is also known (but is relatively rare); much more common, however, are families of columnar phases in which the molecules stack one upon another. The example shown in Fig. 3*e* is the columnar hexagonal (Col_h) phase. The organization of these molecules gives some hint to one application in OLED displays.

Liquid crystals for OLEDs—improved charge transport. If electron hopping is now reconsidered, then

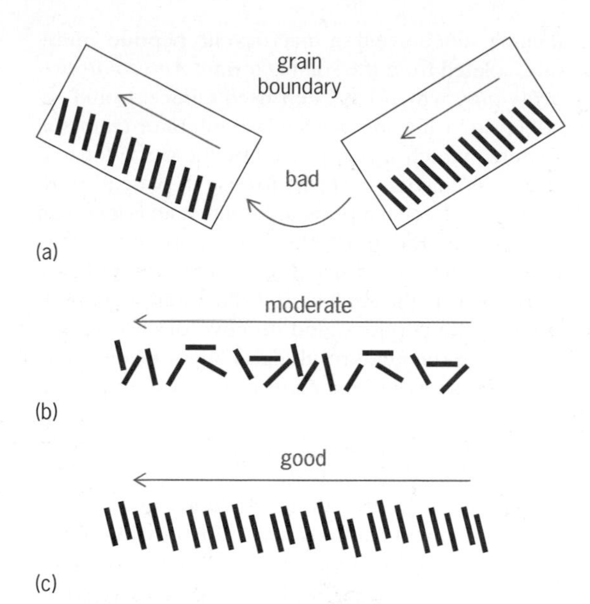

Fig. 4. Liquid-crystalline materials offer better organization of molecules for charge transport.

it should be clear that the correct orientation of molecules can make this process easier, which will then reduce the power required to drive the device and save energy. The ultimate way to order molecules over long length scales is by growing perfect single crystals. However, this is a very difficult process and such thin single crystals would be incredibly fragile, rendering them useless. The situation would not be improved by using a layer of crushed single crystals, for while the charge mobility in the individual crystallites would be very good, the resistance where the crystals meet would be very high indeed (**Fig. 4*a***). To get around this problem, the layers in the device can be deposited in such a way that they form an amorphous arrangement, which has the advantage that there are no grain boundaries. However, as shown in Fig. 4*b*, the relative organization of the molecules is random and this slows the charge transport.

Returning now to Fig. 3, it can be seen that liquid-crystal phases represent a way of ordering molecules at the molecular level with the possibility of having this order persist over relatively macroscopic length scales. In this way, good charge transport (lower power requirements) can be achieved (Fig. 4*c*).

One issue remains, in that liquid crystals are fluids and ought not to be suitable for application in solid-state devices, such as OLEDs, because they will not make the required robust thin-film layer. How can this problem be addressed? One answer is to use reactive liquid crystals. Such liquid crystals contain reactive groups, and once alignment is achieved, they can be polymerized to lock in the order (**Fig. 5**). This is very similar to the way dentists use a viscous liquid to prepare a filling and then transform it into a solid by the action of ultraviolet light. A second approach could be to use some of the more ordered phases of liquid crystals, something that has been used to produce charge-carrier films of discotic molecules used in xerography.

Liquid crystals for OLEDs for LCDs—polarized emission. The optics of liquid-crystal displays require that the display cell be encased between two polarizers oriented with their polarization directions (in most cases) perpendicular to one another. However, this creates an inefficiency, as each polarizer can absorb up to 50% of the light passing through it. If the LCD of a laptop computer is considered, there is an unpolarized backlight and so, because of the two polarizers alone, the amount of transmitted light can be reduced to 25% of the backlight intensity. However, if the backlight emitted polarized light, then the rear polarizer would no longer be needed and twice as much light would pass through the device. Using the same alignment approach described above, such an approach is, in principle, possible, although its realization is a little way off just yet. Similarly, application of chiral liquid-crystalline materials may result in OLEDs emitting circularly polarized light, which is of interest in holographic displays.

Outlook. Liquid crystals are unique materials that offer advantages over both crystalline and amorphous materials and are currently attracting attention in development of OLEDs. Although OLEDs are considered by some to be superior to LCDs, OLED technology still has materials and manufacturing issues to address before it becomes a low-cost and

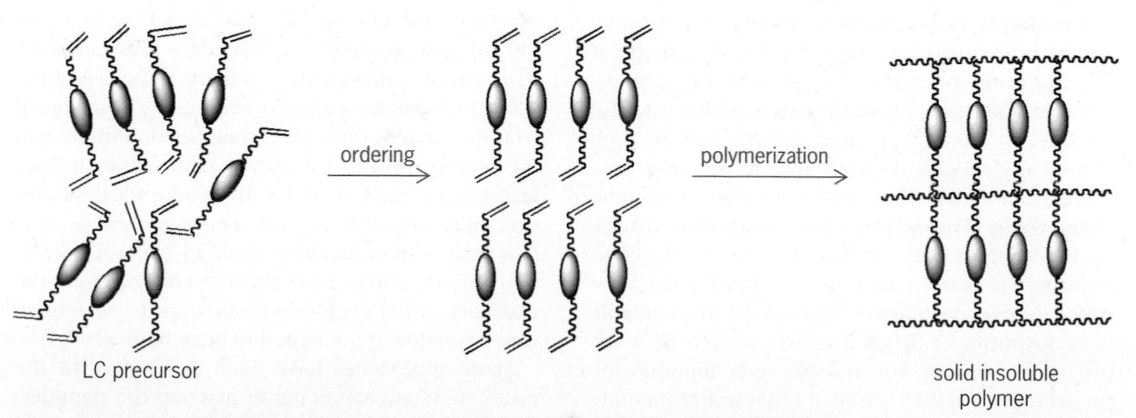

Fig. 5. Application of reactive mesogens to prepare large, well-ordered thin films for OLEDs.

ubiquitous alternative. Maybe one day the most efficient OLED displays will take the best of both worlds and use liquid crystals at their heart.

For background information *see* BAND THEORY OF SOLIDS; CRYSTAL; ELECTRIC FIELD; ELECTRON; ELECTRON-HOLE RECOMBINATION; ELECTRONIC DISPLAY; EXCITON; HOLE STATES IN SOLIDS; LIGHT; LIGHT PANEL; LIGHT-EMITTING DIODE; LIQUID CRYSTALS; MOLECULAR ORBITAL THEORY; POLARIZED LIGHT in the McGraw-Hill Encyclopedia of Science & Technology. Valery N. Kozhevnikov; Duncan W. Bruce

Bibliography. J. H. Burroughes et al., Light-emitting diodes based on conjugated polymers, *Nature*, 347:539–541, 1990, DOI:10.1038/347539a0; R. J. Bushby, S. M. Kelly, and M. O'Neill (eds.), *Liquid Crystalline Semiconductors*, Springer, Dordrecht, Netherlands, 2013, ISBN 978-90-481-2873-0; W. Pisula et al., Liquid crystalline ordering and charge transport in semiconducting materials, *Macromol. Rapid Commun.*, 30(14):1179–1202, 2009, DOI:10.1002/marc.200900251; C. W. Tang and S. A. VanSlyke, Organic electroluminescent diodes, *Appl. Phys. Lett.*, 51(12):913–915, 1987, DOI:10.1063/1.98799.

Macrocyclic peptides used as drugs

Macrocyclic peptides are a remarkably diverse family of peptides found in a range of organisms, including bacteria, plants, fungi, and mammals. The common feature defining macrocyclic peptides is a ring structure, but a range of linkages can be involved, including N-terminus-to-C-terminus (head-to-tail), head-to-side-chain, side-chain-to-tail, and side-chain-to-side-chain linkages. These different linkages result in a wide array of structures. In addition, the peptide sizes vary significantly, with peptides ranging from less than 10 residues to proteins of approximately 70 residues. Nonnative residues, which can enhance biological stability, are often present in the smaller peptides, whereas the larger peptides tend to have predominantly standard residues. This distinction is a reflection of the fact that smaller peptides are often biosynthesized by nonribosomal pathways, whereas larger macrocyclic peptides are generally gene-encoded.

Advantages of macrocyclic peptides. Macrocyclic peptides are being pursued for the development of drug leads, primarily because of a few key features, including a large surface area (which is useful for interactions with targets and can lead to high affinity and specificity), resistance to proteases (enzymes that break down proteins), and membrane permeability. The extent to which individual macrocyclic peptides possess these features varies enormously, but such characteristics have been illustrated in several examples. Perhaps the most notable is cyclosporine, a fungal macrocyclic peptide comprising 11 residues, which is used as an immunosuppressant drug in the treatment of some autoimmune diseases and organ transplantation. Other examples include vancomycin, a macrocyclic peptide antibiotic isolated from the soil bacterium *Amycolatopsis orientalis*, which has been used clinically, and romidepsin, a histone deacetylase inhibitor from soil bacteria, which was approved by the U.S. Food and Drug Administration (FDA) for use in patients with cutaneous T-cell lymphoma. Romidepsin is less than 600 daltons (Da) and is therefore more like a small molecule in terms of size. Importantly, advances are being made in the engineering of naturally occurring macrocyclic peptides, and de novo discoveries are expanding the field, resulting in larger macrocyclic peptides showing exciting results in preclinical studies.

Potential therapeutic applications of macrocyclic peptides. Examples of larger macrocyclic peptides with potential in drug design include those isolated from venomous creatures, which are providing interesting starting points for the development of therapeutic agents. The macrocyclic rings in these structures are formed by side-chain-to-side-chain bonds, and the applications that are being pursued primarily involve the treatment of pain and their use as imaging agents. Cone snails produce numerous peptides in their venom. These venom peptides are termed conotoxins, and one (ziconotide) has been approved for the treatment of chronic pain. However, ziconotide requires intrathecal administration (that is, injection into the spinal fluid), and therefore conotoxins with oral activity are being developed, including an analog that has been head-to-tail cyclized. A scorpion venom peptide, chlorotoxin, is a particularly exciting venom peptide that has the potential to assist surgeons in removing tumors because it selectively binds to tumor cells. The cyclotides, which constitute another class of macrocyclic peptides, have been used in rational drug design approaches to develop peptides with potent bioactivities and interesting pharmacokinetic properties. Collectively, the cyclotides make up one of the largest families of macrocyclic peptides, with more than 200 different sequences being characterized. They are isolated from plants and have a head-to-tail cyclic backbone and three disulfide bonds arranged in a cystine-knot motif. The combination of the cyclic backbone and disulfide bond connectivity makes for a very stable framework. Researchers have used kalata B1, the prototypic member of the cyclotide family, as a scaffold to graft bioactive sequences with bradykinin B1-receptor antagonistic activity. The resulting cyclic peptides have shown oral activity in an animal pain model and, thus, have therapeutic potential in the treatment of inflammatory pain. Other investigators have used a similar approach with another cyclotide, MCoTI-I, to engineer a grafted peptide that targets an intracellular protein–protein interaction important in cancer. These studies highlight the plasticity of the cyclotide framework to accommodate nonnative sequences and bioactivities.

Many approaches have been developed for the production and screening of macrocyclic peptides. Chemical production was initially the only viable method for producing cyclic peptides for engineering and structure activity relationship studies. Several methods are available for the chemical synthesis of cyclic peptides, including sulfur-mediated and metal-ion-assisted cyclization. For disulfide-rich macrocyclic peptides, thioester chemistry has been used extensively to facilitate the cyclization reaction. Biological approaches for the synthesis of macrocyclic peptides are emerging as powerful tools for the development and discovery of novel peptides with therapeutic potential. Recombinant methods based on a modified protein-splicing unit (intein) have been used, and the protease sortase has been used to mediate cyclization and can be applied to either expressed or chemically produced peptides. However, one of the most promising fields of study for the development of novel macrocyclic peptides is the use of genetic reprogramming. This approach has expanded rapidly in recent years and has allowed the introduction of nonnative residues into the sequences. When combined with screening methods, genetic reprogramming has been used in the evolution of nonnatural cyclic peptides that potently inhibit the protease thrombin, and libraries of macrocyclic peptides have been generated that potently target proteases and protein–protein interactions.

It is clear that the potential therapeutic applications of macrocyclic peptides are as diverse as the structures themselves, with the cyclotide examples highlighting the potential for the treatment of pain and cancer. Major excitement lies in the development of macrocyclic peptides that are able to inhibit protein–protein interactions. Traditionally, these targets have been difficult to modulate with small molecules because of the larger surface area required. However, as more protein–protein interactions become valid targets for drug design, macrocyclic peptides are likely to play a major role in designing effective inhibitors with druglike properties. The diversity of cyclic peptides in terms of structure and function means that not all will have characteristics suitable as drug leads, and a greater understanding of properties such as membrane permeability and those important for oral activity needs to be further explored. Still, with several examples now in the clinic, as well as exciting advances in screening technologies and examples showing intracellular activity and oral availability, it is likely that significant research effort will continue to be placed on the development of macrocyclic peptides as drug leads.

For background information *see* CANCER (MEDICINE); MACROCYCLIC COMPOUND; PAIN; PEPTIDE; PHARMACEUTICAL CHEMISTRY; PHARMACOLOGY; PHARMACY; TOXIN in the McGraw-Hill Encyclopedia of Science & Technology.

Norelle L. Daly

Bibliography. V. Baeriswyl and C. Heinis, Polycyclic peptide therapeutics, *ChemMedChem*, 8:377–384, 2013, DOI:10.1002/cmdc.201200513; Y. Ji et al., In vivo activation of the p53 tumor suppressor pathway by an engineered cyclotide, *J. Am. Chem. Soc.*, 135(31):11623–11633, 2013, DOI:10.1021/ja405108p; T. Passioura et al., Selection-based discovery of druglike macrocyclic peptides, *Annu. Rev. Biochem.*, 83:727–752, 2014, DOI:10.1146/annurev-biochem-060713-035456; Y. Schlippe et al., In vitro selection of highly modified cyclic peptides that act as tight binding inhibitors, *J. Am. Chem. Soc.*, 134(25):10469–10477, 2012, DOI:10.1021/ja301017y; C. T. Wong et al., Orally active peptidic bradykinin B1 receptor antagonists engineered from a cyclotide scaffold for inflammatory pain treatment, *Angew. Chem. Int. Ed. Engl.*, 51:5620–5624, 2012, DOI:10.1002/anie.201200984.

Maya civilization

The Maya civilization prospered in an area comprising the present-day locations of southern Mexico, Guatemala, Belize, Honduras, and El Salvador (**Fig. 1**). It encompassed a diversity of environmental zones, including the tropical lowlands of the Yucatán Peninsula and the highlands located to the south of it. The beginning of the Maya society was surprisingly late. Sedentary villages and ceramics did not appear in the Maya lowlands until the onset of the Middle Preclassic period (1000–400 BC), whereas many other groups in areas surrounding the Maya area had adopted a sedentary way of life centuries earlier. In particular, the so-called Olmec civilization on the southern Gulf of Mexico coast had developed the large center of San Lorenzo (in present-day Mexico), which is known for its colossal head sculptures that are thought to have depicted

Fig. 1. Map of the Maya area and the surrounding regions.

Fig. 2. Cache of greenstone axes dating to approximately 1000 BC found at Ceibal (in present-day Guatemala).

rulers. A central question concerning the origins of the Maya civilization has been whether it emerged under the influence of the Olmecs or whether it developed more independently.

Early Maya society and political organization. Recent investigations at the Maya center of Ceibal (in present-day Guatemala) have indicated that the residents suddenly built a formalized ceremonial complex at approximately 1000 BC and began elaborate rituals involving the deposits of greenstone axe caches (**Fig. 2**). Chronological examination suggests that this new development occurred during the transitional period after the decline of San Lorenzo and before the rise of the next Olmec center, La Venta (in present-day Mexico). If this is true, the direct influence of the Olmec civilization was probably not a primary factor in the initial development of the sedentary Maya communities. However, this does not mean that the Maya civilization emerged independently. Formal ceremonial complex remains and caches similar to those found at Ceibal are also found at Chiapa de Corzo and other contemporaneous centers in Chiapas, Mexico. Therefore, close interactions with other regions must have been critical. Another possible factor is an increase in the productivity of maize around 1000 BC.

In comparison to the Olmec centers, which probably formed hierarchical polities (political organizations) with rulers, the degree of social inequality in the Maya lowlands was not prominent. Although the emergent elite groups of the Middle Preclassic period possibly organized the construction of ceremonial buildings, they did not make elaborate tombs or stone monuments glorifying themselves. More centralized polities involving rulerships probably developed during the Late Preclassic period (400 BC to AD 250). Excavations at San Bartolo (in modern-day Guatemala) have revealed evidence of hieroglyphic writing dating to around 250 BC and elaborate murals, including a scene of a ruler's coronation, painted around 150 BC. About the same time, the center of El Mirador (in present-day Guatemala) developed into a large city with enormous pyramids and other massive structures. It has been debated when stelae (carved stone pillars or slabs) depicting rulers began to be erected in the Maya lowlands. There is no unambiguous evidence of such monuments before 100 BC, although some scholars suggest the possibility of earlier dates. In the Maya highlands, the center of Kaminaljuyu (located near the modern site of Guatemala City) flourished during the Verbena-Arenal period, which was marked by elaborate royal tombs and stelae depicting rulers. Although the Verbena-Arenal period was commonly thought to have begun around 400 BC, a recent reevaluation of the available data places it between 100 BC and AD 150. In other words, fully established rulerships with rich tombs and depictions of rulers on monuments may have emerged roughly contemporaneously in the Maya lowlands and highlands around 100 BC.

Decline of Preclassic centers. Toward the end of the Preclassic period, around AD 150, many Maya centers experienced a period of decline. These centers included El Mirador in the lowlands and Kaminaljuyu in the highlands. This phenomenon of decline is sometimes called the Preclassic Maya collapse. It has been suggested that droughts occurred at this time, and these may have severely affected agricultural production in the region. However, the magnitude and frequency of any droughts and their social effects need to be further investigated. The effect of humans on the environment may also have been an important factor. Studies suggest that extensive deforestations caused substantial soil erosions, which may have transformed perennial lakes into seasonal swamps. In addition, social problems may have led to political instability. Construction of fortifications increased around this time, which implies that warfare among competing groups intensified.

Recovery and the Classic period. Maya society recovered from the Preclassic collapse and enjoyed

Fig. 3. Plaza of Tikal (in modern-day Guatemala), where rulers probably presided over public ceremonies.

its heyday during the Classic period (AD 250–900). Political and cultural developments during this period occurred primarily in the Maya lowlands. The Maya further refined their writing system at this time by combining logographic signs (those representing meanings) and phonetic ones. The results of hieroglyphic decipherment, as well as the analysis of the archeological data, indicate that the Maya area was never politically unified. Many centers, such as Copan (in present-day Honduras) and Palenque (located in modern southern Mexico), had their own dynasties and royal courts and maintained some level of autonomy, although certain powerful cities, particularly Tikal (**Fig. 3**) and Calakmul in the Maya lowlands, exerted strong influences over other dynasties. The interactions among the various dynasties often combined diplomacy, marriage alliances, and wars. Military campaigns rarely led to territorial conquests, and defeated dynasties typically persisted.

Rulers served not only as political and military leaders, but also as the symbolic center of each community and as ritual performers. Many stelae depict their dances and other performances carried out in public plazas. Other hereditary elites also played important roles in carrying out various aspects of administration, war, and rituals. Many of these elite persons were highly skilled scribes and artists, and their production of stone sculptures, painted ceramics, and other artistic objects was closely tied to their power and prestige. Farmers constituted a large part of the nonelites, and they developed specific strategies for adapting to the tropical lowlands. Although heavily emphasizing maize cultivation, they also combined other cultigens (cultivated plants) and various farming methods, such as raised fields in wetlands and agricultural terraces, as adaptations to specific microecological zones.

Classic Maya collapse. Around the ninth century AD, many Maya centers went into a period of decline, often termed the Classic Maya collapse. Multiple studies of lake sediments and speleothems (secondary mineral deposits created by the action of water in a cave; also known as cave formations) indicate that a series of droughts occurred at this time. However, the nature of their effects on Maya society is debated. An important issue is that an early wave of collapse around AD 810 was particularly notable in the southern lowlands, which are typically characterized by high precipitation. In the drier northern lowlands, which should have been more vulnerable to any droughts, powerful centers such as Uxmal and Chichen Itza (both located on the Yucatán Peninsula) prospered through the period of droughts. Therefore, some scholars suggest that the collapse was caused primarily by social problems, including political problems and human-induced environmental degradation. Investigations at Aguateca (in modern-day Guatemala; **Fig. 4**) indicate that the center was destroyed by invading enemies and then abandoned.

Fig. 4. Figurine depicting a Maya noble dating to approximately AD 810 found at Aguateca (in present-day Guatemala).

The Maya society continued through the Classic collapse and the later Spanish conquest, and several million Maya people live in the region today. Some recent studies have focused on the society's ability to withstand natural and social disasters, as well as the relationship between archeology and the modern Maya people.

For background information *see* AGRICULTURAL SCIENCE (PLANT); AGRICULTURE; ARCHEOLOGY; CERAMICS; CIVIL ENGINEERING; CLIMATE MODIFICATION; CORN; DROUGHT; LINGUISTICS; NORTH AMERICA; TERRACING (AGRICULTURE) in the McGraw-Hill Encyclopedia of Science & Technology.

Takeshi Inomata

Bibliography. M. Coe, *The Maya*, 8th ed., Thames & Hudson, London, 2011; S. Houston and T. Inomata, *The Classic Maya*, Cambridge University Press, Cambridge, UK, 2009; T. Inomata et al., Early ceremonial constructions at Ceibal, Guatemala, and the origins of lowland Maya civilization, *Science*, 340:467–471, 2013, DOI:10.1126/science.1234493; S. Martin and N. Grube, *Chronicle of the Maya Kings and Queens: Deciphering the Dynasties of the Ancient Maya*, 2d ed., Thames & Hudson, London, 2008; W. A. Saturno, D. Stuart, and B. Beltrán, Early Maya writing at San Bartolo, Guatemala, *Science*, 311:1281–1283, 2006, DOI:10.1126/science.1121745.

Measuring formaldehyde emissions from wood-based panels

Formaldehyde is the common name for methanal. It is the smallest of the aldehydes and is a gas at room temperature with a pungent odor. In commercial processes, it is normally used as an aqueous solution (formalin) or in a solid polymerized form called paraformaldehyde. Products like wood-based panels that are made with formaldehyde-based adhesives have the potential to emit formaldehyde, which is classified as a carcinogen. Therefore, methods for measuring these emissions have been developed.

Production of formaldehyde. Formaldehyde is most commonly made by vaporizing methanol, mixing it with air, and passing the mixture over a silver catalyst at temperatures up to 650°C (1200°F). The formaldehyde gas is then dissolved in water to a concentration of up to 60%. Often, however, the concentration used is much lower, so as to increase its storage life. A common concentration of formalin found in laboratory stores is 37%, and solutions of this type normally include some methanol for stability.

Uses of formaldehyde. Formaldehyde has a great many uses. Most of the millions of tons of formaldehyde that are made each year are used to make thermosetting resins, such as urea formaldehyde, melamine-urea formaldehyde, and phenol formaldehyde, which are the most commonly used adhesives for the manufacture of wood-based panel (WBP) products.

Types and levels of formaldehyde emissions. Formaldehyde is used in many products. Thus, there are also many sources of formaldehyde emissions. Emissions can be direct via burning (for example, cigarette smoke and emissions from motor engines) or indirect by the oxidative degradation of volatile organic compounds (VOCs) [for example, the oxidation of ethane and propene]. In fact, indirect emissions have been determined to account for more than 90% of the formaldehyde found in outdoor air.

The outdoor background levels of formaldehyde have been found to be approximately 5.8 ppb (parts per billion). The average human can detect the presence of formaldehyde at levels of 50 ppb. At concentrations above 100 ppb (1.2 mg/m^3), it becomes an irritant to the eyes, nose, and throat. Indoor formaldehyde concentrations are typically an order of magnitude higher than outdoor levels because of the presence of formaldehyde-emitting products.

The International Agency for Research on Cancer (IARC) classified formaldehyde as carcinogenic to humans in 2004, and this classification was confirmed in 2009. Therefore, the World Health Organization (WHO) recommends an exposure limit of 0.1 mg/m^3. Limits on formaldehyde concentration in indoor air have been established by various regulatory bodies. Thus, WBPs must be tested to verify whether they conform to accepted emission rates.

Measuring and testing of formaldehyde. There are two approaches to measuring formaldehyde. One approach measures the formaldehyde level in a WBP, whereas the other measures the concentration of formaldehyde in the air surrounding a WBP. Most methods belong to the latter category.

Fig. 1. One of the 1-m^3 formaldehyde emission chambers at École Supérieure du Bois (Nantes, France).

The European (EN) formaldehyde extraction method, known as the perforator method (EN 120), uses toluene to extract formaldehyde from approximately 100 g (3.5 oz) of small cubes [25 × 25 mm (1 × 1 in.)] of WBPs. This method is used extensively for quality-control purposes in manufacturing plants. However, there are concerns over the accuracy of this method for today's low-emitting products.

In the majority of countries, the reference method for determining the emission rate of a WBP is via the use of a controlled environment chamber. Chamber volumes vary considerably, but the 1-m^3 chamber (**Fig. 1**) is probably the most common in laboratories and test houses. Samples are placed in the chamber and then clean air is continually pumped into the chamber at a precise rate, known as the air exchange rate. Pumping air into the chamber generates a slight pressure within it, and this prevents the possibility of air from the laboratory entering the chamber. The temperature, relative humidity, and speed of the air in the chamber are controlled and are kept constant during a test (**Table 1**). Thus, the WBP samples in the chamber will reach equilibrium with these conditions, and the formaldehyde emission rate is determined at this point.

The decay curve of a typical test result is shown in **Fig. 2**. Tests take quite a long time to complete, usually between 2 and 4 weeks. Consequently, these methods are not well suited to product development. Instead, they are better for product verification.

TABLE 1. A comparison of European, American, and ISO chamber methods

Properties	EN 717-1	ISO 12460-1	ASTM D6007 (ISO 12460-2)	ASTM E1333
Chamber volume (m^3)	0.225, 1, or 12	1	0.02–1	≥22
Chamber loading (m^2/m^3)	1	1	0.13–0.95*	0.13–0.95*
Temperature (°C)	23	23	25	25
Relative humidity (%)	45	50	50	50
Air exchange rate (number/h)	1	1	≈0.5	≈0.5
Speed of the air (m/s)	0.2	0.1–0.3	Not specified	Not specified
Unsealed edge:surface ratio (m/m^2)	1.5	1.5	Thickness dependent	Thickness dependent

*Varies with panel type.

At various points during the test, known volumes of air (between 30 and 120 L, depending on the test method) are bubbled through water [International Organization for Standardization (ISO) and EN standards] or sodium bisulfite solutions [American Society for Testing and Materials (ASTM) standards] to capture the formaldehyde in the air. In the case of EN and ISO standards, the concentration of formaldehyde in the water is subsequently measured using the Hantzsch reaction. This involves the addition of acetylacetone and ammonium acetate to the water containing the formaldehyde. These compounds react with formaldehyde to produce diacetyldihydrolutidine (DDL). The concentration of DDL is easily determined by light spectroscopy because DDL absorbs light at 412 nm. In contrast, ASTM tests use another spectrophotometric method (the chromotropic acid method), with measurements at 580 nm. A previously prepared calibration curve is then used to determine the concentration of formaldehyde in both the solution and the original sample of air.

Over the course of 1 or 2 weeks, data points are collected to provide an indication of the steady-state formaldehyde emission of the WBP. Each of the standards specifies a different method for determining when a steady-state condition is achieved. However, all accept a change in formaldehyde emission of less than 5% over a given period as representing a quasi-steady-state condition. In addition, the test is stopped after 28 days, even if the steady-state condition has not been reached.

The gas analysis method (EN 717-2, equivalent to ISO 12460-3) uses very small chambers to measure formaldehyde emissions. It is a production control method for plywood. The specimens used are relatively small, being only 400 mm (16 in.) in length and 50 mm (2 in.) in width, whereas WBPs in use tend to be large and are often more than 2 m (6.6 ft) in length and 1 m (3.3 ft) in width. Consequently, the permeability of the WBP faces has a great influence on emission rates. The edges of the gas analysis specimens are therefore sealed, most commonly with aluminum tape, so only the formaldehyde emission through the faces is measured.

The gas analyzer itself is a small cylinder, measuring 555 mm (22.2 in.) in length and 96 mm (3.84 in.) in diameter. This gives a chamber volume of just over 4 L. The chamber is heated to 60°C (140°F), and clean, dry air is pumped through it at 60 L per hour. The air exits the chamber and passes through two impingers in series (each containing about 30 mL of distilled water), which capture any formaldehyde in the air. The test lasts for 4 h. At the end of each hour, the air is automatically diverted to a new pair of impingers. After the test, the formaldehyde concentration in the water is determined via the Hantzsch reaction (as described previously).

Overall, the gas analysis method is relatively quick, taking 4 h, plus another 1.5 h for the analysis of the water. Consequently, the method is used by laboratories for product development.

Formaldehyde emissions are dependent on the moisture content of the samples. The use of 60°C (140°F) and dry air in the gas analysis method causes the specimen to dry during the test. Thus, formaldehyde emissions fall during the test (**Table 2**). The falling emission values during the course of the test have caused people to assume that EN 717-2 is an extraction method. This is a false assumption because similar quantities of formaldehyde are emitted as were observed during the first test if the samples are reconditioned and retested.

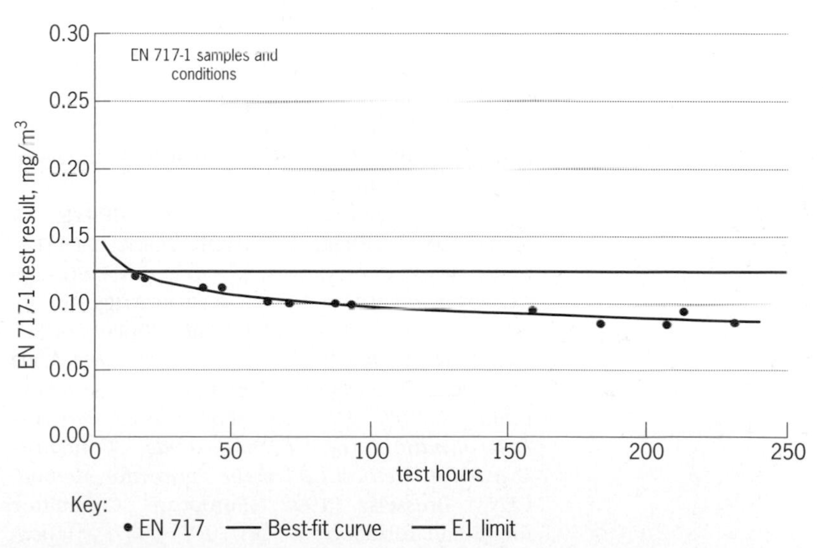

Fig. 2. A typical test result following the EN 717-1 chamber method.

TABLE 2. Typical formaldehyde emission levels observed from E1 grade, thin medium-density fiberboard during a test conducted following EN 717-2

Emission period	Value observed, $mg/m^2/h$
First hour	2.69
Second hour	1.81
Third hour	0.78
Fourth hour	0.25
Average	1.38

Another type of test used to measure formaldehyde emissions from WBPs is the sealed chamber method. In this method, the air is not changed during the test. The two most-popular types employing this method are the European flask method (EN 717-3) and the Japanese desiccator method (JIS A 1460). The principle of both types is the same: namely, a quantity of WBP is sealed in a vessel containing a quantity of water, for a set period, at a specific temperature. The formaldehyde concentration in the water is measured at the end of the test. The results are expressed in mg/L, so it is not obvious how these results relate to the values provided by other emission tests that express their results in mg/m^3. Therefore, correlations between test methods are being investigated.

The correlations are often good for a specific panel. Unfortunately, these correlations cannot be used to accurately predict the behavior of other panels. This is probably because the emission of formaldehyde from a WBP is dependent on a complex interaction of a wide range of panel properties. For example, the moisture content of specimens during a test is dependent on the test being conducted, and the moisture content will affect many panel properties, including surface permeability, diffusion coefficients, and adhesive hydrolysis. Consequently, it is not surprising that a correlation is not generally applicable to all panel types.

Conclusions. There are a number of different tests and ways to measure the levels of formaldehyde emitted from WBPs. Various standards are required in different countries, with regulatory bodies limiting the concentrations of formaldehyde in indoor air. Thus, WBPs need to be tested properly and accurately so that they conform to accepted rates of formaldehyde emission.

For background information *see* ADHESIVE; AIR POLLUTION, INDOOR; ALDEHYDE; FORMALDEHYDE; RESIN; WOOD COMPOSITES; WOOD ENGINEERING DESIGN; WOOD PRODUCTS; WOOD PROPERTIES; WOODWORKING in the McGraw-Hill Encyclopedia of Science & Technology.

Mark A. Irle

Bibliography. European Committee for Standardization (CEN), *EN 120: Wood-based Panels—Determination of Formaldehyde Content—Extraction Method Called the Perforator Method*, CEN, Brussels, 1992; European Committee for Standardization (CEN), *EN 717-1: Wood-based Panels—Determination of Formaldehyde Release—Part 1: Formaldehyde Emission by the Chamber Method*, CEN, 2005; European Committee for Standardization (CEN), *EN 717-2: Wood-based Panels—Determination of Formaldehyde Release—Part 2: Formaldehyde Release by the Gas Analysis Method*, CEN, 1995; European Committee for Standardization (CEN), *EN 717-3: Wood-based Panels—Determination of Formaldehyde Release—Part 3: Formaldehyde Release by the Flask Method*, CEN, 1996; Japanese Industrial Standards Committee, *JIS A 1460: Building Boards Determination of Formaldehyde Emission—Desiccator Method*, Japanese Standards Association, Tokyo, 2001; D. D. Parrish et al., Primary and secondary sources of formaldehyde in urban atmospheres: Houston Texas region, *Atmos. Chem. Phys.*, 12:3273–3288, 2012, DOI:10.5194/acp-12-3273-2012; M. Risholm-Sundmana et al., Formaldehyde emission—Comparison of different standard methods, *Atmos. Environ.*, 41:3193–3202, 2007, DOI:10.1016/j.atmosenv.2006.10.079; M. Z. M. Salem et al., Evaluation of formaldehyde emission from different types of wood-based panels and flooring materials using different standard test methods, *Building Environ.*, 49:86–96, 2012, DOI: 10.1016/j.buildenv.2011.09.011; T. Salthammer, S. Mentese, and R. Marutzky, Formaldehyde in the indoor environment, *Chem. Rev.*, 110:2536–2572, 2010, DOI:10.1021/cr800399g; United States Environmental Protection Agency, *Locating and Estimating Air Emissions from Sources of Formaldehyde*, Report EPA-450/4-91-012, Office of Air Quality Planning and Standards, U.S. EPA, Research Triangle Park, NC, 1991; World Health Organization, *WHO Guidelines for Indoor Air Quality: Selected Pollutants*, WHO, Bonn, Germany, 2010.

MESSENGER mission results

The MErcury Surface, Space ENvironment, GEochemistry, and Ranging (MESSENGER) mission is the first spacecraft to orbit Mercury and has greatly extended our knowledge of the innermost planet. *MESSENGER* was launched in 2004 and subsequently executed flybys of Earth, Venus (twice), and Mercury (three times). The primary motivation for these flybys was to reduce the required propulsion for Mercury orbit insertion on March 18, 2011. In addition, the three *MESSENGER* flybys of Mercury allowed its camera to image much of the planet's surface that had never before been seen. Once *MESSENGER* was in orbit, an imaging campaign was undertaken for its entire surface, expanding this coverage even further. A complete global image mosaic of the surface was finished on December 30, 2012 (**Fig. 1**).

This global image mosaic and measurements from *MESSENGER*'s other instruments revealed a heavily cratered planet that is different in important ways from the other terrestrial planets. Much of Mercury's

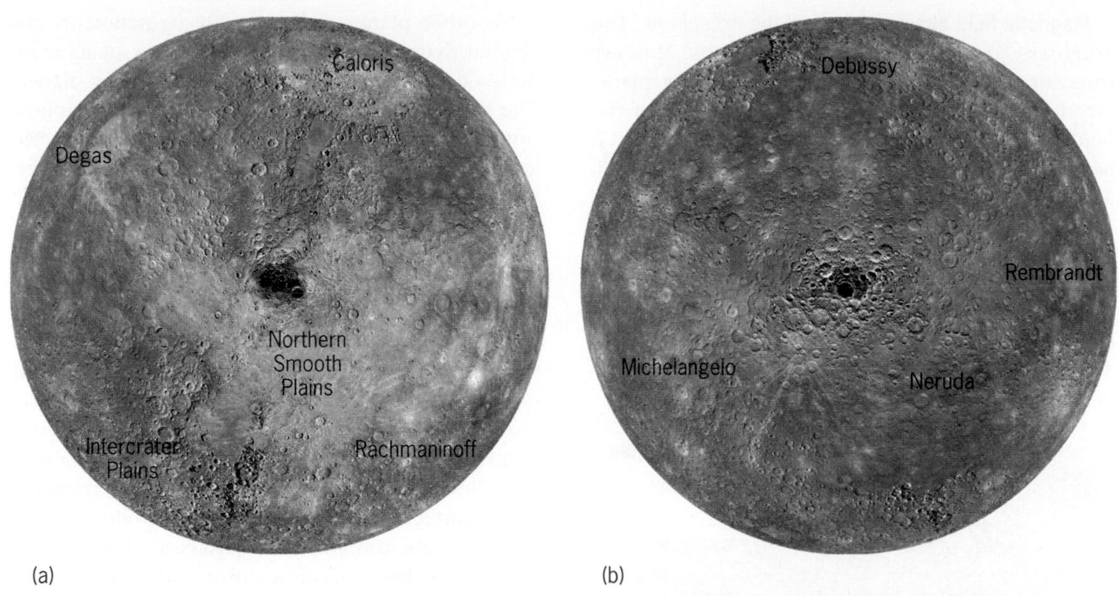

Fig. 1. Global views of Mercury's surface from *MESSENGER*'s Mercury Dual Imaging System camera. (*a*) Northern hemisphere (looking down on the north pole). (*b*) Southern hemisphere (looking up on the south pole).

surface has been resurfaced by volcanism. The youngest large region of volcanic smooth plains is at high northern latitudes, surrounding Mercury's north pole in a broad, topographically low expanse. Expansive plains are also found within and adjacent to large impact basins such as Caloris, approximately 1550 km (963 mi) in diameter, and Rembrandt, approximately 715 km (444 mi) in diameter. These volcanic plains are broadly similar in appearance to the lunar maria. Unlike on the Moon, however, even the most heavily cratered portion of Mercury's crust (which is confusingly called the intercrater plains) is unlikely to be primordial or date to the earliest part of the solar system. Instead, its cratering record implies that volcanic resurfacing was essentially global on Mercury early in its history, and its oldest surfaces are approximately 4.0–4.1 billion years old.

Both the young, smooth plains, and the older, more heavily cratered intercrater plains are tectonically deformed, mainly by large lobate scarps and smaller, plains-cutting wrinkle ridges that are the result of compressional deformation (**Fig. 2**). The predominance of compressional faulting on Mercury has been interpreted to a result of planetary contraction—in other words, a decrease in Mercury's radius—as it cooled, causing it to shrink up to approximately 10–20 km (6–12 mi or 0.4–0.8%).

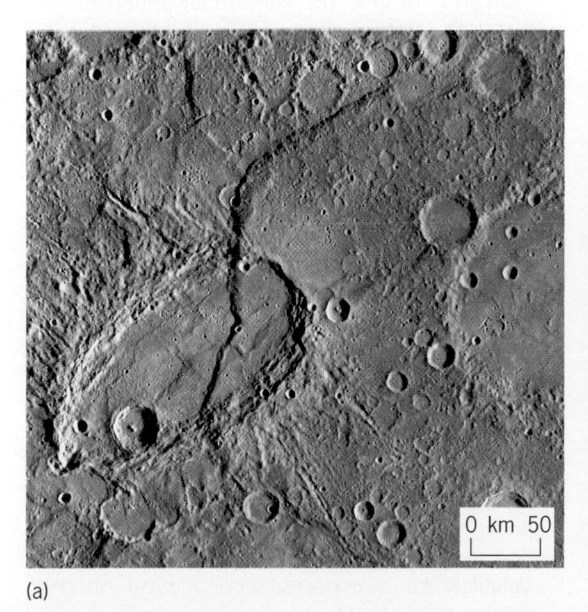

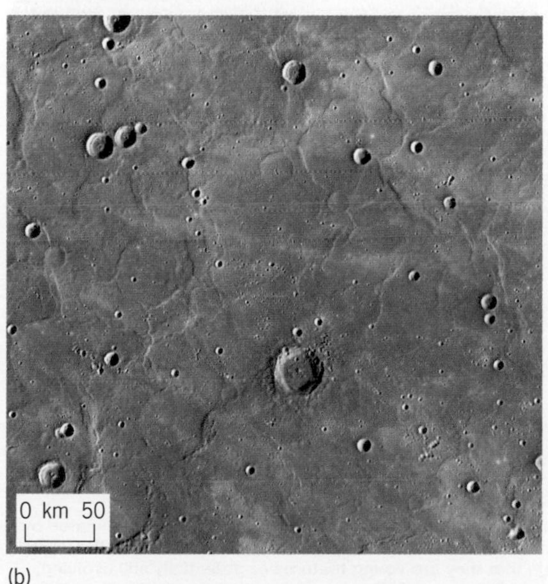

Fig. 2. Examples of common tectonic features on Mercury. (*a*) Beagle Rupes, a large lobate scarp. (*b*) Wrinkle ridges in the northern plains of Mercury. 50 km = 31 mi.

Magnetic field observations and the exosphere. The *Mariner 10* spacecraft, which flew by Mercury three times in 1974–1975, detected a weak planetary magnetic field, but its characterization of the field was hampered by limited observations and a lack of concurrent measurements of the solar wind. MESSENGER data have demonstrated that Mercury's internal field is a spin-aligned dipole with a magnetic moment approximately 0.1% as strong as Earth. In addition, the dipole's magnetic equator is offset north of the planet's equator by 479 ± 6 km (298 ± 4 mi), or approximately 20% of the radius of Mercury.

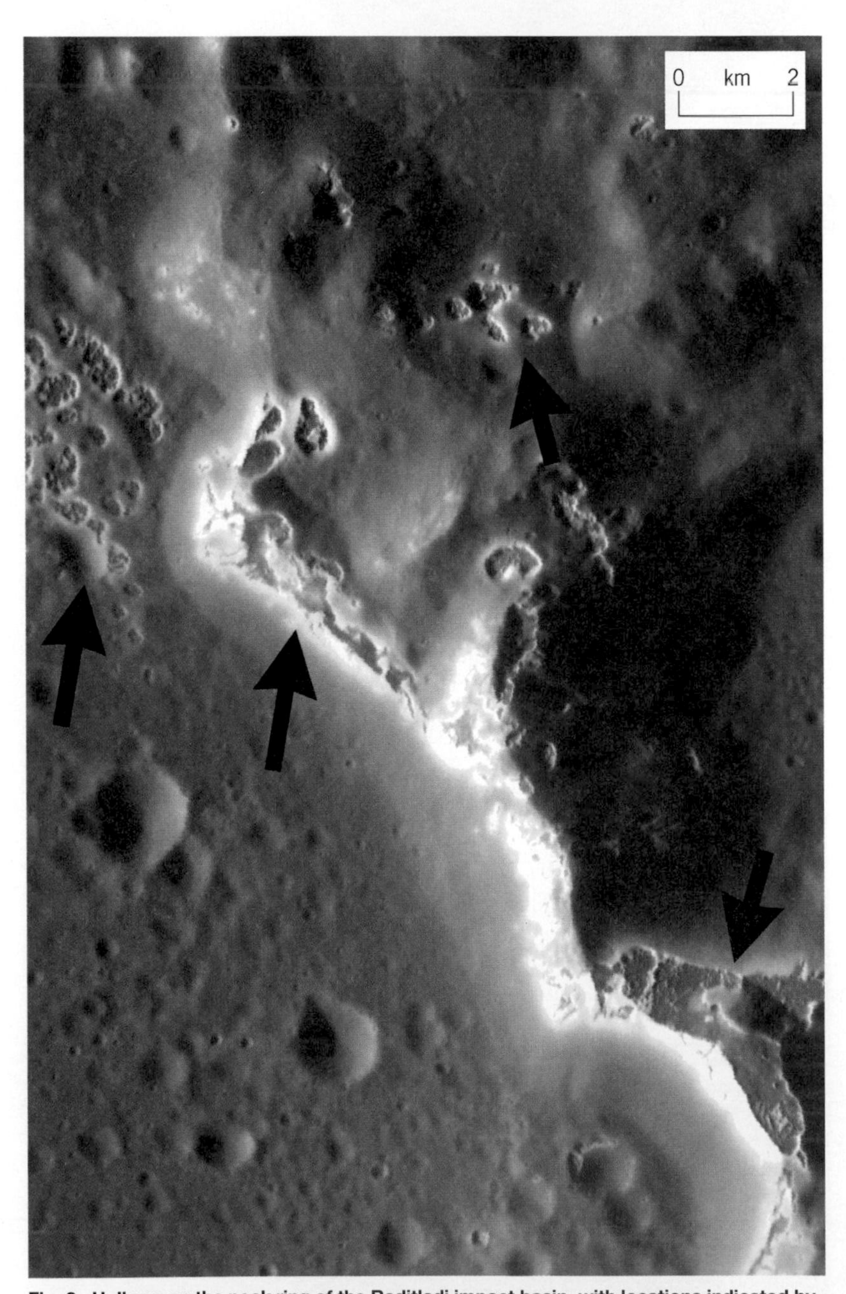

Fig. 3. Hollows on the peak ring of the Raditladi impact basin, with locations indicated by the four arrows. The hollows have sharp boundaries (especially the hollow near the lower rightmost arrow). This suggests that they are young features or potentially still evolving, because, if they were older, the hollows would have been smoothed out and softened over time (like the impact craters in the image), probably from the constant bombardment of Mercury's surface with very small impactors. 2 km = 1.24 mi.

No other planetary body with a rotationally coaligned dynamo has such a large offset of its magnetic equator relative to the equator of the planet. The magnitude of this offset has important implications for the strength of the magnetic field at the surface, which is a factor of 3.4 times stronger at the north pole than at the south pole. The offset dipole also leads to hemispheric asymmetry in the size of the magnetic cusps, regions near to the magnetic poles where currents from the magnetopause nearly cancel the internal field. Because of the weak resulting fields within these regions, high-energy particles from the solar wind reach the surface with ease. The offset in the dipole results in a much larger cusp region in the southern hemisphere than the northern hemisphere, resulting in an enhanced flux (about four times higher intensity) of high-energy particles to the surface in the south compared to the north.

The bombardment of the surface by high-energy particles at these magnetic cusps and in other areas where the solar wind is able to impinge on the surface provide a likely source for Mercury's tenuous exosphere. MESSENGER has measured the composition of the exosphere, which includes sodium, magnesium, calcium, potassium, sulfur, and silica. The abundance of these elements is higher in Mercury's exosphere than in the solar wind, which implicates the surface as their likely source.

Interior structure, geochemistry, and surface mineralogy. Mercury has the highest uncompressed bulk density of any terrestrial planet, 5.3 g/cm^3, and geophysical constraints from MESSENGER and terrestrial radar observations are consistent with a core radius of 2020 ± 30 km (1255 ± 19 mi), which is a very large fraction of the radius of the planet (2440 km or 1516 mi). This implies that iron makes up 60–75% of Mercury's mass. The silicate crust and mantle are thus thin, with a thickness of only 420 ± 30 km (261 ± 19 mi).

However, despite the substantial abundance of iron on the planet as a whole, visible and near-infrared spectroscopy and gamma- and x-ray spectroscopy from MESSENGER suggest that surface materials contain little iron (less than a few percent by weight). Because iron is a common cation (positively charged atom) in silicate minerals in the lunar, Martian, and asteroid crust, the mineralogy of Mercury's surface must be distinct from those bodies. Mercury's surface is also rich in magnesium and depleted in aluminum and calcium compared to the other terrestrial planets. This elemental chemistry is consistent with a surface mineralogy with a high abundance of magnesium-rich silicates, such as enstatite (a magnesium-rich pyroxene) and forsterite (a magnesium-rich olivine), as well as some amount of plagioclase feldspar and sulfides. Taken together, these data suggest that Mercury formed from highly reduced precursor materials, compositionally akin to the enstatite chondrite meteorites.

Volatiles. An unexpected geochemical observation from MESSENGER is that the abundance of moderately volatile elements, such as sodium, potassium, and sulfur, are higher on Mercury than expected. The presence of up to approximately 4% sulfur by weight is particularly striking, as it is a factor of 10 times higher on Mercury than on the Moon or Earth. These volatile abundances seem inconsistent with some of the explanations that have been put forward for Mercury's high metal-to-silicate ratio and large core, which have invoked periods of very high temperature early in its history.

Volatiles appear to have influenced the geology of Mercury's surface, too. During the first *MESSENGER* flyby in 2008, rimless pit structures were observed on the surface, associated with and surrounded by deposits with relatively high albedo. These pits are interpreted to be loci of past pyroclastic volcanism. The size of their associated deposits—up to approximately 70 km (43 mi) in radius—would require up to approximately 1–2% by weight of a volatile such as carbon monoxide (CO), hydrogen sulfide (H_2S), or water (H_2O) if the eruptions were produced in a Hawaiian-style fire fountain.

An additional distinctive class of surface features on Mercury that also appears to relate to volatiles is called hollows (**Fig. 3**). These are small, scarp-bounded depressions of tens to thousands of meters across that commonly have high-reflectance interiors and halos. The process hypothesized to form these features is loss of volatiles, such as by sublimation. Taken together, the results of the MESSENGER mission imply that Mercury's surface and interior have a greater abundance of volatiles than has traditionally been predicted for the innermost planet.

Polar volatiles. Earth-based radar observations during the 1990s led to the discovery that some craters in Mercury's polar regions were anomalously bright at radar wavelengths, consistent with the signature of water ice or another volatile. Given the proximity of Mercury to the Sun and its peak temperature of more than 700 K (800°F), the presence of ice on its surface might seem surprising. However, these radar-bright deposits are areas of permanent shadow, which is possible because of Mercury's near-zero rotational tilt (obliquity). Models of the thermal environment of these permanently shadowed regions suggest that the temperature in the shallow subsurface remains at approximately 100 K (−280°F) or less, allowing water ice to be stable. Data from *MESSENGER*'s neutron spectrometer confirm the presence of high concentrations of hydrogen—almost certainly in the form of water ice—in the upper meter in many permanently shadowed crater floors.

MESSENGER has also made measurements of the surface reflectance of these deposits using laser altimetry. The amount of laser light reflected from the surface of these deposits surprisingly reveals that they are anomalously darker than is typical for Mercury's surface. This suggests that a thin layer of organic materials may overlie the ice deposits. Only a few deposits in craters at the highest northern latitudes (greater than 84.9°N) have high reflectance, consistent with ice actually exposed at Mercury's surface. As is the case for similar deposits at the lunar poles, the source of water ice in these deposits is unknown, although delivery from comets is a plausible mechanism.

Prospects. Plans call for *MESSENGER* to continue making observations until it impacts Mercury in March 2015. During its last year of operations, the spacecraft's altitude at periapsis is progressively decreasing, bringing it closer to Mercury's surface in the northern hemisphere. This will potentially allow surface imaging at the scale of a few meters, at least locally, and geophysical observations at substantially higher fidelity. These data have the potential to improve the characterization of polar volatiles, as well as to provide an unprecedented look at the morphology of hollows, fresh craters, tectonics, and volcanic features.

For background information *see* ENSTATITE; FELDSPAR; MAGNETOSPHERE; MERCURY (PLANET); METEORITE; OLIVINE; PLANET; PYROXENE; RADAR ASTRONOMY; SOLAR SYSTEM; SOLAR WIND; SPACE PROBE in the McGraw-Hill Encyclopedia of Science & Technology. Caleb I. Fassett

Bibliography. B. Denevi and C. Ernst, MESSENGER: Revealing Mercury's secrets, *Lunar and Planetary Information Bulletin*, 127:2–5, 2011; S. C. Solomon, A new look at the planet Mercury, *Phys. Today*, 64:50–55, 2011, DOI:10.1063/1.3541945; S. Z. Weider and L. R. Nittler, The surface composition of Mercury as seen from MESSENGER, *Elements*, 9(2):90–91, 2013; A. Witze, Mercury shrinking more than thought, *Nature News*, December 9, 2013, DOI:10.1038/nature.2013.14331.

Metamemory

Metamemory refers to our knowledge and awareness of our own memory processes. Knowledge in this case means self-knowledge about our memory processes. For example, when a person asserts that he or she is good at remembering faces, but poor at remembering names, that person is making a statement concerning metamemory knowledge. Metamemory awareness refers to our feelings or experiences of our own memory. For example, if a person feels certain that he or she will remember later something just learned now, that person is having a metamemory experience. Metamemory experience has been extensively studied by cognitive psychologists and will be the focus of this discussion.

Focus and models of metamemory research. The current focus of research on metamemory experience derives from two historical trends within cognitive psychology. On the one hand, Endel Tulving and Stephen A. Madigan proposed in 1970 that we need to understand the phenomenological experiences that accompany memory in order to understand human memory. In particular, they emphasized that one of the unique aspects of human memory is its self-knowledge. On the other hand, metamemory research also arose from theories within cognitive development stating that improvements in memory as children grow older are likely the result of children acquiring more awareness and knowledge about how memory works. These two trends have contributed to how we conceptualize metamemory today.

However, it was the work of Thomas O. Nelson and his colleagues that brought metamemory research into the mainstream of cognitive psychology in the 1990s. Nelson introduced an important model of metamemory, accounting for both metamemory process and metamemory function. For Nelson, metamemory served the function of monitoring ongoing memory processes, which then could be used to control memory processes. Nelson proposed that metamemory processes are themselves cognitive processes, but serve at a meta-level, monitoring the activity of object-level processes. In this way, monitoring is the awareness of ongoing memory activity. It can be measured by asking participants to make judgments about their learning or retrieval. The outcome of these judgments can affect how people decide to learn and remember—that is, how they control their own memory.

In this view, metamemory needs to be accurate. Specifically, monitoring must accurately reflect the underlying object-level memory process in order to direct the control function in an adaptive fashion. That is, for an individual to make good control decisions, that individual must be able to accurately monitor the relevant memory process. For example, tip-of-the-tongue states are metacognitive experiences concerning the likelihood of recalling a currently inaccessible word. Tip-of-the-tongue states must indeed be correlated with actual retrieval if they are going to help us determine what we know but cannot recall. Indeed, research shows strong correlations between tip-of-the-tongue states and subsequent memory retrieval. In general, metamemory is accurate at predicting memory performance, but exceptions can be found.

Assessing metamemory. Assessing metamemory accuracy involves two components, which are called resolution and calibration. Resolution refers to the ability of people to discriminate easy items from hard items. In terms of metamemory judgments, if higher judgments are followed by successful memory performance and lower judgments are followed by unsuccessful metamemory performance, then the resolution scores will be high. In metamemory judgments, such as feeling-of-knowing judgments and judgments of learning, resolution is usually accurate; that is, there is a strong correlation between the level of judgment and the likelihood of successful retrieval. Resolution is typically measured by the gamma correlation, which measures accuracy on a scale of −1 to 0 to 1. Calibration refers to the correspondence between the absolute level of the judgment and the percent likelihood of remembering. Thus, if a person is making judgments of learning on a scale of 0 to 100, it will be necessary to determine how closely his or her average judgment corresponds to his or her actual memory score. Moreover, calibration can refer to the percent correct as plotted against the predicted correct. For example, a person may predict that he or she will recall 80% of items committed to memory; however, will that person actually recall 80%? Calibration is usually measured by calibration curves that examine whether the judgments are overconfident or underconfident relative to actual performance.

Metamemory judgments. Judgments of learning are made during study and are predictions of future memory performance. For example, consider a student who is studying for an exam in his or her French language class. The student must decide if he or she has learned a specific translation (for example, "the job" = "le boulot") well enough to remember it for a later test. In an experiment, the judgment is often made on a scale of 0 to 100 so that the calibration can be assessed. Research shows that the judgment-of-learning resolution is particularly good, especially if the judgment of learning is delayed and done when only the cue is presented. However, research also shows that judgments of learning are subject to spurious influences, such as the fluency of reading the word pair. For example, it has been found that judgments of learning were higher for cue-target pairs in larger fonts, even though font size affected neither memory for the target words nor the resolution of the judgments of learning.

Feeling-of-knowing judgments are predictions of the future recognition performance of a particular item made at the time of retrieval. Typically, feeling of-knowing judgments are made when a person fails to recall a target answer but predicts the likelihood of recognizing that item. For example, after failing to recall the name of the first person to walk on the moon, a participant judges the likelihood of recognizing that name from a list of names of astronauts. Resolution can then be determined by correlating the magnitude of the feeling-of-knowing judgment with performance on the actual recognition test. Typically, feeling-of-knowing judgments are accurate when applied to both episodic memory (a person's memory of an episode or event) and semantic memory (the recollection of facts, concepts, and information). However, similar to the case with judgments of learning, feeling-of-knowing judgments are sometimes based on spurious factors, such as whether the cue term is familiar or not.

Because it is a cognitive process, it is possible to look for the neural correlates of metamemory. There are now a handful of studies that have examined both judgments of learning and feeling-of-knowing judgments using neuroimaging technology. Typically, these studies show that areas in the prefrontal lobe of the brain are correlated with making metamemory judgments. In particular, the anterior cingulate cortex is associated with identifying a mismatch between metamemory and memory (as occurs in a tip-of-the-tongue state), and the medial frontal cortex is associated with various aspects of monitoring and control. The dorsolateral prefrontal cortex also has been found to be associated with aspects of control.

Conclusions. Metamemory is a thriving area of research. It may be thought of in terms of a system in which underlying object-level processes give rise to self-relevant phenomenal experiences that comment on these underlying processes. The fact that these metamemory states are consciously perceived by the individual is important in allowing the ensuing action to be freely determined and to potentially produce a change in the individual's current knowledge or behavior. Thus, metamemory stands at the crossroads of cognitive psychology, philosophy, and what it is to be human.

For background information *see* BRAIN; COGNITION; INFORMATION PROCESSING (PSYCHOLOGY); LEARNING MECHANISMS; MEMORY; NEUROBIOLOGY; PERCEPTION; PROBLEM SOLVING (PSYCHOLOGY) in the McGraw-Hill Encyclopedia of Science & Technology. Bennett L. Schwartz

Bibliography. E. F. Chua, D. Pergolizzi, and R. R. Weintraub, The cognitive neuroscience of metamemory monitoring: Understanding metamemory processes, subjective levels expressed, and metacognitive accuracy, pp. 267–291, in S. M. Fleming and C. D. Frith (eds.), *The Cognitive Neuroscience of Metacognition*, Springer, New York, 2014; J. Dunlosky and J. Metcalfe, *Metacognition*, Sage Publications, Thousand Oaks, CA, 2009; S. M. Fleming and R. J. Dolan, The neural basis of metacognitive ability, pp. 245–265, in S. M. Fleming and C. D. Frith (eds.), *The Cognitive Neuroscience of Metacognition*, Springer, New York, 2014; T. O. Nelson and L. Narens, Metamemory: A theoretical framework and new findings, pp. 125–141, in G. H. Bower (ed.), *The Psychology of Learning and Motivation*, vol. 26, Academic Press, San Diego, 1990; T. O. Nelson and L. Narens, Why investigate metacognition?, pp. 1–25, in J. Metcalfe and A. P. Shimamura (eds.), *Metacognition: Knowing about Knowing*, MIT Press, Cambridge, MA, 1994; M. G. Rhodes and A. D. Castel, Memory predictions are influenced by perceptual information: Evidence for metacognitive illusions, *J. Exp. Psychol. Gen.*, 137:615–625, 2008, DOI:10.1037/a0013684; B. L. Schwartz and J. Metcalfe, Tip-of-the-tongue (TOT) states: Mechanisms and metacognitive control, in B. L. Schwartz and A. S. Brown (eds.), *Tip-of-the-Tongue States and Related Phenomena*, Cambridge University Press, Cambridge, UK, 2014; A. K. Thomas, J. B. Bulevich, and S. J. Dubois, Context affects feeling-of-knowing accuracy in younger and older adults, *J. Exp. Psychol. Learn. Mem. Cogn.*, 37(1):96–108, 2011, DOI:10.1037/a0021612; E. Tulving and S. A. Madigan, Memory and verbal learning, *Annu. Rev. Psychol.*, 21:437–484, 1970, DOI:10.1146/annurev.ps.21.020170.002253.

Microbial forensics

The anthrax biological attacks that occurred in the United States in the weeks following the terrorist attacks on September 11, 2001, thrust the field of microbial forensics into hyperdrive. Investigators used DNA analyses of the anthrax spores to determine the laboratory strain and thus were able to narrow down the suspects responsible for the biological attacks. Since that time, the field of microbial forensics has expanded to include the study of bioterrorism, foodborne illness, personal identification, postmortem death interval (an estimation of the time of death), and toxicology. This article will focus on exciting, recent advances in the use of microbes for personal identification and postmortem analysis. Although these practices are not yet standard in forensic investigations, the science is promising and lends itself to being standardized in much the same way that DNA fingerprinting was standardized for use in criminal cases.

Human microbiome. The human body is teeming with bacterial life, with nearly 10 bacterial cells per 1 human cell. These bacteria typically have a symbiotic relationship with the human body, providing needed metabolic functions, which human cells are incapable of, such as the synthesis of the vitamins folate and biotin. In an effort to understand the health implications of microbial communities within the human body (the microbiome), the U.S. National Institutes of Health launched an initiative called the Human Microbiome Project in 2007. Phase 1 of this project defined the microbiota (microbial flora) of four distinct body sites—the skin, mouth, gut, and vagina—using high-throughput DNA sequencing and computational methods. Results from nearly 300 healthy volunteers revealed a conservation of microbial metabolic pathways within a given body site from person to person. However, the diversity and prevalence of microbes are highly variable from person to person within each body site. For example, only 13% of the microbes found on the palms of the hands are shared between any two individuals. The finding that microbe composition was unique to each person set up the possible use of microbes as personal identifiers. In other words, the microbe composition could be used as a microbial fingerprint (see **illustration**).

Microbial fingerprints. Two promising areas of interest in developing standardized microbial fingerprint practices are the use of skin microbes, which could be detected on touched surfaces, and salivary microbes, which could be detected in bite marks. In some crime-scene situations, microbial fingerprints could serve as tools more accurate than traditional fingerprinting and DNA fingerprinting.

Research has been carried out to demonstrate the viability of using skin microbiota as a forensic identification tool. For example, investigators have shown that bacterial DNA could be isolated from computer keyboards and computer mouse devices in

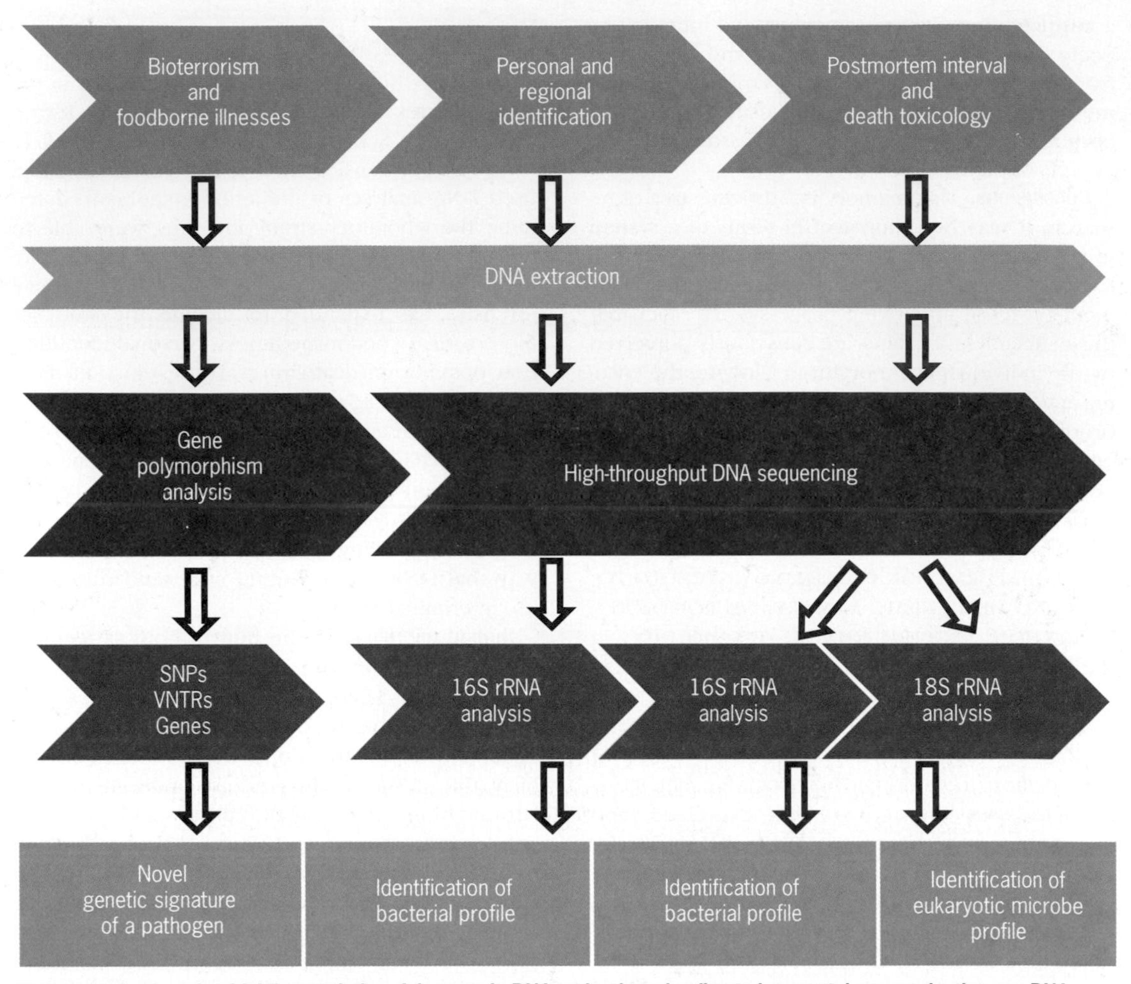

Methodologies for microbial fingerprinting. Advances in DNA technology, leading to low-cost, large-scale, de novo DNA sequencing (DNA sequencing with no prior knowledge of the actual nucleotide sequence), and computational tools have paved the way for the use of microbial profiles in forensics. To determine the identification of a criminal perpetrator, microbial DNA can be isolated from samples as small as fingerprints. The isolated DNA is sequenced using high-throughput DNA sequencing, which is also known as next-generation sequencing. The 16S gene sequences are then compared to a database of known microbial sequences to establish the diversity and prevalence of microbial species, thus yielding the microbial profile of an individual. Similar methods can be used to determine the time of death of an individual or to gather toxicology information. DNA can be isolated from either the corpse or surrounding soil. In addition to the use of 16S gene sequences, which are specific to prokaryotes, 18S gene sequences are utilized to define eukaryotic microbial profiles. In the case of bioterrorism or foodborne illness, the pathogen is often known, but the origin or source of the pathogen is unknown. Small variations in the DNA sequence, single nucleotide polymorphisms (SNPs), variable number of tandem repeats (VNTRs), and the presence or absence of genes can provide the information needed to identify a particular laboratory strain of a bacteria or the source of food contamination. For example, these techniques were used to identify the Colorado farm responsible for the 2011 *Listeria* outbreak that was traced to contaminated cantaloupe, which killed 33 people.

analyzable quantities up to two weeks after being touched. Using high-throughput DNA sequencing and analyses of specific 16S gene sequences to establish microbial profiles, surfaces touched by an individual could be matched to that individual's fingertip. Microbial fingerprints could prove to be a more powerful tool than standard forensic practices in the case of objects touched by a criminal perpetrator because traditional fingerprints are frequently smudged and often there is not enough nuclear DNA present on an object for analysis. It also should be noted that skin microbial fingerprints are relatively stable because the washing of hands by an individual only temporarily alters the microbial content, which returns within a few hours.

In addition, other researchers have demonstrated that streptococcal DNA isolated from bite marks can be matched to an individual's salivary microbes by comparing three specific genomic sequences: the 16S gene, the 16S–23S intergenic spacer, and the RNA polymerase beta subunit. This is particularly important because many violent crimes involve bite marks, and nuclear DNA is difficult to obtain from saliva as a result of the presence of salivary nucleases, which cause the exposed nuclear DNA to be degraded. These nucleases are less likely to affect the integrity of microbial DNA because the genetic material of the microbes is encased in a protective cell envelope. In general, sufficient quantities of bacterial DNA can be obtained up to 24 h after a bite.

Postmortem analysis. For centuries, insects and other arthropods have been used by investigators to predict the time of death of a person (postmortem interval) using knowledge of the known life cycles of these invertebrates. However, an estimation of the time of death using forensic entomology can be highly variable, ranging from weeks to months. In addition, forensic entomology cannot be used for all corpse situations, especially extreme conditions involving freezing temperatures or in a desert, where few to no arthropods colonize a corpse. Given that microbes are capable of thriving in all known earthly environments, including extreme environments, microbial colonizers could serve as better predictors of the time of death. In fact, researchers have performed experiments with mice demonstrating that microbial profiles can be used to estimate the time of death of a mouse within 3 days of the date of death. In these experiments, the microbial profiles were determined for the abdominal cavity and skin of mouse corpses, as well as for the soil surrounding these corpses, over a 48-day period (the methods are presented in the illustration). Within the 48-day duration of the experiments, mouse corpses progressed through all major stages of decomposition. During the bloat stage (approximately 6 to 9 days following death), endogenous anaerobic gut microbes, such as Firmicutes and Bacteroidetes, increased in numbers in the abdominal cavity. After rupture (about 9 days following death), these gut microbe numbers decreased dramatically, and both the aerobic bacteria Rhizobiales (Alphaproteobacteria) and the facultative anaerobic bacteria Gammaproteobacteria became the predominant microbes. Also following rupture, there was an increase in pH from 6.0 to 8.5 in the grave soil, a decline in Acidobacteria, and an increase in Rhizobiales. Overall, consistent and reproducible bacterial profile changes occurred in the skin, abdominal cavity, and grave soil throughout each of the major stages of decomposition, allowing for the time of death to be predicted much more accurately than forensic entomology techniques. These same methods are presently being applied to the study of human corpse decomposition to develop a set of bacterial profiles to accurately predict the time of death in people. Thus, with further standardization, the use of microbial forensics will be an important tool in human postmortem analyses and criminal investigations.

For background information *see* ANTHRAX; BACTERIA; BACTERIOLOGY; CRIMINALISTICS; DEATH; DEOXYRIBONUCLEIC ACID (DNA); FORENSIC BIOLOGY; FORENSIC ENTOMOLOGY; FORENSIC EVIDENCE; FORENSIC MEDICINE; MEDICAL BACTERIOLOGY; MICROBIAL ECOLOGY; MICROBIOLOGY; MICROBIOTA (HUMAN) in the McGraw-Hill Encyclopedia of Science & Technology.

Rebekah L. Waikel; Rebecca M. Lynch

Bibliography. D. Gevers et al., The Human Microbiome Project: A community resource for the healthy human microbiome, *PLoS Biol.*, 10(8):e1001377, 2012, DOI:10.1371/journal.pbio.1001377; D. M. Kennedy et al., Microbial analysis of bite marks by sequence comparison of streptococcal DNA, *PLoS One*, 7(12):e51757, 2012, DOI:10.1371/journal.pone.0051757; S. L. Leake, Is human DNA enough?—Potential for bacterial DNA, *Front. Genet.*, 4:282, 2013, DOI:10.3389/fgene.2013.00282; J. L. Metcalf et al., A microbial clock provides an accurate estimate of the postmortem interval in a mouse model system, *eLife*, 2:e01104, 2013, DOI:10.7554/eLife.01104; S. J. Song et al., Cohabiting family members share microbiota with one another and with their dogs, *eLife*, 2:e00458, 2013, DOI:10.7554/elife00458.001; R. Yang and P. Keim, Microbial forensics: A powerful tool for pursuing bioterrorism perpetrators and the need for an international database, *J. Bioterror. Biodef.*, S3:007, 2012, DOI:10.4172/2157-2526.S3-007.

Middle East respiratory syndrome (MERS)

The severe acute respiratory syndrome (SARS) pandemic that occurred in 2002–2003 was caused by a coronavirus (SARS-CoV). The name coronavirus (Latin *corona*, meaning crown or a halo appearance) comes from the shape of the virus when it is observed using an electron microscope. Coronaviruses typically infect mammals and birds. Only a handful of coronaviruses cause disease in humans, with the most publicized being the SARS-CoV. The SARS-CoV caused a contagious and sometimes fatal pneumonia that spread throughout the world after its first appearance in China in November 2002. There have been no known cases of SARS since 2004. However, in 2012, another new human coronavirus emerged in Saudi Arabia. This new coronavirus is responsible for a viral disease known as Middle East respiratory syndrome (MERS).

Reporting of a new human virus. In September 2012, Ali Mohamed Zaki, a virologist at the Dr. Soliman Fakeeh Hospital in Jeddah, Saudi Arabia, announced the discovery of a new human virus on ProMED-mail (an online disease reporting system). He isolated the virus from a sample of sputum (mucus coughed up from a patient's lungs) taken from a 60-year-old male patient who died in June 2012 from atypical pneumonia and kidney failure. The term "atypical" is applied when diagnostic tests for known expected pathogens are negative. After isolating the virus, Zaki shipped the genetic material (RNA) of the virus to Ron Fouchier, a virologist at the National Influenza Center and Department of Virology, Erasmus Medical Center, located in Rotterdam, the Netherlands, for identification and sequencing.

Laboratory investigation. Fouchier's research team focuses on the identification and characterization of newly discovered viruses, including the human metapneumovirus, new human coronaviruses [such as the SARS-CoV and a human coronavirus isolated from hospitalized Dutch children (hCoV-NL)], and a new influenza A virus (H16N3) isolated from black-headed gulls found in urban city parks in the Netherlands. Asking for Fouchier's expertise was a logical action for Zaki to take. Fouchier's team confirmed that Zaki's isolated virus was a new human coronavirus.

Public awareness of a new virus threat. Zaki's ProMED-mail report caught the attention of doctors in a London hospital who were puzzled by the case of a 49-year-old man from Qatar suffering from atypical pneumonia. Eight days prior to Zaki's initial post, the Qatari man had been airlifted to the London hospital. The patient's symptoms began while he was in Qatar, but he had visited Saudi Arabia in August 2012. While trying to solve the Qatari man's case, a physician and a scientist independently noticed the ProMED-mail announcement of the newly discovered virus. Immediately, they decided to test for the new virus and subsequently determined that the man had been infected with a human coronavirus. Within a day, the viral genome was sequenced and confirmed to be nearly identical to the new coronavirus discovered by Zaki.

Zaki's post ignited a global response and alert. By 2012, laboratories around the world had been equipped with the means to diagnose new cases of severe respiratory illnesses. Therefore, the world was much more prepared than it had been in 2002. When the SARS outbreak occurred in 2002, the outbreak came first and the SARS-CoV was discovered later. In 2012, the reverse situation occurred. The novel coronavirus was discovered first, and virologists and health-care workers watched and waited for any possible epidemic.

Epidemiology of MERS. As of May 2014, 571 confirmed cases from 18 different countries, including 171 deaths (34%), had been reported to the World Health Organization (WHO). The initial cases originated in Jordan, Qatar, Saudi Arabia, and the United Arab Emirates. Therefore, the new coronavirus was named the Middle East respiratory syndrome coronavirus (MERS-CoV). The average age of the laboratory-confirmed cases was 60 years. Forty-two (21%) of the cases were health-care workers (**Fig. 1**). Despite this evidence of person-to-person transmission, the number of contacts infected by persons with laboratory-confirmed infections has been very low. The WHO was confident enough in the global response to MERS that no travel restrictions were imposed on pilgrims traveling to Saudi Arabia for the Hajj, which begins during mid-October of each year. The Hajj is one of the largest mass gatherings of Muslim people in the world (**Fig. 2**).

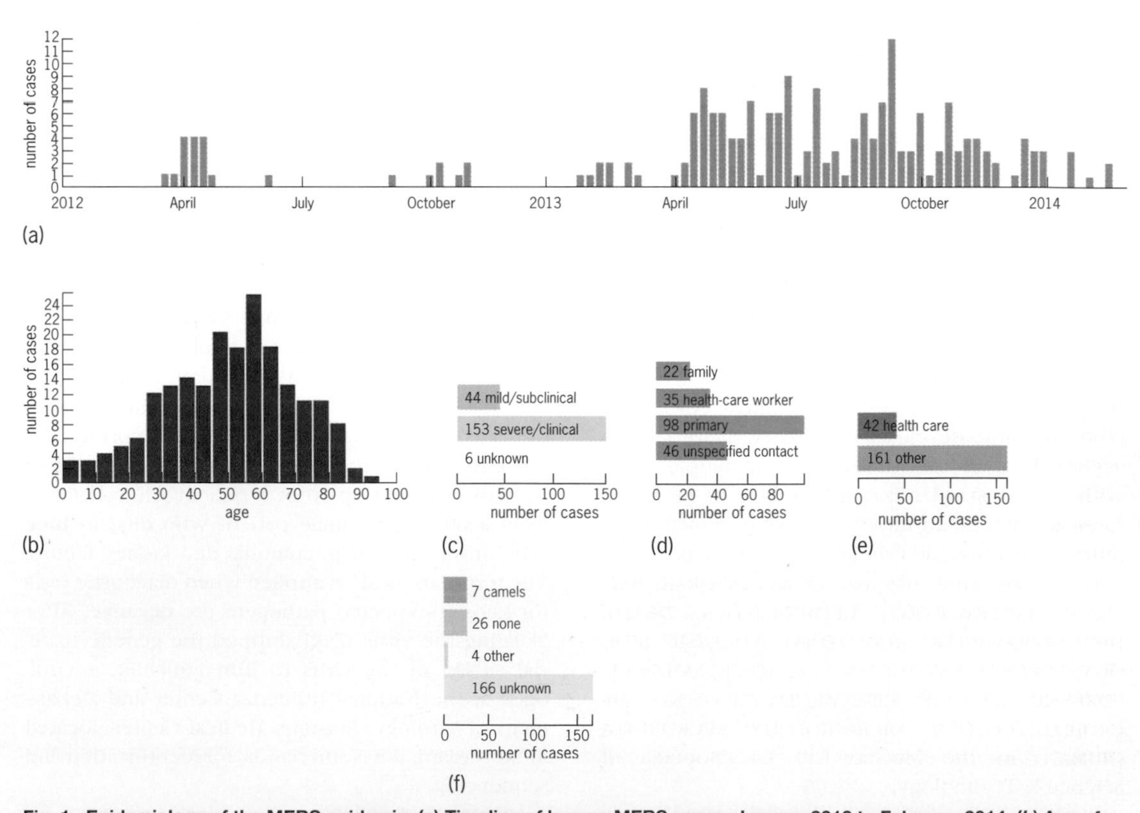

Fig. 1. Epidemiology of the MERS epidemic. (*a*) Timeline of human MERS cases: January 2012 to February 2014. (*b*) Age of human cases. (*c*) Cases categorized by severity of illness. (*d*) Cases categorized by contact with other sick individuals. (*e*) Cases categorized by occupation. (*f*) Cases categorized by contact with animals. (*Adapted from Epidemic: Molecular Epidemiology and Evolution of Viral Pathogens—Coronaviruses; Andrew Rambaut, Institute of Evolutionary Biology, University of Edinburgh; http://epidemic.bio.ed.ac.uk/coronavirus_background*)

Fig. 2. The Hajj is an Islamic pilgrimage to Mecca that has been going on for centuries. It is the largest gathering of Muslim people in the world every year. This photograph depicts people praying at al-Masjid al-Harm, the holiest site of Islam, during the Hajj in 1889. Today, the crowds are much larger, posing an even greater risk for human-to-human transmission of infectious diseases. (*Image courtesy of the Library of Congress; http://www.loc.gov/pictures/item/2013646214/*)

Investigations have continued to search for an animal reservoir of the MERS-CoV. The MERS outbreak was different from the SARS outbreak, which was associated with live-animal marketplaces in China. So far, researchers have found that the MERS-CoV could replicate in various bat cell lines. In addition, seven teams of researchers have collaborated to collect blood samples from camels (**Fig. 3**), goats, sheep, and cattle in the Middle East (Oman), Spain, the Netherlands, and Chile. They discovered that 100% (50 of 50) of the blood samples from camels in Oman and 14% (15 of 105) of the blood samples from Spanish camels contained neutralizing antibodies against the MERS-CoV, indicating infection with or exposure to the new virus. Blood samples from European sheep, goats, and cattle did not contain antibodies. Additional studies have found antibodies against the MERS-CoV in camels in Egypt, Jordan, Saudi Arabia, and the Canary Islands. These results suggest that the MERS-CoV or a related virus has infected the camel populations, making them a possible animal reservoir of the virus. Of the 203 human cases of MERS, 7 had confirmed contact with camels.

A more recent study on archived camel blood samples originally collected from 1992 to 2010 in Saudi Arabia has determined that the MERS coronaviruses have been circulating in camels countrywide since at least 1992. The research team also screened blood samples from sheep and goats, but found no evidence of MERS-CoV infections in these animals. It is not known whether human infections with the

Fig. 3. A camel relaxing. The MERS-CoV is posited to be a zoonosis (a disease transmitted from animals to humans). Recent studies suggest that contact with camels may play a role in the transmission of the MERS-CoV. (*Image courtesy of Alex Langmuir, Centers for Disease Control and Prevention, Atlanta, GA*)

MERS-CoV occurred before 2012 because diagnostic tests were not yet available. The only way to answer this question would be to screen archived samples of human blood to determine when a cross-species zoonotic transmission from camels to humans occurred.

Controversy over the rights to the MERS-CoV. The growth of the MERS epidemic has been slow, but it has not been without controversy. When Zaki provided the coronavirus (which was isolated in the laboratory in Saudi Arabia) to Fouchier for testing, he handed over the sovereign and intellectual property rights to the first diagnostic tests or treatments based on the viral sequencing results to the Erasmus Medical Center in the Netherlands. The Erasmus Medical Center thus had control over requests by researchers to access samples of the virus through material transfer agreements (MTAs) related to patent applications. This created tensions among Zaki, the Saudi Ministry of Health authorities, and other researchers. The hospital authorities terminated Zaki's contract, and he left to work as a microbiologist in Cairo, Egypt. Dutch researchers have fulfilled the MTA requests by other laboratories, but they have placed certain restrictions on experiments with and applications of the new coronavirus. Unfortunately, this controversy has led to conflicts of political and commercial interests that may hamper the global community in its quest to develop a suitable vaccine or treatment.

Outlook. Regardless of the legal controversy involving the MERS-CoV, the main goal of scientists has been to stop the spread of the MERS illness. So far, the disease has been confined to areas of the Arabian Peninsula. A few travel-associated cases occurred in Europe and the United States. Most recently, two health-care workers who traveled to Indiana (April 2014) and Florida (May 2014) from Saudi Arabia became ill. A health-care worker taking care of the patient with MERS in the Florida hospital also became ill. Coordinated efforts were in place to identify contacts and prevent the spread of MERS. After the hospital worker in Florida became ill, the WHO held an emergency meeting on MERS to discuss public health response efforts.

For background information *see* CAMEL; CLINICAL MICROBIOLOGY; DISEASE ECOLOGY; EPIDEMIC; EPIDEMIOLOGY; INFECTIOUS DISEASE; RESPIRATORY SYSTEM; RESPIRATORY SYSTEM DISORDERS; VIRUS; VIRUS CLASSIFICATION; ZOONOSES in the McGraw-Hill Encyclopedia of Science & Technology.

Teri Shors

Bibliography. A. N. Alagaili et al., Middle East respiratory syndrome coronavirus infection in dromedary camels in Saudi Arabia, *mBio*, 5(2):e00884-14, 2014, DOI:10.1128/mBio.00884-14; D. Butler, Progress stalled on coronavirus, *Nature*, 501:294–295, 2013, DOI:10.1038/501294a; B. I. Haagmans et al., Middle East respiratory syndrome coronavirus in dromedary camels: An outbreak investigation, *Lancet Infect. Dis.*, 14:140–145, 2014, DOI:10.1016/S1473-3099(13)70690-X; R. Hilgenfeld and M. Peiris, From SARS to MERS: 10 years of research on highly pathogenic human coronaviruses, *Antivir. Res.*, 100:286–295, 2013, DOI:10.1016/j.antiviral.2013.08.015; K. V. Holmes, Adaptation of SARS coronavirus to humans, *Science*, 309:1822–1868, 2005, DOI:10.1126/science.1118817; L. Lu et al., Middle East respiratory syndrome coronavirus (MERS-CoV): Challenges in identifying its source and controlling its spread, *Microb. Infect.*, 15:625–629, 2013, DOI:10.1016/j.micinf.2013.06.003; A. M. Zaki et al., Isolation of a novel coronavirus from a man with pneumonia in Saudi Arabia, *N. Engl. J. Med.*, 367:1814–1820, 2012, DOI:10.1056/NEJMoa1211721.

Molecular magnets

Over the past 2 decades, considerable worldwide attention has focused on the development of superparamagnetic complexes, known collectively as single-molecule magnets (SMMs). These molecular species belong to a class of magnetic materials that can be easily prepared, tuned, and studied via a range of spectroscopic and magnetic characterization techniques. The most celebrated of these are those containing transition-metal centers linked via oxo- and carboxylate bridges, with $\{Mn_{12}O_{12}(O_2CMe)_{16}(OH_2)_4\}$ (**1**) receiving the most attention (**Fig. 1**). These soluble, single-domain clusters generally display large spin ground states ($S_T = 10$, 20 unpaired electrons for **1**) due to efficient antiferromagnetic interactions between the Mn^{IV} ($S = {}^3/_2$, •) and Mn^{III} ($S = 2$, •) spin centers. Owing to a nearly parallel alignment of the Jahn-Teller axes (elongated) at the Mn^{III} sites, magnetic anisotropy is generated (where $D < 0$ with small E for the Hamiltonian: $H = DS^2_{T,z}$), which leads to the observation of slow magnetic relaxation at cryogenic temperatures (for example, below about 10 K; Fig. 1). Under ideal circumstances, these and other magnetic complexes

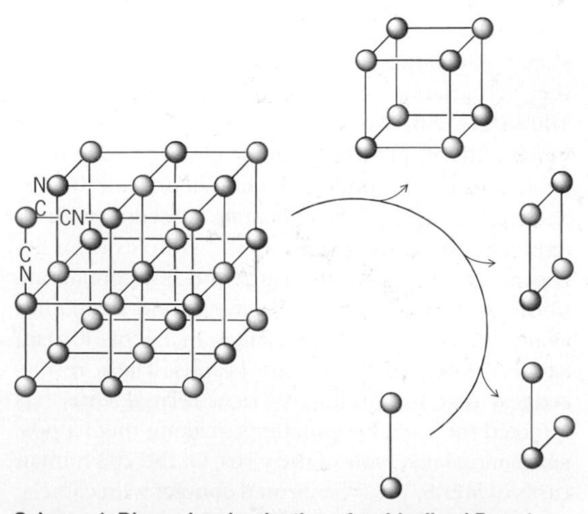

Scheme 1. Dimensional reduction of an idealized Prussian blue structure into discrete polynuclear complexes. *(Adapted from P. Ferko and S. M. Holmes, Curr. Inorg. Chem., 3:172–193, 2013)*

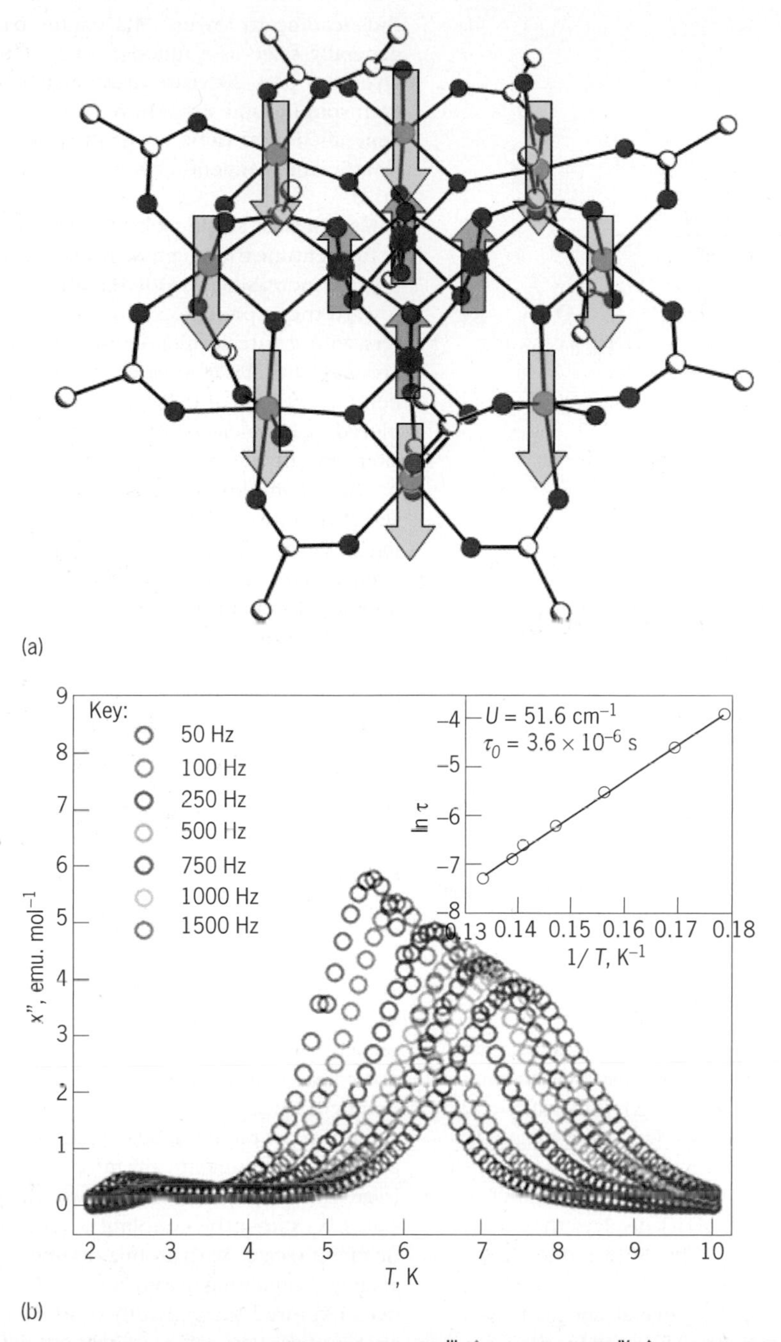

Fig. 1. *(a)* **X-ray structure and magnetic interactions between Mn^{III} (↑, green) and Mn^{IV} (↑, purple) unpaired spins.** *(b)* **Structure and magnetic data for $\{Mn_{12}O_{12}(OAc)_{16}(OH_2)_4\}$. Inset: Arrhenius data show that magnetic relaxation is thermally activated.** ***(Adapted from D. Ruiz et al., Angew. Chem., Int. Ed., 37:300–302, 1998 and R. Sessoli and D. Gatteschi, Angew. Chem., Int. Ed., 42:268–297, 2003)***

exhibit sizable energy barriers to magnetization reversal, where $\Delta \sim |D|S_T^2$ or $\Delta = |D|(S_T^2 - {}^1/_4)$ for integer or half-integer S_T values, respectively.

In SMMs the creation of an energy barrier (Δ) between two thermodynamically equivalent $m_S = \pm S$ configurations allows the possibility that slow relaxation dynamics may be seen (**Fig. 2*a***). Using a variety of magnetization measurements, below a critical temperature (the so-called blocking temperature, T_B,), the available thermal energy is insufficient to overcome Δ and the spin becomes trapped in one of two possible configurations. Application of large external magnetic fields (*H*) saturate the magnetization (*M*) of the sample and, upon removal of this dc field ($H_{dc} = 0$), *M* slowly decays toward zero (becomes demagnetized) with a characteristic relaxation time (τ). The relaxation time usually exhibits temperature-dependent (or thermally activated) behavior and can be measured by following magnetization (*M*) as a function of time or the frequency (ν) dependence of the ac susceptibility. At very low temperatures (Fig. 2*b*), quantum tunneling of the magnetization (QTM, also known as short-cut) can be seen, where magnetic relaxation occurs at a faster

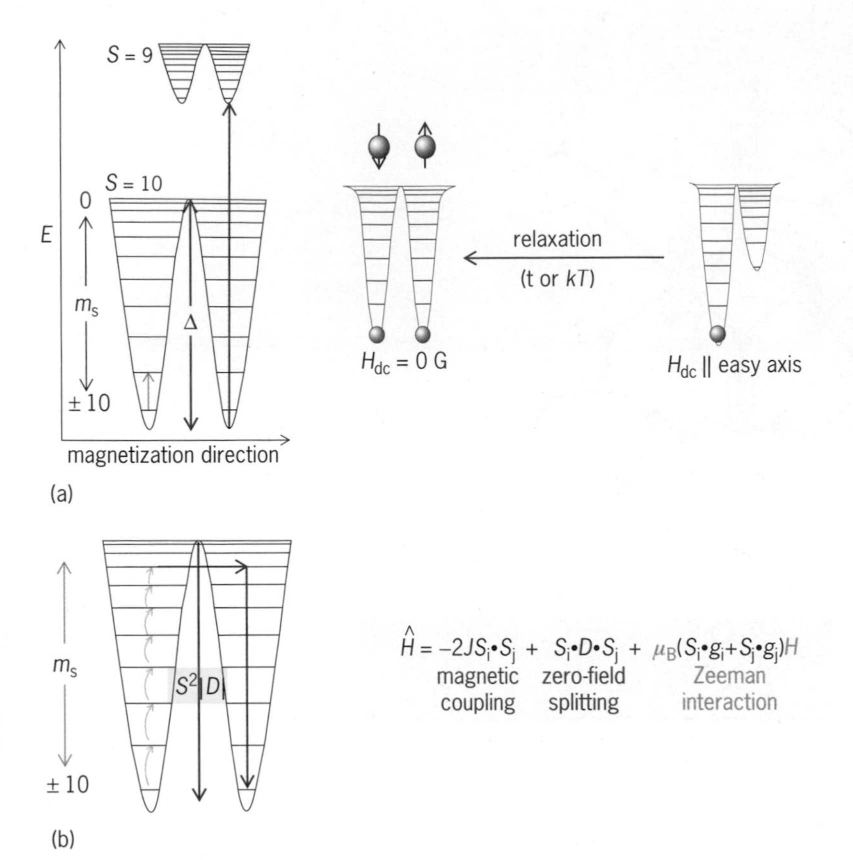

Fig. 2. *(a, left)* **Idealized energy-level diagram for** $\{Mn_{12}O_{12}(OAc)_{16}(OH_2)_4\}$. *(a, right)* **Idealized magnetic relaxation from saturated to equilibrium state with time and/or thermal energy.** *(b, left)* **Pictorial representation of quantum tunneling of the magnetization.** *(b, right)* **General design considerations for tuning magnetic properties in polynuclear SMMs.** *(Adapted from R. Sessoli and D. Gatteschi, Angew. Chem., Int. Ed., 42:268–297, 2003)*

rate than thermally activated pathways allow, thus bypassing the energetic costs associated with climbing the potential-energy barrier (Δ). With sufficiently large spin-reversal energy barriers, slow magnetization relaxation can be observed at low temperatures when QTM is slow. Of known analogs, $[Mn^{III}{}_6O_2(Et\text{-}sao)_6(O_2CPhMe_2)_2(EtOH)_6]$ displays the highest spin-reversal barrier ($\Delta = 86$ K) for this class of SMMs.

Over the past 20 years several successful synthetic strategies have been developed for the preparation of polynuclear complexes with higher spin ground states. The majority of oxo- and carboxylate SMMs contain first-row transition-metal ions and generally exhibit small zero-field splitting terms ($D \sim -0.1$ to $0.2\ \text{cm}^{-1}$) because orbital angular momentum contributions to their magnetic ground states are nearly quenched by the low-symmetry environment of the paramagnetic metal ions. Several research groups have attempted to tune this behavior by systematically inserting a variety of organic molecules (ligands) into a given structure, so that the total spin state and zero-field splitting may be maximized (and thus Δ; Fig. 2). Unfortunately, as a general trend, when the overall spin ground state of polynuclear complexes increases, the uniaxial magnetic anisotropy $|D|$ becomes proportional to S_T^{-2}, thus leading to lower SMM energy barriers which generally scale as a function of S_T^0 rather than S_T^2. Over the past 20 years many analogs that display high spin ground states have been known, but Δ is generally found to be lower than predicted, due to insufficient magnetic anisotropy control and rapid QTM.

Mononuclear single-molecule magnets. In an effort to further tune the magnetic properties of these complexes, increasing worldwide attention has turned toward the exploitation and insertion of metal centers with greater orbital contributions to their spin ground states. Significant efforts to control the magnetic properties of these complexes have been explored, with most concerned with the systematic alteration of the zero-field splitting term exhibited by these complexes; the general approach exploits selective substitution of transition metals in existing structural archetypes in favor of those with even greater single-ion anisotropy (**Fig. 3**). These interactions fall into two general categories, in-state and out-of-state orbital contributions, where the former concern symmetry-allowed mixing of degenerate spin states (for example, that are related by rotation, substantial first-order orbital angular momentum contributions) or by the latter, where low-energy excited states with orbital contributions are allowed to mix electronically with the lower-energy ones; simultaneously and preferentially orienting the resulting single-ion anisotropy tensors is highly important and can often be controlled by choosing the right ligand(s). For example, through the use of sterically demanding polydenate ligands, preferential orientation of the Jahn-Teller axes and/or single-ion anisotropy tensor orientations during the self-assembly can give complexes that display dramatically different magnetic properties for a given number of paramagnetic centers present.

More recent approaches exploit heavy-metal ions as a route for systematically introducing centers with large orbital contributions to their magnetic ground states. As spin–orbit coupling is generally a relativistic effect (scales with atomic number), several SMM analogs containing paramagnetic *f* elements have been explored, given that their orbital contributions are significant and often remain unquenched by the ligand field. Several lanthanide and actinide derivatives are known, and many of these polynuclear complexes afford degenerate spin ground states (and high magnetic anisotropy). These hybrid *d*–*f* compounds can, under ideal circumstances, display higher spin-reversal energy barriers than their 3*d* analogs, but systematic control of their single-ion tensors still remains a difficult synthetic chemistry challenge.

Given that control of single-ion anisotropy tensors is a difficult feature inherent to many SMM systems, and that the energy barriers for spin reversal are not dramatically changed with increasing spin values, a logical approach has investigated how reducing the numbers of paramagnetic ions to a few or a single anisotropic metal center may circumvent

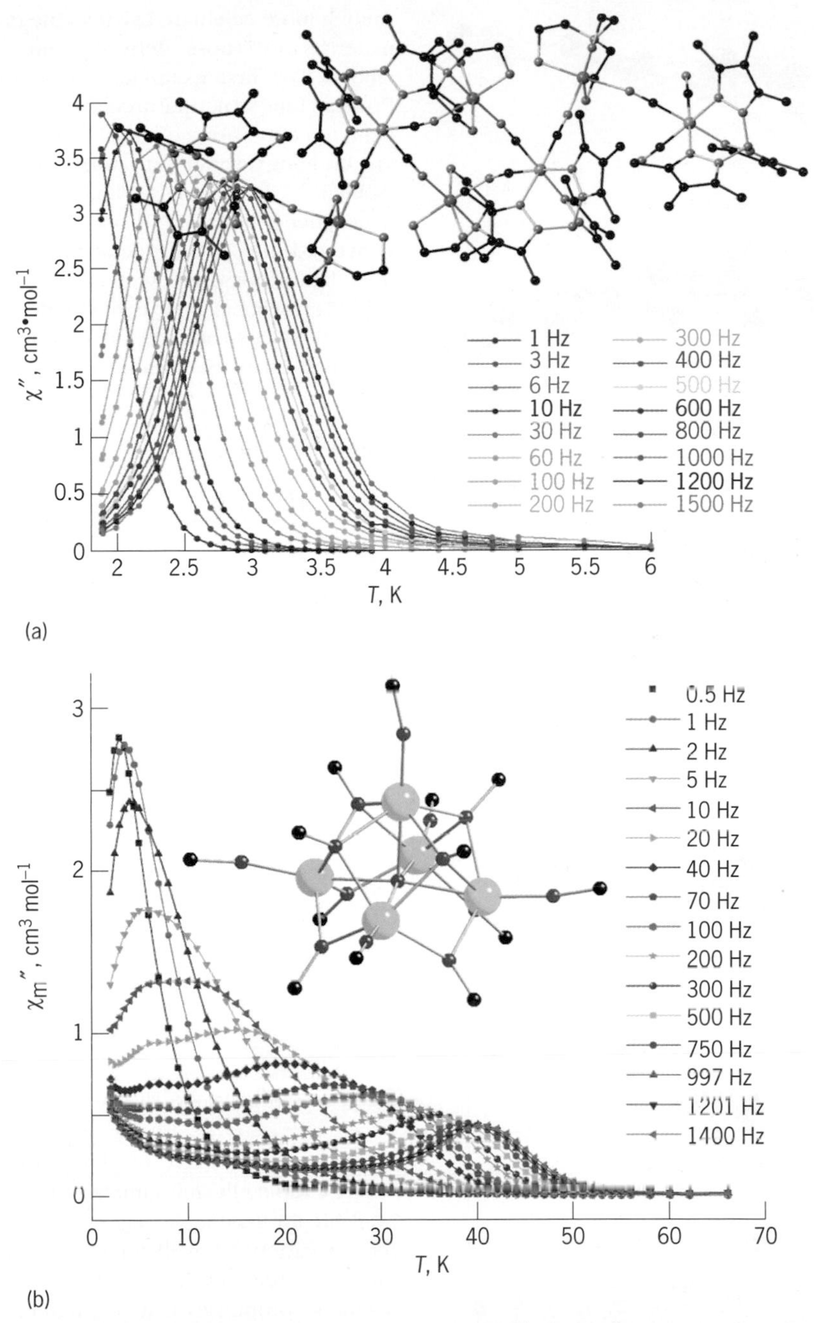

Fig. 3. X-ray structures and magnetic data for (*a*) octanuclear $[(Tp^{*}Me)Fe^{III}(CN)_3]_4[Ni^{II}(tren)]_4$ and (*b*) pentanuclear $[Dy_5(\mu\text{-}O)(OiPr)_{12}]$ complexes. (*Adapted from Y.-Z. Zhang et al., Chem. Commun., 46:4953–4955, 2010 and R. J. Blagg et al., Angew. Chem., Int. Ed., 50:6530–6533, 2011*)

these apparent engineering limitations. The smallest members contain $4f$ ions whose complexes of general $[LnPc_2]^n$ stoichiometry, where Ln = Tb, Dy, Ho (for $n = \pm 1, 0$) and Pc = phthalocyanine; alternative strategies explore structurally related mononuclear low-coordinate $3d$ complexes (**Fig. 4**). Astonishingly, owing to unquenched orbital angular momentum and well-isolated states, several of these two-coordinate $3d$ complexes display exceptionally high thermal barriers to magnetization reversal [$\Delta \sim 325(4)$ K, for Fe^{I}] and slow magnetic relaxation (up to about 29 K) is found. In these mononuclear iron complexes, spin-orbit interactions give rise to significant magnetic anisotropy, in addition to efficient coupling of the molecular vibrations to those of lattice phonon modes. Consequently, rapid QTM occurs even under an applied magnetic field.

Thermo- and photochromic materials. The systematic engineering of molecular complexes with atom-economical efficiency is an exciting area of worldwide research activity, as realization of this goal may afford a diverse array of materials suitable for molecule-based electronic applications. These materials are of interest owing to their potential use in a variety of high-density information storage, sensor, display, and device switching applications.

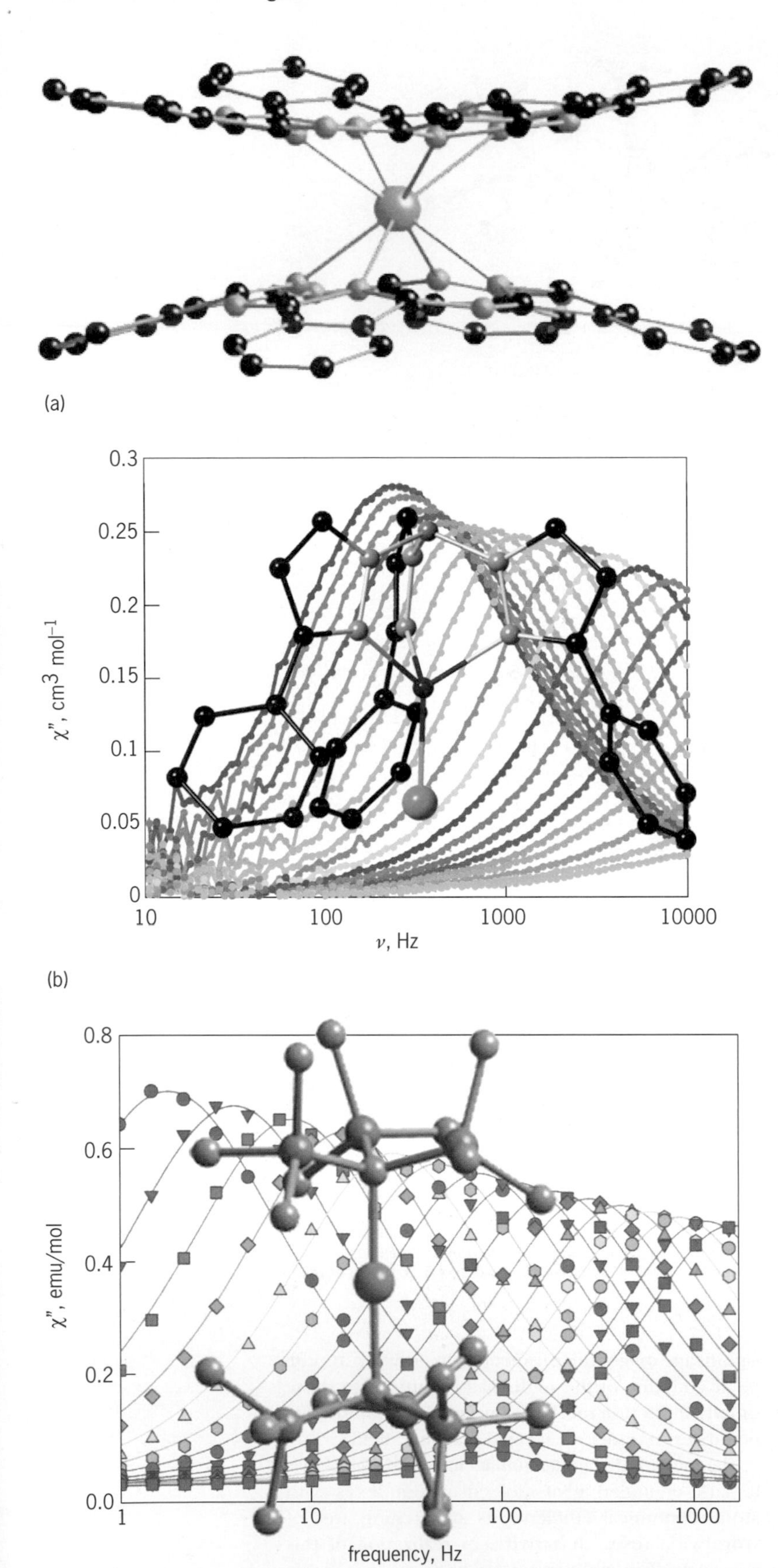

Fig. 4. Recent examples of mononuclear *(a)* $[NBu_4][TbPc_2]$, *(b)* [K(222-crypt)] $[Fe^{I}\{C(SiMe_3)_3\}_2]$, and *(c)* $(Tp^{Ph})Co^{II}Cl$ single-molecule magnets. *(Adapted from N. Ishikawa et al., J. Am. Chem. Soc., 125:8694–8695, 2003; P. J. Ferko et al., submitted; and J. M. Zadrozny et al., Nat. Chem., 5:577–581, 2013)*

Among more celebrated switchable molecule-based materials are those derived from cyanometalate anions. The first example, which belongs to the Prussian blue structural archetype, consists of a face-centered cubic array of metal centers that are linked via bridging cyanides to form Fe(μ-CN)Co units (scheme 1). In these three-dimensional-defect solids, diamagnetic $Fe^{II}_{LS}(\mu\text{-CN})Co^{II}_{LS}$ units are reversibly converted into paramagnetic $Fe^{III}_{LS}(\mu\text{-CN})Co^{II}_{HS}$ ones upon light exposure or with changes in temperature (**Fig. 5**); their behavior is highly dependent on the ligand-field environment (at Co) and presence of interstitial alkali metal cations.

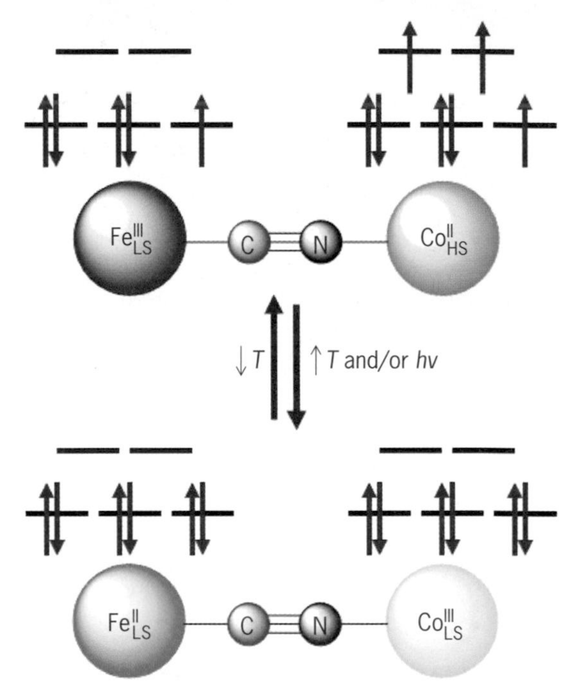

Fig. 5. Thermally and photoinduced electron transfer.

Using the concept of dimensional reduction, where capping ligands limit the number of cyanides available for forming linkages to adjacent metal centers, a variety of well-defined molecular clusters may be systematically prepared, whose optical and magnetic properties can be engineered to resemble those of Co/Fe Prussian blues (scheme 1). The optical and magnetic behavior of these Co/Fe valence tautomers may be systematically controlled as a function of ancillary ligand steric demand [various ligands such as 2,2′-bipyridines (bpy^R), pyrazolylborates (Tp^R), etc.], electron-density donation ability, and internuclear separation in the solid state (**Fig. 6**). It appears that tri-, tetra-, and hexanuclear complexes exhibit a wide temperature range over which their magnetic and optical properties may be switched; we were the first to demonstrate conclusively that the interconversion of the $Fe^{II}_{LS}(\mu\text{-CN})Co^{III}_{LS} \leftrightarrow Fe^{III}_{LS}(\mu\text{-CN})Co^{II}_{HS}$ states proceeds through a common intermediate and that their relaxation follows Arrhenius behavior. Current efforts are aimed at understanding structural and electronic factors responsible for these changes.

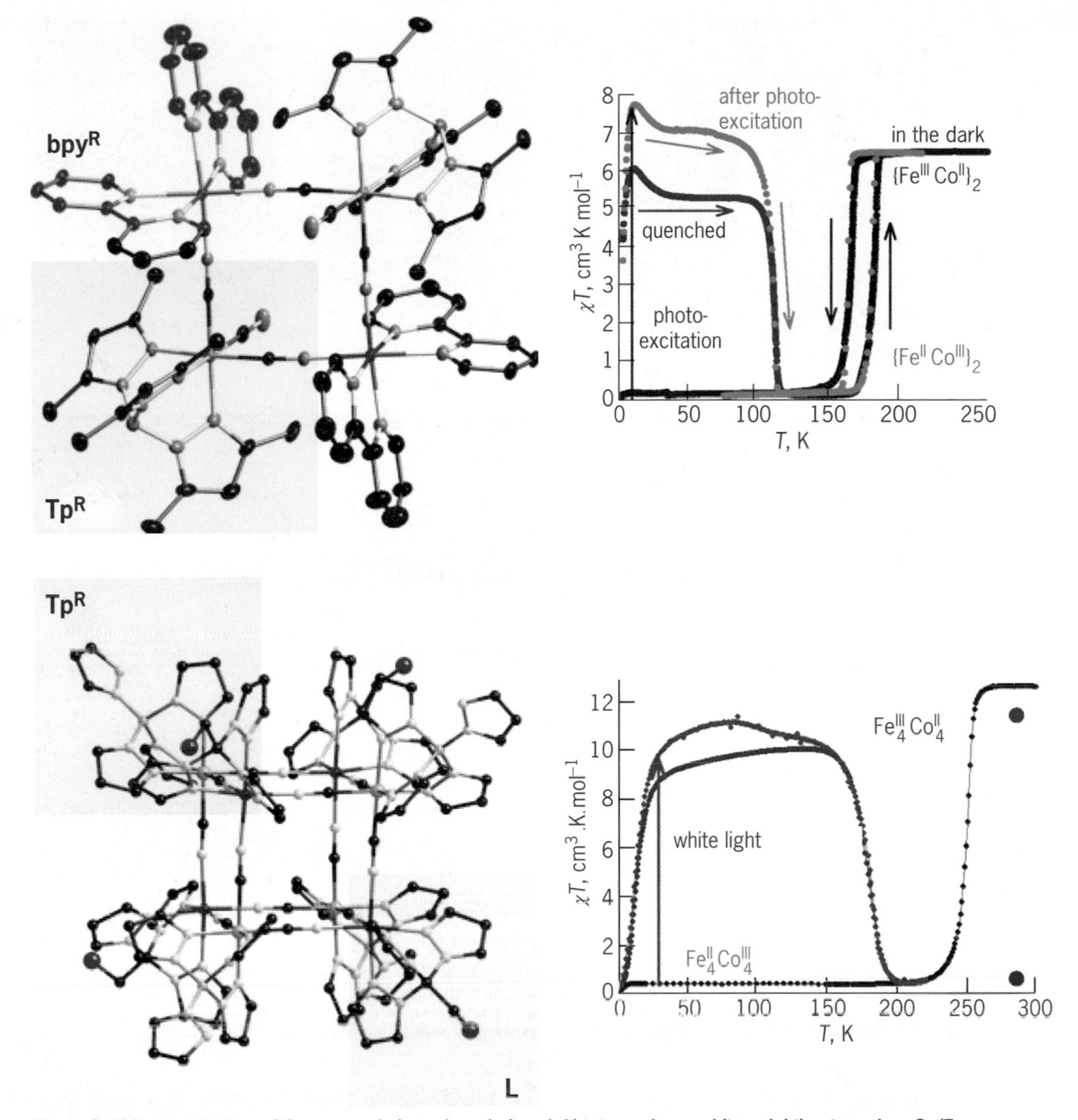

Fig. 6. (left) X-ray structure of thermo- and photochromic (top, left) tetranuclear and (top, right) octanuclear Co/Fe complexes. (bottom) Variable temperature magnetic data collected for both complexes in the presence and absence of light. (*Adapted from D.-F. Li et al., J. Am. Chem. Soc., 128:252–258, 2008 and Y.-Z. Zhang et al., Angew. Chem., Int. Ed., 49:3752–3756, 2010*)

For background information *see* ANTIFERROMAGNETISM; ATOM ECONOMY; CRITICAL PHENOMENA; CYANIDE; JAHN-TELLER EFFECT; LIGAND; MAGNETIC RELAXATION; MAGNETISM; MAGNETOCHEMISTRY; PARAMAGNETISM; PRUSSIAN BLUE in the McGraw-Hill Encyclopedia of Science & Technology.

Stephen M. Holmes

Bibliography. L. Bogani et al., Molecular spintronics using single-molecule magnets, *Nat. Mater.*, 7:179–186, 2008, DOI:10.1038/nmat2133; P. Ferko and S. M. Holmes, Pyrazolylborate cyanometalate single-molecule magnets, *Curr. Inorg. Chem.*, 3:172–193, 2013, DOI:10.2174/1877944111303020009; M. A. Halcrow, *Charge-Transfer-Induced Spin-Transitions in Cyanometalate Materials*, Wiley, New York, 2013; T. N. Hooper et al., Clusters for magnetic refrigeration, *Angew. Chem.*, Int. Ed., 51:4633–4636, 2012, DOI:10.1002/anie.201200072; R. A. Layfield, Organometallic single-molecule magnets, *Organometallics*, 33:1084–1099, 2014, DOI: 10.1021/om401107f; D. N. Woodruff et al., Lanthanide single-molecule magnets, *Chem. Rev.*, 113:5110–5148, 2013, DOI:10.1021/cr400018q.

Nanostructured thermoelectric energy scavenging

Approximately 50–60% of the energy input that our society uses is eventually wasted as heat. Automobiles use only approximately 20–25% of the heat output of combustion, with about one-third of the remaining energy being rejected as heat through the exhaust pipe and another third being lost through

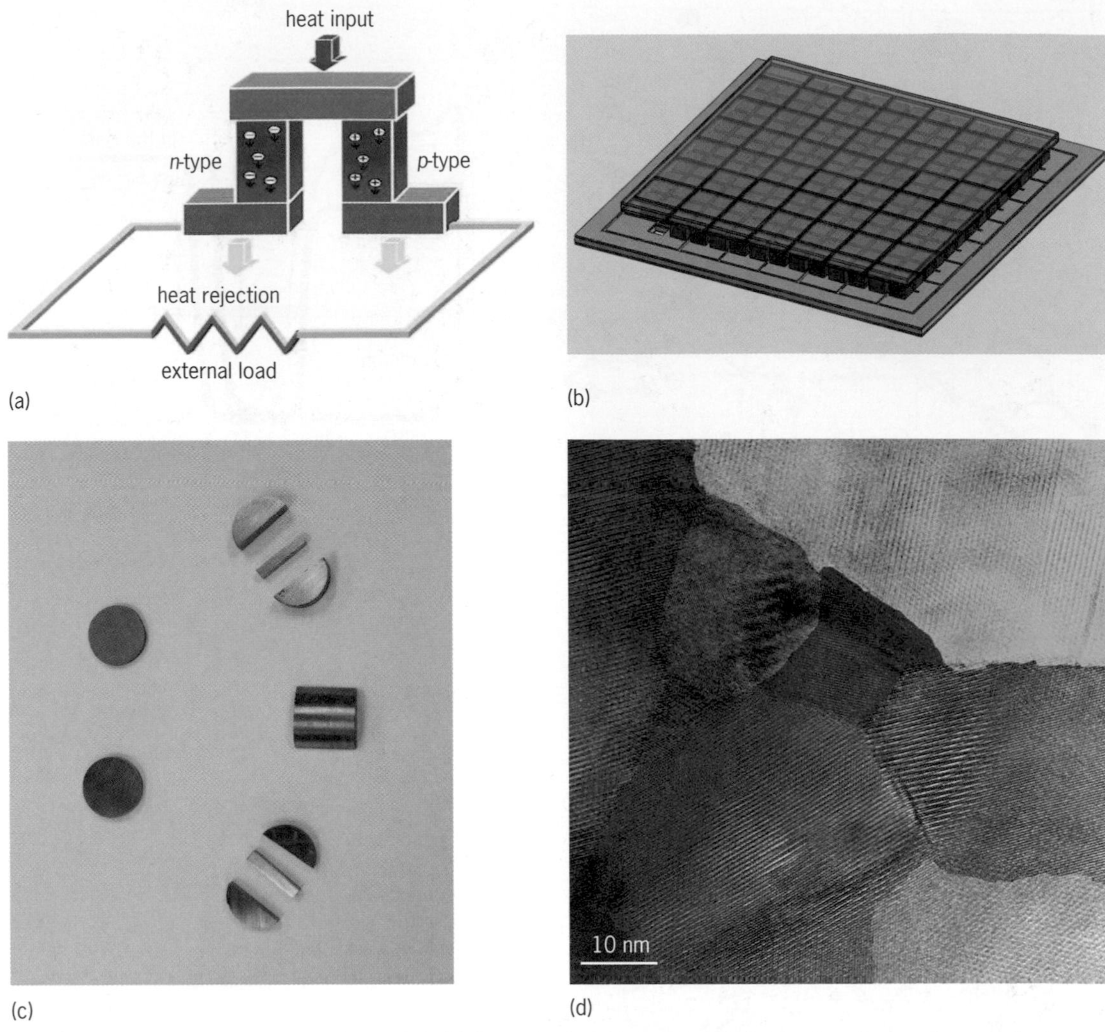

Fig. 1. Thermoelectric generators and materials. (*a*) One pair of *p-n* thermoelectric legs under a temperature gradient, showing electrons and holes diffusing from the hot and cold sides, driving an external load. (*b*) Many pairs connected electrically in series and thermally in parallel in a device. (*c*) Bulk bismuth telluride (Bi_2Te_3)–based materials used to make thermoelectric devices. (*d*) Transmission electron micrograph showing nanoscale grains in such materials.

the radiator. Even electricity is eventually dissipated as heat, a fact that is made evident by the whirring of fans cooling computers and other consumer electronics. Recovering a fraction of this wasted heat and converting it into usable electricity will reduce our overall energy needs and benefit our environment. Capturing this waste heat and converting it into useful energy, however, is difficult because of the distributed nature of the sources of the waste heat and other constraints such as space, weight, and cost. Nevertheless, many people are working on different approaches to utilize the waste heat, one of which is based on thermoelectric energy conversion. This article discusses the principles of thermoelectric energy generators, the need to increase the efficiency of thermoelectric devices if they are to be widely used, the search for thermoelectric materials with higher efficiencies, and finally, the use of nanostructures to improve thermoelectric efficiency. In conclusion, challenges to thermoelectric conversion of waste heat and the present status of the field are discussed.

Thermoelectric effects. In 1821, Thomas Johann Seebeck discovered that an electric voltage is generated when an electrical conductor is subjected to a temperature difference. This phenomenon is now called the Seebeck effect. Correspondingly, the Seebeck coefficient of a material, S, is defined as the voltage generated per degree of temperature difference. The most familiar application of the Seebeck effect is in temperature measurements using thermocouples made of two different metallic wires. Employing the same effect, thermoelectric power generators have been used for most past NASA deep space missions, since spacecraft moving away from the Sun cannot rely on diminishing solar energy. Power is generated by a radioisotope heat source coupled to a thermoelectric generator. In addition to the Seebeck effect, there are also other thermoelectric effects, especially the Peltier effect, that have been used to make small-scale solid-state refrigerators, air conditioners, semiconductor laser wavelength stabilizers, and automobile seat air conditioners and heaters.

A typical thermoelectric generator (**Fig. 1***a* and *b*) uses many pairs of *n*-type and *p*-type semiconductors, connected electrically in series, but thermally in parallel. The thermoelectric legs are electrically connected in series because the Seebeck voltage is typically small, of the order of 200 μV/K. The *n*-type and *p*-type semiconductors have opposite Seebeck coefficients, and hence the total voltage between a *p-n* pair adds up. When the same temperature difference is applied across several *p-n* pairs, a thermally parallel configuration, the resulting total voltage is the product of the per-pair voltage and the number of pairs.

Efficiency of thermoelectric devices. Thermoelectric devices have all the attractive features of solid-state devices: they are quiet, scalable, and environmentally friendly. However, despite these advantages, they are not widely used because their efficiencies are low compared to the efficiencies of thermal-mechanical energy conversion devices. For both power generation and cooling, the efficiency of a thermoelectric device depends on a combination of three properties of materials as $ZT = S^2\sigma T/k$, where ZT is termed the nondimensional figure of merit, σ is the electrical conductivity, k is the thermal conductivity, and T is the absolute temperature. Thus, a good thermoelectric material should have a large Seebeck coefficient and be a good electrical conductor, so that the numerator $S^2\sigma$, also called the power factor, is large. At the same time, the material should be a good thermal insulator with a low thermal conductivity to prevent heat leakage from the hot to the cold side. Good thermoelectric materials, such as the bismuth telluride (Bi_2Te_3)-based materials (Fig. 1*c*) discovered in the 1950s, have a peak ZT of about 1.

Search for good thermoelectric materials. Finding materials with a high ZT is challenging. Fundamentally, the Seebeck coefficient is a measure of the average thermal energy per electron relative to the chemical potential, or Fermi level, of the electrons. In metals, roughly half of the mobile electrons have energies above the Fermi level and the other half have energies below it, producing a cancellation effect and, consequently, a low Seebeck coefficient. In addition, the high electrical conductivities of metals lead to high thermal conductivities, making metals unsuitable thermoelectric materials. The best thermoelectric materials are usually heavily doped semiconductors whose Fermi levels are brought close to the band edge through the optimization of the dopant concentration (**Fig. 2***a*). This leads to a large asymmetry of carriers with energies above and below the Fermi level, and consequently a large Seebeck coefficient. Meanwhile, there are still enough carriers around the Fermi level to move heat around. Hence, the search for good thermoelectric materials involves the control of carriers in the material by external dopants, stoichiometry control, or defects in the material. Optimized thermoelectric materials usually have a carrier concentration of the order of 10^{19}–10^{21} cm^{-3}, leading to an electrical conductivity of the order of 10^5 S/m. The electronic contribution to thermal conductivity is around 0.3–1 W/(m · K), depending on the carrier concentration.

For most semiconductors, the electronic thermal conductivity mentioned above is much smaller than the thermal conductivity due to phonons. To the first order, phonons do not contribute to thermoelectric energy conversion (leaving out scattering and phonon drag effects), while their finite thermal

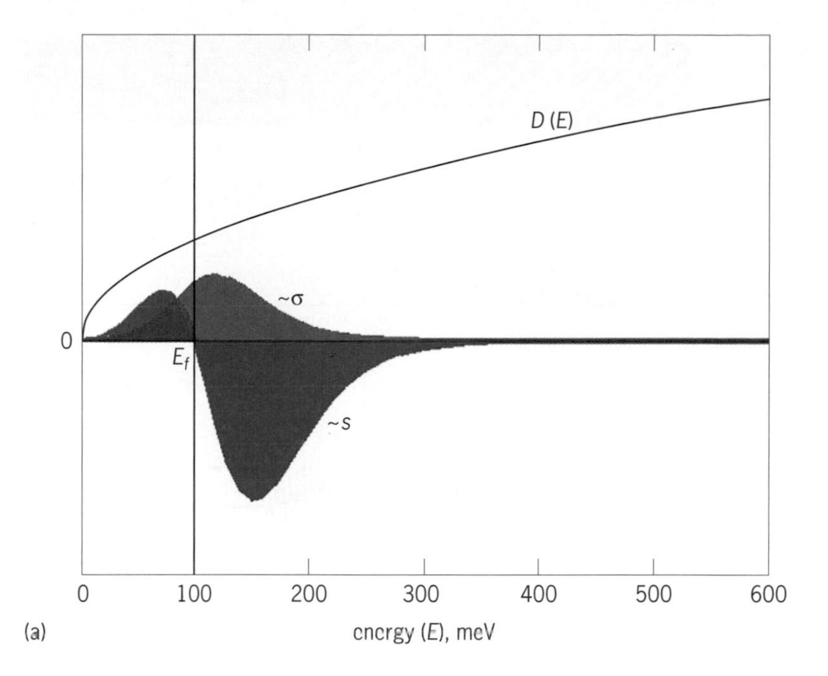

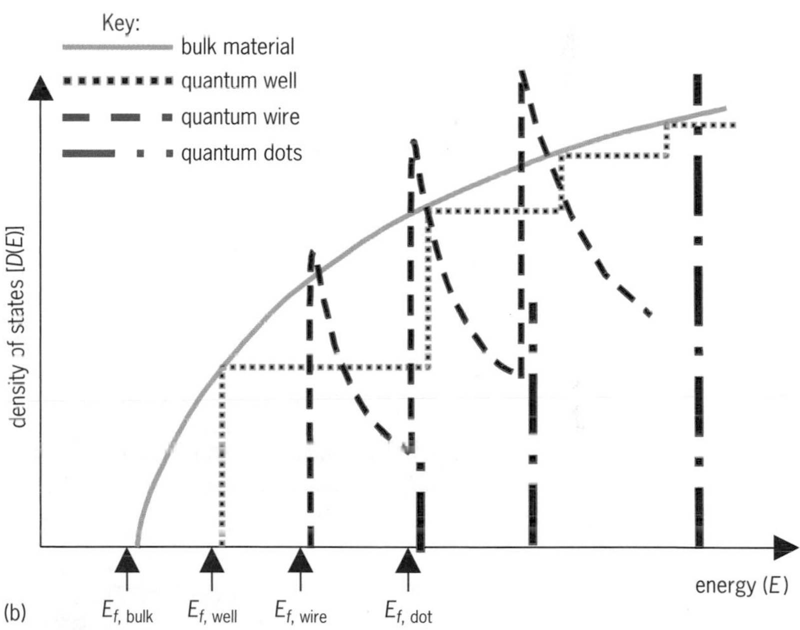

Fig. 2. Electronic density of state functions. (*a*) Density of states, *D*(*E*), in a typical semiconductor. For a given chemical potential determined by the dopant concentration, the electrical conductivity, σ, is proportional to the blue area, while the Seebeck coefficient, *S*, is proportional to the red area. Partial cancellation of *S* may be noted for electrons above and below the Fermi level, E_f. (*b*) Density of states in low-dimensional structures (quantum wells, quantum wires, and quantum dots), compared with that in the bulk material. Sharp features in the density of states make it possible to achieve a larger power factor, $S^2\sigma$, by optimally placing the chemical potential (the Fermi level, $E_{f,bulk}$, $E_{f,well}$, $E_{f,wire}$, or $E_{f,dot}$) through doping.

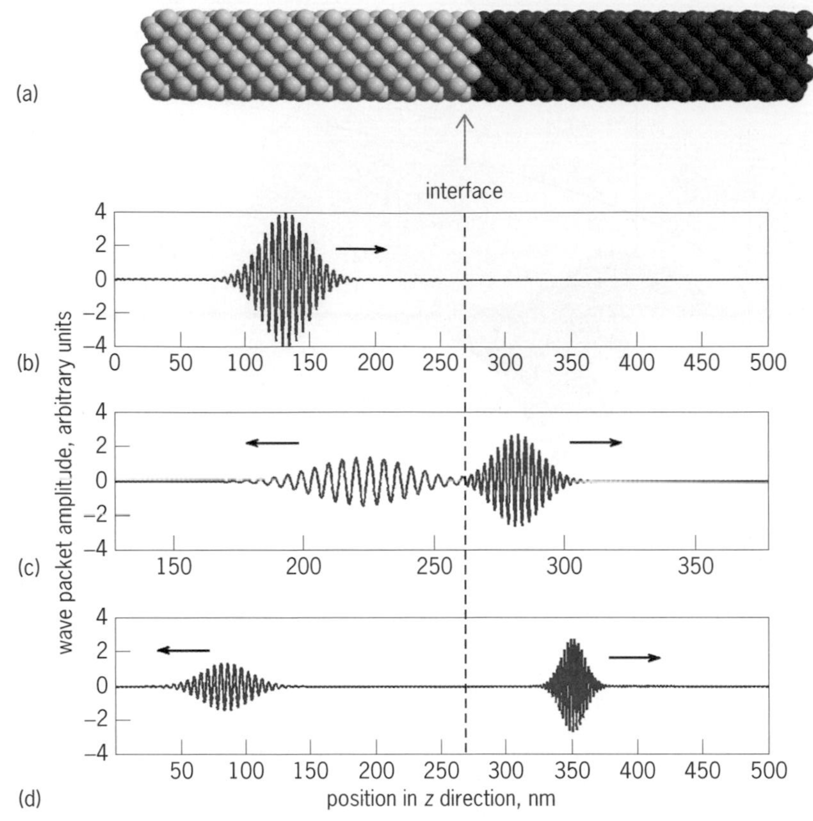

Fig. 3. Phonon reflection at an interface. (*a*) Lattice of atoms near the interface. (*b–d*) A series of screen shots of a phonon wave packet (*b*) launched from one side and (*c*) passing through the interface to form (*d*) reflected and transmitted waves. High interface density in nanostructures creates frequent interface reflection, leading to reduced phonon thermal conductivity.

conductivity acts as a channel for heat loss and should be minimized. One direction for reducing phonon thermal conductivity is through the use of heavy materials, which usually have low phonon frequencies and small phonon group velocities, and consequently have a low thermal conductivity.

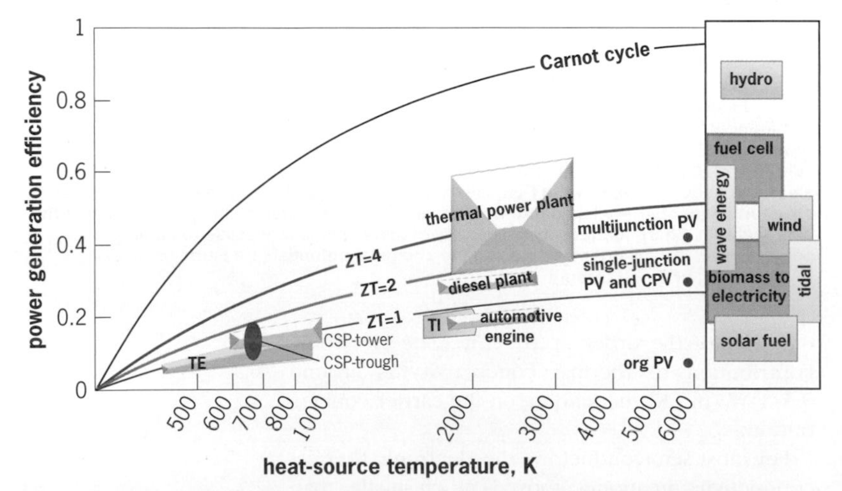

Fig. 4. Theoretical efficiencies of thermoelectric devices as functions of the hot-side temperature, assuming that the cold side is at 50°C (323 K or 122°F). Also plotted in the figure are the efficiency ranges of other energy conversion technologies. CSP, concentrated solar power via thermomechanical engines; TE, reported range of thermoelectric devices; TI, thermionic engines; PV, photovoltaic; CPV, concentrated PV; org PV, organic PV.

Bi_2Te_3-based thermoelectric materials used in commercial thermoelectric coolers are good examples. Another avenue for reducing the transport of heat by phonons is to increase their scattering rates. For example, silicon (Si) and germanium (Ge) have thermal conductivities of 150 W/(m · K) and 60 W/(m · K), respectively. However, because mass fluctuations create more phonon-scattering channels, the thermal conductivity of a $Si_{0.2}Ge_{0.8}$ alloy is only about 8 W/(m · K), significantly lower than the parent materials Si and Ge. This mass mixing strategy is used in most thermoelectric materials, and has enabled the deployment of SiGe alloys for power generation in deep space missions.

Use of nanostructures. Thermoelectric materials such as Bi_2Te_3 and their alloys were developed in the 1950s, and progress in finding new materials has been slow, although commercial devices using these traditional materials have found increasing niche applications, mainly in the area of cooling. In the 1990s, new ideas emerged on how to improve the *ZT* of existing materials and search for new thermoelectric materials. One fruitful idea is the use of nanostructures. Nanostructures bring several potential benefits to thermoelectrics. For example, nanostructuring can change electron energy levels through quantum mechanical size effects. This occurs when the electron wavelengths become comparable to the characteristic lengths of the nanostructures. Quantum size effects lead to sharp features in the electron density of states (Fig. 2*b*). By bringing the Fermi level close to these sharp features, enhanced asymmetries in the carrier concentration can lead to an increase in the power factor. This idea of creating sharp features in the electronic density of states has lately been extended to bulk crystalline, polycrystalline, and nanocrystalline materials, in which resonant states, *d*-level or *f*-level electron states, can also exhibit similar effects.

Another benefit of nanostructures is that interfaces scatter phonons and thus can lead to reduced thermal conductivities. Mass differences in alloys more strongly scatter short-wavelength phonons, similarly to the way molecular fluctuations in the sky scatter blue light more strongly than red light. Longer-wavelength phonons, on the other hand, are not effectively scattered by mass differences. Interfaces between two different materials and at the grain boundaries within a single material (Fig. 1*d*) reflect long-wavelength phonons the way echoes are reflected off the walls of a room (**Fig. 3**). Significantly reduced phonon thermal conductivities were first observed in thin films, especially in periodic thin films called superlattices. Later, similar effects were observed in nanowires and bulk nanocomposites. As most device applications require large quantities of thermoelectric materials with thicknesses above tens of micrometers, the bulk nanocomposite approach is now widely pursued worldwide.

The nanostructuring approach, in tandem with other ideas employed in the search for bulk crystals with high *ZT* values, has elevated the *ZT*s of many materials to values larger than 1, with the largest value being around 2. However, continued *ZT* improvements are still needed, as *ZT* is temperature dependent and large average *ZT* values over the entire range of temperatures in which a device operates are desirable for high device efficiency. **Figure** 4 shows the maximum device efficiency as a function of the hot-side temperature using materials with different values of *ZT*, and compares this efficiency with typical efficiency ranges of other technologies.

Challenges. There are many challenges to overcome in converting waste heat into useful electricity and converting thermoelectric materials into devices. First, for specific applications, one must determine the most appropriate *n*-type and *p*-type materials to use. Decisions are made based not only on *ZT* values, but also on considerations of mechanical properties, since there are large thermal stresses generated in the device when a temperature gradient of a few hundred degrees is applied across a distance of a few millimeters, and also on other factors such as cost and the abundance of the materials. Once a material is chosen, one must solve issues of how to join *p*-type and *n*-type materials to form devices, how to minimize the electrical and thermal resistances of the contacts, how to form strong, reliable bonds among these materials to withstand high temperatures and cyclic operation, and how to couple the devices to heat sources and heat sinks. Despite these difficulties, many companies and research institutions are developing thermoelectric devices and systems for a variety of applications, one of which is the recovery of waste heat from the exhaust pipes of trucks and automobiles. In addition to recovering waste heat, thermoelectric power generators can also be coupled to solar energy converters, thus turning solar heat directly into electricity and other renewable heat sources.

For background information *see* ARTIFICIALLY LAYERED STRUCTURES; CONDUCTION (ELECTRICITY); CONDUCTION (HEAT); GRAIN BOUNDARIES; NANOSTRUCTURE; PELTIER EFFECT; PHONON; QUANTIZED ELECTRONIC STRUCTURE (QUEST); SEEBECK EFFECT; SEMICONDUCTOR; SOLAR ENERGY; THERMOELECTRIC POWER GENERATOR; THERMOELECTRICITY in the McGraw-Hill Encyclopedia of Science & Technology.

[Acknowledgments: The author would like to thank A. Muto, M. Luckyanova, and Z. T. Tian, X. W. Wang, and M. Zebarjadi for contributions to the figures and proofreading, and Professors M. S. Dresselhaus and Z. F. Ren for long-term collaboration. This work is supported as part of the S^3TEC, an Energy Frontier Research Center funded by the U.S. Department of Energy, Office of Science, Office of Basic Energy Sciences under Award Number DE-FG02-09ER46577.] Gang Chen

Bibliography. G. Chen, *Nanoscale Energy Transport and Conversion: A Parallel Treatment of Electrons, Molecules, Phonons, and Photons*, Oxford University Press, New York, 2005; H. J. Goldsmid, *Introduction to Thermoelectricity*, Springer Series in Materials Science, vol. 121, Springer, Berlin, 2010; L. D. Hicks and M. S. Dresselhaus, Effect of quantum-well structures on the thermoelectric figure of merit, *Phys. Rev. B*, 47:12,727–12,731, 1993, DOI:10.1103/PhysRevB.47.12727; B. Poudel et al., High-thermoelectric performance of nanostructured bismuth antimony telluride bulk alloys, *Science*, 320:634–638, 2008, DOI:10.1126/science.1156446; M. Zebarjadi et al., Perspectives on thermoelectrics: From fundamentals to device applications, *Energ. Environ. Sci.*, 5:5147–5162, 2012, DOI:10.1039/C1EE02497C.

Nearby supernova SN 2014J

SN 2014J is the name given to a nearby supernova in the galaxy M82. On January 21, 2014, Dr. Steve Fossey was teaching a group of undergraduate students how to observe with a telescope when they decided to observe M82. They noticed a bright, star-like object that did not appear in previously taken online images of the galaxy. It was officially designated SN 2014J. Supernovae (the plural form of the Latin-based "nova") are named based on the year the supernova was first observed and given a letter based on the order in which it is announced. Thus, SN 2014A was the first supernova observed in 2014 to be given an official name, and SN 2014J was the tenth. After 26 supernovae, two letters are used, continuing with the pattern aa, ab, through az, and then continuing with ba, bb, and so on. Hundreds of supernovae are now discovered each year.

Early images. Interestingly, it was soon determined that SN 2014J was more than a week old when it was first discovered. Supernovae in the nearest galaxies are often found within a few days of explosion because they are imaged so regularly by amateur and professional astronomers alike. In this case, the supernova had been imaged many times before discovery but not noticed because of the complex appearance of the host galaxy. Many such images were announced after the supernova discovery was publicized. These images were used to accurately determine when the explosion first began to shine. They also show a rapid brightening that was faster than predicted by theoretical models.

Host galaxy, M82. The host galaxy of SN 2014J is M82, the M indicating it came from a list of interesting objects identified by Charles Messier in the eighteenth century. M82 is located about 3.3 megaparsecs (10.8 million light-years) away. It features a dusty disk that we view edge-on and filaments of gas and dust blown out perpendicular to the disk (see **illustration**). M82 is known as a starburst galaxy, in which a large number of stars are currently being formed. The high rate of star formation in M82 also results in a high supernova rate, particularly explosions that come from short-lived, massive stars. It is estimated that a supernova occurs every 10 years in M82, compared to one every 100–300 years in

(a)

(b)

Images from the *Swift* Ultraviolet-Optical Telescope (UVOT), showing the host galaxy M82 (*a*) before and (*b*) after SN 2014J exploded. The images are made by combining an optical image (shown in red), a near-ultraviolet image (green), and a mid-ultraviolet image (blue). The main disk of the galaxy is viewed edge-on and appears red with many dark dust lanes blocking the light behind it. The energy of many previous supernova explosions has blown out gas and dust perpendicular to the disk, scattering ultraviolet light that is seen as blue in the image.

our Milky Way Galaxy. The energy from these supernovae in M82 is responsible for blowing out gas and dust from the disk, whose light shows up as blue in the images in the illustration.

Classification of supernovae. SN 2014J was identified as a type Ia supernova. Supernovae are classified based on their spectra, the brightness at different wavelengths (or colors). Different chemical elements absorb or emit light at specific wavelengths, allowing the chemical composition of the explosions to be studied. Supernovae are most broadly separated into type I and II, based on the appearance of hydrogen spectral lines (for type II) or the lack thereof (for type I). Type I supernovae are further subclassified, based on the presence of strong silicon (Ia), helium (Ib), or neither helium nor strong silicon (Ic). Understanding the explosion process itself, supernovae are more appropriately grouped as thermonuclear explosions (Ia) or core-collapse explosions (Ib, Ic, II). Core-collapse supernovae are more common in galaxies with a lot of recent star formation because they result from the deaths of massive stars with short life spans. Type Ia supernovae can occur in galaxies with or without recent star formation. Thus, a new supernova in a starburst galaxy might be expected to be of the core-collapse variety, but a type Ia supernova is still possible. M82 previously hosted a type IIP (hydrogen-rich, core-collapse) supernova in 2004 and a radio supernova in 2008. Other supernovae in this galaxy could have occurred in recent times but have not been visible because of the thick clouds of dust in M82.

Origin of type Ia supernovae. Type Ia supernovae are thought to arise from the explosions of white dwarf stars. These are the end states of low-mass stars that have completed their lifetime of nuclear fusion of hydrogen and helium, leaving a core of carbon and oxygen. In low-mass stars, the pressure and temperature are not high enough to trigger further nuclear fusion processes. In some cases, however, the mass of the white dwarf can be increased by mass transferred from a companion star or by two white dwarfs merging. The increase of mass causes the white dwarf to compress, increasing the temperature and pressure to the point, just over a specific mass limit, that nuclear fusion resumes in a runaway fashion, burning through the star and releasing enough energy to blow it apart.

Type Ia supernovae as standard candles. The explosion of type Ia supernovae at a similar mass results in each having a similar energy and intrinsic brightness. By comparing the brightness we observe to the intrinsic brightness we know they have, we can use them as standard candles to measure how far away they are. Type Ia supernovae are important for measuring distances in the universe. By measuring the brightness of supernovae at larger and larger distances, we can measure the rate at which the universe is expanding. Type Ia supernovae were used to discover that the expansion of the universe is accelerating because of an unknown force termed "dark energy."

Effect of dust. One complication of measuring distances with a "standard candle" is that dust between the supernova and the observer causes it to appear fainter. The amount of dust can be measured by its effect on the different colors. This effect is called reddening, because shorter wavelengths are absorbed and scattered more, resulting in a redder color. Ultraviolet light is especially sensitive to the effects of dust. Dust is very effective at scattering ultraviolet photons, but the dust grain size and geometry leaves a distinctive imprint on the ultraviolet light. However, because the ultraviolet light is strongly suppressed, it is very hard to observe dusty supernovae in the ultraviolet. A supernova with a lot of dust would have to be extremely close to detect the important ultraviolet light. Thus, it was fortuitous that the closest type Ia supernova in 40 years was very dusty, as this allows unique science to be done that is impossible for more distant supernovae. Two other type Ia supernovae have been discovered at comparable distances, SN 1972E and SN 1986G.

Observations. Two space telescopes with ultraviolet capabilities, the *Hubble Space Telescope* (*HST*) and the *Swift* satellite's Ultraviolet-Optical Telescope (UVOT), both obtained ultraviolet images and spectra (see illustration). One of the *Swift* satellite's key strengths is the ability to respond quickly to new objects. *Swift* began observing SN 2014J less than 15 h after discovery. This quickness is remarkable for a space-based observatory. *Swift* was able to follow the rise and fall of the light with daily observations. Simultaneous to the ultraviolet observations, *Swift* observed with the X-ray Telescope and gamma-ray Burst Alert Telescope. No x-rays were detected with *Swift*'s X-ray Telescope or the *Chandra* telescope, nor was radio emission detected from various ground-based facilities. The first detection of a type Ia supernova in gamma rays was reported by scientists working with observations of SN 2014J with the *Integral* satellite.

Origin of reddening. Because it was so close, SN 2014J was incredibly bright in optical light despite the dust. Observations by *HST* and *Swift*'s UVOT show that it is much fainter at shorter wavelengths. Based on maps of the infrared emission of dust in our own Milky Way Galaxy, we know that only a small portion of the light is dimmed because of Milky Way dust. The rest of the dimming must come from dust right around the supernova or elsewhere in the host galaxy, M82. By studying the different colors of SN 2014J, we can determine the location and properties of that dust. The flux ratios at different wavelengths are inconsistent with extinction from Milky Way–type dust. The intervening dust is probably made of smaller grain sizes. The change in the colors is not consistent with light being scattered back into our line of sight by dust behind the supernova. Thus, the reddening is probably caused by dust in M82 that just happens to be in the line of sight and is not physically associated with the supernova system. SN 2014J in M82 is still being studied and promises to be an important object in our understanding of type Ia supernovae and the dust around them.

For background information *see* ACCELERATING UNIVERSE COSMOLOGY; ASTRONOMICAL SPECTROSCOPY; COSMOLOGY; DARK ENERGY; GALAXY, EXTERNAL; GAMMA-RAY ASTRONOMY; GAMMA-RAY BURSTS; HUBBLE CONSTANT; HUBBLE SPACE TELESCOPE; INTERSTELLAR EXTINCTION; INTERSTELLAR MATTER; MESSIER CATALOG; MILKY WAY GALAXY; SCATTERING OF ELECTROMAGNETIC RADIATION; STARBURST GALAXY; SUPERNOVA; ULTRAVIOLET ASTRONOMY; WHITE DWARF STAR in the McGraw-Hill Encyclopedia of Science & Technology.

Peter J. Brown

Bibliography. R. Amanullah et al., The peculiar extinction law of SN2014J measured with *The Hubble Space Telescope*, *Astrophys. J. Lett.*, 788:L21 (15 pp.), 2014, DOI:10.1088/2041-8205/788/2/L21; D. Branch, Type Ia supernovae and the Hubble constant, *Annu. Rev. Astron. Astrophys.*, 36:17–55, 1998, DOI:10.1146/annurev.astro.36.1.17; A. V. Filippenko, Optical spectra of supernovae, *Annu. Rev. Astron. Astrophys.*, 35:309–355, 1997, DOI:10.1146/annurev.astro.35.1.309; R. Kirshner, *The Extravagant Universe: Exploding Stars, Dark Energy, and the Accelerating Cosmos*, Princeton University Press, Princeton, NJ, 2002; S. Webb, *Measuring the Universe: The Cosmological Distance Ladder*, Springer-Praxis, Chichester, UK, 1999; W. Zheng et al., Estimating the first-light time of the type Ia supernova 2014J in M82, *Astrophys. J. Lett.*, 783:L24 (6 pp.), 2014, DOI:10.1088/2041-8205/783/1/L24.

New wind power technology: INVELOX

The use of wind energy to propel boats dates as far back as 5000 BC, and wind-energy conversion systems in the form of windmills have a recorded history from 200 BC. The first horizontal-axis windmill for electricity production was built in Glasgow, Scotland, in 1887 by Professor James Blyth of Anderson's College, Glasgow (now Strathclyde University). In 1931 in the United States, George Darrieus invented and patented the first vertical-axis wind turbine, known as the Darrieus turbine. The 1930s became a turning point for the wind power generation industry as several wind turbines with high electric power-generating capacity were built. In 1941 the world's first megawatt wind turbine was built in Vermont, United States, and was connected to the power grid. This turbine had 23-m (75-ft)

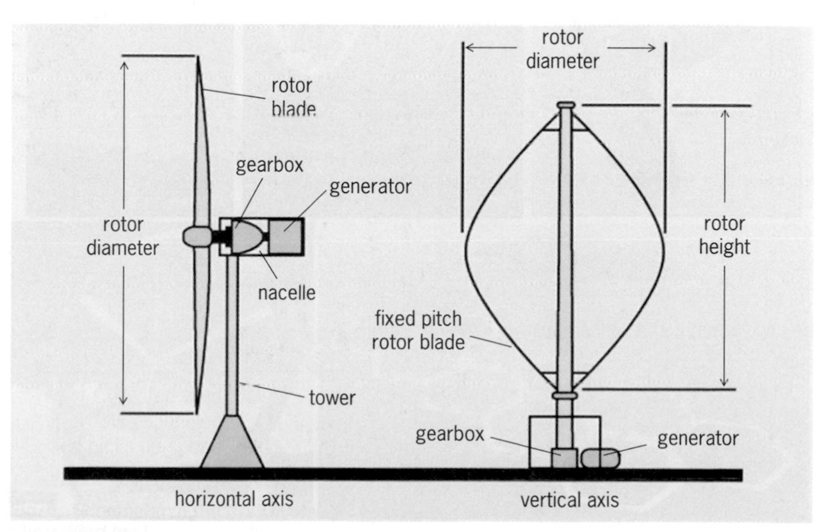

Fig. 1. Horizontal-axis and vertical-axis turbines.

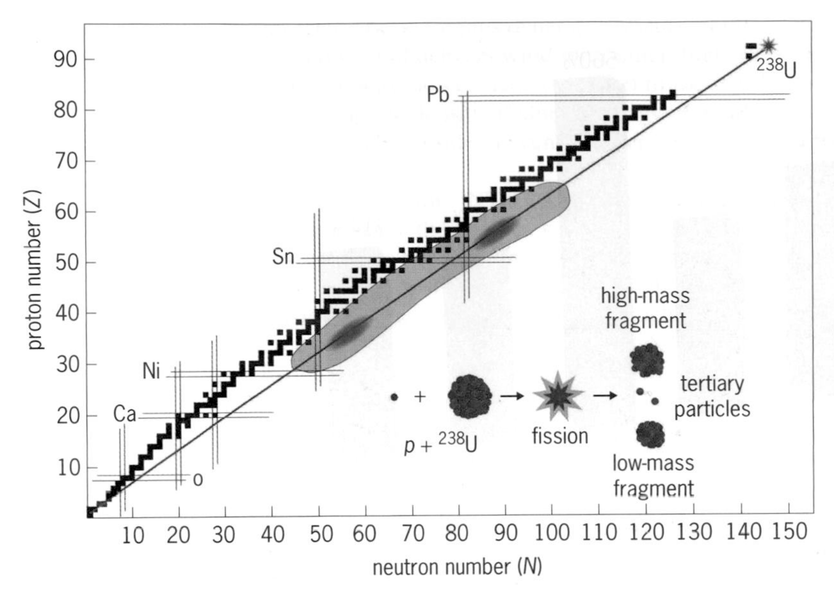

Fig. 1. The chart of nuclei. With the fission of ^{238}U, a distribution of neutron-rich nuclei (blue-shaded area) is produced. The production yields peaks in two regions, commonly called the low- and high-mass regions, which are illustrated with a deeper shade of blue. The red line identifies the nuclei with the same ratio of protons to neutrons as ^{238}U.

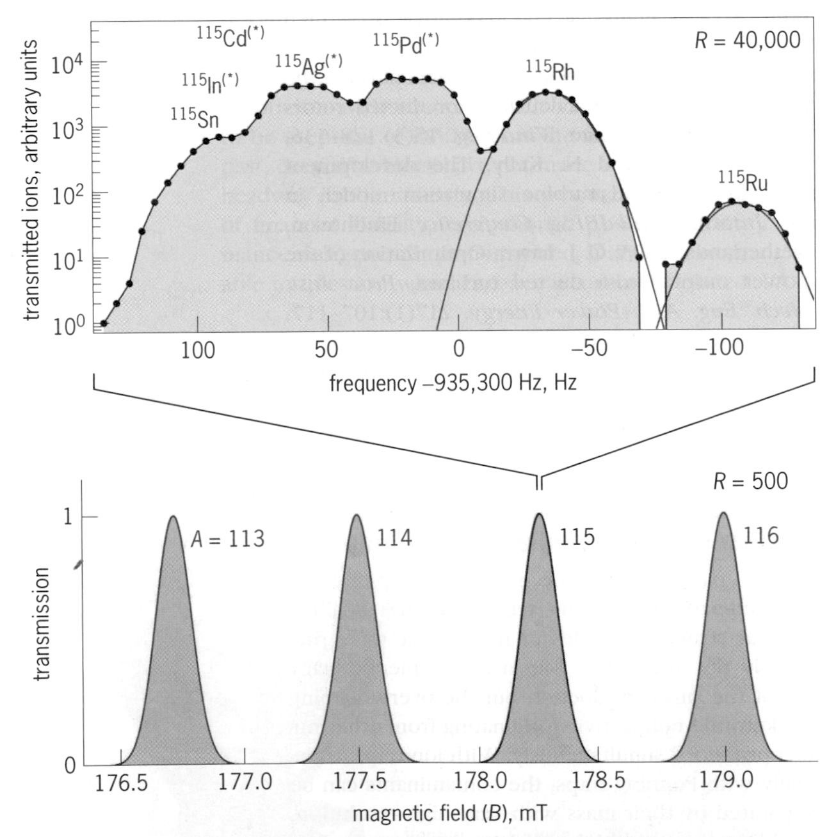

Fig. 2. Transmission spectra. The lower panel is a schematic transmission spectrum through a conventional dipole magnet as a function of the applied magnetic field. Here, a typical mass resolving power ($R = M/\Delta M$) of 500 is reached, which is enough to separate isotopes, here depicted with mass numbers $A = 113$ up to $A = 116$. The upper panel shows the transmission of isobars with $A = 115$ through a Penning trap. A modest mass resolving power of 40,000 is used, which is enough to separate, for example, ^{115}Rh from ^{115}Ru. The nuclei marked with asterisks [such as ^{115}Pd$^{(*)}$] have close-lying isomeric states. The transmission is plotted on a logarithmic scale, whereby it can be seen that the highest peak (^{115}Pd) suffers from detector saturation.

measurement of atomic masses. In recent decades, nuclear structure studies further away from the valley of stability against nuclear (beta) decay have vastly expanded, reaching out to both the neutron-rich and neutron-deficient sides of the nuclide chart. Production mechanisms vary considerably for the two sides of the nuclide chart. Because heavier stable nuclei have more neutrons than protons, one can produce neutron-deficient nuclei by fusing two stable nuclei, and the resulting product will have fewer neutrons than the stable nuclei with the same number of nucleons.

For the neutron-rich side, the common reaction for production is nuclear fission induced in a heavy nucleus such as uranium-238 (^{238}U) or californium-252 (^{252}Cf) [**Fig. 1**]. The products, usually two medium-heavy nuclei, with one being somewhat heavier than the other, are neutron-rich. Production distributions are quite broad, and neutron-rich nuclei from nickel (Ni) to well above tin (Sn) are produced.

Commonly, mass separators—magnetic or other type—can get rid of most of the coproduced contaminants, but what usually remains, in addition to the nuclei of interest, are the radioactive isobars that have the same atomic mass number.

In this article, the development of cleaning techniques will be demonstrated through studies of the neutron-rich ruthenium-115 (^{115}Ru) nucleus, which was discovered at the IGISOL (Ion Guide Isotope Separator On-Line) facility at the University of Jyväskylä, Finland, in 1991 and reported in 1992 by J. Äystö and colleagues. IGISOL is based on the ion guide (IG) method developed in the mid-1980s. In an IG-method reaction, products recoiling out from the target are stopped in a buffer gas, where they end up being transformed into singly-charged ions through charge-exchange reactions. This method is chemically independent and thus equal in efficiency for all elements, unlike the conventional ion source, where different elements have widely varying extraction efficiencies, usually totally inhibiting the extraction of refractory elements. Consequently, the majority of decay data about exotic, refractory isotopes produced in fission originate from the studies done at IGISOL.

Techniques. Traps for charged particles have proved to be extremely useful in the mass measurements of stable particles. The pioneering experiment for mass measurements of radionuclei was carried out at the ISOLDE facility at CERN, where it was demonstrated in the mid-1980s that radioactive ions can be stored and their masses can be measured at relatively short time scales of less than one second. The setup that was later to be known as ISOLTRAP has performed mass measurements of hundreds of radionuclei and is still in operation.

Two main types of traps are in use. One, usually referred to as a Paul trap, is a simple radio-frequency (RF) trap that confines particles merely with time-varying electric fields. A more sophisticated type of trap is a Penning trap, which confines particles with a homogenous magnetic field and a quadrupolar electrostatic potential. The former devices are typically used as ion-beam improvement and storage devices, while the latter are used mostly for high-precision mass measurements by linking the ion's revolution frequency in the magnetic field to its mass.

A typical trap installation in a radioactive ion-beam provider facility consists of an ion-beam buncher (Paul-type trap) and one or more Penning traps for ion-beam purification or atomic mass measurements. ISOLTRAP was set up for atomic mass measurements. The first trap setup with beam purification as the main function, consisting of a buncher and two Penning traps, was JYFLTRAP (JYFLTRAP is a trap setup at the Physics Department of the University of Jyväskylä) at the IGISOL mass separator. A Penning trap can provide a resolution in excess of one part in 10^5, which surpasses the resolving power of an ordinary dipole magnet by a factor of at least 100, enabling the separation of nuclear isobars such as ruthenium-115 (^{115}Ru) and rhodium-115 (^{115}Rh) [**Fig. 2**], and in some cases even the separation of nuclear isomers.

In studies of fission fragments, it is of the essence to have mass separation capabilities far beyond those of an ordinary mass separator, which reaches a mass resolving power ($M/\Delta M$) of perhaps 1000. In a beam consisting of all the ions with the atomic mass number $A = 115$, there is very little ^{115}Ru. In this case, more than 99% of the ion beam contributes to the background, and ultimately detectors will saturate from background radiation.

In 1991, when the ^{115}Ru isotope was first discovered, the data allowed only an identification of a single gamma-ray line and determination of its beta-decay half-life. Despite the increasing beam intensities that became available over the years, no progress was made in the spectroscopy because of the proportionally increasing isobaric background. Progress came only after the introduction of trap-assisted purification, which made possible the isolation of ^{115}Ru from the more abundant isobaric contaminants (Fig. 2). The resulting sensitivity gain of the spectroscopy is impressive (**Fig. 3**), and dozens of new levels were reported in 2011.

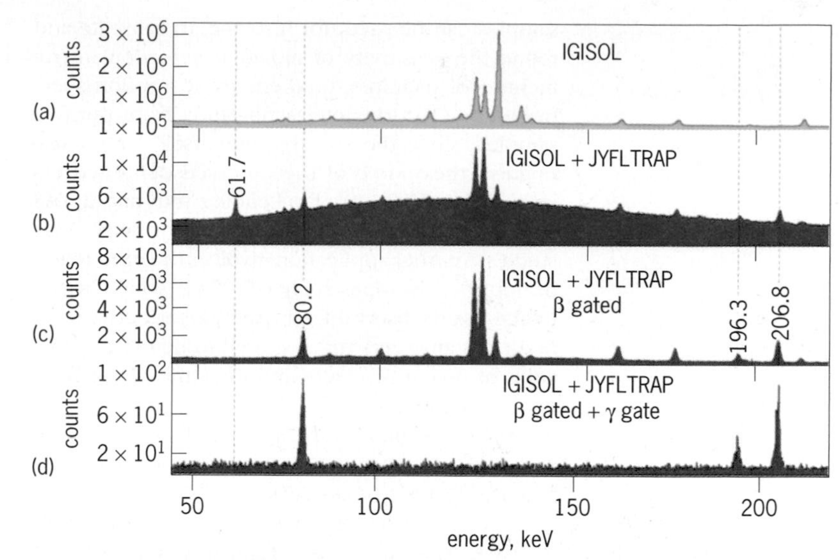

Fig. 3. An illustration of the enhancement of spectroscopic sensitivity with purification in a Penning trap. (*a*) The green spectrum shows data collected when only IGISOL mass separation is used. The few lines observed originate from the most abundant isobaric background in the mass chain $A = 115$. (*b*) Additional purification in the trap has been utilized. With a factor-of-100 reduction in the amount of data (mostly through the removal of contaminants), gamma transitions belonging to the beta decay of the ground state of ^{115}Ru become visible as well as the line corresponding to the decay of an isomeric state in ^{115}Ru (61.7 keV). Further filtering (c) with beta coincidence and (*d*) with gamma coincidence as well results in spectra whose dominant lines are related to the beta decay of ^{115}Ru.

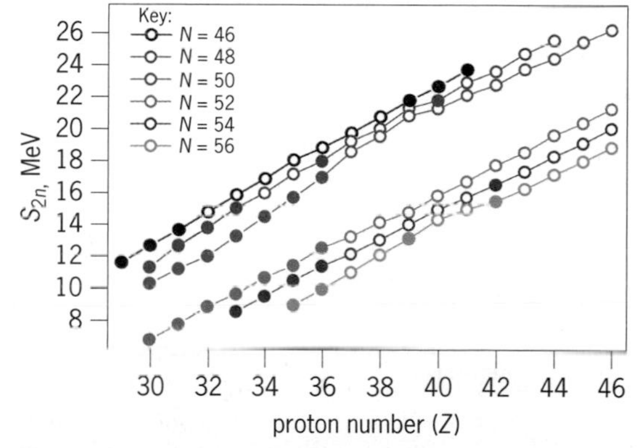

Fig. 4. Evolution of two-neutron separation energies for even-*N* isotones for element numbers from $Z = 29$ (copper) to $Z = 46$ (palladium). A large gap between the isotones with neutron numbers $N = 50$ and $N = 52$ is related to the closed shell $N = 50$. Open symbols represent available data prior to Penning trap measurements from JYFLTRAP and ISOLTRAP. Filled symbols extend the systematics very close to the doubly magic nickel-78 (^{78}Ni; $Z = 28$, $N = 50$) nucleus and into the region of the expected astrophysical rapid-proton capture process path, which dominates the nucleosynthesis of heavy elements.

Nuclear structure studies through atomic masses. While ion traps are excellent tools for providing clean samples of ions for decay spectroscopy studies, Penning traps can also be utilized directly to measure the masses of radioactive ions. Precise masses allow the determination of nuclear binding energies. Comparing differences in binding energies and other energy-related quantities such as two-nucleon separation energies between neighboring nuclei, one can see subtle effects such as shell closures and onsets of nuclear deformations. **Figure 4** shows a typical shell closure, which is seen as a sudden decrease in two-neutron separation energy (S_{2n}) between the isotones with neutron numbers $N = 50$ and $N = 52$. [The two-neutron separation energy is defined as a mass difference between two isotopes having a neutron number separation of two units, $S_{2n} = -M(A, Z) + M(A - 2, Z) + 2M(n)$, where $M(A, Z)$ is the mass of a nucleus with mass number A and proton number Z, and $M(n)$ is the neutron mass.] Similarly, one can study, for example, two-proton and one-nucleon separation energies to reveal nuclear structure changes.

Applications. Ultraclean samples of radionuclides have spun interest in various applications. Clean samples can be used, for instance, to calibrate and refine the sensitivity of radiation detection instruments. For example, total energy absorption spectrometers (TAS) benefit significantly from purified samples, since the spectrometer itself cannot distinguish the source of the observed deposited energy. A combination of a Penning trap and a TAS also has been used recently for reactor decay heat studies. Another application involves the production of noble-gas isotopes such as ^{133m}Xe, which is one of the main observables in the global network of noble-gas analyzers that are used to detect possible spills of fission products into the atmosphere from nuclear detonations.

For background information *see* EXOTIC NUCLEI; ISOBAR (NUCLEAR PHYSICS); ISOTONE; ISOTOPE; MASS SPECTROSCOPE; NUCLEAR BINDING ENERGY; NUCLEAR FISSION; NUCLEAR ISOMERISM; NUCLEAR SHELL MODEL AND MAGIC NUMBERS; NUCLEAR SPECTRA; NUCLEAR STRUCTURE; NUCLEOSYNTHESIS; PARTICLE TRAP; RADIOACTIVITY in the McGraw-Hill Encyclopedia of Science & Technology. Ari Jokinen; Tommi Eronen

Bibliography. K. Blaum, J. Dilling, and W. Nörtershauser, Precision atomic physics techniques for nuclear physics with radioactive beams, *Phys. Scr.*, T152:014017 (32 pp.), 2013, DOI:10.1088/0031-8949/2013/T152/014017; K. Blaum, Yu. N. Novikov, and G. Werth, Penning traps as a versatile tool for precise experiments in fundamental physics, *Contemp. Phys.*, 51:149–175, 2013, DOI:10.1080/00107510903387652; A. Kankainen, J. Äystö, and A. Jokinen, High-accuracy mass spectrometry of fission products with Penning traps, *J. Phys. G Nucl. Part. Phys.*, 39:093101 (37 pp.), 2012, DOI:10.1088/0954-3899/39/9/093101.

Nutraceuticals

Nutraceuticals are compounds or products isolated or purified from different food sources that are sold for their medical or health benefits of preventing or treating chronic diseases. The word "nutraceutical" is a combination of nutrition and pharmaceutical and was coined in 1989 by Stephen DeFelice, founder of the Foundation for Innovation in Medicine (originally located in Crawford, New Jersey). A large number of nutraceutical and functional food products with reported health benefits are now available in various supermarkets and grocery stores. In 1991, Japan was the first country to establish regulations for functional foods by introducing "Foods for Specified Health Use" (FOSHU). Since then, the rapid growth in nutraceutical research and production has led to regulations in the United States, Canada, Europe, and Asia. This article will review selected nutrients as major sources of nutraceuticals.

Carbohydrates. Carbohydrates account for approximately 55% of our daily caloric needs and include oligosaccharides, starches, fibers, and sugars. Of particular importance are the complex carbohydrates or polysaccharides that cannot be digested in the upper part of the gastrointestinal tract. Instead, they pass through to the colon, where the microbial flora (microflora) will convert them to bioactive compounds. Dietary fiber is one such bioactive compound and is composed of undigested polysaccharides, cellulose, hemicelluloses, beta-glucans, lignins, pectins, and gums. Dietary fiber has a reduced transit time in the gut, and it also maintains healthy microorganisms in the colon. Among the beneficial gut bacteria are lactic acid bacteria, which ferment the complex polysaccharides, increasing the acidity of the colon and reducing the number of pathogenic organisms. Dietary carbohydrates, including resistance starch, insoluble fiber, and soluble fiber, enhance the growth of healthy and beneficial bacteria known as bifidobacteria or probiotics. When these microorganisms, that is, bifidobacteria and lactobacilli, become dominant, they displace the unhealthy pathogenic bacteria in the gut. Unlike the pathogenic bacteria, which require proteins for growth, the probiotics require complex polysaccharides for their metabolism and proliferation. This illustrates the importance of having these polysaccharides, referred to as prebiotics, in the diet to enhance the growth of the beneficial probiotic bacteria. They also provide antioxidants and anticancer substances, and they reduce the risk of cardiovascular diseases. Other prebiotics include inulin, which is found in a wide range of fruits and vegetables, and an oligofructose enzymatically produced from inulin (known as Raftilose®).

An important part of the carbohydrates found in fruits and vegetables is dietary fiber. This material cannot be fully digested and is either soluble or insoluble fiber. Whereas soluble fiber can dissolve in water, insoluble fiber is insoluble in water and goes directly through the digestive system. Moreover, whereas high-fiber foods require longer digestion, their slow and steady digestion in the gut controls blood sugar and maintains weight control. Whole-grain products and vegetables, which are rich sources of dietary fiber, lower blood cholesterol and reduce the risk of cardiovascular and diverticular (colon) diseases.

Proteins and peptides. Some large proteins, such as buckwheat and soybean proteins, contain large parts that are indigestible. When these proteins are consumed, the indigestible parts trap and remove toxins (such as natural toxins in foods or carcinogens formed during processing). They also can reduce the reabsorption of cholesterol from the intestine. Digestion or hydrolysis of the digestible parts of these proteins produces short-chain molecules or peptides, some of which may be bioactive. This can occur during the normal digestion of proteins, with the peptides transported into the blood circulatory system. These bioactive peptides exhibit antioxidant activity, and they also have antimicrobial, antifungal, blood-lowering, and cholesterol-lowering properties. In addition, they exhibit antithrombotic and immunomodulatory effects and increase mineral absorption. Bioactive peptides, produced by the enzymatic digestion of plant and animal proteins, have been shown to reduce blood pressure when fed to hypertensive individuals. Commercial products containing bioactive peptides in a sour milk are available in Japan and Europe, and a milk powder containing bioactive peptides is sold in the United States.

Lipids and fatty acids. Extensive research has confirmed the important health benefits accrued from polyunsaturated fatty acids, particularly omega-3 (ω3) fatty acids. Some of the most noteworthy omega-3 fatty acids are linolenic acid (C18:3 ω3, LN), docosahexaenoic acid (C22:6 ω3, DHA), and eicosapentaenoic acid (C20:5 ω3, EPA). DHA and EPA are important components in the brain and appear to play a role in memory functions. Current evidence suggests that these fatty acids enhance brain development, and they also may be beneficial in individuals suffering from Alzheimer's disease. Fish oils are very rich sources of DHA and EPA, which (together with linoleic and linolenic acids) contribute important cardiovascular benefits as well. In addition, omega-3 fatty acids have exhibited anticancer effects. For example, prostate tumor growth has been shown to be reduced by omega-3 fatty acids, resulting in increased survival. Other studies have indicated that the type and the amount of fatty acids in the diet, particularly omega-3 fatty acids, were critical in modulating inflammatory bowel disease (IBD), which can lead to colorectal cancer.

Another important fatty acid is found in large amounts in canola and olive oils. This is the monounsaturated fatty acid, oleic acid, which is very beneficial in reducing blood cholesterol. Together with linoleic and linolenic acids, it also has a beneficial effect in reducing the risk of cardiovascular diseases. Conjugated linoleic acid (CLA), a unique fatty acid found in dairy and meat products, is a mixture of positional and geometric isomers of linoleic acid produced in the cow's stomach by rumen microorganisms. A number of health benefits are associated with CLA, including the prevention of the start of adult-onset diabetes (by lowering insulin) and the inhibition of cancer tumor growth in experimental animals.

Future research. This article has highlighted just a few of the dietary components in our diet that provide important nutraceuticals. Extensive research is being conducted around the world to identify new and better sources of nutraceuticals that will further reduce the risk of chronic diseases, including cardiovascular diseases, cancers, and Alzheimer's disease.

For background information *see* ANTIOXIDANT; CARBOHYDRATE; CARBOHYDRATE METABOLISM; FAT AND OIL (FOOD); FOOD; FOOD ENGINEERING; FOOD SCIENCE; FRUIT; LIPID; LIPID METABOLISM; NUTRITION; PEPTIDE; PROTEIN in the McGraw-Hill Encyclopedia of Science & Technology. N. A. Michael Eskin

Bibliography. R. E. Aluko, *Functional Foods and Nutraceuticals*, Food Science Text Series, Springer, New York, 2012; S. A. El-Sohaimey, Functional foods and nutraceuticals—modern approach to food science, *World Appl. Sci. J.*, 20(5):691–708, 2012, DOI:10.5829/idosi.wasj.2012.20.05.66119; N. A. M. Eskin and T. Snait, *Dictionary of Nutraceuticals and Functional Foods*, Taylor and Francis, Boca Raton, FL, 2005.

Obesity as a factor of gut microbiota

Obesity is an increasing epidemic in the United States and in many other countries worldwide. This increase in obesity has been attributed to a variety of causes, including poor diet choices, lack of exercise, and food-eating disorders. There is now, however, scientific evidence that microorganisms found in the intestinal microbiota (microbial flora) may actually play a role in the development of obesity. Many articles in the popular press are touting the benefits of "good bacteria" and the disadvantages of "bad bacteria" in terms of human health. The latest evidence indicates that these microbes may play a greater role in our overall health than previously understood.

Obesity and overweight categories. The Centers for Disease Control (CDC) in the United States reported that, in 2009–2010, 69.2% of the adult population over 20 years of age was overweight. Moreover, 35.9% of these individuals were obese. Overweight and obesity are terms used to describe weight ranges that are greater than what is healthy for a given height. These terms are also closely linked to the development of certain diseases and other health problems, including type 2 diabetes and heart disease. The CDC uses the body mass index (BMI) to determine whether an individual is at a healthy weight, overweight, or obese. BMI is calculated as shown in **Table 1**. An adult who has a BMI between 25 and 29.9 is considered overweight, whereas an adult whose BMI is 30 or higher is considered obese. (Although BMI is a good indicator of a person's general health, athletes who are heavily muscled will be labeled as "overweight" because the BMI does not measure the percentage of body fat. It is simply a ratio of weight to height.) An example of how BMI is used for a person who is 5′ 9″ [5 ft 9 in. = 69 in. (1.75 m)] in height is shown in **Table 2**. Body fat can also be more directly measured in other ways,

TABLE 1. Calculation of BMI*

Measurement units	Formulas and calculations
Kilograms and meters	Formula: weight (kg)/[height (m)]^2 Example: weight = 68 kg; height = 1.65 m Calculation: 68 ÷ (1.65)^2 = 24.98
Pounds and inches	Formula: {weight (lb)/[height (in.)]^2} × 703 Example: weight = 150 lb; height = 65 in. Calculation: [150 ÷ (65)^2] × 703 = 24.96

*BMI is calculated the same way for both adults and children.
SOURCE: http://www.cdc.gov/healthyweight/assessing/bmi/adult_bmi/index.html; http://www.cdc.gov/obesity/adult/defining.html.

TABLE 2. BMI examples for a typical individual

Height	Weight range	BMI	Considered
5′ 9″	124 lb or less	Below 18.5	Underweight
	125–168 lb	18.5 to 24.9	Healthy weight
	169–202 lb	25.0 to 29.9	Overweight
	203 lb or more	30 or higher	Obese

SOURCE: http://www.cdc.gov/obesity/adult/defining.html.

such as magnetic resonance imaging (MRI), ultrasound, skinfold thickness, and waist circumference measurements.

Focus on microbiota. The role of microbiota in human health and disease has been the focus of a number of recently published articles in microbiology. Breast-feeding by an infant in the first nine months of life appears to play a role in the development of a healthy immune system because this activity increases the prevalence of *Lactobacillus* and *Bifidobacterium* organisms in the gastrointestinal tract. The female breast has been the subject of studies examining its normal microbiota, which has been shown to be unique when compared to the rest of the body. The focus of this article, however, is the role that microbiota may play in the development of obesity.

Evidence points to a bacterial role. Determining the role of gut microbiota in obesity development is a difficult task for scientists. Separating the effects of gut microbiota from those of diet, lifestyle, and genetic factors is nearly impossible. However, an innovative research group came up with the idea of using human twins, one obese and one of healthy weight (both female), as their source of microbiota. Fecal samples from each twin were first transplanted into separate groups of germ-free (gnotobiotic) mice, referred to as "lean" (Ln) or "obese" (Ob). Members from both groups were fed an unlimited, low-fat (4%) diet that was high in plant polysaccharides. The mice were monitored for their weight and general health. Samples were taken from their fecal matter (pellets) at periodic intervals (1, 3, 7, 10, and 15 days postcolonization), and their genetic material was analyzed to determine the taxonomic groups of the microbiota. In order to determine the taxonomy (species) of the microbes present in the gut, 16S ribosomal RNA (rRNA) profiling was performed. Of the greatest significance, perhaps, was the analysis of the metabolic activities of the microbiota after transplantation, which differed significantly between the Ln and Ob mice. In particular, the mice were measured for increases in adipose (fat) mass over time following the transplantation. It was found that the Ob mice gained significantly more adipose mass than the Ln mice fed an identical diet. This increase occurred despite the fact that the Ob mice did not consume significantly more food than their Ln counterparts.

Overall, the genetic analysis showed that there were significant differences in the metabolic activities of the microbes found in each type of mouse. Ob mice had a higher expression of microbial genes involved in producing branched-chain amino acids, such as valine and leucine, as well as elevated levels of these amino acids in their bloodstream. Elevated valine and leucine levels are also seen in individuals who are obese and insulin-resistant when compared to lean and insulin-sensitive individuals. Therefore, the possibility exists that the microbiota of the intestinal tract can influence the metabolites that are present in obese individuals. Ln mice, conversely, had microbiota that expressed higher levels of genes involved in digestion and breakdown of plant-derived polysaccharides, as well as those involved in fermenting the resulting monosaccharides and disaccharides. This seems to indicate that the microbiota found in these Ln mice is better able to break down and ferment polysaccharides than the microbiota in their Ob counterparts, thus avoiding the buildup of fat deposits.

An additional experiment performed by the aforementioned research team examined the effect of housing the two types of mice together. After the transplantation was performed, the scientists sought to determine, by fecal analysis, if the population was stable in the gut of each type of mice. The mice were then housed together in a 1:1 (Ln:Ob) ratio. Because mice are coprophagic (that is, they will eat their own fecal matter), it was assumed that they would each consume material produced by their respective cage mate. The mice were monitored for adipose deposition, and their gain (or lack thereof) was compared to separately housed Ob mice. It was found that the Ob mice housed with Ln mice gained significantly less adipose mass when compared to the control Ob mice, and the Ln mice did not gain when compared to Ln mice controls. It also was found that the microbiota of the cohoused Ob mice became altered over time to resemble that of the Ln mice. However, the microbiota of the Ln mice did not change when exposed to the Ob mice, indicating that this population of microbes easily becomes stabilized within the host and is not influenced or overwhelmed by the Ob microbiota.

Conclusions and outlook. What does this mean in terms of the human population? Currently, there are a number of different supplements on the market that contain microbes prepared for human consumption (prebiotics, probiotics, and combinations of both). In the future, will the obese individual be able to take a pill that will change his or her microbiota to a lean phenotype? It is definitely a question of significance in a population that is becoming more and more obese. Therefore, more studies are required to investigate the effects of microbiota on human health and wellness.

For background information *see* ADIPOSE TISSUE; BACTERIA; DISEASE; FOOD; METABOLISM; MICROBIAL ECOLOGY; MICROBIOTA (HUMAN); NUTRITION; OBESITY; PUBLIC HEALTH in the McGraw-Hill Encyclopedia of Science & Technology. Marcia M. Pierce

Bibliography. M. K. Cowan, *Microbiology: A Systems Approach*, 4th ed., McGraw-Hill Education, New York, 2015; R. Francavilla et al., Salivary microbiota and metabolome associated with celiac disease, *Appl. Environ. Microbiol.*, 80(11):3416–3425, 2014, DOI:10.1128/AEM.00362-14; P. R. Murray, K. S. Rosenthal, and M. A. Pfaller, *Medical Microbiology*, 7th ed., Mosby, St. Louis, 2013; E. Nester, D. Anderson, and C. E. Roberts, Jr., *Microbiology: A Human Perspective*, 7th ed., McGraw-Hill, New York, 2012; V. K. Ridaura et al., Gut microbiota from twins discordant for obesity modulate metabolism in mice, *Science*, 341:1079, 2013, DOI:10.1126/science.1241214; C. Urbaniak et al., Bacterial microbiota of human breast tissue, *Appl. Environ. Microbiol.*, 80(10):3007–3014, 2014, DOI:10.1128/AEM.00242-14; E. P. Widmaier, H. Raff, and K. T. Strang, *Vander's Human Physiology: The Mechanisms of Body Function*, 13th ed., McGraw-Hill Education, New York, 2014.

Offshore wind energy conversion

Offshore wind energy conversion is one of the most effective, renewable, and sustainable power generation technologies. These advantages arise from both the overall efficiency of wind power conversion, which is in the range of 50%, and from the advantageous wind conditions out at sea.

The offshore wind farms can be located either close to a shoreline, which is more effective with respect to maintenance and interconnection, or far out at sea, where wind conditions are better and energy yields can therefore be higher. Thus, the corresponding number of full-load hours (FLH), defined as the ratio of annual energy yield divided by the nominal power of the wind turbine, reaches up to 4500 FLH at far-offshore locations, while in the onshore regions it lies in the range of about 1600–2500 FLH.

Just over 2% of worldwide installed capacity for wind power is in offshore projects. In 1991, Denmark installed the world's first offshore wind farm, a 5-MW project in the Baltic Sea. Denmark already gets more than 30% of its electricity from wind—onshore and offshore—and aims to increase that share to 50% by 2020.

Figure 1 shows the cumulative installed offshore wind power capacity by country at the end of 2013. More than half of the 7 GW of offshore capacity belongs to the United Kingdom. According to the

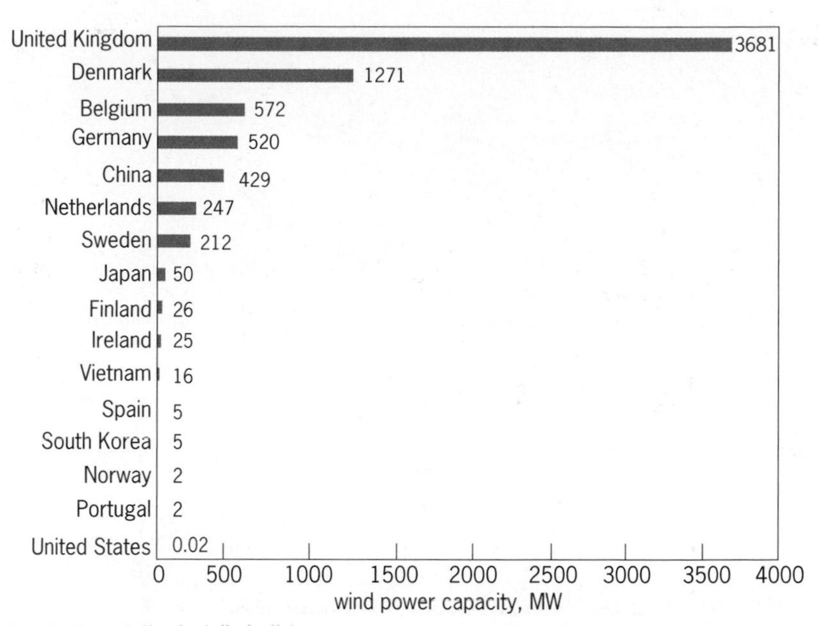

Fig. 1. Cumulative installed offshore wind power capacity by country, 2013. (*Courtesy of J. Matthew Roney, Plan B Updates, 2013 to be Record Year for Offshore Wind, Earth Policy, October 30, 2013, http://www.earthpolicy.org/plan_b_updates/2013/update117, and Global Wind Energy Council, Global Wind Report: Annual Market Update 2013, Brussels, 2014*)

U.S. Department of Energy, shallow waters along the eastern seaboard could host 530,000 MW of wind power, which would be capable of covering more than 40% of current U.S. electricity generation needs. Adding in deeper waters and the other U.S. coastal regions boosts the potential to more than 4.1 million MW.

Offshore wind power has become a major component of the European electricity mix. The European Wind Energy Association (EWEA) expects 148 TWh of electricity to be produced in 2020, meeting more than 4% of the European Union's (EU's) total demand. EWEA expects 40 GW of offshore wind

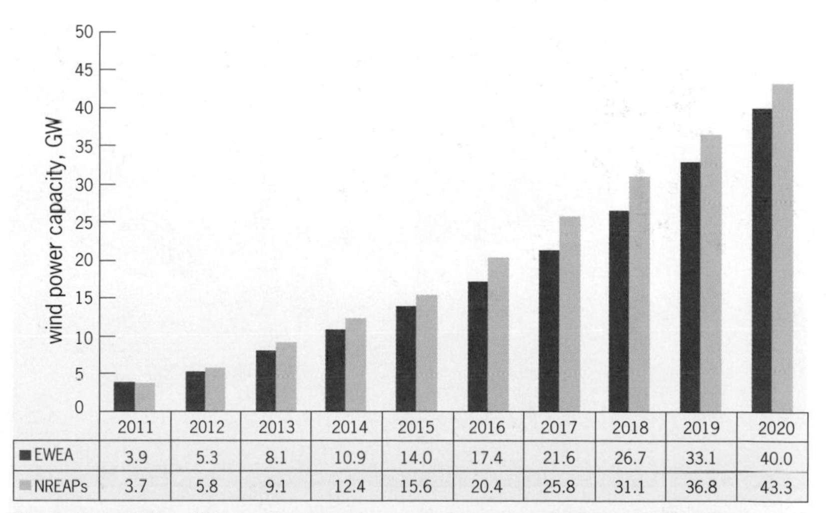

	2011	2012	2013	2014	2015	2016	2017	2018	2019	2020
EWEA	3.9	5.3	8.1	10.9	14.0	17.4	21.6	26.7	33.1	40.0
NREAPs	3.7	5.8	9.1	12.4	15.6	20.4	25.8	31.1	36.8	43.3

Fig. 2. Cumulative offshore wind capacity across Europe, projected in 2010 by the European Wind Energy Association (EWEA) and by National Renewable Energy Action Plans (NREAPs). (*Courtesy of European Wind Energy Association, Wind in our sails—The coming of Europe's offshore wind energy industry, EWEA, Brussels, 2011, http://www.ewea.org/fileadmin/ewea_documents/documents/publications/reports/23420_Offshore_report_web.pdf*)

Fig. 3. Wind farm in the Baltic Sea. (*Courtesy of Siemens AG, Germany; photograph by Paul Langrock*)

capacity to be installed across Europe by 2020. By 2030, that capacity would grow to 150 GW, producing 562 TWh, or enough to cover 14% of the EU's electricity demand. **Figure 2** shows cumulative offshore wind capacity as projected in 2010 by the EWEA and the National Renewable Energy Action Plans (NREAPs) of member states of the EU.

Technological issues. Generally, offshore wind turbines are installed in groups referred to as wind farms (**Fig. 3**).

Electrical system infrastructure. The electrical system for an offshore wind farm consists of a medium-voltage electrical collection grid within the wind farm and a high-voltage electrical transmission system capable of delivering power to an onshore transmission line.

Fig. 4. Part of Burbo Bank Offshore Wind Farm in the Irish Sea, showing wind turbine order. (*Courtesy of Siemens AG, Germany; photograph by Paul Langrock*)

Transformers at each wind turbine, usually in the base of the tower, step up the electrical output from its low generation voltage, typically 690 V, to a medium voltage of typically 25–40 kV. The preference for this voltage range represents a pragmatic compromise: standardized equipment for delivering it is available at competitive prices and higher voltage transformers would be too large to fit readily into the tower cross sections. A grid of medium-voltage submarine cables, typically buried 1–2 m (3–7 ft) beneath the seabed, connects the wind turbines to an offshore substation. The substation steps up the voltage to 130–150 kV, the highest voltage currently used for ac submarine cables. The higher voltage allows much smaller-diameter and lower-cost submarine cables to be used for the long run to shore. A high-voltage submarine cable (which is also buried in the seabed for protection) carries the power from the offshore substation to land, and then either underground or overhead to an onshore substation for connection to a transmission line. An additional transformer in this substation may step up the voltage to a higher level to match the transmission grid.

Two technology options are available for this transmission system: high-voltage ac (HVAC) and high-voltage dc (HVDC). HVAC is the most economical option for distances shorter than 50 km (31 mi). Between 50 and 80 km (31–50 mi), HVAC and HVDC are expected to be similar in cost. For farther than 80 km (50 mi), HVDC systems will likely cost less, mainly because the capacity of a given HVAC cable drops off with distance because of the capacitive and inductive characteristics of the cable.

The number of wind turbines in the wind farm is limited by the maximal connection capacity, which corresponds currently to about 500 MW. Taking into account the usual rating of the offshore wind turbines in the range of 2–6 MW, the number of wind turbines in the wind farm can reach 100 or more (**Fig. 4**).

The structure of the offshore wind farm has to be optimized with respect to the placement and distance between individual wind turbines. The distances must be minimized in order to have a shorter collector cable system and to lower both investment costs and power losses. Yet, the distances between individual wind turbines should also be sufficient to avoid the extensive shadowing effects that individual wind turbines can have on each other by altering the available wind flow. In the case of shadowing effects, the energy yield of the wind farm could be significantly decreased.

Grid interconnection and power capture. In contrast to low-capacity wind generation, which does not need to meet stringent electric-grid-code requirements, wind farms can contribute hundreds of megawatts to municipal power grids, with a profound impact on the grids' stability, regulation, and fault-ride-through (FRT) capability. Wind farms are therefore required to follow connection requirements regarding scheduling and reserve availability, reactive power support, transient recovery, system stability and voltage/frequency regulation, and power quality.

Offshore wind farms are different from onshore wind farms with respect to grid interconnection. If the wind power connection points incorporated converter systems, many of the network connection issues would be eliminated and the stability of the ac system might even be enhanced. However, interconnection with the ac network of the onshore grid would still require the use of additional equipment for reactive power and voltage support. Issues related to low inertia, power, and frequency control and stabilization of the ac system, which are significantly important with large wind farm interconnection, would not be resolved. Recently, some European countries, such as Denmark, Germany, and Spain, have issued grid interconnection codes addressed to wind farms, both at the transmission and distribution levels.

A number of offshore wind farm configurations for collecting power and feeding into the grid have been suggested. The produced electricity would go through many conversion stages, both via ac transformers and via ac/dc and dc/ac converters. A configuration that minimizes the number of conversion stages is advantageous both from an initial cost and system efficiency point of view.

Environmental issues. Wind farms may adversely affect the local environment, but offshore wind farms located far from shore have a lower noise and visual impact.

Impacts on birds. More knowledge is required about the overall effects of offshore wind farms on bird populations. Nevertheless, it seems clear that offshore wind farms should not be placed in protection areas for birds. Potentially detrimental effects of offshore wind farms on birds include short-term or long-term habitat losses due to human disturbance at the stages of construction or maintenance, collision fatalities, and barriers to the movement of birds. These disturbances can lead directly to expulsion and thus a loss of territory for certain species of birds. Displacement of birds from favorable habitats into less suitable ones may adversely affect the sustainability of their populations. Conversely, some studies have indicated that species such as gulls and terns show a preference for wind farm areas.

Larger rotor diameters and taller towers increase the risk of birds colliding with the turbines or being injured in the associated turbulence vortices. Particular bird species susceptible to collisions may become endangered or have declining populations. Information about bird mortality due to offshore wind farms is still limited because of the difficulty of detecting collisions and the bodies of dead birds at sea. Some studies conducted at existing offshore wind farms have suggested that birds can avoid the facilities by flying around the perimeter of the wind farm rather than between the turbines. Collision rates per turbine may therefore be relatively low during days of good visibility.

Impacts on marine mammals and fish. Most marine mammals depend to some extent on hearing for communication, navigation, foraging, and predator detection, or other activities. Consequently, human-caused excessive underwater sound can adversely affect these marine mammals over very large distances and cause temporary or permanent hearing impairment, behavioral disturbance, and possible stress.

Construction of an offshore wind farm can generate intense noise for limited periods, such as hydraulic hammers pile-driving monopile foundations for shallow water turbines. Such high-intensity sound could cause hearing loss in marine mammals such as seals. Moreover, even after an offshore wind farm is put into operation, its operational wind turbines and associated maintenance activities can continuously produce underwater noise for the lifetime of the wind farm.

Noise may also have negative effects on fish, such as hearing damage, interference with acoustic communication, forced changes in habitat, and disturbances in routines and behaviors. Because various fish species react very differently to sound, the lack of adequate knowledge and long-term monitoring data makes it difficult to draw firm conclusions as to the likelihood of health effects on fish from exposure to noise emitted by offshore wind farms.

Some fish use natural magnetic fields like a compass to navigate and detect their prey. Thus artificial electromagnetic fields may cause abnormal migration or even changes in the fish's physiological mechanisms. Some research has suggested that the submarine cables could retard migration of some fish species or slow down the speed of migrating silver eels.

On the other hand, offshore wind farm foundations may also have the potential to benefit marine life by acting as artificial reefs and aggregators for fish, increasing marine biomass, and attracting some fish species.

Aesthetics, economics, and other concerns. The aesthetics of offshore wind is a common cause for concern. The costs of current offshore wind technology limit the placement of wind turbines to waters within view from the shore. Public attitudes toward offshore wind vary from community to community, depending on the culture, economy, coastal development, and recreational uses of the coast.

Two of the first large offshore wind farms were constructed off the coast of Denmark in 2002 and 2003. People in two coastal areas of Denmark reacted differently.

The Nysted wind farm was built 10 km (6 mi) offshore from the quiet, scenic coastal area of Lolland, Denmark. People who opposed the development were worried that the farm would spoil their unique natural scenery, and their concerns remained unchanged after construction.

The Horns Rev wind farm was built 13 km (8 mi) offshore from an area with a vibrant tourism industry. People who opposed the wind development

Fig. 5. Offshore wind conversion devices in the Burbo Bank Wind Farm during a calm period. Each turbine is positioned in a different direction. Elements: rotor blades, nacelle, and wind speed measuring device. (*Courtesy of Siemens AG, Germany; photograph by Paul Langrock*)

were concerned about its impacts on business and property values. After construction, tourists still visited the area, and the fear of a decrease in the summer house prices has proved to be groundless. Two years after construction, 89 percent of residents in that area supported new offshore developments in Denmark.

Cape Wind is the first offshore wind farm proposed in Nantucket Sound between Cape Cod and Nantucket Island, Massachusetts. In a 2005 survey, a majority (55%) of Cape Cod residents were opposed to the proposed Cape Wind offshore wind farm. Cape residents expected more negative outcomes (on aesthetics, community harmony, fishing, boating, property values, and tourism) than positive ones (on job creation, electricity rates, and air quality) from the proposed project. Many respondents indicated they would increase their support if Cape Cod received the electricity, if electricity rates decreased, if local fishing was helped, and if air quality improved.

The health effects of air pollution from fossil-fuel-based electric generation also deserve consideration in evaluating the value of offshore wind farms. For example, the Cape Wind project could result in avoiding approximately 11 premature deaths per year due to particulate-matter emissions alone from two Massachusetts-based coal-fired plants. Taking the effect of worker deaths from construction and maintenance into account suggests a net of 10 human lives per year saved as a result of the project.

The question of scale needs to be addressed. Discussions about the future of offshore wind farm development almost invariably focus on the prospects of individual projects rather than on how to develop the huge wind power resources off a region such as the Eastern coast of the United States. To evaluate the claims that offshore wind technology can deal with the large-scale problems of fossil-fuel power generation and import dependence, the prospect of offshore wind power at comparably large scales needs to be evaluated.

The vulnerability of coastal areas to the effects of climate change such as sea-level rise and more severe storms is another issue to consider. An area's proximity to the ocean, its beaches and other environmental features near sea level, and the high tourist and recreational value of its shoreline all make it valuable. Offshore wind is the only U.S. wind resource strong enough and close enough to Eastern cities to displace significant greenhouse gas emissions in the near term, and it is evident that mitigation of climate change is a benefit of offshore wind projects.

Support structure technologies. The towers and foundations that constitute the support structures of offshore wind turbines need special design attention. Strong offshore winds and severe wave and sea conditions pose hazards distinct from those facing onshore support structures.

Tower. The turbine tower is fixed to the foundation on the seafloor and carries the nacelle and turbine rotor. The towers must be able to withstand cyclonic wind gusts, and must be rigid so as not to damage the wind turbine generator system or itself through vibration and bending. They must also have high resistance to fatigue stress and resonant vibrations. To increase the strength of the tower, the most commonly used shapes are lattice and tubular conical designs. A lattice design is lighter than a corresponding tubular steel tower would be and the building materials for truss towers cost much less. Truss towers also have the advantages of experiencing less force from waves, a smaller foundation, and a flexible design.

The increasing generating capacity of offshore wind turbines makes concrete a competitive material. A concrete tower is durable, reliable, and needs less maintenance than a steel tower would. A concrete/steel hybrid tower could be economically viable if the whole-life cost for a wind farm were considered.

Foundation. Offshore wind turbine foundation types and design depend on seafloor soil properties, water depth, wave heights, and currents. Offshore wind farms built in shallow waters less than 30 m (100 ft) deep and close to shore use relatively simple types of foundations such as monopole and gravity-based foundations.

Far-offshore and deeper water areas require technologies such as suction caissons installed into the seabed either by pushing or by pumping water out of the caisson to create a pressure differential across its top. Floating foundation prototypes are being developed.

Protection measures for offshore wind turbines. Protection against damage from corrosion is very important for offshore wind turbines. Anticorrosion protection measures should involve a simple construction method and a long-term service life that is not dependent on maintenance and management work. Measures must prevent corrosion on both the interior and the exterior of the turbine, which mainly consist of its nacelle and support structure. The anticorrosion protection on the inside of the nacelle can be achieved by keeping the air dry, and by using a watertight tower and a local air conditioning system to ensure low relative humidity. Alternatively, the whole machine can be sealed and the gear and generator cooled by heat exchangers that recycle the air in the air-cooling system. (These can take the place of conventional air-cooled components on earlier turbines.) Methods used to protect the exteriors of offshore wind turbines from corrosion consist mainly of strengthening the anticorrosion, cathodic protection, environmentally friendly painted and multifunctional coating, and spray and nano-based layer systems.

The design of offshore wind turbines must comply with anti-typhoon requirements in areas with a high frequency of hurricanes or typhoons. Extreme wind speed, unusual turbulence, and sudden changes in wind direction are the main characteristics of a hurricane or a typhoon and also the main causes of damage to the vital components of wind turbines—destroying turbine blades, nacelles, and even towers.

Future trends. The European Union has set an ambitious master plan for offshore wind energy deployment that assumes a very high increase in offshore wind capacity. China and the United States are also trying to use the high potential of this clean and sustainable energy source.

The main obstacle in the way to reaching more offshore energy is the reliability of supply, which is not as high as in the onshore power system.

First, the intermittent character of wind generation is a factor. The energy generation is quite unstable. Some wind power forecasting tools allow for the operation of the wind power system according to a schedule; nevertheless, the forecast error can be high. Furthermore, at sea, periods of no wind (calm periods) lasting for up to a week can sometimes unexpectedly occur (**Fig. 5**). In these calm periods, the offshore wind farms are out of operation due to the missing driving force (wind) and no power can be provided to the onshore power system.

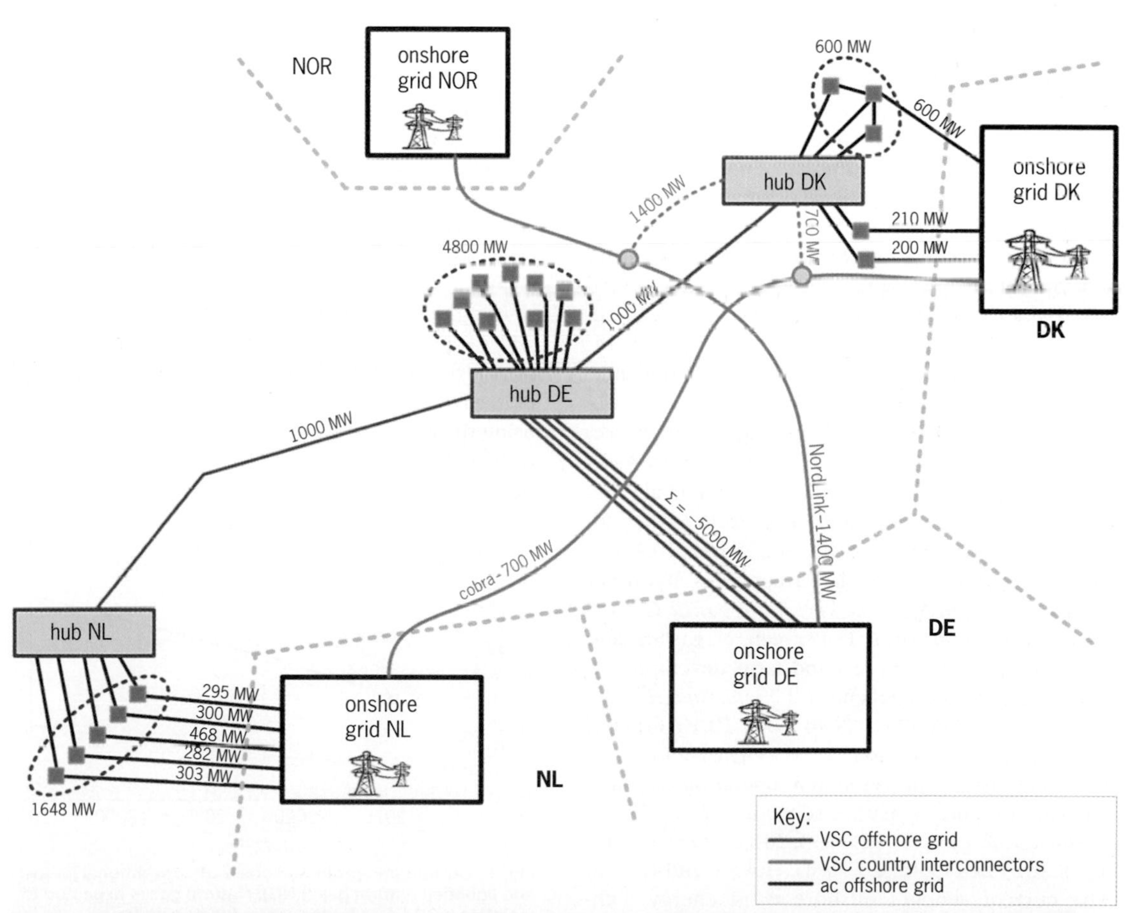

Fig. 6. General structure of an offshore test power system for part of the North Sea. VSC, voltage-source converter; NOR, Norway; DK, Denmark; DE, Germany; NL, the Netherlands. (*Courtesy of K. Rudion, A. G. Orths, and P. B. Eriksen, Offshore power system operation planning considering energy market schedules, IEEE Trans. Sustain. Energy, 4(3):725–733, 2013, DOI:10.1109/TSTE.2013.2245350*)

Maintenance of the offshore infrastructure is a second challenge. Winter storms and other rough sea conditions periodically make it impossible to conduct any maintenance. During these times, neither boats nor helicopters can reach the faulted infrastructures for many days or even weeks.

Third, the single-cable connections between a big wind farm and the onshore power system are critical points. If the connection experiences an outage, none of the energy from the wind farm can be transported to the onshore power system. The onshore power system is usually meshed, so if one connection is on outage, the power flows through the other parallel paths. A similar principle has been developed for the offshore power system. In **Fig. 6** a schematic structure of a test offshore power system in the North Sea is presented. Such measures can essentially increase the reliability of the offshore power system with relatively small expenses.

For background information *see* AIR POLLUTION; DIRECT-CURRENT TRANSMISSION; ELECTRIC POWER GENERATION; ELECTRIC POWER SYSTEMS; ELECTRIC POWER TRANSMISSION; GLOBAL CLIMATE CHANGE; SUBMARINE POWER CABLE; UNDERWATER SOUND; WIND POWER in the McGraw-Hill Encyclopedia of Science & Technology. Zbigniew A. Styczynski; Krzysztof Rudion; Mohamed E. El-Hawary

Bibliography. T. Ackermann (ed.), *Wind Power in Power Systems*, 2d ed., Wiley, Chichester, UK, 2012; A. Athanasiaa, G. Anne-Bénédicteb, and M. Jacopo, The offshore wind market deployment: Forecasts for 2020, 2030 and impacts on the European supply chain development, *Energy Procedia*, 24:2–10, 2012, DOI:10.1016/j.egypro.2012.06.080; European Wind Energy Association, *Wind in our sails—The coming of Europe's offshore wind energy industry*, EWEA, Brussels, 2011; Global Wind Energy Council, *Global Wind Report: Annual Market Update 2013*, Brussels, April 2014; J. Green et al., Electrical collection and transmission systems for offshore wind power, pp. 2215–2221, *2007 Offshore Technology Conference*, Houston, TX, April 30–May 3, 2007, Curran Associates, Red Hook, NY, 2007; W. Kempton et al., The offshore wind power debate: Views from Cape Cod, *Coast. Manag.*, 33:119–149, 2005, DOI:10.1080/08920750590917530; S. M. Muyeen (ed.), *Wind Energy Conversion Systems: Technology and Trends*, Springer, London, 2012; R. Perveen, N. Kishor, and S. R. Mohanty, Off-shore wind farm development: Present status and challenges, *Renew. Sustain. Energ, Rev.*, 29:780–792, 2014, DOI:10.1016/j.rser.2013.08.108; K. Rudion, A. G. Orths, and P. B. Eriksen, Offshore power system operation planning considering energy market schedules, *IEEE Trans. Sustain. Energy*, 4(3):725–733, 2013, DOI:10.1109/TSTE.2013.2245350; X. Sun, D. Huang, and G. Wu, The current state of offshore wind energy technology development, *Energy*, 41:298–312, 2012, DOI:10.1016/j.energy.2012.02.054.

Options for future methane- and LNG-fueled aircraft

In recent years, natural gas has become increasingly plentiful and has been shown to have a cost advantage on an energy basis compared to jet fuel refined from oil or other feedstocks. This is especially true in North America, but technological advances in natural gas production are likely to be deployed worldwide, leading to an extended era of abundant and relatively low-cost natural gas. Natural gas enjoys a significant cost advantage over conventional jet fuel, even when the cost of processing natural gas to liquefied natural gas (LNG) is considered (**Fig. 1**).

Natural gas and hydrogen as aircraft fuels. Natural gas is composed mostly of methane and is generally cleaner burning than conventional jet fuel, producing less carbon dioxide, NO_x, and other pollutants. Previous studies have looked at hydrogen-fueled aircraft, and hydrogen has great promise as an aircraft fuel. It has high specific energy (energy/weight), but low energy density (energy/volume). The properties of natural gas fall between those of jet fuel and those of hydrogen (**Table 1**). From an environmental perspective, because it is clean burning and produces no carbon dioxide when combusted, hydrogen has a significant advantage over conventional jet fuel. However, the production of hydrogen, either by cracking natural gas or by electrolysis, is currently an inefficient and energy-intensive process that tends to eliminate hydrogen's environmental advantages when the entire life cycle is considered and results in a high fuel production cost. Both natural gas and hydrogen need to be liquefied to low cryogenic temperatures to increase their storage density, so that additional energy is required to liquefy the fuel. In addition, cryogenic fuel storage requires special fuel tanks and fuel system insulation, significantly increasing fuel system complexity and weight, and raising safety concerns that must be addressed through careful leak monitoring and extensive demonstration. Another consideration in using these alternative fuels is that a new airport infrastructure would need to be developed to deliver the fuel to the aircraft.

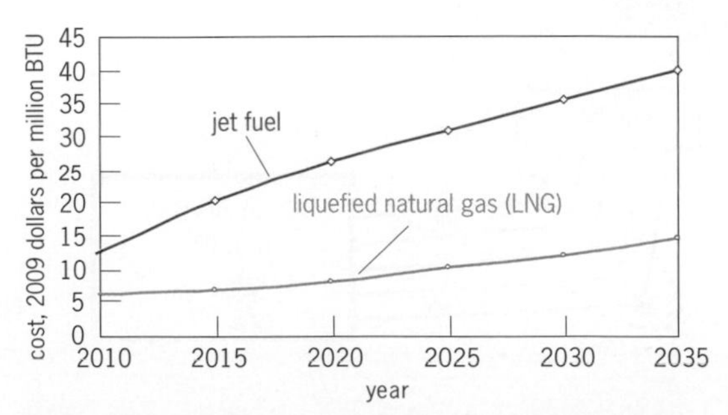

Fig. 1. Current and projected costs of conventional jet fuel and liquefied natural gas (LNG). Natural gas is expected to maintain a 2-to-1 or better cost advantage over conventional jet fuel out to 2035 and beyond. *(Copyright © 2012 Boeing; all rights reserved)*

TABLE 1. Comparison of fuel properties for jet fuel, LNG, and hydrogen

Property	Conventional jet fuel	LNG (mostly methane)	Hydrogen (liquid)
Energy/weight, MJ/kg	43.2 (100%)	50 (116%)	120 (278%)
Energy/volume, MJ/L	34.9 (100%)	21.2 (61%)	8.4 (24%)
Storage temperature, °C	−40 or higher	−163	−253

SOURCE: G. Hemighaus et al., *Alternative Jet Fuels*, Chevron Corp., Houston, TX, 2006.

Natural gas has advantages over hydrogen in addressing these issues. It does not have to be manufactured, and it can be readily transported in pipelines. Methane is a larger molecule than hydrogen, making it much less prone to leakage, and it is more compatible with typical tank and fuel system materials. Natural gas is liquefied to a higher cryogenic temperature (−163°C versus −253°C for hydrogen), so it requires less energy to liquefy. Being stored at a higher temperature means that less fuel system and tank insulation is required, reducing weight and insulation volume. A special challenge for natural gas is that methane is itself a potent greenhouse gas that can be released during natural gas extraction, transportation, storage, and fueling, and by boil-off from the aircraft tank. These releases must be controlled to prevent the erosion of the environmental advantages of using natural gas. *See* GREEN AVIATION.

NASA study. In a research study conducted for NASA in 2012, several alternative energy possibilities for aviation were investigated. One of the most promising was judged to be using natural gas in its liquid form (LNG). In the study, a variety of engine cycles that could be deployed in the 2040–2050 time frame were evaluated for use in commercial aircraft. Many other advanced technologies that could reduce fuel burn and increase energy efficiency were postulated. These technologies included full implementation of advanced air traffic management improvements; advanced materials and structural design methods; higher engine temperatures and higher component efficiencies; improved aerodynamics, including significant regions of laminar flow; and more efficient and lighter aircraft subsystems. These technologies result in up to a 54.5% reduction in jet fuel burned compared to a contemporary aircraft such as the Boeing 737. However, the goal of the study was greater than a 60% fuel burn reduction, so several alternative propulsion system architectures were investigated:

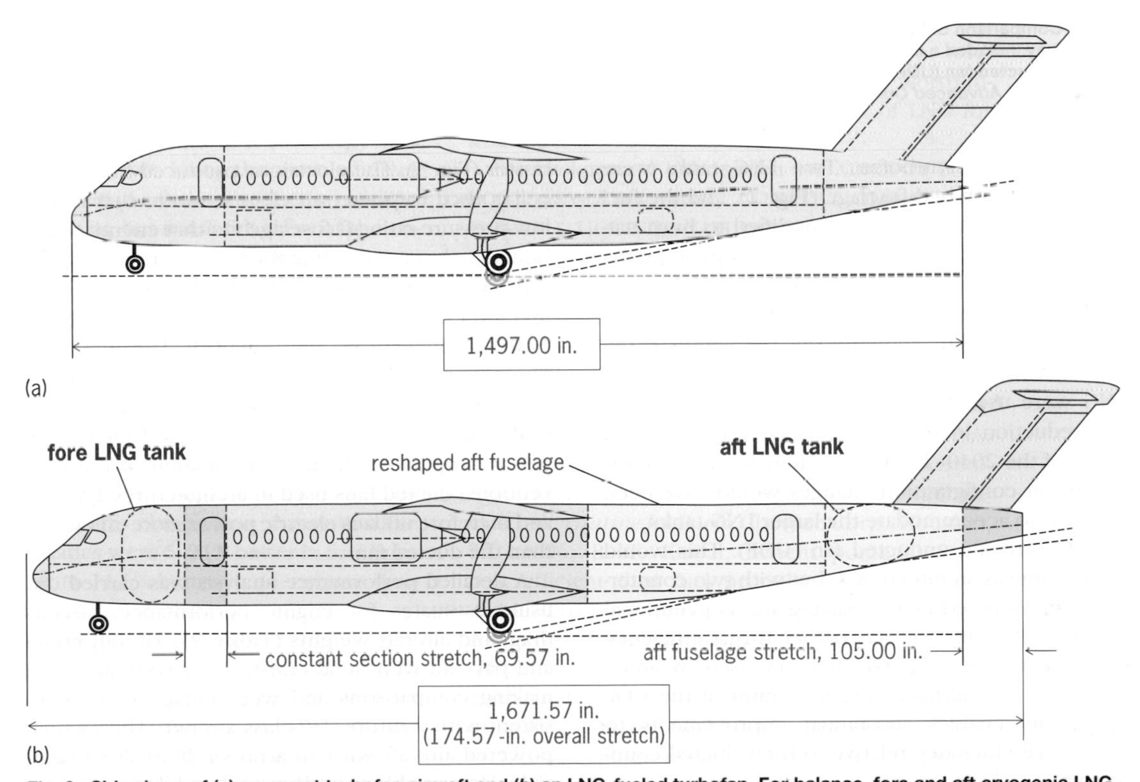

Fig. 2. Side views of (*a*) a current turbofan aircraft and (*b*) an LNG-fueled turbofan. For balance, fore and aft cryogenic LNG tanks were added to a stretched fuselage. Other integrations are possible, such as a cylindrical tank installed above the circular fuselage or possibly wing-mounted tanks. Because of other technologies developed by 2040–2050 to reduce fuel burn, the tank volume is reduced to less than half the size that would be required for an aircraft of today. (*Copyright © 2012 Boeing; all rights reserved*)

Origin of vertebrates

The fossil record for vertebrates is arguably one of the best of any major group of organisms and enables the evolutionary development from primitive jawless fish through more advanced fish to tetrapods (and ultimately humans) to be delineated. Despite this, it is often not clear when major groups diverged from each other, particularly when the first vertebrates separated from other chordates. Because the earliest vertebrates do not appear to have possessed mineralized skeletons, their preservation potential is low and their fossil record is very sparse. However, in recent years, material from the Cambrian Konservat-Lagerstätten (cases of exceptional soft-tissue preservation), particularly that of Chengjiang, South China, has pushed the fossil record of vertebrates down into the Early Cambrian Period [525 million years ago (MYA)], indicating that vertebrate origins must lie even earlier. This fits with divergence times based on molecular data, which suggest that chordates had evolved in the Upper Precambrian (Ediacaran Period, 600–545 MYA) and that vertebrates may have diverged not long afterward.

Chordates and early vertebrates. Vertebrates belong to the phylum Chordata, which also includes the marine cephalochordates and tunicates. Vertebrates have the basic chordate characters of a notochord (a tough flexible rod that runs down the back), a dorsal longitudinal nerve cord, pharyngeal slits or gills, and sigmoidal (S-shaped) muscle blocks or myomeres along the length of their bodies. In addition, they have elements of the vertebral column and the ability to secrete phosphatic hard tissue (bone), which are characteristics not shared by the other chordates. The tunicates are small, attached filter feeders that have a mobile larval stage in which they closely resemble primitive fish. In this stage, the animal has a notochord and a longitudinal nerve cord, but these are lost once it anchors itself and metamorphoses into the adult form. The cephalochordates, or lancelets, are small eel-like filter feeders and are exemplified by amphioxus. They retain the chordate features as adults and live in sediments as mobile filter feeders. It is generally considered that the cephalochordates are the closest relatives of the vertebrates, whereas the tunicates are less closely related. Within the vertebrates, the significance of the development of the jaw is recognized by the classification of vertebrates into gnathostomes ("jaw mouths") and agnathans ("without jaws"). The jawless vertebrates are considered to be the most primitive and have been separated as the Agnatha, which includes modern lampreys, modern hagfishes, and a number of fossil groups often termed ostracoderms ("shell-skinned"). The modern forms share a similarity in their primitive eel-like appearance and their feeding habits.

The earliest fossil chordates are known from the Chengjiang and Burgess Shale Lagerstätten, which are Early Cambrian and Middle Cambrian in age, respectively (525 and 505 MYA). Of these, *Pikaia gracilens* has been recently redescribed in detail, showing it to be a leaf-shaped organism with clearly defined sigmoidal myomeres and a notochord, indicating that it is a chordate. However, there is some doubt about its cephalochordate affinities because there is no evidence of pharyngeal slits. A similar animal from the Chengjiang fauna, *Cathaymyrus*, is thought to be a cephalochordate because it does have pharyngeal slits, and both *P. gracilens* and *Cathaymyrus* have been compared with the modern cephalochordate amphioxus. Thus, there is evidence of the closest relatives of vertebrates as far down as the Early Cambrian Period.

Two additional species have been reported from Chengjiang, *Myllokunmingia* and *Haikouichthys*, both described as vertebrates. These two species show sigmoidal myomeres, a relatively complex and presumably cartilaginous skull, gill arches, a heart, and fin supports, and they have been compared to lampreys. However, the apparent presence of serially arranged gonads and a dorsal fin with forwardly tilted radials has been cited to cast doubt on their inclusion within the vertebrates. Very recently, an agnathan, *Metaspriggina*, has been described from the Burgess Shale. Although known from only two specimens, it shows complex cranial anatomy and sigmoidal myomeres.

Prior to the discovery of the Chengjiang and Burgess Shale taxa, the evidence for the presence of vertebrates in the Cambrian Period rested on the preservation of fragments of phosphatic hard tissue. Of these, *Anatolepis* has been reported from the Late Cambrian (although initially described from the Early Ordovician of Spitsbergen). It consists of microscopic plates and spines with scalelike ornamentation and a layered internal structure. Although there has been some controversy over the vertebrate attribution of this material, recent histological studies show the presence of the characteristic vertebrate hard tissue known as dentine. Although fragments of purported vertebrate hard tissue have also been reported from the Late Cambrian of Australia, there is some doubt as to their affinity because thin sections show a resemblance to some arthropod cuticles (outer skeletons).

Thus, both the soft-tissue fossil evidence and the hard-tissue fossil evidence point to the presence of vertebrates within the Cambrian and possibly as far down as the Early Cambrian.

Ediacaran fauna. The occurrence of possible fossil vertebrates as far down as the Early Cambrian has resulted in speculation about their presence in even older rocks and has focused attention on the Upper Precambrian Ediacaran Fauna. This fauna is a Precambrian (Neoproterozoic) assemblage of organisms, which existed from approximately 600 to 545 MYA (the base of the Cambrian). It was originally recognized in the Flinders Range of South Australia, but is now known from localities worldwide, and the fossils represent both shallow- and deep-water ecosystems and evoke a marine life very different from that of today's oceans. These representatives of the first metazoans or of complex life had no hard parts and are preserved mostly as disk- and frond-shaped impressions that may be up to several feet in length. They were originally thought to represent the precursors to animal families present today; that is, disk shapes were thought to represent ancestral jellyfish, and frond shapes were thought to represent sea pens (relatives of sea anemones). However, it has been proposed more recently that they were not relatives of modern organisms, but instead were an initial experiment in which a hydraulic architecture that could be swollen with fluid was used. This fauna was thought to have become extinct at the end of the Precambrian, replaced by the proliferation of organisms in the basal Cambrian that is referred to as the "Cambrian Explosion." Recent work, however, has shown that some Ediacaran organisms did survive into the Cambrian and may indeed have been ancestral to some modern forms. Recently, a possible chordate was reported from the Ediacaran of the Flinders Range, but it now appears likely that it represents a distorted specimen of a known invertebrate organism. Regardless of whether this is evidence of a Precambrian chordate, it is possible to explore the timing of the origin of chordates and the divergence of vertebrates from them by utilizing molecular data.

Molecular data. Although fossil evidence continues to refine the picture for the divergence of fossil groups, the lack of fossil data for the earliest history of chordates in general and vertebrates in particular requires that other techniques be tried. Studies on organic molecules have shown that the amino-acid sequences in DNA or proteins are accurately reproduced through the generations, with only minor changes (mutations) caused by natural selection or genetic drift, and that comparison of these sequences among a range of living taxa can be used to provide evidence of interrelationships. In addition, molecular distances (a measurement of cumulative mutations) can be used to date divergence events between taxa if the "molecular clock" is first calibrated using organisms that have well-dated divergence events (that is, organisms that have a good fossil record). Recent studies on the divergence time for cephalochordates and vertebrates using molecular data estimate that this event may have happened as far back as 750 MYA (see **illustration**), whereas the divergence between agnathans and gnathostomes may have occurred approximately 500 MYA. These divergence dates await corroboration from the fossil record. However, it is interesting to note that the date for the origin of chordates is about the same as that for the first major Neoproterozoic glaciation event, which some suggest may have led to extensive evolutionary changes as a result of the contraction of ranges and genetic isolation.

Future analyses. Given the pace at which new information about the earliest vertebrates has appeared recently, it is to be expected that new material will come to light in the next few years that will substantially clarify our understanding

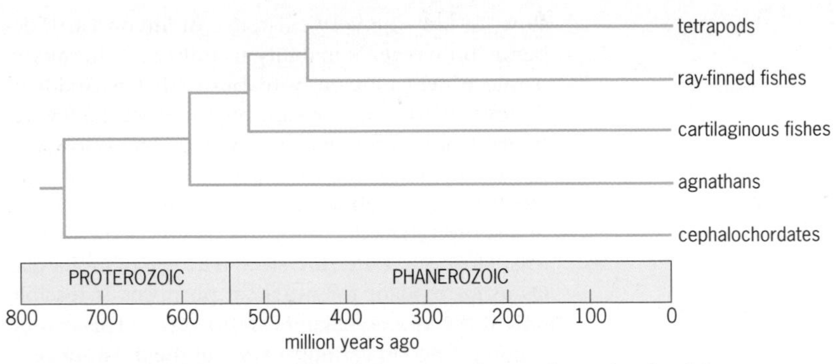

Vertebrate relationships and estimated divergence times using molecular data.

of both the timing and the causes of vertebrate origins.

For background information *see* ANIMAL EVOLUTION; CAMBRIAN; CEPHALOCHORDATA; CHORDATA; EDIACARAN BIOTA; FOSSIL; JAWLESS VERTEBRATES; MACROEVOLUTION; PALEONTOLOGY; PHYLOGENY; PRECAMBRIAN; TAPHONOMY; VERTEBRATA in the McGraw-Hill Encyclopedia of Science & Technology.

David K. Elliott

Bibliography. M. J. Benton, *Vertebrate Palaeontology*, 3d ed., Blackwell Publishing, Malden, MA, 2005; J-Y. Chen, D-Y. Huang, and C-W. Li, An early Cambrian craniate-like chordate, *Nature*, 402:518–522, 1999, DOI:10.1038/990080; S. Conway Morris, A redescription of a rare chordate, *Metaspriggina walcotti* Simonetta and Insom, from the Burgess Shale (Middle Cambrian), British Columbia, Canada, *J. Paleontol.*, 82:424–430, 2008, DOI:10.1666/06-130.1; S. Conway Morris and J-B. Caron, *Pikaia gracilens* Walcott, a stem-group chordate from the Middle Cambrian of British Columbia, *Biol. Rev. Camb. Philos. Soc.*, 87:480–512, 2012, DOI:10.1111/j.1469-185X.2012.00220.x; D. H. Erwin and J. W. Valentine, *The Cambrian Explosion: The Construction of Animal Biodiversity*, Roberts and Company, Greenwood Village, CO, 2013; X-G. Hou et al., *The Cambrian Fossils of Chengjiang China: The Flowering of Early Animal Life*, Blackwell Publishing, Malden, MA, 2007; P. Janvier, *Early Vertebrates*, Clarendon Press, Oxford, UK, 2003; P. Janvier, Vertebrate characters and the Cambrian vertebrates, *C. R. Palevol.*, 2:523–531, 2003, DOI:10.1016/j.crpv.2003.09.002; J. A. Long, *The Rise of Fishes: 500 Million Years of Evolution*, The Johns Hopkins University Press, Baltimore, 2010; D-G. Shu et al., Head and backbone of the Early Cambrian vertebrate *Haikouichthys*, *Nature*, 421:526–529, 2003, DOI:10.1038/nature01264.

Oxygen activation by metal complexes

Molecular oxygen (or dioxygen, O_2) is a powerful oxidant. It is also safe and abundant and would therefore seem to be a perfect candidate as an oxidant in both industry and the laboratory. However, oxidations of organic substrates by O_2 are extremely slow. In fact, the very existence of life on Earth depends on oxygen's inability to utilize its thermodynamic power efficiently to burn (that is, oxidize) organic matter. The reason for the discrepancy between large thermodynamic potential and slowness of the reaction lies in the fact that the overall four-electron reduction of O_2 to water must take place in a series of smaller steps, the first of which is the least facile and therefore slow. The use of molecular O_2 as an oxidant for practical purposes is feasible only if the kinetic barriers are lowered. This can be accomplished in various ways, but the most successful approach is based on the use of transition-metal complexes (TMCs) as O_2 activators. In the process of activation, TMCs interact with O_2 and generate new compounds that are more reactive than O_2 itself. TMCs cover an enormous range of substitution rates and reduction potentials, and many have multiple oxidation states that can be utilized in the reaction with oxygen or with various intermediates generated along the way to final products, as described below.

Another large area that depends critically on oxygen activation involves biological oxidations that drive metabolism, biosynthesis, and other processes. Oxygen cannot accomplish such oxidations unless it is activated by metalloenzymes, which have transition-metal ions, most frequently copper or iron, at their active sites. Activation by metalloenzymes enables the oxidation of even the most inert substrates such as hydrocarbons, including methane. The oxidations are catalytic, that is, enzymes facilitate and speed up the reaction but are not consumed in the process.

Superoxo complexes. The key components of oxygen activation by TMCs are coordination and partial reduction of O_2. For example, the reaction between O_2 and a mononuclear chromium(II) complex $[Cr^{II}]^{2+}$ can be broken down into two steps: the binding of O_2 to $[Cr^{II}]^{2+}$ in reaction (1) and the

$$[Cr^{II}]^{2+} + O_2 \rightarrow [Cr^{II}(O_2)]^{2+} \quad (1)$$

intramolecular electron transfer from chromium to O_2 in reaction (2). Roman superscripts represent the

$$[Cr^{II}(O_2)]^{2+} \rightarrow [Cr^{III}(O_2^-)]^{2+} \quad (2)$$

oxidation state of the metal. Several molecules of water (solvent) are also bound to chromium, but they are not involved in oxygen activation and will not be shown. The result is a compound in which the coordinated O_2 is reduced by one electron, and the metal is oxidized to 3+ state. This superoxo-chromium(III) ion is an example of metal-activated oxygen. The single negative charge on the O_2 unit is shown in reaction (2) to illustrate the direction of electron flow, but will be omitted hereafter. The charge can be calculated for every form of activated oxygen from the oxidation state of the metal (III in the above case) and overall charge on the compound (2+).

Superoxo complexes can oxidize organic substrates that are otherwise unreactive toward O_2, for example, phenols, reaction (3).

$$[Cr^{III}(O_2)]^{2+} + C_6H_5OH \rightarrow [Cr^{III}(HO_2)]^{2+} + C_6H_5O \quad (3)$$

Another signature reaction is that with radicals, that is, with species having an unpaired electron. Of major importance in biology and environment is the coupling with an inorganic radical, nitrogen monoxide (NO), which leads to strongly oxidizing and harmful species such as metal peroxynitrite $[Cr^{III}(OONO)]^{2+}$, nitrogen dioxide (NO_2), and metal oxo compounds such as $[Cr^{IV}(O)]^{2+}$, reaction (4).

$$[Cr^{III}(O_2)]^{2+} + NO \rightarrow [Cr^{III}(O_2NO)]^{2+} \rightarrow [Cr^{IV}(O)]^{2+} + NO_2 \quad (4)$$

Hydroperoxo complexes. $[Cr^{III}(O_2)]^{2+}$ also reacts with inorganic reductants, including $[Cr^{II}]^{2+}$, reaction (5), which leads to the hydroperoxo-

$$[Cr^{III}(O_2)]^{2+} + [Cr^{II}]^{2+} \xrightarrow{H^+} [Cr^{III}(HO_2)]^{2+} + [Cr^{III}]^{3+} \quad (5)$$

chromium(III) ion, $[Cr^{III}(HO_2)]^{2+}$, the same species that is generated in the reactions with organic reductants, such as phenol in reaction (3). Hydroperoxo compounds such as $[Cr^{III}(HO_2)]^{2+}$ also represent transition-metal-activated oxygen that reacts readily with reducing species. A noteworthy reaction for hydroperoxides is the transfer of an oxygen atom to acceptors such as phosphines to generate phosphine oxides, reaction (6). This reaction type is also

$$[Cr^{III}(HO_2)]^{2+} + PPh_3 \rightarrow [CrIII(HO)]^{2+} + OPPh_3 \quad (6)$$

observed with H_2O_2 from which hydroperoxides are formally derived by replacement of one of the protons with a reduced metal complex, for example, $[Cr^{II}]^{2+}$. As one should expect for activated oxygen, the hydroperoxides react faster than H_2O_2. Another important reaction of metal hydroperoxides is the spontaneous cleavage of the peroxo O-O bond. In the example at hand, the transformation generates a chromium(V) compound, $[CrV(O)(OH)]^{2+}$, a strong oxidant for organic and inorganic substrates, reaction (7). This type of O-O cleavage, which increases

$$[Cr^{III}(HO_2)]^{2+} \rightarrow [CrV(O)(OH)]^{2+} \quad (7)$$

the oxidation state of the metal by 2, is known as heterolytic cleavage.

A different type of O-O cleavage has been observed with some iron(III) hydroperoxo complexes. Here the cleavage is homolytic; that is, the oxidation state of the metal increases by 1 so that $Fe^{IV}(O)$ is produced. The remaining oxidizing equivalent appears as a very reactive hydroxyl radical. Both $Fe^{IV}(O)$ and hydroxyl radicals can engage in oxidation of organic substrates, reaction (8).

$$[Fe^{III}(O_2H)] \rightarrow [Fe^{IV}(O)] + HO \xrightarrow{\text{Substrate}} \text{Products} \quad (8)$$

High-valent oxo complexes. In addition to homolytic and heterolytic cleavage of metal hydroperoxo complexes described earlier, high-valent metal oxo species are also produced by reduction of hydroperoxo compounds. For example, one-electron reduction of $[Cr^{III}(HO_2)]^{2+}$ by $[Cr^{II}]^{2+}$ or by other one-electron reductants causes the O-O bond to cleave, leading to tetravalent oxochromium ion, a powerful oxidant, reaction (9). Interestingly, oxidations of

$$[Cr^{III}(HO_2)]^{2+} + [Cr^{II}]^{2+} \rightarrow [Cr^{IV}(O)]^{2+} + [Cr^{III}(OH)]^{2} \quad (9)$$

organic substrates by $[Cr^{IV}(O)]^{2+}$ often take place in two-electron steps that regenerate $[Cr^{II}]^{2+}$ and provide the basis for catalysis, as shown in **Scheme 1**. The efficiency of catalysis is, however, quite low because of destructive side reactions between various intermediates. For example, $[Cr^{II}]^{2+}$ will react with $[Cr^{IV}(O)]^{2+}$ to generate two molecules of $[Cr^{III}]^{3+}$, which is unreactive toward both O_2 and organic substrates.

Oxygen activation by other mononuclear transition-metal complexes follows the same general principles and pathways as those described for the $O_2/[Cr^{II}]^{2+}$ reaction. An overview is presented in **Scheme 2** for a generic $[LM^{II}]$ complex, where M = metal, ML = metal complex, and L = ligand system (unreactive atoms or groups bound to the metal, such as molecules of water). Electrons (e^-) can be supplied either by excess $[LM^{II}]$ or by other one-electron reductants present in the reaction mixture. RH represents organic and inorganic substrates that can be oxidized by high-valent metal oxo species.

The top two lines in Scheme 2 illustrate the binding of O_2 and subsequent electron-transfer steps to generate hydroperoxo and oxo species, followed by substrate oxidation and regeneration of $[LM^{II}]$, as shown for $[Cr^{II}]^{2+}$ in Scheme 1. The bottom branch shows heterolytic cleavage of the peroxo bond and formation of $[LM^{V}(O)]$, which in the next step oxidizes the substrate. The metal-derived product, $[LM^{III}]$, requires an additional electron to generate $[LM^{II}]$ and close the catalytic cycle.

In several well-documented examples, including the $[Cr^{II}]^{2+}$ reaction, all of the key intermediates and transformations in Scheme 2 have been observed directly. More frequently, however, some or all of the intermediates are too short-lived to be observed, and their involvement is surmised on the basis of indirect observations and final reaction outcome.

Strikingly, the interaction of dioxygen with simple transition-metal complexes involves the same fundamental chemistry that is observed with metalloenzymes, which feature transition-metal ions at their active sites. This point is illustrated, for example, by the chemistry of $[Cr^{II}]^{2+}$ (molecular weight 160) described earlier and a family of cytochrome (Cyt) P450 enzymes (molecular weight of about 3×10^5),

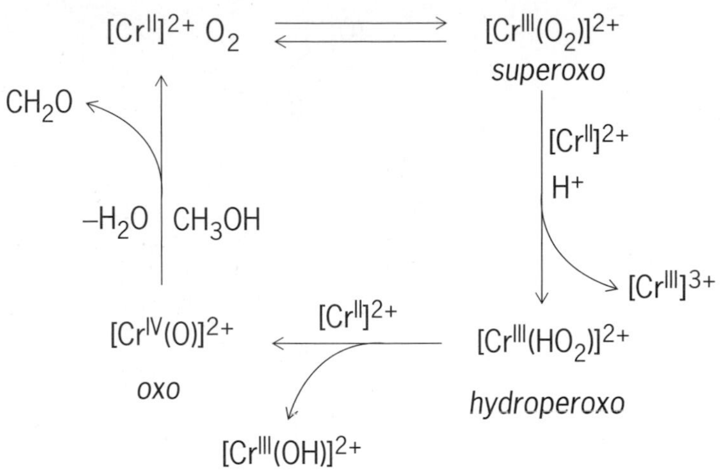

Scheme 1. $[Cr^{II}]^{2+}$-catalyzed oxidation of methanol to formaldehyde with O_2.

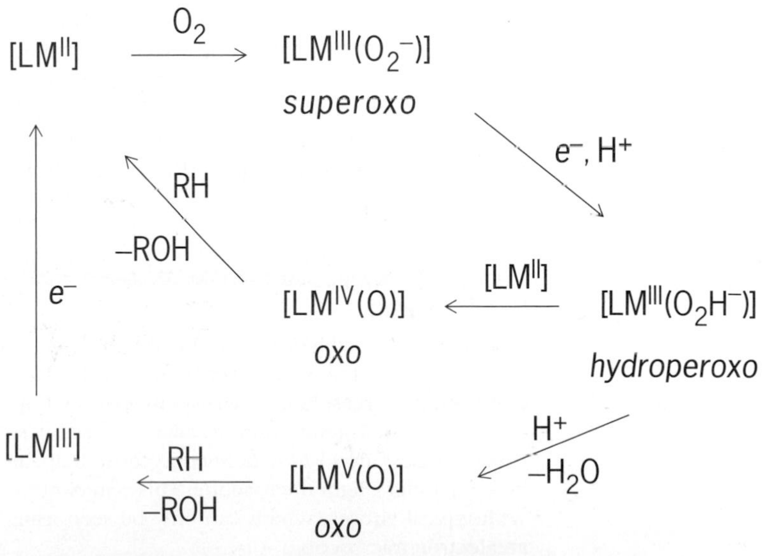

which have an iron porphyrin complex at the active site. (Porphyrins are large aromatic molecules that bind to metals through four nitrogen atoms.) Cyt P450 enzymes catalyze oxidations with molecular oxygen in biological processes, including detoxification, biosynthesis, and metabolism. Catalytic oxidations with O_2/Cyt P450, starting at the Fe^{II} state, follow Scheme 2, just as $[Cr^{II}]^{2+}$ does. Nonetheless, the catalytic performance of Cyt P450 is infinitely better than that of $[Cr^{II}]^{2+}$. Obviously, the complex structure of the enzyme, and the chemical composition and geometry around iron, serve to optimize the rates and thermodynamics of the individual steps of O_2 reduction and to control accessibility of active sites; however, they do not introduce novel chemical reactivity.

As expected, activation of oxygen by binuclear and larger metal complexes involves a greater number and different types of intermediates, but the

Scheme 2. Simplified reaction scheme for oxygen activation by a generic mononuclear $[LM]^{II}$ complex leading to catalytic oxidation of substrate. Roman numerals in superscript denote formal oxidation states of the metal.

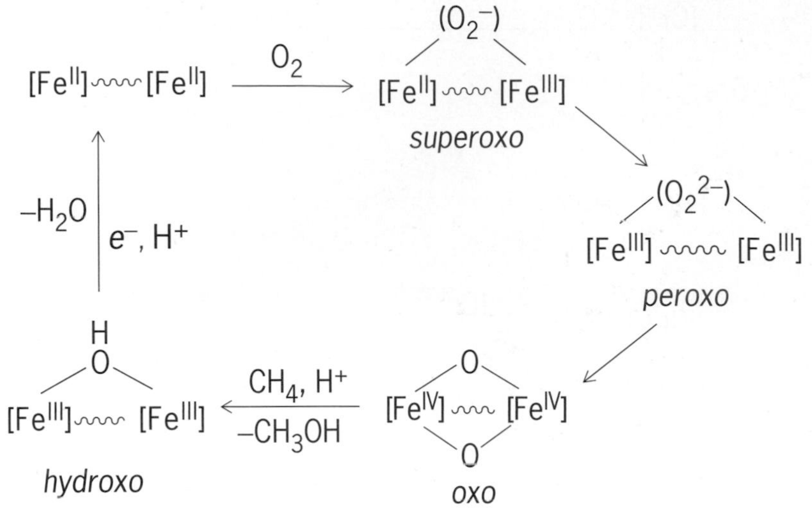

$[Fe^{II}] \sim [Fe^{II}] \xrightarrow{O_2} [Fe^{II}] \sim [Fe^{III}]$ (O_2^-) *superoxo* $\rightarrow$ $[Fe^{III}] \sim [Fe^{III}]$ (O_2^{2-}) *peroxo* $\rightarrow$ $[Fe^{IV}] \sim [Fe^{IV}]$ (O, O) *oxo* $\xrightarrow[-CH_3OH]{CH_4,\ H^+}$ $[Fe^{III}] \sim [Fe^{III}]$ (OH) *hydroxo* $\xrightarrow[-H_2O]{e^-,\ H^+}$ $[Fe^{II}] \sim [Fe^{II}]$

Scheme 3. Simplified reaction scheme for oxygen activation by a dinuclear Fe^{II} complex as a model for an active site in soluble methane monooxygenase (MMO). The two iron atoms at the active site in MMO are bridged by carboxylate and hydroxyl groups shown as wavy lines.

fundamental chemistry remains quite similar to that at mononuclear metal centers. This is illustrated in **Scheme 3** for a dinuclear iron complex serving as a model for an active site in soluble methane monooxygenase (MMO), an enzyme that catalyzes oxidation of methane to methanol in bacteria.

For background information *see* BIOLOGICAL OXIDATION; BIOINORGANIC CHEMISTRY; CATALYSIS; ELECTRON-TRANSFER REACTION; ENZYME; FREE RADICAL; HYDROXYL; LIGAND; NITROGEN OXIDES; ORGANOMETALLIC COMPOUND; OXIDATION PROCESS; OXIDATION-REDUCTION; OXYGEN; PEROXIDE; PORPHYRIN in the McGraw-Hill Encyclopedia of Science & Technology. Andreja Bakac

Bibliography. D. R. Barton et al., *The Activation of Dioxygen and Homogeneous Catalytic Oxidation*, Springer, New York, 1993; I. Bertini et al., *Biological Inorganic Chemistry, Structure & Reactivity*, University Science Books, Sausalito, CA, 2007; S. J. Lippard and J. M. Berg, *Principles of Bioinorganic Chemistry*, University Science Books, Sausalito, CA, 1994; L. I. Simandi, *Catalytic Activation of Dioxygen by Metal Complexes*, Springer, New York, 1992.

Pandoravirus

Pandoraviruses constitute a family of ameba-infecting viruses that were discovered in 2013. They comprise the largest known viruses in terms of both particle size and gene content. Giant viruses (also called giruses) are loosely defined as those that can be seen under a light microscope. This is in contrast with typical viruses, which can only be seen using an electron microscope.

History of giant viruses. The viral nature of the first giant virus, named *Mimivirus*, was discovered in 2003 in Marseille, France, after being first isolated in 1992 in Bradford, United Kingdom. However, it was mistaken for a bacterium for 10 years. *Mimivirus*, like all other known giant viruses, infects a special genus of amebas, called *Acanthamoeba*, which is ubiquitous in humid environments. The entire icosahedral virus particle (virion) was approximately 0.7 micrometers (μm) in diameter (which is about seven times larger than typical viruses) and contained a double-stranded DNA genome of 1.18 million nucleotides coding for 979 proteins. This was the first known occurrence of a virus possessing more genes than small bacteria, leading to an evolutionary paradox because many viruses only carry a dozen genes that are sufficient to infect cells and multiply, causing disease. In 2010, another giant virus relative, called *Megavirus chilensis*, was discovered off the coast of Chile. This new giant virus was found to carry a genome encoding 1120 proteins. As additional *Mimivirus* and *Megavirus* relatives of similar dimensions were subsequently found in various environments, there was a growing consensus that no virus much bigger than *Megavirus* would ever be found.

Pandoravirus particle structure. An entirely new type of giant virus was revealed in 2013, when the two first representatives of the genus *Pandoravirus*, named *P. salinus* and *P. dulcis*, were simultaneously described in the same publication. *Pandoravirus salinus* was isolated in the Pacific Ocean from a sediment sample taken at the mouth of the Tunquen River, Chile, whereas *P. dulcis* was isolated from a shallow freshwater pond located in the middle of the La Trobe University campus in Melbourne, Australia. The two virions (entire virus particles) are morphologically identical, exhibiting a distinctive amphora-like shape that is 1.2 μm in length and 0.5 μm in diameter, making them easy to see under a light microscope (**Fig. 1**). At one apex, the particle is closed by a structure similar to a plug that is removed at the time of infection to allow the interior of the particle (including the DNA genome) to be released into the ameba cell (**Fig. 2**). Pandoraviruses are named after Pandora, who was given a "box" full of mysteries by the gods according to

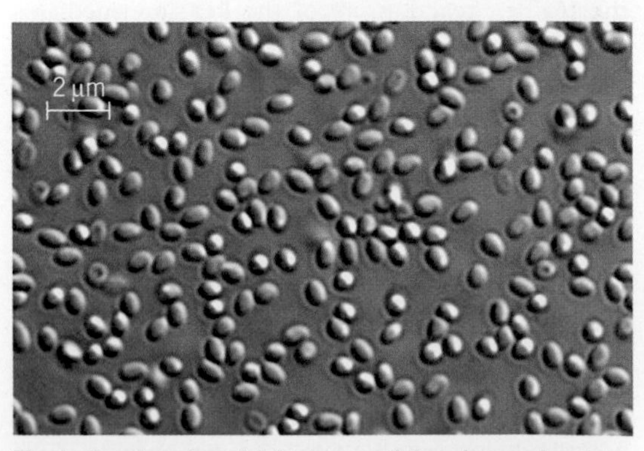

Fig. 1. ***Pandoravirus dulcis*** **virus particles observed under a light microscope (differential interference contrast).**

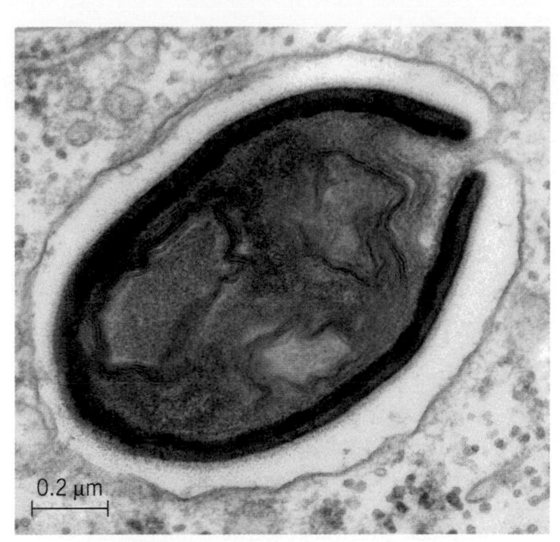

Fig. 2. A *Pandoravirus salinus* particle releasing its contents into the *Acanthamoeba* cytoplasm (thin section, electron microscopy).

Greek mythology. (Although mistranslated as a box, the container in the original myth was in fact a large, human-sized amphora.) A proteomic analysis of the virus particles has indicated that they are made of more than 200 gene products, as for *Mimivirus* particles. However, in contrast with *Mimivirus*, no component of the virus-encoded transcription machinery was found in the *Pandoravirus* particles (see the section on *Pandoravirus* replication below).

Genome size and gene content. In addition to their unique amphora-like shape (note that most large DNA viruses are pseudo-icosahedral or nearly spherical in shape), the pandoraviruses have provided a number of other surprises. The genomes of *P. salinus* and *P. dulcis* were found to be double-stranded DNA molecules of 2.77 million base pairs and 1.95 million base pairs, respectively. These genomes were predicted to encode 2556 proteins for *P. salinus* and 1502 proteins for *P. dulcis*. These gene contents are not only much bigger than that of *Megavirus*, and are comparable to those of many bacteria, but they exceed the number of genes of parasitic eukaryotic microorganisms such as *Encephalitozoon* species. Another surprising feature of pandoraviruses is the extremely low proportion of their proteins (16%) sharing a significant sequence similarity with proteins of any other organism, including other viruses. Together with their unique particle shape and size and their huge gene content, this lack of sequence similarity is making the pandoraviruses into one of the most alien microorganisms ever discovered.

Pandoravirus replication. In contrast with the previously discovered giant viruses of the *Mimivirus* and *Megavirus* family, in which replication entirely occurs in the *Acanthamoeba* cytoplasm leaving the cell nucleus intact, the pandoraviruses appear to be highly dependent on the host's nuclear functions for their replication. Approximately 2–4 h after the content of the particle has been released in the cytoplasm, the host nucleus undergoes a drastic reorganization, losing its spherical appearance and then its nucleolus, before dissolving entirely. Then, 8–10 h postinfection, as the ameba cells become round and lose their adherence, new pandoravirus particles appear at the periphery of the region formerly occupied by the nucleus. These observations suggest that the pandoraviruses, despite their much larger genomes, are more dependent on the host for their cellular and nuclear functions than mimiviruses and megaviruses. Data from proteome analyses confirmed this view by showing the presence of spliceosomal introns (spliced-out intervening genomic sequences) in an estimated 10% of *Pandoravirus* genes. The importation of the *Pandoravirus* genome within the cell nucleus is thus likely to be an obligatory step for its successful expression. The known pandoraviruses are nonpathogenic to animals or humans and are unable to replicate in animal macrophages in vitro.

Pandoravirus relationship with other giant DNA viruses. As already suggested by the small number of genes encoding proteins with recognizable homologues in other organisms, the phylogenetic tree computed from DNA polymerases (enzymes that link nucleotides together to form polynucleotide chains) places the pandoraviruses in their own clade, outside of all the previously defined families of large DNA viruses infecting eukaryotes. The use of a handful of less ubiquitously conserved viral proteins also found in pandoraviruses generates a faint phylogenetic signal, suggesting a slight affinity with an isolated group of large viruses infecting a specific type of unicellular algae, called the coccolithophores. The analysis of more *Pandoravirus* relatives will eventually shed further light on their evolutionary origin and their relationship with the rest of the tree of life (phylogenetic tree).

For background information *see* AMEBA; CELL (BIOLOGY); CELL NUCLEUS; CYTOPLASM; DEOXYRIBONUCLEIC ACID (DNA); GENE; GENOMICS; PHYLOGENY; VIRUS; VIRUS CLASSIFICATION in the McGraw-Hill Encyclopedia of Science & Technology.

Jean-Michel Claverie; Chantal Abergel

Bibliography. D. Arslan et al., Distant *Mimivirus* relative with a larger genome highlights the fundamental features of Megaviridae, *Proc. Natl. Acad. Sci. USA*, 108(42):17486–17491, 2011, DOI:10.1073/pnas.1110889108; J-M. Claverie and C. Abergel, *Mimivirus* and its virophage, *Annu. Rev. Genet.*, 43:49–66, 2009, DOI:10.1146/annurev-genet-102108-134255; J-M. Claverie and C. Abergel, Open questions about giant viruses, *Adv. Virus Res.*, 85:25–56, 2013, DOI:10.1016/B978-0-12-408116-1.00002-1; B. La Scola et al., A giant virus in amoebae, *Science*, 299(5615):2033, 2003, DOI:10.1126/science.1081867; N. Philippe et al., Pandoraviruses: Amoeba viruses with genomes up to 2.5 Mb reaching that of parasitic eukaryotes, *Science*, 341(6143):281–286, 2013, DOI:10.1126/science.1239181.

Papaya

Papaya (*Carica papaya*) is a semiwoody, usually single-stemmed, herbaceous plant (**Fig. 1**). It originated along the Caribbean lowlands of Mesoamerica and is now extensively and exclusively distributed in tropical and subtropical regions because of its sensitivity to chilling temperatures. Papaya is mainly known for its climacteric fruits (that is, they continue to ripen after harvest), which are produced constantly during the plant's adult life and broadly commercialized; and also, to a lesser degree, for its latex, which contains enzymes having industrial uses (for example, papain, which is employed as an ingredient in meat-tenderizing, brewing, pharmaceuticals, and cosmetics), and many other important compounds.

Papaya production. Papaya production ranks fourth among tropical fruits worldwide, after banana, mango, and pineapple, with significant growth in recent years, particularly in India. This country is the leading producer of papaya (nearly 40% of the world's production), followed by Brazil (17.5%), Indonesia (6%), Nigeria (6%), Mexico (6%), Ethiopia (2%), the Democratic Republic of the Congo (2%), Colombia (2%), Thailand (2%), and Guatemala (2%).

Despite the increase in the international trade of this fruit observed in recent years, only a small fraction of the total world production (approximately 3%) is exported. Mexico, Brazil, and Belize accounted for 63% of the global trade between 2007 and 2009, with Mexico providing approximately 41% of the trade and Brazil and Belize providing approximately 11% each. Because of the extremely high domestic demand for papaya, India exported less than 1% of its total production in 2009.

Fig. 1. Adult papaya plant of the "Pococí" F1 hybrid.

Fig. 2. Ripe fruit and seeds of the "Pococí" red-fleshed papaya F1 hybrid.

Plant development and cultivation. The development of the papaya plant occurs very quickly. The juvenile phase (from seed germination to flowering) lasts between 3 and 8 months, and harvesting requires between 9 and 15 months. In natural conditions, the plant can live up to 20 years. However, the commercial life span of cultivated papaya is usually restricted to 2–3 years so as to avoid excessive plant height and because of pathological constraints.

Cultivated papaya plants have three possible sexual forms: female, male, and hermaphroditic. Hermaphroditic and female plants produce their flowers in groups of 2 or 3 to 15, whereas the male plants produce very long inflorescences [flower clusters segregated from other flowers on the same plant, together with their associated stems and bracts (reduced leaves)], containing dozens or even hundreds of flowers. Although recent evidence suggests that hawkmoths are mainly responsible for pollination, papaya flowers are also visited by several beetles, skipper butterflies, bees, flies, and hummingbirds. Wind pollination has also been cited.

Fruit characteristics. Papaya fruits are very diverse in size and shape. Fruit weight can range from less than 100 g (3.5 oz) to more than 10 kg (22 lb) in extreme cases. The small-sized Solo-type papayas, also known as Hawaiian papayas, have a weight range of 0.5–1.0 kg (1.1–2.2 lb) per fruit, whereas the large-sized papayas, also denominated Mexican papayas, can weigh up to 4.5 kg (9.9 lb) each. Fruits from hermaphroditic plants are elongated (varying from cylindrical to pear-shaped), and female fruits are spherical. Hermaphroditic fruits are usually preferred over female fruits for commercialization in international markets because the cavity of the former is usually smaller and their shape allows more efficient arrangements in packaging. The large cavity hosts the seeds, making up most of the fruit volume (**Fig. 2**). Well-pollinated fruits can bear more than 600 black seeds.

As they ripen, papaya fruits change in color. The color of ripe fruits may vary from yellow to salmon red, according to the carotenoid profile of the fruits. Carotenoids are common pigments of yellow, orange, or red hues that serve in both light absorption and the protection against too much light in photosynthesis. The most important carotenoids found in papaya fruits are β-carotene and β-cryptoxanthin (in red and yellow papayas), and lycopene (exclusively in red papayas). These carotenoids have a high bioavailability (that is, a high level of absorption into the body's systemic circulation after consumption) that is between 2.6 and 3 times greater than that supplied by carrots and tomatoes (the traditional sources of these compounds). Higher bioavailability could be related to the liquid-crystalline deposition of β-carotene and the storage of lycopene in very small crystalloids in papayas. Because of the former, papaya may be a readily available dietary source of provitamin A, providing a means to diminish the frequency of vitamin A insufficiency in many individuals in subtropical and tropical developing countries.

The nutritional requirements of papaya plants are high. Nitrogen, phosphorus, and potassium are extracted from the soil in high amounts and are taken up mainly by the fruits (which require 20–30% of these extracted nutrients). This is reflected when considering the nutritional contribution of this fruit to the human diet. Although the nutritional composition of a papaya fruit can vary widely, depending on the type of cultivar (cultivated variety), maturity, and ripening, fresh papaya fruits [100 g (3.5 oz)] on average supply 179 kJ of energy [mainly from 11 g (0.385 oz) carbohydrates (mostly sucrose, glucose, and fructose), 1.7 g (0.06 oz) fiber, and less than 1 g (0.035 oz) protein or fat], and they supply 9–12% of the Dietary Reference Intake (DRI) for folate, 3–7% of the DRI for potassium, 6–8% of the DRI for magnesium, and 9% of the DRI for copper.

Pathology. Probably the main threat to global papaya production is the papaya ringspot virus (PRSV). This disease has drastically reduced fruit yield, size, and quality in several countries, causing, in extreme cases, devastation of entire papaya plantations. Because of this disease, production of this fruit in Hawaii almost vanished in the 1990s. Countermeasures to combat this disease included the development of the genetically modified (GM) papaya cultivars, SunUp and Rainbow, introduced in 1998. Additional GM cultivars have been released or are in the process of being released in China, Jamaica, Taiwan, Thailand, Australia, the Philippines, Malaysia, and Vietnam.

Postharvest losses, of up to 75%, are mainly caused by fungal diseases (primarily anthracnose and stem-end rot), physiological disorders, mechanical damages, and chilling injuries (low-temperature injuries in the absence of freezing). Most of these detrimental factors, including diseases, can be overcome by cultural practices, including adequate harvesting time, careful oversight throughout the handling chain, and proper management to avoid exposure to chilling environments [below 10°C (50°F)], low relative humidity (which can lead to excessive weight loss), and direct sunlight [which can cause fruit scald (skin damage)]. In addition, postharvest diseases can be efficiently managed with hot water and fungicide treatments.

Future outlook. Because of the papaya's high yield, nutritional value, functional properties, and year-round fruit production, the importance of this crop around the world is undeniable. In particular, fruit consumption is likely to increase in the future, especially in developed countries, as a result of the latest findings regarding its functional properties.

For background information *see* AGRICULTURAL SCIENCE (PLANT); BIOTECHNOLOGY; CAROTENOIDS; ENZYME; FOOD MANUFACTURING; FRUIT; FRUIT, TREE; GENETICALLY ENGINEERED PLANTS; HORTICULTURAL CROPS; PHYSIOLOGICAL ECOLOGY (PLANT); PLANT PATHOLOGY; POLLINATION; SEED in the McGraw-Hill Encyclopedia of Science & Technology.

Víctor M. Jiménez

Bibliography. E. A. Evans and F. H. Ballen, *An overview of global papaya production, trade, and consumption* (FE 913), Food and Resource Economics Department, Florida Cooperative Extension Service, Institute of Food and Agricultural Sciences, University of Florida, Gainesville, 2012; V. M. Jiménez, E. Mora-Newcomer, and M. V. Gutiérrez-Soto, Biology of the papaya plant, pp. 17–33, in R. Ming and P. H. Moore (eds.), *Genetics and Genomics of Papaya*, Springer, New York, 2014; R. E. Paull and N. J. Chen, Recent advances in postharvest management of papaya, *Acta Horticulturae (ISHS)*, 1024:321–327, 2014; R. M. Schweiggert et al., Carotenoids are more bioavailable from papaya than from tomato and carrot in humans: A randomised cross-over study, *Br. J. Nutr.*, 111(3):490–498, 2014, DOI:10.1017/S0007114513002596; M. M. Wall and S. Tripathi, Papaya nutritional analysis, pp. 377–390, in R. Ming and P. H. Moore (eds.), *Genetics and Genomics of Papaya*, Springer, New York, 2014.

Phorid flies

Phorids, also known as humpbacked flies or scuttle flies, are small insects [having body lengths of 0.4–6 mm (0.016–0.24 in.)] and are found worldwide. There are approximately 4000 described species of the family Phoridae in the world, but this is estimated to be only 10% of the true diversity. Revisionary work on phorids continually encounters new species, even in relatively well-sampled regions such as Europe. The tropics, especially the New World tropics, are the most species-rich areas for these flies, but they are found in virtually all terrestrial habitats, including above the tree line in the Arctic. The larvae of some phorid flies are aquatic.

History and diversity. Phorids have a long fossil history, extending back 125 million years to the Cretaceous Period in Burmese and other ambers. They are extremely diverse in the younger (44 million-year-old) European Baltic amber and the Miocene (20 million-year-old) Dominican amber. All Cretaceous phorid flies appear to be from extinct genera, whereas the Baltic amber fauna is a mixture of extinct and extant genera. The Miocene fauna appears to be completely modern at the genus level.

Phorids are commonly referred to as the most biologically diverse family of insects because their larvae are known to be scavengers, herbivores, fungivores, predators, parasites, and parasitoids. The parasitoid larvae live as parasites that eventually kill their hosts. One species, *Megaselia scalaris*, is found worldwide and is one of the most versatile scavengers in the insect world. Because of this species, many entomologists mistake all phorids as being similarly undiscerning feeders. Actually, the lifestyle of only about 10% of the described phorid flies is known, and more than half of these are parasitoids, with most of them attacking ants, but others attacking bees, millipedes, beetles, other dipterans, scale insects, and a variety of other hosts.

Parasitoid lifestyle. The most species-rich group of parasitoids is the genus *Apocephalus*, commonly referred to as the "ant-decapitating flies." These flies dart down to their hosts, usually injecting a single egg into the host's body by piercing the membrane between segments with their sharp ovipositors (specialized egg-depositing structures). Often, the site of egg laying is the head; however, if this is not the case, the larva migrates to the head internally through the ant's body. Once it is located in the head, the larva feeds on the huge mandibular muscles that dominate the head capsule of the ant host. Eventually, this feeding leads to the loss of the host ant's head, sometimes while the ant's body still continues moving.

Many other phorid fly genera also pursue this lifestyle, including those that are parasitoids of invasive ants. Most notable are the species of the genus *Pseudacteon*, which parasitize small ants such as fire ants and Argentine ants. Because their hosts can be problems in areas where they are accidentally introduced, some species of *Pseudacteon* are being used to attempt to control the populations of these ants.

Other phorid species are parasitoids of leafcutter ants in South America and Central America. These ants can be agricultural pests, so there have been some studies on the behavior of parasitoid flies such as *Eibesfeldtphora*, *Myrmosicarius*, and *Apocephalus* that attack these ants. However, it is unclear whether the effects of these native flies can be augmented to provide some relief from the activities of these pests.

Variety of body forms in adult female Phoridae. Clockwise from upper left: a *Thaumatoxena* species; a termitoxeniine; and a *Vestigipoda* species. (*Images courtesy of the Entomological Society of America*)

The smallest fly in the world is an ant-parasitizing phorid. This species is known as *Euryplatea nanaknihali* and is found in Thailand. Adult females are only 0.4 mm (0.016 in.) in body length. Another (larger) species of this genus is known from Africa, where it attacks ants of the genus *Crematogaster*.

One species of parasitoid phorid fly has become of recent concern to apiculture (beekeeping) because it has been found to attack domesticated honeybees in North America. This species, *Apocephalus borealis*, has been dubbed the "zombie fly" because parasitized hosts appear to be disoriented and are unable to fly properly. They are also attracted to lights at night, which is an unusual behavior for honeybees. The native hosts of *A. borealis* appear to be bumblebees and yellow jacket wasps. In South America and Central America, beekeepers are further plagued by the phorid fly, *Pseudohypocera kerteszi*, which can be a pest in both honeybee colonies and those of native stingless bees that are reared for their honey. Some species of bee-killing flies, genus *Melaloncha*, also attack honeybees in the tropical parts of South America and Central America.

Although phorid flies are small, their parasitic behavior can have strong effects on the populations of their hosts. Parasitism rates can be as high as 80% in some firefly populations in eastern North America, causing a late-season collapse in their numbers. Mating aggregations of male stingless bees are attacked by a species of *Apocephalus*, with parasitism rates reaching nearly 50%.

Other phorid types. Some phorids can cause economically significant damage. Species have been reared from corn, chickpeas, and mushrooms, but they normally do not achieve pest status.

In addition to their biological diversity, phorid flies exhibit considerable structural diversity (see **illustration**). For example, many female phorids lack wings and have reduced eyes and reduced sclerotized structures, thus appearing more membranous than most flies. Other phorids have become extremely defensive in their body form, looking like small beetles rather than flies (for example, *Thaumatoxena*; see illustration).

One exceptionally unusual group is the Termitoxeniinae (see illustration), whose members live in the nests of fungus-gardening Old World termites. The females emerge with shriveled abdomens; however, over a period of hours, they develop greatly enlarged structures that resemble termite nymphs on their backs. The termites interact with the structures, allowing the flies to integrate into the colony.

The most structurally unusual phorids, and perhaps the most unusual flies in the world, are the wingless, legless females of the genus *Vestigipoda* (see illustration) that live in army ant colonies in Southeast Asia. The body of the female flies closely resembles that of the ant larvae, seemingly with a phorid head and thorax attached. The ants care for the female flies as they do for their own larvae, feeding them and moving them as the colony goes about its migrations.

Classification. The classification of phorid flies is currently in flux. Based on new research, using both molecular and new morphological characters, it appears that the earlier hypotheses about a sister-group relationship between the southern hemisphere group Sciadocerinae and the rest of the Phoridae were correct. Genus *Chonocephalus* appears to be a sister group to the rest of the family, with Termitoxeniinae being the next branch recognized. All other phorids are placed in the large subfamilies Metopininae and Phorinae.

For background information *see* ARTHROPODA; DIPTERA; ECOLOGY; ENTOMOLOGY, ECONOMIC; HYMENOPTERA; INSECT CONTROL, BIOLOGICAL; INSECT PHYSIOLOGY; INSECTA; PARASITOLOGY; PREDATOR-PREY INTERACTIONS in the McGraw-Hill Encyclopedia of Science & Technology. Brian V. Brown

Bibliography. B. V. Brown, Phoridae, in B. V. Brown et al. (eds.), *Manual of Central American Diptera*, vol. 2, pp. 725–761, NRC Research Press, Ottawa, 2010; R. H. L. Disney, Natural history of the scuttle fly, *Megaselia scalaris*, *Annu. Rev. Entomol.*, 53:39–60, 2008, DOI:10.1146/annurev.ento.53.103106.093415; R. H. L. Disney, *Scuttle Flies: The Phoridae*, Chapman and Hall, London, 1994; L. W. Morrison, Biological control of *Solenopsis* fire ants by *Pseudacteon* parasitoids: Theory and practice, *Psyche*, vol. 2012, no. 424817, 2012, DOI:10.1155/2012/424817; L. W. Morrison, Biology of *Pseudacteon* (Diptera: Phoridae) ant parasitoids and their potential to control imported *Solenopsis* fire ants, *Recent Res. Dev. Entomol.*, 3:1-13, 2000.

Photoacoustic imaging in biomedicine

The photoacoustic effect, or, as it is frequently called, the optoacoustic effect, refers to the production of sound by the absorption of light. The mechanism of sound generation is based on the principle that virtually all materials, whether in the gas, liquid, or solid phase, expand when heated. The expansion initiated by the absorption of light is a mechanical motion that exerts a force on the surrounding medium, with the result that a sound wave that can be detected with high sensitivity is launched. There are numerous applications of the photoacoustic effect; however, in this article, the application of this effect to imaging human and animal tissue and, in particular, applications in biomedicine, such as the detection of cancers, will be discussed.

Photoacoustic effect. In most cases, the properties of the photoacoustic effect are described by the wave equation (1), where p is the photoacoustic

$$\left(\nabla^2 - \frac{1}{c^2}\frac{\partial^2}{\partial t^2}\right) p = -\frac{\beta}{C_P}\frac{\partial H(\mathbf{x}, t)}{\partial t} \quad (1)$$

pressure, c is the sound speed, β is the thermal expansion coefficient, C_P is the specific heat capacity, $\mathbf{x}$ is the coordinate, t is the time, and H is the energy per unit volume and time deposited in the absorbing medium. For a radiation source, H is given by Eq. (2),

$$H = \tilde{\alpha}(\mathbf{x}) I(\mathbf{x}, t) \quad (2)$$

where $\tilde{\alpha}(\mathbf{x})$ is the absorption coefficient and $I(\mathbf{x}, t)$ is the intensity of the light beam. It can be seen from Eq. (1) that any variation in the light intensity of a beam of radiation directed at an absorbing object gives a nonzero forcing term that results in sound production.

Alexander Graham Bell discovered the photoacoustic effect in 1881. The method he used for producing the photoacoustic effect was based on the modulation of a continuous beam of sunlight by a mechanical chopping wheel, with the modulated light being directed onto a solid or liquid surface. Sound was recorded at first by using the human ear, and later by using sensitive microphones. The invention of infrared lasers in the 1970s resulted in wide interest in the photoacoustic effect, since the high power typical of laser sources gives a far larger photoacoustic signal than sunlight or other broadband light sources, making the method a sensitive detector of gases. For trace gas detection, a laser beam with a wavelength, typically in the infrared, tuned to an absorption line of the molecule of interest is modulated by a chopping wheel and directed into an acoustic resonator to produce sound. The chopping wheel gives a time dependence to $I(\mathbf{x}, t)$ and, in turn, to $H(\mathbf{x}, t)$, so that the forcing term in Eq. (1) [the right-hand side of this equation] is nonzero and a sound wave is produced. Typically, the modulation frequency of the laser beam is between 10 Hz and 10 kHz. The output power of a typical infrared gas laser can be a few watts, with a linewidth of only 10 Hz. Thus, owing to both the high spectral intensity of the laser (that is, essentially its high power in a narrow wavelength region) and the remarkable sensitivity of the microphone, the photoacoustic effect has been shown to be capable of detecting trace gases at the parts-per-billion level.

It is obvious from the forcing term in Eq. (1) that other means of exciting the photoacoustic effect with, for instance, a pulsed laser beam will also give a time dependence to $H(\mathbf{x}, t)$. In the case of fluids, a pulsed laser beam can be directed into an object inside a fluid or turbid medium and the sound detected with a hydrophone. With pulsed lasers, it can be shown from the solution to Eq. (1) that the photoacoustic effect gives information about the irradiated object; that is, recordings of the temporal profiles of ultrasonic waves emitted from bodies with simple geometries give information about the bodies, such as their geometries, dimensions, and acoustic properties.

Photoacoustic imaging. If an absorbing object is located within tissue irradiated by a pulsed laser, the photoacoustic effect can act as a probe of the spatial dependence of the absorption of the object. In an experiment, if the sound speed of the transparent tissue is known and the time of arrival of the ultrasonic wave is recorded following the firing of the laser, it is possible to determine the distance between the absorber and the transducer. Of course, when imaging objects within turbid media or tissue, the optical radiation is strongly scattered. However, the directionality of the optical radiation is not relevant to sound production, so that even though the light beam is diffuse when it reaches the absorbing object, a photoacoustic effect is nevertheless generated. As the photoacoustic wave propagates out from the absorbing region of the irradiated body, it experiences far less scattering and attenuation than does the optical radiation, at least for acoustic waves with frequencies below a few megahertz. Thus, the acoustic wave calculated from Eq. (1) will be relatively unchanged as it passes through tissue to be recorded in an externally placed pressure transducer.

It is possible to take advantage of what is essentially a mapping of the time profile of the acoustic wave into the space dependence of the absorber to design a three-dimensional imaging system by placing transducers in an array around the irradiated body or moving the transducer to a number of positions around the body as a laser is fired, in order to record the emitted ultrasonic waveform at numerous angles from the irradiated body. After the information from the transducers is recorded and stored in a computer, a construction of the image of the optical absorption in space can be carried out based on a mathematical inversion of Eq. (1). The inversion gives the spatial dependence of the region probed, that is, $\tilde{\alpha}(\mathbf{x})$, from the time profile of the pressure wave. **Figure 1** presents three different images of a mouse taken with two different laser wavelengths and different image processing, demonstrating the unique capability to display and emphasize certain tissue structures that are of interest and to hide other tissues that are not of interest.

Photoacoustic imaging is currently being evaluated as a modality for detecting breast tumors. A characteristic of these tumors is that they force the body to grow a network of blood vessels to supply nutrients to the rapidly growing cells within the tumor. As a result of the presence of these blood vessels, blood becomes a marker for the presence of tumors. Fortunately, breast tissue has an absorption minimum in the red to near infrared spectral range, so that there is significant contrast between the blood at the site of the tumor and the surrounding tissue. In this wavelength range, tissue has a minimum in its absorption, so the penetration depth of the radiation can be several centimeters. The method for breast cancer detection involves placing the patient so that the breast lies at the top of a water-filled chamber containing the transducer array. The breast is irradiated by a pulsed laser, and the ultrasound generated through the photoacoustic effect is recorded by the transducer array. The optical radiation dose is quite small and is, of course, far safer than x-ray mammography, which requires

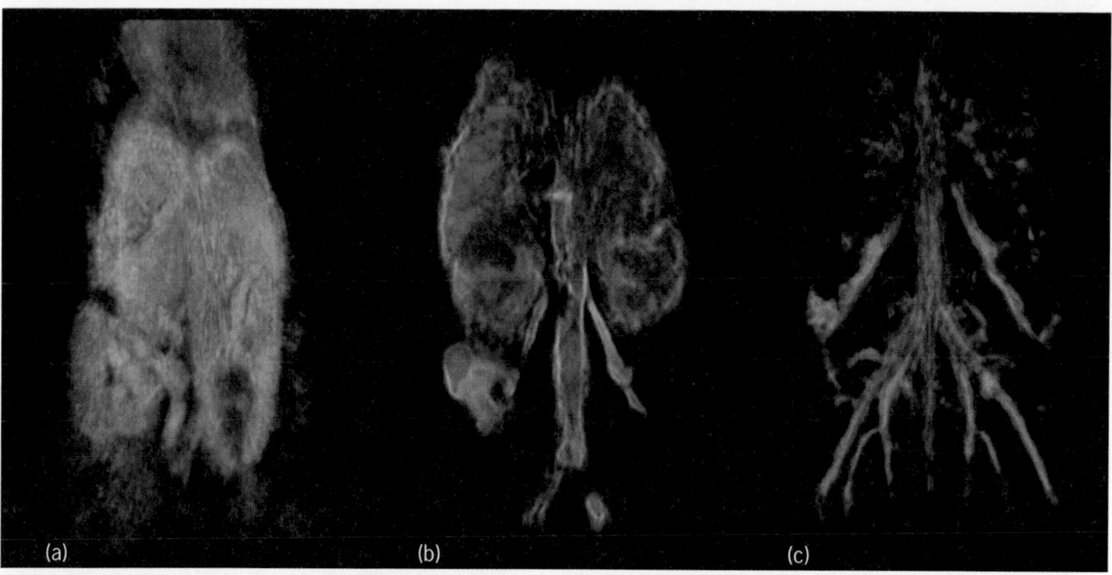

Fig. 1. Images of a live mouse taken with different laser excitation wavelengths. (*a*) Raw 3D image, acquired using a wavelength of 700 nm. (*b*) A slice of image (*a*), showing the kidneys. (*c*) An image acquired with a wavelength of 1064 nm to enhance the visibility of arteries, and further processed with a filter based on a curvelet transform to remove the low frequencies associated with organs, so that only the vasculature is visible. (*From R. Su et al., Laser optoacoustic tomography: Towards new technology for biomedical diagnostics, Nucl. Instrum. Meth. Phys. Res. A, 720:58–61, 2013, DOI:10.1016/j.nima.2012.12.035*)

the use of ionizing radiation. The present stage of experimentation with breast tumor detection is that clinical trials have been initiated. Images presented by researchers in this field show spectacular contrast in comparison with x-ray images in nonfatty breast tissue.

The salient feature of photoacoustic imaging is that its contrast mechanism is based on optical absorption. This feature has been used to probe the degree of oxygen saturation of blood, which has a strong dependence of its absorption coefficient on wavelength as a result of the color change of hemoglobin as the oxygen concentration varies. Thus, a photoacoustic imaging system can be configured to give images of the relative degree of oxygen saturation of arteries or blood vessels that lie within tissue. Typically, two experiments with laser beams having two different wavelengths are carried out to determine a differential absorption. In the case of tumors, the degree of oxygenation can be used to help distinguish malignant from benign tumors, as the former are typically hypoxic.

It is possible to devise a tomographic imaging modality by repeatedly firing a laser at the body and rotating the transducer to obtain images at a number of angles. By using the appropriate inversion scheme, the absorption profile of the body that is a direct analog of x-ray tomography can be found. **Figure** 2*a* shows images of a rat brain that were taken with a live animal placed under water with a breathing tube and a transducer that rotated around the head of the rat recording photoacoustic waves each time the laser fired. The skull of the rat is so thin that light can penetrate into the interior of the brain and sound can be transmitted out to the transducer with insignificant distortion. A conventional optical image of the rat brain is shown in Fig. 2*b* after the rat had been sacrificed and the top of the skull removed.

Contrast agents. An important development in the field of photoacoustic imaging is the use of nanoparticles to act as contrast agents. Colloidal gold, which to date appears to be inert in live tissue, has a strong plasmon absorption near 500 nm. It is possible to modify the particles by synthesizing them to be elliptical or rodlike so that the normal absorption in the green is shifted into the near-infrared region, where tissue has an absorption minimum. The particles are strong optical absorbers, so that, upon irradiation, the heating of the nanoparticles results in steam formation, giving a large photoacoustic effect. **Figure** 3 shows images of nanorods in a live mouse taken with a commercially available instrument from Tomowave Laboratories. The accumulation of the nanorods in different organs as a function of time is clearly visible in the three images in Fig. 3. With regard to tumor detection, in some cases the vascularization around tumors is leaky, so that gold nanoparticles accumulate naturally at the sites of tumors. It is thus possible to locate a tumor by noting the enhancement in the photoacoustic amplitude. A recent development in the use of contrast agents is that internal temperature can be measured by noting the magnitude of the signal from the nanoparticles in an image. The method is based on the fact that, for water, the thermal expansion coefficient β has a significant temperature dependence. A goal of many researchers has been to attach antibodies specific to tumors to nanoparticles in order to concentrate the nanoparticles on the tumors, thereby providing a straightforward and sensitive method for tumor detection. Development of this technology is certain to

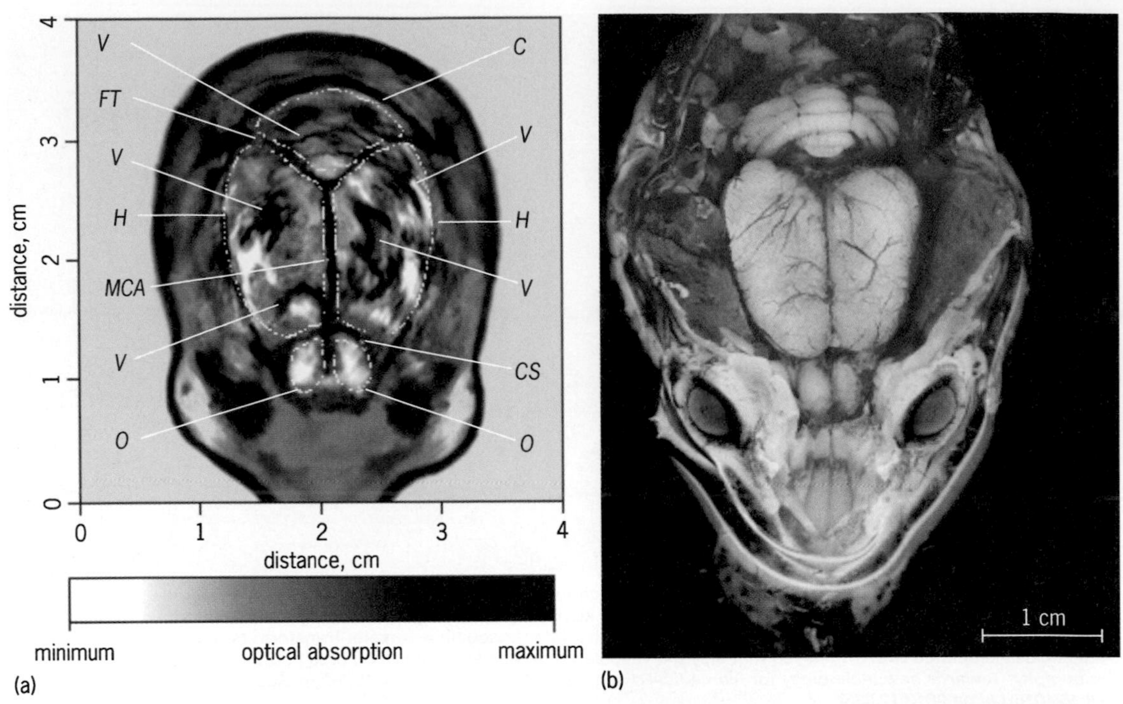

Fig. 2. Images of a rat brain. (*a*) Photoacoustic image of the brain of a live rat using 532-nm radiation, with the dark parts of the image indicating a large photoacoustic amplitude. The letters mark the positions of various parts of the brain: C, cerebellum; H, cerebral hemispheres; O, olfactory bulbs; MCA, middle cerebral artery; CS, curciate sulcus; FT, fissure transversa; V, blood vessels. (*b*) Optical image (photograph) of the same brain, but with the skull removed after the rat has been killed. 1 cm = 0.4 in. (*Reprinted by permission from Macmillan Publishers Ltd., from X. Wang et al., Noninvasive laser-induced photoacoustic tomography for structural and functional in vivo imaging of the brain, Nat. Biotechnol., 21:803–806, 2003, DOI:10.1038/nbt839*)

give photoacoustic imaging far more applications to medicine.

Other developments in photoacoustics. Although the majority of researchers have used pulsed laser radiation to excite the photoacoustic effect in tissue, it has been shown recently that modulation of a continuous laser at a high frequency produces photoacoustic waves that can be detected and processed

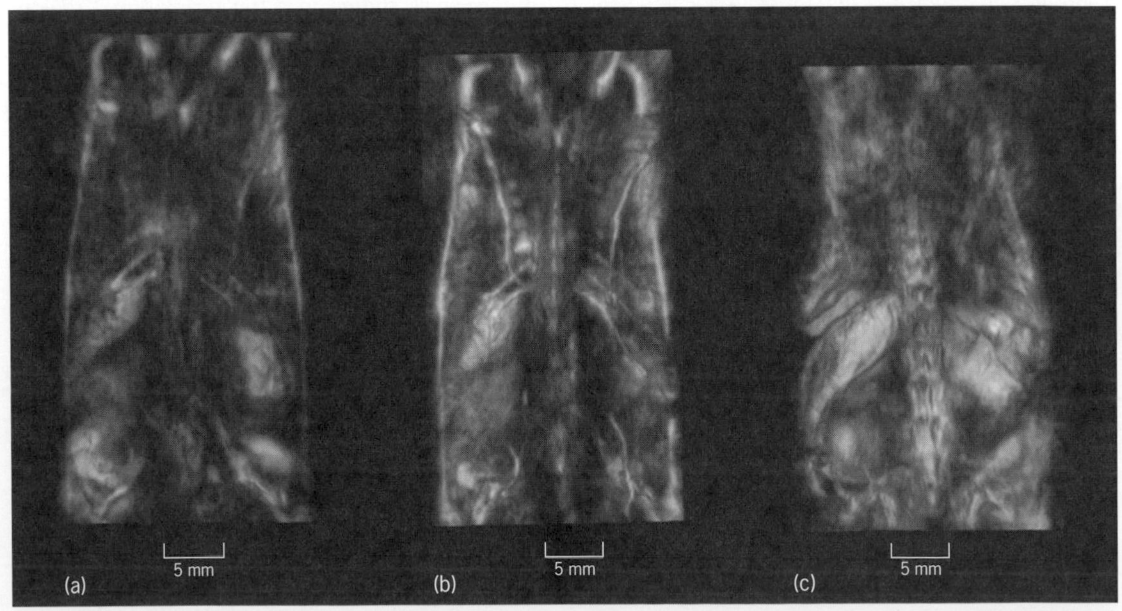

Fig. 3. Distribution of gold nanorods in a live mouse using 3D photoacoustic imaging at 765 nm. (*a*) Image prior to the injection of gold nanorods. (*b*) Image one hour after the injection. (*c*) Image two days after the injection. The migration of the nanorods to different organs is evident from the photoacoustic amplitude in the image. 5 mm = 0.2 in. (*From R. Su et al., Laser optoacoustic tomography: Towards new technology for biomedical diagnostics, Nucl. Instrum. Meth. Phys. Res. A, 720:58–61, 2013, DOI:10.1016/j.nima.2012.12.035*)

using sophisticated mathematical methods that cannot be applied to the waves resulting from pulsed radiation. This method is still in its infancy, but may show enhanced contrast. Although much of the focus on imaging has been directed toward detection of cancers, it has been shown that the photoacoustic effect can detect blood clots that migrate around the body, using the absence of absorption by the clot. Another finding is that a single melanoma cell can be detected in the lymph using photoacoustics. A totally different approach to generating the photoacoustic effect uses pulsed microwaves. Here, contrast is not dependent on the simple optical absorption provided by blood, but has an altogether different mechanism, and promises the possibility of new kinds of contrast agents that may contribute to superior imaging.

For background information *see* ABSORPTION OF ELECTROMAGNETIC RADIATION; BIOMEDICAL ULTRASONICS; COMPUTERIZED TOMOGRAPHY; LASER; MEDICAL IMAGING; PHOTOACOUSTIC SPECTROSCOPY; PLASMON; SOUND in the McGraw-Hill Encyclopedia of Science & Technology. Gerald J. Diebold

Bibliography. G. J. Diebold, M. I. Khan, and S. M. Park, Photoacoustic "signatures" of particulate matter: Optical production of acoustic monopole radiation, *Science*, 250:101–104, 1990, DOI:101126/science.250.4977.101; R. Su et al., Laser optoacoustic tomography: Towards new technology for biomedical diagnostics, *Nucl. Instrum. Meth. Phys. Res. A*, 720:58–61, 2013, DOI:10.1016/j.nima.2012.12.035; L. V. Wang ed., *Photoacoustic Imaging and Spectroscopy*, CRC Press, Boca Raton, FL, Optical Science and Engineering, vol. 144, 2009; X. Wang et al., Noninvasive laser-induced photoacoustic tomography for structural and functional *in vivo* imaging of the brain, *Nat. Biotechnol.*, 21:803–806, 2003, DOI:10.1038/nbt839.

PINK1-Parkin pathway

PINK1 [PTEN (phosphatase and tensin homolog)-induced kinase-1] and Parkin are the prototypical protein factors for which single-gene human mutations were first linked to rare, hereditary forms of Parkinson's disease. Parkinson's disease is a multifactorial progressive neurodegenerative condition characterized by the loss of nerve cells (neurons), especially in a part of the midbrain called the substantia nigra. Because the affected nigrostriatal neurons have high levels of the neurotransmitter dopamine, Parkinson's disease has been treated traditionally with drugs, such as L-dopa or dopamine agonists, which enhance neuronal dopamine signaling. However, such treatments do not address the cause of the disease, which is neuronal death. Indeed, it was genetic studies that identified the mutations (first in Parkin and subsequently in PINK1) that ultimately revealed the mechanism for neuronal toxicity underlying the inherited forms of Parkinson's disease, specifically, a defect in the mitochondrial quality control.

Mitochondria and mitophagy. Mitochondria are ubiquitous and specialized organelles that consume oxygen and nutrients through the process of respiration to generate cellular energy in the form of adenosine triphosphate (ATP). Biochemically, ATP synthesis is driven by the sequential transport of electrons along the mitochondrial respiratory chain and ultimately to molecular oxygen, which then combines with hydrogen to produce harmless water. However, mitochondrial damage can uncouple the electron transport from ATP synthesis, resulting in the formation of chemically reactive oxygen species (ROS). The promiscuous oxidation of mitochondrial and cellular DNA, proteins, and lipids by ROS is damaging (this condition is commonly referred to as oxidative stress), and it contributes to senescence and many chronic degenerative conditions, including Alzheimer's disease and Parkinson's disease. Accordingly, cells have evolved a mechanism termed mitophagy (literally "eating mitochondria") to identify, sequester, and remove the damaged and damaging mitochondria. PINK1 and Parkin are two central effectors of mitophagy.

Parkin and PINK1. Parkin is a ubiquitin ligase, that is, a protein that facilitates the attachment of small ubiquitin peptides to damaged or misfolded substrate proteins, tagging them for degradation via the ubiquitin-proteasome system. The process of ubiquitin attachment is known as ubiquitylation (or ubiquitination). Parkin is normally cytosolic, which means that it is found in the fluid component of the cytoplasm. However, it specifically relocates to damaged mitochondria, where it ubiquitylates outer membrane proteins that attract autophagosomes. These autophagosomes are structures that engulf damaged cellular components and transfer them to lysosomes for degradation. Cells lacking Parkin fail to invoke mitophagy. Thus, in colloquial terms, Parkin acts as a traffic cop, identifying and ticketing (by ubiquitylation) damaged mitochondria and directing them to the cellular junkyard for dismantling and component recycling.

PINK1 is a mitochondrial kinase, which is a protein that modifies other proteins by adding phosphate groups to serine or threonine amino acids. It was not readily apparent how the mutations in a mitochondrial kinase (PINK1) and a cytosolic ubiquitin ligase (Parkin) could cause the same human disease. However, a functional relationship between PINK1 and Parkin was uncovered by genetically suppressing the two proteins in *Drosophila* fruit flies. A deficiency of either PINK1 or Parkin evoked similar muscular and neurological degenerations accompanied by comparable mitochondrial defects. Strikingly, these abnormalities were not aggravated in flies by the combined absence of both factors; that is, the consequences of PINK1 and Parkin deficiency were not additive. This indicates that these factors function sequentially within a common pathway.

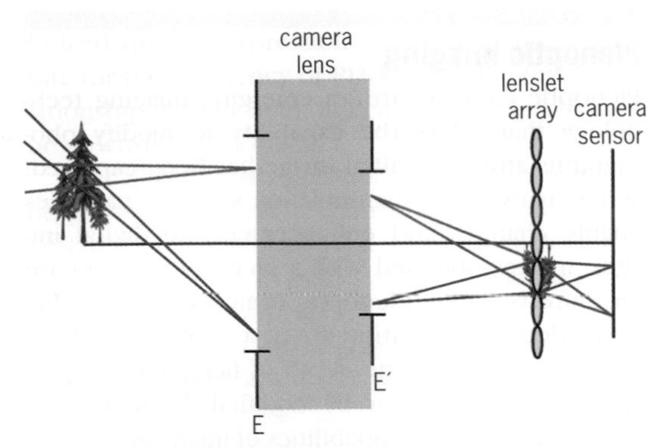

Fig. 2. A plenoptic camera records a different portion of the plenoptic function than a conventional camera. The image is formed on a lenslet array. The camera elements are arranged so that the camera sensor records many tiny images of the exit pupil E′. A pixel on the camera sensor records rays from a patch on the object over a narrow range of angles.

information in the ultraviolet. Finally, polarization information is typically recorded by means of filters that reject some polarization states and pass others. Consequently, information regarding the rejected states is lost. Cameras, in general, capture a reduced portion or projection of the plenoptic function and reproduce the scene as best as possible from this reduced information.

Fig. 3. A small section from the raw image from a plenoptic camera. Each circle is an image of the exit pupil formed by an individual lenslet.

Conventional cameras. Conventional cameras have become ubiquitous. Cell phone, point-and-shoot, and professional digital single-lens reflex (DSLR) cameras each represent a standard geometry for capturing a particular projection of the plenoptic function. The projection is modified slightly by changing the focal length of the lens or adding external filters, but in general these cameras use essentially the same technique to record a projection of the plenoptic function. A conventional camera consists of a camera lens and a camera sensor (**Fig. 1**). To understand the imaging properties of the camera lens, it is useful to consider just the entrance pupil E and the exit pupil E′ of the lens. The entrance pupil is just the image of the aperture stop that is seen when looking into the front of the camera lens. Light rays from the scene must pass through the entrance pupil to get into the camera. Similarly, the exit pupil is the image of the aperture stop when looking into the rear of the camera lens. For a well-designed camera lens, rays that enter the entrance pupil will emerge from the exit pupil.

For a conventional camera, the focus of the camera lens is adjusted so that a sharp image of the scene is formed on the camera sensor. In Fig. 1, a point on the base of the tree is illuminated by the Sun and any other light sources nearby. In addition, scattered light from the sky, the ground, and other surrounding objects may also fall on this point. The incident light rays from the illumination, in turn, are absorbed, scattered, or reflected from this point. The scattered and reflected rays appear as if they are leaving the base of the tree in all directions. A small fraction of these rays will enter the entrance pupil of the conventional camera and be focused onto a pixel in the sensor. The pixel will record the color and intensity associated with the light rays leaving the point at the base of the tree and entering the entrance pupil. Conventional cameras typically ignore all polarization information, although polarization filters are sometimes used to reduce glare. Thus, the sensor pixels record a reduced version of the plenoptic function that captures only a subset of the rays leaving a point in the scene; compress the wavelength information into red, green, and blue components; and ignore polarization. Furthermore, short exposure times usually freeze any changes in these variables. Finally, the trajectory information of the incident rays is lost as all of these rays are combined on the pixel. From the information recorded by the camera sensor, a snapshot is created that mimics the original scene, but mimics it imperfectly.

Plenoptic cameras. Plenoptic cameras operate differently from a conventional camera. The camera lens focuses the scene onto a lenslet array (**Fig. 2**). In turn, each lenslet forms an image of the exit pupil on the camera sensor. In this configuration, each lenslet corresponds to a patch on the objects in the scene, and each pixel on the sensor records that patch from different angles. Examining the raw image from a plenoptic camera shows a representation of the scene consisting of many small circular regions (**Fig. 3**). Pixels within each circular region correspond to the same patch on the object, but from a different vantage point. Plenoptic cameras still capture a reduced version of the plenoptic function, but this projection is different from that of a conventional camera and enables novel images to be generated from the original raw image. Because plenoptic cameras are examining patches in the scene from different vantage points, they retain trajectory information about the rays entering the entrance pupil. Because plenoptic cameras examine patches and not points on the object, some spatial detail regarding the scene is lost. Plenoptic cameras trade spatial resolution for knowledge of the ray trajectories.

Plenoptic image possibilities. The raw image from a plenoptic camera needs to be processed by software to create snapshots of the scene. The advantage of retaining information regarding the ray trajectories that made up the scene is that many images, each with different characteristics, can be generated from the same raw image. Refocusing is one example of a characteristic that can be modified. The data from the raw plenoptic image can be used to simulate conventional cameras with lenses of different focal lengths. In doing so, snapshots can be created in which the focus is set to different distances (**Fig. 4**). Other possibilities include modifying the depth of field. Each circular region in the raw image is a miniature reproduction of the exit pupil. If the diameter of these regions is reduced by software, the effect is equivalent to stopping down the aperture in a conventional camera. Small apertures create increased depth of field in the image. Furthermore, moving the reduced-diameter region around within the bounds of the original circular region leads to images from slightly different perspectives. Finally, the distance to various objects within the scene can be estimated, creating a three-dimensional representation of the scene from a single snapshot.

Plenoptic cameras trade spatial resolution for the capability to create all these different possible images. A 10-megapixel sensor in a conventional camera creates a high-resolution static reproduction of a scene. The same sensor in a plenoptic camera may provide only a 1-megapixel reproduction of the scene, but a multitude of images with different planes of focus, different depths of field, and different perspectives can be generated. Currently, the spatial detail in these low-resolution plenoptic images is just marginally acceptable for electronic display and small prints. However, a second-generation consumer plenoptic camera, based on a 40-megapixel sensor that provides improved image detail, was released in 2014. The resolution of camera sensors will continue to increase rapidly, so the quality of plenoptic images is expected to improve steadily, leading to more detailed and versatile images.

(a)

(b)

Fig. 4. Raw images processed to focus on (*a*) the sphere and (*b*) the background.

For background information *see* CAMERA; FOCAL LENGTH; GEOMETRICAL OPTICS; IMAGE PROCESSING; LENS (OPTICS); PHOTOGRAPHY; POLARIZED LIGHT in the McGraw-Hill Encyclopedia of Science & Technology. Jim Schwiegerling

Bibliography. T. Adelson and J. Y. A. Wang, Single lens stereo with a plenoptic camera, *IEEE Trans. Pattern Anal. Mach. Intell.*, 14:99–106, 1992; R. Ng et al., *Light Field Photography with a Hand-Held Plenoptic Camera*, Stanford Tech Rep. CTSR 2005-02, 2005; G. Wetzstein et al., Computational plenoptic imaging, *Comput. Graph. Forum*, 30:2397–2426, 2011, DOI:10.1111/j.1467-8659.2011.02073.x.

Polariton laser

Since their first demonstration in 1960, lasers have been tremendously influential in almost all aspects of scientific research, industrial manufacturing, and in the medical, military, and engineering fields. Whereas an optical laser generates coherent light beams amplified through stimulated emission from a cavity filled with various gain media, a polariton laser emits coherent electromagnetic waves from a cavity with embedded quantum wells. Coherent polaritons at the system ground state are amplified by stimulated scattering processes to undergo Bose-Einstein condensation. This article explains basic structures, fundamental operating principles, and several types of polariton-based light sources.

Polaritons. What are polaritons? The term "polariton" refers generally to a quasiparticle that results as a consequence of the strong interaction between photons and electric- or magnetic-dipole-carrying particles. Well-known examples are phonon-polaritons, surface plasmon polaritons, and exciton-polaritons. We use "polariton" as a short name for a microcavity exciton-polariton, the hybrid mixture of a cavity photon and an exciton in single or multiple quantum wells embedded inside the cavity (**Fig. 1***a*). In 1992, Claude Weisbuch and his colleagues reported for the first time the direct signature of polariton states in a gallium arsenide (GaAs)–based semiconductor microcavity, whose physical size was submicrometer length, comparable to the emitted optical wavelengths, a few hundred nanometers. This was the outset of the polariton research field.

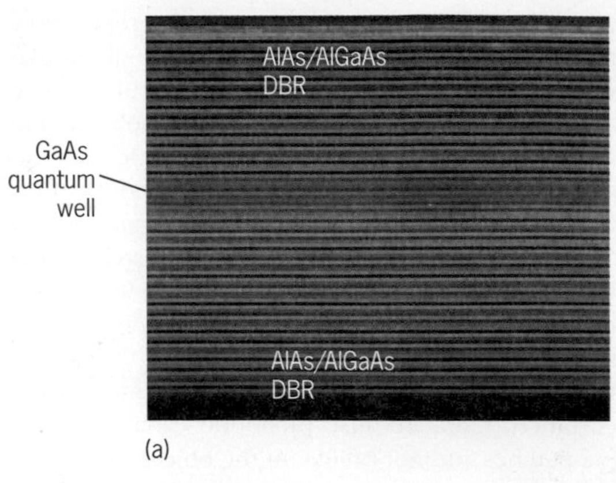

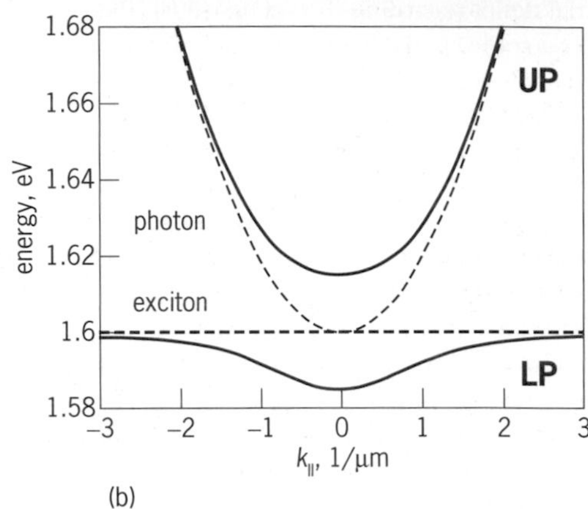

Fig. 1. Microcavity exciton-polaritons. (*a*) A scanning electron microscope image of a submicrometer-wavelength-sized microcavity formed by a pair of AlAs/AlGaAs distributed Bragg reflectors (DBRs). A single GaAs quantum well is at the center of the microcavity. (*b*) Relations of energy and in-plane momentum ($k_{\|}$) values for a cavity photon and a quantum-well exciton (broken blue lines), and for anticrossed upper and lower polaritons (UP and LP, solid red lines) arising from the strong coupling between a cavity photon and a quantum-well exciton.

Device structures of polaritons. For the last half century, we have acquired powerful and versatile techniques in semiconductor band-gap and advanced growth engineering, which enable us to design and fabricate high-quality versatile structures in zero, one, two, and three dimensions. The choice of different elements and their concentrations results in an infinite set of semiconductor alloys, and the precise control of the layers by molecular-beam epitaxy and metal-organic vapor-phase epitaxy is possible even down to single-atom thickness. Repeated alternation of two semiconductor alloys with different refractive index values, whose layer thickness is a quarter of an optical wavelength, can produce a dielectric mirror, a so-called distributed Bragg reflector. The reflectance of the mirror increases with the number of constituent layers. Figure 1*a* shows the cross-sectional image of a Fabry-Perot microcavity formed by two distributed Bragg reflectors, wherein electromagnetic waves are confined. Photons bounce back and forth between the reflectors during their lifetime. In a high-quality cavity, photons reside for a relatively long time before their release from the cavity. A quality factor Q is a quantitative measure relating to cavity photon lifetime. Photons last a few picoseconds inside a $Q \sim 2000$ microcavity, but they can remain up to hundreds of picoseconds in a $Q \sim 100{,}000$ microcavity.

Single or multiple quantum wells, where a small-band-gap semiconductor is sandwiched by large-band-gap semiconductors, are placed at the locations of the maximum field amplitudes of electromagnetic waves inside the microcavity. When lasers shine on two-dimensional quantum wells, or the wells are biased electrically, a negatively charged electron in the valence bands is excited to the conduction bands, leaving behind a positively charged hole in the valence bands. The oppositely charged electron and hole are attracted to each other, and a bound state, an exciton, is formed. It is a primary excitation of quantum wells, having a finite lifetime, roughly 100 ps to 1 ns, in a GaAs quantum well.

In contrast to the weak-coupling case, where cavity photons and excitons do not interact, cavity photons in a high-quality cavity can exchange energy with excitons in a reversible manner, namely, a cavity photon excites an exciton, which decays to a cavity photon. When the rate of this reversible energy exchange, proportional to the mutual interaction strength, is much faster than photon and exciton decay rates, the system enters into a strong-coupling regime, exhibiting the distinct feature of two anticrossed modes. These two emerging particles are called microcavity exciton-polaritons, in short, polaritons. The relations between energy and in-plane momentum, $k_{||}$, for polaritons and their constituent particles are plotted in Fig. 1*b*. It is important to note that polaritons escape the cavity in the form of light radiation. The in-plane momentum, wavelength (energy), and polarization of the escaped photons are in a one-to-one correspondence to those of the polaritons in quantum wells, as is guaranteed by translational invariance in two dimensions. This correspondence provides a powerful advantage for directly accessing quantitative information about polaritons.

Operating principle of the polariton laser. A key concept for understanding the polariton laser is the condensation of boson particles. According to Bose-Einstein statistics, below a critical temperature, indistinguishable bosons have the propensity to accumulate limitlessly in the system ground state. The transition temperature is inversely proportional to the effective mass of participating bosons. Inheriting from the bosonic nature of their constituent particles, photons and excitons, polaritons are considered to be bosons in a small-density limit. Typically, the effective mass of polaritons is roughly 10^{-5} times the bare electron mass, and only 10^{-8}–10^{-9} times the hydrogen atomic mass. Therefore, extremely light polaritons condense at temperatures around 1–300 K (**Fig. 2*a***), 10^{8}–10^{9} times higher than atomic condensation critical temperatures, which are around tens to hundreds of nanokelvins.

How are polaritons created in a cavity? Normally we shine a laser at higher energy values on a quantum-well-microcavity sample at either normal incidence or a specific angle. At low pump-power values, thermal polaritons are distributed along the whole lower polariton branch via polariton-phonon scattering. As the number of polaritons at $k_{||} = 0$ reaches the quantum degeneracy point (average particle number $\sim$ 1) at higher pump-power values, a polariton-polariton scattering process turns on. This two-particle elastic scattering conserves both energy and momentum so that one polariton relaxes down to the $k_{||} = 0$ state, while the other moves up to a higher energy state. The probability that polaritons are scattered into the $k_{||} = 0$ state is enhanced by $N + 1$, where N is the number of polaritons in the $k_{||} = 0$ state (bosonic final-state stimulation). This stimulated scattering builds up coherence among condensed polaritons spontaneously. Accumulated coherent polaritons in the $k_{||} = 0$ state are naturally leaked out of the cavity in the form of photons.

Figure 2*b* captures the characteristic behavior of polaritons in the $k_{||} = 0$ state as a function of the laser pump-power values. A threshold behavior is clearly seen in the population of polaritons (blue) and the energy linewidth (red). The polariton energy values go up (they are blue shifted) as the laser pump-power values increase, as shown in the inset of Fig. 2*b*. It is understood that the mean-field energy of the condensate is increased by the polariton-polariton scattering energy. Figure 2*c* shows that polaritons in the condensates interfere, and an intensity modulation in Young's double-slit experiments confirms coherence. A polariton laser emits intense, monochromatic, directional light beams with a well-defined phase and a polarization. It is worthwhile remarking that a polariton laser is superior to a photon laser in that coherent light beams are generated at much lower pump-power values in a polariton laser by a factor of $\sim$10 than those of a photon laser. This is because the particle density for a population inversion as a gain process in a photon laser is higher by a factor of $\sim$10 than the particle density for quantum degeneracy in a polariton laser.

Types of polariton light sources. A phase diagram of light sources in GaAs is mapped in Fig. 2*a*, showing polariton-based light sources in a strong-coupling regime (within the rectangle enclosed by broken lines) and photons in a weak-coupling regime. These devices can be made not only by optical pumping, but polaritons can also be created electrically in semiconductors by doping the top and bottom distributed Bragg reflectors with *n*- and *p*-type elements as electron and hole reservoirs, respectively. The device is basically a *p-i-n* diode, where quantum wells are located in an intrinsic (*i*-, or undoped) domain. Under electrical bias, electrons and holes from the *n*- and *p*-doped sections propagate to the central intrinsic area and recombine to form excitons, and consequently polaritons, in the quantum wells. Polariton light-emitting diodes have been demonstrated in InGaAs/GaAs-based semiconductors, and they display the ability to operate even above room temperature, a property that is very promising with regard to the prospects for constructing a practical polaritonic apparatus. Light-emitting diodes are distinguished from lasers in that photons outside of the cavity do not interfere; that is, they are incoherent. Polaritons in light-emitting diodes remain below the threshold of condensation.

A long-sought dream since the discovery of polaritons is to build electrically pumped polariton lasers. They are attractive and advantageous in terms of minimum operating power values, compact device footprint, and integration with other electronic and optoelectronic components. The first polariton laser diode, using InGaAs/GaAs semiconductors,

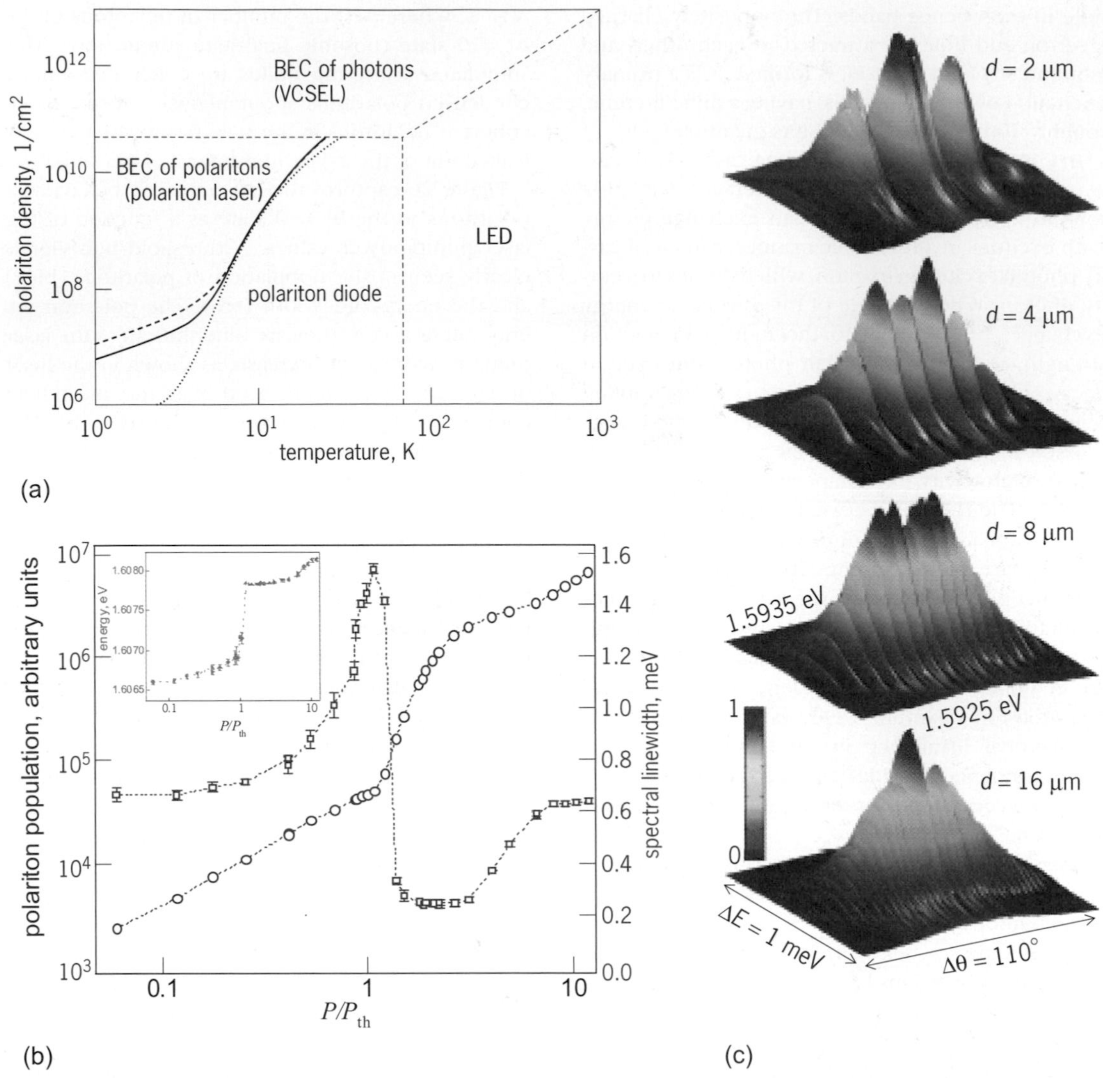

Fig. 2. Polariton condensation. (*a*) A calculated phase map of exciton-polaritons in a GaAs-based microcavity. Within the rectangle enclosed by broken lines, polaritons remain in a strong coupling, but excitons no longer exist outside this area. BEC = Bose-Einstein condensate; VCSEL = vertical-cavity surface-emitting laser; LED = light-emitting diode (*from G. Malpuech et al., Polariton laser: Thermodynamics and quantum kinetic theory, Semicond. Sci. Tech., 18:S395–S404, 2003, DOI:10.1088/0268-1242/18/10/314; copyright and reproduced by permission, courtesy of IOP Publishing, all rights reserved*). (*b*) Three characteristics of polariton condensation in terms of occupancy of the lowest energy level (blue), energy linewidth (red), and energy (green). A threshold for the condensed phase appears near a nonlinear increase of the polariton occupancy and a sharp reduction of energy linewidth. The horizontal scale is the ratio of the laser pump power *P* to its threshold value P_{th}. (*c*) Young's double-slit interference images from polaritons above threshold at different slit distances, *d*. The visibility of the interference patterns confirms coherence in condensed polaritons. The color scale indicates the normalized intensity of the interferogram (1 represents maximum intensity, 0 represents minimum intensity). (*Courtesy of C. W. Lai et al., Coherent zero-state and condensed polexciton-polariton condensate array, Nature, 450:529–532, 2007, DOI:10.1038/nature06334.*)

was demonstrated in 2013 (**Fig. 3*a***). Below threshold, the same structure works as a polariton light-emitting diode, but above threshold it behaves as a polariton laser. A collection of evidence from a series of experiments supports the existence of coherent light beams from polariton states in the strong-coupling regime. This GaAs-based polariton laser diode still operates only at temperatures of approximately 10 K, since thermal energy at room temperature may dissociate excitons, destroying the strong coupling. However, it is strongly anticipated that in the future a room-temperature polariton laser diode can be made of large-band-gap semiconductors such as gallium nitride (GaN), zinc oxide (ZnO), and organic semiconductors.

Perspectives. In the last two decades, our understanding of polariton physics has become solid and firm through numerous theoretical and experimental studies. Furthermore, a solid understanding of polariton physics also enables us to conceive and demonstrate functional polariton light sources: polariton lasers, polariton light-emitting diodes, and polariton laser diodes. Polaritons also can be useful as fast optical switches and modulators, which would potentially be important in classical communications, and it is envisioned that the light-matter duality of polaritons can provide us with an ingenious trick in quantum information processing. We have just embarked on an exciting journey to explore fundamental polariton physics and to

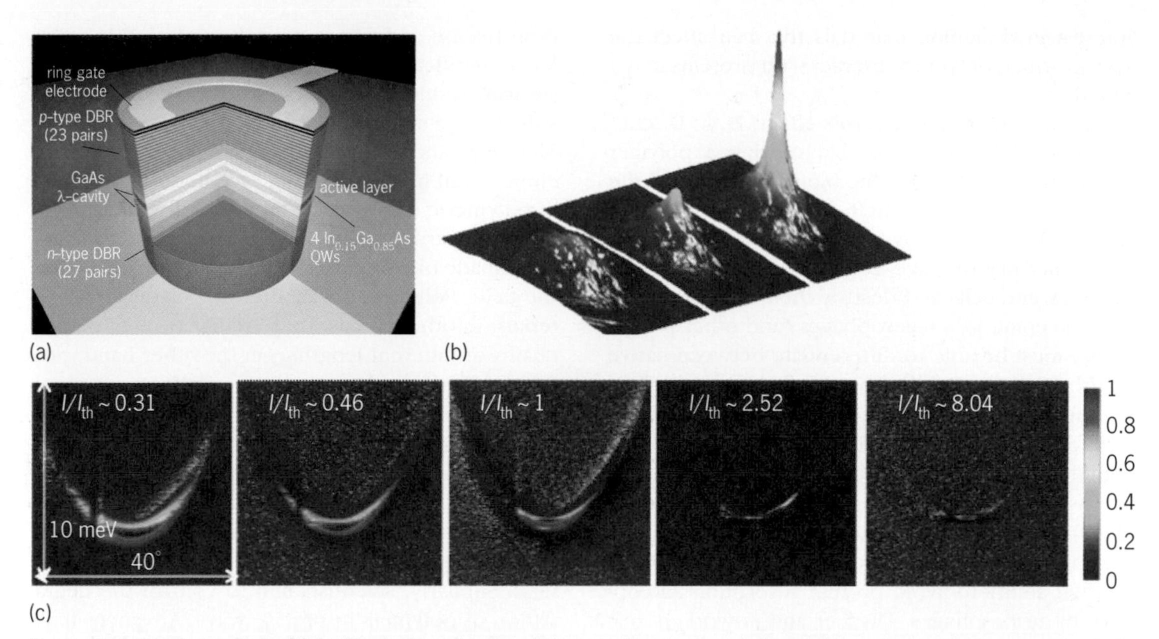

Fig. 3. Polariton laser diode. (*a*) An illustration of a doped microcavity-quantum-well structure, where polaritons are created by electrical fields. DBR = distributed Bragg reflector; QW = quantum well. (*b*) Polariton distributions in momentum spaces from lower (left) to higher current density to excite excitons. (*c*) Evolution of energy-momentum relations as a function of injected current values (*I*) and current density (*J*). The threshold current I_{th} is approximately 1.73 mA, and the threshold current density J_{th} is approximately 22 A/cm^2. The color scale indicates the normalized intensity of the interferogram.

invent pragmatic and efficient polaritonic devices, and more exciting news in the field of polaritons is anticipated.

For background information *see* ARTIFICIALLY LAYERED STRUCTURES; BOSE-EINSTEIN CONDENSATION; CAVITY RESONATOR; COHERENCE; EXCITON; INTERFERENCE OF WAVES; INTERFEROMETRY; LASER; LIGHT-EMITTING DIODE; PHOTON; Q (ELECTRICITY); QUANTIZED ELECTRONIC STRUCTURE (QUEST); SEMICONDUCTOR HETEROSTRUCTURES in the McGraw-Hill Encyclopedia of Science & Technology.

Na Young Kim

Bibliography. B. Deveaud-Pledran, Solid-state physics: Polaritonics in view, *Nature*, 453:297–298, 2008, DOI:10.1038/453297a; A. Kavokin and G. Malpuech, *Cavity Polaritons*, Elsevier, San Diego, 2003; D. Sanvitto and V. Timofeev (eds.), *Exciton Polaritons in Microcavites: New Frontiers*, Springer-Verlag, Berlin, Germany, 2012, DOI:10.1007/978-3-642-24186-4; D. Snoke and P. Littlewood, Polariton condensates, *Phys. Today*, 63(8):42–47, 2010, DOI:10.1063/1.3480075.

Polymers for drug delivery

The field of drug delivery has been through a few eras of advancement. Early efforts were simply directed at using polymers to enable the slow release of drugs. This was done to mitigate side effects, enable prolonged drug action over time, and increase the bioavailability of drugs in the body. The next phase focused on achieving zero-order (steady) drug-release profiles. It was thought that having a constant concentration of a drug would provide an optimal therapeutic effect. Most likely, a system that enables the concentration of a drug to remain above the minimum effective dose and below the maximum toxic dose (within the so-called therapeutic window) will improve the efficacy and duration of therapy. Scientists are currently in a third era of "smart" drug delivery, where researchers are seeking to design drug delivery systems that can release drugs on demand, in response to specific stimuli. This can include pH, temperature, specific enzymes, and other forms of triggers. Scientists would also like to have better control of the interactions of polymer nanoparticles with biological fluids and proteins in the body. Therefore, polymers for drug delivery are now being called upon to perform many functions. These include increasing the time that drugs are available (in blood circulation or in the tissue), mediating the biodistribution (the location of the drug in the body), improving the solubility of drugs (for example, hydrophobic, water-insoluble small-molecule drugs), improving the stability of drugs (for example, nucleic acid or protein drugs), providing active or passive transport of the drug to targeted tissues, enabling controlled release of the drug in response to stimuli, and controlling the specific interaction of the drug with biomolecules in the body. Recently, a number of advancements have been made that will have a positive effect on the future of drug delivery using polymers. A few of them are highlighted in this article.

Upon intravenous administration, polymer-based drug delivery systems immediately come into contact with the biological environment. The exposure to biological fluids often results in a corona of proteins and other biomolecules that coat the surface of the polymer and change its chemical and physical properties. Given this reality, there is significant interest in designing materials that can affect the way in which polymers interact with proteins in the blood.

As one way to mitigate this effect, D. E. Discher and coworkers reasoned that coating a polymer nanoparticle with a specific biomolecule to tell the body that this nanoparticle "belongs" would be a powerful way to improve drug delivery. In the blood, cells called macrophages recognize foreign particles, bacteria, and cells and destroy them. But to do this housekeeping job, macrophages (and other phagocytes) must be able to differentiate between native healthy cells and cells from foreign species. They do this through the recognition of protein signals. One of these markers is CD47, a glycoprotein that signifies "self" and tells macrophages not to engulf it. Passive polymers, such as poly(ethylene glycol) [PEG], have long been used to delay the immune clearance of foreign particles, but there are limits to their ability to avoid protein adsorption and uptake by macrophages. Discher and coworkers reasoned that attachment of CD47, or even a minimal "self" peptide version, to the surface of a nanoparticle would cause the host macrophage to recognize that nanoparticle as being part of itself by signaling through the phagocyte receptor CD172a. They found that self peptides delay macrophage-mediated clearance of nanoparticles. This resulted in a significant increase in the persistent circulation time in the blood. In turn, these polymer nanoparticles with attached self peptides accumulated in tumors in mice to a much greater extent than nonfunctionalized nanoparticles. Thus, controlling the surface chemistry of polymer-based nanoparticles for drug delivery is an important direction for the future of nanomedicine. The use of peptides and other markers that can tell the host that the particle belongs is a creative and effective way to increase blood circulation time and increase the delivery of drugs to tumors. Ultimately, this may lead to more effective tumor imaging and chemotherapies using "smart" polymers that control specific interactions with the biological environment.

In the clinic, polymers have been used for many years as excipients in tablet and capsule formulations, as enhancers of blood circulation time (for example, antibody PEG conjugates), and in the creation of slow-release depots. As we move into an era in which polymers must also actively control the multitude of requirements listed previously, clinical translation of these new systems must be carefully studied. In one such study, J. Hrkach and coworkers evaluated a platform of block copolymers of either poly(D,L-lactide) [PLA] or poly(D,L-lactide-*co*-glycolide) [PLGA] and PEG, loaded with the anticancer drug docetaxel. The nanoparticles were targeted to the extracellular domain of prostate-specific membrane antigen (PSMA), using a small-molecule ligand displayed on the nanoparticle surface. Although targeted nanoparticles have been studied for more than 30 years, there have been few trials done in humans. This study reported controlled drug release and tumor growth inhibition in mice. More significantly, interim Phase 1 data in humans showed results comparable to those of the animal studies, and also showed signs of tumor shrinkage. More time and studies are obviously needed, but this clinical trial may be a key bellwether for the future of polymeric nanoparticle-mediated drug delivery.

With regard to synthetic chemistry, advances have been made on a number of important challenges in the field. Polymers are inherently polydisperse materials; in other words, they are always mixtures of chains of different lengths. On the other hand, proteins, nature's polymers, have a uniform composition that results in a higher-order self-assembly ability and defined 3D structure. Therefore, scientists are focused on developing synthetic strategies to control the sequence of units within polymer chains. It is anticipated that this will improve drug delivery by providing greater control over "smart" responsiveness. Similarly, scientists aim to control the degradation of polymers in precise ways. Advances have been made in this area as well, including polymers that respond to shear stress, light, ATP (adenosine triphosphate), and toxins. In order to understand how polymers can advance drug delivery, we must also develop improved ways to analyze and characterize these new and exciting materials of the future. Polymer science is in an exciting era—one with increasing demands on polymer capabilities, but also one with brave ideas for stimuli-sensitive and controllable materials for the future of drug delivery.

For background information *see* ANTIBODY; CANCER (MEDICINE); CHEMOTHERAPY AND OTHER ANTINEOPLASTIC DRUGS; COPOLYMER; DRUG DELIVERY SYSTEMS; GLYCOPROTEIN; LIGAND; NANOPARTICLES; PHAGOCYTOSIS; PHARMACOLOGY; POLY(ETHYLENE GLYCOL); POLYMER; PROTEIN in the McGraw-Hill Encyclopedia of Science & Technology.

Daniel J. Siegwart

Bibliography. S. Biswas et al., Biomolecular robotics for chemomechanically driven guest delivery fuelled by intracellular ATP, *Nat. Chem.*, 5:613–620, 2013, DOI:10.1038/nchem.1681; J. Hrkach et al., Preclinical development and clinical translation of a PSMA-targeted docetaxel nanoparticle with a differentiated pharmacological profile, *Sci. Transl. Med.*, 4:128ra139, 2012, DOI:10.1126/scitranslmed.3003651; C. M. J. Hu et al., A biomimetic nanosponge that absorbs pore-forming toxins, *Nat. Nanotechnol.*, 8:336–340, 2013, DOI:10.1038/nnano.2013.54; N. Korin et al., Shear-activated nanotherapeutics for drug targeting to obstructed blood vessels, *Science*, 337:738–742, 2012, DOI: 10.1126/science.1217815; J. F. Lutz et al., Sequence-controlled polymers, *Science*, 341:628–636, 2013, DOI:10.1126/science.1238149; M. P. Monopoli et al., Biomolecular coronas provide the biological identity of nanosized materials, *Nat. Nanotechnol.*, 7:779–786, 2012, DOI:10.1038/nnano.2012.207; J. P. Patterson et al., The analysis of solution self-assembled polymeric nanomaterials, *Chem. Soc. Rev.*, 43:2412–2425, 2014, DOI:10.1039/

C3CS60454C; P. L. Rodriguez et al., Minimal "self" peptides that inhibit phagocytic clearance and enhance delivery of nanoparticles, *Science*, 339:971–975, 2013, DOI:10.1126/science.1229568; R. Tong, H. H. Chiang, and D. S. Kohane, Photoswitchable nanoparticles for in vivo cancer chemotherapy, *Proc. Natl. Acad. Sci. USA*, 110:19048–19053, 2013, DOI:10.1073/pnas.1315336110; Y. F. Zhang et al., Chain-shattering polymeric therapeutics with on-demand drug-release capability, *Angew. Chem. Int. Ed.*, 52:6435–6439, 2013, DOI:10.1002/anie.201300497.

Prairie restoration on the American Prairie Reserve

The overarching goal of the American Prairie Reserve (APR) is to restore and conserve the biodiversity (species, habitats, and ecological processes) that is native to this region of the Great Plains (**Fig. 1**). Since Euro-Americans came to the Great Plains, prairie habitats and several wildlife species have declined or disappeared as a result of agriculture, overhunting, and other environmental impacts. By purchasing private land and through cooperation with the managers of adjacent public lands, especially the Bureau of Land Management and the Charles M. Russell National Wildlife Refuge, the APR's goal is to create a reserve of 5000 mi^2 (12,950 km^2)—an area twice the size of Delaware—that will once again harbor the diversity and abundance of wildlife that so stirred Meriwether Lewis and William Clark on their journey through this region in the early 1800s.

Prairie restoration on the APR involves three major interconnected strategies: (1) restoring natural habitat, (2) restoring ecological processes that sustain habitats and wildlife, and (3) bringing back populations of imperiled and extirpated wildlife.

Restoring habitat. Most of the APR region is native prairie that has never been plowed (**Fig. 2**). However, establishing native plant cover on APR land that has been plowed is important because invasive nonnative plants quickly take root without it. Therefore, the APR has begun to reseed plowed ground with native plants and is researching what methods work best for reseeding and for controlling nonnative plants.

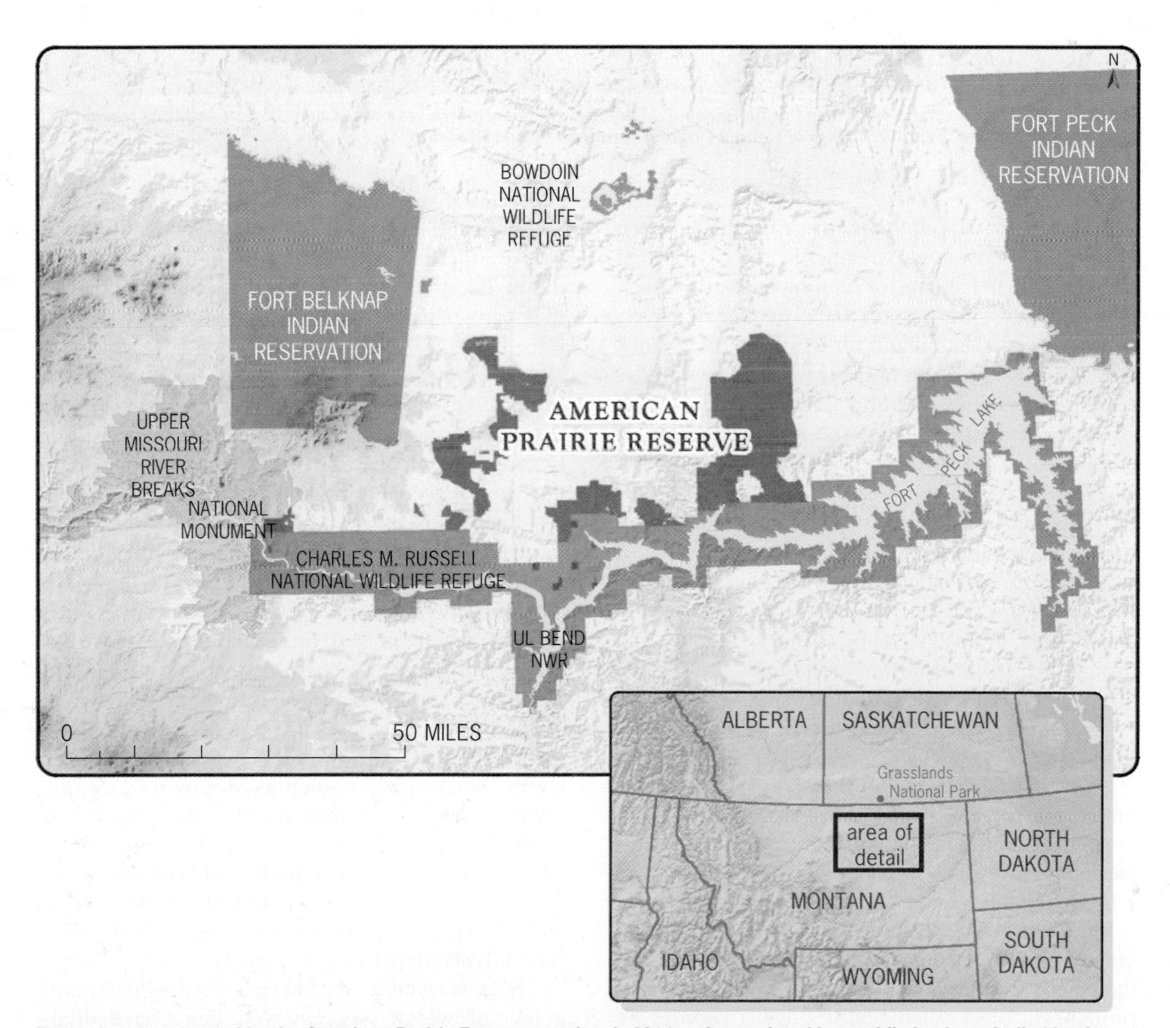

Fig. 1. Lands managed by the American Prairie Reserve as of early 2014 and associated key public lands and tribal lands. The checkerboard ownership pattern of private, state of Montana, and Bureau of Land Management lands in the region is not shown. (*Map courtesy of Liz Juers/American Prairie Reserve*)

Fig. 2. **Native prairie on the American Prairie Reserve. Most lands in the region have never been plowed and thus remain as largely native prairie. (*Photo courtesy of Dennis Lingohr/American Prairie Reserve*)**

Prairie streams and riparian areas (moist, often forested habitats along streams) in the APR region provide critical habitats for many species. However, the streams in the region have been extensively dammed for irrigation and for the creation of stock ponds for livestock. Because these dams greatly alter the water flow, leading to the degradation of streams and riparian areas, the APR has begun removing dams to restore natural flows.

Another crucial step in restoring the prairie habitat is to reduce "habitat fragmentation." Fences, roads, power lines, buildings, and windbreaks of trees fragment the habitat in ways that harm wildlife. Fencing, for example, impedes the wide-ranging movements of bison (*Bison bison*) [**Fig. 3**] and pronghorn (*Antilocapra americana*). Pronghorn often get hung up and die in their attempts to cross fences. Greater sage-grouse (*Centrocercus urophasianus*) are killed when they collide with fences while flying low over the prairie. Many prairie birds avoid nesting near vertical structures such as fences and buildings. Every year, the APR restores natural prairie by removing miles of fencing and scores of buildings and other artificial structures.

Restoring ecological processes. Fire and grazing are the two most important ecological processes that shape the prairie ecosystem. Grazing by bison, pronghorn, black-tailed prairie dogs (*Cynomys ludovicianus*) [**Fig. 4**], and many other species has strongly influenced the evolution of prairie plants and animals. Lightning can cause prairie fires, and Native Americans periodically lit prairie fires. In addition, livestock ranching has caused major changes in grazing patterns and has led to the suppression of fires.

Restoring fire to the prairie is important for two reasons. First, periodic fires kill young tree saplings that invade the prairies. Without fire, prairies often become woodlands. The other reason is that nutrients are released into the soil when the prairie burns, resulting in the growth of nutritious plants, which in turn attract bison and other grazers. Thus, under natural conditions, fire and grazing interact to create a shifting mosaic of habitats—recently burned sites that are heavily grazed, unburned sites that are not grazed and have tall but not very nutritious vegetation, and variations between these two extremes. These different habitats are crucial for supporting the diversity of prairie wildlife. Among birds, for example, some species, such as McCown's longspur (*Rhynchophanes mccownii*), like short vegetation, whereas other species, such as Baird's sparrow (*Ammodramus bairdii*), prefer tall vegetation. The APR is restoring the interactions of fire and grazing and is working with researchers to better understand their effects on prairie biodiversity.

Restoring wildlife. Several species of wildlife, only a few of which are reviewed here, have disappeared or been greatly reduced in number because of human activity in the APR region over the last

Fig. 3. Bison on the American Prairie Reserve. Restoring wild bison and their natural grazing patterns is important for restoring and maintaining the diversity of prairie habitats and species. (*Photo courtesy of Dennis Lingohr/American Prairie Reserve*)

200 years. Bison were extirpated from the region by hunting in the late 1800s, and truly wild bison are still rare in North America. With animals supplied by the U.S. and Canadian park services, the APR began reintroducing wild bison in 2005. By 2014, the herd had grown to approximately 400, with a long-term goal of thousands of bison roaming the prairie.

Fig. 4. Black-tailed prairie dogs on the American Prairie Reserve. Restoration of the black-tailed prairie dog populations is crucial for the recovery of the highly endangered black-footed ferret and for multiple other effects on prairie habitats and species. (*Photo courtesy of Dennis Lingohr/American Prairie Reserve*)

Fig. 3. Lion (*Panthera leo*) feeding on a zebra (*Equus burchelli*), Serengeti, Tanzania. (*Photo courtesy of Anne Hilborn, Virginia Tech*)

variations related to major threats that felids are facing, including habitat loss, habitat fragmentation, and climate change.

For background information *see* BEHAVIOR GENETICS; CAT; ENDANGERED SPECIES; EXTINCTION (BIOLOGY); GENE; GENETICS MAPPING; GENETICS; GENOMICS; PHYLOGENY; POPULATION ECOLOGY; POPULATION GENETICS; PREDATOR-PREY INTERACTIONS in the McGraw-Hill Encyclopedia of Science & Technology.

Claudia Wultsch

Bibliography. W. L. Allen et al., Why the leopard got its spots: Relating pattern development to ecology in felids, *Proc. R. Soc. Biol. Sci. B*, 278:1373–1380, 2011, DOI:10.1098/rspb.2010.1734; Y. S. Cho et al., The tiger genome and comparative analysis with lion and snow leopard genomes, *Nat. Comm.*, 4:2433, 2013, DOI:10.1038/ncomms3433; P. Christiansen, Canine morphology in the larger Felidae: Implications for feeding ecology, *Biol. J. Linn. Soc.*, 91:573–592, 2007, DOI:10.1111/j.1095-8312.2007.00819.x; P. Christiansen, Evolution of skull and mandible shape in cats (Carnivora: Felidae), *PLoS ONE*, 3(7):e2807, 2008, DOI:10.1371/journal.pone.0002807; R. Diogo et al., The head and neck muscles of the serval and tiger: Homologies, evolution, and proposal of a mammalian and a veterinary muscle ontology, *Evol. Biol.*, 295:2157–2178, 2012, DOI:10.1002/ar.22589; E. Eizirik et al., Defining and mapping mammalian coat pattern genes: Multiple genomic regions implicated in domestic cat stripes and spots, *Genetics*, 184:267–275, 2010, DOI:10.1534/genetics.109.109629; S. Lovari et al., Common and snow leopards share prey, but not habitats: Competition avoidance by large predators?, *J. Zool.*, 291:127–135, 2013, DOI:10.1111/jzo.12053; S. Lyngdoh et al., Prey preferences of the snow leopard (*Panthera uncia*): Regional diet specificity holds global significance for conservation, *PLoS ONE*, 9(2):e88349, 2014, DOI:10.1371/journal.pone.0088349; T. M. McCarthy and G. Chapron, *Snow Leopard Survival Strategy*, International Snow Leopard Trust and Snow Leopard Network, Seattle, WA, 2003; A. Schneider et al., How the leopard hides its spots: *ASIP* mutations and melanism in wild cats, *PLoS ONE*, 7(12):e50386, 2012, DOI:10.1371/journal.pone.0050386; M. Sunquist, What is a tiger?: Ecology and behavior, pp. 19–33, in R. Tilson and P. J. Nyhus (eds.), *Tigers of the World: The Science, Politics, and Conservation of Panthera tigris*, Academic Press, London, 2010.

Pressure-gain combustion

Pressure-gain combustion (PGC) represents a novel approach for significantly increasing the efficiency of aerospace propulsion systems and ground-based power systems. PGC is defined as a combustion process whereby there is a rise (gain) in the averaged total pressure across the device in which the combustion is taking place. This pressure rise is in contrast to the more typical 4–8% total pressure drop associated with the combustion processes that are currently utilized in aircraft propulsion and electric power generation systems. The pressure rise can be used to produce increased thrust in a ram propulsion application, where the combustor is followed by a nozzle. Alternatively, it can be used to extract additional shaft work from the downstream turbine in a gas turbine application. The latter application is particularly significant given the ubiquitous use of gas turbines in the United States. In 2011, the operation of these turbines alone was responsible for approximately 13% of fossil fuel consumption. Thus, even small improvements in their efficiency can have a substantial positive impact on energy consumption and the environment. For example, each 1% improvement in the thermal efficiency of the U.S. gas turbine fleet saves as much energy as is produced by 12 average-size wind farms, each consisting of 140 average-size wind turbines operating at full capacity.

Thermodynamics. From an ideal (that is, no component losses) thermodynamic perspective, PGC can be thought of as implementing an Atkinson or Humphrey cycle in a propulsion or power system rather than the currently implemented Brayton cycle. The Atkinson/Humphrey and Brayton cycles differ in the manner by which heat is added via combustion. In the Brayton cycle, combustion occurs at constant pressure. (Idealization assumes no pressure loss.) In the Atkinson/Humphrey cycle, it occurs at constant volume. The relative impact of these two combustion modes can easily be seen when the cycles are represented on a so-called temperature-entropy diagram (**Fig. 1**). Here, both cycles undergo the same mechanical compression of 25:1. This is a representative "middle of the road" value for modern gas turbines. Both cycles also have the same amount of heat added per unit mass of flow (that is, the same

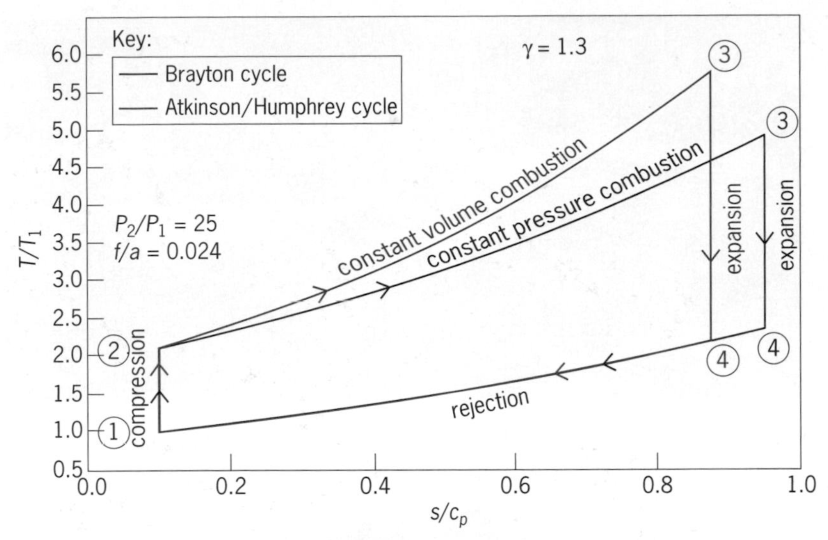

Fig. 1. Temperature-entropy diagram showing the ideal Brayton and Atkinson/Humphrey cycles. Both cycles have the same mechanical compression and fuel-to-air ratio. Subscripts on P (pressure) and T (temperature) refer to circled points shown on the diagram; s/c_p is the ratio of the specific entropy (entropy per unit mass) to the specific heat capacity (heat capacity per unit mass) at constant pressure.

fuel-to-air ratio, f/a). The working fluid is air for both cycles, as is the single value for the ratio of specific heats, γ. It is seen that constant-volume combustion produces less entropy than constant-pressure combustion for the same amount of chemical energy reacted. As a result, there is more energy available for the expansion phase of the cycle, which is represented by the length of the vertical line running from state 3 to state 4. Expansion occurs in the turbine (or nozzle) and produces the useful work of the cycle. With efficiency defined as net work (that is, turbine work extracted minus compression work input) divided by fuel energy added, the Atkinson/Humphrey cycle shown is 6% more efficient than the Brayton cycle.

Basic operation. A key feature of PGC is that it is fundamentally unsteady. By this is meant that periodic unsteadiness is necessary in order to achieve a combustion-driven pressure gain. This may be understood by considering a notional PGC device (**Fig. 2**). The device consists of a chamber with an ideal valve at the inlet end and a nozzle at the exhaust end. The working fluid that passes through the device undergoes a cyclic process that roughly consists of three phases:

1. The inlet valve opens and the chamber fills with a fuel-and-air mixture supplied by an upstream compression system.
2. The inlet valve closes and a rapid, confined combustion or chemical heat release follows that raises both temperature and pressure.
3. The chamber exhausts or blows down, sending the effluent through a turbine or nozzle. When the chamber pressure drops below the inlet pressure, the inlet valve reopens and the process repeats.

During the blowdown phase, the pressure (and temperature) in the chamber is continually dropping. As a result, there is a time-varying distribution of exhaust states and mass-flow rates from which

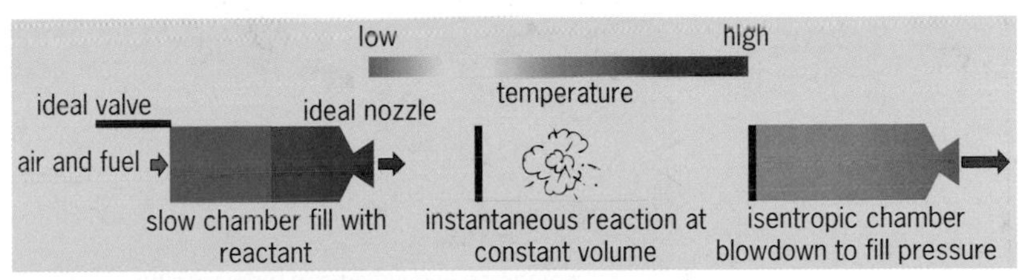

Fig. 2. Notional PGC device executing a three-phase cycle.

work can be extracted by a downstream component. However, the average value of this continuum of states yields a higher pressure than the inlet pressure (that is, pressure gain). The average temperature is precisely the same as that of a conventional steady-flow combustor that is normally used in propulsion and power systems. This latter outcome is important, since the average exit temperature of a combustor is an important performance parameter (higher is generally better); however, it is also typically limited by available materials. Thus, although PGC yields higher instantaneous exit temperatures than conventional combustion (for the same fuel addition), it does not yield higher average temperatures.

Implementation. PGC devices have been constructed and tested in a laboratory setting (**Fig. 3**). Generally, they are made up of a multitude of tubes, each of which (in a sequential fashion) undergoes some semblance of the process described earlier. The required inlet valving action is typically accomplished in one of two ways: the tubes are fixed and some type of valve regulates the inlet (Fig. 3*a*), or there is a fixed end plate with ducted openings, and the tubes rotate to regulate inflow (Fig. 3*b*). In the latter scenario, often referred to as a wave rotor implementation, there is also the possibility of regulating the exhaust flow.

(a)

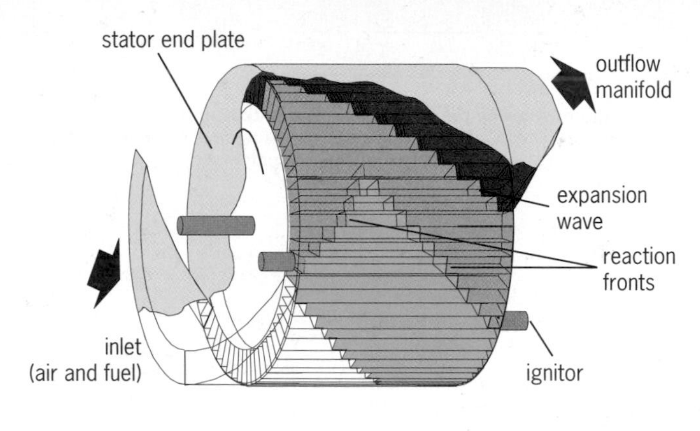

(b)

Fig. 3. Photograph and schematic representation of actual PGC devices. (*a*) Bank of stationary tubes (chambers) with dynamic inlet valves (*used by permission of the Air Force Research Laboratory, Dayton, OH*). (*b*) Annular ring of rotating tubes with fixed end plates for valving.

Few of these devices execute the notional process just described precisely, since instantaneous constant-volume combustion is not possible. Instead, combustion modes that produce very fast-moving combustion waves (for example, detonation) are utilized. The waves travel supersonically (or nearly so) and pass through all the reactive contents in the tube before any can escape from the exhaust end. The post-combustion products are in states very similar to that of actual constant-volume combustion.

Because of this wave style of combustion and the relatively long, narrow geometry of the chambers, the actual blowdown process described earlier is accomplished through a series of gas-dynamic waves that traverse the tubes rapidly (Fig. 3*b*).

Published results. Measured and predicted performance results available in the open literature are limited. The left-hand scale of the plot in **Fig.** 4*a* pertains to a particularly simple PGC device that utilizes only a partially confined (as opposed to constant-volume) combustion mode, often called resonant pulsed combustion (Fig. 4*b*). The simplicity of this device limits its performance potential; however, it also produces a more benign flow field (for example, lower material stresses on the confining walls). The plot shows total pressure as a function of total temperature across the device. It can be seen that a modest pressure gain is measured at all operating points. This magnitude of pressure gain yields a 1.7% improvement in thermal efficiency for the 25:1 mechanical pressure ratio engine described previously.

The right-hand scale of the plot refers to computational fluid dynamic (CFD) simulations of an idealized detonative device. The performance potential is clearly higher, as detonative PGC represents a much closer approximation to a constant-volume process.

Technology challenges. Like any significant advancement in technology, PGC faces challenges. These include the development of long-lasting, low-loss inlet valves; effective thermal management strategies; techniques for integration into the constrained space of a gas turbine; and effective operation of nominally steady-state turbomachinery in the presence of fundamentally unsteady PGC flows. Research is ongoing to address all of these areas. If it proves to be successful, PGC offers substantial efficiency improvements for future aerospace propulsion and ground-based power systems.

Disclaimer. This article was written as part of the author's official duties as an employee of the NASA John H. Glenn Research Center. However, the contents do not represent an endorsement by NASA of the technology concepts presented.

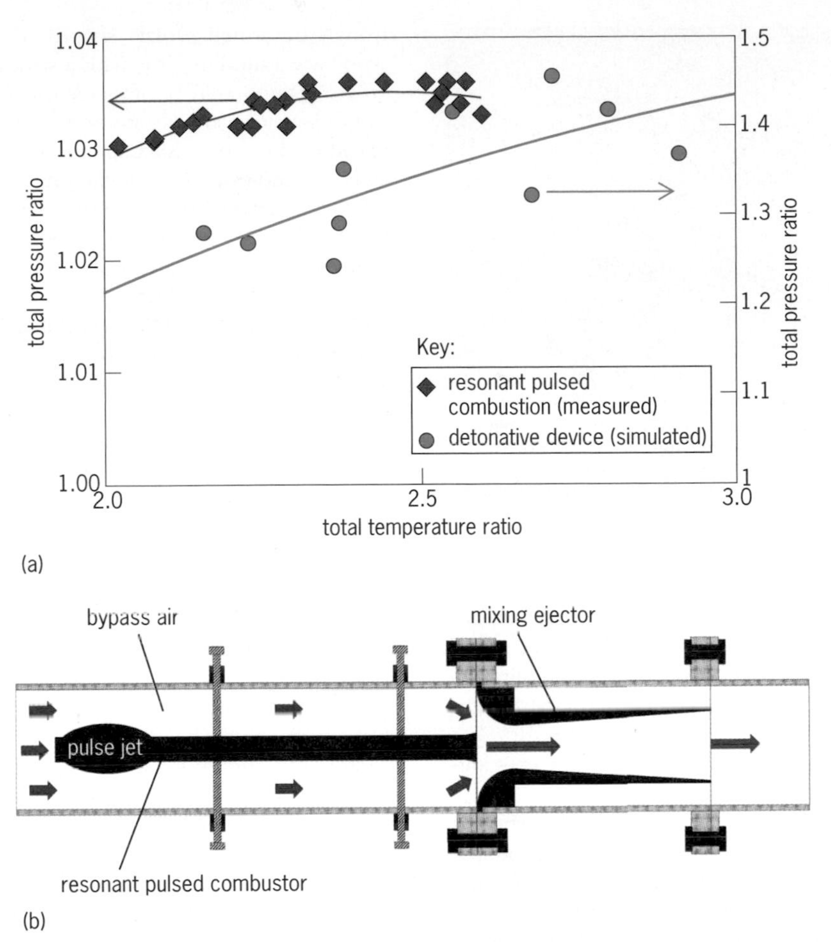

Fig. 4. PGC performance as shown by total pressure ratio versus total temperature ratio. (*a*) Measured performance of a resonant pulsed combustion device (red) and predicted performance of an ideal detonative device (green). (*b*) Schematic of the resonant pulsed combustion device on which measurements were made. The resonant pulsed combustor is also known as a pulse jet.

For background information *see* AIRCRAFT ENGINE PERFORMANCE; BRAYTON CYCLE; COMBUSTION; COMBUSTION CHAMBER; COMPUTATIONAL FLUID DYNAMICS; ENTROPY; GAS TURBINE; SCRAMJET; THERMODYNAMIC CYCLE; THERMODYNAMIC PROCESSES; TURBINE PROPULSION in the McGraw-Hill Encyclopedia of Science & Technology. Daniel E. Paxson

Bibliography. D. E. Paxson and K. Dougherty, Ejector enhanced pulsejet based pressure gain combustors: An old idea with a new twist, American Institute of Aeronautics and Astronautics, AIAA 2005-4216, in *41st AIAA/ASME/SAE/ASEE Joint Propulsion Conference & Exhibit*, Tucson, AZ, July 10–13, 2005, AIAA, Reston, VA, 2005, DOI:10.2514/6.2005-4216; D. E. Paxson and T. H. Kaemming, Influence of unsteadiness on the analysis of pressure gain combustion devices, *J. Propul. Power*, 30:377–383, 2014, DOI:10.2514/1.B34913; K. P. Rouser et al., Experimental performance evaluation of a turbine driven by pulsed detonations, American Institute of Aeronautics and Astronautics, AIAA 2013-1212, pp. 17950–17978, in *51st AIAA Aerospace Sciences Meeting including the New Horizons Forum and Aerospace Exposition*, Grapevine, TX, January 7–10, 2013, Curran Associates, Red Hook, NY, 2013, DOI:10.2514/6.2013-1212; P. H. Snyder et al., Pressure gain combustor component viability assessment based on initial testing, American Institute of Aeronautics and Astronautics, AIAA 2011-5749, pp. 2465–2477, in *47th AIAA/ASME/SAE/ASEE Joint Propulsion Conference & Exhibit*, San Diego, CA, July 31–August 2, 2011, Curran Associates, Red Hook, NY, 2011, DOI:10.2514/6.2011-5749.

Prolate-oblate shape transitions in nuclei

The shape of an atomic nucleus determines the allowed excited energy levels of a nucleus and other nuclear properties. Current and future developments of more sophisticated nuclear models to explain the behavior of nuclear matter inside a nucleus depend on knowing nuclear shapes. Nuclear shapes can be determined by measuring the properties of the excited states in a nucleus. Most atomic nuclei have either spherical shapes or prolate shapes (like an American or rugby football) for their ground states. In a few regions on the chart of the nuclides, for a given atomic number (*Z*), nuclei are found to change their ground-state shapes in a complex way from prolate shapes to oblate shapes (like a discus or thick pancake), or vice versa, as their neutron

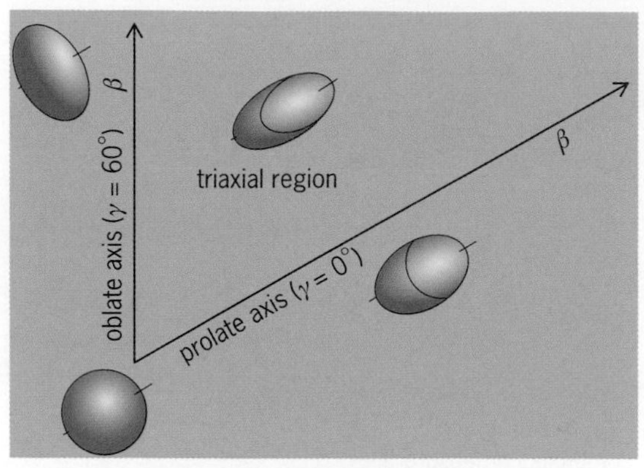

Fig. 1. Various nuclear shapes are shown in the β-γ plane between $\gamma = 0°$ on the prolate shape axis and $\gamma = 60°$ on the oblate shape axis (for $\beta > 0$). In-between nuclei are triaxial. At the origin, a nucleus is spherical. (*From F. Yang and J. H. Hamilton, Modern Atomic and Nuclear Physics, rev. ed., World Scientific, Singapore, 2010*)

number increases. Both prolate and oblate shapes are axially symmetric, with two axes being equal. Ground-state shape transitions with changing neutron number and coexistence of different shapes in one nucleus provide very sensitive tests to develop and refine microscopic models of the shapes and structure of nuclei.

Nuclear shapes. Besides prolate and oblate shapes, nuclei can have shapes where all three principal axes of the nuclei have different lengths, so that the nuclei are no longer axially symmetric. These are called triaxial shapes, and can be based on either prolate or oblate shapes. However, all nuclei were initially considered to have spherical shapes. When well-deformed ground states were observed in rare-earth and actinide nuclei, they all were found to have prolate shapes. Early searches for nuclear ground states with oblate or triaxial shapes had little if any success. Why oblate and triaxial shapes were not observed was a major problem.

Two parameters, β (beta) and γ (gamma), describe nuclear shapes (**Fig. 1**). Here, β is a measure of the total deformation of the nucleus from a spherical shape. (As beta increases, the long axis of a prolate nucleus increases while the total volume is constant.) The parameter γ is a measure of the degree of triaxiality of a nucleus (a measure of how different are the lengths of the three axes), with $\gamma = 0°$ corresponding to a prolate ellipsoid (for $\beta > 0$); with $\gamma = 60°$ or $-60°$ (the sign depends on the choice of definition) corresponding to an oblate ellipsoid; and with triaxiality increasing as γ approaches 30° or $-30°$, where maximum triaxiality occurs.

Shape transitions in selenium and krypton nuclei. Early inklings of the existence of oblate shapes came from nuclei around selenium-72 (^{72}Se, $Z = 34$) and krypton-74 (^{74}Kr, $Z = 36$), where nuclear shape coexistence was found in 1974 and 1981, respectively. In ^{72}Se, the ground state and states built on them were found to have near-spherical or oblate shapes, while other excited states were built on strongly deformed prolate shapes. The reverse behavior was found in ^{74}Kr, with a strongly deformed prolate ground state. Comparisons of the most recent theoretical calculations with experimental data reproduced the two coexisting shapes in these two nuclei. In addition, the calculations stressed the importance of triaxial shapes in the coexistence of the two shapes, and predicted a ground-state shape transition from oblate in ^{72}Kr, as was just established in 2014, to prolate in ^{74,76}Kr.

Shape transitions in ruthenium and palladium nuclei. Various theoretical calculations were carried out in the 1990s, predicting a similar, but reverse, prolate-to-oblate ground-state shape transition as the neutron number increased in neutron-rich ruthenium (Ru, $Z = 44$) and palladium (Pd, $Z = 46$) nuclei through the nuclides $^{107-114}$Ru and $^{108-116}$Pd.

Triaxial shape transitions in Ru nuclides. In the next decade, the results from studies of high-spin states in $^{109-112}$Ru were interpreted as being due to a triaxial shape transition from prolate to oblate occurring in ^{111}Ru. Theoretical studies of nuclear shapes throughout the chart of the nuclides indicated that ^{108}Ru is the center of the triaxial region on the nuclear chart, which has the largest lowering of the ground-state energy for a rigid triaxial shape compared to an axially symmetric shape. Then, new experimental studies of ^{108,110,112,114}Ru established ^{112}Ru instead of ^{108}Ru as the center of the region with a rigid triaxial shape. The case that ^{112}Ru acts as this center was based on the strong evidence for two sets of right- and left-handed chiral doublet bands of energy levels, with energy degeneracy and similar properties, and a staggering with spin of the energy levels in the gamma-type vibrational band (where the gamma vibration of a prolate-type nucleus is like pushing on the sides of the football, so that the shape parameter γ is no longer zero) to form bands of energy levels built on a wobbling motion of the angular momentum about a nuclear axis. Both of these phenomena as found in ^{112}Ru can occur only for rigid triaxial nuclei, but they are not found in ^{108}Ru.

Energy levels in Pd nuclides. In 2013, energy levels in $^{112-118}$Pd were investigated in greater detail both experimentally and theoretically to study their predicted shape transitions. **Figure 2** shows the changes in the energies (described by the rotational transition energy parameter, $\hbar\omega$, calculated as one half the transition energy, where $\hbar$ is Planck's constant divided by 2π and ω is the angular frequency of rotation) of various rotational bands as a function of the nuclear spins of the levels: the bands built on the ground states of ^{112,114,116,118}Pd; bands 1 in ^{115,117}Pd, built on an $h_{11/2}$ neutron (band-head energies of 177 and 267 keV, respectively, and spin and parity $11/2^-$); and excited bands in ^{115}Pd.

For $^{112-118}$Pd, one may note the first sudden changes in $\hbar\omega$ with spin around $\hbar\omega \sim 0.39$ MeV. These sudden changes arise from the rotational alignment of a pair of $h_{11/2}$ neutrons, due to the Coriolis force, to form a new rotational band. In ^{115,117}Pd,

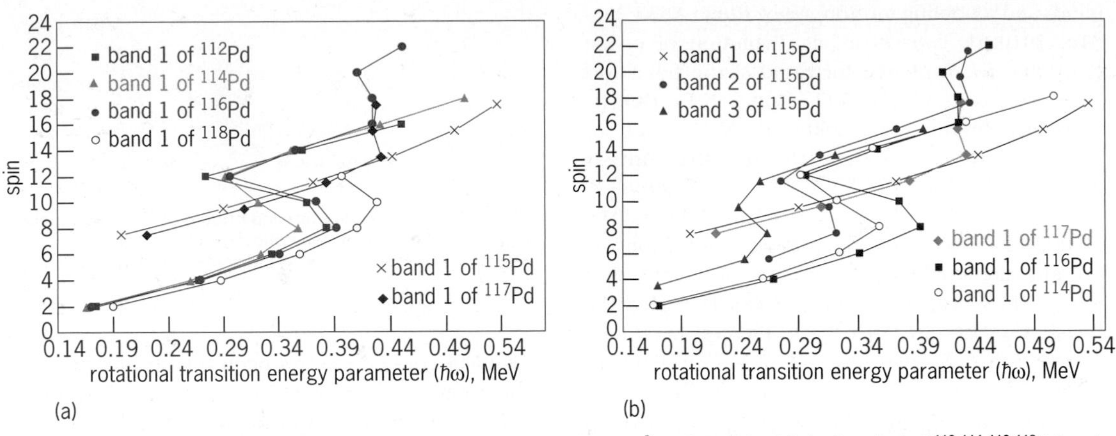

Fig. 2. Graphs of spin versus the rotational transition energy parameter, $\hbar\omega$, plotted for (*a*) the bands 1 of $^{112,114,116,118}Pd$ and (*b*) all the bands in ^{115}Pd and band 1 of ^{117}Pd. For comparison, bands 1 of $^{115,117}Pd$ are also shown in part *a*, and bands 1 of $^{114,116}Pd$ are also shown in part *b*. (*After Y. X. Luo et al., New insights into the nuclear structure in neutron-rich $^{112,114,115,116,117,118}Pd$, Nucl. Phys. A, 919:67–98, 2013, DOI:10.1016/j.nuclphysa.2013.10.002*)

bands 1, built on an $h_{11/2}$ neutron, do not exhibit these band crossings, which are blocked by the presence of the $h_{11/2}$ neutron that forms the band head.

One may also note that ^{116}Pd exhibits a second sudden change at $\hbar\omega \sim 0.42$ MeV, as does the $h_{11/2}$ band in ^{117}Pd. This second change arises from the alignment of a pair of $g_{9/2}$ protons to form a different new rotational band, and is accompanied by a dramatic change of their triaxial deformations. However, the second change is not seen in ^{114}Pd nor in band 1 of ^{115}Pd in the available region of $\hbar\omega$. On the other hand, the next higher energy in ^{115}Pd, band 2, built on a $g_{7/2}$ neutron (with band-head en ergy of 355 keV and spin and parity $7/2^+$) exhibits the same sudden change at the same value of $\hbar\omega$ as ^{117}Pd (Fig. 2*b*). These data indicate that bands 1 and 2 in ^{115}Pd have different shapes, and thus manifest prolate-oblate shape coexistence. The second change in ^{114}Pd is calculated to be delayed to a higher energy by the dramatic change of γ accompanying the alignment of a pair of $g_{9/2}$ protons.

Triaxial deformation parameter in Pd nuclides. Theoretical calculations traced the behavior of the triaxial deformation parameter γ in $^{110-118}Pd$ as a function of $\hbar\omega$, where $\hbar\omega = 0$ corresponds to zero angular frequency ω of the ground state. Recall that for $\gamma = 0°$ nuclei are prolate, and for $\gamma = -60°$ nuclei are oblate. As seen in **Fig. 3**, for $^{114,116}Pd$, the ground states and those states with $\hbar\omega$ lower than ~ 0.2 MeV have $\gamma \approx -60°$, for an oblate shape, while ^{110}Pd with $\gamma \approx -20°$ has a triaxial prolate shape, and $^{112,118}Pd$ with $\gamma \approx -40°$ are triaxial oblate. The calculations thus show a shape transition from triaxial prolate in ^{110}Pd, via triaxial oblate in ^{112}Pd, to nearly oblate in $^{114,116}Pd$, and back to triaxial oblate in ^{118}Pd, with shape-coexisting structures in ^{115}Pd. A triaxial oblate shape is calculated in ^{114}Pd for $\hbar\omega \approx 0.3–0.45$ MeV. Furthermore, the changes in γ as $\hbar\omega$ increases (Fig. 3) reproduce well the sudden changes of $\hbar\omega$ with spin observed in all the palladium isotopes (Fig. 2), again revealing the important role of the triaxial deformations.

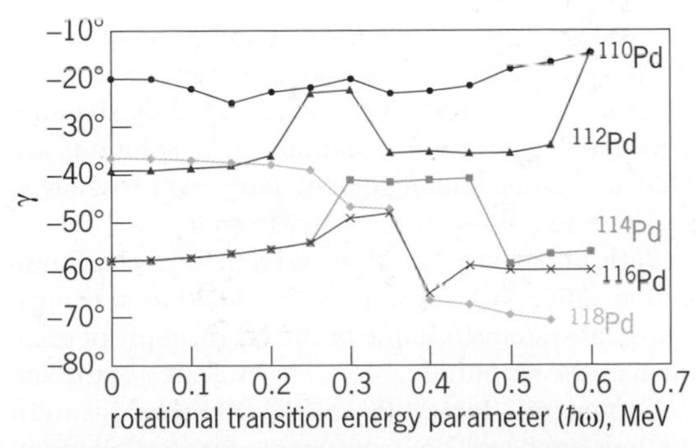

Fig. 3. Theoretically calculated values of the triaxial deformation parameters, γ, for the palladium isotopes plotted against the rotational transition energy parameter, $\hbar\omega$. (*After Y. X. Luo et al., New insights into the nuclear structure in neutron-rich $^{112,114,115,116,117,118}Pd$, Nucl. Phys. A, 919:67–98, 2013, DOI:10.1016/j.nuclphysa.2013.10.002*)

In summary, all the studies of nuclear shape changes from prolate to oblate or oblate to prolate in selenium, krypton, rubidium, and palladium nuclei now show clear pictures of complex, not simple, ground-state shape changes and shape coexistences, both with increasing neutron number and as a function of energy, and these phenomena are found to be strongly influenced by the triaxial degree of freedom in recent, more sophisticated theoretical calculations. Both the complexity of these prolate-oblate shifts and the central role of triaxiality in the shifts are significant advances in our understanding of the behavior of nuclear matter.

For background information *see* NUCLEAR SHAPE COEXISTENCE; NUCLEAR SHELL MODEL AND MAGIC NUMBERS; NUCLEAR STRUCTURE in McGraw-Hill Encyclopedia of Science & Technology.

Joseph H. Hamilton; Yixiao Luo; John O. Rasmussen

Bibliography. J. H. Hamilton et al., Evidence for coexistence of spherical and deformed shapes in ^{72}Se, *Phys. Rev. Lett.*, 32:239–243, 1974; J. H. Hamilton et al., Super deformation to maximum triaxiality in $A = 100$-112; superdeformation, chiral bands and wobbling motion, *Nucl. Phys.*, A834:28c–31c, 2010; H. Iwasaki et al., Evolution of collectivity in ^{72}Kr: Evidence for rapid shape transition, *Phys. Rev. Lett.*, 112:142502 (5 pp.), 2014, DOI:10.1103/PhysRevLett.112.142502; Y. X. Luo et al., New insights into the nuclear structure in neutron-rich ^{112,114,115,116,117,118}Pd, *Nucl. Phys. A*, 919:67–98, 2013, DOI:10.1016/j.nuclphysa.2013.10.002; P. Möller et al., Global calculations of ground-state axial shape symmetry of nuclei, *Phys. Rev. Lett.*, 97:162502 (4 pp.), 2006, DOI:10.1103/PhysRevLett.103.212501; R. B. Piercey et al., Evidence for deformed ground states in light Kr isotopes, *Phys. Rev. Lett.*, 47:1514–1517, 1981; K. Sato and N. Hinohara, Shape mixing dynamics in the low-lying states of proton-rich Kr isotopes, *Nucl. Phys.*, A849:53–71, 2011, DOI:10.1016/j.nuclphysa.2010.11.003; F. Yang and J. H. Hamilton, *Modern Atomic and Nuclear Physics*, rev. ed., World Scientific, Singapore, 2010.

Quantum communications

Quantum communications exploits the quantum mechanical properties of elementary particles such as photons to encode and decode information in order to transmit it between distant locations more securely and efficiently than classical communication systems can. This article will review these phenomena and techniques, and the enabling ideas that underlie them. In addition to the currently available solutions, the challenges involved will also be surveyed.

More than 100 years after its introduction, quantum mechanics continues to strain the human imagination. Among physical theories, it is particularly counterintuitive because it challenges assumptions that seem self-evident from our daily experience. An important example of this is the concept of superposition, whereby a quantum bit contains both of the logical values 0 and 1 at the same time, whereas a classical bit can contain only either one of them.

EPR paradox. As a way of explaining a very special quantum phenomenon called entanglement, we can perform a thought experiment that may be difficult to implement in practice, but that can be analyzed and understood directly in the context of quantum mechanics. A famous example is the so-called EPR paradox, which was introduced by Albert Einstein, Boris Podolsky, and Nathan Rosen in 1935; a twenty-first-century version of it is presented here.

Our assistants, Alice and Bob, send a laser beam through a piece of potassium dihydrogen phosphate (KDP) crystal. KDP is a nonlinear optical material that causes some of the photons in the laser beam to undergo a process called spontaneous down conversion, in which they split into photon pairs with the same polarization. The photons leaving the crystal thus form two beams, and a special quantum mechanical connection has been established between the photons of the two different beams (**Fig. 1**).

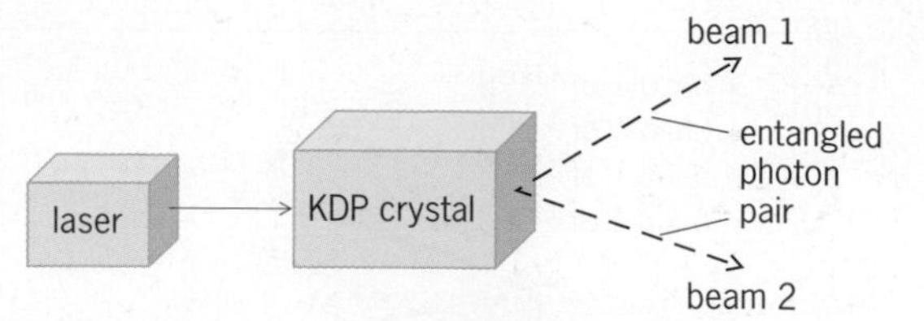

Fig. 1. Generation of entangled photon pairs. Spontaneous down conversion splits the photons into mutually entangled photon pairs with the same polarization.

Now, Alice picks up a photon from beam 1 and Bob catches its connected photon from beam 2. Next, Alice travels to the North Pole with her photon in her pocket, while Bob flies to the South Pole with his own particle. Exactly at noon, Alice puts her photon into a device that turns on a red bulb if the photon is vertically polarized or, alternatively, a green one if it has horizontal polarization. One femtosecond later, Bob measures his photon, too. Because of the special connection, called entanglement, between the two photons, the results of the two measurements will be the same; that is, Bob will obtain the same color as Alice did. One femtosecond is a very short period of time, less time than that required for light to travel from Alice to Bob. Einstein called this apparent contradiction (paradox) to the relativity theory "spooky action at a distance" and never accepted it. Experimental results eventually confirmed the existence of this strong connection between specially prepared elementary particles. For Einstein's sake, nature does not allow Alice and Bob to set the bulbs intentionally. The operation remains random. No information is transmitted faster than the speed of light; only the coordination between the bulbs determines the color of the second bulb.

Entanglement is one of the most efficient and potentially dangerous tools of quantum communications and could, for example, enable teleportation, communication over zero-capacity channels, and tremendously fast algorithms for computer scientists. In fact, it is often mentioned as a basic resource. On the other hand, unintentional entanglement with the physical environment is a major obstruction, since it cannot be avoided by shading. Fortunately, quantum communication solutions are able to handle entanglement fairly well, whereas computing architectures suffer from it.

Basic principles. Classical and quantum communication differ in their applied modulation schemes. Classical systems use one of the classical properties, such as the amplitude, phase, or presence of the signal, to encode logical 0s and 1s, whereas quantum systems are based on quantum mechanical phenomena, for example, various types of polarization of photons.

The simplest and best-studied quantum communication protocol is teleportation, in which quantum information is transmitted using an entangled photon pair and a classical communication channel (**Fig. 2**). To transmit quantum information, Alice and Bob share the entangled photon pair. Alice entangles her half-pair with the unknown quantum state to be transmitted and converts the result to classical bits. These bits are then sent to Bob via a classical channel. Bob controls his receiver by using the received classical bits to modify his half-pair to make a copy of the original quantum state. Is this system, then, a copying machine, violating the no-copy theorem (discussed later)? No, because the first quantum state disappears from Alice's transmitter before Bob produces his own state. Interestingly, no quantum channel is needed; a preshared entanglement supported by classical communication is enough.

However, when discussing the possibilities of quantum communication, we should never forget that human beings are able to understand information only in classical form—for example, a sequence of 0s and 1s. Therefore, typical quantum communication systems must accept and emit classical bits, although the information is delivered via superpositions or even by means of preshared entangled photon pairs.

Communication channels suffer from the effects of noise and interference. To overcome these natural and artificial phenomena, error correction codes for classical systems have been developed. Fortunately, their generalization to quantum channels has been successful in spite of the fact that quantum channels typically have rather different behavior.

No-copy theorem. Classical computers often use the COPY command to make several copies of a register or object. We would like to generalize the COPY command to quantum objects, but when we attempt to do this, we fail. Interestingly, only special quantum states (for example, classical states) can be copied without error.

Secure quantum communication. When quantum computers spread all over the world, any public key based cryptography solution will fall victim to them. However, even without this threat, malicious hackers regularly attack classical communication networks. The restriction provided by the no-copy theorem can be exploited in a rather straightforward way to open new techniques for strong secure communication (**Fig. 3**). Alice encodes her information into carefully selected photon polarizations. Eve the eavesdropper catches these photons by inserting a copying machine in the quantum channel. In order to remain undiscovered, she tries to make one copy for herself and one to forward to Bob. However, because of the no-copy theorem, both copies will differ from the original state, and thus Bob will detect her intervention. In order to recognize Eve's action, Alice and Bob need a public classical channel that allows them to exchange and compare several randomly selected bits. However, eavesdropping on these bits will not help Eve, since they are not connected to the secure information and are simply discarded.

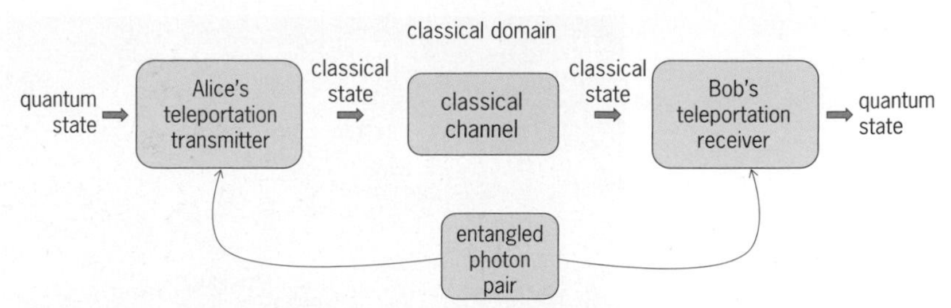

Fig. 2. Teleportation, a quantum communication protocol.

Alice and Bob can share encryption keys in a very secure manner that can be utilized for communication, even over classical channels. Secure quantum communication requires quantum key distribution (QKD). First-generation QKD protocol proposals, such as BB84 or B92, used single-photon sources and single-photon detectors, which made their implementation expensive, and the effective communication distance was rather limited. Currently available second-generation QKD solutions apply weak photon pulses comprising a few dozen photons to carry the quantum information (**Fig. 4**).

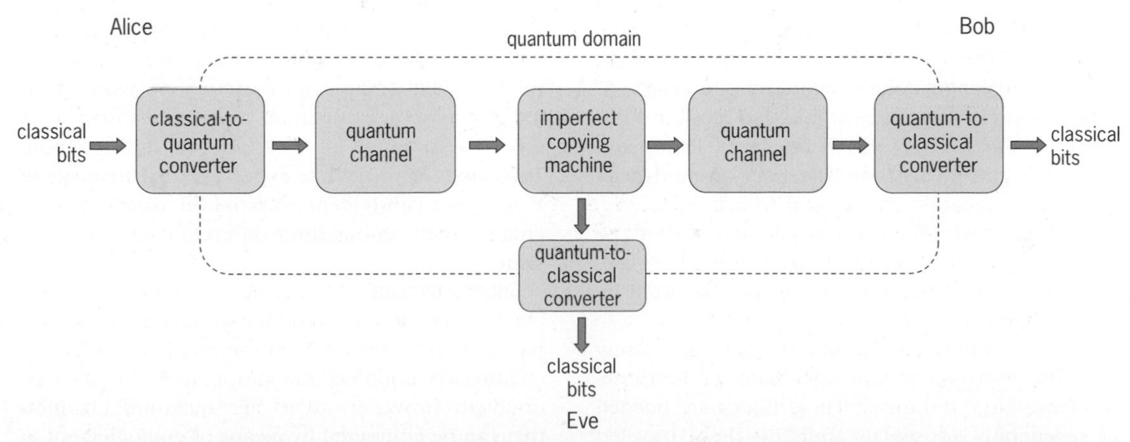

Fig. 3. Basic scheme of quantum key distribution. Alice encodes her classical bits into quantum properties of photons by means of a classical-to-quantum converter. Since arbitrary quantum states cannot be copied without error, Eve's interaction will modify the classical bits detected by Bob, and thus her eavesdropping action will be revealed.

Fig. 4. Second-generation QKD. This experimental QKD realization at the Budapest University of Technology and Economics (BME) applies photon packets made from coherent states to carry quantum information between Alice and Bob.

This makes it possible to produce QKD devices for the public market.

As with classical communication systems, one of the primary challenges of free-space quantum communications is handling the disturbing effects of the very noisy environment compared with the wired case. Since the introduction of QKD by Charles H. Bennett and Gilles Brassard in 1984, several groups have demonstrated quantum communications and quantum key distribution over multikilometer distances using both optical-fiber and free-space transmission. The first free-space QKD experiment was performed in 1991. In 1998, Los Alamos National Laboratory, New Mexico, performed a free-space QKD over outdoor optical paths up to 950 m (3117 ft) in length under nighttime conditions. In 2006, an international research group reached a distance of 144 km (89 mi) between La Palma and Tenerife in the Canary Islands. A space-ground quantum communication experiment from the *International Space Station (ISS)* is planned within the framework of the Space-QUEST (Quantum Entanglement in Space Experiments) project supported by the European Space Agency.

Quantum communication networks. Global communication networks require amplifiers or repeaters to bridge large distances. These nodes are needed to regenerate the signals after they have traveled a certain distance because they suffer from attenuation and noise in the communication channel. Amplification can be regarded as a form of copy function. The repeater receives several photons and emits a great number of copies of them. Unfortunately, the no-copy theorem, while allowing for secure quantum communication, makes the design of quantum repeaters very difficult. Another challenge that was solved recently is using the same optical fiber for classical and quantum communications in a parallel manner. Furthermore, in order to exploit the already deployed infrastructure, these combined classical-quantum protocols should be capable of operating over commercial optical fibers. For general-purpose quantum networks, customers will have to wait several years; however, special applications like QKD are much closer. Researchers at the Los Alamos National Laboratory are working on extending secure quantum communications from the current range of 100 km (62 mi) to about 1000 km (621 mi), while experts at the University of Waterloo recently demonstrated the distribution of entanglement among three different locations at the same time.

Superactivation. Classical communication channels—like water conduits—have a certain capacity. Furthermore, when the number of classical channels is doubled, the throughput will also be doubled. However, there are quantum channels that can be combined by means of entanglement in such a way that the joint capacity will be more than twice the individual capacity. This phenomenon is called superadditivity. This strange effect operates even in the case of quantum channels with zero throughput. In other words, two fully choked pipes that are equipped with entanglement can deliver information between the input and the output; these channels are said to have been superactivated.

For background information *see* CRYPTOGRAPHY; DATA COMMUNICATIONS; ELECTRICAL COMMUNICATIONS; ELECTRICAL INTERFERENCE; ELECTRICAL NOISE; INFORMATION THEORY; MODULATION; NONLINEAR OPTICS; OPTICAL COMMUNICATIONS; OPTICAL FIBERS; PHOTON; QUANTUM COMPUTATION; QUANTUM MECHANICS; QUANTUM TELEPORTATION; SUPERPOSITION PRINCIPLE in the McGraw-Hill Encyclopedia of Science & Technology. Sándor Imre

Bibliography. C. H. Bennett and G. Brassard, Quantum public key distribution system, *IBM Tech. Disclosure Bull.*, 28:3153–3163, 1985; S. Imre, Quantum communications: Explained for communication engineers, *IEEE Comm. Mag.*, 51(8):28–35, 2013, DOI:10.1109/MCOM.2013.6576335; R. Van Meter and J. Touch, Designing quantum repeater networks, *IEEE Comm. Mag.*, 51(8):64–71, 2013, DOI:10.1109/MCOM.2013.6576340

Radio and television broadcasting during disasters

For many decades, radio and television broadcasters have been the primary source of critical information to the public in the event of disasters such as tornadoes, hurricanes, tropical storms, floods, snowstorms, earthquakes, tsunamis, solar storms, terrorist violence, mass transportation accidents, and industrial or technological catastrophes. Broadcasters can play this important role both before an impending event and also during and after an event. On these occasions, radio and television broadcasting provides reliable point-to-everywhere delivery of essential information and safety advice to the public, to first responders, and to others via widely available consumer receivers, both mobile and fixed.

By their nature, natural and human-made disasters, whether impending or immediate, quickly capture the attention of a very large majority of the entire population in an affected area. In disaster situations, members of the public seek at first to be informed, so that they may understand what is happening, and to assess whether and how they and their family and friends may be affected. The typical individual reflex is to tune to radio or television broadcast stations that are known to have a strong record of serving viewers by reporting and interpreting emergency situations. The broadcast listening and viewing public is aware that, in such circumstances, scheduled radio and television programming is quickly interrupted by broadcast station news personnel who report information they have collected from many sources throughout the emergency. As coverage continues, broadcasters include information from reporters at various scenes; police and fire departments; relevant federal, state, and local government agencies; weather and geological bureaus; and the like.

Following the initial need for information, personal communication is then attempted as citizens seek rapid contact with family members and friends. At this point, communication networks can suffer connectivity failures caused by blocking or traffic congestion, and often loss of power at key network centers, cellular transmission towers, fiber links, or other intermediate processing points. Failed elements can include wired and wireless telephone and mobile data networks, cable television networks, and in cases of severe weather, direct-to-home satellite services. Although these nonbroadcast media often suffer infrastructure failure within a disaster area, broadcasting's architecture is uniquely simple and powerful. If the main transmitter and the radio or television studios that feed it remain on the air, reception is available wherever there are working receivers.

Importance. A particular attribute of information provided by many broadcasters is the professional quality of the compilation and analysis of local facts and guidance by experienced broadcast station news personnel, who often report around the clock. Broadcasters gather and convey information and video from reporters deployed at various locations including police and fire departments; relevant national and local government agencies; and weather, geological, scientific, and medical bureaus.

Radio and television broadcasters have an expert ability to interpret information and impact for their viewers in the local broadcast coverage area. Viewers are offered comfort by hearing or seeing well-known, trusted local news reporters and anchors interpret the situation and provide advice and guidance for viewers' safety.

All but the smallest radio and television stations have an important ability to gather and summarize information for the public by bringing to bear their electronic news, traffic, and meteorological personnel with special knowledge of the local area, and their field audio and video reporting capabilities, as well as sophisticated graphics, mapping, and weather radar systems. Most broadcasters have disaster plans that include the presence of backup generator power at key studio and transmitter locations, and associated long-term fuel storage, as well as backup facilities in secondary locations where information gathering and studio work can be moved if the primary location is disabled because of catastrophic conditions.

It is also common for radio and television receivers to be available in critical locations, such as police and fire stations, hospitals, government buildings, auditoriums, indoor stadiums, and public shelters, often with backup generator power. Thus, both citizens and emergency responders will often benefit from the distribution of key information by local broadcasters. In the United States, emergency broadcasts are provided free of charge to the public and are planned, executed, and fully paid for by

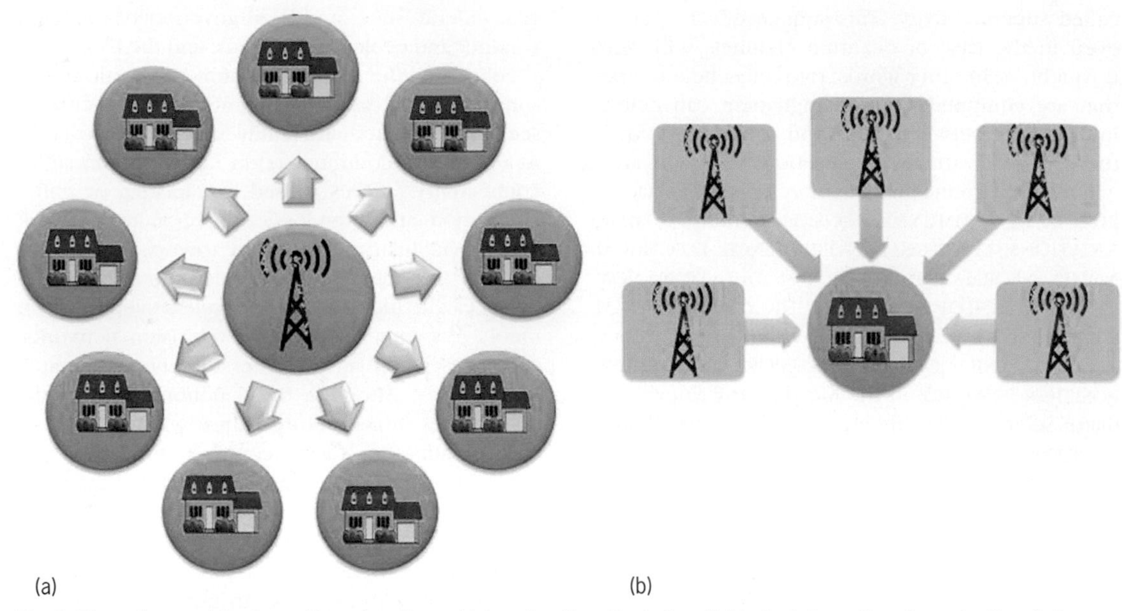

Fig. 1. The unique architecture of broadcasting, which makes it particularly suitable for information dissemination during disasters. (*a*) For the broadcaster: one-to-many architecture, providing reach and scale. (*b*) For the household: many-to-one communication, providing survivability.

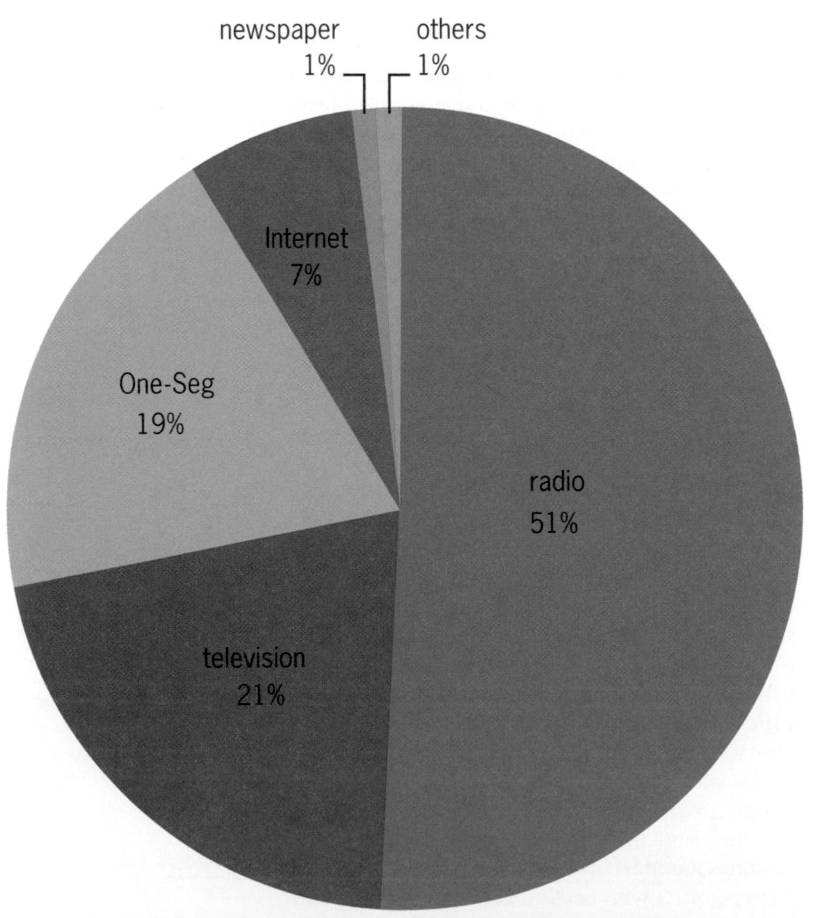

Fig. 2. Media use in Japan immediately following the Great East Japan Earthquake in March 2011. (*After NHK Broadcasting Culture Research Institute, The Great East Japan Earthquake: How disaster survivors used the media (Japanese), The NHK Monthly Report on Broadcast Research, Japan Broadcasting Corporation, September 2011, http://www.nhk.or.jp/bunken/summary/research/report/2011_09/20110902.pdf*)

commercial broadcasters as part of their business mission, which includes the responsibility to serve their communities in times of special need.

The architecture of broadcasting is particularly suited to maintaining operations during disasters (**Fig. 1**). Operating as a point-to-multipoint network, with high-power transmissions from a single tall tower, a broadcast transmitter can theoretically reach an unlimited number of receivers within its service area, as adding receivers has no effect on the network load, unlike the mobile telephony network, which can easily overload in times of crisis because of unusually high demand.

From the household perspective, not only is the broadcast network highly reliable because of the one-to-many nature of the architecture, but there are typically multiple broadcast transmission towers that are geographically located in different places. There is consequently high likelihood that at least one broadcast transmission will continue to be available during an emergency as a result of this geographic diversity.

Infrastructure and procedures. The broadcast imperative is to be on the air and available at all times, especially during emergency situations. Most facilities possess redundant capabilities and signal paths in order to maintain their over-the-air signals. In larger markets, more robust measures are employed. These are usually "case-hardened" facilities that include multiple power feeds from diverse power generation stations, full backup power generators at the studio and transmitter sites, multiple signal paths from studio to transmitter sites, redundant transmitters and antennas, and direct feeds to cable and satellite operators. All of these minimize the number of single points of failure that could keep vital information from being broadcast.

Television and radio broadcasting's commitment to providing local news and information for many hours a day has created established in-house procedures to deal with the dissemination of all types of news. These same procedures are easily and quickly adaptable to provide life and safety information to the public. Stations are linked via emergency alert systems to state and national emergency information channels and can repeat messages from civil and governmental authorities very quickly. Electronic news gathering and satellite uplink vehicles are quickly deployed to be on the scene with live pictures and sound. Closed-captioning systems, along with full-screen graphical displays, news "tickers," and lower-third-screen text information, make sure that those who are hearing impaired are also provided with emergency information.

As society becomes more mobile, there is increased appreciation by broadcasters for including broadcast reception capability in mobile devices. In Japan, for example, mobile digital television service is available in a large majority of mobile phones. In some parts of the world, such as Europe, FM radio reception capability in mobile phones is commonplace, whereas in the United States and some other countries, this feature is less prevalent. Active programs are under way in the United States to encourage mobile network providers and phone manufacturers to include broadcast signal reception in more products.

A number of emergency warning systems exist that allow broadcast networks to alert people of impending disasters and enable them to prepare for emergencies. The emergency warning systems use special alert signals to issue an emergency bulletin, alerting people to an impending or existing critical situation.

In the United States, the Emergency Alert System (EAS) is a public alert and warning system that uses the communications assets of EAS participants, including terrestrial broadcasters, cable television systems, wireless cable systems, Satellite Digital Audio Radio Service (SDARS) providers, direct broadcast satellite (DBS) services, and wireline video service providers, to allow the President of the United States the capability to address the American public during a national emergency. This system is designed to be available under all conditions. The system is also used by state and local authorities to deliver important emergency information, such as AMBER (America's Missing: Broadcast Emergency Response) alerts. The National Weather Service (NWS) also regularly uses the system to disseminate emergency weather alerts and advisories.

Case study of earthquake in Japan. An earthquake with a magnitude of 9.0 hit Japan on March 11, 2011. Some 30 min to 1 h after the earthquake, tsunami waves at least 10 m (33 ft) high hit various parts of the Japanese archipelago, causing serious casualties and devastating damage in coastal areas. Strong shaking and the tsunami triggered a nuclear crisis at Tokyo Electric Power Company's Fukushima Daiichi Nuclear Power Station. The disaster left more than 15,000 people dead and almost 3000 people missing. In all, 126,483 houses were destroyed and 272,287 others were damaged.

According to investigations by NHK Broadcasting Culture Research Institute, the most popular media used for getting information soon after the Great East Japan Earthquake were radio (51%), television receiver (21%), and One-Seg (19%) [**Fig. 2**]. One-Seg is one of the features of the Japanese digital television standard, which enables television programs to be received on mobile phones. In the disaster-stricken area, the information could not be acquired from television receivers because of power outages. Mobile phone calls and text messages were not connected because of traffic congestion and damage to infrastructure, and this meant that most people were isolated from information sources. A broadcast radio receiver, which can operate for long periods with a portable battery, was the most vital device to obtain information in this situation. Only about 1 in 14 people used the Internet during this critical time after the disaster.

For background information *see* CABLE TELEVISION SYSTEM; CLOSED-CAPTION TELEVISION; INTERNET; MOBILE COMMUNICATIONS; OPTICAL COMMUNICATIONS; RADIO; RADIO BROADCASTING; SATELLITE TELEVISION BROADCASTING; TELEPHONE SERVICE; TELEVISION; TELEVISION TRANSMITTER; VOICE OVER IP in the McGraw-Hill Encyclopedia of Science & Technology.

Lynn Claudy

Bibliography. Federal Emergency Management Agency, *An Emergency Alert System Best Practices Guide-Version 1.0*, update, 2012; International Telecommunication Union, Broadcasting for public warning, disaster mitigation and relief, Report ITU-R BT.2299-0, 2014; International Telecommunication Union, Use of satellite and terrestrial broadcast infrastructures for public warning, disaster mitigation and relief, Recommendation ITU-R BT.1774, 2007; NHK Broadcasting Culture Research Institute, The Great East Japan Earthquake: How disaster survivors used the media, *The NHK Monthly Report on Broadcast Research*, Japan Broadcasting Corporation, September 2011; D. E. Wilson, Emergency alert system, pp. 247–277, in E. A. Williams et al. (eds.), *NAB Engineering Handbook*, 10th ed., Elsevier, Burlington, MA, 2007.

Remote sensing with small unmanned aircraft systems

In science, remote sensing refers to the process of observing the Earth and other terrestrial planets and moons from above. It is used in many practical applications and day-to-day operations, including mapping, weather forecasting, and as an aid for measuring the state and trends of the environment. In order to deliver effective tools that meet the needs of researchers and professionals, remote sensing is continually evolving. Over the past several decades, major advances have occurred in sensor technology, software, and public access to remote sensing data. Yet the conventional remote sensing platforms, upon which most sensors are placed, have not changed considerably. Most remote sensing data is still acquired with manned aircraft and satellites, platforms that have several trade-offs for some applications in terms of their high cost and limited operational capacity. However, this may be changing with the relatively recent introduction of remote sensing with small unmanned aircraft systems (UASs).

Small UASs are a type of autonomous aerial robot that consist of an aircraft (also known as drone), a ground control system (GCS), a pilot or navigator who operates the aircraft from the GCS, and one or more spotters who monitor other aircraft and hazards in the survey area. The term "small" refers to a certain UAS class (<25 kg). The number of small UASs for civil, commercial, and scientific applications is expected to increase rapidly because they are less expensive and more versatile than larger UASs. Already, small UASs equipped with cameras are being used for nonmilitary purposes, such as precision agriculture, pipeline inspection, utilities management, mapping, wildlife enumeration, search and rescue, and hazard assessment.

A number of advantages make small UASs appealing for remote sensing. Small UASs (1) are relatively inexpensive to purchase and operate compared to manned aircraft; (2) perform missions and acquire remote sensing data autonomously so that human operation is minimized; (3) are highly maneuverable; (4) can fly at low altitude, which increases the spatial resolution of the imagery; (5) can fly below cloud cover, which often limits the usefulness of some form of satellite remote sensing; and (6) reduce the risk to pilots. However, research has shown that there are also some important challenges in remote sensing with small UASs, including (1) an immature regulatory landscape, which may slow or prevent certain remote sensing applications from being realized; (2) distortions that result from the low-altitude images and the process of combining a large number of overlapping images into one; (3) difficulty in adapting image processing techniques developed for conventional satellite and airborne imagery; and (4) limited areal coverage due to regulations, small power systems (for example, batteries), and limited fuel-carrying capacity.

Small UASs for remote sensing. Small UASs used for remote sensing normally may be fixed-wing or vertical take-off and landing (VTOL) aircraft (**Fig. 1**). In many ways, they resemble conventional radio-controlled or model aircraft in terms of their airframes and engines. But what makes small UASs stand apart is that they also have integrated autopilot systems that enable autonomous flight and remote sensing. Typical autopilots for remote sensing are designed for passive sensors such as cameras.

In general, small UASs have several integrated components: (1) a Global Navigation Satellite System [GNSS, such as GPS (United States), GLONASS (Russian), and Galileo (European)] receiver to measure aircraft position and airspeed; (2) an inertial measurement unit (IMU) that measures aircraft attitude; (3) a microprocessor for flight control and interfacing with the remote sensing payload; and (4) a flight data recorder that logs aircraft position and attitude parameters during image acquisition.

Each aerial survey is pre-programmed with flight-planning software in the GCS. The software calculates the waypoints for the images based on user-specified parameters, such as the amount of image overlap, the desired ground resolution, and the camera focal length. During flight, the position of the aircraft is relayed to the GCS by telemetry, which allows the pilot or navigator to monitor its progress and make adjustments if needed.

(a)

(b)

Fig. 1. Examples of two types of small unmanned aircraft systems (UASs) used for remote sensing: (*a*) a vertical take-off and landing (VTOL) UAS (quadcopter) used for a river survey and (*b*) a fixed-wing UAS used for mapping a glacier.

In some ways, small UASs are just small-scale versions of a manned aircraft. However, there are two notable differences in the context of remote sensing. First, the payload weight capacity of a small UAS is typically much smaller than for a manned aircraft. This restricts the size and type of remote sensing system that can be supported by the aircraft, and has so far limited most applications to those involving passive imaging sensors such as digital cameras. Laser- and radar-based sensors, as well as other advanced passive sensors that are currently used on manned aircraft, may be more commonly used on small UASs in the future as the technology improves and is scaled down to meet the payload weight restriction. Second, in most countries, aviation regulations limit the flying height of the aircraft and require that it is operated within line of sight (LOS). In most cases, small UASs are required to operate within a few hundred meters above the surface, although this may vary from one country to another. While this provides the opportunity to acquire imagery with high spatial resolution, it can also require a large number of images in order to cover the survey area. The combination of high spatial resolution and numerous (perhaps hundreds) of images results in long processing times and can result in radiometric and geometric distortions in the final mosaic (composite) image. The LOS restriction limits the size of the survey area and necessitates a series of overlapping surveys for larger areas.

Evan Thomas Creek flood. A case study from an aerial survey after a major flood in June 2013 along Evan Thomas Creek in southern Alberta, Canada, is presented to help illustrate the opportunity and challenges of remote sensing with a small UAS (**Fig. 2**). The main purpose of the survey was to develop a topographic map of the creek and to test the accuracy of the data relative to ground-based measurements from a survey-grade GNSS. The survey was performed with a small fixed-wing UAS that used a consumer-grade camera to acquire images. A Special Flight Operation Certificate was issued for the survey. Prior to the flight, 22 ground control targets visible in the imagery were distributed throughout the survey area and the map coordinates of each were estimated with the GNSS to within a few centimeters, both horizontally and vertically. The purpose of the targets was to transform the imagery into the map coordinates during the processing stage (orthorectification). The survey was flown in September 2013

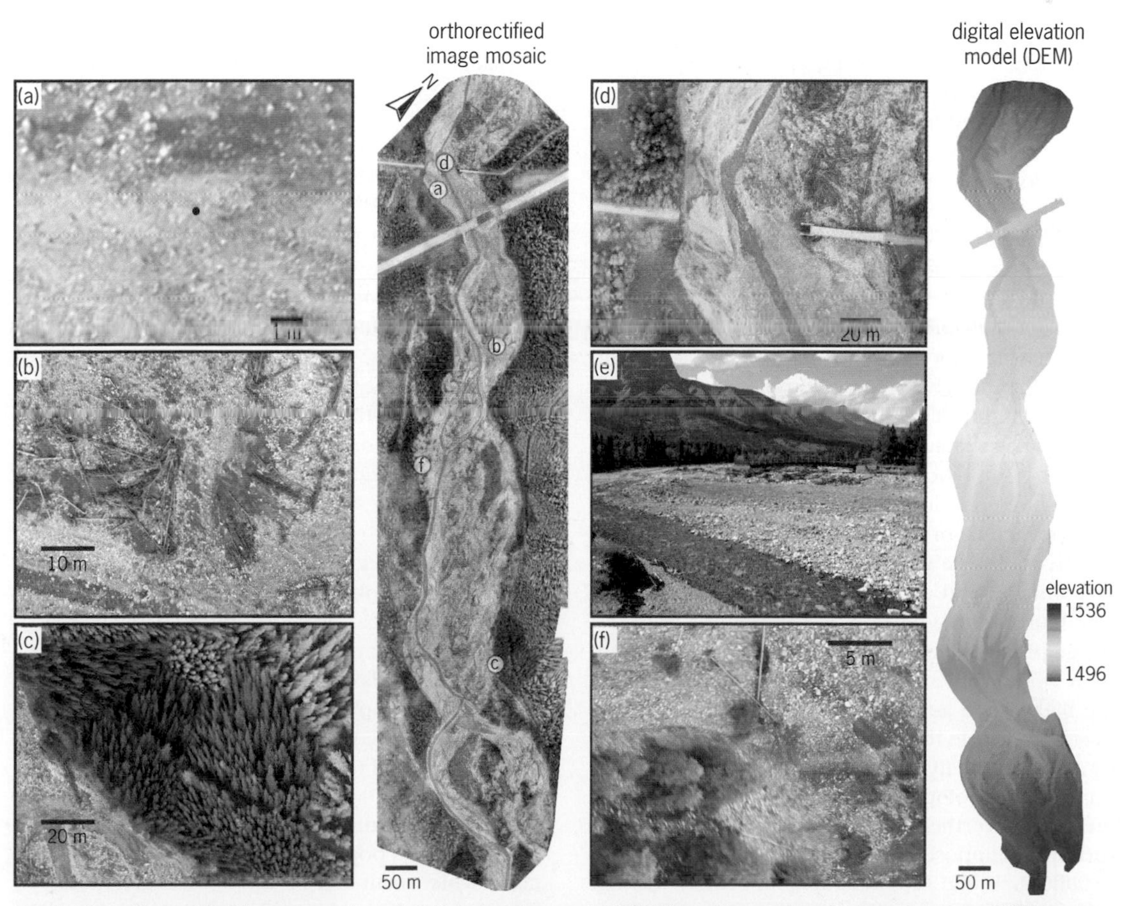

Fig. 2. The orthorectified image mosaic and digital elevation model (DEM) from the unmanned aircraft system (UAS) survey of Evan Thomas Creek, Alberta, Canada. This creek flooded in June 2013, wiping out two bridges and completely altering the riverbed. Some of the features resolved in the imagery include: (*a*) a ground control target, (*b*) trees and woody debris, (*c*) geometric and radiometric distortions, (*d*) damage to a pedestrian bridge with the corresponding ground photo in (*e*), and (*f*) blurring in portions of the image.

and consisted of 188 images from five northwest-southeast trending flight lines flown at an approximate height of 122 m above the ground. By applying photogrammetric processing techniques—the process of extracting the 3D world from 2D images—a digital elevation model (DEM) and an othorectified image mosaic were produced (Fig. 2).

The imagery and maps generated from the aerial survey provide unprecedented spatial resolution. Each pixel in the DEM and image mosaic is equivalent to about 3.5 cm on the ground, which is about 10 times the resolution typically available from conventional platforms. This means that features in the image that are about 10 cm (that is, three or more pixels) or larger can be resolved, including individual boulders, trees and woody debris, bank erosion, and a whole host of sedimentary features generated by the flood.

In addition to documenting the state of the creek post-flood, an accuracy assessment of the DEM and image mosaic shows that the horizontal error of the mosaic is only 3 and 4 cm in the east and north directions, and the vertical error of the DEM is only 5 cm. This means that on average, the coordinates of features in the image mosaic and DEM are very close to the coordinates that would be measured using ground-based survey methods. For sake of comparison, the horizontal and vertical error in imagery and DEMs acquired with conventional platforms are typically 10–100 times greater. So, overall, the mapping accuracy achievable with the small UAS is remarkable.

On closer examination, there are notable distortions in the mosaic image that highlight some of the challenges of remote sensing with small UASs. The distortions are either geometric or radiometric (brightness). An example of a geometric distortion, called relief displacement, is the apparent and inconsistent lean of trees in the forest bordering the river (Fig. 2*c*). This occurs because each image is taken from different locations, and objects near the edge of each image appear to lean away from the center. Thus, when a large number of overlapping images are combined into one, some features have the appearance of leaning in various directions, like the trees. Two examples of radiometric distortion are blurring and differences in image brightness. The blurring may be from the camera or from the image processing. Variations in brightness are common when ambient lighting conditions change during flight, such as when a cloud blocks the Sun. It may be possible to reduce distortions in the future by adjusting the flying height and by developing new imaging technology and procedures to correct the imagery. Nevertheless, the case study shows that distortions do impose certain limitations.

Outlook. While the limitations of remote sensing with small UASs are still not well understood, there is growing evidence of their advantages in situations where the cost, efficiency, and resolution favor their use over conventional remote sensing platforms. In many countries, the proliferation of small UASs for remote sensing applications ultimately depends on how civil aviation regulations are adapted to permit their use.

For background information *see* AERIAL PHOTOGRAPHY; AUTOPILOT; CAMERA; CARTOGRAPHY; IMAGE PROCESSING; INERTIAL NAVIGATION SYSTEM; REMOTE SENSING; SATELLITE NAVIGATION SYSTEMS; TOPOGRAPHIC SURVEYING AND MAPPING; UNINHABITED AERIAL VEHICLE UAV in the McGraw-Hill Encyclopedia of Science & Technology.

Christopher Hugenholtz

Bibliography. C. H. Hugenholtz et al., Small unmanned aircraft systems for remote sensing and Earth science research, *EOS*, 93(25):236–237, 2012, DOI:10.1029/2012EO250005; D. Turner, A. Lucieer, and L. Walace, Direct georeferencing of ultrahigh-resolution UAV imagery, *IEEE Trans. Geosci. Rem. Sens.*, 52(5):2738–2745, 2013, DOI:10.1109/TGRS.2013.2265295; C. Vermeulen et al., Unmanned aerial survey of elephants, *PLoS ONE*, 8(2):e54700, 2013, DOI:10.1371/journal.pone.0054700; A. C. Watts, V. G. Ambrosia, and E. A. Hinkley, Unmanned aircraft systems in remote sensing and scientific research: Classification and considerations of use, *Remote Sens.*, 4(6):1671–1692, 2012, DOI:10.3390/rs4061671; K. Whitehead, B. J. Moorman, and C. H. Hugenholtz, Low-cost, on-demand aerial photogrammetry for glaciological measurement, *The Cryosphere*, 7:1879–1884, 2013, DOI:10.5194/tcd-7-3043-2013.

Restoring regeneration

Many human individuals suffer from diseases related to the body's gradual failure in restoring or repairing nonfunctional tissues. These age-related ailments, including those of the cardiovascular and nervous systems, are among the leading causes of death worldwide and are significant public-health concerns. Thus, the expansion of our knowledge in tissue regeneration would provide a great benefit.

The field of regenerative medicine aims to repair or replace tissues that are damaged as a result of trauma, disease, or aging. Most fully differentiated (functionally specialized) cells within the body enter a tangential phase of the cell division cycle and are difficult to coax as contributors to in vivo regeneration. In some tissues, such as the human liver's hepatocytes, cells can naturally reenter the cell cycle to aid in tissue repair. This partially explains our ability to expand the mass of preexisting hepatic tissue following partial liver amputation. The possibilities of regenerative therapies must take into account the complexity of other tissues. At the heart of success lies a powerful population of adult cells, called stem cells, which maintain or drive the regeneration of these tissues. For example, in certain animal models, such as salamanders, limb regeneration after injury (amputation) is possible. An injury initiates a cascade of events that coax stem cells to multiply, migrate to the wound, and subsequently differentiate into the desired cell types. A regeneration blastema, a required compartment of regenerating tissue, develops at the amputation site during this process. Our understanding of the molecular signals involved with this process is weak, but extraordinary efforts are being made to unlock these secrets. By uncovering the molecular components of this cascade, the expectation is that it will be possible to establish more effective therapies that target human tissues.

Modern model system approaches. Regenerative medicine uses multiple approaches. Whereas one approach studies the mechanisms by which stem cells are stimulated in vitro, another approach studies natural stem cell growth within the context of model organisms. Many cases of regeneration of which humans are incapable occur naturally in the animal kingdom. Recently, two model systems—specifically, the zebrafish and planarian model systems—have driven some of the most exciting regeneration studies. What makes these organisms more attractive models is that, while they exhibit amazing regenerative capacity in one region of the body, other regions fail to exhibit these capabilities. Recent studies have utilized extremely advanced "whole transcriptome" approaches to compare gene activities in regions of differing regenerative potentials. In these studies, scientists have taken advantage of these lists of gene function by transforming a regeneration-deficient animal into one with great regenerative strength.

Sex-specific regeneration differences in fish. The zebrafish is an excellent model for studying regeneration because of its capacity to regenerate portions of its heart, kidney, and fins following ablation (surgical removal). A recent study unveiled that zebrafish are sexually dimorphic (distinct) for fin regeneration, where male zebrafish exhibited sex-specific regenerative defects in amputated pectoral fins. This regenerative failure coincides with the onset of male sexual maturity and androgen production within the testes. Androgens are hormones that direct the development and maintenance of male characteristics, making their presence or absence an excellent switch for controlling male and female differences in regenerative potential. This model is supported by the rescue of pectoral fin regeneration in males who have undergone castration or treatment with androgen-inhibiting drugs.

To unveil the downstream target pathways of the androgen signal, a microarray screen was undertaken. This whole transcriptome approach allowed for a high-throughput comparison between the collection of active fin genes between male and female fish. Four hundred regeneration-specific genes were shown to exhibit sexually distinct expression patterns. Among these, two genes have been found to encode proteins that "shut down" a particular pathway known to be a promoter of fin regeneration. This pathway is called the Wnt pathway. The model put forth was that male fish demonstrate a defect in pectoral fin regeneration because circulating androgen induces the expression of inhibitors of the proregenerative Wnt pathway. Investigators were able to completely rescue pectoral fin regeneration by simply turning off the inhibitor-based switch. These findings could explain human male–female differences in tissue maintenance. In addition, because of lifetime hormonal variances, it could shed light on age-related defects in tissue repair.

Restoring regeneration in planarians. One of the most prodigious of regenerative model organisms is the freshwater planarian (a nonparasitic flatworm belonging to the Tricladida order of turbellarians). The planarian is a champion of whole-organism regeneration because it possesses the ability to regenerate fully from tiny amputated fragments. This power, along with the design of molecular tools to assay gene expression and to knock down (reduce) a specific gene's function, has made planarian studies highly prevalent over the last decade. The two main species driving this resurgence are *Dugesia japonica* and *Schmidtea mediterranea* because of their strong regenerative capabilities along nearly any amputation plane across the anterior–posterior (A–P) axis. Scientists have used these model species to determine the heterogeneous nature of the planarian stem cell (neoblast) pool and have uncovered the molecular mechanisms driving neoblasts to support tissue replacement after severe injury.

Three recent studies have utilized planarian species (*Dendrocoelum lacteum*, *Phagocata kawakatsui*, and *Procotyla fluviatilis*) with severely limited regenerative capacities. Investigators have characterized this failure within the posterior region of the worm. Whereas tissue fragments cut from anterior "trunk" positions regenerate all missing head structures, more posterior "tail" fragments fail to regenerate a head and eventually die. These variations in potential were used in comparative studies whereby modern whole transcriptome RNA sequencing was done on trunk and tail fragments. The motivation was to uncover a list of genes exhibiting different levels of activity to find the molecular switches determining the regeneration of head tissues. Analyses showed that the genes encoding stimulators of the Wnt signaling pathway (ligands and receptors) were expressed inappropriately in regeneration-deficient fragments and were turned off in highly regenerative fragments. As observed with the zebrafish pectoral fin, the ability or inability to regenerate head structures in these planarian species hinges on the activation of the Wnt pathway. More importantly, investigators were able to restore full regenerative capacity in these regeneration-deficient animals by knocking down Wnt activity.

Moving forward. In many ways, animal models such as the zebrafish and planarians reflect the human context because some tissues regenerate, whereas others do not. Gaining a better understanding of the molecular switches controlling the variation of regenerative potential between tissues and species is an important step in understanding human tissue maintenance, aging, and disease. As the key factors involved in regeneration are uncovered, regenerative medicine moves forward as well.

For background information *see* ANDROGENS; CELL CYCLE; CELL DIFFERENTIATION; CELL DIVISION; CELL FATE DETERMINATION; DEVELOPMENTAL BIOLOGY; GENE; MORPHOGENESIS; PATTERN FORMATION (BIOLOGY); REGENERATIVE BIOLOGY; SEXUAL DIMORPHISM; STEM CELLS; TRICLADIDA; TURBELLARIA in the McGraw-Hill Encyclopedia of Science & Technology.

Daniela I. Alarcón; Robert J. Major

Bibliography. S. Y. Liu et al., Reactivating head regrowth in a regeneration-deficient planarian species, *Nature*, 500(7460):81–84, 2013, DOI:10.1038/nature12414; G. Nachtrab, M. Czerwinski, and K. D. Poss, Sexually dimorphic fin regeneration in zebrafish controlled by androgen/GSK3 signaling, *Curr. Biol.*, 21(22):1912–1917, 2011, DOI:10.1016/j.cub.2011.09.050; J. M. Sikes and P. A. Newmark, Restoration of anterior regeneration in a planarian with limited regenerative ability, *Nature*, 500(7460):77–80, 2013, DOI:10.1038/nature12403; E. M. Tanaka and P. W. Reddien, The cellular basis for animal regeneration, *Dev. Cell*, 21(1):172–185, 2011, DOI:10.1016/j.devcel.2011.06.016; Y. Umesono et al., The molecular logic for planarian regeneration along the anterior–posterior axis, *Nature*, 500(7460):73–76, 2013, DOI:10.1038/nature12359.

S-MIM radio interface for mobile satellite services

The use of spacecraft for data relay to provide radio coverage to the entire Earth dates back to 1945, when Arthur C. Clarke, at that time serving in the Royal Air Force as officer and radar specialist, published the visionary article in *Wireless World* entitled "Extra-Terrestrial Relays. Can Rocket Stations Give World-wide Radio Coverage?" Twelve years later, the first communications satellite, named *Sputnik 1*, was launched by the Russians. It carried two radio transmitters at 20.005 and 40.002 MHz, with batteries lasting 22 days. Today, the number of communications satellites listed in the National Space Science Data Center (NSSDC) Master Catalog is greater than 2000; the number of active communication satellites in geosynchronous orbits, characterized by an orbital period equal to one sidereal day, is almost 400; and the lifetimes of such satellites are typically around 15 years. Television and radio broadcasting are by far the most important markets for satellite manufacturers and operators, accounting today for almost 80% of the total satellite service revenue. In contrast, for many years, bidirectional communication services via satellite remained a relatively small, although important, market niche, mainly addressing institutional or professional applications. This was chiefly due to the high cost of satellite airtime and terminals and to the high latency due to the propagation delay. Only recently, affordable broadband-access satellite networks have been emerging in the United States, complementing the terrestrial networks' coverage in locations where wireline solutions are not economically viable. At the time of this writing (2014), Google has announced plans to invest in a fleet of satellites that will expand Internet access to unconnected regions of the world.

Use of satellite communication for M2M services. Moreover, an increasing number of emerging applications are based on the sporadic transmission of short messages from and to remote sensors or mobile devices used to track specific events or monitor automatic systems, and they might radically change the current picture. These applications are typically labelled under the broad term "machine-to-machine" (M2M), and include, for example, industrial supervisory control and data acquisition (SCADA) systems; fleet management or containers' tracking; public systems, such as automatic highway tollgates and traffic-light controllers; and energy systems, such as current sensors in a solar panel array and water-level sensors in a dam. Although terrestrial wireless networks are expected to be capable of serving most of the traffic generated by such applications, the intrinsic cross-border nature of satellite communications makes it the ideal complement to provide M2M services over a wide area to remote and sparsely populated locations.

This article describes a solution for M2M applications called S-band Mobile Interactive Multimedia (S-MIM), recently standardized by the European Telecommunications Standards Institute (ETSI). S-MIM (**Fig. 1**) is an integrated satellite-terrestrial mobile system capable of providing interactive broadcasts and multicasts as well as packet data acquisition services to subscribers. The technical solution adopted for its satellite uplink channel (that is, from terminal to gateway) is particularly innovative and offers a low-cost yet spectrally efficient solution to send short messages with modest power requirements on the terminal side via a geosynchronous satellite acting as relay. Remarkably, the consequent reduction of capital and operational expenditures is expected to significantly contribute to the enlargement of market share for satellite-based M2M services. The rest of this article focuses on the uplink transmission scheme.

Multiple-access protocols. Whenever a certain number of terminals share a physical wireless communication channel (that is, a portion of the available radio spectrum), a suitable multiple-access protocol is required. Existing multiple-access protocols can be grouped into two main categories, namely reservation-based and contention-based. In the first case, often referred to as demand assignment multiple access (DAMA), coordination among all terminals is required to ensure that each one has access to an exclusive portion of the available spectral and time resources. In the simplest cases, this can be achieved, for example, by partitioning the available radio spectrum into smaller frequency sub-channels or different time slots. A central network control unit will dynamically allocate available time

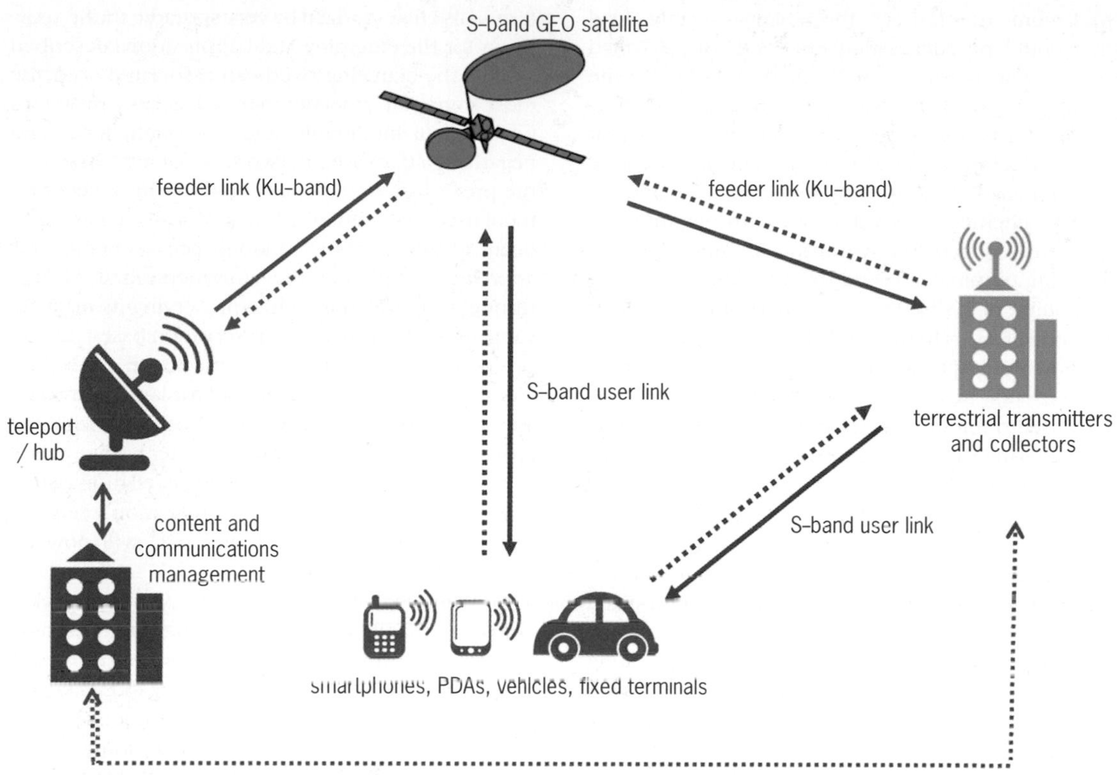

Fig. 1. S-MIM system architecture assuming a transparent S-band payload. (*Courtesy of S. Scalise et al., S-MIM: A novel radio interface for efficient messaging services over satellite, IEEE Comm. Mag., 51(3):119–125, March 2013, DOI:10.1109/MCOM.2013.6476875*)

and frequency resources to the requesting terminals. In contrast to this procedure, contention-based access protocols assume that no, or very little, coordination exists among terminals. This decentralized approach does not guarantee that data packets transmitted by different terminals will not overlap in time or frequency, thus leading to mutual interference (**Fig. 2**). The main advantage is that the network does not require the extra burden (in time and signaling overhead) of centralized resource management.

The simplest type of contention-based protocol is named ALOHA. This was a pioneering computer networking system developed at the University of Hawaii in the 1960s to interconnect by radio links computer terminals located in the Hawaiian Islands to the university campus mainframe. The basic version of the ALOHA protocol consists in letting each terminal transmit at arbitrary time instants, regardless of what other terminals sharing the same channel will do. Communication theory tells us that, when the power of the received packets is balanced (that is, when each terminal sets its transmitted power to a level that ensures that all packets are received with the same power level), the

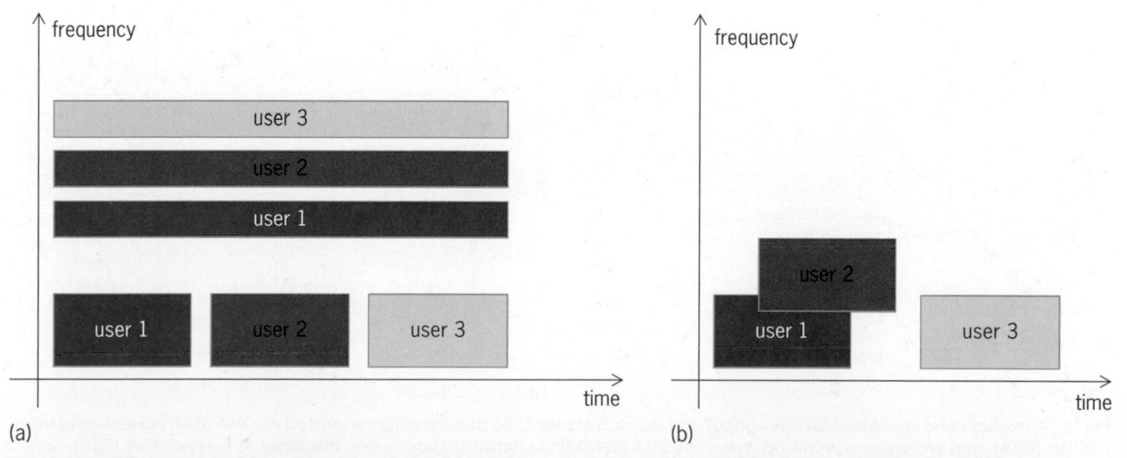

Fig. 2. Comparison of (*a*) reservation-based and (*b*) contention-based access. In the first case, the three users transmit either in dedicated time slots (lowermost) or in dedicated-frequency subchannels (uppermost). In the second case, data packets might partly or totally overlap in time or frequency (for example, users 1 and 2).

maximum fraction of the channel traffic load that could be successfully received, the so-called throughput, is roughly 0.184 with ALOHA. [In the following, we will measure the throughput as the ratio of information bits per signaling unit (chip), that is, in bits/chip. This is a more general and accurate way to measure the spectral efficiency of a random-access scheme.] This value of the throughput can be increased to 0.368 by introducing time slots and allowing terminals to send their packets only at the beginning of a slot; the resulting protocol is called slotted ALOHA. (In practice, the achievable random-access throughput is much lower than its peak value; in fact, it may be two orders of magnitude less. This is because, to minimize retransmission over the large-latency satellite channel, the packet loss ratio, defined as the ratio of total lost packets to total transmitted packets, should be kept below 10^{-3}.) It should be noted that this modification already implies a higher level of coordination among all the terminals, in order to keep the transmitted packets synchronized to the common network timing reference for aligning each packet to be transmitted at the beginning of a time slot. In general, any type of coordination among terminals requires exchanging signaling information over the communication network and, in many cases, the presence of a central entity in charge of gathering, processing, and distributing this information is also required. In a properly designed communication network, the amount of exchanged data signaling should be minimized to avoid subtracting precious spectral resources from the transmission of real data traffic. The consequence of poor throughput performances of ALOHA-like protocols is that, today, they are mainly used only during satellite terminal logon to exchange initial signaling, whereas reservation-based access protocols are preferred for traffic data exchange.

Use of SSA with iSIC for M2M applications. However, in the case of networks with a very large number of terminals characterized by very sporadic traffic activity, as for the emerging M2M applications described earlier, the signaling overhead associated with the most common reservation-based access protocols becomes an intolerable source of inefficiency and negatively affects the network scalability. Based on the previously stated consideration, the access protocol used in the uplink of the S-MIM system is based on an advanced type of random-access scheme. This approach overcomes the aforementioned ALOHA throughput limitation, while preserving its main advantage with respect to reservation-based access protocols, namely, a very simple logon procedure and very low signaling overhead for large networks. Spread-spectrum ALOHA (SSA), together with the application at the demodulator of a technique called iterative successive interference cancellation (iSIC), provides the best fit to this application scenario. The basic concepts of SSA and iSIC will now be described.

Conventional SSA allows the resolution of packet collisions thanks to the terminal-unique spreading sequence discrimination characteristics. (In reality, each terminal is reusing the same spreading sequence but with a different time, frequency, and power offset so as to make it appear unique at the gateway demodulator.) Relatively high throughput with low packet loss ratio can be achieved when the power of the received packets is balanced. Deviation from this condition leads to a rapid performance degradation. Balancing the packet power at the gateway demodulator input is not straightforward and, again, it requires some level of coordination, resulting in the aforementioned need to exchange signaling information. Otherwise, because of the fact that the gain of the satellite receiving antenna within its coverage area is not uniform and that packets sent by terminals located in different positions will undergo different levels of link attenuation caused by different propagation conditions, even if two

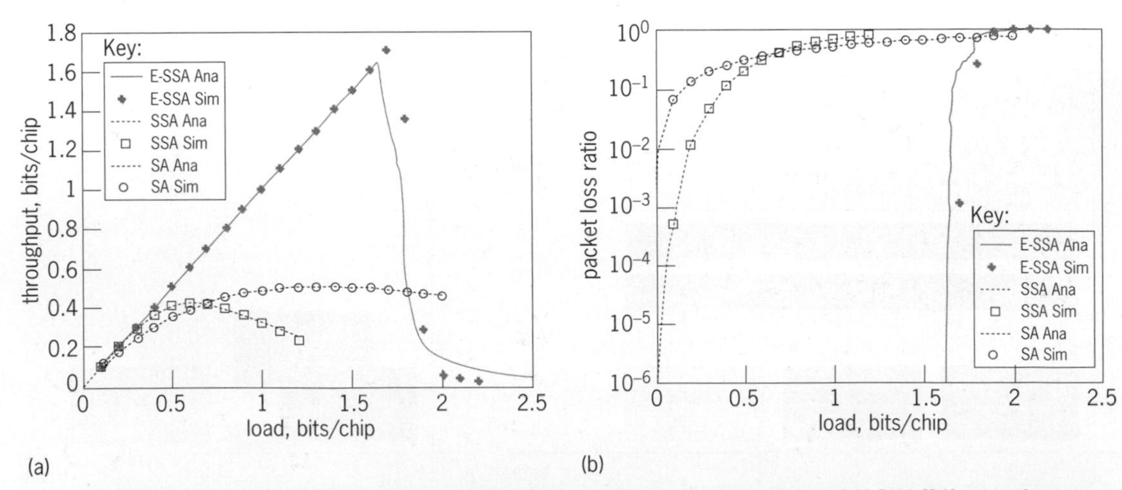

Fig. 3. Analytical and simulated (*a*) throughput and (*b*) packet loss ratio of conventional slotted ALOHA (SA), spread-spectrum ALOHA (SSA), and enhanced spread-spectrum ALOHA (E-SSA) via iterative successive interference cancellation (iSIC), as used in the uplink of the S-MIM system. It is assumed that the received power level is unbalanced and follows a lognormal distribution with null mean, standard deviation $\sigma = 2$ dB, and a signal-to-noise ratio (SNR) equal to 10 dB. Ana = analytical; Sim = simulated.

terminals transmit with the same power, their packets will have different power levels when they reach the gateway demodulator.

A change of paradigm was therefore introduced in the design of the S-MIM uplink demodulator by exploiting the inherent presence of power unbalance at the demodulator input to enhance the system throughput. This is achieved through iSIC, which extends the conventional successive interference cancellation (SIC) technique, adapting it to the asynchronous and bursty nature of the transmitted packets. The demodulator stores in its memory a portion of the received signal, which is the superposition of many incoming packets, and, in the first iteration, attempts to detect and decode the packet with the highest received power. Then, this packet is cancelled from the received signal stored in the demodulator memory. During the second iteration, the packet with the next highest received power is detected and decoded. The process is repeated several times to resolve as many collisions as possible. Finally, the content of the demodulator memory is partly updated by storing a new portion of the received signal according to a sliding window approach and the process is repeated. As shown in **Fig. 3**, SSA combined with iSIC can allow, with truly affordable implementation complexity, a maximum throughput close to 1.8 bits/chip, about 3 dB below what is achievable with code-division multiple access (CDMA) with random spreading capacity. Further improvements can be obtained by optimizing the statistical distribution of the received power level at the demodulator, since it has been proved that a uniform distribution in decibels maximizes the performance of the iSIC algorithm, whereas the results in Fig. 3 correspond to a lognormal distribution. An open-loop power control algorithm can be used by all terminals to set their power so as to obtain an effective power distribution at the demodulator as close as possible to the optimum one.

For background information see COMMUNICATIONS SATELLITE; DATA COMMUNICATIONS; MOBILE COMMUNICATIONS; MULTIPLEXING AND MULTIPLE ACCESS; PACKET SWITCHING; SPREAD SPECTRUM COMMUNICATION in the McGraw-Hill Encyclopedia of Science & Technology.

Sandro Scalise; Riccardo De Gaudenzi

Bibliography. N. Abramson, The ALOHA System: Another alternative for computer communications, pp. 281–285, *Proceedings of the AFIPS '70 Fall Joint Computer Conference*, November 17–19, Houston, AFIPS/ACM, New York, 1970; O. Del Río Herrero and R. De Gaudenzi, High efficiency satellite multiple access scheme for machine-to-machine communications, *IEEE Trans. Aero. Electron. Syst.*, 48:2961–2989, 2012, DOI:10.1109/TAES.2012.6324672; European Telecommunications Standards Institute, *Satellite Earth Stations and Systems (SES): Air Interface for S-band Mobile Interactive Multimedia (S-MIM)*, ETSI TS 102 721, V1.2.1, 2013; S. Scalise et al., S-MIM: A novel radio interface for efficient messaging services over satellite, *IEEE Comm. Mag.*, 51(3):119–125, March 2013, DOI:10.1109/MCOM.2013.6476875; S. Verdu and S. Shamai, Spectral efficiency of CDMA with random spreading, *IEEE Trans. Inform. Theor.*, 45: 622–640, 1999, DOI:10.1109/18.749007.

Schmallenberg virus

In August–September 2011, dairy cows on farms in Germany and the Netherlands were sick with fever [40°C (104°F)], loss of appetite, and diarrhea, and they had up to 50% reductions in milk production. The symptoms were similar to those of the bluetongue virus infections that caused a major epizooty from 2006 to 2009 in Europe. An epizooty is an epidemic among large numbers of animals in a defined location. Farmers feared that this was the reemergence of a bluetongue virus outbreak. To their surprise, the cattle tested negative for endemic or emerging viruses such as bluetongue virus, bovine herpesvirus type-1, malignant catarrhal fever virus, bovine diarrhea virus, foot-and-mouth disease virus, epizootic hemorrhagic disease virus, Rift Valley fever virus, and bovine ephemeral fever virus.

Investigation of the mysterious cattle disease. Blood samples were collected and pooled from three sick cows on a farm near the city of Schmallenberg (North Rhine-Westphalia, Germany) in October 2011. These samples, along with a blood sample for comparison from a healthy cow located on a different farm, were sent to a team of scientists at the Friedrich-Loeffler-Institut in Greifswald, Insel Riems, Germany. The blood samples from the sick cows contained sequences of a novel virus related to a Shamonda-like virus that was detected in cattle in Japan in 2002. The new viral sequences were not detected in the blood sample of the healthy cow. Previously, the Shamonda virus was isolated from cattle and *Culicoides* species of biting midges in Nigeria in the 1960s. The transmission of this virus was known to occur mainly through *Culicoides* biting midges and mosquitoes. Biting midges are extremely small flies [wingspans of 1–2 mm (0.04–0.08 in.)] that pester and bite humans and animals. They lay their eggs on plants in water, in mud, or on the surface of water. More than 65 viruses have been isolated from *Culicoides* midges, but only a few viruses cause serious infections in humans or livestock.

The new Shamonda-like virus was isolated from the blood samples of the diseased cows. In addition, genetic material (RNA) of the new Shamonda-like virus was detected in a high proportion of different species of biting midges, thereby helping to explain its transmission. Biting midges trapped in Denmark, Norway, Poland, and Sweden also tested positive for the virus. Field-trapped common European mosquito species tested negative for the virus. Scientists screened archival blood samples collected from livestock in Europe prior to 2011; none of these samples tested positive for the new virus. This new or emerging virus was named the

Schmallenberg virus, after the place of its discovery in the sick dairy cows.

A new symptom raises concerns. During the following months after the initial outbreak of the disease, more severe symptoms were observed. Large numbers of malformed lambs (**Fig. 1**) that were born dead or died after birth, with defects such as malformations of the skull, fused joints, and twisted spinal columns, were reported from farms in the Netherlands in November and December of 2011. This raised concerns that the Schmallenberg virus might cause congenital defects because related viruses were known to cause arthrogryposis-hydranencephaly syndrome (AHS). AHS results in stillbirths and congenital defects in fetuses of cattle, sheep, and goats following infection during pregnancy. Notably, the brain of a malformed lamb in the Netherlands tested positive for the Schmallenberg virus. Therefore, surveillance was stepped up across Northern Europe, raising awareness among veterinarians.

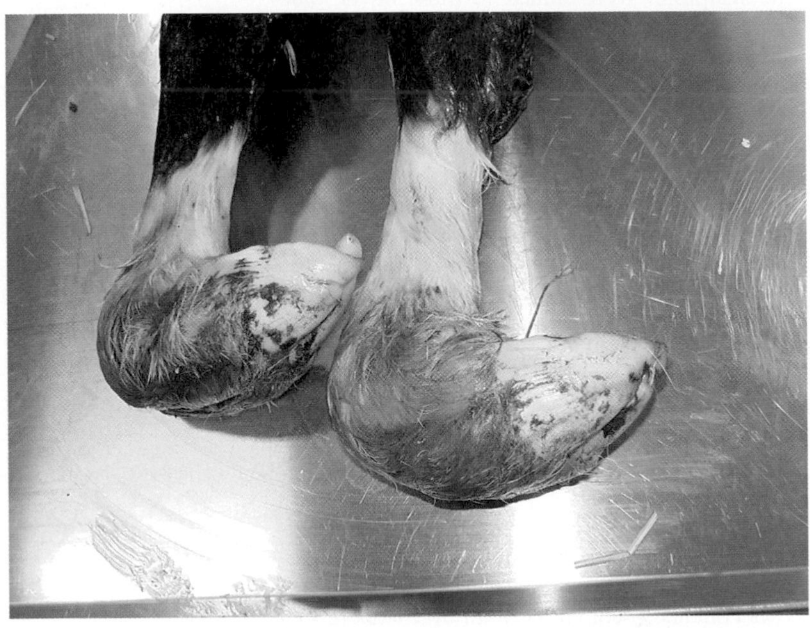

(a)

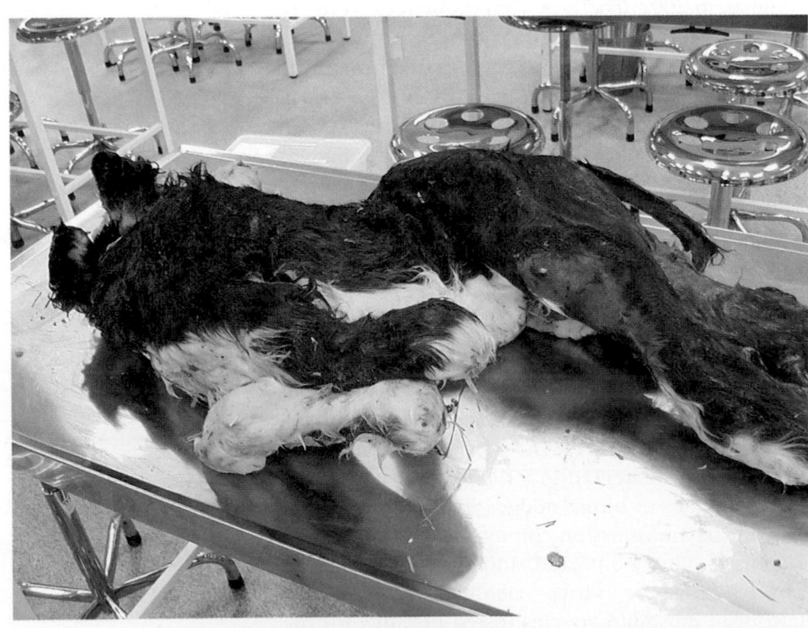

(b)

Fig. 1. Malformed lambs associated with Schmallenberg virus infection. (*a*) The lamb has fused joints (arthrogryposis). (*b*) The whole calf has fused joints and twisting of the spine (torticollis). (*Images courtesy of Rachael Tarlinton, School of Veterinary Medicine and Science, University of Nottingham, Loughborough, UK*)

Schmallenberg virus spreads through Europe. The Schmallenberg virus spread rapidly in cattle, sheep, and goats in vast parts of Europe from its first identification in September 2011 through April 2013. Within a few months, it was observed in large areas in Belgium, France, Germany, Luxembourg, the Netherlands, the United Kingdom (south and east of England), and Switzerland. Sporadic infections were also reported from Italy, Spain, and Denmark. By October 2012, goats, sheep, and cattle in Poland, Austria, Scotland, Norway, the Czech Republic, Slovenia, and Estonia tested positive for the Schmallenberg virus. To date, the Schmallenberg virus has spread to virtually all countries of Europe (**Fig. 2**).

Reasons for the rapid spread. Serum samples from livestock, farmers, and wildlife were screened for the presence of antibodies against the Schmallenberg virus. Antibodies were found in red and roe deer, mouflon (wild sheep), and alpacas, indicating their exposure to the Schmallenberg virus. However, there was no evidence of outbreaks or congenital malformations of animals in the wild. In addition, there was no evidence of antibodies in the serum from farmers working near or with sick livestock.

The geographical locations affected by the Schmallenberg virus in 2011 overlap many of the same areas affected by the bluetongue virus in 2006. Some scientists think that changes in climate and land use may favor transmission. The influence of wind may also be a factor. For example, midges can be blown by wind for hundreds of kilometers or miles. One study has suggested that midges arriving by downwind movements during dusk (when the midges are most active) could explain 62% of the Schmallenberg infections, while a mixture of downwind and random movements could explain 38% of the Schmallenberg infections. Another consideration is the presence of midges in the shipments of plants. For example, the Netherlands is a major center for international flower imports (including those from sub-Saharan Africa). These flowers are typically cut and packed at night under bright lights. Midges are attracted to bright lights and could be trapped within these shipments, which then could be released upon unpacking.

Economic impact of the Schmallenberg virus. The United Kingdom is the largest producer of sheep meat in Europe and the fifth largest producer worldwide. Therefore, outbreaks caused by the Schmallenberg virus have the potential for considerable impact (economic and otherwise) on farmers, the meat industry, and the food supply, both in the United Kingdom and on mainland Europe. Furthermore,

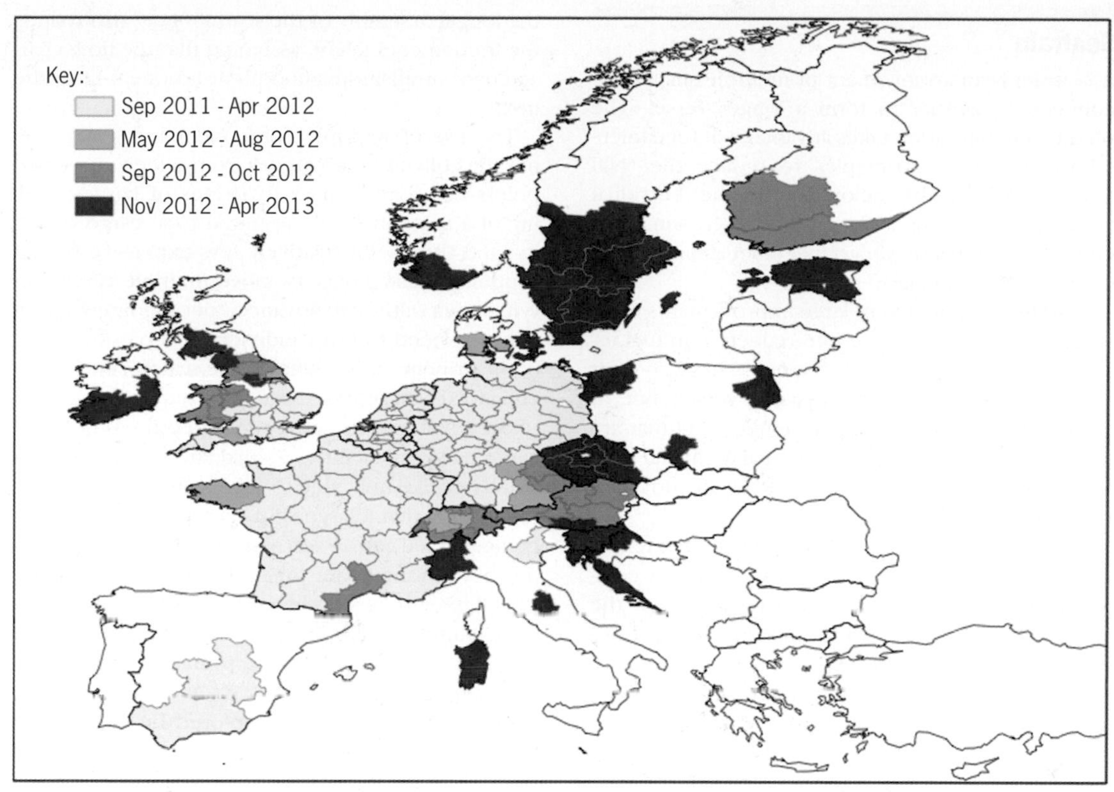

Fig. 2. Mapped laboratory-confirmed cases of the Schmallenberg virus. The virus spread rapidly through livestock herds on European farms from September 2011 through April 2013. Most countries in Europe reported Schmallenberg virus cases. (*Adapted from Fig. 4, p. 14, "Schmallenberg" Virus: Analysis of the Epidemiological Data, European Food Safety Authority, Supporting Publications 2012:EN-429, May 2013, http://www.efsa.europa.eu/en/search/doc/429e.pdf*)

the semen of naturally infected bulls contains the Schmallenberg virus, raising the concern that the semen could be infectious and may have an impact on fertility. More than 15 countries, including the United States, Mexico, and Japan, have imposed restrictions on the imports of cattle and animal products (such as semen and embryos) from the European Union.

Controlling future outbreaks. Midges are one of the most difficult insects to control. In addition, chemicals to control midges may harm aquatic environments and nontarget organisms. The breeding sites for midges are also large, making chemical control impractical. The best method to prevent bluetongue virus outbreaks has been through vaccination. Based on the similarities between the bluetongue virus and the Schmallenberg virus regarding the insect vector involved and the affected hosts (livestock), vaccination against the Schmallenberg virus is considered to be one of the most important ways to control diseases on the farm. Two inactivated vaccines for cattle and sheep became available in 2013: SBVvax (Merial) developed in France and Bovilis SBV (MSD) developed in the United Kingdom. The efficacy of these vaccines has yet to be shown. However, these vaccines will hopefully protect livestock from Schmallenberg virus outbreaks in the future.

For background information *see* ANIMAL VIRUS; BLUETONGUE; DIPTERA; DISEASE ECOLOGY; EPIDEMIOLOGY; SHEEP; VIRUS; VIRUS CLASSIFICATION in the McGraw-Hill Encyclopedia of Science & Technology.

Teri Shors

Bibliography. M. Beer, F. J. Conraths, and W. H. van der Poel, "Schmallenberg virus"—a novel orthobunyavirus emerging in Europe, *Epidemiol. Infect.*, 141:1–8, 2012, DOI:10.1017/S0950268812002245; V. Doceul et al., Epidemiology, molecular virology and diagnostics of Schmallenberg virus, an emerging orthobunyavirus in Europe, *Vet. Res.*, 44:31, 2013, DOI:10.1186/1297-9716-44-31; M. Ganter, R. Eibach, and C. Helmer, Update on Schmallenberg virus infections in small ruminants, *Small Ruminant Res.*, 18:63–68, 2014, DOI:10.1016/j.smallrumres.2013.12.012; R. Lühken, E. Kiel, and S. Steinke, Impact of mechanical disturbance on the emergence of *Culicoides* from cowpats, *Parasitol. Res.*, 113(4):1283–1287, 2014, DOI:10.1007/s00436-014-3766-3; P. Mertens et al., Schmallenberg virus: Emergence of a novel pathogen in Europe, *Microbiologist*, 14(3):8–12, 2013; F. Ruiz-Fons et al., The role of wildlife in bluetongue virus maintenance in Europe: Lessons learned after the natural infection in Spain, *Virus Res.*, 182:50–58, 2014, DOI:10.1016/j.virusres.2013.12.031; L. Sedda and D. J. Rogers, The influence of the wind in the Schmallenberg virus outbreak in Europe, *Sci. Rep.*, 3:3361, 2013, DOI:10.1038/srep03361; R. Tarlinton et al., The challenge of Schmallenberg virus emergence in Europe, *Vet. J.*, 194:10–18, 2012, DOI:10.1016/j.tvjl.2012.08.017.

Seatrain

A seatrain is an arrangement of multiple small hulls connected together to form a longer vessel. The seatrain configuration takes advantage of fundamental hydrodynamic principles to reduce the total drag of the assembly below that of the individual components if separately propelled. In some circumstances, the seatrain arrangement can also offer operational advantages.

Seatrains can offer very large improvements in resistance that will translate into reduction in fuel usage and potentially increased range. They can offer the economy associated with large vessels but, at the same time, may be able to enter ports that are too small for big ships. The principal technical challenge arises from the forces and loads occurring at the interfaces between units.

Principles of operation. The drag force that resists a ship's forward motion through the water is composed of two major parts: frictional resistance of the water passing over the ship's hull, and energy dissipated in the waves of the ship's wake. Wave-making resistance is a function of the ratio of a ship's speed to its length on the waterline, usually expressed as the nondimensional Froude number, $\frac{V}{\sqrt{gL}}$, where V is the ship's speed, g is the gravitational acceleration, and L is the length in any consistent system of units. From the definition of the Froude number, it follows that, for a given speed, a longer ship operates at a lower Froude number and hence with lower wave-making resistance.

Wave-making becomes significant at a Froude number of approximately 0.2, and dominant at a Froude number of 0.45. The fundamental innovation of the seatrain is to assemble multiple smaller units to increase effective length, thus reducing the Froude number and the wave-making energy required to propel them. In addition, relative to a single unit, the longer assembly of the seatrain may also reduce the friction coefficient, assuming that the flow characteristics transition smoothly from one unit to the next.

The use of multiple component hulls may permit the collection of cargoes from widely dispersed points of origin, similar to strings of barges; scaling of a seatrain to the required total cargo capacity; and the use of relatively few expensive sets of propulsion machinery to move multiple cargo carrying units (although advanced oceangoing seatrain concepts tend to favor individual propulsion). One of the principal advantages of the seatrain concept is enhanced port access. Large ships require large areas of open water to maneuver and specialized port and cargo handling facilities to load and unload. A short and barge-like hull, able to access congested or shallow waters that a large ship cannot, potentially can be assembled into a seatrain to gain the energy efficiency advantage for long-distance transportation. This flexibility has potential for military operations and commercial shipping in areas without extensive fixed maritime infrastructure, sometimes referred to as austere ports.

The concept has been explored by naval architects since the 1950s. Early model tests associated with the concept focused on the seatrain as a means of increasing the waterborne speed of military amphibious vehicles. As single units, these vehicles encountered the traditional speed limitation on displacement-type craft caused by the generation of the bow wave. Their formation into a seatrain, consisting of as many as eight units, decreased the resistance per unit and some speed increase was achieved. However, because all units including the lead unit were boxy amphibious vehicles, the seatrain speed tended to be limited by the bow unit pitching (tilting about the athwartships axis) downward and taking water over the bow.

Fig. 1. The Improved Navy Lighterage System demonstrates the principles of the seatrain concept, here shown in a two-unit assembly. (*Image courtesy of United States Navy*)

Fig. 2. Model tests of a proposed purpose-built high-speed Transoceanic Seatrain Concept. (*Images courtesy of Naval Surface Warfare Center, Carderock Division*)

Improved Navy Lighterage System. A relatively new practical application of the concept is the U.S. Navy's Improved Navy Lighterage System, which uses a system of barges propelled by small powered units called warping tugs to construct floating piers (**Fig. 1**). Each unit is about 90 ft (27 m) long and 24 ft (7 m) wide. The front module is a beaching unit with a ramp, and all power and steering is in the aft power module. One or two unpowered load units can be connected between the beaching and powered modules to form three- and four-unit assemblies. Each unit is connected by a bull-nose and socket arrangement that has a tension member called a flexor. This connection mechanism allows each unit to pitch relative to the adjacent unit. Cargo can be driven from one unit to the next. Each component unit can be transported on a large barge or ship to the theater of action.

Transoceanic Seatrain Concept. The U.S. Navy Transoceanic Seatrain Concept is an extensive design study that demonstrates the principles for large-scale commercial or military use of purpose-built oceangoing units. The concept is to cross the ocean as a seatrain taking advantage of the lower wave resistance offered by the length increase. Upon arrival at the coast, the units are to disengage and proceed individually into a port lacking the facilities to handle a large commercial or military cargo vessel.

Each unit of this high-speed seatrain would have a sharp bow that would have a waterline with a 10° half angle of entrance (the angle that the unit's bow forms in the plane of the water surface) to accommodate the high speed. Each bow would fit into a similarly shaped notch in the stern of the unit in front. The notch length would be 30% of the ship length, and the bow and stern geometry details were chosen to minimize the section-area discontinuity between units. Each unit would be equipped with

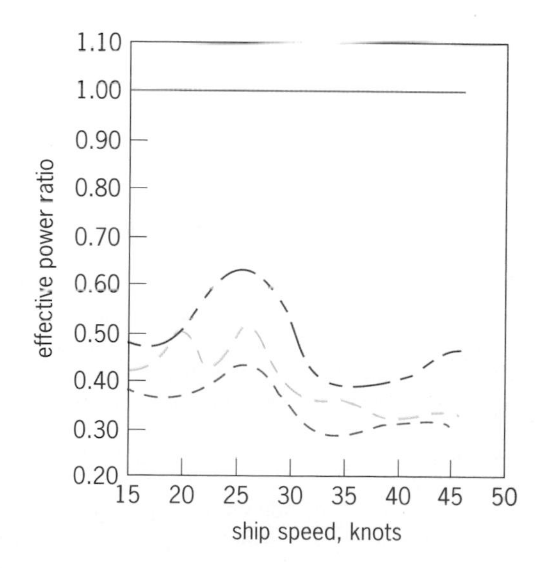

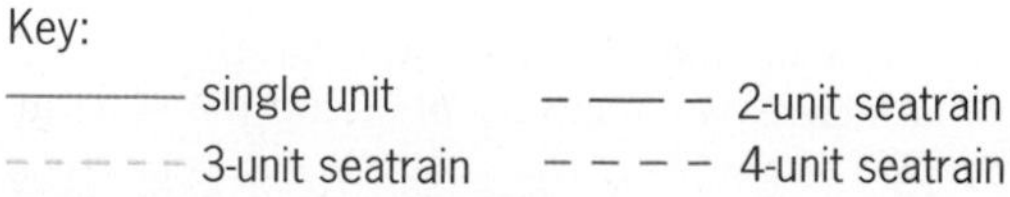

Fig. 3. Demonstrated effective power reductions in a seatrain. 1 knot = 0.514 m/s.

(a) (b)

Fig. 4. Mechmath LLC seatrain with air-cavity drag reduction. (*a*) Concept. (*b*) Model test for 24 knots (12 m/s), sea state 4. (*Image courtesy of Naval Surface Warfare Center, Carderock Division, and Mechmath LLC*)

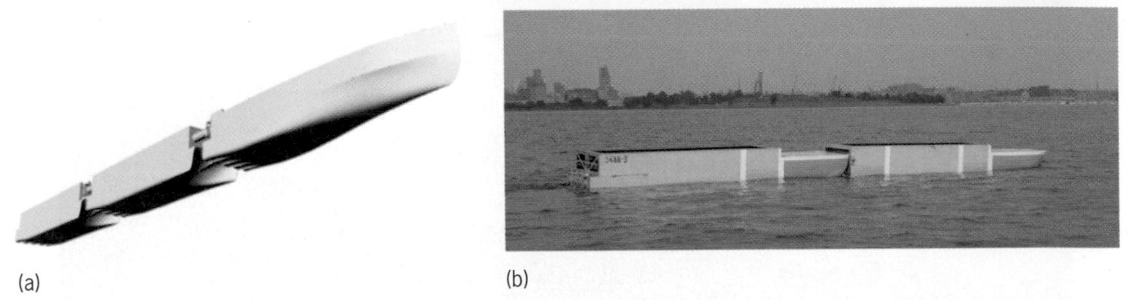

(a) (b)

Fig. 5. MAPC seatrain. (*a*) Initial concept with three units. (*b*) Outdoor coupling test with two units. (*Image courtesy of Naval Surface Warfare Center, Carderock Division, and Maritime Applied Physics Corporation*)

water jets at the stern, one on each side of the notch. The water-jet flow would exit above the waterline into the air, and would be splayed outward in order to avoid impingement on the following unit.

The model test of such a high-speed design with 7.5-ft-long (2.3-m) individual models is shown in **Fig. 2**. The design is representative of a four-unit seatrain at a speed corresponding to 38.3 knots (19.7 m/s) full scale, with each unit 400 ft (122 m) long, with 9050 long tons (9195 metric tons) displacement. The units are towed in close proximity to each other, but not touching. They are free to move relative to each other. The drag of each individual unit was measured.

Measurements taken during model testing of this concept demonstrated the hydrodynamic advantages of the seatrain concept. **Figure 3** shows the effective power requirement per unit to propel a seatrain as a percentage of the power required to propel a single unit operating by itself; that is, Fig. 3 shows the effective power ratio given by the equation below. The dramatic savings shown are

$$\text{effective power ratio} = \frac{\text{total seatrain effective power}}{(\text{number of units}) \times (\text{single-unit effective power})}$$

the practical demonstration of the hydrodynamic effects of increased effective length and the seatrain concept.

Connection between units. An engineering challenge in the design of a seatrain is the physical connection between the units. A ship experiences longitudinal bending as it moves in waves, and other forces (vertical, torsional) vary along its length. In a conventional ship, those forces are taken up by the ship structure. In a seatrain, the individual units will potentially move relative to each other. In some respects this may be advantageous, since bending loads are proportional to ship length, and allowing units to flex relative to each other will reduce bending stress, but mechanical joint mechanisms capable of appropriate articulation and of resisting the potentially large forces that could develop between units are a challenging structural design issue. Mechanical connectors of very large capacity have been developed by the articulated tug barge industry and may form a practical solution for a physically connected design.

Small Business Innovative Research (SBIR) projects. Subsequent to the favorable resistance results of Fig. 3, the U.S. Navy further pursued the seatrain concept with SBIR projects awarded to Mechmath

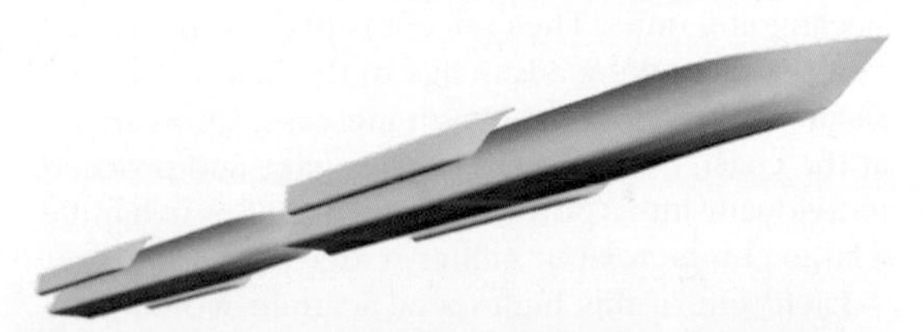

Fig. 6. Art Anderson seatrain concept with two units that was model tested at NSWCCD. (*Image courtesy of Naval Surface Warfare Center, Carderock Division, and Art Anderson Associates*)

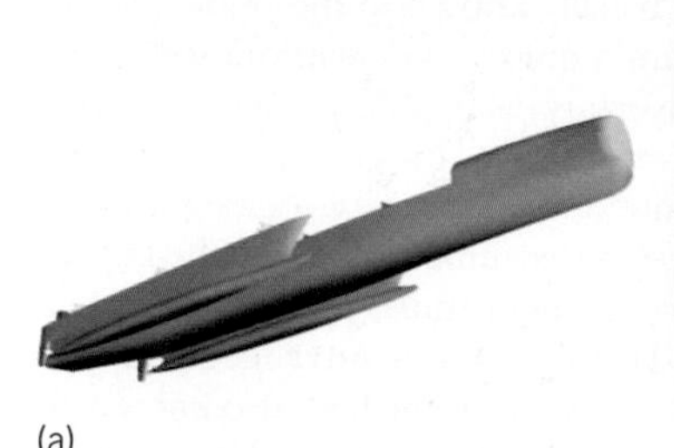

(a) (b)

Fig. 7. AMH trimaran seatrain. (*a*) Concept. (*b*) Model test outdoors for maneuvering performance. *(Images courtesy of Computer Sciences Corporation)*

LLC, Maritime Applied Physics Corporation (MAPC), and Art Anderson Associates. All these seatrain designs were model tested at the Naval Surface Warfare Center, Carderock Division (NSWCCD). These efforts and some related independent work are briefly summarized below to provide an overview of the range of applications and technologies being considered.

Mechmath LLC concept. A two-unit seatrain concept with the use of forced-air cavities beneath the hulls to reduce frictional drag was designed by Mechmath LLC (**Fig. 4**). The depth of the air cavity was sized to operate in ocean waves as well as in calm water. Model resistance tests confirmed the effectiveness of the air cavity drag reduction in both calm water and in oncoming seas. Loads between units were measured in calm and irregular seas for a gimbaled connector decoupling all vertical bending moments.

MAPC concepts. MAPC performed several seatrain design iterations ranging in size from an assembly of existing landing craft with a 180-long-ton (177-metric-ton) payload to an eight-module transport with 30,000-long-ton (29,500-metric-ton) payload per module, conceptually designed to provide flight decks for large aircraft. A final seatrain concept with sizing for austere port access was designed to carry one army heavy battalion per train. The design was model tested and the hull form was refined to reduce hydrodynamic interaction between units. Model tests measured resistance and connector loads. A two-unit train was tested outdoors to demonstrate the self-connect/disconnect capability (**Fig. 5**).

Art Anderson Associates concepts. Several seatrain concepts varying in size from a seatrain of landing craft units to a seatrain that formed a runway for aircraft operations were investigated at Art Anderson Associates. Special emphasis was placed on connector designs. Model resistance testing was conducted at NSWCCD on a two-unit seatrain made up of individual trimaran units (**Fig. 6**). The tests determined the drag with several alternative separation distances between the units, with alternative fairings between the units, and with a coupling that restricted relative vertical motion between the lead-unit stern and the bow of the following unit.

America's Marine Highway (AMH) concept. The motivation for this concept was to provide a mode of U. S. coastal transport, especially on the East Coast, that relieved highway congestion, principally along the I-95 corridor. (Interstate 95 is the main highway along the East Coast of the United States.) Fast ship loading and unloading combined with good fuel economy and regularly scheduled sailing times were considered to be very important to the economic viability of a sea transport system. A seatrain carrier for rolling stock (trucks and trailers) was envisioned along with the possibility of continuous coastal service with units dropping off and other units joining up to deliver cargo.

All of the previously explored seatrain concepts resulted in a very long assembly of units, and the concepts with no freedom to yaw (turn) between units were expected to have very large turning diameters. Therefore a bulbous above-water cylindrical nose and socket was designed to allow yaw articulation. The trimaran configuration was adopted with an attractively large deck area for quick cargo movement and for resistance benefits. The design was prepared by Computer Sciences Corporation (CSC) with assistance from NSWCCD.

Several outrigger longitudinal positions were tested to lower the resistance by cancelling a portion of the wave system generated by the main hull. Resistance tests up to a speed of 30 knots showed that a four-unit seatrain has about 25% less resistance than that of four units operating individually, even though the individual units are quite fine compared to the more barge-like seatrain component hulls. A three-unit seatrain was tested and showed a turning diameter of two seatrain lengths, which is considered very good (**Fig. 7**). Additional testing measured connector loads and the influence of the propeller on a subsequent unit.

For background information *see* FROUDE NUMBER; INLAND WATERWAYS TRANSPORTATION; MERCHANT SHIP; NAVAL SURFACE SHIP; SHIP DESIGN; SHIP POWERING, MANEUVERING, AND SEAKEEPING; TOWING TANK in the McGraw-Hill Encyclopedia of Science & Technology. Gabor Karafiath; Jonathan Slutsky

Bibliography. E. L. Amromin, G. Karafiath, and B. Metcalf, Viscous drag reduction by air cavity as an inverse ideal fluid problem: Achievements and issues, pp. 961–976, in *28th Symposium on Naval Hydrodynamics*, Pasadena, CA, Sept. 12–17, 2010, Curran Associates, Red Hook, NY, 2011;

E. L. Amromin, B. Metcalf, and G. Karafiath, Synergy of resistance reduction effects for a ship with bottom air cavity, *J. Fluids Eng.*, 133:021302 (7 pp.), 2011, DOI:10.1115/1.4003422; G. Karafiath, B. Metcalf, and J. Geisbert, Seatrain for high-capacity, high-speed, ocean transport, in *FAST 2009: 10th International Conference on Fast Sea Transportation*, Oct. 5–8, 2009, Athens, Greece, National Technical University of Athens, 2009.

Secure internet communication

Communication over the Internet is inherently insecure unless various schemes of authentication, integrity checking, and confidentiality protection are combined to achieve the desired level of security. Secure internet communication is the set of protocols that provides a certain level of security to any parties who use the inherently insecure Internet to exchange data and information. Because of its ubiquity, the Internet has become the most frequently used medium for exchanging information and data across the globe. Its application spans fields as diverse as e-commerce, education, and the financial sector. However, the Internet was not designed with security in mind. It was not until recently that new protocols and add-on services were introduced to establish and strengthen security on the Internet.

Threats to internet-based communication. Internet traffic may cross several autonomous networks that are managed by different independent Internet service providers (ISPs). Such traffic could easily be accessed by rogue individuals, organizations, or governments. There are numerous ways to threaten the traffic. The **table** enumerates some of these threats and the security services that help mitigate such threats. For there to be trust in the security of any communication channel, such as the Internet, the parties involved typically require many of the security services listed in the table. This article provides an overview of how these services provide security for internet communication. Communicating parties may choose all or some of the security services, depending on their need and the cost-versus-risk assessment.

Encryption. To hide its content from unauthorized access, a message or a document is encrypted by taking it through a sophisticated, yet reversible, process of alteration. The output of the encryption algorithm is known as the ciphertext equivalent of the original document or plaintext. The encryption algorithm is usually known to the public. However, the algorithm requires a key without which an unauthorized adversary is unable to decrypt (that is, reverse the encryption process) and retrieve the original plaintext. There are two categories of encryption algorithms: symmetric and public-key algorithms. Symmetric algorithms, such as the Data Encryption Standard (DES) and Advanced Encryption Standard (AES), use the same key, also known as the session key, for both encryption and decryption of a document. Therefore, entities using symmetric encryption must first securely exchange the session key, using either physical delivery to all parties or electronic delivery by applying a key-exchange protocol.

In contrast, public-key encryption algorithms require two keys: one is made public, while the other is kept private and must be guarded with the utmost secrecy. This pair of keys is created by each entity, using a procedure such as the Rivest-Shamir-Adleman (RSA) algorithm. For example, Ganna and Yusuf, two friends who wish to use public-key encryption, independently create the key pairs $(\boldsymbol{PUB}_{\mathbf{Ganna}}, \boldsymbol{PRV}_{\mathbf{Ganna}})$ and $(\boldsymbol{PUB}_{\mathbf{Yusuf}}, \boldsymbol{PRV}_{\mathbf{Yusuf}})$, respectively. They make the ***PUB*** component available to the general public. To send a secret message **M** to Yusuf, Ganna first encrypts it using Yusuf's public key as follows: $\mathbf{M}_{\mathbf{cipher}} = \textbf{RSA-Encrypt}(\boldsymbol{PUB}_{\mathbf{Yusuf}}, \mathbf{M})$. Without loss of generality, we assume that the parties are using the RSA public-key algorithm. $\mathbf{M}_{\mathbf{cipher}}$ is now transmitted through the Internet to Yusuf. Only Yusuf can retrieve **M,** using $\mathbf{M} = \textbf{RSA-Decrypt}(\boldsymbol{PRV}_{\mathbf{Yusuf}}, \mathbf{M}_{\mathbf{cipher}})$, because no one else has access to $\boldsymbol{PRV}_{\mathbf{Yusuf}}$. It is extremely difficult and prohibitively time-consuming for an adversary to recover $\boldsymbol{PRV}_{\mathbf{Yusuf}}$ from $\boldsymbol{PUB}_{\mathbf{Yusuf}}$ and $\mathbf{M}_{\mathbf{cipher}}$. This is because both components of the key pair are very large numbers (for example, 300 decimal digits), and recovery would require factorizing these very large numbers. However, if Ganna transmits $\mathbf{M}_{\mathbf{cipher}} = \textbf{RSA-Encrypt}(\boldsymbol{PRV}_{\mathbf{Ganna}}, \mathbf{M})$ (that is, she encrypts **M** using her own private key), anyone can recover **M** as follows: $\mathbf{M} = \textbf{RSA-Decrypt}(\boldsymbol{PUB}_{\mathbf{Ganna}}, \mathbf{M}_{\mathbf{cipher}})$. This feature of public-key encryption algorithms is used in digital signature schemes.

Authentication. A party wishing to provide evidence of its identity to the Internet community applies to some publicly trusted certification authority (CA) for a "certificate," which is a digital data

Threats and mitigations for secure communication

Threats	Mitigation
Identity Theft: impersonating either party of the communication	Authenticate: validate the identity claimed by the other party.
Malicious or Accidental Corruption of Content	Enforce Integrity Checking.
Disclosure of Secrets	Use Confidentiality/Encryption Services.
Replication of Content to Confuse Recipient(s)	Enforce Message Uniqueness/Freshness.
Repudiation: falsely denying ever sending the information	Use Digital Signatures: An arbiter validates the signature of the repudiating party.

structure containing public information, such as name, address, and public key. The certificate is endorsed (that is, digitally signed) by the authority and is presented by its owner to the public. Any alteration to a certificate can easily be detected. The signature of the CA on the certificate is inspected by the recipients, and if determined to be valid, the identity claimed by the party presenting the certificate must exactly match the one specified in the certificate. Certificates are freely transmitted on the Internet, however, and can be intercepted by an adversary. To prevent identity theft, when Yusuf receives a certificate presented by someone claiming to be Ganna, Yusuf first validates the certificate and learns Ganna's public key from it. He then challenges the presenting party by creating a large random number **N**, say 200 decimal digits long, which he then encrypts using **RSA-Encrypt(*PUB*$_{\mathbf{Ganna}}$, N)** and transmits the result to the party being authenticated. He challenges her to decrypt his message and respond with 1+ the value she recovers. If Yusuf receives the expected N + 1 value, he is then assured that the other party is indeed Ganna. After all, she is the only one able to decrypt Yusuf's original message with her private key ***PRV***$_{\mathbf{Ganna}}$. This authentication scheme fails only if someone manages to steal ***PRV***$_{\mathbf{Ganna}}$ from Ganna.

Secure hashing. To create the equivalent of a physical signature on an electronic document **M**, a fingerprint, also known as a message digest, of the document, **h = H(M)**, is first created by the sender using some secure hashing function H(•) such as SHA-512. A hashing function accepts as input an arbitrarily long message **M** and produces a fixed-length digest **H(M)**, usually hundreds of bits long. The hashing function is secure if and only if:

1. An adversary who intercepts a hash value ***b*** is unable to reverse the process and recover the original message **M**; that is, the algorithm $\mathbf{M = H^{-1}(\mathit{b})}$ does not exist. This is important because the digest of a message is easily accessible to the public and is sometimes attached to the encrypted version of that message while in transit.
2. For any given message $\mathbf{M_1}$, it is computationally prohibitive to find a different message $\mathbf{M_2 \neq M_1}$ such that $\mathbf{H(M_1) = H(M_2)}$.
3. An even stronger version of requirement 2 is that it is computationally prohibitive to locate any pair of messages $\mathbf{M_1 \neq M_2}$ such that $\mathbf{H(M_1) = H(M_2)}$.

Constraints 2 and 3 are essential for preventing an adversary from having you digitally sign a check M_1 and then attach your signature to another check M_2. This is known as the "birthday attack" and is illustrated in **Fig. 1**. Secure hash functions are used in combination with public-key encryption to assure that the sender has read and approved the content of the message, and to guarantee the integrity of the message's content.

Digital signature. For certain transactions, it is vital that the entity executing the transaction gets an undisputed approval, in the form of a digital signature, from the entity requesting that specific transaction. For any digital signature scheme to be secure, the signer, and only the signer, must deliberately produce his signature on the document. Moreover, the signer's signature on some document $\mathbf{M_1}$ cannot be used on a different document $\mathbf{M_2}$, even one with very minor changes. The scheme should make it easy for any entity to verify the authenticity of the signature. Several digital signature schemes are available, for example, the Digital Signature Algorithm (DSA). An RSA-based digital signature scheme is shown in **Fig. 2**, where the signature is **RSA-Encrypt(*PRV*$_{\mathbf{Signer}}$, H(M))**, that is, the digest of the message is encrypted with the signer's private key. The message itself can be transmitted either as clear text or after being encrypted with a previously exchanged symmetric session key. Figure 2 also shows the process of validating the signature at the recipient's end. By verifying the sender's digital signature on a document, not only is the recipient assured of the integrity of the message, but also of the signer's deliberate approval of the document's content. This in turn disqualifies any repudiation claims by the signer in the future.

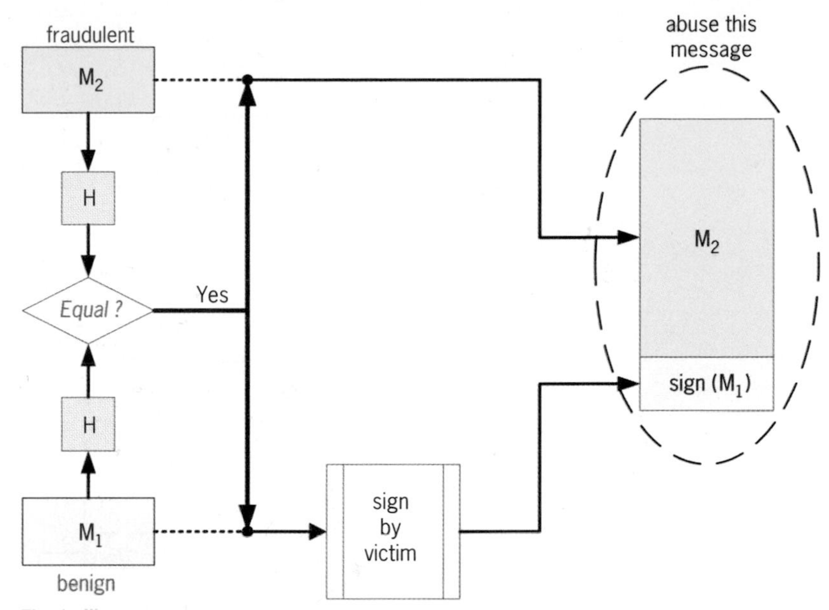

Fig. 1. Illustration of the birthday attack.

Confidentiality. Symmetric encryption is much faster to execute on large documents than public-key encryption. However, symmetric encryption requires the exchange of a session key among the parties involved in the communication. A simple scheme calls for one party to locally generate a random session key and then send a copy of the key, encrypted with each recipient's public key, to that recipient. For example, Ganna generates the session key **K** and sends **RSA-Encrypt(*PUB*$_{\mathbf{Yusuf}}$, K)** to Yusuf, who can easily recover **K** using his private key, ***PRV***$_{\mathbf{Yusuf}}$, as illustrated at the top of Fig. 2. Subsequently, they both can encrypt/decrypt any communication using **K**. However, it is very important that they generate a new key for every communication session, or after a certain use-time or

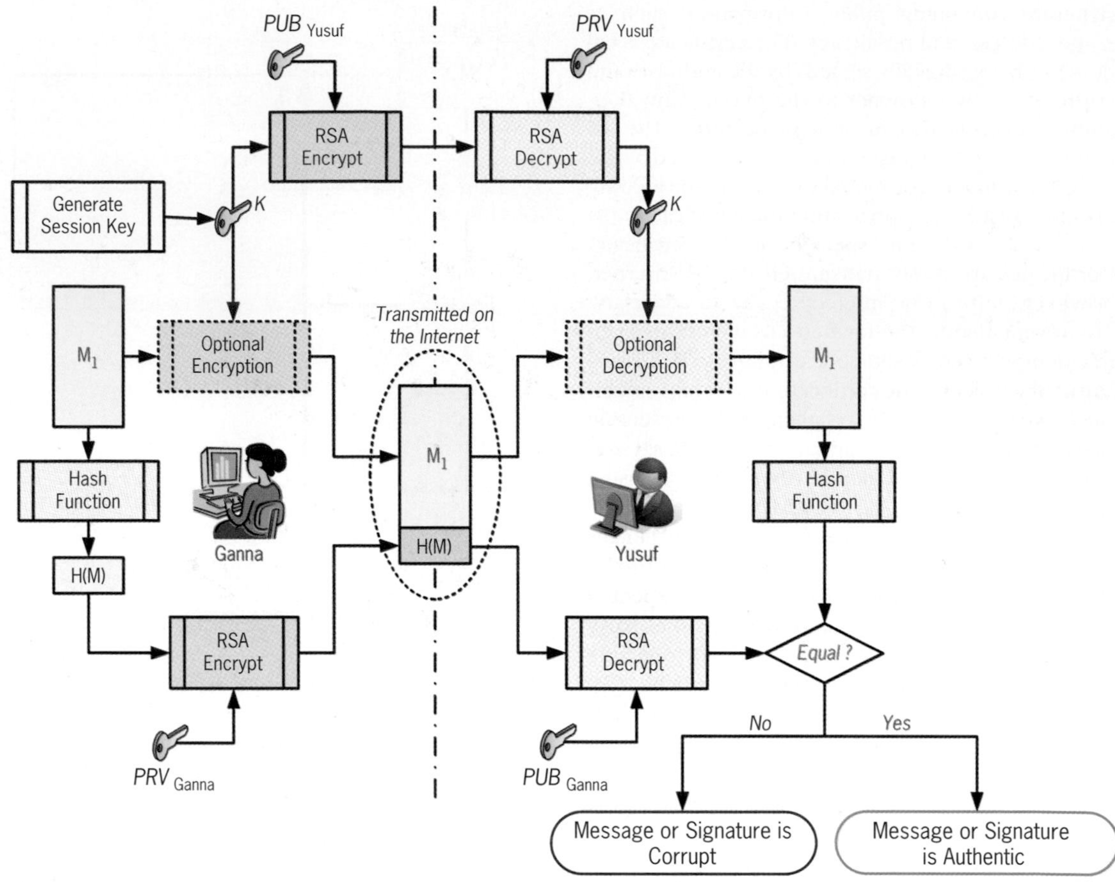

Fig. 2. RSA-based digital signature with optional confidentiality.

data amount, whichever comes first. Using the same key for an extended period of time, or for encrypting a huge amount of data, enhances an adversary's chance to break the code.

Mitigating playback attacks. Replicas of an old digitally signed message confuse the recipient, thus they must be detected and discarded. Two approaches exist: a time-stamp-based scheme, and a random-number-based scheme. The signer can include the current time as part of the message before producing her signature. The recipient then rejects messages that arrive after a long time from being signed. This requires both the sender and the recipient to maintain tightly synchronized clocks, which has proved to be a globally challenging problem. Another approach requires the recipient to initiate the request for the signed message. In his request, Yusuf sends a large (that is, 200–300 decimal digit) random number, referred to as a nonce, to Ganna. Yusuf must never recycle a previously used nonce. Because of its randomness, no adversary can predict the nonce generated by Yusuf. Ganna then includes the nonce, or a function of it such as the nonce +1, into the message before producing her signature. On arrival, Yusuf discards any message that does not include a recently generated nonce of his own.

For background information *see* ALGORITHM; COMMUNICATIONS SCRAMBLING; COMPUTER SECURITY; CRYPTOGRAPHY; DATA COMMUNICATIONS; INTERNET; NUMBER THEORY in the McGraw-Hill Encyclopedia of Science & Technology.

Mohamed S. Aboutabl

Bibliography. M. Bishop, *Introduction to Computer Security*, Addison-Wesley, Indianapolis, IN, 2004; M. Boyd, *Protocols for Authentication and Key Establishment*, Springer, New York, 2003; B. A. Forouzan, *Cryptography and Network Security*, McGraw-Hill, New York, 2008; W. Stallings, *Cryptography and Network Security: Principles and Practices*, 6th ed., Prentice Hall, Upper Saddle River, NJ, 2014.

Self-adaptive design techniques

Self-adaptive design techniques are techniques for designing integrated circuits that can adjust the value of a circuit parameter during run time by feedback control, or allow the values of some key run-time parameters to be set after design time, by the use of fuses, registers, memory, or some similar means.

Moore's law is the observation that, over the history of computing hardware, the number of transistors on integrated circuits doubles approximately every 18 months. The law is named after Gordon E. Moore, cofounder of Intel Corporation, who described the trend in 1965. Today, the semiconductor industry is able to manufacture chips that have a capacity of a few billion transistors, which allows for integrating complete systems comprising multiple processing cores, memories, and various peripheral components onto a single chip (resulting in a system-on-chip, abbreviated SoC). However, the ongoing process of scaling down the physical transistor parameters is becoming an increasingly challenging issue for future SoC design. The margin between nominal device operation and device failure is steadily shrinking, and the impending physical limits of the underlying complementary metal-oxide-semiconductor (CMOS) technology lead to increased vulnerability of a chip to radiation and degradation effects and to growing process variabilities during manufacturing. Moreover, the ongoing miniaturization is accompanied by increasing power density on the chip, which requires efficient heat removal and thermal management techniques to reduce the occurrence of "hot spots" and thermomechanical stress. This situation is likely to get worse if 3D stacking of multiple chips is applied to reduce the interconnect delay and the package size.

These problems have been identified as the major challenge to integrated-circuit design to be solved within the next decade. One solution is to evolve from today's deterministic design to probabilistic and statistical design for the future. These new design methods and tools have to guarantee reliable and robust SoCs composed of less reliable components. Looking at design methodology, for the last 40 years, most chips have been produced based on fixed and rigid designs, where every uncertainty about the final product and its dynamics and operating conditions has been addressed during design and by applying higher design margins. However, with increasing variability, variation, and degradation, this approach becomes less favorable, because the worst case design approach based on higher margins would swallow up the advantages of new technology nodes, threatening to make them economically untenable. Therefore, future SoCs cannot be statically designed for all worst-case conditions and design margins, but they have to adapt to changing operating conditions. A solution to this problem is to integrate additional circuits that enable the chip to self-adapt individually to its actual manifestation in order to reduce the underlying uncertainties and their dynamics. The self-adapting circuits allow the chip to reach its individually maximal performance and functionality. Essentially, self-adaptation moves some decisions that have formerly been taken at design time to the run time of a chip, because the particular optimal settings of the chip are unknown at design time.

Components of self-adaptive systems. Generally, self-adaptive systems comprise three essential components to implement a self-adaptive design technique: a monitor, an evaluator, and an actuator. The monitor reads the current state of the chip (for example, the current timing-error rate or the values of thermal sensors), the evaluator deduces an appropriate response (for example, to increase the supply voltage), and the actuator carries out the response (for example, by addressing the integrated voltage regulators). Self-adaptive design techniques are known by various names, for example, the IBM MAPE (monitor, analyze, plan, execute) model.

DVFS. Many self-adaptive systems use dynamic voltage and frequency scaling (DVFS) to reduce the overall energy consumption of a computer system. DVFS can be used either to push the operating frequency of a circuit beyond the nominal operating frequency to improve the performance, or to reduce the voltage and frequency to lower the energy consumption at times when the full performance of the hardware is not required by the application to be executed.

Examples. The Turbo Boost technique, a Core i7-specific feature from Intel, uses DVFS to dynamically change the operating frequency and operating voltage of a processor core based on current operating conditions. Razor is an approach to dynamic voltage scaling (DVS) that adapts the voltage to the current temperature variations such that the timing-error rate is kept at a predetermined value. ReCycle is an adjustable circuit, which redistributes cycle time across the processor pipeline to compensate process variations; the amount of redistribution is determined by built-in self-tests and is adjusted once after production. Another self-adaptive system is the deskew system of the Intel Itanium 2 processor with code name Montecito. It adapts a delay buffer to nullify clock skew due to power, voltage, and temperature variations at run time; for maximum performance, the deskew system is also adjusted once after production to compensate clock skew due to process variation.

Autonomic SoC platform. Another example of a self-adaptive system is the autonomic SoC (ASoC) platform. It addresses the challenges arising from increased chip capacities by embedding autonomic principles in SoCs. The concept of this platform is inspired by the way nature deals with complexity. For example, the human organism has an autonomic nervous system that fulfills life-critical control and supervision functions (such as heart-beat control, breathing, and temperature control through skin functions) without the active and conscious awareness of the human being. The ASoC architecture platform (**illus.** *a*) is split into two logical layers: The functional layer contains all the functional elements (FEs), which are general-purpose central processing units (CPUs), memories, on-chip buses, special-purpose processing units (PUs), or interfaces, as in a conventional, nonautonomic design. The autonomic layer consists of autonomic elements (AEs) and an interconnection structure among the AEs. Each AE is composed of a monitor, an evaluator, and an actuator (Illus. *b*). The monitor senses signal or state information from the associated FE. The state information of an ASoC is manifold: It includes inputs from temperature sensors, counts of observed errors, measurements of performance and power, information on defective or degraded units, and much

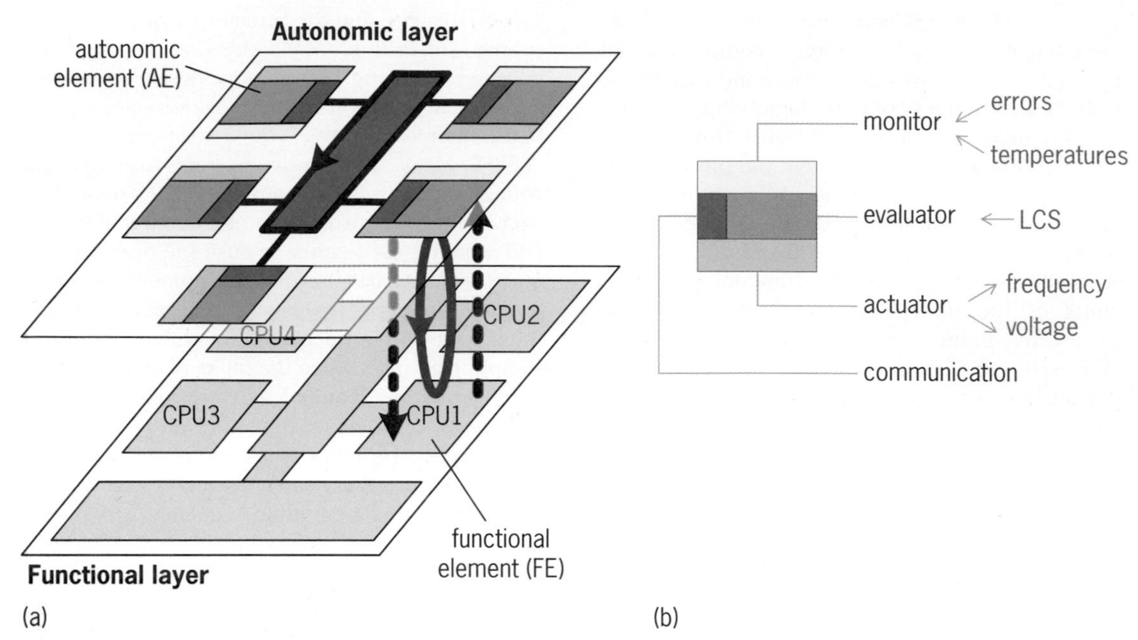

Autonomic system-on-chip (SoC) platform. (*a*) Platform with two logical layers. (*b*) Autonomic element (AE). CPU = central processing unit; LCS = learning classifier system.

more. The evaluator merges and processes the locally obtained information with the state information obtained from other AEs or memorized local knowledge.

In self-adaptive systems such as Razor, specially trained engineers who are faced with a self-adaptation problem must craft an individual solution while considering all possible events that may affect it. In the ASoC architecture platform, an evaluator is used that automatically learns the optimal actions for a given situation by itself. The evaluator uses a learning classifier system (LCS) based on reinforcement learning techniques to guide the self-adaptation. An LCS is an adaptive system that learns to perform the best action given its input. The learning process can be divided into a design-time phase and a run-time phase. During design time, an initial rule set is determined that is used at run time to support a correct working system immediately after the SoC starts operation. Afterwards, run-time learning is applied for self-adaptation of the SoC with respect to changing operating and environmental conditions or malfunctioning components of the chip.

Comparison to evolvable hardware. In contrast to evolvable hardware, self-adaptive systems adjust to changing system and operating conditions to maintain their intended functionality, whereas the functionality of evolvable hardware is the result of an emergent process during run time.

For background information *see* ADAPTIVE CONTROL; AUTONOMIC NERVOUS SYSTEM; DIGITAL COMPUTER; FEEDBACK CIRCUIT; INTEGRATED CIRCUITS in the McGraw-Hill Encyclopedia of Science & Technology. Oliver Bringmann; Thomas Schweizer

Bibliography. A. Bouajila et al., Autonomic system on chip platform, in C. Müller-Schloer, H. Schmeck, and T. Ungerer (eds.), *Organic Computing—A Paradigm Shift for Complex Systems*, Springer, Basel, Switzerland, pp. 413–425, 2011; D. Ernst et al., Razor: A low-power pipeline based on circuit-level timing speculation, in *Proceedings of the 36th annual IEEE/ACM International Symposium on Microarchitecture (MICRO 36)*, San Diego, CA, December 3–5, 2003, IEEE Computer Society, Washington, DC, pp. 7–18, 2003, DOI:10.1109/MICRO.2003.1253179; S. Naffziger et al., The implementation of a 2-core, multi-threaded itanium family processor, *Solid-State Circuits*, 41:197–209, 2006, DOI:10.1109/JSSC.2005.859894; E. Rotem et al., Power-management architecture of the Intel microarchitecture code-named Sandy Bridge, *IEEE Micro*, 32(2):20–27, March 2012, DOI:10.1109/MM.2012.12; A. Tiwari, S. R. Sarangi, and J. Torrellas, ReCycle: Pipeline adaptation to tolerate process variation, in *Proceedings of the 34th Annual International Symposium on Computer Architecture (ISCA)*, San Diego, CA, June 9–11, 2007, Association for Computing Machinery, New York, pp. 323–334, 2007, DOI:10.1145/1250662.1250703.

Self-assembled organic optical materials

Self-assembly refers to the spontaneous organization of a predesigned functional molecule (building block) under appropriate conditions by using the noncovalent forces of interactions between the molecules. The noncovalent interactions commonly used for the self-assembly of functional organic molecules rely on the use of π-stacking, hydrogen bonding, electrostatic, metal-coordination, hydrophobic, and van der Waals forces of interaction. Self-assembly, therefore, can be a very useful tool to generate structures much larger than the

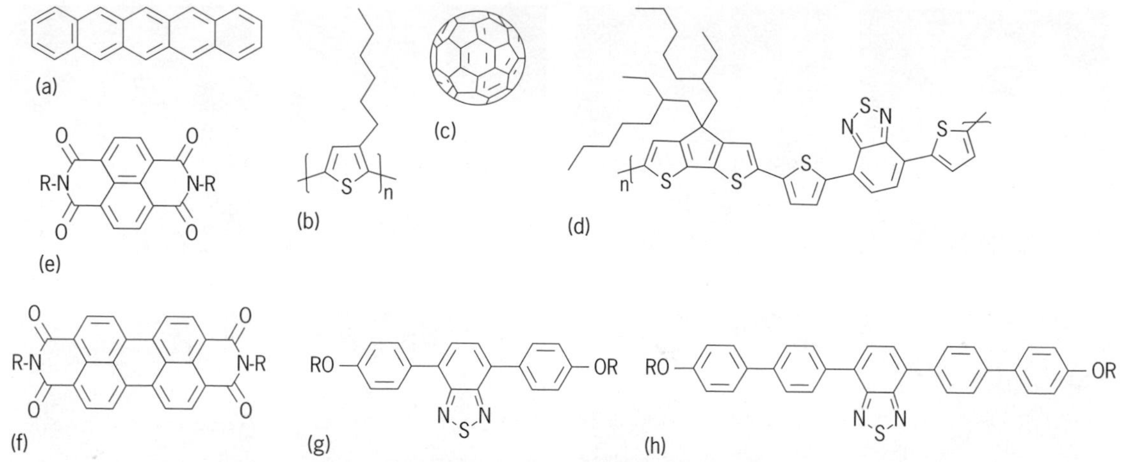

Fig. 1. Chemical structures of some of the commonly employed π-conjugated molecules and polymeric structures in self-assembly and device applications. (*a*) Pentacene. (*b*) Poly(3-hexylthiophene-2,5-diyl) [P3HT]. (*c*) C_{60} (fullerene). (*d*) PCPDTTBTT. (*e*) Naphthalene diimide (NTCDI). (*f*) Perylene diimide (PTCDI). (*g*) 4,7-bis(alkyloxyphenyl)-2,1,3-benzothiadiazole (BTD-1). (*h*) 4,7-bis(alkyloxybiphenyl)-2,1,3-benzothiadiazole (BTD-2).

components (molecules) themselves (for example, large-scale crystals, nanostructures, and microstructures). The resulting structures are also referred to as supramolecular structures because of the large number of individual molecular building block units present in the final assembled structures. The vast array of π-conjugated molecules and polymers (some of the molecular structures are shown in **Fig. 1**) are particularly attractive for advancing both the materials and device applications because of the ability to tailor the molecular and self-assembled properties by predesigned synthesis. The π-conjugated molecules contain alternating single and double bonds. The double bonds are comprised of the sigma bonds (σ-) and three π-orbitals (p_x, p_y, and p_z). The p_z orbitals lie perpendicular to the coplanar carbon-to-carbon double bond (C═C) and interact with neighboring orbitals, leading to additional bonding and delocalization of the electron density (both below and above the molecule). The interactions of the π-orbital lead to many stacking (π-stacking) possibilities and are evident in abundance in π-conjugated systems (such as dimers of benzene and stacking of DNA).

Role of size and dimension. In comparison to their inorganic counterparts, the organic molecules can be tailored specifically to suit the applications. Moreover, by controlling the functionalization (often, side chains), a high degree of solubility can be achieved in various solvents, enabling ease of processing. The molecular interactions, such as π-stacking, interdigitation of side chains, and hydrogen-bonding interactions to impact direction control, also allow for controlling the morphology; that is, they provide size and dimension control. This ability to tailor dimension and size has significant importance for device applications since both properties and miniaturization can be addressed. Essentially, the packing of the molecules within the organized structures allows for the determination of the final properties and therefore their characteristic device performance. The dimension- and size-controlled self-assembled structures from some of the functional organic molecules are shown in **Fig. 2**.

The physical vapor deposition method has been widely employed for crystallization and controlled deposition of materials over desired substrates. In this method, there is no chemical transformation. The organic material to be deposited or crystallized is carried in an inert vapor stream at much below the decomposition temperatures (unlike chemical vapor deposition). Additionally, a gradient in flow temperature allows for controlled deposition of crystalline thin films from organic materials. This approach has now been extended to the production of dimension- and size-controlled nanostructures such as nanowires.

Although vapor-based methods are popular, solution processing approaches have come to the forefront particularly for organic molecules, since the dimension and size can be controlled by optimizing the growth conditions such as solvent and temperature. The most widely studied self-assembled structures belong to the 1D morphology, and elegant strategies have been developed toward control of size (lengths and widths) and features necessary for optimizing the ensuing properties (such as chemical, mechanical, optical, and electrical). In the 1D morphology, molecules have been self-assembled into fibers, wires, rods, and tubules. More recently, 2D self-assembled structures (such as sheets) have also gained attention, owing to the realization of 2D nanomaterials such as graphene, and possess remarkable optical properties for integration in optical and electronic devices. The ability of the dimension- and size-controlled self-assembled structures to show unique optical and electrical properties is critical for advancing their use in next-generation photonic components such as waveguides, light-emitting diodes (LEDs), solar cells, display, polarized-light emitters, photodetectors, phototransistors, and lasers. The self-assembly process is a bottom-up method for the

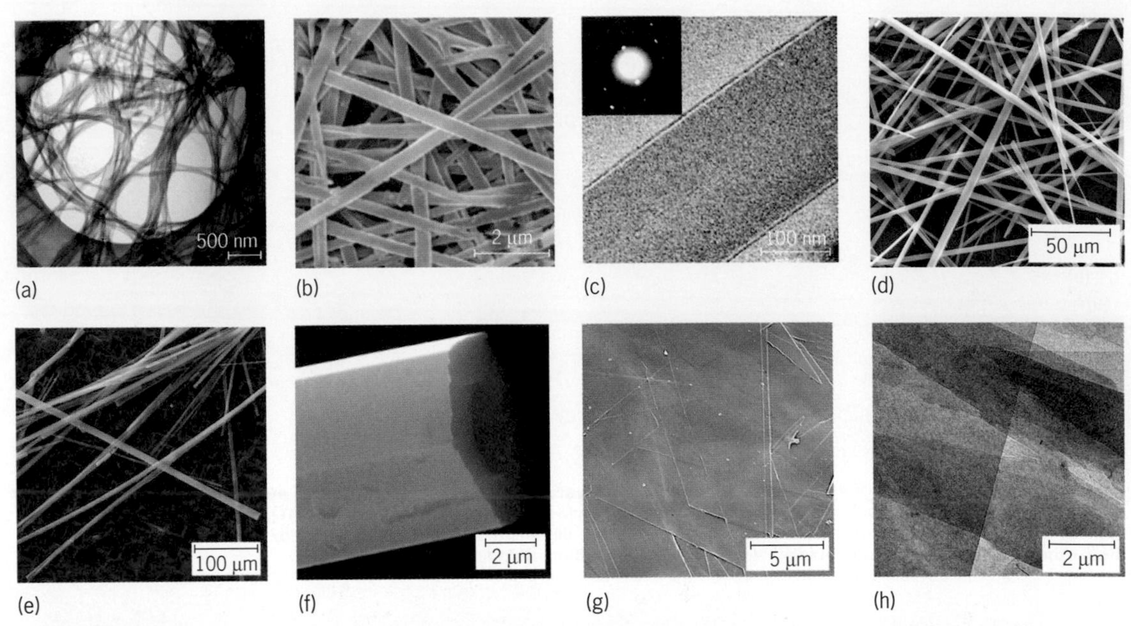

Fig. 2. Morphology of structures self-assembled from PTCDI and BTD molecules. (*a*) One-dimensional (1D) nanofibers (*reprinted with permission from K. Balakrishnan et al., Effect of side-chain substituents on self-assembly of perylene diimide molecules: Morphology control, J. Am. Chem. Soc., 128:7930–7938, 2006, copyright © 2006, American Chemical Society, DOI:10.1021/ja061810z*). (*b, c*) Nanobelts (*reprinted with permission from K. Balakrishnan et al., Nanobelt self-assembly from an organic n-type semiconductor: Propoxyethyl-PTCDI, J. Am. Chem. Soc., 127:10496–10497, 2005, copyright © 2005, American Chemical Society, DOI:10.1021/ja052940v*). Inset of part (c) shows diffraction pattern. (*d*) Microwires. (*e, f*) Microtubules (*from K. Balakrishnan et al., Multi-mode waveguides from ultra-long self-assembled hexagonal faceted microtubules of a benzothiadiazole molecule, Chem. Commun., 48:11668–11670, 2012, DOI:10.1039/C2CC36327E*). (*g, h*) Two-dimensional (2D) sheets.

realization of ordered structures and complements traditional fabrication techniques (such as top-down methods including e-beam and photolithography) for device fabrication. Some of the critical features of the self-assembled materials and the resulting optical components and devices will be discussed.

Polarized-light emission and waveguides. Both 1D and 2D self-assembled structures have been studied for achieving polarized-light emission. The anisotropic arrangement of molecular entities within the self-assembled structures is essential for achieving such emission. The strong packing of the molecules often leads to quenching of emission because of the forbidden transitions. This is often described by an empirical observation called Kasha's rule, which governs the optical transitions and fluorescence emission yield from the molecular structures and aggregates. However, in some self-assembled structures the favorable packing of the molecules enables strong light emission because of the distortions in the molecular arrangement. The molecular distortions within the stacks allow electronic transitions that would otherwise be forbidden in an ideal stack. Recently, strong fluorescence emission was observed from self-assembled nanowires and microtubules made from PTCDI and BTD-2 molecules (Figs. 1*f* and 1*b*) because of the distorted packing of the molecules. Such self-assembled structures show unique ability for light emission to be polarized along the preferred direction of molecular arrangement. Examples of polarized-light emission from 1D self-assembled structures are illustrated in **Fig. 3**. The self-assembled structures with appropriate size, dimension, and refractive index also can be used as optical waveguides. The combination of top-down fabrication techniques and self-assembly of 1D structures with controlled widths, lengths, and unique morphologies shows enormous potential for photonics device applications.

Phototransistors and photodetectors. The dynamic modulation of charge carrier density by light intensity at specific wavelengths leads to photodetection systems. Many of these are based on the operating principles of the field-effect transistor, except that the charge carrier mobility or density is changed as a function of the incident light—the device is a phototransistor. Such detectors are normally fabricated from complex inorganic materials. However, recent advances toward improving the materials quality and the new know-how with dimension- and size-controlled self-assembled structures now enable high-performance phototransistors and highly sensitive photodetection over a range of wavelengths. The initial studies on the synthesis of new building blocks and their assemblies were important steps toward understanding critical material and device parameters. The subsequent utilization of highly crystalline and single-crystal materials not only has improved our understanding of operating mechanisms but also has led to significantly higher values of carrier mobility (10–40 $cm^2V^{-1}s^{-1}$), an important performance metric. Today, there are more than 50 organic systems with charge carrier mobilities higher than 1 $cm^2V^{-1}s^{-1}$. The emergence of

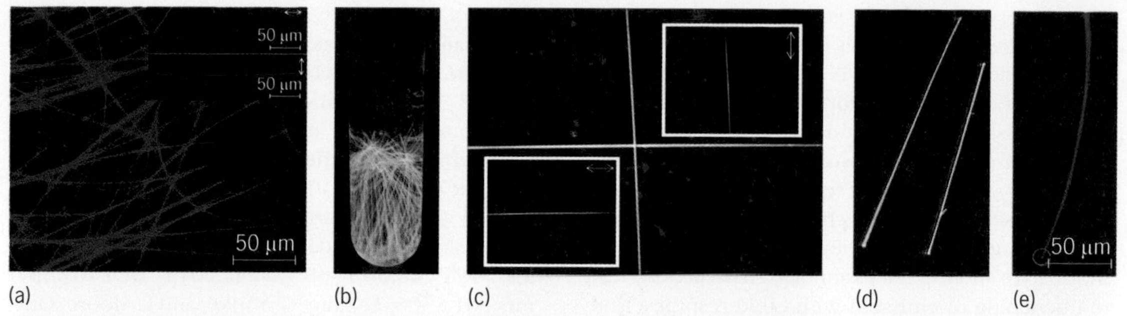

Fig. 3. Photoluminescence from (*a*) 1D nanowires of PTCDI and (*b, c*) microtubules. The fluorescence of 1D self-assembled structures also exhibits polarization along the favorable packing direction of the molecules, as seen in the insets of parts *a* and *c*. Arrows in the insets of part *a* indicate the direction of the linear polarizer with respect to the 1D wires. Arrows in the insets of part *c* indicate the direction of linear polarization of light emission. Such 1D self-assembled structures coupled with appropriate size and refractive index can exhibit waveguiding behaviors, as seen in (*d*) BTD and (*e*) PTCDI, characterized by the confinement of light within the self-assembled structures and the ability to guide light of different wavelengths and polarization along the elongated direction. (*Parts a and e reprinted with permission from Y. Che et al., Highly polarized and self-waveguided emission from single-crystalline organic nanobelts, Chem. Mater., 21:2930–2934, 2009, copyright © 2009, American Chemical Society, DOI: 10.1021/cm9007409; parts b–d from K. Balakrishnan et al., Multi-mode waveguides from ultra-long self-assembled hexagonal faceted microtubules of a benzothiadiazole molecule, Chem. Commun., 48:11668–11670, 2012, DOI:10.1039/C2CC36327E*)

self-assembled materials with both highly controlled structural assemblies and unique optical and electrical characteristics is anticipated to be critical in many of the photonics technologies such as photodetection and imaging.

Solar cells. The ability of π-conjugated optoelectronic molecules and polymers to absorb light and convert the incident radiation into electrical charges (collected at electrodes) has allowed for evolving paradigms for inexpensive organic solar cells, given the low cost of processing and materials. On photoexcitation in organic systems, excitons (bound electron-hole pairs) are produced, unlike the free charge carriers that often result in inorganic or silicon-based photovoltaic materials. These excitons must be broken into the corresponding free charge carriers (charge separation) followed by rapid charge migration (or transport) within the material so that the carriers can be collected at respective electrodes. Since the introduction of the prototype light-harvesting demonstrations using a bilayer structure of an electron donor and electron acceptor, significant strides have been made using combinations of organic materials and new device architectures. Several materials and device design paradigms have been established that have now allowed for achieving ~6–8% efficiency in single-junction binary structures and ~10% efficiency using tandem (multijunction) solar cells.

Presently, exciton diffusion and charge migration are both major issues in thin-film organic solar cells and represent major bottlenecks. Only recently have self-assembled structures made inroads into this arena, and already high-quality crystalline assemblies are providing valuable insight into the mechanistic details. Additionally, in organic solar cells, the conversion of light into electrical energy occurs when an electron donor (excited by the incident photon), instead of relaxing back to its ground state, transfers an electron to another material (electron acceptor), creating free carriers at the interface that migrate to the cathode and anode. The carrier mobility (that is, the speed at which the charges are transported) is often low in amorphous or polycrystalline organic materials, but improves significantly in single-crystalline systems. Therefore, in organic solar cells, the use of single-crystalline systems, particularly cocrystalline systems (systems comprised of electron donor-acceptors that are made of two or more different types of molecules) using crystal growth and self-assembly paradigms, can in part alleviate the problem of charge migration and is rapidly evolving into a critical strategy for high-performance organic solar cells. *See* SENSITIZED MESOSCOPIC SOLAR CELLS.

Organic light-emitting diodes (OLEDs). These devices are based on the principle of electroluminescence and are important in lighting and display applications. An OLED consists of a transparent conductor (anode), hole-transport and electron-transport layers, and a cathode electrode. Upon application of voltage, the charge carriers (holes and electrons) mix and produce excitons. The relaxation of the exciton produces the desired light emission. The process is the opposite of that in an organic solar cell, where the exciton produced upon photoexcitation is broken into charge carriers. Significant advances in performance have occurred recently whereby record light emissions from organic materials are now competing fiercely for commercial lighting applications. Additionally, large-scale low-power displays have also been realized and are now expected to be widely employed in mobile devices. The recent adaptations of self-assembled structures in such applications also promote enhanced movement of charges leading to strong emission at the interfaces and are likely to play a critical role in future devices.

Lasers. The π-conjugated optoelectronic organic materials have strong absorption and tunable emission that make them attractive for the fabrication of lasers. Recently, strong stimulated emission has been observed in 1D self-assembled nanowires and shows high gain. Advances in the control over morphology and properties will therefore be critical for next-generation high-performance lasers fabricated from self-assembled structures.

Chemical sensors. The strong fluorescence emission from π-conjugated self-assembled structures can be used directly in chemical-sensing applications. In such a device, the binding event between the analyte and the self-assembled structure can lead to quenching of emission with rapid response time and high sensitivity. Such sensors are now available for detection of trace amounts of explosives and are already a critical technology in defense-related applications.

Nonlinear optics. The recent nonlinear optics characterization of several molecular crystals has revealed new design principles for synthesis of tailored molecules. Large values of the nonlinear susceptibilities $\chi^{(2)}$ (second order) and $\chi^{(3)}$ (third order) are key to achieving high nonlinear conversion efficiency and light modulation, and their values depend greatly on the crystal symmetry and the molecular hyperpolarizability tensor. Such factors lead to intensive consideration of the molecular packing capabilities that arise from predesigned molecular systems, particularly for the cocrystal design.

Status and prospects. The self-assembly of molecular materials and polymers into dimension- and size-controlled structures is a continually changing field supported by advances in the synthesis of new molecules, polymers, hybrids, and growth processes, along with the characterization of properties. In turn, a wide range of optical materials and technology has been developed based on the ability to use light as a source, as a detector, or for energy harvesting. The integration of top-down (lithographic process) and bottom-up (self-assembly) technologies has further enabled realization of a range of optical devices based on self-assembled nanostructures, which is crucial for miniaturization and the widespread adaptation of materials technology. The innumerable combinations of self-assembled materials and the exploitation of these materials in photonic devices will be critical for achieving new milestones in the figures of merit for device technologies. The future is poised for reaping the rewards from dimension- and size-controlled self-assembled organic optical materials.

For background information *see* CHEMICAL BONDING; CRYSTAL GROWTH; EXCITON; GRAPHENE; INTERMOLECULAR FORCES; LASER; LIGHT-EMITTING DIODE; MOLECULAR ORBITAL THEORY; NONLINEAR OPTICS; OPTICAL MATERIALS; PHOTOELECTRIC DEVICES; PHOTONICS; PHOTOTRANSISTOR; POLARIZED LIGHT; SEMICONDUCTOR; SOLAR CELL; TRANSISTOR; WAVEGUIDE in the McGraw-Hill Encyclopedia of Science & Technology.

Kaushik Balakrishnan; Stanley Pau

Bibliography. K. Balakrishnan et al., Multi-mode waveguides from ultra-long self-assembled hexagonal faceted microtubules of a benzothiadiazole molecule, *Chem. Commun.*, 48:11668–11670, 2012, DOI:10.1039/C2CC36327E; B. Kippelen, Optical materials: Self-assembly reaches new heights, *Nat. Mater.*, 3:841–843, 2004, DOI:10.1038/nmat1273; T. Lei and J. Pei, Solution-processed organic nano- and micro-materials: Design strategy, growth mechanism and applications, *J. Mater. Chem.*, 22(3):785–798, 2012, DOI:10.1039/c1jm14599a; V. Podzorov, Organic single crystals: Addressing the fundamentals of organic electronics, *MRS Bulletin*, 38:15–24, 2013, DOI:10.1557/mrs.2012.306; L. Zang, C. Yanke, and L. Moore, One-dimensional self-assembly of planar π-conjugated molecules: Adaptable building blocks for organic nanodevices, *Accounts Chem. Res.*, 41:1596–1608, 2008, DOI:10.1021/ar800030w.

Sensitized mesoscopic solar cells

A sensitized mesoscopic solar cell is based on a "mesoporous" oxide film, where mesopores with diameters of 2–50 nm are formed by the packing of oxide nanoparticles. A mesoscopic solar cell is usually composed of the nanoparticle titanium dioxide (TiO_2) film on a transparent conductive oxide (TCO) substrate, the light-harvesting sensitizer that is adsorbed on the TiO_2 surface, and the redox electrolyte for a liquid cell or the hole-transporting material for a solid-state cell. The TCO substrate with the sensitizer-adsorbed TiO_2 film serves as the working electrode or negative electrode. A platinum-coated conductive substrate is used as a counter electrode in the liquid electrolyte cell, while metal thin films, such as gold or silver, are used as the counter electrode in the solid-state cell. The mesoscopic solar cell was first developed by Brian O'Regan and Michael Grätzel in 1991, and is called the Grätzel cell.

The power conversion efficiency (PCE) was about 8% in 1991, which was improved to 10% in 1993. The high conversion efficiency was ascribed mainly to the enlarged surface area of the TiO_2 film providing numerous sites for adsorption of the sensitizer. Since 1991, intensive research has been performed on the sensitized mesoscopic solar cell. As a result, a PCE as high as 12% was achieved in 2011 using porphyrin dye and cobalt redox electrolyte. Cobalt redox electrolyte has a higher photovoltage than iodide redox electrolyte, and the 12% efficiency was mainly due to the high photovoltage of cobalt electrolyte, nearly 1 V.

The cheap material and low-cost process are advantages of the mesoscopic sensitized solar cell. However, a PCE much higher than 12% is required to fulfill grid parity. In order to meet this criterion, a novel technology must be developed to achieve a PCE comparable to or higher than that of the silicon solar cell. The achievement in 2013 of a PCE of 15% using an organolead halide perovskite sensitizer seems to promise a bright future.

Types of cells. The sensitized mesoscopic solar cell, originally termed the dye-sensitized solar cell, can be fabricated as either of two different types, the electrochemical junction type and the all-solid-state type. The former is based on a liquid redox electrolyte and the latter is based on a solid hole conductor (**Fig. 1**). The electrochemical junction type can be prepared by coating the TiO_2 paste, which is annealed at around 500°C (932°F) in air, on the fluorine-doped tin oxide (FTO)-coated substrate. The annealed mesoporous TiO_2 film is immersed in a solution containing dye molecules to adsorb dye on the TiO_2 surface. This working electrode is sandwiched with the platinum-coated counter electrode, and then liquid electrolyte is injected between the working and counter electrodes.

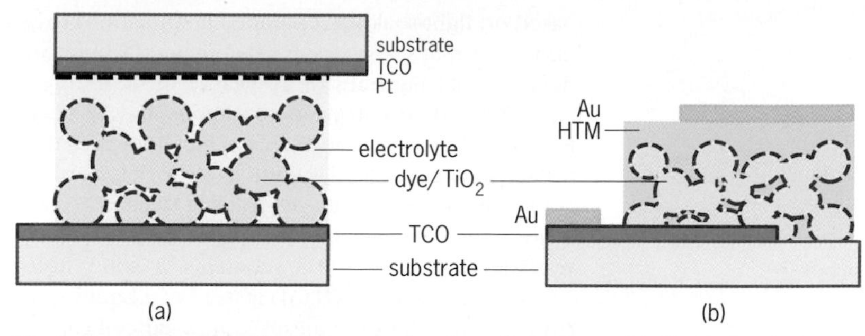

Fig. 1. Two different sensitized mesoscopic solar cell structures. (*a*) Electrochemical junction structure based on a liquid redox electrolyte. (*b*) Solid-state pseudo *p-i-n* junction structure based on a hole-transporting material (HTM).

Working principle of the electrochemical junction type. When a semiconductor is in contact with a liquid electrolyte, an electrochemical junction is formed. In the electrochemical junction structure, the entire oxide surface should be in contact with the redox electrolyte. Thus, an open pore structure is important for a semiconductor-electrolyte contact. How does the sensitized mesoscopic solar cell work? The working principle of the electrochemical junction type of sensitized solar cell is schematically depicted in **Fig. 2**. When the sensitizer absorbs incoming solar light, electrons in the ground state [or the highest occupied molecular orbital (HOMO) level for a molecular sensitizer, or the valence band for an inorganic semiconductor sensitizer] are excited to the excited state [or the lowest unoccupied molecular orbital (LUMO) level for a molecular sensitizer, or the conduction band for an inorganic semiconductor sensitizer]. Thus the role of the dye (that is, the sensitizer) is to harvest light and generate electron-hole pairs. Therefore, the use of the dye molecule distinguishes between the sensitized mesoscopic solar cell and the conventional semiconductor *pn*-junction solar cell without a sensitizer. In order to adsorb a lot of dye, a high-surface area, nanostructured TiO_2 film is required, which is also quite different in nature from the low-surface-area, single-crystalline characteristics required for the conventional semiconductor solar cell.

The photoexcited electrons should be injected into the conduction band of the *n*-type oxide; otherwise, those excited electrons cannot be collected at the TCO substrate. Once the excited electrons are injected, they are transported through the porous oxide network by a diffusion mechanism, collected at the TCO substrate, and flow in an external circuit to generate photocurrent. The oxidized sensitizer is regenerated by oxidation of iodide when using iodide and tri-iodide as a redox couple. The electron movement in the electrochemical junction type of sensitized mesoscopic solar cell is shown in Fig. 2. The solar cell voltage is determined by the difference between the Fermi energy (E_F) of TiO_2 and the redox potential of the electrolyte. To show both the electron flow and the open-circuit voltage (V_{oc}), the energy diagram in Fig. 2 is sketched based on the open-circuit condition. A TiO_2 film thickness of 10–15 μm is recommended for high efficiency in the liquid-type device.

In preparing the liquid-based dye-sensitized solar cell, the handling of the liquid electrolyte is carefully managed since it contains a volatile solvent. Thus, the sealing process affects performance and long-term stability as well. After the sandwiching of the two electrodes using a hot-melt sealant called Surlyn®, liquid electrolyte is introduced in the gap between the electrodes by capillary force through a hole in the TCO glass, a process which is followed by sealing the holes. If the sealing of the device is not perfect, it will have a short life. However, perfect sealing will guarantee the long life of the device. The volatility of the solvent is also a critical issue. The highly volatile acetonitrile that was used in the early stages of development was replaced by less volatile solvents such as valeronitrile and metoxypropiontrile or by ionic liquids. It was reported in 2011,

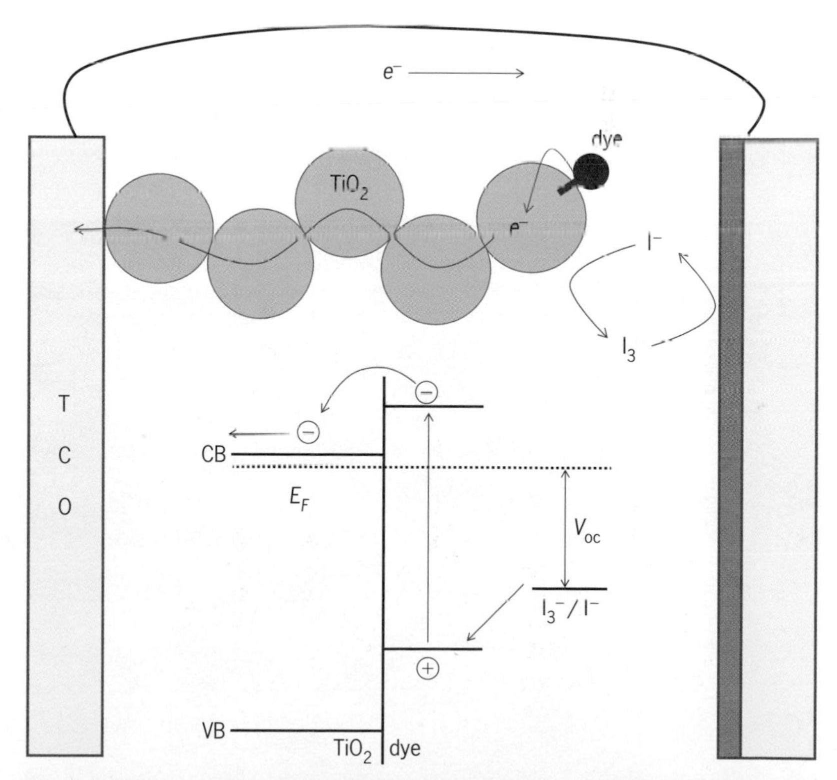

Fig. 2. Electron flow and energy levels for the electrochemical junction type of sensitized mesoscopic solar cell. Open-circuit voltage (V_{oc}) is determined by the difference between the Fermi level (E_F) of TiO_2 and the redox potential of the electrolyte. CB, conduction band; VB, valence band; TCO, transparent conductive oxide.

based on light-soaking (extended illumination) tests under thermal stress of solvent-contained laboratory devices, that lifetimes of 40 years can be extrapolated for Middle European conditions and 25 years for South European conditions, which indicates that the sensitized solar cell is quite stable for a long time.

All-solid-state pseudo-*pin*-junction type. The all-solid-state sensitized mesoscopic solar cell, which was developed in 1998, contains a solid hole-transporting material (HTM) instead of a liquid electrolyte. The solid-state mesoscopic solar cell can be classified to be a pseudo-*pin*-junction type, where the intrinsic (*i*) light harvester adsorbed on an *n*-type (*n*) TiO_2 surface is in contact with the *p*-type (*p*) HTM. **Figure 3** shows the cross-sectional scanning electron microscope (SEM) image of a real solid-state mesoscopic solar cell. A compact thin layer is required to avoid direct contact between the transparent conductive substrate and the HTM. The mesoporous oxide film, which is decorated with sensitizer, is deposited on the compact layer. The HTM is infiltrated into the pores in the oxide film. A metal layer with a work function higher than the HOMO of the HTM is deposited on top of the HTM overlayer.

Ordinarily, 2,2′,7,7′-tetrakis(*N,N-p*-dimethoxyphenylamino)-9,9′-spirobifluorene (spiro-MeOTAD) has been used as the organic HTM. However, the first demonstrated device showed poor performance, with a PCE of less than 1%. Little attention was given to the solid-state mesoscopic solar cell in its early stages of development because of its extremely low efficiency. The low efficiency was overcome by incorporating butylpyridine and lithium salt in the spiro-MeOTAD, which improved the PCE to 2.56%. The TiO_2 film thickness may be restricted in the solid-state device, unlike the liquid-type device, because of the limited mean free path of hole transport in the HTM. Therefore, a film thickness of less than 2 μm has been proposed for the solid-state device. For such a thin film, high-absorption-coefficient sensitizers are required. Replacement of the low-absorption-coefficient ruthenium complex coded N719 with the higher-absorption-coefficient organic sensitizer coded D102 increased the PCE to 4%. The structural and optical design of organic dye led to a better organic sensitizer, coded C220, which further improved the photovoltaic performance to 6%. The low conductivity of the pristine spiro-MeOTAD was found to be improved by doping with a small amount of the hole dopant tris(2-(1H-pyrazol-1-yl)pyridine)cobalt(III), which resulted in a higher PCE of 7.2%.

Striking progress in solid-state devices was not achieved until nanoscale inorganic semiconductors were applied as light harvesters. A breakthrough in solid-state mesoscopic solar cells in 2012 was based on the organometal halide sensitizer $CH_3NH_3PbI_3$ with perovskite structure. In general, perovskite has the formula ABX_3 (X = halogen or oxygen), where the A and B cations have 12 and 6 coordinates with X anions, respectively. Lead halide perovskites form two- and three-dimensional (2D and 3D) structures with the chemical formulas $(RNH_3)_2PbX_4$ and $(RNH_3)PbX_3$ (X = halogen), respectively. The 3D lead halide perovskite can be a better candidate than the 2D one for solar cell application because of its smaller band-gap energy with lower exciton binding energy. The submicrometer-thick mesoporous TiO_2 film adsorbed with perovskite $CH_3NH_3PbI_3$ nanocrystals demonstrated a PCE of 9.7%. This device showed good long-term stability, retaining its photovoltaic performance when stored for 500 h in air at room temperature without encapsulation.

Soon after this achievement, a higher efficiency of 10.9% was attained using perovskite-adsorbed aluminum oxide (Al_2O_3). Since the Al_2O_3 film used in this device acts simply as a scaffold layer, not as an electron accepting layer, this structure is called a superstructured mesoscopic solar cell. One year later, the efficiency was increased to 15% using a two-step process for the $CH_3NH_3PbI_3$ deposition.

The remarkable achievements using perovskite materials promise further breakthroughs in this field of study. The progress in solid-state mesoscopic solar cells is plotted in **Fig. 4**. There was no noticeable achievement from 1998 to 2011, but there was very rapid progress in a very short period from 2012 to 2013, and perovskite sensitizers played a central role in this quantum leap.

Perspective and prospects. Sensitized mesoscopic solar cells have been studied for more than 2 decades. Fundamental and applied research has been intensively carried out and, in consequence, a high efficiency of approximately 15% has been attained and long-term-stability technologies have been developed. Some companies have launched commercially available products. Applications now extend from portable solar power for small electronic devices to a façade solar window for powering a building. Thus, the sensitized mesoscopic

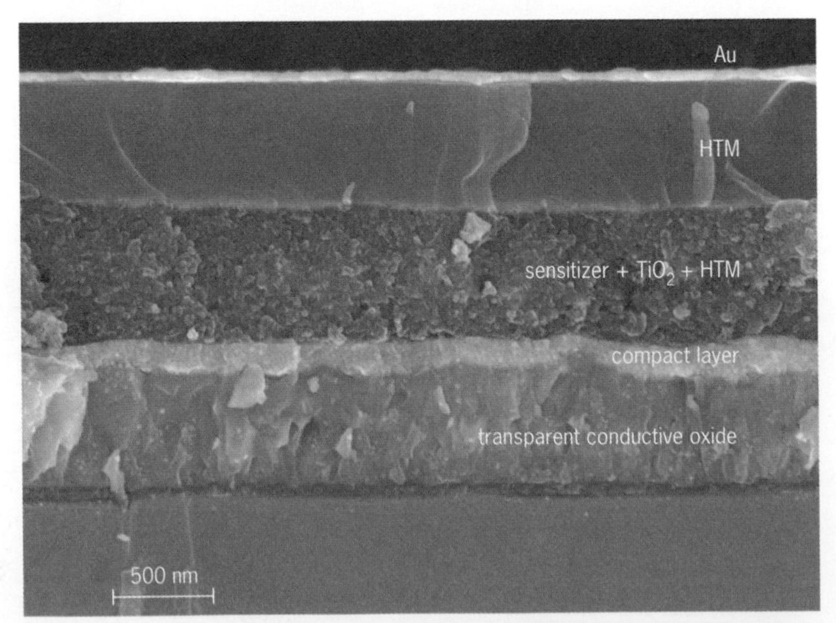

Fig. 3. Cross-sectional scanning electron microscope (SEM) image of a real solid-state mesoscopic solar cell.

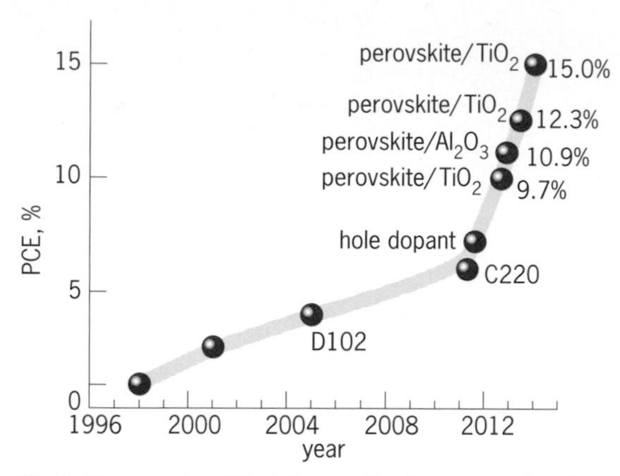

Fig. 4. Progress in solid-state sensitized mesoscopic solar cells. The PCE has reached 15% thanks to the $CH_3NH_3PbI_3$ perovskite organic-inorganic hybrid sensitizer.

solar cell is expected to find a potential market in the very near future.

Nevertheless, the efficiency of the sensitized mesoscopic solar cell is still lower than that of the silicon solar cell. Solar cell conversion efficiency is the product of photocurrent density, voltage, and fill factor. A maximum photocurrent density of 28 mA/cm^2 is possible by converting photons in the wavelength range of 280–820 nm into electrons. A light-harvesting material with a band-gap energy of around 1.5 eV utilizes incoming light with wavelengths up to 820 nm. The perovskite $CH_3NH_3PbI_3$ has a band-gap energy of about 1.5 eV, while the typical organic dye coded N719 has a band gap of around 1.7 eV. Taking into consideration the 20% light reflection at the TCO glass substrate, approximately 22 mA/cm^2 is a realistic photocurrent density for a 1.5-eV band-gap material. Therefore, a PCE of approximately 18% is realistic for a $CH_3NH_3PbI_3$-based, solid-state, sensitized solar cell because a photocurrent density of 22 mA/cm^2, a photovoltage of 1.1 V, and a fill factor of 0.7–0.8 are achievable. A photovoltage of 1.1 V is derived by subtracting the driving force energies of 0.4 eV required for electron and hole separation from the band-gap energy of 1.5 eV. Further improvement is possible by considering that perovskite plays dual roles in both charge transportation and light harvesting. More than 20% efficiency is expected because a maximum voltage of 1.3 V is available. Thus, the thinly coated and highly efficient mesoscopic sensitized solar cell is believed to be a very promising technology for harnessing the power of solar light to generate electricity.

For background information *see* DYE; ELECTROCHEMISTRY; ELECTROLYTE; HOLE STATES IN SOLIDS; MOLECULAR ORBITAL THEORY; PEROVSKITE; PHOTOCHEMISTRY; PHOTOVOLTAIC EFFECT; PORPHYRIN; SOLAR CELL; SOLAR ENERGY; TITANIUM OXIDES in the McGraw-Hill Encyclopedia of Science & Technology.

Nam-Gyu Park

Bibliography. U. Bach et al., Solid-state dye-sensitized mesoporous TiO_2 solar cells with high photon-to-electron conversion efficiency, *Nature*, 395:583–585, 1998, DOI:10.1038/26936; J. Burschka et al., Sequential deposition as a route to high-performance perovskite-sensitized solar cells, *Nature*, 499:316–319, 2013, DOI:10.1038/nature12340; J.-H. Im et al., 6.5% efficient perovskite quantum-dot-sensitized solar cell, *Nanoscale*, 3:4088–4093, 2011, DOI:10.1039/c1nr10867k; H.-S. Kim et al., Lead iodide perovskite sensitized all-solid-state submicron thin film mesoscopic solar cell with efficiency exceeding 9%, *Sci. Rep.*, 2:591, 2012, DOI:10.1038/srep00591; A. Kojima et al., Organometal halide perovskites as visible-light sensitizers for photovoltaic cells, *J. Am. Chem. Soc.*, 131:6050–6051, 2009, DOI:10.1021/ja809598r; M. M. Lee et al., Efficient hybrid solar cells based on meso-superstructured organometal halide perovskites, *Science*, 338:643–647, 2012, DOI:10.1126/science.1228604; J. H. Noh et al., Chemical management for colorful, efficient, and stable inorganic-organic hybrid nanostructured solar cells, *Nano Lett.*, 13:1764–1769, 2013, DOI:10.1021/nl400349b; B. O'Regan and M. Grätzel, A low-cost, high-efficiency solar cell based on dye-sensitized colloidal TiO_2 films, *Nature*, 353:737–740, 1991, DOI:10.1038/353737a0, A. Yella et al., Porphyrin-sensitized solar cells with cobalt (II/III)-based redox electrolyte exceed 12 percent efficiency, *Science*, 334:629–634, 2011, DOI:10.1126/science.1209688.

Sensorless electric drives

Sensorless electric drives are closed-loop variable-speed drives that use, instead of motion (speed) sensors, an online speed calculator (called the state observer) that is based on the equations that govern the operation of the electric machines and on measurements of current, voltage, or both.

Variable-speed drives. An electric drive consists of one or more electric machines, usually operating as a motor, and the associated electric control equipment that is intended to govern the performance of the motors. The term variable-speed drive (VSD) refers to an electric drive whose design permits the motor speed to vary through a considerable range as a function of load. Specifically, a VSD consists of an electric motor (in general, a brushless one), a static power converter [an ensemble of electronic (static)

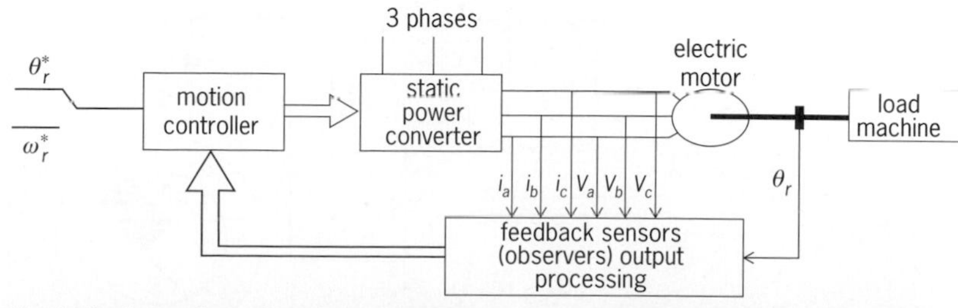

Fig. 1. Basic topology of modern electric drives. In this and following figures, an asterisk next to a letter denotes a reference (desired) value of a quantity; i_a, i_b, and i_c are phase currents; V_a, V_b, and V_c are voltages; and θ_r and ω_r are the rotor's position and speed (in mechanical terms).

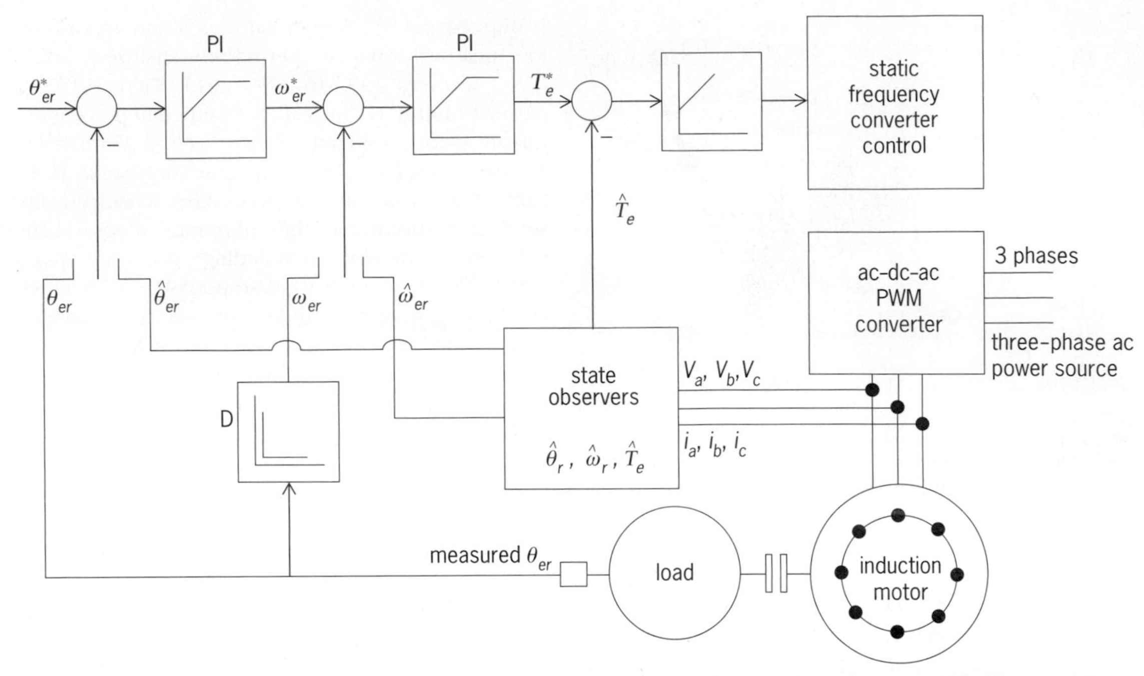

Fig. 2. Motoring (motion) control in an electric drive. In this and the following figure, a circumflex accent ("hat") over a letter denotes a value of a quantity that is estimated by the state observers; θ_{er} and ω_{er} are the rotor's position and speed in electrical terms ($\theta_{er} = p_1\theta_r$, where p_1 is the number of motor pole pairs). PWM = pulse-width modulation; PI = proportional integral regulator; and D = derivative operator.

power switches that vary the amplitude, phase, and frequency of electric power voltages], a section processing the output of feedback sensors and state observers, and the motion or power controller (**Fig. 1**). Variable-speed electric drives achieve considerable energy savings when the electric machine is operating as a motor driving industrial loads, such as pumps, ventilators, and compressors.

The term motion (or power) control describes closed-loop (feedback) control of the motor's shaft position, speed, or torque when the motor drives the load, and active and reactive power control when the electric machine is used as a generator (such as in wind generator control). Conventional motion control employs sensors to measure the rotor speed, while sensorless control estimates (infers or constructs) the speed online, using measurements of voltage, current, or both (**Fig. 2**). In all cases, knowledge of the position of the motor's rotor θ_{er} (in electrical terms; $\theta_{er} = p_1\theta_r$, where p_1 is the number of motor pole pairs, and θ_r is the position in mechanical terms) is required.

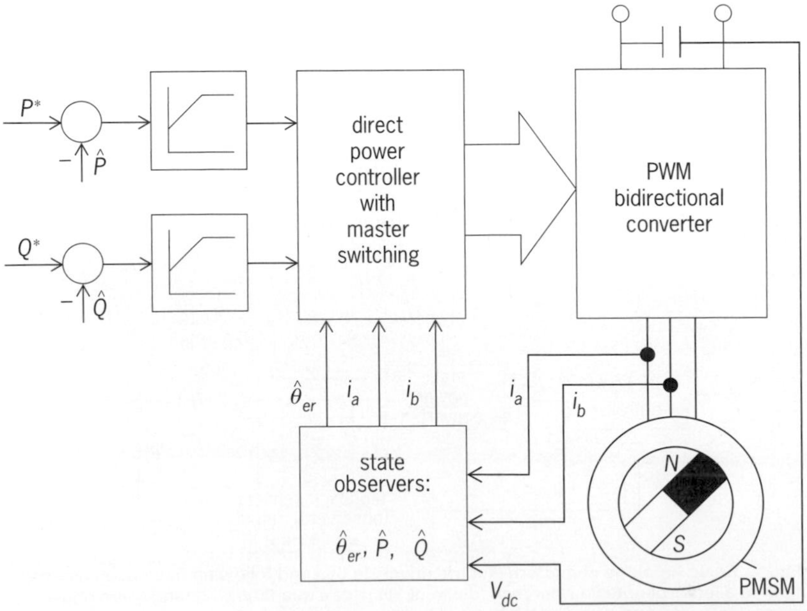

For electric power generation, active power (P) and reactive power (Q) regulators are added (**Fig. 3**).

Fig. 3. Generating control of a permanent-magnet synchronous motor (PMSM).

Reconstruction of phase voltages and currents. In the standard inverter in **Fig. 4**, measurements of the dc-link voltage and current, V_{dc} and i_{dc}, which involve low-frequency-band (low-cost) sensors, can be used to reconstruct simply the phase voltages, V_{an}, V_{bn}, and V_{cn}, where n is the null point of the machine's three-phase (a, b, c) windings (Fig. 4*a*), using Eqs. (1).

$$V_{an} = \frac{V_{dc}}{3}(2S_a - S_b - S_c)$$
$$V_{bn} = \frac{V_{dc}}{3}(2S_b - S_a - S_c) \quad (1)$$
$$V_{cn} = \frac{V_{dc}}{3}(2S_c - S_a - S_b)$$

The values or states of the switches S_a, S_b, and S_c vary with time (according to the six-pulse operation sequence dictated by Fig. 4*b*), but take on the values of either 1 or 0, corresponding to upper- and lower-branch (leg) switching in conduction.

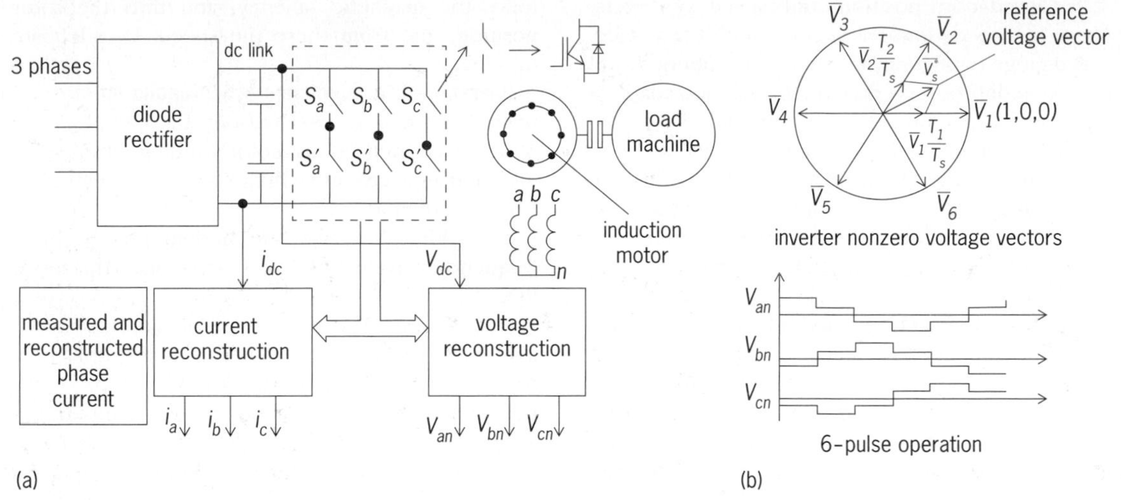

Fig. 4. Standard inverter. (a) Reconstruction of phase voltages (V_{an}, V_{bn}, V_{cn}) and currents (i_a, i_b, i_c). (b) Six-pulse operation technique.

(To avoid short circuiting the converter, S_a and $S_{a'}$ are not allowed to conduct simultaneously.) For the voltage vector $\bar{V}_1(1, 0, 0)$, $S_a = 1$ and $S_b = S_c = 0$.

The current reconstruction is less obvious because, when all upper-leg switches or all lower-leg switches are on (closed, with $S = 1$), the machine appears to be short circuited, and so the dc current does not correspond to the phase currents. Zero voltage is required at low speeds when low fundamental voltage is required. To estimate the current when the voltage is zero, the latter is replaced by a sequence of fast nonzero voltages whose sum is zero (Fig. 4*b*).

In space-vector pulse-width modulation (PWM, Fig. 4*b*), the reference (desired) stator voltage vector required for drive control, $\bar{V}_s{}^*$, is composed by simple trigonometric rules from two nonzero adjacent voltage vectors and the zero voltage ($\bar{V}_0$), with sample time T_s and the durations T_1, T_2, and T_0 for $\bar{V}_1$, $\bar{V}_2$, and $\bar{V}_0$, respectively, as in Eq. (2).

$$\bar{V}_s^* = \bar{V}_1\frac{T_1}{T_s} + \bar{V}_2\frac{T_2}{T_s} + \bar{V}_0\frac{T_0}{T_s},\ T_s = T_1 + T_2 + T_0 \tag{2}$$

In the reconstruction of the phase currents, i_a, i_b, and i_c, $\bar{V}_0$ is replaced by three nonzero voltage vectors, as in Eq. (3).

$$\bar{V}_0\frac{T_0}{T_s} = (\bar{V}_1 + \bar{V}_3 + \bar{V}_5)\frac{T_0}{3T_s} \tag{3}$$

Now $i_{dc} = i_a$ for $\bar{V}_1$, $i_{dc} = -i_a$ for $\bar{V}_4$, $i_{dc} = -i_b$ for $\bar{V}_2$, $i_{dc} = i_b$ for $\bar{V}_5$, $i_{dc} = i_c$ for $\bar{V}_3$, and $i_{dc} = -i_c$ for $\bar{V}_6$. And thus the phase currents may be reconstructed over the entire time span, but at the cost of additional nonzero voltages, which create some additional losses.

Motion state observers and sensorless control. Brushless induction, permanent-magnet, and reluctance synchronous ac motors and generators are the current leaders of electric drive technology. A popular approach to sensorless control is based on the position-and-speed motion state observer, whose complexity depends on the application.

The terms "state variable" and "state vector" refer to system variables, which in electric circuits are the voltage across the terminals of a capacitor and the current through an inductor (coil). In most control situations, however, the state vector is not available for direct measurement. In this case, a reasonable substitute for the state vector must be found. In 1964, David G. Luenberger introduced the concept of a state observer, which shows how the available system inputs and outputs may be used to construct an estimate of the system states. The device that reconstructs the state vector from measurements of the input and output of the real system is called an observer (typically computer-implemented). The observer itself is a time-invariant linear system driven by the inputs and outputs of the system that it observes.

Two principal classes of application are:

1. General variable-speed drives, with 0.4–1 p.u. (per unit, that is, relative to rated value) speed range, slow torque disturbance rejection response (±100% torque variation in the 30–300-ms range), and speed control precision of around 2–3% at maximum speed.

2. Variable-speed servo-drives, where the speed range in sensorless control is 0.01–1 p.u., with ±100% torque response in less than 5 ms, and speed control precision less than 1% at maximum speed and less than 2 revolutions per minute (rpm) when 2 rpm minimum speed is required.

Motion state observers are classified as:

1. Rotor speed and position online open-loop observers. These are fast, but do not guarantee stability and convergence when voltage and current amplitudes versus frequency are given a priori. Alternating-current drives without or with stabilizing loops belong to this category.

2. Closed-loop position and speed observers. These have been implemented in quite a few wide-speed-range configurations such as Luenberger (full motor model) observers, model reference adaptive observers, and Kalman filter observers.

For ac machines, the combined voltage and current method proved to be reliable down to 3–50 rpm for a reasonable on-line computation time and for mild calibration efforts. The lower speed implies the use of sliding modes in the state observers and current loops.

The signal injection methods use the high-frequency model of ac machines and complex filtering to extract from the machine leakage flux (in the case of induction motors), or from the main flux (in the case of the permanent-magnet synchronous motors and reluctance synchronous motors), the magnetic saliency, and thus the rotor position, and from there the speed, by adequate filtering.

Operation of a reluctance synchronous motor. **Figures 5** and **6** show results from the operation of a reluctance synchronous motor with high-frequency-injection-assisted active-flux-based sensorless control implementation.

Also, **Fig. 7** shows the hodographs of high-frequency traveling voltages, currents, and active flux at zero speed.

Note that the flux components of Fig. 7 are shown as alpha (α) and beta (β) values using the alpha-beta ($\alpha/\beta/\gamma$) transformation (also known as the Clarke transformation), which is employed to simplify the analysis of three-phase circuits. (It is conceptually similar to the dq0 or dqo transformation and the

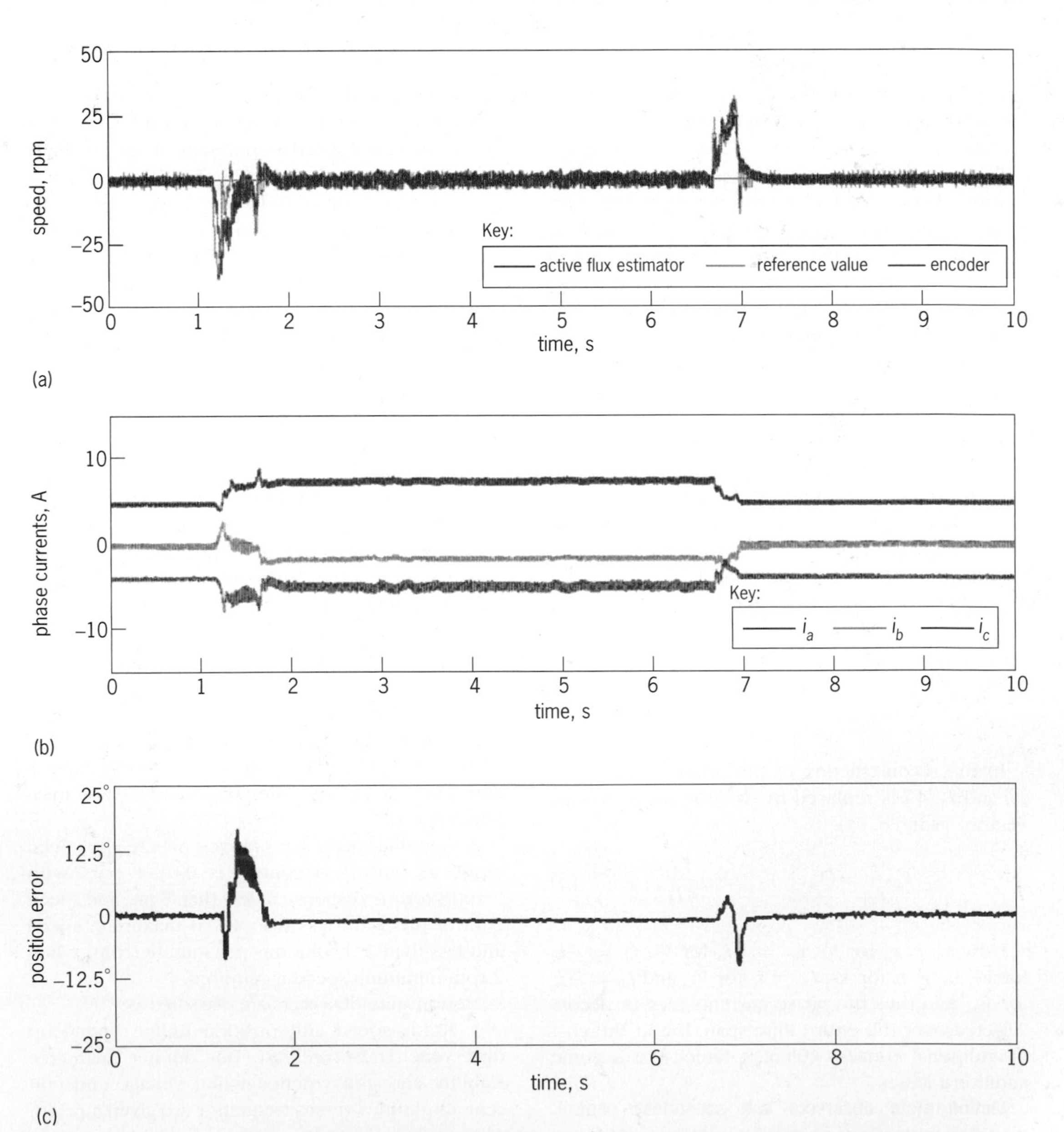

Fig. 5. Zero-speed operation of a reluctance synchronous motor with ± full-step torque perturbation. Graphs show variation of (*a*) rotor speed, (*b*) phase currents, and (*c*) rotor position error.

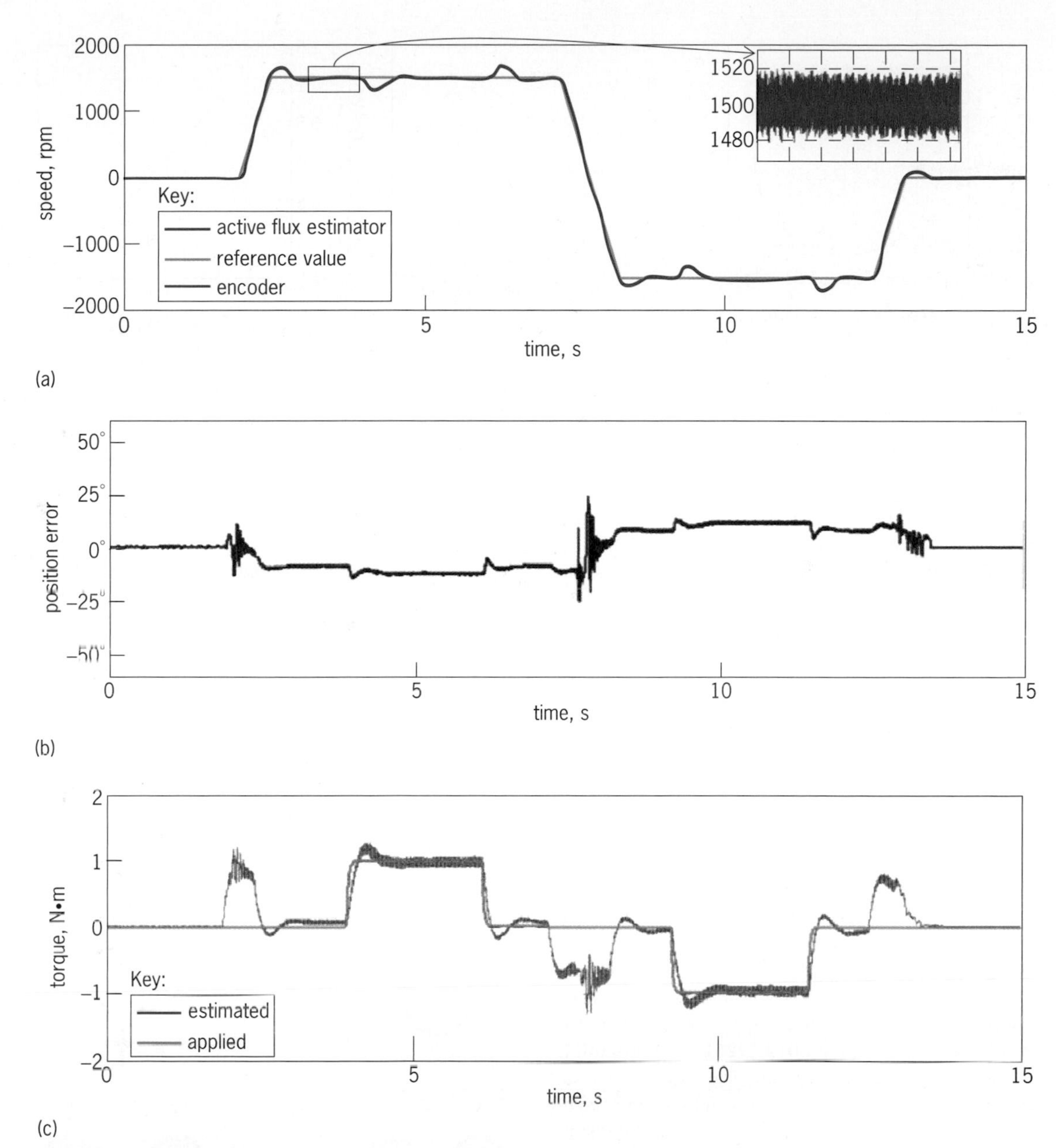

Fig. 6. Acceleration of a reluctance synchronous motor to 1500 rpm with loading cycle and speed reversal. Graphs show variation of (*a*) rotor speed, (*b*) rotor position error, and (c) applied and estimated torque.

symmetrical component transformation.) A useful application of the $\alpha/\beta/\gamma$ transformation is to generate the reference signal used for modulation control of three-phase inverters.

It may be noted that

1. The operation at zero speed with full-step torque perturbation has been feasible only with signal injection contribution. It shows good performance in precision and stability.

2. The running speed and torque transients (Fig. 6) also show good dynamic performance with fast torque response and less than 1% speed steady-state error at 1500 rpm.

3. Voltage traveling-wave injection results (Fig. 7) show clearly that the stator flux is traveling (as the voltage). The current hodograph is an ellipse due to machine saliency, but the active flux stays aligned to rotor position (without complex and error-prone filtering).

In conclusion, sensorless electric drives are based on reconstructing phase voltages and currents (from the dc link voltage V_{dc} and current i_{dc}) to estimate rotor position, speed, and torque. Applying this approach reduces the number and cost of voltage and current sensors, and hence, reduces cost and increases reliability. The drives, with combined signal injection and model-based state observers, permit very good position precision (1–2 mechanical degrees), speeds less than 1 rpm, full-step torque rejection, and a few-millisecond slope of ±100% torque response. The subject area continues to grow with further algorithmic enhancements.

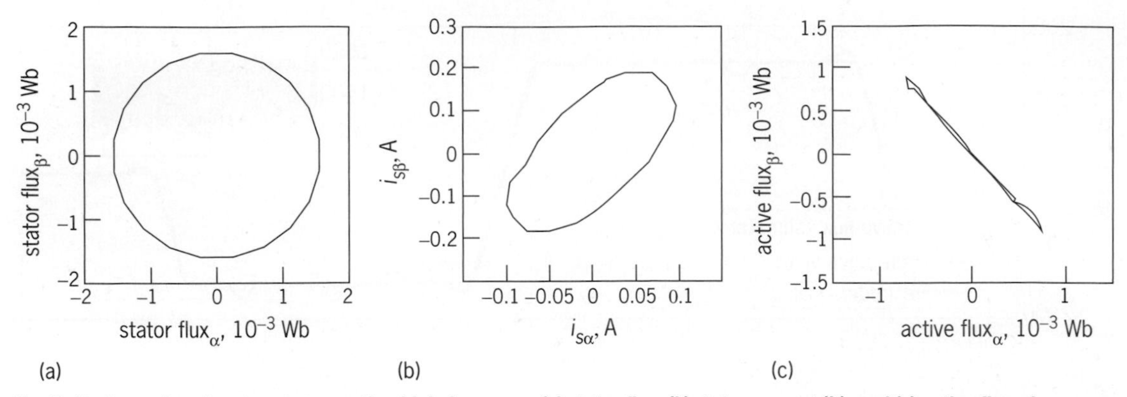

Fig. 7. Hodographs of vectors representing high-frequency (*a*) stator flux, (*b*) stator current (i_s), and (*c*) active flux of a reluctance synchronous motor at zero speed.

[Acknowledgment. This work was partially supported by the strategic grant POSDRU/89/1.5/S/57649, Project ID 57649 (PERFORM-ERA), co-financed by the European Social Fund—Investing in People, within the Sectoral Operational Programme Human Resources Development 2007–2013.]

For background information *see* ALTERNATING CURRENT; ALTERNATING-CURRENT GENERATOR; ALTERNATING-CURRENT MOTOR; CONTROL SYSTEMS; ELECTRIC ROTATING MACHINERY; GENERATOR; INDUCTION MOTOR; MOTOR; OPTIMAL CONTROL (LINEAR SYSTEMS); RELUCTANCE MOTOR; SYNCHRONOUS MOTOR in the McGraw-Hill Encyclopedia of Science & Technology.

Ion Boldea; Sorin Cristian Agarlita

Bibliography. S. C. Agarlita, I. Boldea, and F. Blaabjerg, High-frequency-injection-assisted "active-flux"-based sensorless vector control of reluctance synchronous motors, with experiments from zero speed, *IEEE Trans. Ind. Appl.*, 48:1931–1939, 2012, DOI:10.1109/TIA.2012.2226133; R. Ancuti, I. Boldea, and G. D. Andreescu, Sensorless V/f control of high-speed surface permanent magnet synchronous motor drives with two novel stabilising loops for fast dynamics and robustness, *IET Elec. Power Appl.*, 4:149–157, 2010, DOI:10.1049/iet-epa.2009.0077; I. Boldea, M. C. Paicu, and G. D. Andreescu, Active flux concept for motion-sensorless unified AC drives, *IEEE Trans. Power Electron.*, PE-23:2612–2618, 2008, DOI:10.1109/TPEL.2008.2002394; R. W. Hejny and R. D. Lorenz, Evaluating the practical low-speed limits for back-EMF tracking-based sensorless speed control using drive stiffness as a key metric, *IEEE Trans. Ind. Appl.*, 47:1337–1343, 2011, DOI:10.1109/TIA.2011.2126013; H. Kim and T. M. Jahns, Phase current reconstruction for AC motor drives using a DC-link single current sensor and measurement voltage vectors, *IEEE Trans. Power Electron.*, 21:1413–1419, 2006, DOI:10.1109/TPEL.2006.880262; P. D. C. Perera et al., A sensorless, stable V/f control method for permanent-magnet synchronous motor drives, *IEEE Trans. Ind. Appl.*, 39:783–791, 2003, DOI:10.1109/TIA.2003.810624.

Sex chromosomes

The genome of every mammal consists of the autosomal chromosomes (or autosomes), of which all individuals normally carry two copies, and the sex chromosomes, which vary in number and type with the individual's sex. In mammals, the sex chromosomes are called the X and Y chromosomes. In placental and marsupial mammals, females have two X chromosomes, whereas males have one X chromosome and one Y chromosome (see **illustration**). (Note that monotreme mammals have a different X/Y system, which will be discussed briefly in this article.)

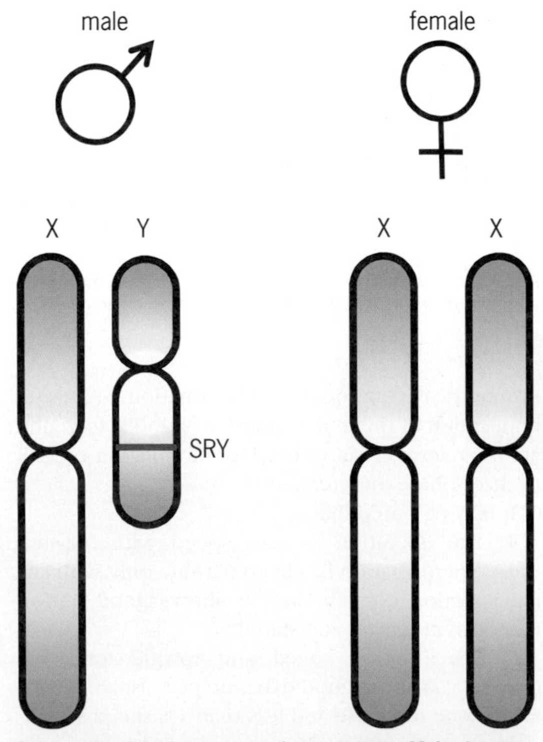

Schematic of the human sex chromosomes. Males have an X chromosome and a Y chromosome, whereas females have two X chromosomes. On the Y chromosome, it is the SRY (sex-determining region on the Y) gene that triggers male development.

Some other types of animals determine sex through different chromosome systems (see the final section of this article), but the mammalian X/Y system is particularly well studied. The difference in the distribution of the sex chromosomes between males and females is a unique feature of the mammalian genome. Because all the mammalian autosomes exist in duplicate, there are two copies of all autosomal genes. This means that there is, in effect, a backup of each autosomal gene. If one gene copy is mutated, these duplicate chromosomes can undergo recombination to repair the damaged DNA.

The X chromosome. The X chromosome contains approximately 2000 genes, or almost 8% of the entire human gene set. The X chromosome can undergo recombination repair of these genes only in females because only females have two X copies. In contrast, in males, any mutations on the X chromosome cannot be repaired or compensated for by a backup copy and often lead to disease (see **table**). A common example of an X-linked disease is red-green color blindness, which affects almost 10% of men but less than 1% of women. This is because males have only a single X chromosome, and a male will be color-blind if this particular gene is defective. Females must have two mutant copies of the gene if they are to be color-blind, which is much less likely to occur. The single X chromosome in males poses a problem with regard to disease, but it also places a unique set of evolutionary pressures on this chromosome that have shaped its gene content. Although many of the gene mutations that are harbored in the genome are masked by duplicate gene copies, any mutation or change in a gene on the X chromosome will be exposed in males. Thus, if the mutation causes a male reproductive advantage, it can immediately be selected for. As a result, the mammalian X chromosome contains a disproportionately high number of genes associated with male reproduction. The lack of recombination repair in males also means that genes on the X chromosome tend to evolve faster than genes on the autosomes. Despite the fact that the X chromosome is a relatively high-risk home for a gene required for reproduction, the advantage of being able to immediately select for genes that improve male reproductive success is far stronger.

The mammalian X chromosome is also unique in that it undergoes a process known as dosage compensation in females. All the chromosomes in the genome exist in a tightly regulated balance of copies. Extra or missing chromosomes lead to severe disease phenotypes, such as Down syndrome, which results from an extra copy of chromosome 21. Similarly, the difference in X-copy number between males and females poses a gene balance problem for the cell. Mammals rectify this problem by silencing one of the X chromosomes in females through a process known as X-inactivation. The X chromosome to be silenced is selected at random in the cell and is packaged up into an inactive state. Through this mechanism, both males and females have only a single active X chromosome in each cell.

Just as with the autosomes, it is occasionally possible to inherit extra or missing copies of the sex chromosomes. Klinefelter syndrome occurs when a male inherits two X chromosomes in addition to the Y chromosome, leading to an XXY chromosome complement. Klinefelter syndrome is actually the most common chromosomal abnormality seen in humans. It occurs in 1 in 500 male births, almost double the incidence of Down syndrome. Many XXY males do not have any adverse symptoms, but individuals with Klinefelter syndrome generally are taller, show reduced masculinization, and have increased rates of infertility. A more serious condition results from the inheritance of a single X chromosome (in the absence of the Y chromosome), termed XO or Turner syndrome. Individuals with Turner syndrome are female and have many characteristic physical features, including short stature. They are also infertile.

The Y chromosome. The Y chromosome is a unique chromosome that exists only in the male sex. It is very small in size and contains only a few genes (less than 80), most of which are duplicate copies of one another. Because the Y chromosome is always present as a single copy in the normal population, it cannot undergo recombination repair. As a result, genes on the Y chromosome are rapidly evolving and often lost. The only genes that can be retained on the Y chromosome are those that are essential for male function. As a result, most Y chromosome genes are associated with sperm production, which is a biological process that occurs only in males and is essential for fertility.

The Y chromosome also determines the sex of a developing mammalian embryo, based solely on the presence or absence of a single gene, SRY (sex-determining region on the Y). If a developing mammalian embryo has a Y chromosome with an intact SRY gene, testes will develop, and it will become male. If there is no Y chromosome, or if the SRY gene on the Y chromosome is mutated, ovaries will develop, and it will become female. Despite its critical importance for sex, the SRY gene is rapidly changing and is barely recognizable among mammal species. In fact, in the monotremes (platypus and echidna), the gene does not exist at all.

Other sex chromosome systems. X and Y chromosomes, like those that exist in humans, are found only in live-bearing mammals. The egg-laying monotremes have five X and five Y chromosomes

Common X-linked diseases*
Red–green color blindness
Hemophilia A
Hemophilia B
Duchenne muscular dystrophy
Becker muscular dystrophy
X-linked ichthyosis
X-linked agammaglobulinemia
Glucose-6-phosphate dehydrogenase deficiency

*Each of these diseases affects males disproportionately because they have only a single X chromosome.

that join up end to end during cellular division and determine sex through a gene other than SRY. Many species of birds and reptiles have Z and W sex chromosomes, where the males are ZZ and the females are ZW. In these species, sex can be determined by a gene on the W that triggers female development, or it can be determined by the dosage of a gene on the Z such that female development is triggered by one dose and male development is triggered by two doses.

For background information *see* CELL (BIOLOGY); CELL BIOLOGY; CHROMOSOME; CHROMOSOME ABERRATION; GENE; GENETICS; HUMAN GENETICS; MAMMALIA; MUTATION; RECOMBINATION (GENETICS); SEX DETERMINATION; SEX-LINKED INHERITANCE in the McGraw-Hill Encyclopedia of Science & Technology.

Andrew J. Pask

Bibliography. D. W. Bellott and D. C. Page, Reconstructing the evolution of vertebrate sex chromosomes, *Cold Spring Harbor Symp. Quant. Biol.*, 74:345–353, 2009, DOI:10.1101/sqb.2009.74.048; J. A. Marshall Graves, Weird animal genomes and the evolution of vertebrate sex and sex chromosomes, *Annu. Rev. Genet.*, 42:565–586, 2008, DOI:10.1146/annurev.genet.42.110807.091714; A. J. Pask and J. A. Marshall Graves, Sex chromosomes and sex-determining genes: Insights from marsupials and monotremes, *Cell. Mol. Life Sci.*, 55:864–875, 1999, DOI:10.1007/s000180050340.

Shark history

Sharks belong to a group of fishes known as the chondrichthyans ("cartilaginous fishes," referring to their unossified internal skeleton). This group also includes the rays and chimaeras (ratfish). The popular perception of sharks as primitive relatives of bony fishes (osteichthyans) is undeserved because both have equal antiquity, and traces of bonelike tissue in some fossils suggest that early sharks evolved from bony ancestors. Modern and extinct chondrichthyans uniquely possess a type of endoskeletal mineralization known as tessellated calcified cartilage. In addition, modern sharks are anatomically advanced over earlier ones in the structure of their cranium, jaws, gill arches, fin skeletons, vertebral column, teeth, skin denticles (small toothlike projections), and fin spines. Today, sharks are among the top modern marine predators, and the fossil record reveals that they have held this distinction for more than 400 million years, despite occasional challenges from Paleozoic armored placoderms and lobe-finned fishes, Mesozoic marine reptiles, and modern marine mammals.

Chondrichthyan fossils. The oldest fossils attributed to the chondrichthyans are from the Paleozoic Era. The most ancient of these are isolated skin denticles from the Late Ordovician Harding Sandstone of Colorado, dating back to approximately 450 million years ago (MYA). The earliest shark teeth are from the Early Devonian Period (*Leonodus*, 418 MYA). The oldest complete shark body fossil is *Doliodus* (409 MYA). *Doliodus* resembles modern chondrichthyans in several important respects: Its cartilage endoskeleton has a thin, apparently tessellated calcified outer layer instead of bone; its head (and apparently the rest of its body) is covered by small dermal denticles instead of large bony plates; and it has batteries of teeth arranged as in modern sharks (although its teeth were replaced very slowly). However, *Doliodus* differs from modern chondrichthyans in a number of features, including odd, double-pronged teeth and large paired spines projecting from its front (pectoral) fins. This last feature is of particular interest because it is reminiscent of extinct bony acanthodian fishes, which may be close relatives of the chondrichthyans.

Cladoselache (**Fig. 1*a***) is one of the best known sharks from the Devonian Period. It was approximately 1.5 m (5 ft) in length and probably was a fast-swimming predator, with stiff, hydrofoil-like paired fins and a deeply forked tail. The chondrichthyans underwent major diversifications in the Devonian, Mississippian, and Pennsylvanian Periods (360–300 MYA), with more than 20 recognized families of sharks and a dozen or so families of chimaera-like fish. Many Paleozoic sharks achieved worldwide distribution, occupying many marine and freshwater environments that are dominated today by bony ray-finned fishes (actinopterygians). Some of these Paleozoic sharks had rodlike opercular cartilages (for example, *Tristychius*, some xenacanths, and some symmoriiforms), suggesting that their gills were covered by a fleshy operculum (a lidlike flap) as in modern chimaeroids. In other fossil sharks (including *Cladoselache* and probably *Doliodus*), these rods were either uncalcified or absent, and the gill openings may have been exposed, as in modern sharks and rays.

Many Paleozoic sharks were small [measuring less than 1 m (3.3 ft)], but some exceeded 6 m (20 ft) in length (as large as modern great white sharks). Among these giants were ctenacanths (sharks with stout fin spines projecting from the dorsal fins, and with teeth specialized for grasping, stabbing, or cutting prey). Symmoriiform sharks (including forms

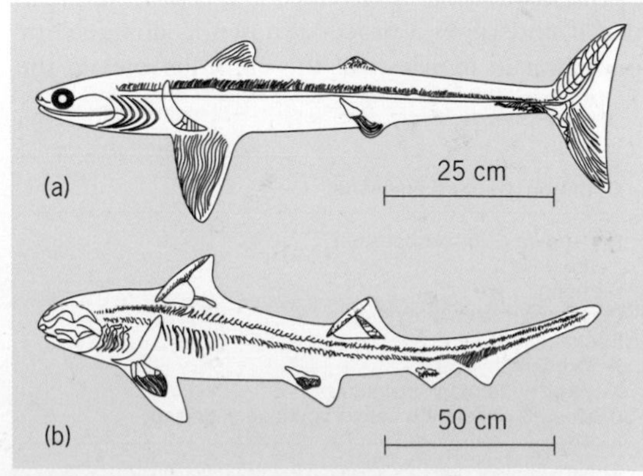

Fig. 1. Lateral views of fossil sharks: (*a*) *Cladoselache*, a primitive shark; (*b*) *Hybodus*, a hybodont.

such as *Akmonistion* and *Stethacanthus*, with a recurved bladelike dorsal spine just behind the head) were only slightly smaller. Late Paleozoic freshwater environments were dominated by two groups of sharks: eel-like xenacanths [up to 2 m (6.6 ft) in length, with double-pronged teeth resembling those of *Doliodus*] and hybodonts (with posteriorly barbed fin spines; note that male hybodonts also had odd hooklike spines on the head).

Many chondrichthyan families disappeared in the Pennsylvanian and Permian Periods. However, a few survived the Permian mass extinction (250 MYA), including the hybodonts and the early relatives of the chimaeras and modern ("neoselachian") sharks and rays, plus the last xenacanths and edestoids (large marine sharks with stiff paired fins like those of cladoselachians and having bizarre tooth whorls located centrally in the mouth). Hybodonts shared some advanced anatomical features with the neoselachians, including fast tooth replacement, a constricted and flexible shoulder joint for the pectoral fins (features that they also shared with xenacanths), a robust pelvic girdle, a highly specialized inner ear morphology capable of low-frequency sound detection, and more kinetic (movable) jaws than most Paleozoic sharks. Hybodonts and neoselachians quickly adapted themselves to a wide range of marine and freshwater ecologies in the Mesozoic Era, resulting in considerable taxonomic diversity. In contrast, Mesozoic chimaeroids were highly specialized, exclusively marine fishes that evolved to fill a somewhat restricted ecological role; their survival and low modern diversity may reflect the long-term stability of their marine environment over the past 200 million years.

Hybodonts were the dominant sharks of the Mesozoic Era (250–65 MYA). Although hybodonts were never as diverse as neoselachians, they nevertheless evolved a variety of tooth morphologies, including sharp, multicusped teeth adapted for grasping prey [*Hybodus* (Fig. 1b) and *Egertonodus*], serrated cutting teeth (*Priohybodus*), crushing teeth (*Acrodus*, *Asteracanthus*, and *Lissodus*), and flat, raylike interlocking teeth (*Tribodus*). Hybodonts included freshwater sharks as small as 15 cm (6 in.) in length and marine sharks measuring more than 4 m (13 ft) in length, with distinctive endemic faunas in Europe, North America, southeast Asia, and perhaps western Gondwana (South America and Africa; note that Gondwana was the ancient landmass that later fragmented and eventually drifted apart to form the present continents). In the Cretaceous Period, hybodonts were largely replaced by modern sharks and rays, with one notable exception: the small hybodont *Lonchidion* persisted almost until the end of the Cretaceous Period in North America, surviving in freshwater habitats that few other chondrichthyans were able to colonize. Apparently no hybodonts survived beyond the end of the Cretaceous Period (65 MYA).

Modern sharks. Modern sharks (and rays) are distinguishable anatomically from hybodonts and other extinct sharks by the presence of heavily calcified vertebrae and specialized features in their teeth, skin denticles, and fin spines. These modern attributes were probably not all acquired at once, but their evolutionary history is poorly understood because the fossil record is fragmentary. Triassic neoselachians include *Nemacanthus*, *Ostenoselache*, *Reifia*, *Rhomphaiodon*, and *Grozonodon*. Older (Mississippian) fragmentary neoselachian-like fossils include *Amelacanthus* and *Cooleyella*.

The first evidence of modern shark diversity appears in the Early Jurassic Period (200 MYA), with fossils attributed to orectolobiforms (*Annea*), heterodontiforms (*Paracestracion*), hexanchiforms (*Hexanchus arzoensis*), and rays (*Doliobatis*). Many extant neoselachian lineages arose successively in the Mesozoic Era: for example, squalids, squatinids, lamniforms, and carcharhiniforms appeared by the Late Jurassic Period (150 MYA); rhinobatids, sclerorhynchids (early relatives of sawfish), "dasyatids" (generalized stingrays), echinorhinids, pristiophorids, hemiscyllids, orectolobids, brachaelurids, and mitsukurinids appeared by the Early Cretaceous Period (110 MYA); and platyrhinids, rajids (skates), gymnurids, myliobatids, chlamydoselachids (frilled sharks), centrophorids, etmopterids, somniosids, oxynotids, dalatiids, parascyllids, odontaspids, scyliorhinids, and triakids appeared by the Late Cretaceous Period (100 MYA). Modern representatives of these groups occupy habitats ranging from shallow freshwaters and marginal shelf seas to the ocean depths. Large seagoing "mega-sharks" resembling modern mako sharks were abundant in the Cretaceous Period, but they may have belonged to one or more extinct lineages rather than being directly ancestral to modern forms.

Although the Cretaceous/Paleogene boundary marks an important worldwide extinction event, its impact upon sharklike fishes is unclear. For example, "sclerorhynchid" sawfishes apparently disappear at the Cretaceous/Paleogene boundary. However, did they become extinct, or did some "sclerorhynchids" evolve into modern pristid sawfish? Similar doubts surround the apparent demise of other Cretaceous elasmobranchs, including cardabiodontids, cretoxyrhinids, and anacoracids, as well as the Cenozoic "odontids" (including the notable supershark, "megalodon").

Several neoselachian lineages appear by the Eocene Period, including rhynchobatids, "modern" pristids, torpedinids, narcinids, hexatrygonids, urolophids, megascyliorhinids, hemigaleids, carcharhinids (**Fig. 2**), rhincodontids, alopiids, and cetorhinids. A few modern lineages have appeared even more recently, including megachasmids ("megamouth sharks") in the Oligocene Period (33–23 MYA) and hammerhead and bonnethead sharks (sphyrnids) in the Miocene Period (23–6 MYA).

Although modern shark diversity is low compared to that of bony fishes, scientists have doubled the number of living shark and ray species known from a decade ago. Today, this number totals more than

Fig. 2. Blacktip reef shark (*Carcharhinus melanopterus*). (*Photo courtesy of NOAA's Coral Kingdom Collection: Mariana Islands, Guam; photo by David Burdick*)

1220, but it probably still underestimates their actual diversity. Many of these newly recognized species have been distinguished genetically, but still await description, and some may be morphologically identical to their sibling species. This possibility has important implications for shark conservation policies and marine resource management because many more species may be endangered through human activities than was previously supposed.

Phylogeny (evolutionary history). There are competing views on the evolution of modern sharks and rays: Either rays evolved from within the sharks, and are therefore more closely related to some sharks than to others; or modern sharks and rays may share a common ancestor and be of equal antiquity. Phylogenetic analyses of their morphological features suggest that the closest relatives of rays are sawsharks and angel sharks. However, phylogenetic analyses of the molecular data mostly support the "equal antiquity" hypothesis, which is more in accordance with the fossil record than the morphologically based one. The earliest ray fossils are from the Early Jurassic Period (200 MYA, which is about the same age as the oldest hexanchoid, heterodontid, and orectolobiform fossils), whereas the oldest known angel sharks are from the Late Jurassic Period (150 MYA) and the oldest known sawsharks, bramble sharks, and squaleans are from the Late Cretaceous Period (approximately 75 MYA). In addition, the morphology-based phylogeny suggests that rays have lost essentially all the basal characters used in the analyses, and imply that the stratigraphic range of modern sharks (mostly documented from isolated teeth) is seriously underrepresented in the Triassic and Jurassic periods. Moreover, a still-undescribed complete ray fossil from the Lower Jurassic of Germany (200 MYA) displays features that contradict some of the morphological similarities shared by modern sawsharks, angel sharks, and rays, suggesting that there might be anatomical convergence in the modern forms (related perhaps to shared benthic lifestyles), which could be obscuring their evolutionary relationships.

For background information *see* ANIMAL EVOLUTION; CLADOSELACHE; ECHINORHINIFORMES; ELASMOBRANCHII; EXTINCTION (BIOLOGY); FOSSIL; HETERODONTIFORMES; HEXANCHIFORMES; LAMNIFORMES; ORECTOLOBIFORMES; PHYLOGENY; PRISTIOPHORIFORMES; SELACHII; SQUALIFORMES; SQUATINIFORMES in the McGraw-Hill Encyclopedia of Science & Technology.

John G. Maisey

Bibliography. H. Cappetta, Chondrichthyes—Mesozoic and Cenozoic Elasmobranchii: Teeth, in H. P. Schultze (ed.), *Handbook of Paleoichthyology*, vol. 3E, Verlag Pfeil, Munich, 2012; M. Ginter, O. Hampe, and C. Duffin, Chondrichthyes—Paleozoic Elasmobranchii: Teeth, in H. P. Schultze (ed.), *Handbook of Paleoichthyology*, vol. 3D, Verlag Pfeil, Munich, 2010; J. G. Maisey, What is an "elasmobranch"?: The impact of palaeontology in understanding elasmobranch phylogeny and evolution, *J. Fish Biol.*, 80(5):918–951, 2012, DOI:10.1111/j.1095-8649.2012.03245.x; J. G. Maisey et al., Dental patterning in the earliest sharks: Implications for tooth evolution, *J. Morphol.*, 275:586–596, 2014, DOI:10.1002/jmor.20242; G. J. P. Naylor et al., A DNA sequence-based approach to the identification of shark and ray species and its implication for global elasmobranch diversity and parasitology, *Bull. Am. Mus. Nat. Hist.*, 367:1–262, 2012, DOI:10.1206/754.1; A. Pradel et al., A Palaeozoic shark with osteichthyan-like branchial arches, *Nature*, 509:608–611, 2014, DOI:10.1038/nature13195; M. Zhu et al., A Silurian placoderm with osteichthyan-like marginal jaw bones, *Nature*, 502:188–193, 2013, DOI:10.1038/nature12617.

Ship modernization and conversion

Ships have always represented large capital investments for their purchasers. Until the advent of aircraft they were the most complex structures built by humans, and even today they are the largest mobile objects on Earth. Because of their complexity and level of investment, there is a strong desire to maximize the period of return by extending the useful life of a given ship as long as possible. As technological and economic conditions change, a ship may no longer be competitive in its original mission or market, but it will have the potential for modernization or conversion to prolong its useful life.

In the last 200 years, marine propulsion has advanced from sail to steam to internal combustion (to say nothing of nuclear and other unusual forms of propulsion). Similarly, cargoes have gone from bulk and simply packaged (break-bulk) solids to liquids, trucks and other vehicles, and standardized intermodal containers. Over the same period, navigation and control have become computerized, reliant on satellite communications. For warships, the technological change in weapons systems has been even more acute. Many shipowners have preserved their vessels through one or more of these transitions by some combination of incorporating new technologies via modernization or by changing the type of service via conversion.

Engineering considerations and limiting factors. In an engineering sense, a modernization or conversion project is in many ways as complex as a new design. Because the ship already exists, many aspects of the design are constrained and cannot easily be changed. All aspects of naval architecture must be considered in a conversion, but structural strength and stability are the dominant factors in feasibility and safety. Regulatory requirements are also affected by the degree of conversion. Where the extent of modification is significant, constraints that would normally apply to the construction of new ships (for example, fuel-tank arrangements or ballast-water systems) may apply here as well.

Because it is designed to carry the structural loads imposed by the ocean, the hull structure is both the strongest part of the ship and the most integral element of the overall design. In one sense, this underlies the suitability of ships for conversion or modernization; a well-built hull can support a wide variety of potential facilities and cargoes. On the other hand, it imposes certain ultimate limitations on the useful life of a ship and its suitability for some activities.

Ship structures are designed to a set of cargo and sea loads associated with planned operations. In the modern world, that process entails an analysis of the expected maximum structural loading over the design life of the ship based on a statistical analysis of sea conditions. Before the advent of wave observation technology (buoys and satellites) and computer-aided structural design, structural design was a product of the craftsmanship and experience of the maritime community using rules-based approaches to provide adequate strength.

Whereas large vessels are generally designed for open-ocean service anywhere in the world, smaller vessels may be designed for more benign or locally specific conditions. Using a ship in an area for which it was not designed has been a contributing factor in several marine casualties.

A ship's cargo can be a significant factor in its overall structural load, particularly for large cargo carriers such as oil tankers and bulk carriers. In these cases, the structural design needs to take the quantity and type of cargo into account, and a mid-life change could compromise the overall design.

The repeated flexing action of a ship in waves leads to cyclic loading, causing fatigue in its structure. Although appropriately designed steel structures are relatively resistant to fatigue, it is still a concern and can be a particularly limiting factor in aluminum structures.

When ships are lengthened (**Fig. 1**), the designer must accommodate the shape of the hull and must make sure that the resulting structural configuration suits the resulting wave load distribution. For many ships, there is a portion of "parallel midbody" where the section is constant. This makes it somewhat easier to insert new material to make the ship longer. The arrangement of the bulkheads and machinery are also factors to consider. The new ship configuration will have different loads, and the structure must suit those loads.

Fig. 1. A Stena Line RoPax vessel is lengthened in a floating drydock at Lloyd Werft Bremerhaven. (*Photo courtesy of Lloyd Werft Bremerhaven AG*)

Although many ships have a significant structural capacity to absorb change, hydrodynamic stability is usually a more limiting factor. The relationship of the centers of gravity and buoyancy is the key parameter, and conversion can very easily raise the center of gravity either by adding structure or changing assumed cargo loading. This can quickly lead to a dangerously unstable ship if not monitored.

The saltwater environment is also extremely corrosive. Even with regular maintenance, corrosion eventually wears away plating and compromises the local strength of a ship. Before conversion of an existing ship, it must be surveyed to confirm the condition of the structure so that the design takes it into consideration and repairs are made where necessary.

Types of modernization. Modernization is, to some extent, inevitable in the routine process of maintaining a ship in service. When a component breaks, whether a sail, piece of deck equipment, or electronics gear, it will generally be replaced with an available equivalent, leading to a piecemeal modernization over the life of the ship. In many cases, though, a more intentional and complete process is undertaken. In some cases, the engineering documentation for an operating ship is not kept current at the level of detail necessary to engineer a conversion design. The starting point for any modification is documenting the existing arrangement and configuration so that the cumulative small changes are taken into account.

The propulsive machinery of a ship is usually the most complex and substantial system aboard, and its efficiency is a major determinant of the economics of ship operation. Motivation for changing propulsion systems might include achieving higher speed, reducing operating costs through improved efficiency or reduced maintenance, decreasing labor costs through automation, or meeting new environmental regulations. Modernization of propulsion machinery and controls is a common but substantial alteration. Historically, this has included major changes of technology, such as the retrofit of steam propulsion to sailing vessels or the conversion from coal to oil fuel. Recently, the use of liquefied natural gas (LNG) as a clean propulsion fuel has driven an increasing number of repowering projects. New regulations for emissions are coming into force, and LNG is a comparatively clean fuel. Many present-day modernizations have focused on computerized and automated control for existing power plants. The capacity to monitor and control engine performance can reduce the number of crew necessary to stand watch, tailor the performance of the plant to reduce fuel consumption, and provide data to identify maintenance needs before a failure, all of which make the economics of this type of modernization attractive. *See* LIQUEFIED NATURAL GAS AS A MARINE FUEL.

Similarly, upgraded propellers can be economically attractive (**Fig. 2**). The investment to understand the hydrodynamics of flow around the hull and design an optimized propeller makes it more attractive in situations in which the new design can be applied on multiple ships and the engineering investment distributed accordingly.

The pace of advancement in electronics is many times faster than the cycle of replacement for ships. In most civilian vessels, the electronics suites are focused on navigation and communications functions. Computerized charts, broadband satellite data connections, and the ubiquitous Global Positioning System (GPS) have all become virtually universal through piecemeal updates to existing vessels as well as more deliberate integrated electronic bridge system designs.

Modernizations of military vessels are a result of the same physical and technological processes that affect commercial ships, but with a more complex blend of economic and mission-effectiveness considerations. Warships' physical architecture and infrastructure, often referred to as hull, mechanical, and electrical (HM&E) systems, are generally robust compared to commercial vessels and feature considerable redundancy. This provides a durable platform for modernization, although a warship's service tends to involve irregular and often harsh operating conditions compared to commercial vessels, which increases physical wear and tear. In some cases, such as the Iowa class battleships, the same ship might undergo multiple cycles of modernization over the course of decades.

The driving force in the modernization of warships is the continuing effectiveness of the weapons and sensors that make up their combat systems. Even in the nineteenth century, the main batteries of wooden warships were sometimes changed to feature fewer, larger cannon as shell-firing guns came into service. After World War I, naval arms-limitation treaties (agreed to in large part because of the large costs of building new warships)

Fig. 2. A propeller with winglet tips designed to improve efficiency and save fuel, retrofitted to a Maersk Line, Limited, container ship. (*Photo courtesy of Maersk Line, Limited*)

Fig. 3. Models of *USS Lexington* as a battle cruiser (above) and an aircraft carrier (below) being examined by U.S. Navy and Congressional leaders in 1922. The naval arms-limitations treaties of the interwar era drove a boom in the modernization and conversion of warships. (*U.S. Navy photo*)

restricted the building of new capital ships. This led to a boom in extensive modernization and conversion efforts as nations attempted to match the capabilities of their perceived adversaries without building "new" ships (**Fig. 3**). World War II led to another round of major modernization programs as radar and early integrated control systems entered the fleet and a prewar focus on surface gunfire engagements gave way to the new reality of threats from the air.

Modern combat systems are composed of sophisticated integrated networks of sensors, processors, and weapons. The need for modernization can be driven by any element of this triad. New threats may require improved sensors to track them or improved weapons to kill them, but equally important is the computer system necessary to control the function of sensors and systems. A modern warship is a major information technology installation in its own right, incorporating millions of lines of software code running on hundreds of processors throughout the ship. A major modernization effort throughout the U.S. Navy in recent decades has been the transition from custom-built military computing equipment to commercial processors in order to take advantage of the increasing speed and decreasing cost of computing in the larger information technology industry.

The U.S. Navy designs combatant warships with the expectation of periodic minor modernizations and at least one major midlife modernization effort over the course of their service life. Aircraft carriers receive even more extensive, multibillion-dollar refurbishments.

Conversions. Changes to ships lie on a continuum of complexity. Although there are no clear dividing lines, some changes can be so extensive that they fundamentally alter the mission and capabilities of a ship. Conversion design can carefully take advantage of the inherent characteristics of the original design or it can simply try to extract some additional use out of the initial investment in the ship.

A driving force in major conversion programs has been the aftermath of wartime building programs. Both world wars left the United States with very large numbers of ships, often lightly used. Because the major first-cost investment had already been made, there was strong incentive to use those hulls in the future. Many ships were sold to other nations (usually entailing a modernization or conversion in the process), but many saw extensive conversion for new missions.

A number of World War I–era destroyers, though too small for the demands of frontline service in the Pacific theater during World War II, were converted into high-speed transports by the removal of some weaponry and one boiler room for troop berthing and landing craft. Much faster than conventional troop transports of the time, these converted ships filled a niche for the transport of smaller detachments in dangerous conditions.

Similarly, as surface-to-air missiles entered service, they were initially fielded on World War II cruisers

(a)

(b)

Fig. 4. ***USS Chicago* (*a*) before and (*b*) after her conversion from a heavy cruiser to a guided-missile cruiser. This conversion and modernization effort is among the most substantial undertaken. (*U.S. Navy photos*)**

(**Fig. 4**). These conversions were extensive, requiring a complete rebuilding of the ship from the deck up. Cargo ships from the wars were sold into civilian service and were converted for a variety of passenger and cargo trades.

For the military, ship conversion often serves as a relatively cost-effective way of testing a new capability at sea before committing to the design and construction of new ship classes. This approach was followed by navies around the world in the initial development of the aircraft carrier. Incentivized by the reduction in capital ship tonnage mandated by treaties, navies converted battleships and battle cruisers into aircraft carriers (Fig. 3), many of which served as important assets through World War II, and in some cases proved superior to early purpose-built carriers. Conversions of merchant ships to serve specialized government purposes, such as strategic sealift, have provided low-cost and rapidly fielded capability.

In the commercial world, conversions are more strictly based on economics. Fluctuations in global trade patterns can lead to periodic surpluses of ships that cannot be employed in their designed trade. Some of these may be held inactive until economic conditions change, but others may be attractive candidates for conversion to another trade. Some commercial conversions can approach the scope and complexity of major military upgrades, including major alterations such as lengthening a ship by inserting additional hull structure (Fig. 1).

Although modernization almost always increases the utility and sophistication of a vessel, conversions can also represent a shift to a less sophisticated or lucrative form of trade as a vessel ages, wears out, or otherwise becomes unsuitable for its designed employment. Commercial cargo carriers may have their machinery removed and serve as barges. Old warships have served as warehouses, barracks, and prisons. These conversions represent the extraction of the last utility from the initial investment in the ship. After a ship is no longer capable, it is sold for scrap, providing raw materials for other purposes and, hopefully, one final source of income to be invested in new ventures.

For background information *see* LIQUEFIED NATURAL GAS (LNG); MARINE ENGINE; MARINE MACHINERY; MARINE NAVIGATION; MERCHANT SHIP; NAVAL SURFACE SHIP; PROPELLER (MARINE RAFT); SHIP DESIGN; SHIP POWERING, MANEUVERING, AND SEAKEEPING in the McGraw-Hill Encyclopedia of Science & Technology.

Jonathan Slutsky

Bibliography. N. Friedman, *U.S. Destroyers: An Illustrated Design History*, Naval Institute Press, Annapolis, MD, 2004; A. Mansour, D. Liu, and J. R. Paulling (eds.), *Principles of Naval Architecture Series: Strength of Ships and Ocean Structures*, Society of Naval Architects and Marine Engineers, Jersey City, NJ, 2008; C. S. Moore and J. R. Paulling (eds.), *Principles of Naval Architecture Series: Intact Stability*, Society of Naval Architects and Marine Engineers, Jersey City, NJ, 2010; R. O'Rourke, *Navy Aegis Cruiser and Destroyer Modernization: Background and Issues for Congress*, Congressional Research Service Rep. RS22595, 2010; R. Storch et al., *Ship Production*, Society of Naval Architects and Marine Engineers, Jersey City, NJ, 1995.

Social class and poverty: missing elements in psychological research

It is interesting that psychological literature often does not include information on issues relating to social class and poverty. This is curious, especially given the large number of psychologists who have been interested in topics of social justice (including race, gender, and sexuality). Issues of class would seem to be a natural area of investigation by researchers interested in these social justice matters. For example, in looking at health disparities between ethnic minorities and the white (Caucasian) majority in the United States, many researchers have included class variables as largely explaining these differences. However, their respective attentions have remained on racial and ethnic differences instead of class differences. Poverty has been related to poorer health, decreased job security, and a host of other kinds of activities and conditions with negative consequences. Even if someone from poor circumstances were to become affluent later, an early exposure to toxins or malnutrition (as a result of poverty), for example, can have long-term negative physical and emotional effects.

There are also other reasons why social class should be a topic of interest for researchers. For example, children who are affluent can develop a sense of entitlement and may have underdeveloped intra- and interpersonal abilities. This can leave them more vulnerable to later stressors in life. In a psychological context, intrapersonal and interpersonal abilities can be considered as forms of intelligence. One might thus speculate that the higher rates of suicide among older white men, many of whom were successful but then suddenly lost their wealth or security (for example, because of bankruptcy or a loss of job), provide evidence of their lack of preparedness for the assaults of life.

Why, then, has psychology been absent in the investigation of social class? According to some researchers, people who are not poor engage in distancing themselves from the poor, making the plight of the poor foreign to such individuals. This distancing results in stereotypes and other forms of prejudice, rendering those in poverty unworthy of investigation because we already "know" about them. This is obviously not the case. Furthermore, psychological theories tend to be preoccupied with people who are similar to those who construct the theories (specifically, middle-class Euro Americans). In addition, because most theorists have not experienced poverty firsthand, it is much more difficult for them to relate to those in poverty and to have much insight into the plight of classism.

New conceptions of social class in the United States. Discussions of low-income families usually take into account the amount of money that a family brings in. However, it is important to consider the psychological factors that go into what sociologists call "socioeconomic status" (SES). SES was an automatic calculation based on household income. Traditionally, income had been divided into five categories: lower class (the bottom 20%), lower middle class (the second 20%), middle class (the third 20%), upper middle class (the fourth 20%), and the upper class (the top 20%). Thus, in this scheme, social class was treated as an externally imposed grouping variable as opposed to a legitimate prism through which one might view the world.

Recently, however, the traditional classification has been discarded in favor of more nebulous terms, such as "those in poverty," "the working class," "the middle class," and "the superrich." Those in poverty or working-class families are either in poverty or just above the poverty line, so nearly every family decision is dominated by the family's ability to afford even the basic needs of the family (for example, a choice between food and medicine or new clothes). There also tends to be a reluctance to categorize these individuals as being "lower class" because the subjective impact of this label suggests that "lower class" means "no class," which is a judgmental label as opposed to a categorical label. On the other hand, those who are superrich are never in a situation of having to think about the cost of basic needs. Instead, they are more inclined to concern themselves with making more money or deciding how to improve their legacies. Granted, many of these legacies are associated with benefits to society, such as donating money for the construction of a new hospital building that will bear the donor's name, establishing foundations in their names that give grants to important projects or scholarships, or having important awards named in their honor. However, this prism through which one views the world is certainly different from the prism of survival. The broadest category is the middle class, which extends from the old concept of lower middle class to the upper class, excluding the superrich. In other words, those who are quite wealthy in this country are quite often seen—even by themselves—as being in the middle class because they are not in the superrich category.

Wealth disparity in the United States. To demonstrate the disparity in wealth in the United States, it is interesting to compare the statistics about the superrich (the top 0.1%) and the rest of the country's earners. For example, the top 0.1% had incomes 220 times greater than the average incomes of the bottom 90%. This means that 50% of the individuals in the traditional SES category of upper class (the top 20%) is included in the bottom 90% of income earners in this country. To state it even more dramatically, the top 0.1% earn in 1.5 days approximately what the bottom 90% earn in 1 year. Such an income disparity must have profound effects on one's psychological outlook on life. Inevitably, it also creates an atmosphere in which it is difficult for the superrich to understand the depth of the problems (including unemployment) faced by the lower-income earners.

Unfortunately, the wealth disparity in the United States has gotten larger over time. The top 1% of earners in this country increased their wealth by 111% from 1979 to 2002, the middle fifth of income earners increased their wealth by 15% during this period, and the bottom fifth increased their wealth by only 5%. Moreover, the economic advantages of the superrich have increased even more in recent years. For example, from 1979 to 2007, the top 1% of earners in this country saw their worth increase by 275%.

It is also interesting to observe the fate of this wealth disparity during and after the major economic downturn that occurred in 2008. During this economic recession, the superrich lost a great deal of their wealth in the stock market. However, they had recovered their wealth by 2010 because of the recovery of the stock market. In contrast, because the wealth of the middle class was tied up predominantly in their homes and because the housing market had still not fully recovered as of 2014, a vast proportion of earners (especially those with large mortgages) have seen their wealth eliminated.

Future inequity. Overall, the true inequity of our economic system is the lack of financial mobility for those in poverty. Today, people in the United States are the least financially mobile of those living in Western societies. Those living in poverty in the United States, as opposed to those living in other Western countries, are the most likely to remain in poverty, and those born into higher economic categories in the United States are more likely to stay in those categories as well. Although there are still very good economic opportunities for those in the middle and upper middle classes to advance, these opportunities have eroded significantly for those in the lowest economic level.

For background information *see* HUMAN ECOLOGY; INFORMATION PROCESSING (PSYCHOLOGY); INTELLIGENCE; MOTIVATION; PERCEPTION; PSYCHOLOGY; SOCIOBIOLOGY; STRESS (PSYCHOLOGY); SUICIDE in the McGraw-Hill Encyclopedia of Science & Technology.

Jeffery Scott Mio

Bibliography. H. E. Bullock and B. Lott, Building a research and advocacy agenda on issues of economic justice, *Anal. Soc. Issues Public Policy*, 1:147–162, 2001, DOI:10.1111/1530-2415.00008; H. Gardner, *Frames of Mind: The Theory of Multiple Intelligences*, Basic Books, New York, 1983; H. Gardner, *Intelligence Reframed*, Basic Books, New York, 1999; H. Gardner, *Multiple Intelligences: The Theory in Practice*, Basic Books, New York, 1993; R. C. Kessler, G. Borges, and E. E. Walters, Prevalence of and risk factors for lifetime suicide attempts in the National Comorbidity Survey, *Arch. Gen. Psychiatry*, 56:617–626, 1999, DOI:10.1001/archpsyc.56.7.617; P. R. Krugman, *End This Depression Now!*, Norton, New York, 2012; W. M. Liu, *Social Class and Classism in the Helping Professions: Research, Theory, and Practice*, Sage, Thousand Oaks, CA, 2011; B. Lott, Cognitive and behavioral distancing from the poor, *Am. Psychologist*, 57:100–110, 2002, DOI:10.1037/0003-066X.57.2.100; B. Lott, Low-income parents and the public schools, *J. Soc. Issues*, 57:247–259, 2001, DOI:10.1111/0022-4537.00211; B. Lott, Recognizing and welcoming the standpoint of low-income parents in the public schools, *J. Educ. Psychol. Consult.*, 14:91–104, 2003, DOI:10.1207/S1532768XJEPC1401_05; B. Lott and H. E. Bullock, *Psychology and Economic Justice: Personal, Professional, and Political Intersections*, American Psychological Association, Washington, DC, 2007; J. E. Stiglitz, *The Price of Inequality: How Today's Divided Society Endangers Our Future*, Norton, New York, 2012; J. M. Stillion and E. E. McDowell, *Suicide across the Life Span: Premature Exits*, 2d ed., Taylor & Francis, Washington, DC, 1996.

Space flight, 2013

Commercial space flight continued to gain momentum in 2013. The Orbital Cygnus commercial resupply module flew its first mission to the *International Space Station (ISS)* in September. SpaceX's Dragon module also continued flights to the *ISS* for resupply and experiment delivery. Several major issues were faced by crew members on the *ISS* during the year. An ammonia leak forced an emergency spacewalk in May, and Italian astronaut Luca Parmitano had his helmet fill with water during a spacewalk in July.

Human space flight. The second SpaceX Dragon resupply vehicle launched on March 2 and docked to the *ISS* on March 4. It carried supplies and experiments to the *ISS*. The Dragon lost three of its four thruster pods shortly after separation from the launch vehicle; however, engineers were able to get the pods back online, allowing for a successful mission.

A *Soyuz TMA-08M* spacecraft carrying three new Expedition 35 crew members docked with the *ISS*'s Poisk module on March 28, completing its trip to the orbiting laboratory in less than 6 h. It carried Pavel Vinogradov and Alexander Misurkin of the Russian Federal Space Agency (Roscosmos) and NASA astronaut Chris Cassidy.

On April 21, Orbital Sciences launched its Antares rocket carrying a mass simulator, a nonfunctional spacecraft mockup, for the Cygnus *ISS* resupply craft. The launch occurred from Wallops Flight Facility in Virginia. The test flight paved the way for launches of the Antares vehicle to the *ISS* carrying a Cygnus spacecraft, offering a second American *ISS* resupply system (along with SpaceX's Dragon).

In May, NASA discovered an ammonia leak on the *ISS*. Ammonia is used to cool the space station, and the leak forced NASA astronauts Tom Marshburn and Cassidy to go out on a spacewalk to replace a leaky ammonia pump (**Fig. 1**).

Space station Commander Chris Hadfield of the Canadian Space Agency, Soyuz Commander Roman Romanenko of the Russian Federal Space Agency, and NASA Flight Engineer Marshburn undocked their *Soyuz TMA-07M* spacecraft from the space station on May 13 and returned to Earth. The crew had spent 146 days in space on board the *ISS* as part of Expedition 35.

On May 28, Commander Fyodor Yurchikhin of the Russian Federal Space Agency, NASA astronaut Karen Nyberg, and European Space Agency (ESA) astronaut Parmitano launched, rendezvoused, and docked their Soyuz to the *ISS* to become Expedition 37. The Soyuz went from launch to docking with *ISS* in less than 6 h.

On June 5, the European Automated Transfer Vehicle (ATV) *4 Einstein*, loaded with more than 7 tons of supplies, launched to the *ISS*. The Ariane 5 vehicle launched the ATV from Kourou, French Guiana.

A Long March 2-F rocket with the *Shenzhou 10* crewed spacecraft, carrying Chinese astronauts Nie Haisheng, Zhang Xiaoguang, and Wang Yaping, lifted off from the Chinese Jiuquan Satellite Launch Center on June 11. It docked with the *Tiangong 1* space laboratory, spent 15 days in space, and landed on June 25 in Inner Mongolia. Wang was the second female Chinese astronaut to fly in space.

ISS astronaut Parmitano's spacesuit helmet began filling up with water during his July 16 spacewalk. The spacewalk was aborted after 1 h 32 min into a $6^{1}/_{2}$-h scheduled spacewalk. There was some fear that he might drown. It was found that aluminum silicate contamination clogged a line in the system, a blockage that had caused water to back up into the astronaut's helmet.

Expedition 36 crew members Vinogradov, Cassidy, and Misurkin landed in Kazakhstan on September 10, completing their $5^{1}/_{2}$-month stay in space.

Expedition 37 crew members Oleg Kotov (Russia), Mike Hopkins (United States), and Sergey Ryazanskiy (Russia) were welcomed aboard the *ISS* after their launch on September 25 onboard *Soyuz TMA-10M*.

The Orbital Sciences Cygnus cargo module berthed to the *ISS* on September 29. Orbital became the second commercial partner, after SpaceX, to send a resupply mission to the *ISS*. The Cygnus module was placed into orbit using the Orbital Antares launch vehicle. Cygnus delivered about 589 kg (1299 lb) of cargo, experiments, food, and clothing to the station. The Cygnus had launched September 18 from the Mid-Atlantic Regional Spaceport Pad-0A at NASA's Wallops Flight Facility in Virginia.

On October 7, original *Mercury* 7 astronaut Scott Carpenter passed away at the age of 88 after complications from a stroke. In 1962, he became the second American to orbit the Earth.

On November 7, *Soyuz TMA-11M* launched NASA Flight Engineer Rick Mastracchio, Soyuz Commander Mikhail Tyurin of the Russian Federal Space Agency, and Flight Engineer Koichi Wakata of the Japan Aerospace Exploration Agency. They took with them the Olympic torch that was later used to start the 2014 Winter Games in Russia. The crew docked with the *ISS* about 6 h later to become part of Expedition 38.

ISS Expedition 37 crew members Yurchikhin, Nyberg, and Parmitano ended their 166-day mission, safely returning in their Soyuz on November 10. The torch to be used as part of the Olympic Winter Games returned from space with them.

Robotic solar system exploration. During 2013, NASA's *Gravity Recovery and Interior Laboratory (GRAIL)* spacecraft pair mapped the sources for gravity fluctuations on the Moon, including the mascons (mass concentrations) that had been known for decades. These are uneven mass distributions in the Moon that cause lunar orbital anomalies. The data indicates that these mass concentrations occurred when large asteroids or comets impacted the Moon when it was younger and hotter. The mapping of the Moon's gravitational field is expected to achieve a much finer resolution than for other terrestrial planets, including Earth. The results also show that the bulk density of the Moon's highlands is lower than previously assumed, implying a high crustal porosity, which is important for the flow of liquids such as magma or water. The bulk composition of refractory elements was found to be about the same as for Earth, in accord with theories of lunar origin. *See* GRAVITY RECOVERY AND INTERIOR LABORATORY (GRAIL) MISSION.

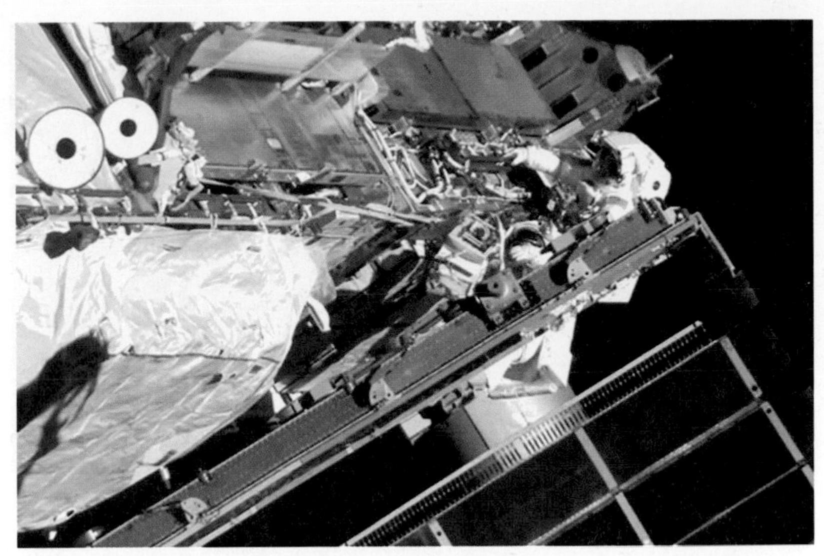

Fig. 1. NASA astronauts replace a leaky ammonia pump on the *International Space Station*. (*NASA*)

Using data collected through 2012, NASA's Mercury orbiter, *MESSENGER*, obtained the final image needed to view 100% of Mercury's surface under daylight conditions. The mosaics cover all of Mercury's surface and were produced using the monochrome mosaic released by NASA's Planetary Data System (PDS) on March 8, 2013 as the base. The full globe of Mercury is seen in **Fig. 2**. This is the first time that the entire surface of Mercury has been mapped. *See* MESSENGER MISSION RESULTS.

Approaching 10 years of exploring Mars, NASA's Mars Exploration Rover *Opportunity* continued to rove across the red planet's surface in 2013. It explored areas of raised segments of the western rim of Endeavour Crater, which is about 22 km (14 mi) in diameter.

Researchers using NASA's *Mars Reconnaissance Orbiter* have found that temperatures in the Martian atmosphere regularly rise and fall not just once each day, but twice. This was the first time that a twice-a-day structure in the temperatures of Mars was observed. Water-ice clouds in the Martian atmosphere absorb sunlight and cause this twice-a-day heating.

On June 27, NASA's *Interface Region Imaging Spectrograph* (*IRIS*) spacecraft launched from

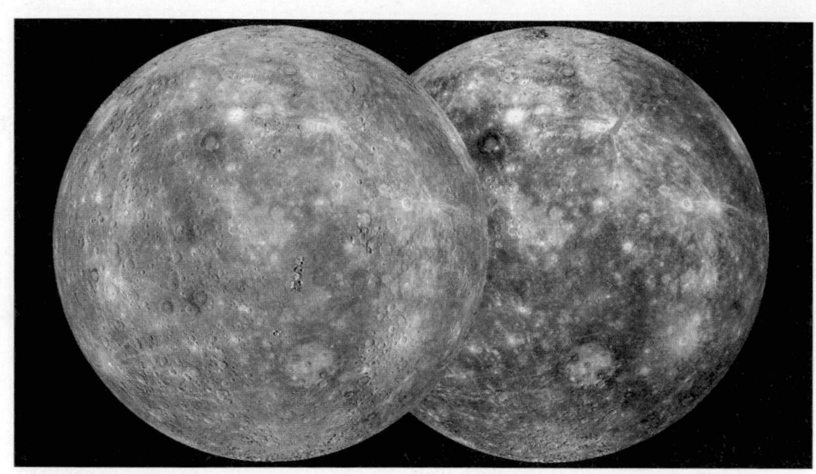

Fig. 2. Full-globe view of Mercury, made from data collected by the *MESSENGER* spacecraft. The globe on the left is a monochromatic view. The one on the right a color imager view of the same surface. (*NASA*)

Vandenberg Air Force Base in California. The mission to study the solar atmosphere was placed in orbit by an Orbital Sciences Corporation Pegasus XL rocket.

In 2013, NASA combined nearly 900 pictures taken by their *Curiosity* rover on the surface of Mars to create a billion-pixel view of the planet. Three cameras were used to take the component images on several different days between October 5 and November 16, 2012 (**Fig. 3**). The panoramic view shows the area where *Curiosity* first sampled the Martian soil.

Using NASA's Moon Mineralogy Mapper (M3) instrument aboard the Indian Space Research Organization's *Chandrayaan 1* spacecraft, scientists remotely detected magmatic water, or water that originates from deep within the Moon's interior, on the surface of the Moon. These results further demonstrate that the Moon is not as dry as once thought. Lava that had erupted from the interior of the Moon exhibited signatures of internal lunar water inside it.

In September, NASA published an atlas of the large asteroid Vesta, created from images taken as NASA's *Dawn* space probe flew around the object. *Dawn* will next explore the icy spherical asteroid Ceres in 2015.

NASA's *Lunar Atmosphere and Dust Environment Explorer* (*LADEE*) was launched to the Moon from Wallops Island, Virginia, on September 7. *LADEE* orbited the Moon and observed the environment around the Moon and dust in its vicinity.

On November 18, *MAVEN* (*Mars Atmosphere and Volatile Evolution*) launched on an Atlas V rocket from Cape Canaveral Air Force Station, Florida, on its way to Mars. The spacecraft is scheduled to arrive at Mars in September 2014 and observe Mars's upper atmosphere, ionosphere, and interactions with the Sun and solar wind.

Other activities. On February 11, NASA launched the *Landsat Data Continuity Mission* (*LDCM*) from Vandenberg Air Force Base in California. Later renamed *Landsat 8*, the spacecraft became the eighth in the Landsat series of satellites that have been continuously observing the Earth since 1972. These remote sensing observations provide information about the Earth and its resources, land use, and processes, as well as a record of changes occurring on the Earth's land surface. The same week, the *Landsat 5* spacecraft set a world record for the longest operating Earth satellite, having been operational in space for 28 years and 10 months.

On July 3, the *Jason-1* spacecraft was decommissioned after collecting data on the Earth's oceans for $11^1/_2$ years. During its lifetime, the joint NASA and French Centre National d'Etudes Spatiales (CNES) satellite charted nearly 4 cm (1.6 in.) of rise in global sea levels, a critical measure of climate change and a direct result of global warming.

On September 29, SpaceX successfully launched its upgraded Falcon 9 v1.1 rocket on its initial flight from the Vandenberg Air Force Base. The rocket was carrying the Canadian *CASSIOPE* satellite as its main payload along with three small nanosats.

On December 19, the European Space Agency launched its Gaia mission from French Guiana. The mission will create the most accurate map yet of the Milky Way Galaxy. By making accurate measurements of the positions and motions of 1% of the total population of stars, the history of our Galaxy can be explored. *See* GAIA MISSION.

Launch summary. In 2013 there were 81 launches, an increase of three over 2012 (see **table**). Of these, 78 were successful. During the year there were 23 commercial launches. Three launches were failures; one was Russian, one was Chinese, and one was one of the multinational Sea Launch's Zenit 3SL launchers. American launch vehicles launched six commercial missions in 2013. Russia had 12 commercial launches. The Sea Launch consortium failed in their commercial launch attempt. Europe had four commercial launches. The realm of small satellites also expanded with a record 92 cubesats launched in 2013.

Fig. 3. NASA combined almost 900 Mars *Curiosity* rover images to create this panoramic view of the planet. Mt. Sharp is in the background. The image processing color balance does not show the actual pink sky of Mars. (*NASA*)

Space launches in 2013

Country of launch	Attempts	Successful
Russia	32	31
China	15	14
United States	19	19
Europe	7	7
Russia–United States–Norway (Zenit-3SL)	1	0
India	3	3
Japan	3	3
South Korea	1	1
Total	81	78

A slew of new launchers appeared on the scene in 2013. The United States introduced the Antares vehicle with a test launch as well as the first Cygnus *ISS* resupply mission. The United States also first flew the Minotaur V, while China first flew the Kuaizhou, Japan first flew the Epsilon, and Russia first flew the Soyuz 2.1V.

Russia once again led in launches in 2013 with 32. Twelve of these were commercial and, of the other 20, eight were devoted to the *ISS*. Four crewless Progress modules were launched on Soyuz launch vehicles on *ISS* supply missions. Four were crewed Soyuz missions. Launch vehicles used by Russia were Proton-M (10), Rockot (4), Soyuz M (4), Soyuz U (4), Soyuz 2.1B (3), Soyuz 2.1A (2), Dnepr (2), Soyuz 2.1V (1), Strela (1), and Zenit 3SLB (1). One Proton-M launch was unsuccessful.

For the United States, the most launches were five with the Atlas V 401. Other vehicles used included the Antares (2), Falcon 9 v1.1 (2), Delta IV-Medium+5,4 (2), Atlas V-501 (1), Atlas V-531 (1), Atlas V-551 (1), Delta IV Heavy (1), Falcon 9 Dragon (1), Minotaur V (1), Minotaur I (1), and Pegasus XL (1). All launches were successful.

The Chinese used a variety of Long March launch vehicles. Four were Long March 4C; two each for the 2C, 2D, 3B, and 4B; and one each for the 2F and 3B. There was also a launch of a new Kuaizhou vehicle. One Long March 4B launch failed. There was one Chinese human space flight in 2013.

Four European launches used the Ariane 5 variants. Two Soyuz 2.1B launches were considered European since they launched from the European French Guiana complex. The Vega rocket also launched for Europe.

Japan launched its HIIA vehicle once as well as the HIIB. The Japanese also introduced a new launcher, the Epsilon Standard.

South Korea joined the nations launching satellites with its Naro-I vehicle. This vehicle had been unsuccessful in launch attempts in 2009 and 2010.

Multinational Sea Launch launched the Zenit 3SL from the Pacific Ocean one time unsuccessfully in 2013. Russia holds a majority share in the Sea Launch company, with the United States and Norway groups having small shares.

For background information *see* ASTEROID; ASTROMETRY; MARS; MERCURY (PLANET); MILKY WAY GALAXY; MOON; SCIENTIFIC AND APPLICATIONS SATELLITES; SPACE FLIGHT; SPACE STATION; SPACE TECHNOLOGY; SUN in the McGraw-Hill Encyclopedia of Science & Technology. Donald Platt

Bibliography. *Aviation Week & Space Technology*, various 2013 issues; ESA Press Releases, 2013; NASA Public Affairs Office, News Releases, 2013; *The Annual Compendium of Commercial Space Transportation: 2013*, Federal Aviation Administration, February 2014.

Spin torque

All electronic devices utilize flows of electrons, carrying with them a negative charge that is used to both store and control the flow of information. Besides their negative charge, electrons also have a number of other physical properties that are transported with them, the most notable being spin. Spin is an intrinsic, quantized form of angular momentum that, for the purposes of this discussion, can be viewed as a direction associated with each electron, often pictured as an arrow. In charged particles, such as electrons, the presence of spin also gives the particle a magnetic moment, and thus a magnetic field oriented along the axis of the spin; in electrons, the magnetic moment points in the direction opposite to that of the spin. In nonmagnetic metals, the spins of the electrons are randomly oriented, and, as a result, currents in such materials typically carry no overall spin. In ferromagnetic metals, on the other hand, the equilibrium electronic state is spin-polarized; that is, the electron population is divided into two subpopulations with opposite spin directions and with more spins pointing in one direction than in the other. As a consequence, currents in ferromagnetic metals are inherently spin-polarized and transmit a flow of spin alongside the flow of charge, allowing spin to play a significant role in device function. A thin ferromagnetic layer can function as a very effective spin polarizer of unpolarized electrons entering from a nonmagnetic metal, a process that is broadly similar to the polarization of sunlight by polarizing lenses. While a growing number of spin-dependent phenomena that such currents can generate, control, and make use of are known, two fundamental effects stand out: (1) the flow of spin-polarized currents can effectively be controlled by manipulating magnetization configurations, and conversely, (2) spin-polarized currents can be used to control the magnetization direction and the dynamic properties of magnetic systems. The second effect is called spin torque, and is the focus of this article.

History. Spin torque, also called spin transfer torque, was first discussed in the late 1970s as a possible mechanism for the observed movement of magnetic domain walls by very large currents of up to 45 A, but did not generate widespread interest.

Instead, the discovery in the late 1980s of interlayer exchange coupling—an interaction between the magnetization of two ferromagnetic layers separated by a nonmagnetic layer—and of giant magnetoresistance (GMR) in such systems would come to define spintronics. In 1989, J. C. Slonczewski predicted that spin currents should tunnel between two ferromagnetic layers separated by an insulator, even in the absence of an external voltage, and provide additional coupling. Experimental techniques available at the time were not capable of testing these predictions; reproducible magnetic tunnel junctions (MTJs) became available only in 1995, and the required high-quality ultrathin magnesium oxide (MgO)-based MTJs only in 2004. In 1996, Slonczewski and L. Berger independently predicted that spin torque could directly affect the magnetization in a multilayer device, either switching it from one orientation to another or causing it to precess continuously. In 1998, this effect was demonstrated experimentally, and so began a rapidly increasing interest in spin torque.

Ferromagnetism and spin polarization. Ferromagnetic materials—such as the most familiar, iron (Fe), cobalt (Co), and nickel (Ni)—display strong responses to applied magnetic fields, as their strong magnetization will try to align with the direction of the applied field. While most metals are nonmagnetic, especially those with relatively simple electronic band structures (copper, for example), ferromagnetic metals have particular features in their electronic band structure that make it energetically favorable for their entire electron population to separate into two antiparallel spin directions that better adjust to these features. In the so-called *s-d* model of ferromagnetic metals, the spin splitting arises almost entirely in the *d* bands (**Fig. 1**). On the microscopic level, all electrons are now spin-polarized. On the macroscopic level, the material is now magnetic, with a maximum magnetization given exactly by the imbalance of the two electron populations.

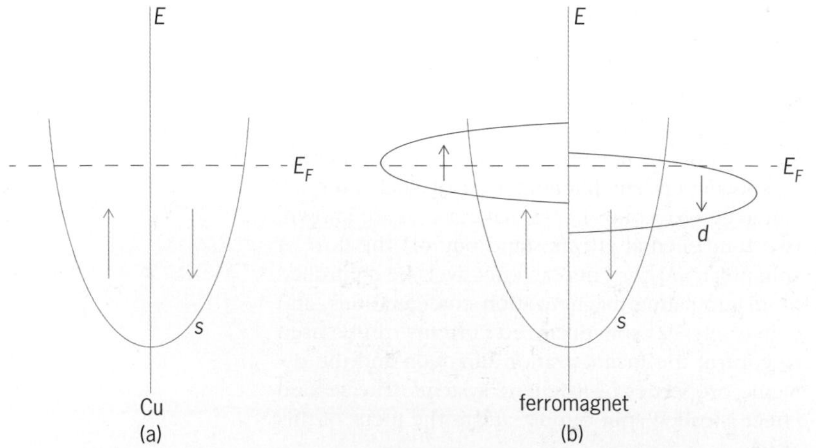

Fig. 1. Densities of states of solids as functions of electron energy *E*. (*a*) Density of states of the conduction band of copper (Cu), showing an equal number of spin-up and spin-down electrons filling the s band up to the Fermi level (E_F). (*b*) Density of states of a typical transition metal ferromagnet, such as cobalt (Co), nickel (Ni), or iron (Fe), where spin-split *d* bands make the material have more electrons with spins pointing down, resulting in a net magnetization, and more electrons at the Fermi level with spins pointing up, creating a net spin polarization with sign opposite to that of the magnetization.

Spin-polarized currents and spin torque. When a voltage is applied to a metal, charge currents start to flow and correspond to slight shifts in the electron distributions, with more electrons having momentum opposite the current flow (since electrons are negatively charged). The only available electronic states for such a redistribution reside at the Fermi surface, that is, the electrons with the highest energies in the metal. Hence, all conduction properties depend on the electronic properties around the Fermi surface, including the spin polarization of the current, and since the band structure of ferromagnetic metals can be quite complex, the spin polarization of the current is usually not the same as the global spin polarization. In Ni, Fe, and Co, the spin polarization even has the sign opposite to that of the total magnetization, as captured by the simple *s-d* model shown in Fig. 1: whereas there are more electrons at the Fermi level (E_F) with spins pointing up ($N_{\uparrow,EF}$), the total number of spin-down electrons is greater than the number of spin-up electrons.

The spin polarization at the Fermi surface, $P = (N_{\uparrow,EF} - N_{\downarrow,EF})/(N_{\uparrow,EF} + N_{\downarrow,EF})$, determines the amount of spin current that is transported along with the charge current. In a uniformly magnetized ferromagnet, this spin current has no net effect. However, since the spins of the flowing electrons must adjust to the local magnetic environment, any nonuniformity in the magnetization will have important consequences for both the electrons and the local magnetization: (1) whenever a spin-polarized current enters an area of magnetization with which it is not aligned, the magnetization must realign the flowing spins by exerting a torque on them, (2) and thus, the electrons exert an equal and opposite torque on the local magnetization; this is what we call spin torque. The effect can be made particularly strong in a GMR trilayer system (**Fig. 2*a***), where two magnetic layers, one that is free to switch (a NiFe free layer) and one that is static (a Co fixed layer), are separated by a nonmagnetic metal spacer (Cu) only a few nanometers thick. The magnetic states of the two layers can then be decoupled, allowing for very large abrupt changes of the magnetization direction and consequently large torques on both electrons and the magnetic layer. Similarly strong effects arise in MTJs, where the metal spacer is replaced by an insulating tunneling barrier.

Spin-torque devices. Spin torque is commonly investigated in GMR and MTJ systems that have two ferromagnetic layers separated by a nonmagnetic spacer (Fig. 2*a*). One layer is thicker or otherwise more resistant to changes in magnetization and is called the fixed layer. The spin current from the fixed layer flows or tunnels through the spacer and exerts its spin torque on the second layer, which is free. Two particular device structures dominate: (1) nanopillars, where all three layers are patterned into circles or ellipses of about 40–400 nm in lateral size and the current is uniform throughout the structure, and (2) nanocontacts, where either only the free layer or all three layers extend laterally to many micrometers and the current is injected through a circular or elliptical nanocontact, again of the order of 40–400 nm (Fig. 2*a* and *b*). In nanopillars, spin torque can be used to switch the entire free layer between two or more well-defined states or alternatively to sustain continuous precession of the free-layer magnetization if a magnetic field is applied in such a way as to reduce the stability of these states. In nanocontacts, the surrounding spins far away from the nanocontact do not experience any spin torque and prevent the free layer from switching. Instead, spin torque can generate continuous local precession of the magnetization underneath the nanocontact, which in turn emits spin waves that can propagate several micrometers away from the nanocontact. This phenomenon is quite remarkable in that a direct current can generate spin waves at microwave frequencies, which otherwise would require large microwave sources (Fig. 2*c*). In this context, one can show, both theoretically and experimentally, that spin torque counteracts the unavoidable intrinsic spin-wave damping in the free layer. Since this reduction in damping is linear in current, the net spin-wave damping can even become negative for large currents; that is, spin-wave damping is replaced by spin-wave amplification (or gain), with the result that the magnetization starts to self-oscillate. Hence, the local area underneath the nanocontact experiences gain from the spin torque, and, as in any system with gain (for example, unwanted feedback in a public announcement system), the system starts to self oscillate at its fundamental frequency. In the free layer, this frequency is the ferromagnetic resonance (FMR), and one can show that the propagating spin waves indeed lie very close to this frequency.

Domain walls and spin torque. If we picture a GMR trilayer in its antiparallel state and then remove the nonmagnetic spacer, we will be left with a magnetic domain wall. Since the exchange energy associated with an abrupt change of the magnetization inside a magnetic material is extremely high, the domain wall will spread out over a much larger thickness range than the original spacer. Electrons passing through a domain wall will again have to adjust their spins to the local magnetization direction and consequently exert spin torque. The net effect is a force on the domain wall that is proportional to the current density and in the direction of the electron flow. Spin torque is hence an effective way of moving domain walls back and forth in magnetic nanowires.

Applications. Commercial uses of spin torque have long been considered, in particular since both GMR and MTJ devices are already finding widespread use in hard drives, magnetoresistive random access memory (MRAM), and field sensors, and are easily integrated into everyday electronics. During the last 10 years, spin torque has become highly relevant in hard-drive read heads, where it causes undesirable magnetic noise because the read-head current

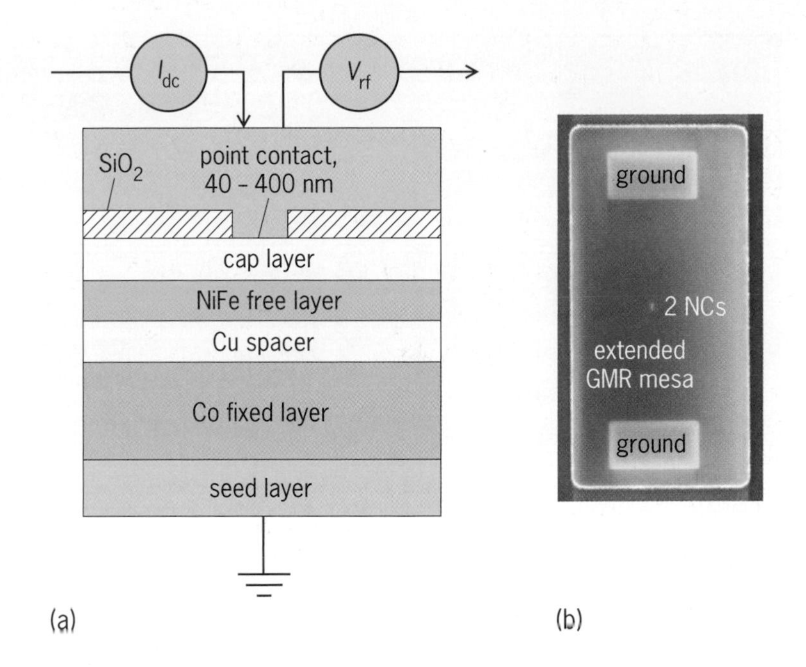

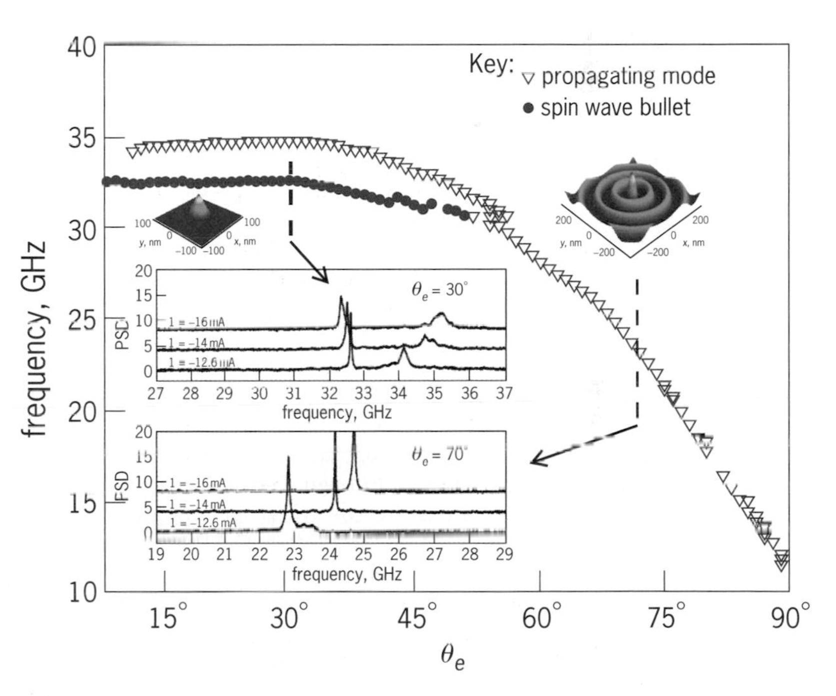

Fig. 2. Nanocontact spin-torque oscillator (NC-STO). (*a*) Schematic cross section, showing an extended spin valve with a fixed Co layer and a free NiFe layer, on top of which a nanocontact is fabricated through a silicon dioxide (SiO_2) insulating layer. (*b*) Top view of an NC-STO in which the extended spin valve is 8 μm × 16 μm wide. Two nanocontacts (NCs) are visible in the center of the structure as well as two much larger ground contacts. (*c*) Microwave output from an NC-STO as a function of applied field angle θ_e at a drive current through the nanocontact of 14 mA and a magnetic field of 1.1 T. (The applied field angle is measured with respect to the film plane; 0° is in the film plane, and 90° is along the film normal.) The insets show the power spectral density (PSD) of the NC-STO (that is, the amount of power in a certain frequency interval, here measured in arbitrary units) as a function of frequency for various drive currents and for applied field angles of 30° and 70°. At high angles, a single mode due to propagating spin waves is observed. The frequency of this mode increases with drive current. At lower angles, another mode starts to dominate, with a frequency that decreases with increasing current. This localized mode confines all spin waves to underneath the nanocontact and is called a spin-wave bullet. (*Main plot of part c adapted from S. Bonetti et al., Experimental evidence of self-localized and propagating spin wave modes in obliquely magnetized current-driven nanocontacts, Phys. Rev. Lett., 105:217204, 2010, DOI:10.1103/PhysRevLett.105.217204; insets of part c from S. Bonetti et al., Power and linewidth of propagating and localized modes in nanocontact spin-torque oscillators, Phys. Rev. B, 85:174427, 2012, DOI:10.1103/PhysRevB.85.174427*)

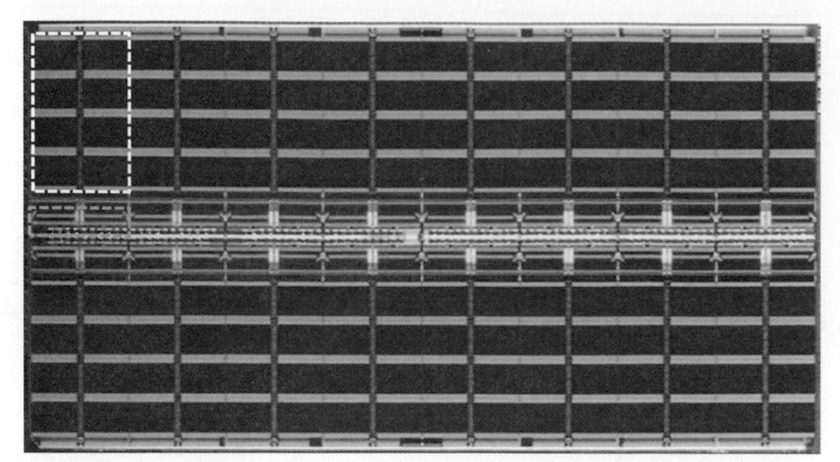

Fig. 3. Commercial 64-Mbit DDR3 spin-torque magnetoresistive random access memory (ST-MRAM). The rectangle with broken white lines outlines one of 16 identical memory areas, each containing close to 5 million magnetic tunnel junctions (including redundancy). The rectangle with broken blue lines outlines one of 16 identical areas for memory programming and read-out electronics. (*Everspin Technologies*)

amplifies thermal excitations through the negative damping. The first commercial spin-torque application was a spin-torque-based MRAM (**Fig. 3**) introduced in 2012. In this device, spin torque enables both smaller memory cells and a much simplified device architecture, since programming can be done by passing a current through the device instead of using two current-carrying lines to create magnetic fields adjacent to the device. A growing number of companies are now trying to catch up and develop the next generation of spin-torque MRAMs, which have the potential to disrupt the memory industry in the long run. There are hopes for a similar disruption of the microwave electronic market by so-called spin-torque oscillators, where the continuous precession of the free layer can generate ultrawideband microwave signals that are easily controllable by the direct current through the device. In contrast to conventional microwave generators, these devices are truly nanoscopic, have an unprecedented tuning range, and can be modulated at very high rates. Their limited output power and undesirably high phase noise are two drawbacks that are currently the subject of intense research.

Prospects. In addition to already being directly relevant for commercial computer memories, spin torque has also very recently opened up novel research directions that are expected to grow rapidly in importance in the next few years. One is magnonics, or spin-wave electronics, where nanocontacts inject, control, and detect spin waves and may use them for information processing at microwave frequencies much higher than those used by today's microprocessors. Another is the recent discovery that spin-torque-generated negative damping can sustain dynamic, solitonic, and nanomagnetic objects, so-called magnetic droplets, which were previously considered only theoretical curiosities. Magnetic droplets not only are highly interesting nanomagnetic objects in themselves, allowing for direct experimental studies of soliton dynamics in magnetic thin films, but also hold great promise for much higher output power in spin-torque oscillators. Just as interest in thin-film magnetism was revitalized by the discovery of GMR about 25 years ago and recognized by a Nobel Prize in Physics in 2007, the prediction of spin torque by Slonczewski and Berger will continue to inspire new discoveries for a long time onward and was awarded the Oliver E. Buckley Prize in Condensed Matter Physics in 2013.

For background information *see* COMPUTER STORAGE TECHNOLOGY; DOMAIN (ELECTRICITY AND MAGNETISM); ELECTRICAL CONDUCTION OF METALS; ELECTRON SPIN; FEEDBACK CIRCUIT; FERMI SURFACE; FERROMAGNETISM; MAGNETISM; MAGNETORESISTANCE; RESONANCE (ALTERNATING-CURRENT CIRCUITS); SOLITON; SPIN (QUANTUM MECHANICS); SPINTRONICS in the McGraw-Hill Encyclopedia of Science & Technology.

Johan Åkerman

Bibliography. L. Berger, Emission of spin waves by a magnetic multilayer traversed by a current, *Phys. Rev. B*, 54:9353–9358, 1996, DOI:10.1103/PhysRevB.54.9353; S. Bonetti and J. Åkerman, Nano-contact spin-torque oscillators as magnonic building blocks, *Top. Appl. Phys.*, 125:177–187, 2013, DOI:10.1007/978-3-642-30247-3_13; S. M. Mohseni et al., Spin torque-generated magnetic droplet solitons, *Science*, 339:1295–1298, 2013, DOI:10.1126/science.1230155; J. C. Slonczewski, Current-driven excitation of magnetic multilayers, *J. Magn. Magn. Mater.*, 159:L1–L7, 1996, DOI:10.1016/0304-8853(96)00062-5; M. Tsoi et al., Excitation of a magnetic multilayer by an electric current, *Phys. Rev. Lett.*, 80:4281–4284, 1998, DOI:10.1103/PhysRevLett.80.4281.

Spring unloader for knee osteoarthritis

Osteoarthritis (OA) is a chronic, degenerative joint disease that affects up to 40 million patients each year in Europe alone. OA is characterized clinically by joint stiffness, decreased mobility, and persistent pain. Although OA can affect any joint in the body, it preferentially targets the large joints and the finger joints, with the knee being the most commonly affected. The worldwide prevalence of OA is increasing and, coupled with aging of the population, is expected to increase by 40% by the year 2025.

Patients with musculoskeletal conditions regularly report among the lowest health-related quality of life scores when compared to other chronic conditions. In fact, knee OA patients report a more significant impact on quality of life than those with other musculoskeletal conditions. In addition to this substantial impact on patients' lives and mobility, OA is also associated with an enormous economic burden, affecting both the individual patient and the healthcare system as a whole. The direct and indirect costs associated with OA in Germany are estimated at nearly €6 billion per year, while costs in France and the United Kingdom are estimated to exceed €3 billion and €5.5 billion per year, respectively.

A summary of the clinical costs and benefits associated with surgical options for treatment of knee osteoarthritis

	KineSpring System	HTO	Prosthesis (joint replacement)
Disease severity	Mild-to-moderate OA	Moderate OA	Severe OA
Surgical invasiveness	Minimal: no bone, ligaments, cartilage removed. Device remains outside of the joint	Moderate: tibia cut and realigned, ligaments resected	Severe: joint surfaces removed, prosthesis implanted within the joint
Reversibility	Reversible	Potential negative effect on future joint replacement	Specialized revision components required
Postsurgical activity expectation	High	High	Low to moderate

HTO, high tibial osteotomy; OA, osteoarthritis.

Per-patient cost estimates range from €1500 in European studies to $4741 per patient, per year, in the United States.

Treatment options. Treatment for OA encompasses a range of options from conservative management to complete surgical replacement of the diseased joint. For the majority of patients, treatment of OA entails a long and difficult course of interventions that too often fail to adequately address the underlying pathology of OA.

Conservative treatment, while by far the most common approach for knee OA, is questionable as a valuable avenue for most OA patients. Studies of effect size (ES)—the standardized mean difference between two given treatments, ranked from 0 (no effect) to 1 (large effect)—have shown conservative care for OA to be clinically ineffective for pain and disease progression and not cost-effective. The most commonly recommended treatments for knee OA, analgesics and non-steroidal anti-inflammatories (NSAIDS), have an estimated effect size of only 0.13, indicating a trivial clinical effect. Viscosupplementation, meanwhile, has been associated with small to moderate treatment effects (ES: 0.37–0.61) in the first 4 weeks after treatment; however, by 13 weeks, clinical effectiveness diminishes to the point where little residual benefit remains and there is an increased risk of serious adverse effects. Paradoxically, pain-relief treatments, which show no significant ability to delay the progression of the disease, may, in fact, accelerate deterioration of the joint. Several authors have noted that patients who feel less pain are more likely to increase their activity level and thus increase the aberrant mechanical stresses placed on an already compromised joint.

When conservative treatments fail, patients often turn to surgery. Surgical treatments for knee OA range from moderate reconstructive procedures designed to unload the affected joint to complex joint replacement procedures. As with any medical procedure, these treatments are associated with benefits and drawbacks, and none is associated with an ideal safety profile. Early surgical interventions include high tibial osteotomy (HTO), which attempts to unload the arthritic compartment but results in irreversible structural changes to the proximal tibia. Partial or unicompartmental knee arthroplasty (UKA) and total knee arthroplasty (TKA) are later-stage, invasive procedures that result in permanent and significant structural modifications to the joint components (see **table**).

Knee replacement procedures, whether total or partial, are major surgical interventions that are often declined by potential patients who desire a more active lifestyle than can be provided by an artificial joint. Although TKA can be transformative for end-stage arthritis sufferers and is associated with large improvements in health-related quality of life, the invasive and irreversible nature of the procedure makes it a last-resort option for most arthritis sufferers. Knee replacement is also associated with notably high revision surgery rates, estimated to approach 20% for total knee replacement and exceed 40% in partial knee replacements after 5 years. The costs associated with these revisions are significant, totaling over €35,000 per case, with a median hospital stay of 5 days. The current annual burden of revision OA surgery is €1.9 billion, an amount that is expected to increase fivefold by 2030. The cost barriers notwithstanding, some estimates suggest that fewer than one-third of patients eligible for knee replacement surgery are even willing to undergo the procedure. Indeed, of the 10 million Americans suffering from arthritis each year, only 500,000 undergo knee replacement procedures, leaving an enormous group of patients searching for an alternative treatment.

Treatment gap. Procedures such as HTO and UKA are mid-stage treatment options, designed to decrease pain, improve mobility, and delay total knee replacement. They are, however, associated with high revision rates when performed in a younger, mid-stage arthritis population. HTO has been a pioneering procedure to unload the diseased joint compartment; however, it is associated with permanent alteration of the knee joint anatomy. This both limits subsequent treatment options and has the potential to increase subsequent treatment costs, as the increased surgical time when TKA is finally performed is associated with increased costs: Each 30-minute increase in surgical time results in a €500–€600 increase in procedure cost.

With the lack of success with conservative treatments and the limitations associated with current surgical interventions, a treatment gap exists, into which fall patients for whom conservative care has proven ineffective but whose disease has not progressed to the point where knee replacement procedures are being actively considered. This treatment gap necessitates a need for an earlier intervention that will address the underlying cause of arthritis, is clinically and economically effective, is reasonably acceptable for patients, and possesses the ability to delay TKA.

KineSpring® Knee Implant System. One treatment option showing promise in addressing the OA treatment gap is the KineSpring® Knee Implant System. The KineSpring System is a minimally invasive device that is implanted outside of the knee joint capsule and unloads the joint by using a unique "absorber" mechanism spanning the height of the joint to decrease the stress placed on the joint itself (**Fig. 1**). The device is both extra-articular and extra-capsular (**Fig. 2**). Similar to HTO, the KineSpring System unloads the diseased joint but, unlike HTO, it does so without disrupting the structural integrity of the joint itself. Device implantation is achieved without resection of bone, muscle, or ligaments and without violation of the joint capsule, thereby leaving the joint in its pretreatment state (**Fig. 3**).

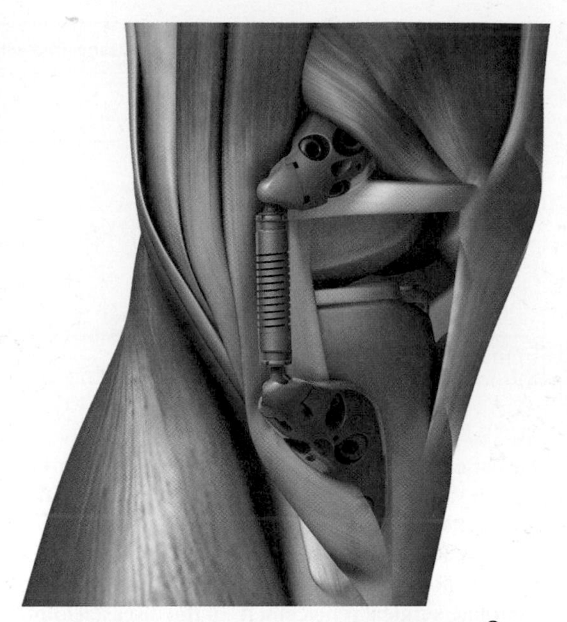

Fig. 2. A schematic representation of the KineSpring® Knee Implant System implanted in the medial compartment of the knee, demonstrating the location of the implant in relation to the surrounding tissues.

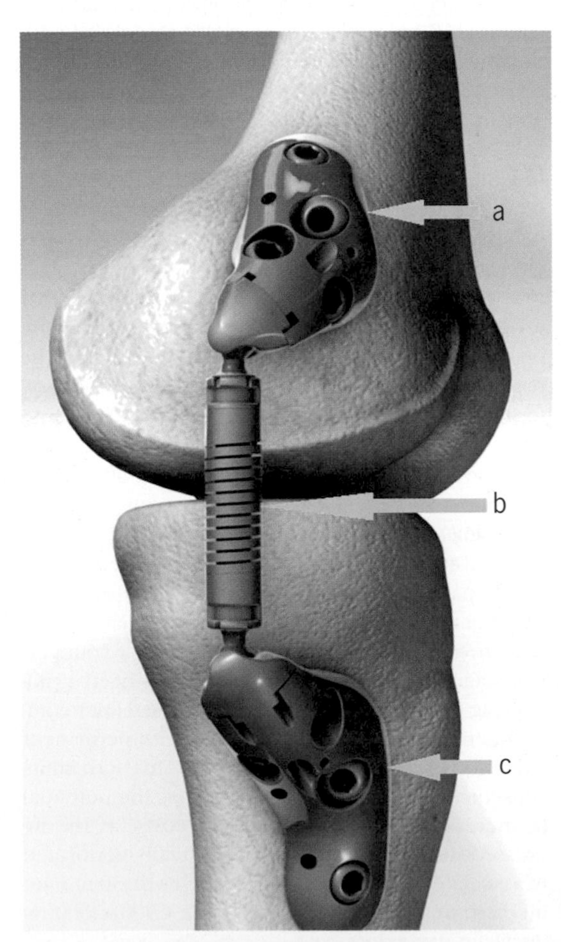

Fig. 1. The KineSpring® Knee Implant System schematically implanted on the medial side of the knee. (*a*) Femoral base. (*b*) Absorber. (*c*) Tibial base.

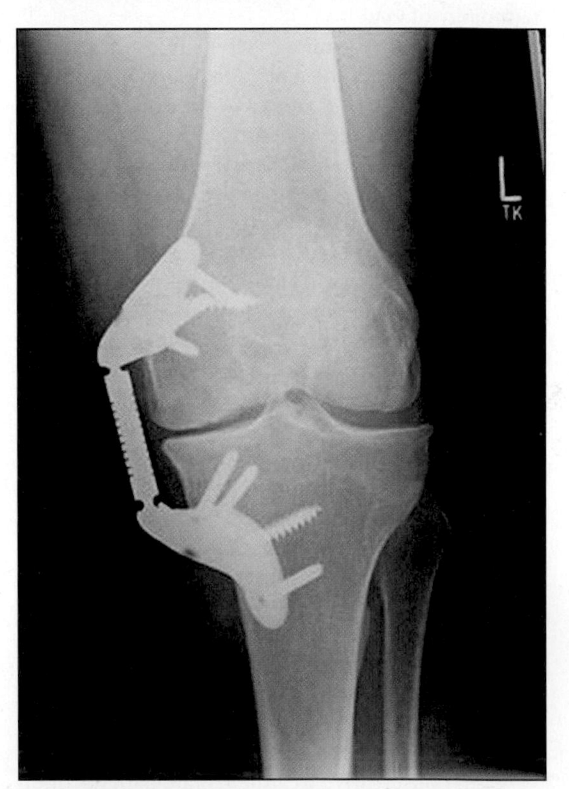

Fig. 3. Radiographic representation of an implanted KineSpring® Knee Implant System.

The KineSpring System consists of a three-piece unit, comprised of the femoral and tibial base units and the bridging absorber unit (Fig. 1). The device is implanted such that the center of rotation of the device on the femur is offset from the center of rotation of the knee. With proper device placement, this design allows the device to compress in extension and lengthen in flexion in excess of 30 degrees. Through device design and correct surgical technique, the absorber is designed to compress and absorb up to 13 kg of joint overload during the stance phase of gait, thus unloading the medial compartment of the knee. The load absorber resides in the subcutaneous tissue on the medial aspect of the knee and is positioned superficial to the medial collateral ligament. The modular design of the device allows for replacement or repair of the system's components, including the absorber itself, without removal of the unit as a whole.

By unloading the knee joint, this spring implant has the unique potential to play an important role in limiting the progression of OA. The KineSpring System is distinctly different from existing surgical options because it addresses the underlying biomechanics of the joint; preserves the native anatomy, including bone and joint surfaces; and targets a pre-joint replacement patient population that is handicapped by pain but is not ideal for joint replacement or osteotomy.

Clinical effectiveness. The KineSpring System has been evaluated in a number of clinical trials. In three studies totaling 100 patients, the KineSpring System was shown to be effective at relieving pain and improving functional ability. Using the Western Ontario and McMaster Universities Arthritis (WOMAC) Index, improvements in pain, function, and stiffness were noted immediately following surgery. These improvements were well maintained, with overall scores improving by as much as 78% postsurgery. Importantly, the improvements in pain and function were noted in a wide range of patients. Gender, age, body mass index, and arthritis severity were found to have no significant impact on the success of the KineSpring System, which was effective at lowering pain scores and improving functioning in all demographics. Indeed, male and female patients both experienced significant improvements in WOMAC scores for pain and function (75.7% and 84.0%, respectively). Similarly, 70% of nonobese patients and 83.4% of obese patients achieved clinical success. Overall success rates were noted to decrease slightly with age; however, even in the oldest category of patients (aged 60 years and older), 70% recorded clinically significant improvements in pain and function, while 80% reported improvements in stiffness.

Cost-effectiveness. Increasing strains on health-care systems and pressure from third-party payers have seen cost-effectiveness become increasingly important when evaluating treatment options. Cost-effectiveness is measured using the quality-adjusted life year (QALY), a gradually decreasing scale in which 1 represents perfect health and 0 represents death. A treatment's cost-effectiveness is expressed as a cost per QALY, and research has shown that patients with knee OA have a much lower willingness to pay than other patients. As such, treatments for OA must be very cost-effective in order to be considered by patients. This poses a significant problem for most current knee OA treatments, as they are generally associated with significant costs. Medications range from €10,800 per QALY for nonsteroidal anti-inflammatories (NSAIDS) to as much as €55,000 per QALY for pain-relief medications such as oxycodone. Even the most clinically effective surgical approach (arthroplasty) is associated with a cost per QALY of up to €13,000.

The KineSpring System presents a cost-effective alternative to the current treatment options. In a multinational study, the KineSpring System was shown to have a cost per QALY of €7,327, which is very favorable in comparison with the current conservative and surgical treatment options. This represents a potential lifetime savings of 13.2 million QALY in the German population alone, assuming conservative estimates that 30% of the OA population is suitable for the KineSpring System.

When compared with a typical treatment protocol of a long course of conservative treatment followed by surgical intervention, the cost-effectiveness of the KineSpring System is apparent. Likewise, with the high and expensive rates of revision surgery associated with knee replacement, the ability of the KineSpring System to provide pain relief, potentially slow disease progression, and postpone more significant surgical procedures is an important consideration.

Strict patient selection criteria will prevent overuse of this procedure. The eligible population for the KineSpring System has been clearly defined by trained surgeons and knee OA experts, resulting in targeted use of the device. The KineSpring System will help to decrease the use of high-cost medications, knee replacement surgeries, and knee revision surgeries, as patients who would otherwise fall into the treatment gap would instead receive effective treatment and would no longer be forced to undergo invasive surgery at a very young age.

For background information *see* ARTHRITIS; BIOMEDICAL ENGINEERING; BONE DISORDERS; BONE; JOINT (ANATOMY); JOINT DISORDERS; PROSTHESIS; SKELETAL SYSTEM; SURGERY in the McGraw-Hill Encyclopedia of Science & Technology.

Joffrey M. Muir; Mohit Bhandari; Jon E. Block

Bibliography. A. Chen et al., The global economic cost of osteoarthritis: How the UK compares, *Arthritis*, 2012:698709, 6 pages, 2012, DOI:10.1155/2012/698709; D. C. Crawford, L. E. Miller, and J. E. Block, Conservative management of symptomatic knee osteoarthritis: A flawed strategy?, *Orthoped. Rev.*, 5:e2, 2013, DOI:10.4081/or.2013.e2; G. Leardini et al., Direct and indirect costs of osteoarthritis of the knee, *Clin. Exp. Rheumatol.*, 22(6):699–706, 2004; C. A. Li et al., Cost-effectiveness and economic impact of the KineSpring Knee Implant System in the treatment for knee osteoarthritis, *Knee Surg., Sports Traumatol., Arthrosc.*, 21(11):2629–2637, 2013, DOI:10.1007/s00167-013-2427-x; N. J. London et al., Midterm outcomes and predictors of clinical success with the KineSpring Knee Implant System, *Clin. Med. Insights: Arthritis Musculoskel. Dis.*, 13(6):19–28, 2013, DOI:10.4137/CMAMD.S11768; A. D. Woolf and B. Pfleger, Burden of major musculoskeletal conditions, *Bull. World Health Org.*, 81(9):46–56, 2003.

Striga: physiology, effects, and genomics

Parasitic plants of the genus *Striga* (witchweed) in the family Orobanchaceae afflict a considerable variety of other plant species, including a number of critical cereal crops. The most economically important *Striga* species include *S. hermonthica*, *S. asiatica*, *S. gesnerioides*, and *S. aspera*. It is thought that *S. hermonthica*, the most problematic *Striga* species, originated in the Nuba Mountains of Sudan and Ethiopia, and then spread to many parts of Africa, Yemen, and Saudi Arabia. The most widespread *Striga* species is *S. asiatica*, which is found in Africa, the Arabian Peninsula, Asia, and the United States. *Striga gesnerioides* is widespread in Africa and also has been reported in the United States. Other *Striga* species of lesser economic importance include *S. angustifolia*, *S. aspera*, and *S. forbesii*.

Parasitic plants can be classified as root or shoot parasites based on their point of attachment to the host. Parasitic plants can also be classified, based on the presence or absence of chlorophyll, as either hemiparasitic plants, that is, those that have the ability to use sunlight to synthesize sugar from carbon dioxide and water (photosynthetic), or holoparasitic plants, which have lost this ability (nonphotosynthetic). Further classification can be done based on whether the parasitic plants can survive without a host (that is, facultative parasites) or whether they are wholly dependent on the host (that is, obligate parasites). For example, *Striga* species are obligate root hemiparasitic plants.

Striga* life cycle.** *Striga* species have evolved several unique survival strategies as parasites. For instance, an adult plant can produce up to 100,000 tiny seeds that can survive in soil for 20 years or more (**Fig. 1a*). Furthermore, seed germination occurs only in response to host root-derived germination stimulants called strigolactones (Fig. 1*b*). After germination, the parasite detects other host-derived factors and develops a multicellular structure called a haustorium to invade the host (Fig. 1*c*). Within 5 days after germination, the parasite establishes a xylem connection with the host vasculature to abstract water and nutrients. All the stages up to this point are below the ground. The parasite eventually grows above the ground to set green leaves and colorful flowers (**Fig. 2***a*).

***Striga* infestations.** *Striga hermonthica* and *S. asiatica* infect crops in the grass family, including maize (corn), rice, and sorghum (Fig. 2*b*). On the other hand, *S. gesnerioides* mainly infects the roots of cowpea (*Vigna unguiculata*). *Striga* species cause severe damage to host plant growth and development. Visible symptoms are plant desiccation, necrosis, and severe stunting. *Striga* species have been estimated to affect 40% of the cereal-producing areas of sub-Saharan Africa, where farmers lose 20–80% of their yields. This affects the livelihoods of approximately 100 million farmers and results in losses of up to $1 billion (U.S. dollars) annually. In some parts

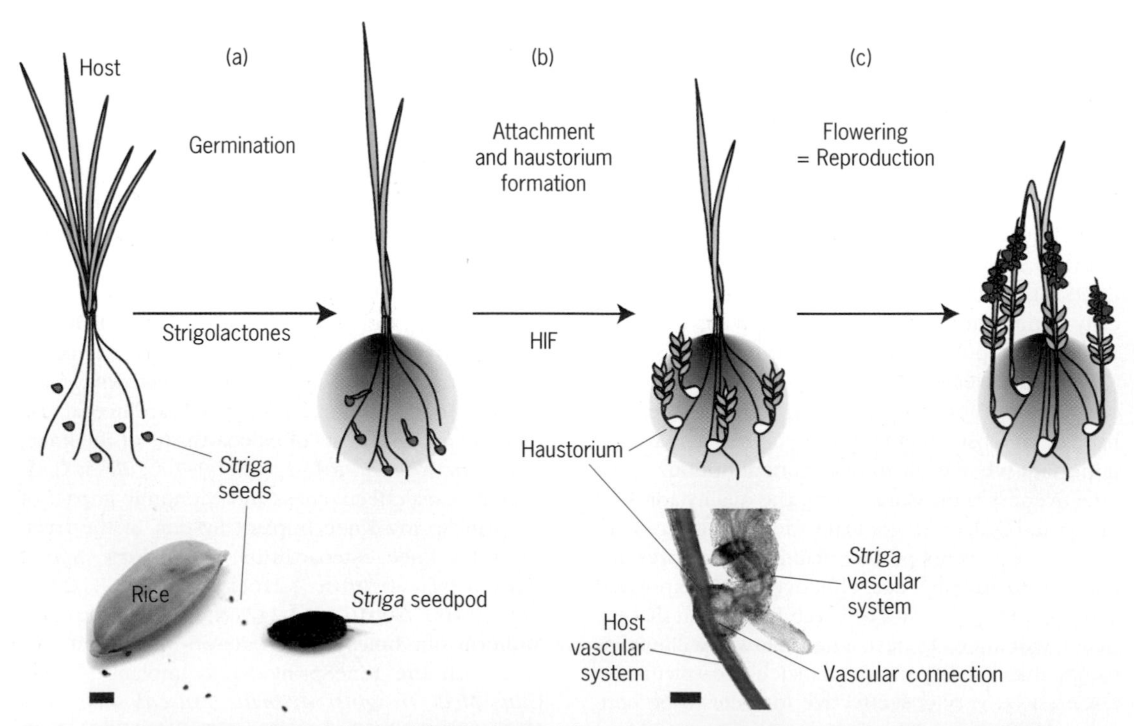

Fig. 1. Critical steps during *Striga* infection. (*a*) Strigolactones induce germination of *Striga* seeds, followed by (*b*) the attachment to host roots and the formation of a haustorium, which is induced by haustorium-inducing factors (HIF). (c) After *Striga* emerges from the soil, new seeds are produced in flowers. (*Lower left*) *Striga* seeds next to a rice grain and a *Striga* seedpod (scale bar = 1 mm). (*Lower right*) Anatomy of a haustorium connected to the vascular system of a host plant (scale bar = 200 μm).

(a) (b)

Fig. 2. *Striga hermonthica*. (*a*) Close-up photo of the purple flowers of *S. hermonthica* in western Kenya. (*b*) *Striga hermonthica*–infested sorghum field in Sudan.

of India, sorghum yield losses of up to 20% occur in infested areas.

***Striga* control.** The main fundamental strategies for controlling *Striga* species are to prevent their reproduction, destroy their soil seed bank, and limit their spread to new uninfested areas. Mechanical methods of control include hand pulling, which targets the parasites before they can reproduce. However, this approach is labor-intensive and best suited to light infestations. Biological control methods use *Fusarium oxysporum*, which is a pathogenic fungus that has been isolated from diseased *Striga* plants. Its application as a bioherbicide has resulted in up to 50% reduction in *Striga* infections in sorghum fields in Burkina Faso in West Africa. Nonetheless, application of bioherbicides as a control method is limited by the lack of efficient delivery systems. Additionally, an intercropping scheme, the so-called push, is used for *Striga* control. For example, intercropping maize and the legume *Desmodium* has been used to inhibit the attachment of *Striga* to maize.

Another successful control method for *Striga* species has been the stimulation of parasite seed germination in the absence of a host. In the United States, ethylene gas has been used to stimulate *S. asiatica* seed germination in the absence of hosts, leading to the eradication of the parasite from farmers' fields in the Carolinas. This effort cost an estimated $250 million (U.S. dollars). Although this method has been successfully applied in the United States, its usefulness for large infested areas in Africa is rather limited because of its cost.

Elucidation of *Striga* resistance. One of the most efficient and cost-effective ways to control *Striga* infestations would be the development of resistance or tolerance in the host species. Resistance is the ability of the host to withstand parasite attack in a manner that prevents the establishment and growth of the parasite. On the other hand, tolerance is the ability of the host to withstand the damage inflicted by the parasite. Cultivars and wild relatives of crop species, including sorghum, maize, rice, and cowpea, that are resistant to *Striga* have been identified. For example, certain cowpea cultivars exhibit the classical gene-for-gene resistance (in which one protein encoded by one host gene detects one protein encoded by a gene of the pathogen) to cognate *S. gesnerioides* isolates. The identified resistance gene encodes a protein that is similar to the receptorlike proteins that are typically required for resistance to fungal or bacterial plant pathogens. Despite the fact that cultivars and wild relatives of several crop species that are resistant to *Striga* parasitization have been identified, there are reports that the resistance tends to break down with the appearance of new parasite races. This implies that the genetic variation in *Striga* populations could be one of the main factors causing the quick breakdown of resistance of the host populations.

Progress in understanding the biology of *Striga* will benefit from genomic resources such as transcriptomes, which are descriptions of all genes actively transcribed under certain conditions. Indeed, a transcriptome analysis of *S. hermonthica* identified a gene that was most likely transferred from sorghum (*Sorghum bicolor*) into the *S. hermonthica* genome in a process called horizontal gene transfer (HGT). Intriguingly, there are studies that have raised the possibility that certain groups of parasitic plants acquire a fitness benefit by host-to-parasite gene transfers. Thus, understanding the unique biology of parasitic plants, undertaking the functional analysis of the genes that are important for *Striga* resistance or parasitism, and determining the genetic variations of *Striga* populations will enable new strategies for their control.

For background information *see* AGRICULTURAL SCIENCE (PLANT); AGRICULTURAL SOIL AND CROP PRACTICES; BOTANY; CEREAL; COWPEA; ECOLOGY; GENE; GENOMICS; GRASS CROPS; PARASITOLOGY;

PLANT GROWTH; PLANT PATHOLOGY; SEED GERMINATION in the McGraw-Hill Encyclopedia of Science & Technology.

Josiah Musembi Mutuku; Thomas Spallek; Ken Shirasu

Bibliography. B. I. G. Haussmann et al., Genomic regions influencing resistance to the parasitic weed *Striga hermonthica* in two recombinant inbred populations of sorghum, *Theor. Appl. Genet.*, 109: 1005-1016, 2004, DOI:10.1007/s00122-004-1706-9; R. Iverson et al., Overview and status of the witchweed (*Striga asiatica*) eradication program in the Carolinas, pp. 51-68, in *Invasive Plant Management Issues and Challenges in the United States: 2011 Overview*, American Chemical Society, Washington, DC, 2011, DOI:10.1021/bk-2011-1073.ch006; L. Musselman and M. Press, Introduction to parasitic plants, pp. 1-11, in M. Press and J. Graves (eds.), *Parasitic Plants*, Chapman and Hall, London, 1995; T. Spallek, J. M. Mutuku, and K. Shirasu, The genus *Striga*: A witch profile, *Mol. Plant Pathol.*, 14:861-869, 2013, DOI:10.1111/mpp.12058; D. Yonli et al., Biological control of witch weed in fields of Burkina Faso using isolates of *Fusarium oxysporum*, *Afr. Crop Sci. J.*, 13:41-47, 2005.

Superconducting quantum interference proximity transistor (SQUIPT)

The superconducting quantum interference proximity transistor (SQUIPT) is a hybrid superconducting interferometer. To achieve high sensitivity to magnetic flux, it exploits the phase dependence of the density of states of a metallic nanowire placed in contact with a superconductor. (The density of states is the physical quantity that refers to the number of states available to be occupied by electrons in the material, per unit energy at each energy and per unit volume.) The operation of a prototype structure based on this principle has recently been reported.

The SQUIPT is similar to the superconducting quantum interference device (SQUID), which is recognized as the most sensitive magnetic-flux detector ever realized, and combines the physical phenomena of the Josephson effect and flux quantization to operate. SQUIDs are nowadays exploited in a variety of physical measurements with applications ranging from pure science to medicine and biology. Recently, the interest in the development of nanoscale SQUIDs has been motivated by the opportunity to exploit these sensors for the investigation of the magnetic properties of isolated dipoles with the ultimate goal of detecting a single atomic spin, that is, one Bohr magneton.

Limited power dissipation together with the opportunity to access single-spin detection makes the SQUIPT interferometer attractive for the investigation of the switching dynamics of individual magnetic nanoparticles. The next section of this article presents the proximity effect as well as a model of the hybrid superconducting magnetometer. The device response in terms of voltage modulation and the transfer function is then discussed. Next, the noise behavior is presented and the feasibility of this structure as a single-spin detector is briefly addressed. The last section of the article discusses the response of a real SQUIPT device.

Proximity effect and device setup. The proximity effect is a phenomenon that can be described as the induction of superconducting-like properties into a normal-type conductor thanks to its contact with a superconductor. The superconductor is altered by the proximity effect as well, displaying a weakening of its typical correlations. The superconducting state is characterized by a macroscopic quantum phase that will affect profoundly the induced correlations. In particular, one relevant consequence of the proximity effect is the modification of the density of states in the normal metal, and the opening of an energy gap (that is, an energy interval with no energy states available to be occupied by electrons) whose amplitude can be controlled by changing the macroscopic quantum phase of the superconducting order parameter. The simplest implementation of a SQUIPT device is shown in **Fig. 1*a*** and consists of a diffusive (that is, with dimensions larger than the electron mean free path) normal metal (N) wire of length L in good electric contact with two superconducting electrodes (S) defining a loop. The contact with S therefore induces superconducting correlations in N through the proximity effect, which is responsible for the modification of the density of states of the wire.

The following calculations are performed in the short-junction limit where the proximity effect in the wire is maximized, therefore optimizing the interferometer performance. Figure 1*b* shows a three-dimensional plot of the amplitude of the N-region zero-temperature density of states [$N_N(\varepsilon,\varphi)$, along the vertical axis], calculated in the middle of the wire as a function of the energy ε and the superconducting phase φ. The density of states N_N shows an energy gap whose magnitude can be controlled through the quantum phase. The energy gap in the density of states turns out to be maximized at $\varphi = 0$, where it attains its largest amplitude Δ_0 (which is also the zero-temperature energy gap of the superconductor), whereas it is gradually reduced by increasing the phase, and it is fully suppressed for $\varphi = \pi$. Thus, it turns out that a proximized metal behaves as a sort of phase-tunable superconductor, meaning that its energy spectrum can be modified at will with the quantum phase, in particular, by closing or opening the gap in its density of states.

One may note that N_N is symmetric with respect to the energy ε since the physical description of the density of states is provided here thanks to the aid of the so-called "semiconductor model." In such a model, the density of states of a normal metal is represented as a continuous distribution of single-particle energy levels that includes states both below (that is, negative) and above (that is, positive) the Fermi level, the latter setting the zero of the energy. Similarly, a superconductor can be represented

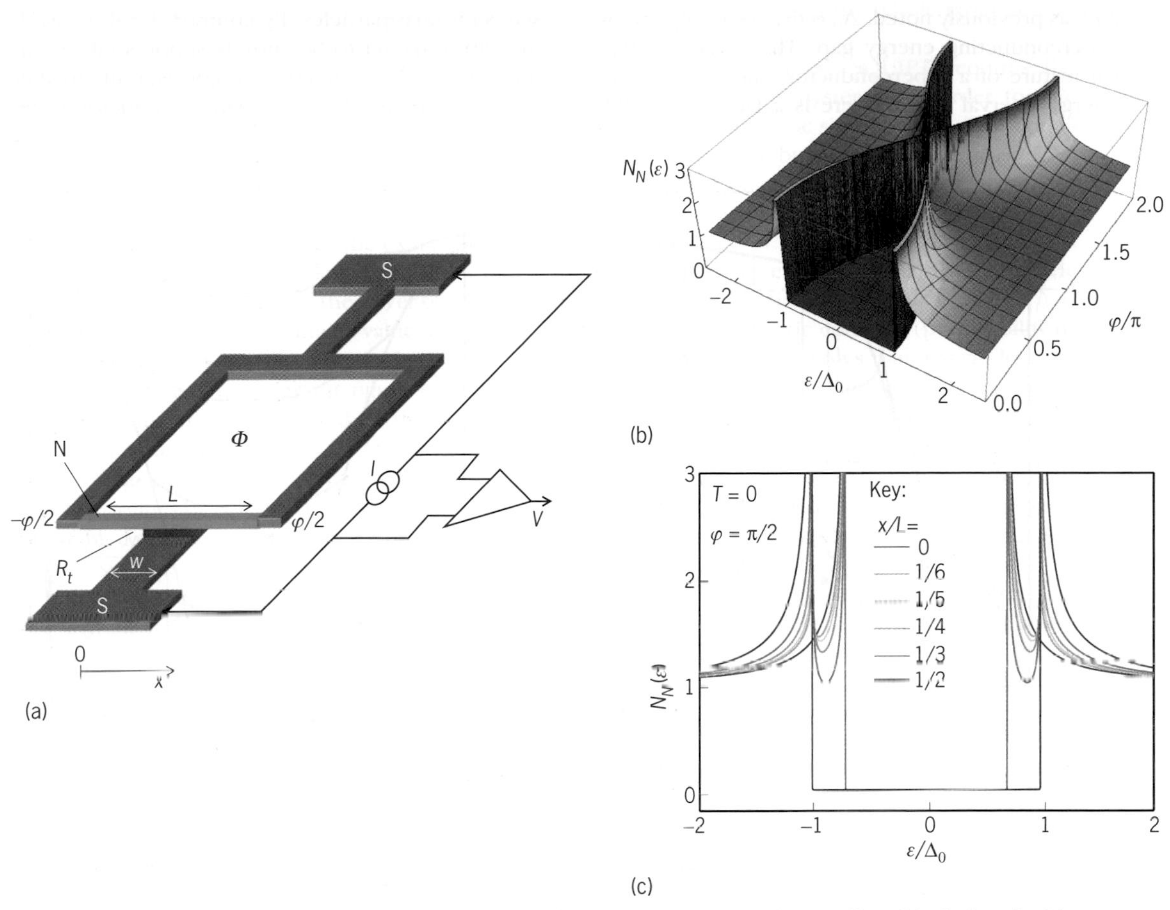

Fig. 1. The SQUIPT and the density of states in the N region. (*a*) Scheme of the interferometer. Here, *L* is the length of the proximized normal metal wire (N), *w* is the width of the superconducting (S) tunnel junction coupled to the middle of the wire, φ is the quantum phase difference in S, and Φ is the externally applied magnetic flux. The tunnel junction normal-state resistance Is denoted by R_t, *I* is the current flowing through the device, and *V* is the voltage drop developed across the junction. The spatial coordinate along the N wire is denoted by *x*. (*b*) Three-dimensional plot showing the evolution of the amplitude (along the vertical axis) of the zero-temperature density of states in the N wire, $N_N(\varepsilon,\varphi)$, normalized to the density of states in the absence of the proximity effect (that is, in the normal state), versus energy ε and phase φ. The calculation is performed at $x = 0$, that is, in the middle of the wire. Here, Δ_0 is the zero-temperature superconducting energy gap, which is a typical feature of a superconductor and indicates the energy interval where there is a lack of available states suitable for quasiparticles. The coloring of the surface has the role of visually emphasizing its three-dimensionality. In particular, the blue color indicates where the amplitude of the density of states approaches that in the normal state, whereas the yellow-orange color emphasizes a strong enhancement of this amplitude. (c) Zero-temperature N_N versus energy calculated for $\varphi = \pi/2$ at different positions along N. Here, *x* is the coordinate along the wire, and $x = \pm L/2$ denotes the N-S interface boundaries.

with the semiconductor model by symmetrizing its density of states with respect to the energy in such a way that, when the energy gap Δ_0 is zero, it tends to the density of states in the normal state. This representation is particularly useful when computing the tunneling current in a system comprising normal metals, superconductor, and proximized metals.

The evolution of N_N also depends on the position x along the N wire, and it is displayed in Fig. 1*c* for $\varphi = \pi/2$ and various values of x. Furthermore, a superconducting junction (S) of width w and normal-state resistance R_t is coupled through a tunnel junction to the middle of the N wire. The loop geometry of the superconducting electrode allows the changing of the phase difference across the normal metal-superconductor boundaries through the application of an external magnetic field that gives rise to a total flux Φ through the ring area. The resulting modification of the wire density of states, therefore, changes electron transport through the tunnel junction. For simplicity, one may suppose that a constant electric current I is fed through the circuit while the voltage drop V developed across the junction is recorded as a function of Φ. In the limit that the kinetic inductance of the superconducting loop is negligible, the magnetic flux fixes a phase difference $\varphi = 2\pi\Phi/\Phi_0$ across the normal metal wire, where $\Phi_0 = 2.0678 \times 10^{-15}$ Wb is the magnetic flux quantum.

Device response. Figure 2*a* shows the low-temperature current-voltage (*I-V*) characteristic of the SQUIPT calculated at a few selected values of Φ. It appears that for $\Phi = 0$, that is, when the gap in the N region is fully developed and maximized, the current-voltage characteristic of the interferometer resembles that of a superconductor-insulator-superconductor (SIS) junction composed of two identical superconductors, where the onset of large quasiparticle current occurs for voltages exceeding that corresponding to the sum of the gaps, that is, for $V \geq 2\Delta_0/e$, where e is the electron charge

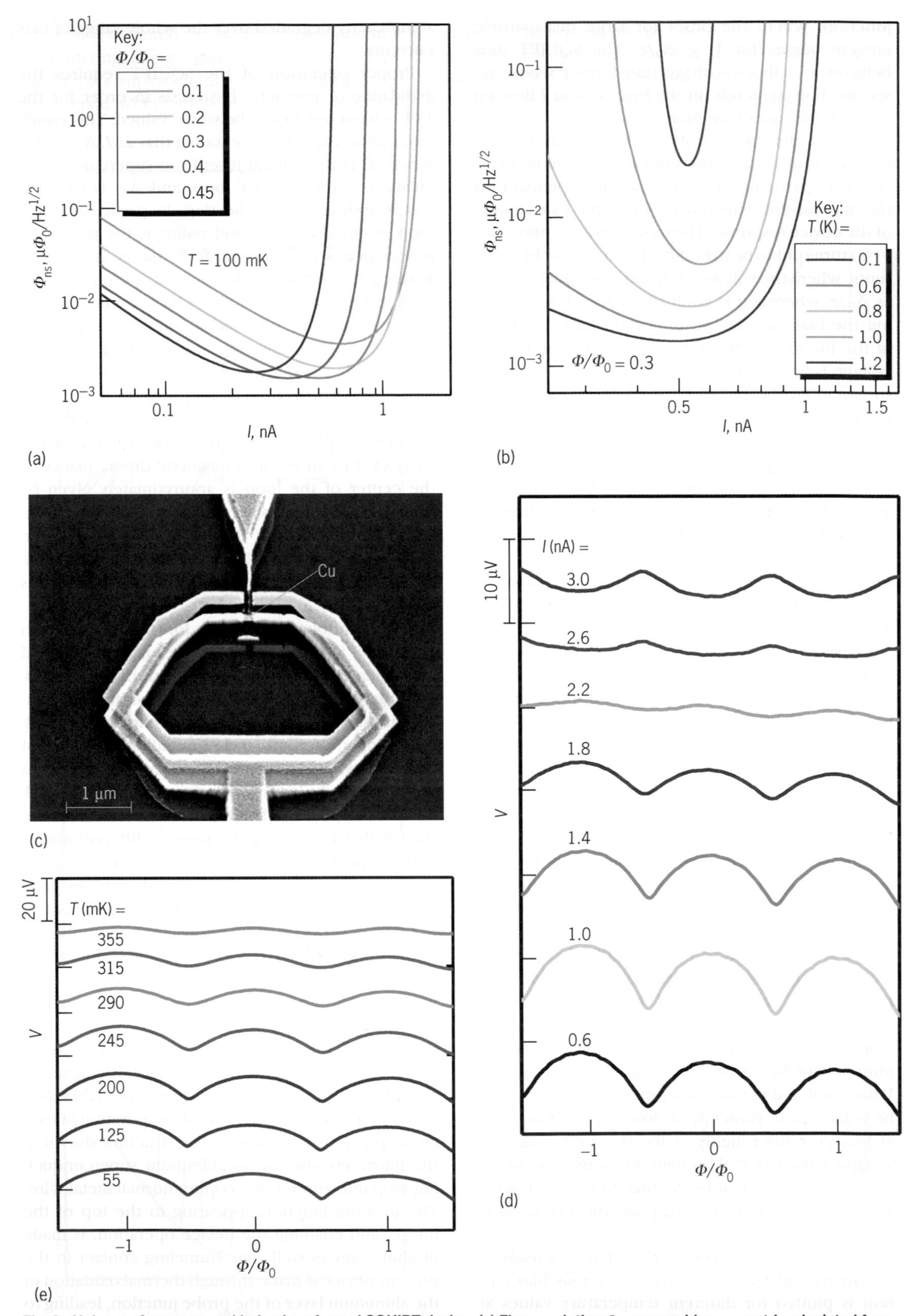

Fig. 3. Noise performance and behavior of a real SQUIPT device. (*a*) Flux resolution Φ_{ns} versus bias current *I*, calculated for a few values of Φ at temperature $T = 100$ mK. (*b*) Flux resolution Φ_{ns} versus *I* calculated at different temperatures for $\Phi = 0.3\Phi_0$. In these calculations, the energy gap is set at $\Delta_0 = 200$ μeV, and $R_t = 200$ kΩ. (*c*) Scanning electron micrograph of a typical aluminum-based SQUIPT. The ring and the probing junction are made of aluminum, whereas the N region is made of copper. The total length of the copper wire is $L \sim 400$ nm, the ring interelectrode spacing is ~150 nm, and the tunnel probe width is $w \sim 60$ nm. (*d*) Voltage modulation $V(\Phi)$ of a typical SQUIPT measured at a temperature of 54 mK for several values of the bias current *I*, given here in nanoamperes (nA). The curves are vertically offset for clarity. (*e*) Voltage modulation $V(\Phi)$ measured at a few bath temperatures, given here in millikelvins (mK), for $I = 1$ nA. The critical temperature of this SQUIPT device is ≅ 1.3 K, and the curves are vertically offset for clarity.

amplitude of $V(\Phi)$ is a nonmonotonic function of I, while $V(\Phi)$ displays changing of concavity for suitable values of the bias current. In this particular sample, the voltage modulation attains values as large as $\sim 7\ \mu V$ at 1 nA. The corresponding transfer function attains values as large as $\sim 30\ \mu V/\Phi_0$ at 1 nA. The impact of temperature is shown in Fig. 3*e*, which displays the modulation amplitude of $V(\Phi)$ measured at 1 nA in another aluminum-based SQUIPT device with critical temperature $T_c \cong 1.3$ K for several increasing bath temperature values. In particular, the modulation amplitude of $V(\Phi)$ monotonically decreases with increasing temperature.

So far, SQUIPTs have demonstrated flux-to-voltage transfer function amplitudes up to $\sim 1.5\ mV/\Phi_0$, leading to flux sensitivities down to $\sim 6 \mu \Phi_0 Hz^{-1/2}$ below 1 K. A large improvement in the intrinsic figures of merit of the interferometer is to be expected through a careful optimization of the structure design parameters as well as through the use of suitable cryogenic readout electronics. These interferometers are currently being developed for the investigation of nanoscale magnetic structures.

Compared to conventional DC SQUIDs, power dissipation (P) is dramatically suppressed in the SQUIPT. In these devices, $P \sim 100$ fW, which can be further reduced by increasing the resistance of the probing junction. Such a power is 4–5 orders of magnitude smaller than that in conventional DC SQUIDs, which makes the SQUIPT ideal for applications where very low dissipation is required. Moreover, there are several peculiarities that make this device attractive for a variety of applications: Only a simple DC readout scheme is required, similar to DC SQUIDs; either current- or voltage-biased measurement is possible, depending on the setup requirements; there is a large degree of flexibility in the fabrication parameters and a wide range of available materials besides normal metals, such as semiconductors, carbon nanotubes, and graphene, to optimize the response and the operating temperature; ultralow dissipation makes the device ideal for nanoscale applications; and the ease of implementation in a series or parallel array (depending on the biasing mode) enables enhanced output.

For background information *see* CARBON NANOTUBES; CRYOGENICS; ELECTRICAL NOISE; GRAPHENE; INDUCTANCE; JOSEPHSON EFFECT; MAGNETISM; MAGNETOMETER; MAGNETON; MICROLITHOGRAPHY; SQUID; SUPERCONDUCTING DEVICES; SUPERCONDUCTIVITY in the McGraw-Hill Encyclopedia of Science & Technology. Francesco Giazotto

Bibliography. J. Clarke and A. I. Braginski (eds.), *The SQUID Handbook*, 2 vols., Wiley-VCH, Weinheim, Germany, 2004, 2006; P. G. de Gennes, *Superconductivity of Metals and Alloys*, W. A. Benjamin, New York, 1966, reprint, Westview Press, Boulder, CO, 2010; F. Giazotto et al., Superconducting quantum interference proximity transistor, *Nat. Phys.*, 6:254–259, 2010, DOI:10.1038/nphys1537; F. Giazotto and F. Taddei, Hybrid superconducting quantum magnetometer, *Phys. Rev. B*, 84:214502 (6 pp.), 2011, DOI:10.1103/PhysRevB.84.214502; M. Meschke et al., Tunnel spectroscopy of a proximity Josephson junction, *Phys. Rev. B*, 84:214514 (5 pp.), 2011, DOI:10.1103/PhysRevB.84.214514; M. Tinkham, *Introduction to Superconductivity*, 2d ed., McGraw-Hill, New York, 1996, reprint, Dover, Mineola, NY, 2004.

Supported amine materials for CO_2 capture

With mounting evidence that anthropogenic carbon dioxide (CO_2) emissions significantly enhance the risk of global climate change, intense efforts have been devoted to developing technologies for alternative energy and CO_2 emissions mitigation. A critical aspect of this effort is to reduce CO_2 emissions from electric power generating plants that burn fossil fuels. For the short term, one practical approach to this is retrofitting current power plants with separation units to remove the CO_2 from the plant's exhaust flue gas, a process referred to as postcombustion CO_2 capture. A schematic of this process is depicted in **Fig. 1**. Currently, this is practiced to a limited extent using aqueous amine gas scrubbers. Here, flue gas leaving the plant, containing 8–15% CO_2 by volume, is contacted by the amine solution so that the mildly acidic CO_2 molecules interact with the basic amines. When the solution is saturated with CO_2, it is heated to regenerate the sorbent and concentrate

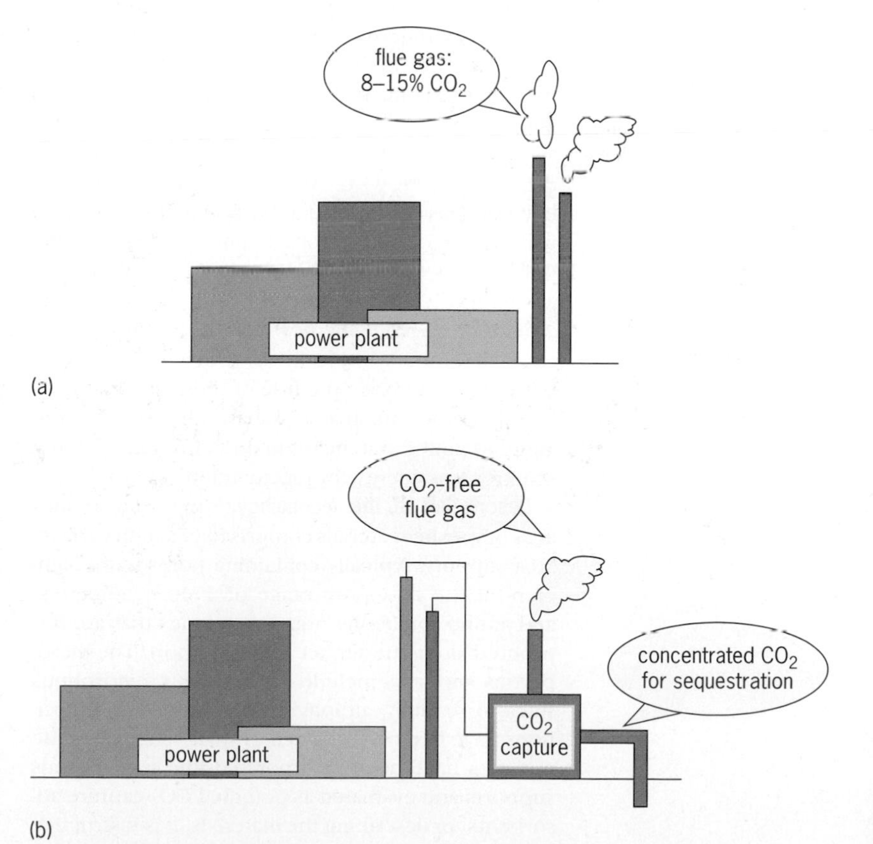

Fig. 1. Electric power generating plant (*a*) without and (*b*) with postcombustion CO_2 capture unit.

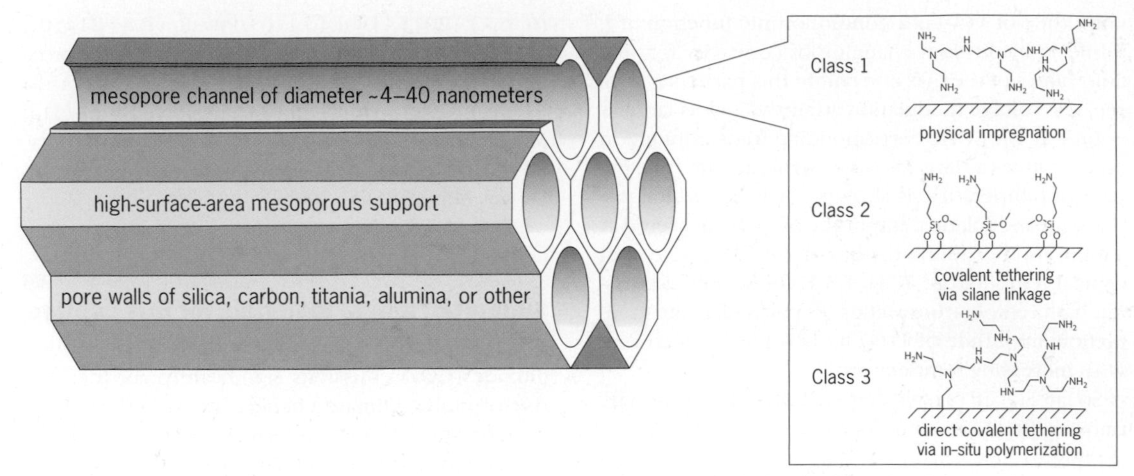

Fig. 2. Pictorial representation of supported amine materials classified by method of amine incorporation.

the CO_2 for downstream sequestration or use. Although these amine solutions remove CO_2 from the flue gas effectively, the process requires high energy input because of the large thermal mass of aqueous solution that must be heated during regeneration, ultimately resulting in a significant cost that is reflected in the price of electricity. Additionally, the amine solutions degrade under some processing conditions and are corrosive to the process equipment.

To overcome some of these shortfalls, the use of solid adsorbents has been suggested. Solids have advantages over aqueous absorbents in that their heat capacities are, in general, much lower, so the thermal energy required for regeneration is potentially reduced. Adsorbents are typically porous materials with high specific surface areas (100 to >2000 m^2/g) and are generally characterized by their pore size and interaction strength with the adsorbate of interest, in this case CO_2. Many solid adsorbents have been proposed for postcombustion flue gas capture, including zeolites, carbons, aluminas, metal-organic frameworks (MOFs), and supported amine materials. Supported amine materials, the subject of this article, are a relatively new class of adsorbent material designed specifically for efficient interaction with CO_2. The basic premise of these adsorbents is to utilize the effective acid-base chemistry of the aqueous amine system but to do so in a material that requires less energy for regeneration.

Description of the technology. Supported amines are composite materials comprised of a high-surface-area support, typically containing pores with diameters in the mesopore range of 4–40 nanometers, and amine-containing organic moieties that are distributed along the surface of the support. The mesoporous supports include metal oxides, amorphous silica, or carbon, although other support compositions have been investigated. A wide range of amine moieties have been incorporated into mesoporous supports and evaluated as potential CO_2-capture adsorbents. In describing the materials, it is instructive to classify them based on the method used to incorporate the amine molecule into the support material, as demonstrated pictorially in **Fig. 2**. Class 1 materials are prepared via the physical impregnation of amines into the pores of the support. Most commonly, amine-containing oligomers and low-molecular-weight polymers are incorporated into supports using this method. Class 1 adsorbents are broadly characterized as having high amine loadings, relative ease of preparation, and weak physical interactions between amine molecules and the support surface. Class 2 materials are prepared via the covalent tethering of amines to the support surface, often via aminosilane grafting. These materials typically contain lower amine loadings than class 1 materials but provide a strong covalent amine–support interaction. Importantly, these materials are conceptually simple, and have provided an excellent model for studying fundamental phenomena. Class 3 materials are prepared via in-situ polymerization of amine-containing monomers directly from the support surface. Class 3 materials combine the high amine densities of class 1 materials and the covalent amine tethering of class 2, but typically at the expense of a lower degree of control of the structure of the amine moiety and relatively difficult laboratory preparation.

Supported amine materials are chemisorbents for CO_2, meaning that the CO_2–amine interaction is relatively strong and often forms a covalent bond. A precise understanding of the chemical bond that is formed is both fundamentally and practically important. To approach this, the established CO_2–amine chemistry from liquid solutions has been extrapolated to the solid system, and has been supplemented by spectroscopic investigation. A reaction scheme is shown in **Fig. 3**, in which it is currently believed that ammonium carbamate ion pairs are the primary reaction product, and are formed via a zwitterionic (dipolar ionic) mechanism. This mechanism requires two base molecules in the immobilization of one CO_2 molecule. In the absence of water, two amines are required to capture one CO_2 molecule, yielding a maximum amine efficiency (moles of CO_2/moles of amine) of one-half at the saturation limit of the

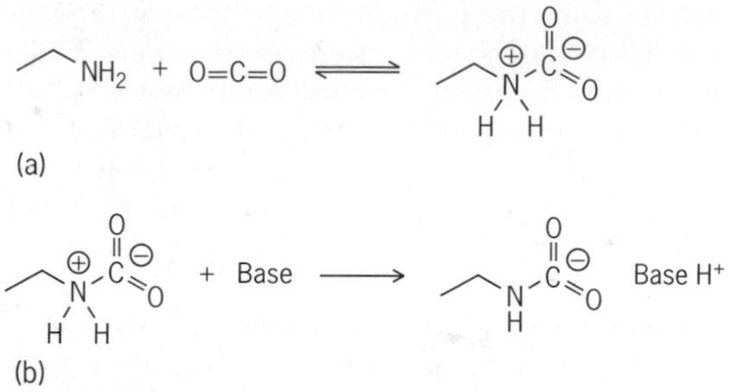

Fig. 3. Zwitterionic reaction mechanism believed to occur during the adsorption of CO_2 onto supported amine materials. (*a*) Formation of the zwitterionic species from a primary amine and CO_2. (*b*) Formation of an alkyl carbamate ion pair via deprotonation of the zwitterion by a free base (either an additional amine or water, if present).

material. However, when water is present and can act as a free base, the theoretical efficiency increases to 1.

Performance as CO_2 adsorbents. The success of any material as a CO_2 separation agent requires superior CO_2 uptake and adsorption rate under relevant process conditions, as well as adequate stability under those conditions. CO_2 uptakes are typically reported as single-point capacities in millimoles CO_2 adsorbed at equilibrium per gram of sorbent, or as isotherms in which capacities are plotted as a function of the partial pressure of CO_2 at a single temperature. The strong CO_2–amine interaction lends these materials very steep CO_2 adsorption isotherms, making them promising not only for flue-gas CO_2 separations but also for separation of CO_2 from ultradilute streams, such as air. The majority of research on supported amine materials has been aimed at developing structure–property relationships and novel material compositions that provide improved CO_2 capacities at relevant CO_2 partial pressures. In general, high CO_2 capacities are reported for materials with large pore volumes, small amine molecules, and optimal loadings of amines in the support. Further modifications, such as support composition and surfactant incorporation, have been shown to enhance CO_2 capacities of class 1 materials. The rates of CO_2 uptake on supported amine materials are generally regarded as being very rapid, as the CO_2 molecules can diffuse freely through the mesopores of the material to the adsorption sites. Data obtained in laboratory testing generally reflects these rapid uptake dynamics, except when the materials are very highly loaded with amines and a restriction on CO_2 diffusion is incurred.

Flue gas is typically at a temperature of 40–75°C, is saturated with water, and contains other contaminants such as oxygen, sulfur oxides (SO_x), and nitrogen oxides (NO_x). Thus, an understanding of how process conditions affect the CO_2 capacity of the material is important in practice. The effect of adsorption temperature on the CO_2 capacity of a material depends on the material composition. Highly loaded materials tend to show optimum capacity at an elevated adsorption temperature because of competing kinetic and thermodynamic effects. Moderately loaded materials tend to show decreasing CO_2 capacities with increasing temperature, as would be thermodynamically expected. As described earlier, water can increase the amine efficiency of a material. Laboratory testing of the materials in the presence of humidity tends to reflect this expected increase, although rarely by the theoretical factor of 2. However, this demonstrated ability to retain high CO_2 uptake in the presence of humidity is a significant practical advantage of these materials compared to other adsorbents such as zeolites and MOFs. Oxygen does not adsorb competitively on these materials, and thus does not impede the CO_2 capacity except when oxidative degradation occurs. However, the negative effects of SO_x and NO_x on the material performance have yet to be fully elucidated.

Degradation of the supported amines under relevant operating conditions remains a primary drawback to their commercial implementation. Degradation is typically incurred under regeneration conditions, such as steam stripping, and temperature, vacuum, and pressure processes. In particular, material degradation at regeneration temperatures (around 100°C) under relevant partial pressures of oxygen, CO_2, and steam have been reported. Exposure to oxygen and CO_2 has resulted in degradation of the amine moieties, while steam exposure has resulted in degradation of the support material, depending in both cases on the chemical composition of the sorbent. Structure–property relationships have been developed for sorbent stability under these different environments, although the engineering of stable and high-performing materials remains an important challenge.

Outlook. Supported amine materials remain promising candidates for incorporation in second-generation postcombustion CO_2 capture technologies because of their high CO_2 uptakes and potential for reduced operating cost, compared to conventional liquid absorbents. However, several barriers must be overcome prior to their large-scale deployment. Process and material development must continue to progress in parallel to reduce the degradation of the materials to provide them with adequate commercial lifetimes. Furthermore, process development and scale-up of the materials will provide better means to estimate the process cost savings over the existing technology. Finally, a comprehensive fundamental understanding of the molecular-level phenomena occurring in these materials, including adsorption, diffusion, and chemical interactions, will provide the means for further material development and optimization.

For background information *see* ABSORPTION; ACID AND BASE; ADSORPTION; AIR POLLUTION; AMINE; CARBON DIOXIDE; ELECTRIC POWER GENERATION; GLOBAL CLIMATE CHANGE; HEAT CAPACITY; ZEOLITE in the McGraw-Hill Encyclopedia of Science & Technology. Miles A. Sakwa-Novak; Christopher W. Jones

Bibliography. P. Bollini et al., Amine-oxide hybrid materials for acid gas separations, *J. Mater. Chem.*, 21:15100–15120, 2011, DOI:10.1039/C1JM12522B;

S. Choi, J. H. Drese, and C. W. Jones, Adsorbent materials for carbon dioxide capture from large anthropogenic point sources, *ChemSusChem*, 2:796–854, 2009, DOI:10.1002/cssc.200900036; A. Goeppert et al., Air as the renewable carbon source of the future: An overview of CO_2 capture from the atmosphere, *Energy Environ. Sci.*, 2:7833–7853, 2012, DOI:10.1039/C2EE21586A; J. C. M. Pires et al., Recent developments on carbon capture and storage: An overview, *Chem. Eng. Res. Des.*, 9:1446–1460, 2011, DOI:10.1016/j.cherd.2011.01.028.

Tandem catalysis

The preparation of chemicals, such as pharmaceuticals, dyes, fuels, or plastics, requires the selective transformation of simple, readily available starting materials into larger, more complex products.

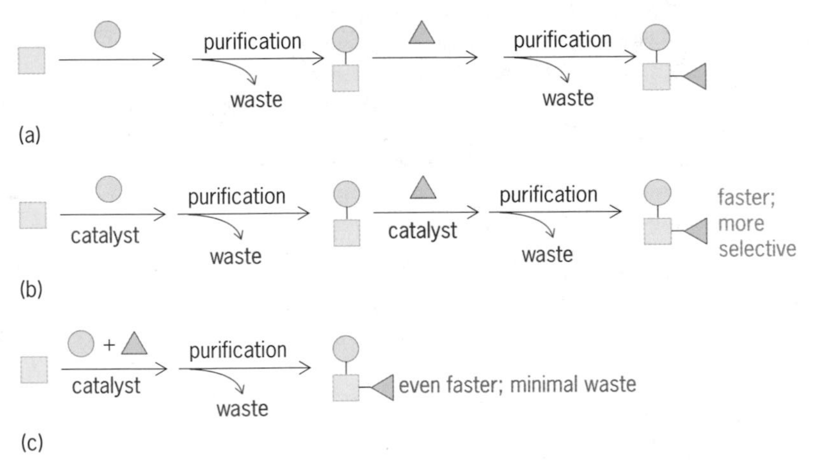

Fig. 1. Synthetic strategies. (*a*) Uncatalyzed stepwise synthesis. (*b*) Catalyzed stepwise synthesis. (*c*) Tandem-catalyzed synthesis.

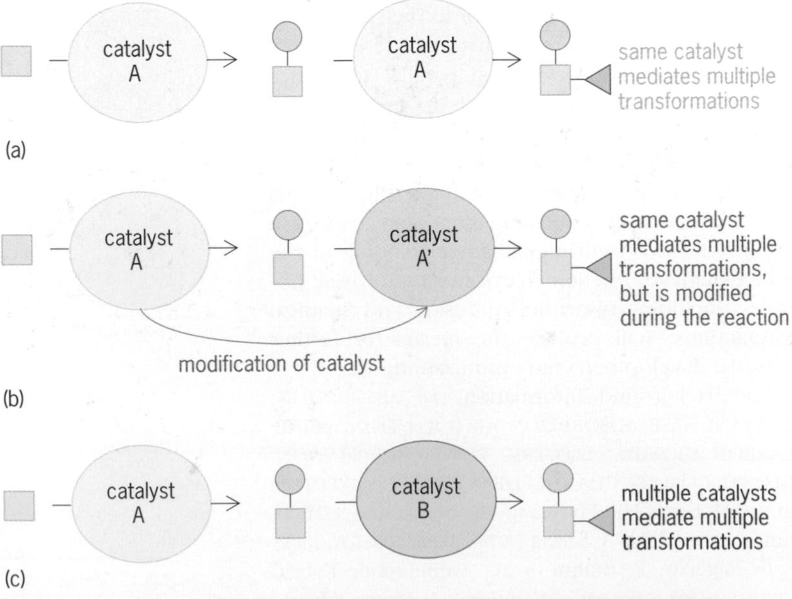

Fig. 2. Types of tandem catalysis. (*a*) Auto-tandem catalysis. (*b*) Assisted tandem catalysis. (*c*) Orthogonal tandem catalysis.

Traditionally, these products are obtained through a series of independent, stepwise reactions. While it provides a straightforward route to the target molecules, this synthetic strategy demands a unique set of reagents, conditions, and purification methods for each step (**Fig. 1***a*). In order to lower costs, catalysts are used in most large-scale modern processes.

Catalysts expedite chemical syntheses by creating new, lower-energy pathways from starting materials to products. The best catalysts result in much faster, more selective, and less wasteful reactions and can be used in small quantities or even reused (Fig. 1*b*). However, while catalysts often reduce the amount of waste generated in an overall transformation, the need for purification between steps still remains. Tandem catalysis is the incorporation of two or more catalytic reactions into a single step. This allows the efficiency and selectivity granted by conventional catalysis, but eliminates the need for individual purification methods (Fig. 1*c*). Although complex, this strategy is a fixture of living systems, where multiple enzymes (nature's catalysts) operate in tandem processes.

Although the term tandem is used in the chemical literature to describe a wide variety of catalytic reactions, the strict definition of tandem catalysis requires the use of one or more catalysts to carry out transformations on a substrate that (1) are distinct, (2) are separated by a discrete intermediate, and (3) occur concurrently in the same reaction vessel. The catalysts can be metal compounds, organic compounds, or enzymes.

Tandem catalysis can be divided into three categories: auto-tandem catalysis, assisted tandem catalysis, and orthogonal tandem catalysis. In auto-tandem and assisted tandem catalysis, a single catalyst accomplishes two distinct transformations. In orthogonal tandem catalysis, two separate catalysts are used, each for its own unique transformation (**Fig. 2**). Each type of tandem-catalyzed reaction has distinct advantages and disadvantages when applied to the synthesis of a molecule. Auto-tandem reactions are simple experimentally, as they require the use of only a single, unmodified catalyst, but they are often limited to a smaller number of transformations. Assisted tandem reactions can achieve a larger number of transformations, yet they require the modification of the catalyst during the reaction. Orthogonal tandem catalysis can combine nearly any two chemical reactions. In practice, it can be difficult to identify catalysts and reagents that will be compatible in the same reaction vessel. In all examples of tandem-catalyzed reactions, unfavorable interactions among catalysts, substrates, and reagents can present challenges.

Auto-tandem catalysis and assisted tandem catalysis. Many catalysts are capable of mediating more than one type of reaction. This ability can be exploited to design auto-tandem systems in which the same catalyst carries out two or more independent transformations on a substrate. With a single catalyst,

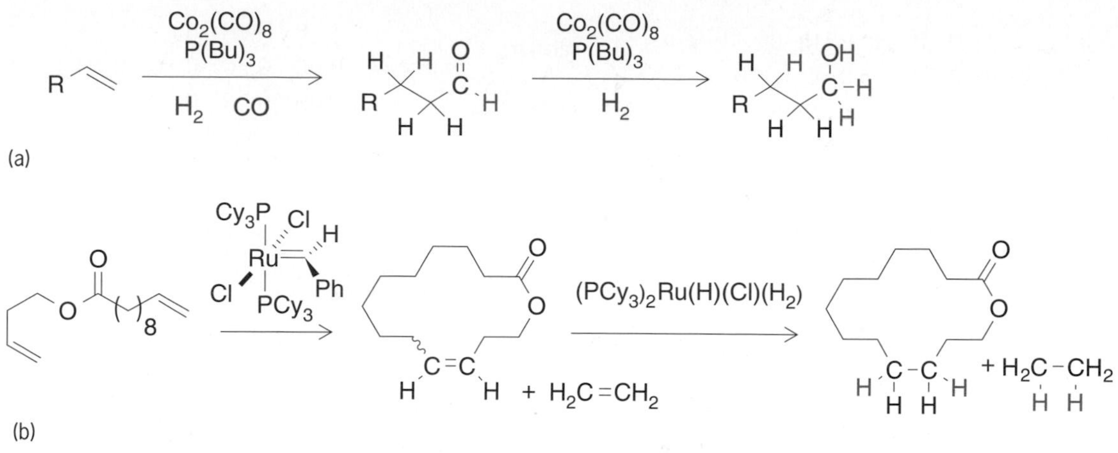

Fig. 3. Examples of (*a*) auto-tandem catalysis (Shell oxo process) and (*b*) assisted tandem catalysis (olefin metathesis/hydrogenation).

these reactions have simple procedures and eliminate the possibility of adverse interactions between different catalysts. A key industrial example of auto-tandem catalysis is the Shell oxo process (**Fig. 3*a***), in which a single cobalt catalyst carries out two different reactions, hydroformylation and hydrogenation, in the same reactor. The process converts petroleum derivatives with carbon-carbon double bonds, which are available at low cost, into industrially valuable alcohols used in detergents and plasticizers.

Unlike the catalyst in the Shell oxo process, other catalysts may require different conditions to carry out two transformations. In these cases, chemists often use assisted tandem catalysis, in which a catalyst that carries out one transformation is modified during the reaction to perform a second, distinct transformation. A common assisted tandem catalysis sequence is ruthenium-catalyzed olefin metathesis followed by ruthenium catalyzed hydrogenation (Fig. 3*b*). Although the ruthenium catalyst promotes the metathesis reaction efficiently in the absence of hydrogen, when it is in the presence of hydrogen, it performs a separate transformation, hydrogenating the double bond of the molecule. Experimentally, the reaction must be monitored to identify when the first transformation is complete, so that hydrogen may be added to induce the second transformation. This required midreaction intervention has led to some debate as to whether these assisted reactions can be accurately classified as tandem catalysis.

Orthogonal, homogeneous tandem catalysis. Homogeneous catalysts, catalysts that are in the same phase as the reagents, are often preferred by chemists because they are much easier to study and improve. In homogeneous tandem catalysis, catalysts are not isolated from each other and can interact freely, increasing the potential for negative interactions.

Despite this possibility, many useful tandem reactions have been developed in the petroleum, pharmaceutical, and polymer industries, such as the synthesis of linear low-density polyethylene (LLDPE). LLDPE is a hydrocarbon polymer with many short side chains branching out from the main chain (**Fig. 4*a***). The branches make the polymer easier to form into plastic bags and packaging materials, but they also introduce complexity into the synthesis, as they generally require an additional, more expensive starting material for the side chains. Using combinations of transition-metal catalysts working in tandem, many varieties of this complicated polymer can be synthesized in a single step from only the more affordable ethylene (Fig. 4*b*). One catalyst converts ethylene into longer-chain olefins that will become the branches, while the other links ethylene with these longer olefins to make the polymer. Examples have even been reported with three catalysts, two that form the longer-chain olefins from ethylene, and one that makes the main polymer chain from ethylene and the longer olefins.

Homogeneous tandem catalysis could also be used in the future for the more efficient conversion of nonpetroleum fuels into diesel fuel, a mixture of hydrocarbon chains about 19 carbon atoms long that comes from oil. While the synthesis of liquid

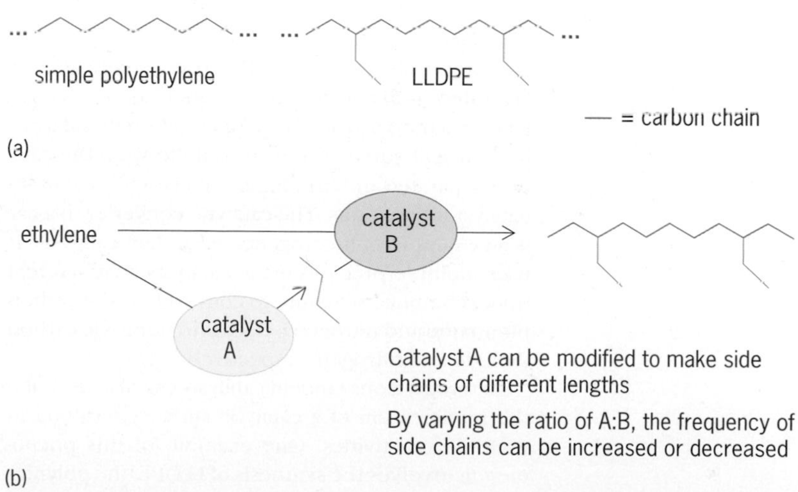

Fig. 4. Tandem synthesis of linear low-density polyethylene. (*a*) Simple polyethylene (unbranched) vs. LLDPE (side chains are shown in green). (*b*) Tandem synthesis of LLDPE.

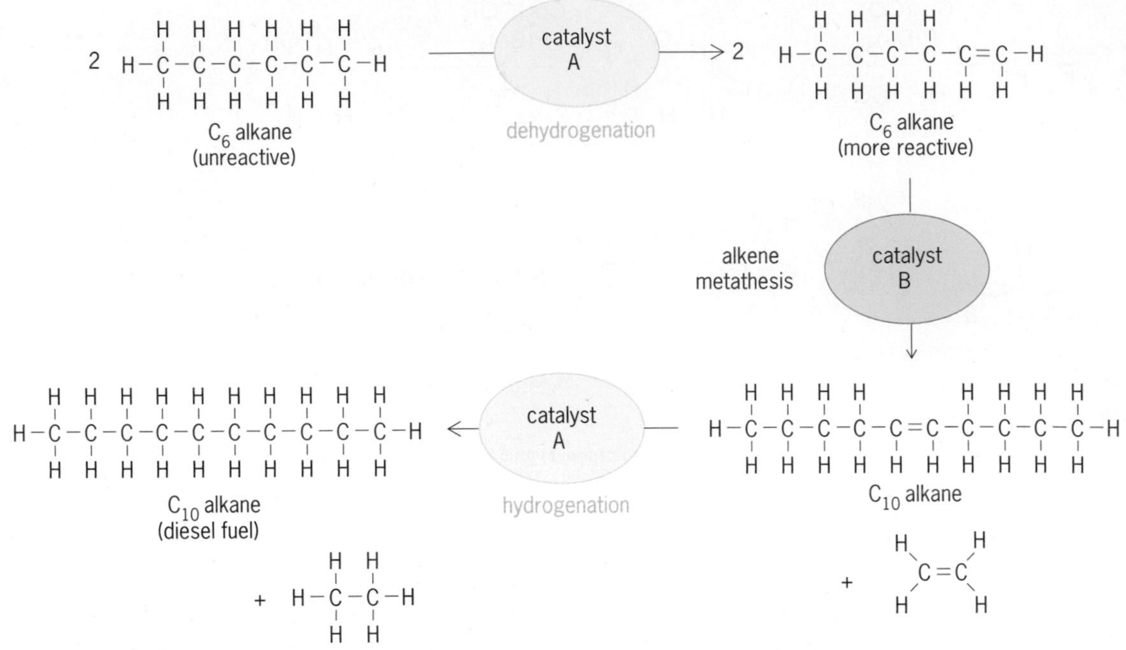

2 C_6 alkane (unreactive) —— catalyst A (dehydrogenation) ——> 2 C_6 alkane (more reactive)

alkene metathesis —— catalyst B

C_{10} alkane + $H_2C{=}CH_2$ —— catalyst A (hydrogenation) ——> C_{10} alkane (diesel fuel) + $H_3C{-}CH_3$

Fig. 5. Alkane metathesis. Catalyst A is iridium-based and catalyst B is molybdenum-based.

hydrocarbon mixtures from other sources, such as natural gas, coal, or biorenewable materials, has long been known, it is not a good route to valuable diesel fuel because there is little control over what hydrocarbons are produced. In addition to diesel, less valuable longer- and shorter-chain hydrocarbons are produced. What is needed is a method to convert these long and short alkanes into more valuable diesel fuel, but alkanes are relatively unreactive. Tandem catalysis offers a solution in the form of a process called alkane metathesis, which uses homogeneous iridium and molybdenum catalysts to first make the alkanes more reactive and then to join them together. The chemistry of the alkane metathesis process is shown in **Fig. 5**.

Orthogonal, heterogeneous tandem catalysis. While homogeneous catalysts are easier to design and improve, heterogeneous catalysts are usually favored in industry because they are robust, can be reused, and are easily recovered. Heterogeneous catalysts are often attached to surfaces such as beads, pellets, or nanoparticles. By placing different catalysts on different surfaces, it is possible to keep the catalysts separated and to minimize unfavorable catalyst-catalyst interactions. The catalytic converter in a car is an example of heterogeneous tandem catalysis. It uses multiple precious-metal catalysts attached to a honeycomblike structure to convert harmful carbon monoxide and nitrogen oxides into nontoxic carbon dioxide and nitrogen, respectively.

Heterogeneous tandem catalysis can also describe the modification of a catalytic surface, resulting in multiple reactivities. One example of this phenomenon involves the synthesis of LLDPE, the polymer mentioned previously, using a chromium catalyst deposited on a surface. Without extra ligands, the chromium catalyst polymerizes ethylene into long, unbranched chains, while the addition of a ligand that sticks to the catalyst causes the chromium to convert ethylene into short chains. By adding sufficient ligand to modify only some of the chromium sites, a system that converts ethylene into longer olefins and then combines those olefins with more ethylene to make LLDPE is possible. The number of side chains branching off from the main polymer can then be adjusted by altering the amount of ligand added.

Heterogeneous tandem catalysis has also benefited from the growing field of nanotechnology. Nanoparticles with different catalytically active surfaces can be easier to make than specialized reactors with catalytic surfaces, and their concentration is easy to adjust. Academic researchers recently developed nanocrystals that can catalyze two different reactions in tandem by using two different types of metal–metal oxide interfaces. Stacked layers of silica,

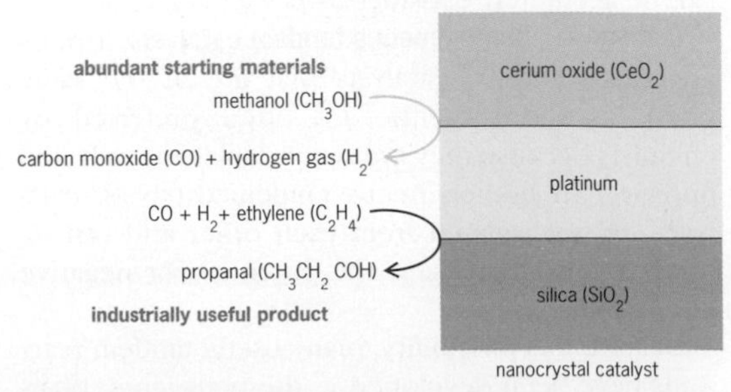

Fig. 6. Catalysis at metal–metal oxide interfaces: synthesis of the industrially useful chemical propanal from methanol and ethylene.

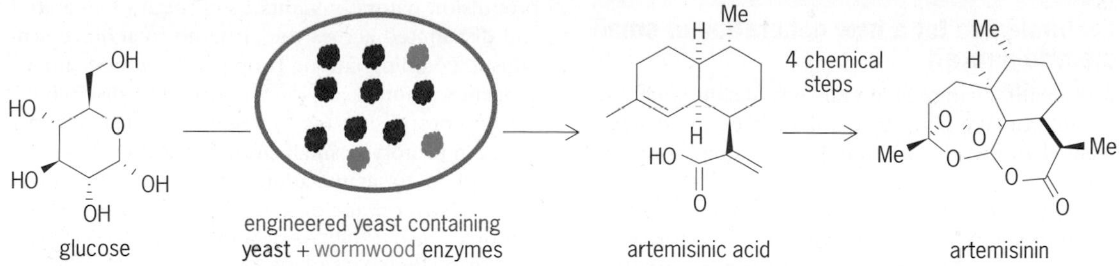

Fig. 7. Semibiological synthesis of artemisinin.

platinum, and cerium oxide create catalytically active silica-platinum and cerium oxide–platinum interfaces in each crystal (**Fig. 6**). These nanoparticles efficiently convert ethylene and methanol into propanal, an industrially important molecule.

Tandem catalysis in "designer" living organisms. Enzymes routinely work in tandem to carry out multiple reactions in a single "flask" (that is, living cells). In the new field of synthetic biology, researchers genetically engineer organisms like yeasts and bacteria to produce enzymes from other plants or animals; these foreign enzymes then work with the organism's own enzymes to synthesize complicated molecules. This could be particularly useful for the manufacture of active pharmaceutical ingredients that are derived from rare or impractical plant sources. By integrating the plant enzymes responsible for making the drug into lab-grown organisms, pharmaceutical manufacturers could cheaply mass-produce drugs that were once rare and expensive.

The semibiological synthesis of artemisinin exemplifies this hybrid of bioengineering and chemical synthesis. Artemisinin is a potent antimalarial drug, but its only source is sweet wormwood, a plant whose long and unreliable growing cycle has led to shortages and price fluctuations. A team of researchers from academia and industry has engineered a new yeast organism in which key enzymes from the sweet wormwood plant work in tandem with native yeast genes to produce artemisinic acid, a close relative of the drug. Artemisinic acid can be converted to artemisinin in just a few chemical steps (**Fig. 7**). Mass production of the drug began in early 2013 using this process.

"Hybrid" types of tandem catalysis. In addition to the types of reactions just described, it is also possible to design tandem reactions that incorporate more than one type of catalyst (homogeneous, heterogeneous, or enzyme). Such combinations are used in polymer synthesis as well as in the synthesis of fine chemicals, and they can combine many of the advantages of each type of catalysis.

The future of tandem catalysis. Tandem catalysis promises to be an effective strategy in the production of chemicals; however, several challenges remain before its synthetic potential can be realized. Predictive models for catalyst compatibility, methods of preventing homogeneous catalysts from interacting with each other (site isolation), and the development of more catalysts that can be combined with enzymes are needed. These challenges are related to the development of an approved understanding of the individual catalysts, a clearer picture of how two catalysts interact with each other, and the synthesis of better catalysts based upon these principles. Similar to the situation in living systems, where multiple enzymes operate in highly complex tandem systems, combining multiple transformations into a single operation will make the movement of chemical processes from the laboratory to large-scale commercial production more effective, economical, and environmentally friendly. The dream transformation for chemists remains the single-step conversion of simple feedstocks into whatever product is needed by society. Tandem catalysis is one of the best strategies we have toward that ultimate goal.

For background information *see* CATALYSIS; ENZYME; HETEROGENEOUS CATALYSIS; HOMOGENEOUS CATALYSIS; HYDROFORMYLATION; HYDROGENATION; NANOPARTICLES in the McGraw-Hill Encyclopedia of Science & Technology.

Daniel J. Weix, Rachel E. Kelemen, Laura K. G. Ackerman

Bibliography. D. E. Fogg and E. N. dos Santos, Tandem catalysis: A taxonomy and illustrative review, *Coord. Chem. Rev.*, 248:2365–2379, 2004, DOI:10.1016/j.ccr.2004.05.012; A. S. Goldman et al., Catalytic alkane metathesis by tandem alkane dehydrogenation-olefin metathesis, *Science*, 312:257–261, 2006, DOI:10.1126/science.1123787; M. C. Haibach et al., Alkane metathesis by tandem alkane-dehydrogenation-olefin-metathesis catalysis and related chemistry, *Accounts Chem. Res.*, 45:947–958, 2012, DOI:10.1021/ar3000713; Z. J. A. Komon et al., Triple tandem catalyst mixtures for the synthesis of polyethylenes with varying structures, *J. Am. Chem. Soc.*, 124:15280–15285, 2002, DOI:10.1021/ja0283551; C. J. Paddon et al., High-level semi-synthetic production of the potent antimalarial artemisinin, *Nature*, 496:528–532, 2013, DOI:10.1038/nature12051; D.-K. Ro et al., Production of the antimalarial drug precursor artemisinic acid in engineered yeast, *Nature*, 440:940–943, 2006, DOI:10.1038/nature04640; J.-C. Wasilke et al., Concurrent tandem catalysis, *Chem. Rev.*, 105:1001–1020, 2005, DOI:10.1021/cr020018n; Y. Yamada et al., Nanocrystal bilayer for tandem catalysis, *Nat. Chem.*, 3:372–376, 2011, DOI:10.1038/nchem.1018.

Technologies for a new generation of small electric aircraft

A dramatic change in aircraft propulsion is just beginning to take place that will likely be more significant than the change from internal combustion to turbine-based engines that took place in the 1950s. Electric propulsion has different characteristics than existing solutions, offering the potential for opening up entirely new types of aircraft configurations with capabilities that have long been desired but were impractical. Most significantly, current aviation propulsion systems have been limited to isolated and centralized integrations, whereas electric propulsion naturally wants to be highly integrated and distributed across the airframe to achieve synergistic coupling among propulsion, aerodynamics, acoustics, control, and even structural disciplines. Electric propulsion technologies are already being tested in prototype small aircraft, but they will soon be applied to many additional aviation products where shorter-range or -endurance missions are acceptable. *See* HYBRID ELECTRIC AND UNIVERSALLY ELECTRIC AIRCRAFT CONCEPTS.

Prototype aircraft. The age of electric propulsion for aviation was ushered in over the past few years through compelling flight demonstrations. In 2011, the Green Flight Challenge (**Fig. 1**) provided the largest aviation prize of all time, and it has been followed by new concepts such as the Airbus E-Fan, which is to go into production as both a two-seat electric trainer and a four-seat hybrid electric general aviation aircraft. Applying this new technology to aviation markets involving smaller aircraft permits more rapid introduction as new certification standards are developed, and the consequences of early-technology shortcomings are less severe. Electric propulsion is also quickly being applied to uninhabited aerial vehicles (UAVs) because of the immediate benefits compared to current small internal combustion engines. Over the coming years, it is likely that, with improved battery and fuel-cell technologies, electric propulsion will be applied to regional jet aircraft that fly shorter routes with a smaller number of passengers than large commercial airliners.

Fig. 1. The Team Pipistrel Taurus G4 purpose-built, dual-fuselage aircraft for the Green Flight Challenge. The $1.5 million prize was won by being able to fly 200 mi (322 km) and achieve an average speed of 108 mi/h (174 km/h) with an equivalent energy use of over 100 mi/gal (42.5 km/L) while carrying a four-passenger payload. (*Photo courtesy of NASA, credit Bill Ingalls*)

Fig. 2. The S2 vertical-takeoff-and-landing (VTOL) personal air vehicle (PAV) concept takes advantage of scale-free electric propulsion to distribute compact electric motors and propellers across the airframe. This permits the aircraft to achieve robust and redundant control in all modes of flight for higher safety than existing helicopters. The distributed propellers operate at low tip speeds at takeoff and landing for dramatically quieter operations, then rotate to provide forward flight, and fold at cruise speeds to achieve an aerodynamic efficiency (lift-to-drag ratio) four times better than that of helicopters. (*Photo art courtesy of Joby Aviation*)

Differences from existing propulsion. Electric propulsion offers improved characteristics over existing solutions in all areas except for one substantial weakness, the specific energy of the available energy storage options (batteries and fuel cells), which are about 50 times heavier than hydrocarbon fuels. However, this is partially compensated by the superior efficiency of electric motors. A typical internal-combustion aviation engine is able to convert about 28% of the stored fuel energy into shaft power, with large turbines near 50% efficiency. However, aviation electric motors are currently able to achieve 95% efficiency, with their digital controllers and battery-management systems only incurring another 5% in losses. This two- to threefold improvement in efficiency means that less energy needs to be stored, and that far lower greenhouse gas emissions are produced. Not only is less energy used, the cost of electricity versus aviation-grade fuel is also 30–70% lower (depending on type of electricity rates and regional location). Electric aircraft also have the potential to achieve lower aerodynamic drag. With both the amount and the relative cost of the energy combined, it is possible to achieve a 10-fold decrease in energy cost by applying electric propulsion to aviation. Because energy cost is approximately 50% of the total operating cost of general aviation aircraft and 50% of the direct operating cost of commercial airliners, achieving this order-of-magnitude reduction in cost is highly desirable. Several other

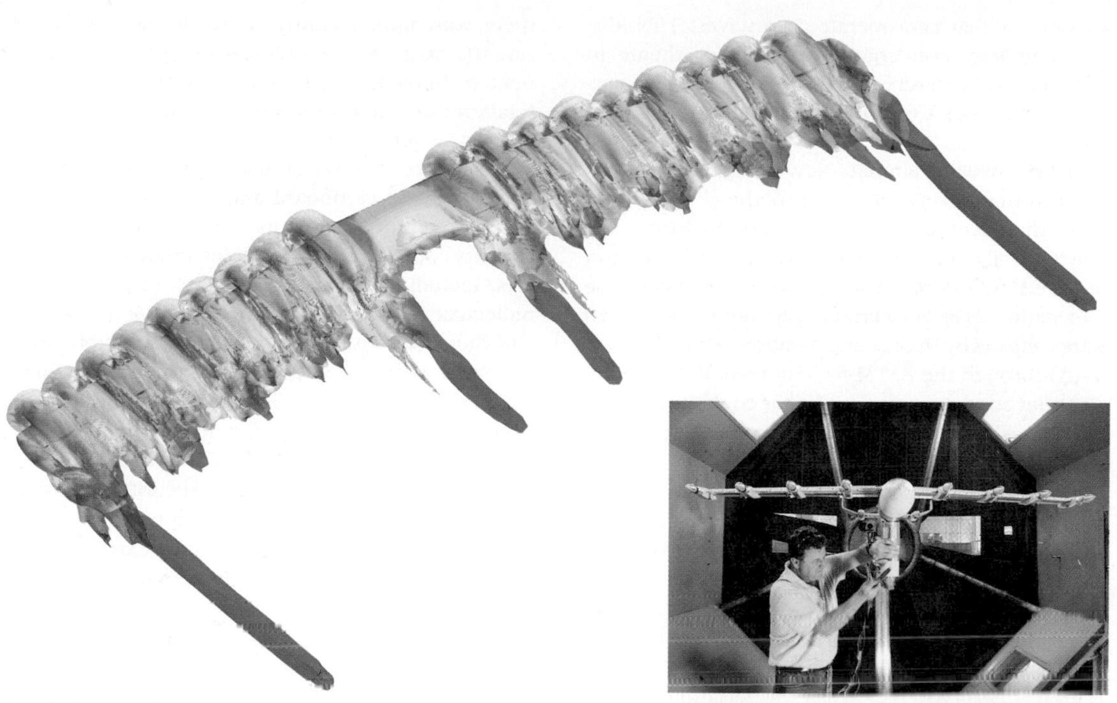

Fig. 3. Computational fluid dynamics (CFD) simulation, showing flow patterns, and wind-tunnel testing (inset) of the NASA LEAPTech general aviation concept, which focuses on achieving high aircraft efficiency as well as improved ride quality. By tightly coupling the wing lift and distributed propulsion thrust, the aircraft is able to achieve the lift required for takeoff and landing with a dramatically smaller wing at low speeds. The threefold reduction in wing area from typical general aviation aircraft means that, if a wind gust is experienced, the occupants avoid the turbulence bumps and experience the same ride comfort as those aboard regional jet aircraft. (*Photo courtesy of NASA and Joby Aviation, credit Alex Stoll*)

unique features make electric propulsion a compelling alternative, such as the fact that there is no decrease in power with altitude, because electric motors are not air-breathing or dependent on the air density. Electric motors are also far quieter and have very high reliability because their only moving parts are the shaft bearings. However, their most intriguing difference is that, unlike other propulsion systems, they are a scale-free technology. *See* GREEN AVIATION.

Scale-free technology. As turbine engines are scaled to smaller sizes, they experience poorer performance as a result of lower Reynolds numbers and larger effective clearances for the same manufacturing tolerances. Internal combustion engines suffer similar tolerance issues, while also experiencing unfavorable piston-chamber surface area-to-volume ratios (cube-square scaling). These issues result in lower fuel efficiency, reliability, and specific power as turbines and internal combustion engines are decreased in size. However, electric motors do not experience these performance penalties at smaller scale, and effectively maintain their same high ratio of power to weight, reliability, and high efficiency with scale independence. Electric motors also provide their full power with high efficiency over a broad revolutions-per-minute (rpm) range (other motors must operate at their maximum rpm). Coupling these characteristics with compactness, distributed propulsion is enabled, without scale penalties or mechanical complexity, to create new degrees of freedom in aircraft design. Distribution across the entire airframe lets the thrust forces be placed in locations where the aircraft drag can be reduced, where control forces can be generated without dependence on airspeed or control surfaces, or where other synergistic benefits can be achieved (**Figs. 2**, **3**).

Emerging vehicles, missions, and technologies. Electric technologies are rapidly progressing, with electric motors already achieving 4 hp/lb (6.6 kW/kg) and lithium batteries achieving 280 W·h/kg. Battery specific power is likely to keep progressing at the long-term rate of 8% per year, creating the opportunity for practical aircraft with 300-mi (483-km) range as pure electric craft and 600-mi (966-km) range as hybrid electric craft before 2025. Because the average distance traveled by general aviation aircraft is less than 300 mi, and for single-aisle commercial transports it is less than 600 mi, the low carbon emissions, low noise, and low operating costs will be applied to shorter-range missions first. However, it is also possible that electric propulsion will create the opportunity for new aviation markets such as on-demand aviation, in which aircraft are used in close-proximity operations with society to achieve improvements in mobility through new transportation options. In the near future, small aircraft may fly packages directly to homes from distribution centers for same-day, or even same-hour, service. The key to achieving such aircraft is that electric propulsion can be integrated with other rapidly progressing technology frontiers such as vehicle autonomy to achieve ultrasafe aircraft that can be operated by practically anyone, or that can operate themselves. This idea of technology convergence is a major theme in new federal research, such as NASA's Transformative Aeronautics Concepts program, announced in 2014.

Future of electric aircraft. New companies have recently formed to take advantage of the new opportunities that electric propulsion offers. However, significant challenges still exist before effective aviation products are offered. The development of certification standards is still a critical gap, but in the United States, efforts by the Federal Aviation Administration (FAA) through the ASTM F44 Subcommittee are developing consensus standards that could be in place as soon as 2016. The two most sensitive technologies for enabling electric aircraft propulsion are a combination of advanced batteries that can permit rapid discharge capability along with rapid charging, and range-extender auxiliary power units (APUs) to supplement battery range.

For background information *see* AIRCRAFT DESIGN; AIRCRAFT ENGINE PERFORMANCE; AIRCRAFT PROPULSION; BATTERY; FUEL CELL; GENERAL AVIATION; INTERNAL COMBUSTION ENGINE; MOTOR; REYNOLDS NUMBER; TURBINE PROPULSION; UNINHABITED AERIAL VEHICLE (UAV); VERTICAL/SHORT TAKEOFF AND LANDING (V/STOL) in the McGraw-Hill Encyclopedia of Science & Technology. Mark D. Moore

Bibliography. A. Gohardani, A synergistic glance at the prospects of distributed propulsion technology and the electric aircraft concept for future unmanned air vehicles and commercial/military aviation, *Progr. Aerospace Sci.*, 57:25–70, 2013, DOI:10.1016/j.paerosci.2012.08.001; M. D. Moore and B. Fredericks, Misconceptions of electric propulsion aircraft and their emergent aviation markets, American Institute of Aeronautics and Astronautics, AIAA 2014-0535, in *52d Aerospace Sciences Meeting*, National Harbor, Maryland, January 13–17, 2014, AIAA, Reston, VA, 2014, DOI:10.2514/6.2014-0535.

Terahertz time-domain spectroscopy and material dynamics

The terahertz region of the electromagnetic spectrum corresponds to frequencies above the microwave and below the infrared (**Fig. 1***a*). This region, alternately termed the far-infrared, spans approximately the frequency range 0.1–30 THz, where 1 terahertz (THz) is 10^{12} Hz, corresponding to a free-space wavelength of 300 μm, a photon energy of 0.004 eV, and a temperature of 47 K. For comparison, microwave ovens operate at approximately 2.5 gigahertz (1 THz = 1000 GHz), and the cosmic microwave background at 2.7 K corresponds to a frequency of approximately 60 GHz. When the universe was approximately 3×10^9 years old, the background radiation in the universe would have been centered at approximately 1 THz.

Development of terahertz technology. While the terahertz region of the spectrum has a long history, there was, until recently, a "terahertz gap," referring to a paucity of sources, detectors, and component technologies required for widespread use of terahertz radiation in science and technology. The interest in terahertz radiation is driven by science and applications. For example, typical packaging materials such as cardboard and Styrofoam are transparent to terahertz radiation. In addition, numerous spectroscopic features reside at terahertz frequencies, including molecular rotations in gases, intermolecular and conformational motions of proteins and biomolecules, and condensed-matter phenomena such as lattice and spin motion, charge transport, and superconductivity. Thus, transparency and spectroscopy have driven the development of terahertz technology because of the considerable potential for applications including security screening, nondestructive real-time evaluation of semiconductors or pharmaceuticals, imaging in extreme environments, and detection of biological or chemical hazards.

Indeed, this scientific interest and technological potential has spurred active research efforts during the past 20 years such that it is reasonable to claim that there is no longer a terahertz gap. Myriad technologies have driven this growth and include large facilities such as free-electron lasers and quantum cascade lasers. However, the greatest advance toward routine laboratory-based terahertz experiments has been the development of femtosecond lasers. This class of lasers includes the ubiquitous titanium-doped sapphire (Ti:sapphire) laser and, more recently, femtosecond fiber lasers. Lasers with pulse durations of less than 100 fs (1 fs = 10^{-15} s) are commercially available and exhibit exceptional long-term stability, thereby facilitating experiments. These lasers have led to the widespread development and adoption of terahertz time-domain spectroscopy, with a typical experimental schematic depicted in Fig. 1*b*.

Experimental setup. The Ti:sapphire laser operates just beyond the visible (in the near-infrared), with a center wavelength of 800 nm, corresponding to a photon energy of 1.55 eV, more than two orders of magnitude greater than that of a 1-THz photon. A portion of the Ti:sapphire laser (labeled A in Fig. 1*b*) is used to generate coherent terahertz radiation, as will be described. A second beam (labeled C) is used to coherently detect the terahertz radiation, subsequent to transmission (or reflection) through a sample of interest. Interestingly, both the Ti:sapphire laser and the generated terahertz radiation are invisible to the eye, which creates a challenge for setting up and aligning terahertz time-domain spectroscopy experiments. These two beams are sufficient for spectroscopic measurements, where changes in the terahertz pulse traversing a sample can be measured accurately and converted to the frequency domain. This technique is very powerful and widespread. However, such a two-beam setup does not enable investigations of the temporal dynamics in a sample.

Ultrafast spectroscopy dynamics requires a third pulse (labeled B in Fig. 1*b*). This pulse can be used to photoexcite, or "pump," a sample to an excited state, with the subsequent dynamics measured with the terahertz pulse used as a "probe." Pump-probe experiments using femtosecond pulses offer considerable potential to interrogate material dynamics at the fundamental timescales of electron and lattice motion. In a simple sense, the terahertz pulses being used as the probe function as a noncontact ohmmeter to monitor the temporal evolution of the conductivity of a given sample with subpicosecond resolution. In fact, probing material dynamics with terahertz pulses has become a standard laboratory technique over the past decade. More recently, as will be described, advances in terahertz generation have also enabled experiments where the terahertz pulse is used to pump, or initiate, dynamics in condensed matter.

Generation and detection of terahertz pulses. Several approaches have been developed to down convert near-infrared Ti:sapphire laser pulses to create terahertz pulses. One ubiquitous technique for coherent terahertz pulse generation is carrier acceleration in a semiconductor using a biased coplanar stripline placed on the semiconductor, typically gallium arsenide. However, nonlinear conversion techniques are more commonly used with amplified Ti:sapphire lasers (regenerative amplifiers), which are the workhorses for pump-probe studies. In a process known as difference frequency generation (DFG), two photons combine in a nonlinear crystal to generate a third photon with an energy that is the difference of the two initial photons. Using femtosecond pulses, coherent DFG is possible over a large fraction of the Ti:sapphire bandwidth, resulting in a broadband terahertz pulse. For example, a 50 fs transform-limited Ti:sapphire laser pulse at 800 nm has a bandwidth of 20 nm, corresponding to 9 THz. Coherent DFG is possible over this bandwidth, although in practice, there is a limit to the highest frequency generated because of absorption and dispersion in the nonlinear crystal. Common crystals used for DFG of terahertz radiation include zinc telluride (ZnTe), gallium selenide (GaSe), gallium phosphide (GaP), and lithium niobate ($LiNbO_3$, sometimes abbreviated LNO). These materials have sufficiently large nonlinear electrooptic coefficients for "efficient" terahertz generation.

These same crystals (usually ZnTe or GaP) are also used for coherent detection of terahertz pulses. This process is coherent in that a portion of the near-infrared beam is used to gate the detector. That is, the terahertz pulse can only be measured when it overlaps temporally and spatially with the gate beam. Most commonly, a terahertz pulse induces birefringence in a crystal (the Pockels effect) that modifies the polarization of the copropagating near-infrared gate beam (C in Fig. 1*b*). Mechanically delaying the relative arrival time of the terahertz transient and the much shorter near-infrared pulse provides a measurement of the terahertz electric field amplitude and phase (**Fig. 2** provides an example).

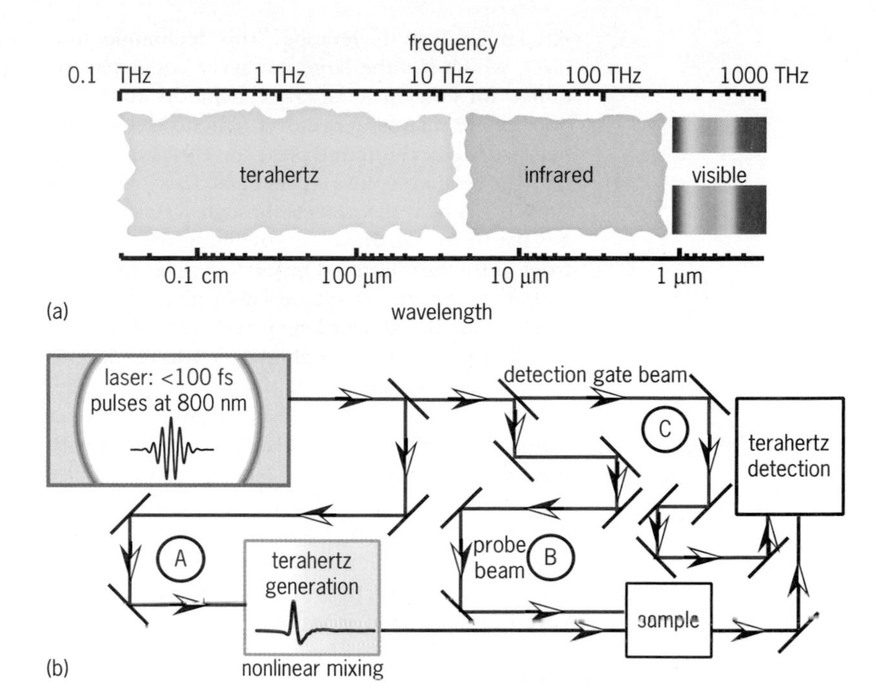

Fig. 1. Terahertz spectroscopy. (*a*) Portion of the electromagnetic spectrum, showing the terahertz region, which spans approximately the range 100 GHz–30 THz. Recently, peak fields approaching 1 MV/cm have been achieved in this range. (*b*) Schematic of terahertz time-domain spectroscopy setup, where terahertz generation, detection, and sample excitation are coherent as they derive from the output of a femtosecond laser. With recent developments in terahertz pulse generation, the terahertz beam (A) and the optical beam (B) can serve as either the pump or the probe to investigate material dynamics, although, as depicted here, the terahertz beam is the pump and the optical beam is the probe.

The previous use of the term "efficient" to describe terahertz pulse generation is relative, with a conversion efficiency of approximately 10^{-6} being quite reasonable, since gated detection enables a signal-to-noise ratio of several thousand. However, if the goal is to use the terahertz pulse to initiate dynamics (that is, to serve as the pump) in materials, it is desirable to have as large a pulse fluence (and corresponding electric field) as possible.

An important advance that has enabled creating terahertz pulses of sufficient intensity to initiate dynamics in condensed matter is a technique called

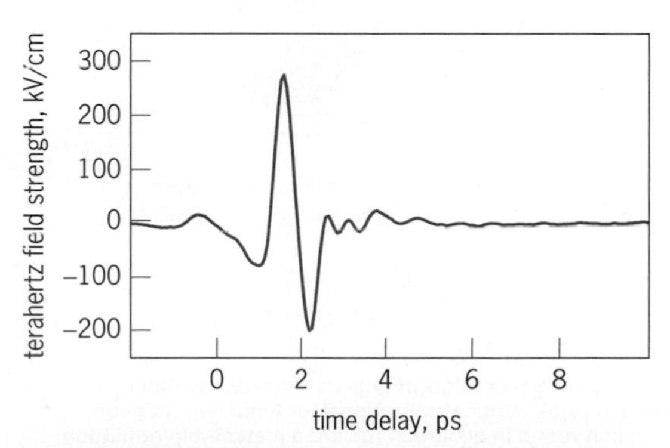

Fig. 2. Terahertz pulse, with the electric field plotted as a function of time. This pulse was generated using tilted-pulse-front generation in lithium niobate ($LiNbO_3$ or LNO), with a peak field of approximately 300 kV/cm.

tilted-pulse-front generation. This technique uses LNO, which has the large nonlinear coefficient required for efficient terahertz generation. However, poor phase matching makes it difficult to use LNO because the near-infrared pulse travels faster in LNO than the terahertz pulse it generates. This behavior is similar to a boat that travels through water creating a wake behind it. However, by tilting the intensity front of the near-infrared generating pulse by an appropriate angle (to match the wake), phase matching can be maintained over long propagation distances, resulting in high-peak terahertz electric fields. This pulse front tilting is accomplished using a grating which, as is well known, leads to angular dispersion. For short pulses, this also leads to pulses with tilted intensity fronts. Figure 2 shows a typical pulse generated using this technique. It has a peak field of approximately 300 kV/cm, with spectral content approximately over the range 0.1–2.0 THz. Fields approaching 1 MV/cm, with a 1-mm beam diameter, have been obtained using this technique. It is certain that this value will continue to increase in the coming years with peak fields of 10 MV/cm in sight. Taken together, the advances in ultrafast lasers and terahertz generation enable routine laboratory use of far-infrared radiation to pump or probe materials with subpicosecond resolution. There is still a great deal to learn with regard to how intense terahertz pulses interact with matter and thus, in the near term, new insights are likely to come from time-resolved nonequilibrium studies.

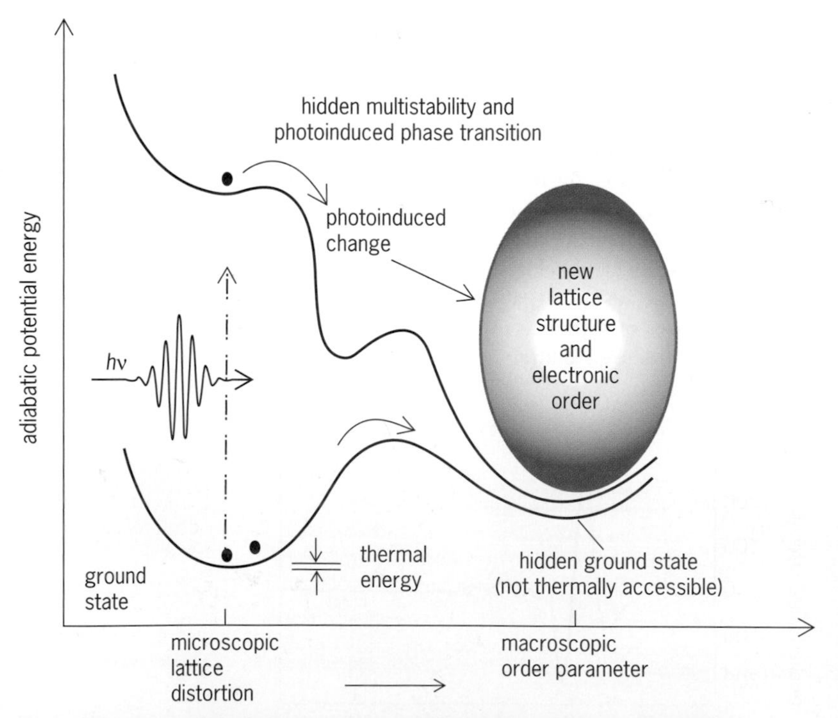

Fig. 3. Diagram of adiabatic energy versus configurational coordinate, indicating nonequilibrium pathways to hidden order. Optical, vibrational, or tunneling-induced excitation in a complex material can result in evolution toward a metastable or hidden phase with properties that are not thermally accessible. (The quantum excitation has energy $h\nu$, where ν is its frequency and h is Planck's constant.) (*Adapted from D. N. Basov et al., Electrodynamics of correlated electron materials, Rev. Mod. Phys., 83:471–541, 2011, DOI:10.1103/RevModPhys.83.47*)

Nonequilibrium dynamics studies. Ultrafast optical spectroscopy studies of nonequilibrium dynamics is a vast field spanning physics, chemistry, and materials science. An emerging area of considerable interest is that of investigations of quantum dynamics in correlated electron materials (CEMs). This class of materials includes, as examples, materials that exhibit superconductivity, insulator-to-metal transitions, or novel magnetic phases. What is common to correlated materials is that the electrons strongly "feel" the presence of the other electrons. For example, the motion of an electron in a material can be suppressed because of strong Coulomb repulsion, leading to insulating behavior. Optical techniques (and, in particular, those involving far-infrared radiation) have much to say about the optical and electronic properties of CEMs because photons probe relevant excitations such as carrier transport, lattice vibrations, spin waves, and gaps that can arise in the electronic excitation spectrum. An exciting aspect of CEMs, over and beyond their fascinating ground-state properties, is the large changes in their optical and electronic properties in response to small external perturbations, such as temperature, magnetic or electric fields, or strain. This sensitivity arises from competition between the charge, lattice, spin, and orbital degrees of freedom, and is ultimately what determines the bulk properties. Controlling this competition is an important step toward selective and deterministic control of the resultant bulk behavior. Along these lines, ultrafast optical spectroscopy plays a particularly important role, since dynamics can be initiated, probed, and controlled at the fundamental timescales of electronic and atomic motion. Technological advances in terahertz pulse generation and, more generally, in ultrafast optical spectroscopy, spanning frequencies from the terahertz region through x-rays, have allowed for increasingly detailed studies using mode-selective excitation and probing of material dynamics.

Mode-selective excitation. Mode-selective excitation refers to the ability to initiate dynamics in a solid through perturbation of a particular degree of freedom. Practically speaking, this simply means choosing the center frequency of the excitation pump pulse. For example, with short pulses in the mid-infrared, it is possible to excite specific vibrational modes to drive novel dynamics. The high-field terahertz pulses previously described also allow for novel mode-selective excitation. One example is impact ionization, whereby electrons are accelerated by the terahertz electric field and acquire sufficient kinetic energy to excite, through collisions, additional carriers on a femtosecond timescale. Terahertz pulses can also induce tunneling of electrons, a process that competes with collisional effects. Finally, the magnetic component of a terahertz pulse can also initiate dynamics. For example, for a pulse with a peak electric field of 1 MV/cm, the corresponding magnetic field is 0.33 tesla. Even with magnetic fields well below this magnitude, dynamics can be initiated by giving a "kick" to spins, leading to

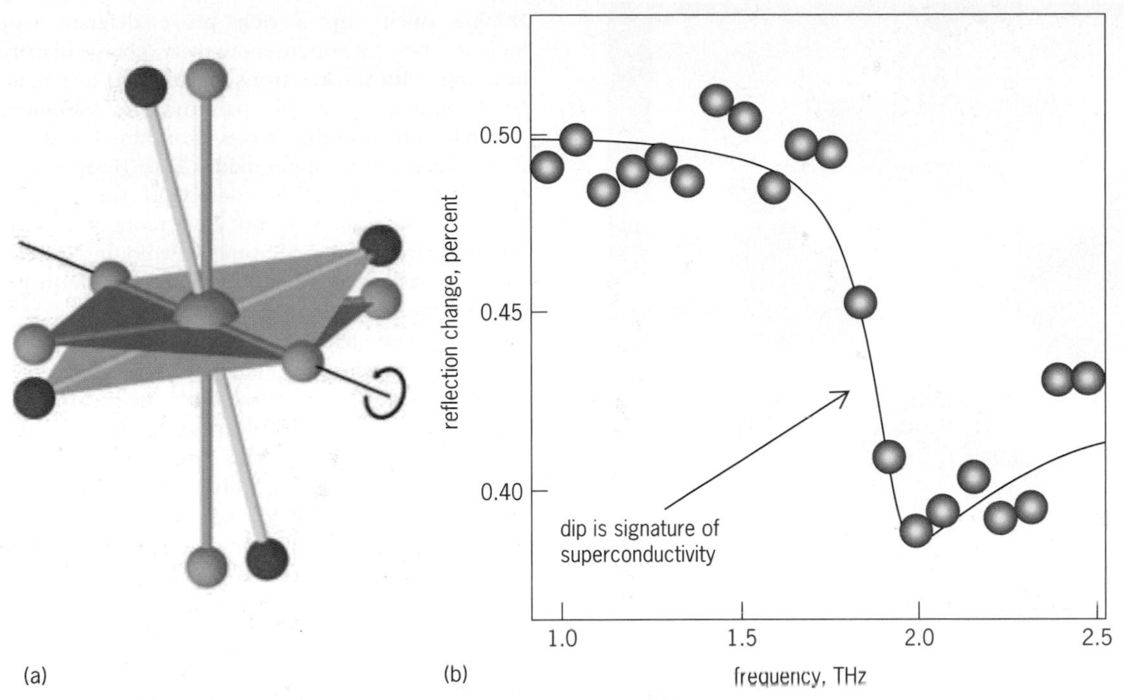

Fig. 4. Excitation by a mid-infrared pulse, leading to transient superconductivity in strontium-doped lanthanum europium copper oxide. (*a*) Depiction of the rocking motion of an elongated oxygen octahedron, which is the phonon mode excited by the pulse. The circular arrow depicts the axis about which the rocking motion occurs. The red and blue balls at the vertices of the octahedron represent oxygen atoms, depicting the extremes of the motion. Copper is at the center of the octahedron, and this rocking motion initiates an electronic transition leading to the superconducting phase. (*b*) Terahertz spectrum obtained 5 ps after the vibrational excitation, with a spectroscopic feature that is indicative of transient superconductivity. (*Adapted from D. Fausti et al., Light-induced superconductivity in a stripe-ordered cuprate, Science, 331:189–191, 2011, DOI:10.1126/science.1197294*)

precession. These phenomena are well established and have been investigated, to some extent, in semiconductors using continuous-wave terahertz sources, but new surprises can be expected with the development of high field, single-cycle terahertz pulses, particularly since the temporal evolution of the dynamics resulting from mode-selective excitation can be followed in real time with femtosecond resolution using probe pulses.

Creation of metastable phases. Investigation of coherent quantum dynamics in CEMs is an emerging area where the goal is to use mode-selective excitation to investigate novel pathways to drive phase transitions, and create metastable phases with bulk properties that are not thermally accessible. Correlated electron materials (and transition-metal oxides in particular) are being extensively studied because of their fascinating properties and aforementioned sensitivity to external perturbation that are manifested dramatically in the optical excitation spectrum. **Figure 3** schematically depicts this rather general idea, showing an energy diagram as a function of a generalized configurational coordinate that could be, for example, a specific lattice distortion. Photo excitation can lead to structural evolution, which in turn drives the material toward a new electronic state with a well-defined order parameter, as depicted on the right side of the figure. These states are hidden in that they originate from a nonequilibrium perturbation and are not thermally accessible. Of course, the excitation need not be to an excited-state manifold, but could be driven from structural excitation or tunneling within the ground state using mid-infrared or terahertz pulses, respectively. The idea of mode-selective control is quite analogous to efforts in quantum chemistry, where the goal is to direct a chemical reaction along an optimized pathway to maximize the overall efficiency. In mode-selective quantum dynamics, the goal is to achieve reversible switching between the parent and product phases. In the following sections, recent examples will be presented that demonstrate some of the progress that has been made during the past several years using terahertz pulses to achieve these goals.

Production of transient superconductivity. As an example, imagine being able to increase the temperature at which superconductivity occurs, or even uncover a hidden superconducting phase (along the lines of Fig. 3), by mode-selective excitation of a given material. Recently, this capability has been demonstrated using mode-selective vibrational excitation of a high-temperature superconductor. High-temperature superconductivity remains one of the most fascinating and enigmatic topics in condensed-matter physics, with the microscopic origin of superconductivity still under active investigation. The undoped parent compounds of the high-temperature superconductors are Mott insulators. This simply means that electron correlations (as previously described) are responsible for the insulating behavior.

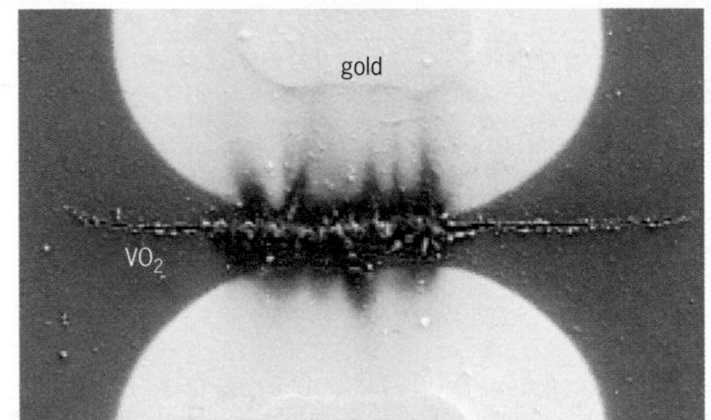

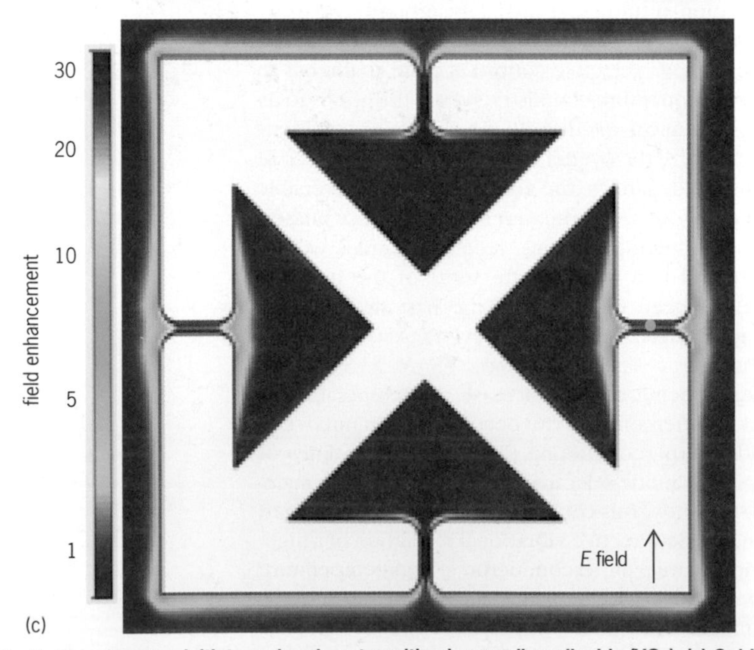

Fig. 5. Experiment to initiate and probe a transition in vanadium dioxide (VO_2). (*a*) Gold metamaterial resonators on VO_2, which are resonant with the terahertz electric field, E_{THz}. (*b*) Enlargement of one of the 1.5-μm gaps in the resonators. The damage in these gaps derives from terahertz-induced excitation where the resonator structure localizes and enhances the electric field. (*c*) Simulation showing the local electric field enhancement, with a peak enhancement of 30 in the horizontal gap (orange). (*Adapted from M. Liu et al., Terahertz-field-induced insulator-to-metal transition in vanadium dioxide metamaterial, Nature, 487:345–348, 2012, DOI:10.1038/nature11231*)

Doping opens up a rich phase diagram that includes, besides superconductivity, charge-density ordering, with the electrons localizing at a specific lattice site in a regular pattern. The particular material studied in **Fig. 4** was strontium-doped lanthanum europium copper oxide ($La_{1.8-x}EuSr_xCuO_4$, with $x = 0.125$). This compound was chosen precisely because it is not superconducting (at the experimental temperatures), exhibiting instead striped charge-density order. However, perturbation of the lattice resulted in "melting" of this ordered phase, leading to spectroscopic features at terahertz frequencies consistent with a transient superconductivity. Figure 4*a* shows the rocking-like vibrational mode of an oxygen octahedron excited with mid-infrared pulses. This rocking motion modulates the electronic coupling between adjacent octahedral units, thereby initiating the transition to the superconducting phase. The subsequent dynamics were monitored with terahertz probe pulses. The terahertz spectrum in Fig. 4*b* was obtained 5 picoseconds after the arrival of the vibrational excitation pulse. The pronounced dip in the reflectivity corresponds to tunneling of superconducting pairs between copper-oxygen planes, the subunits responsible for superconductivity. This data, obtained with time-resolved terahertz techniques, provides a strong indication of transient superconductivity and is an exciting example in line with the idea of photoinduced phase transitions presented in Fig. 3.

Study of a transition in vanadium dioxide. One final example will be considered that highlights future directions in terahertz spectroscopy and dynamics, relying on twenty-first-century advances in high-field terahertz spectroscopy, electromagnetic metamaterials, and high-quality epitaxial growth of correlated oxides. Vanadium dioxide (VO_2) is a correlated transition-metal oxide that exhibits, with increasing temperature, an insulator-to-metal transition at 340 K (67°C or 152°F). Associated with this electronic phase transition is a structural transition. This strong coupling naturally leads to a quandary: Do the structural changes drive the electronic transition or is it the electronic correlations that trigger the structural changes? This interplay between structure and electronic properties has resulted in VO_2 serving as a testbed for ultrafast dynamics aimed at understanding and controlling VO_2. The results of **Fig. 5** are particularly germane, being an example where terahertz pulses were used to initiate and probe the insulator-to-metal transition in VO_2. It was found that terahertz pulses with peak field strengths of approximately 300 kV/cm (similar to those shown in Fig. 2) were not sufficient to initiate the transition to the metallic state. Thus, gold resonators were fabricated on the VO_2 films, serving as subwavelength "lightning rods" that led to localization and enhancement of the terahertz field. The x-shaped resonator structure in Fig. 5*a* is simply an *LC* circuit, where the inductance is given by the loops and the capacitance by the gaps. The outer dimension of this resonator is

75 μm, and the capacitive gaps (Fig. 5*b*) are 1.5 μm across, more than two orders of magnitude less than the wavelength of the incident terahertz radiation. As the simulation in Fig. 5*c* shows, the terahertz electric field is concentrated in the gaps and is enhanced by a factor of approximately 30. This leads to peak fields of several megavolts per centimeter in the gaps. It was experimentally observed that this field was sufficient to initiate the insulator-to-metal transition. Indeed, pronounced damage occurs to the VO_2 at the highest terahertz fields (Fig. 5*b*), something that was unimaginable a few years ago. These results demonstrate that high-field terahertz pulses can initiate quantum dynamics in complex materials and, further, that integration with metamaterial resonators provides a unique route to control and manipulate terahertz fields at microscopic to nanoscopic length scales.

Prospects. The field of terahertz time-domain spectroscopy, and, more generally, ultrafast optical spectroscopy, continues to advance at a rapid pace, taking full advantage of continual improvements in femtosecond laser technology. Numerous exciting possibilities are on the horizon, including nonlinear terahertz imaging and the coupling of terahertz pump sources with time-resolved probes, such as femtosecond x-rays or electron beams. Indeed, efforts are already underway at the Linac Coherent Light Source (LCLS) at SLAC National Accelerator Laboratory, and at the Advanced Photon Source (APS) at Argonne National Laboratory, using terahertz pulses to initiate quantum dynamics that are subsequently probed with x-rays. The future is bright for terahertz time-domain spectroscopy, with scientific and technological achievements in the coming years that are certain to be most impressive.

For background information *see* CHARGE-DENSITY WAVE; COSMIC BACKGROUND RADIATION; DIFFRACTION GRATING; ELECTROOPTICS; INFRARED RADIATION; INFRARED SPECTROSCOPY; LASER; LASER PHOTOCHEMISTRY; LASER SPECTROSCOPY; LATTICE VIBRATIONS; MOLECULAR STRUCTURE AND SPECTRA; NONLINEAR OPTICS; OPTICAL MODULATORS; OPTICAL PULSES; PHASE TRANSITIONS; SPIN-DENSITY WAVE; SUBMILLIMETER-WAVE TECHNOLOGY; SUPERCONDUCTIVITY; SYNCHROTRON RADIATION; ULTRAFAST MOLECULAR PROCESSES in the McGraw-Hill Encyclopedia of Science & Technology.

Richard D. Averitt

Bibliography. D. Fausti et al., Light-induced superconductivity in a stripe-ordered cuprate, *Science*, 331:189–191, 2011, DOI:10.1126/science.1197294; M. C. Hoffmann and J. A. Fülöp, Intense ultrashort terahertz pulses: Generation and applications, *J. Phys. D: Appl. Phys.*, 44:083001 (17 pp.), 2011, DOI:10.1088/0022-3727/44/8/083001; P. U. Jepsen, D. G. Cooke, and M. Koch, *Laser Photon. Rev.*, Terahertz spectroscopy and imaging: Modern techniques and applications, 5:124–166, 2011, DOI:10.1002/lpor.201000011; M. Liu et al., Terahertz-field-induced insulator-to-metal transition in vanadium dioxide metamaterial, *Nature*, 487:345–348, 2012, DOI:10.1038/nature11231; K. Nasu (ed.), *Photoinduced Phase Transitions*, World Scientific, Singapore, 2004; J. Orenstein, Ultrafast spectroscopy of quantum materials, *Phys. Today*, 65(9):44–50, September 2012, DOI:10.1063/PT.3.1717; J. M. Rondinelli, S. J. May, and J. W. Freeland, Control of octahedral connectivity in perovskite oxide heterostructures: An emerging route to multifunctional materials discovery, *MRS Bulletin*, 37:261–270, 2012, DOI:10.1557/mrs.2012.49.

The Dark Energy Survey and Camera

The Dark Energy Survey (DES) Collaboration will study the accelerating expansion of the universe. To perform the 5000-square-degree wide field and 30-square-degree supernova surveys, the collaboration designed and built the Dark Energy Camera (DECam), a 3-square-degree, 570-megapixel charge-coupled-device (CCD) camera. During 2011 and 2012, DECam was installed on the Blanco 4-m telescope at the Cerro Tololo Inter-American Observatory (CTIO) near La Serene, Chile. Funding of the camera construction came primarily from the U.S. Department of Energy, with significant contributions from the collaborating institutions. Funding for operations of CTIO is provided by the National Science Foundation. The survey commenced in August 2013 and is expected to take 525 nights over 5 years to complete. DECam is available for use by the U.S. astronomy community when it is not allocated to the Dark Energy Survey.

The Dark Energy Survey. The 2011 Nobel Prize in physics was awarded to Saul Perlmutter, Adam Reiss, and Brian Schmidt for the discovery that the expansion of the universe is accelerating. That discovery led to our present picture of the cosmos as one in which approximately 5% of the total energy is contained in "normal" matter that we are very familiar with, 25% unseen "dark matter" detectable only through the effects of its gravity, and 70% "dark energy." Dark energy is merely the name that we give to whatever the phenomenon is that is responsible for the acceleration. We do not have a "law of physics" that can explain its cause.

The Dark Energy Survey (DES) is an international collaboration, with over 200 scientists from the United States, the United Kingdom, Spain, Brazil, Switzerland, and Germany. Their common interest is to understand the physics of the accelerating expansion of the universe. They will use four complementary methods for studying dark energy and their collaboration is the first to study them all with data from the same survey. These four techniques are measurements of the population of galaxies and galaxy clusters; the spatial distribution and correlation of the position of galaxies due to baryon acoustic oscillations in the early universe; weak gravitational lensing, which provides the distribution of dark matter as well as normal matter; and the study of the brightness versus redshift of Type Ia supernovae. To make these measurements we will analyze a map of the positions and shapes of 300 million galaxies in a 5000-square-degree "wide-field" area of the southern galactic cap to redshift 1.2, approximately 8 billion light-years distant, and by studying approximately 4000 new Type Ia supernovae discovered in ten 3-square-degree fields. For comparison, the entire sky has 41,253 square degrees and the full moon has an area of about 0.25 square degrees. Note that we avoid observing in the direction of the Milky Way because we cannot detect extragalactic objects through it in our wavelength range. In order to carry out these surveys, the DES Collaboration constructed a new wide-field digital camera named the Dark Energy Camera or DECam (variously pronounced "d-e-cam" in the United States and "dec-am" in the United Kingdom), for the prime focus of the 4-m Victor Blanco Telescope.

The Dark Energy Camera. The Dark Energy Camera was designed and built from 2004 to 2010. Scientists at the Fermi National Accelerator Laboratory led the project, which was primarily funded by the U.S. Department of Energy along with significant contributions from U.S. universities, the United Kingdom, Spain, Brazil, Germany, and Switzerland. Everything about the camera is large-scale (**Fig. 1**). The camera uses five fused-silica optical elements to attain an f/2.7, 2.2-degree-wide image on the focal plane. The first and largest of these lenses is nearly a meter across and weighs 176 kg, making it the largest optical corrector component currently in use in astronomy. The two-blade shutter has a 60-cm diameter. The Dark Energy Survey records images using five filters: g, r, i, z, and Y-band, spanning 400–1080 nm. With a 62-cm diameter, these are the largest in use in the astronomical community. The images themselves are taken on a focal plane that has a 42-cm radius and is populated with 62 2048 × 4096-pixel 250-μm-thick, fully depleted, red-sensitive CCDs. DECam can be "read out" (analog to digital signal conversion) in 20 s, faster than most CCD cameras currently in use.

Fig. 1. The Dark Energy Camera design. The prime focus cage and support "spider" provide the frame that connects the camera to the upper part of the telescope. Focus and alignment is provided by a 6-arm "hexapod" (white) that connects the cage to the "barrel" (dark blue), which houses the corrector optics. The optical corrector is oriented as if the primary mirror was to the left of the camera. The filter changer and shutter are housed in the barrel just to the left of the hexapod. Some filters (green) protrude out of the barrel. The CCD imager dewar in the upper right (also green) is mostly obscured by the readout electronics crates (scarlet).

DECam construction actually started in 2008 and was finished in 2011. DECam was first assembled and tested at Fermilab on a full-scale reproduction of the telescope top rings called the "telescope simulator." The camera and its infrastructure support hardware were delivered to CTIO in a series of shipments from 2010 to late 2011.

The Blanco telescope is an equatorial mount Ritchey-Chrétien telescope with a 15,400-kg primary mirror and prime focus cage supported on a steel Serrurier truss structure. It was commissioned in 1974 at a time when observers operated the shutter and changed photographic plates while riding around in the prime-focus cage. This is relevant because the sturdy telescope structure can support the mass of DECam at the prime focus and the wide-field design of the primary mirror provides one of the few existing platforms for efficient wide-field surveys. CTIO led the camera installation with intense and active participation by the DECam design and construction team. Installation started in January 2011 and was completed in September 2012. **Figure 2** shows a photo of the Dark Energy Camera mounted at the prime focus of the Blanco telescope. The "Official First Light" ceremony was held on September 12, 2012. DECam achieved 0.8 arcsec seeing across the focal plane within 45 min of its first exposure. **Figure 3** is a three-color composite image from the first light night. DECam was commissioned during September and October 2012. A survey and instrument testing period called "Science Verification" was carried out by the DES Collaboration and "Community Astronomers" from November 2012 to February 2013. This period was used to commission the camera, re-commission the telescope, and verify that the camera and telescope could produce the high-quality exposures required for the survey.

Dark energy measurements. To obtain the measurements of positions and shapes of faint, distant galaxies required for the science of the wide-field survey, DES divided the 5000-square-degree area into roughly 1600 slightly overlapping camera "pointings." A "tiling" is comprised of all of the pointings that cover the 5000-square-degree DES field once. Ten such tilings were defined, with each tiling offset from the other tilings by a fraction of the focal-plane diameter. The offsets ensure that the small gaps between the CCDs seen in Fig. 3 will not always coincide with the same area of the sky. The exposures in g-, r-, i-, and z-band filters are of 90-s duration and the exposures in the Y-band filter are of 45-s duration. By taking multiple exposures of a particular place on the sky, rather than one long exposure, we can better identify and compensate for differing observing conditions. After DES has finished with the 10 tilings, DECam will have knitted together a seamless picture of the entire 5000-square-degree survey field. Galaxies will be discovered that are as faint as the 24th magnitude in z-band. The full survey will take at least five seasons to complete.

DES discovers Type Ia supernovae from their appearance in the 10 supernova fields, which are revisited every four to six nights. The stars that undergo supernovae are initially too dim to distinguish from other stars in their host galaxies. After they explode they rise in brightness for 5–10 days and then fade by about a factor of 100 over the next 40–70 days. At peak brightness, they may be brighter than the host galaxy itself. To find the supernovae, DES "subtracts" the images from each exposure of the supernovae fields from "template images" previously made for each of the fields. The new supernovae appear in this "difference image," as do other objects that vary in brightness, including some stars, galaxies, and other types of supernovae. The Type Ia supernovae have a characteristic "brightness curve" over their 35–70 day lifetime that allows them to be distinguished from the other variable objects. The redshifts for the supernovae are determined from the redshift of the host galaxies, not the supernovae themselves, and are determined using spectroscopic instruments on telescopes other than the Blanco, usually well after the supernovae have faded away.

The first observing season, Y1, of the Dark Energy Survey started on August 31, 2013. DES used the whole or parts of 119 nights during that period for a total of 105 nights. The wide-field observations concentrated on four tilings of a 2000-square-degree subset of the full field. More than 14,400 high-quality exposures were recorded. In addition, the supernova fields were visited on 20–28 nights each and 2660 high-quality supernovae exposures were recorded. More than 800 Type Ia supernovae were discovered. Y1 ended on February 10, 2014. The second observing season will start in mid-August 2014.

Between observing seasons, the DES Collaboration will process the Y1 exposures and will produce a catalog of the positions, brightness's, shapes, and redshifts of the galaxies discovered so far. The collaboration is also producing the first scientific results from the Science Verification period in 2012–2013. Answers to the big questions about dark energy are expected to require several years of study to produce.

Meanwhile, DECam is being used by other scientists in the astronomical community. Recently, for instance, DECam was used to discover a new planetoid 450 km in diameter, with an orbit more than 80 times that of the distance of the Earth from the Sun. This is an example of the powerful team-up of the wide field of view of DECam coupled to the 4-m primary mirror of the Blanco telescope.

Fig. 2. The Dark Energy Camera is mounted at the prime focus of the Blanco 4-m telescope at CTIO. The primary mirror is just out of the photo, low and to the left. The camera assembly is approximately 4.0 m long and is secured to the inner telescope ring by the spider. The components of the camera are readily recognizable by comparing this photograph with those parts shown in Fig. 1.

Fig. 3. A three-color composite image of the Fornax galaxy cluster from "first light" night September 12, 2012.

[Acknowledgments: Funding for the DES Projects has been provided by the U.S. Department of Energy, the U.S. National Science Foundation, the Ministry of Science and Education of Spain, the Science and Technology Facilities Council of the United Kingdom, the Higher Education Funding Council for England, the National Center for Supercomputing Applications at the University of Illinois at Urbana-Champaign, the Kavli Institute of Cosmological Physics at the University of Chicago,

Financiadora de Estudos e Projetos, Fundação Carlos Chagas Filho de Amparo à Pesquisa do Estado do Rio de Janeiro, Conselho Nacional de Desenvolvimento Científico e Tecnológico and the Ministério da Ciência e Tecnologia, the Deutsche Forschungsgemeinschaft, and the Collaborating Institutions in the Dark Energy Survey.

The Collaborating Institutions are Argonne National Laboratories, the University of California at Santa Cruz, the University of Cambridge, Centro de Investigaciones Energeticas, Medioambientales y Tecnologicas-Madrid, the University of Chicago, University College London, the DES-Brazil Consortium, the Eidgenoessische Technische Hochschule (ETH) Zurich, Fermi National Accelerator Laboratory, the University of Edinburgh, the University of Illinois at Urbana-Champaign, the Institut de Ciencies de l'Espai (IEEC/CSIC), the Institut de Fisica d'Altes Energies, the Lawrence Berkeley National Laboratory, the Ludwig-Maximilians Universität and the associated Excellence Cluster Universe, the University of Michigan, the National Optical Astronomy Observatory, the University of Nottingham, the Ohio State University, the University of Pennsylvania, the University of Portsmouth, SLAC National Accelerator Laboratory, Stanford University, the University of Sussex, and Texas A&M University.]

For background information *see* ACCELERATING UNIVERSE; ASTRONOMICAL IMAGING; BARYON; CAMERA; CHARGE-COUPLED DEVICES; COSMOLOGY; DARK ENERGY; DARK MATTER; GALAXY, EXTERNAL; GRAVITATIONAL LENS; OBSERVATORY, ASTRONOMICAL; REDSHIFT; SUPERNOVA; TELESCOPE; UNIVERSE in the McGraw-Hill Encyclopedia of Science & Technology.

H. Thomas Diehl; Brenna L. Flaugher

Bibliography. H. T. Diehl, The Dark Energy Survey Camera (DECAM), For the Dark Energy Survey Collaboration, *Phys. Procedia*, 37:1332–1340, 2012, DOI:10.1016/j.phpro.2012.02.472; B. Flaugher, The dark energy survey, *Int. J. Mod. Phys. A*, 20:3121, 2005, DOI:10.1142/S0217751X05025917; B. L. Flaugher et al., Status of the Dark Energy Survey Camera (DECAM) project, *Proc. SPIE*, 8446:844611, 2012, DOI:10.1117/12.926216; J. Frieman, Probing the accelerating universe, *Phys. Today*, 67(4), April 2014, DOI:10.1063/PT.3.2346; E. Gates, *Einstein's Telescope: The Hunt for Dark Matter and Dark Energy in the Universe*, W. W. Norton & Co., New York, 2009; P. Melchior et al., Mass and galaxy distributions of four massive galaxy clusters from Dark Energy Survey Science Verification data, http://arxiv.org/abs/1405.4285; S. Perlmutter et al., Measurements of Ω and Λ from 42 high-redshift supernovae, *Astrophys. J.*, 517:565–586, 1999, DOI:10.1086/307221; A. Riess et al., Observational evidence from supernovae for an accelerating universe and a cosmological constant, *Astron. J.*, 116:1009–1038, 1998, DOI:10.1086/300499; C. A. Trujillo and S. S. Sheppard, A Sedna-like body with a perihelion of 80 astronomical units, *Nature*, 507:471–474, 2014, DOI:10.1038/nature13156.

The next generation of television

Television as a mass medium for public communication is only about 75 years old. Certain earlier experiments (neither viably commercial nor public) helped television to germinate, and without the prior development of wireless audio (radio), the development of television would have been severely constrained. Yet television has risen to its ubiquitous and highly influential current standing in the world with extraordinary speed while also undergoing a series of remarkable technological revolutions. The technology of television is by no means mature, however, and innovations now taking shape should soon catapult it to even greater heights.

A brief history of television technology. The Scottish inventor John Logie Baird is generally credited with the first public demonstration of a true television image—one with 30 lines of resolution, generated mechanically—in the United Kingdom in 1926. However, there were also developments around the world in Germany, Japan, the Soviet Union, and in particular the United States, where electronic television was pioneered by Philo Farnsworth (who in 1927 was first to transmit an electric image without mechanical means) and by Vladimir Zworykin (who developed the cathode-ray tube for both receivers and cameras in the 1920s and 1930s). Farnsworth and Zworykin are often credited as the co-inventors of television, but a debate over the appropriateness of that honor endures today.

This developmental and experimental work continued through the 1930s, although it was severely constrained by World War II. Building on the work of Zworykin, Farnsworth, and others led to the standard specification for an analog monochrome television picture of 525 lines and FM audio, which was adopted in 1941 by the United States and soon thereafter in many other parts of the world. Work by the National Television Systems Committee (NTSC) in conjunction with the U.S. Federal Communications Commission (FCC) resulted in allocations using 6 MHz of bandwidth in the VHF (very high frequency) band. After the war, public and commercial television became a reality around the world. Color television was introduced to the mass market in the 1950s, along with a color specification by the NTSC in 1953 that was compatible with existing black-and-white receivers, and its popularity continued to grow throughout the 1960s and 1970s. Digital television took hold in the 1990s, and high-definition television (HDTV) became a reality in the late 1990s and the first decade of the twenty-first century. (The **illustration** is a functional block diagram describing how HDTV works.)

Keeping within band limitations. It has thus already been a long road from Baird's 30-line mechanical television, and remarkable ingenuity has been required to travel it. (The **table** lists major events in television technology development.) Consider, for example, the significant amount of compression required to fit a signal into 6 MHz in 1941. It became

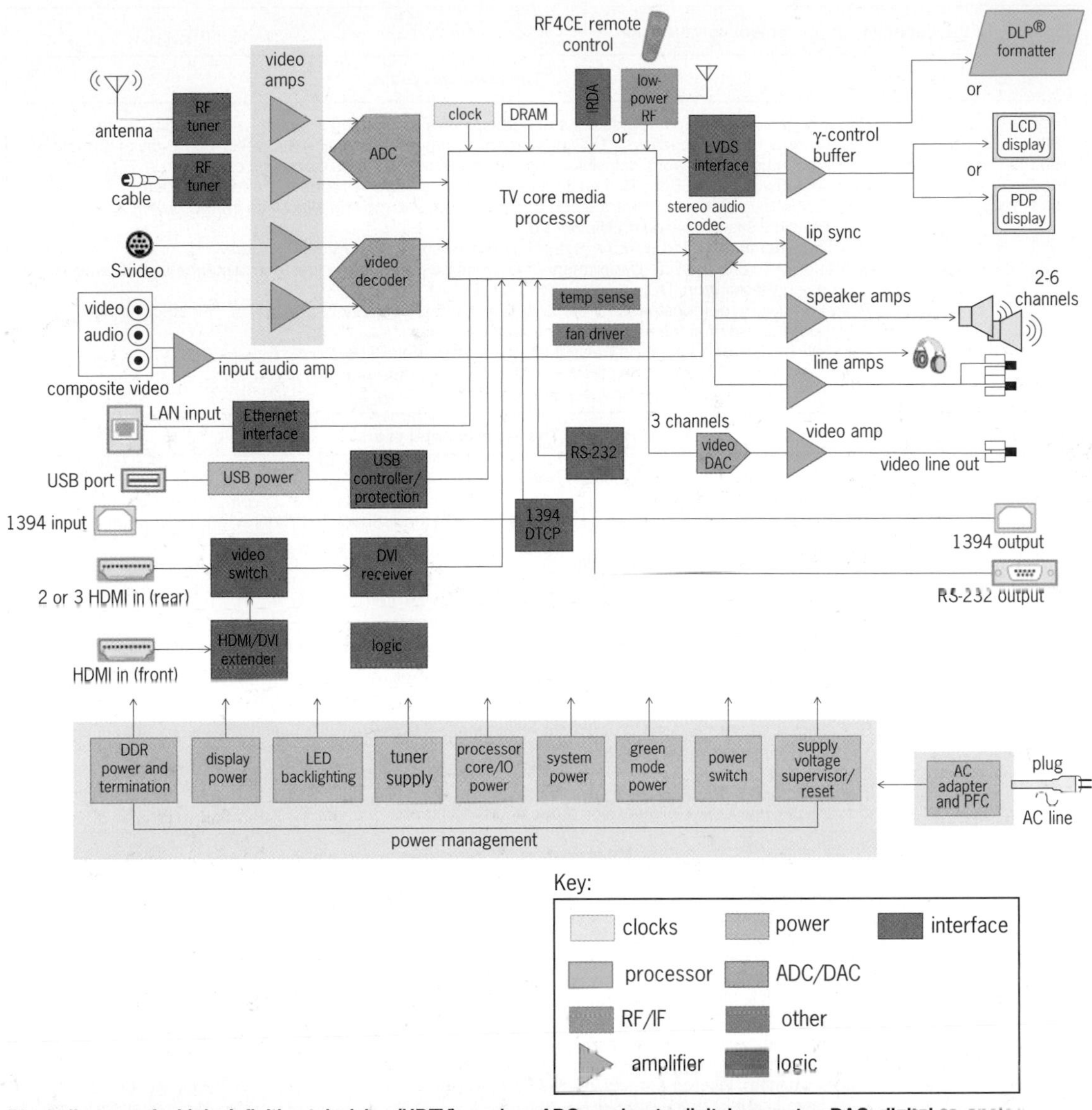

Block diagram of a high-definition television (HDTV) receiver. ADC, analog-to-digital converter; DAC, digital-to-analog converter; DDR, double data rate; DLP, digital light processing (projector technology); DRAM, dynamic random-access memory; DTCP, Digital Transmission Content Protection; DVI, Digital Video Interface (a video display interface), HDMI, High-Definition Multimedia Interface (a compact audio-visual interface); IF, intermediate-frequency; IO, input-output; IRDA, Infrared Data Association (industry standard); LAN, local-area network; LCD, liquid-crystal display; LED, light-emitting diode; LVDS, low-voltage differential signaling (industry standard); PDP, plasma display panel; PFC, power-factor correction; RF, radio-frequency; RS-232, a standard for serial transmission of data; USB, Universal Serial Bus (industry standard). (*Courtesy of Texas Instruments; http://www.ti.com/solution/tv_high_definition_hdtv*)

even more significant in 1953, when almost three times as much information was required to ensure that the color signal performed to the expected standard. Interlaced scanning, which essentially allowed for presentation of two fields of a video frame captured at two different times, enhanced motion perception for the viewer and reduced flicker. This solution was rooted in laboratory work in the 1930s but became a fully functional part of capture and distribution in the 1940s and 1950s. All these enhancements to television technology helped to improve and ensure a quality experience for viewers—and they were all accomplished while fitting signals within the same narrow 6-MHz broadcast channel. Similarly dedicated technical development allowed for further improvements to be made through the 1960s, 1970s, and even the 1980s before the advent of digital television.

Broadcasters have spent hundreds of millions (if not billions) of dollars to improve both the capture and transmission of content over the past three-quarters of a century. They have been efficient in the use of spectrum and have been diligent almost to a fault in trying to implement improvements within the spectrum bandwidth available. Beginning in the 1950s in the United States (and later in Canada and other jurisdictions), with pressure on the VHF band (30–300 MHz), UHF (ultrahigh-frequency, 300 MHz–3 GHz) was opened up for new television stations. In the analog world, however, a 6-MHz UHF channel assigned to a broadcaster resulted in up to 78 MHz of UHF spectrum being affected because

Timeline of the capability improvements of television

	The invention years
1878	William Crookes confirmsthe existence of cathode rays by building a tube in which to display them.
1884	Paul Nipkow sends images with 18 lines of resolution over wires using a rotating-metal-disk technology.
1900	First known use of the word "television" occurs at the International Congress of Electricity.
1907	Campbell Swinton and Boris Rosing suggest using a cathode-ray tube to transmit images.
1924–1925	John Logie Baird (UK) becomes the first person to transmit moving silhouette images using a mechanical system based on Niplow's disk.
1926	Baird is generally credited with the first public demonstration of a true TV image.
1927	Bell Telephone and the U.S. Department of Commerce conduct the first long-distance use of television between Washington, DC, and New York, NY.
	Philo Farnsworth demonstrates for the first time that it is possible to transmit an "electrical image" without the use of any mechanical contrivances.
	Farnsworth files for a patent on the first complete electronic TV system.
	The Radio Act (U.S.) establishes the Federal Radio Commission (FRC, precursor to the FCC), providing radio licenses and that the airwaves are public.
	The experimental years
1928	The FRC issues the first TV station license (W3XK).
1929	Baird opens the first TV studio.
1930	Vladimir K. Zworykin builds on the work of Farnsworth (and others) to produce the iconoscope (the first practical video camera tube).
	The first TV commercial is broadcast.
	BBC begins regular TV transmissions.
1931	CBS begins experimental TV programming.
1932	NBC begins experimental TV programming.
1934	The FCC is established and empowered to regulate all nonfederal use of the radio spectrum, including radio and TV.
1936	About 200 TV sets are in use around the world.
1937	CBS begins TV development.
	BBC begins higher-definition broadcasts in London.
1939	TV is demonstrated at the New York World's Fair.
1940	Paul Goldmark invents a 343-lines-of-resolution color TV system.
1941	The FCC releases the NTSC standard for black-and-white TV.
	Commercial production of television stops until the end of World War II.
	Mass-market TV years (black and white)
1945	Mass-market commercial and public television is introduced.
1948	Cable-TV is introduced in Pennsylvania as a means of bringing TV to rural areas.
	One million homes in the United States now have TV.
1950	The FCC approves the first color TV standard (to be replaced three years later).
1953	Color specification is developed by the NTSC and is compatible with existing black-and-white receivers. It is also designed to fit in the 6-MHz spectrum originally made available by the FCC, through a huge amount of compression.
1954	Just over half of all U.S. households now own a TV set.
1960	The first split-screen broadcast occurs during the Kennedy–Nixon debates.
1962	The All Channel Receiver Act requires that UHF tuners (channels 14–83) be included in all new sets.
	AT&T launches Telstar, the first satellite to carry TV broadcasts.
1963–1964	Most networks now broadcast their schedules partially in color.
	Mass-market TV years (color)
1967	Most TV broadcasts are now in color.
1969	First TV transmission from the Moon; 720 million people watch worldwide.
1972	Half of the TVs in homes in the United States are now color.
	Home Box Office (HBO) becomes the first pay-TV network in the United States.
1976	Sony introduces Betamax, the first home video cassette recorder.
	Turner Broadcasting's WTCG becomes cable TV's first "superstation" as it is beamed across the country via satellite to cable.
1978	PBS becomes the first network to switch to all-satellite delivery of programs.
1981	Japan's NHK demonstrates HDTV with 1125 lines of resolution.
1982	Dolby surround sound for home sets is introduced.
1984	Stereo-TV broadcasts are approved.
1989	The American National Standards Institute gives final approval to a 1125/60 HDTV production standard.
1991	First U.S. testing of HDTV begins.
1993	Closed captioning is required on all new sets.
	Anticipating HDTV, RCA introduces the first widescreen, 16:9-aspect-ratio TV models.
1995	Flat-screen plasma display TVs are introduced at $20,000. By 1997, they are half the price.
1996	One billion TV sets exist worldwide.
	The HDTV years
1996	The FCC approves ATSC's 1.0 Standard, which is adopted as the first HDTV standard (16:9 images, up to 1920 x 1080 pixels in size).
1998	Digital TV sets became available in the United States.
	The first HDTV live broadcast is the launch of the space shuttle *Discovery* in which John Glenn returns to space.

Timeline of the capability improvements of television (*cont.*)

1999	Digital Video Recorders (that is, Tivo, DVRs) enter the market. They are hard-disk–based personal recorders that allow the viewer to pause a live show, record, and skip over advertisements.
2000	85% of U.S. households own VCRs, and 98% own more than one television.
2004	There are now more than 300 cable networks in the United States.
	Raleigh, NC, station WRAL is the first TV station in the United States to provide video news, weather, and traffic information to cell-phone users.
2005	Time-Warner/AOL offers telephone service, in addition to cable television and Internet access.
2006	HDTV penetration hits 35% in the United States.
	The UHDTV and TV everywhere years
2007	Netflix begins streaming Over-the-Top (OTT) content to its paid subscribers via Internet streams.
	Apple releases the iPhone, which instantly revolutionizes the smart-phone market and puts a remote video streaming device into the pockets of millions of users.
2009	The U.S. switch-off of all analog terrestrial TV broadcasts occurs (much of the developed world follows over the next few years). This includes the giveback of channels 52–69.
2010	The 3D-at-home revolution fizzles when consumers fail in any great number to snap up expensive units that require external glasses. Still, 10% of all TVs shipped are 3D-enabled.
2012	The ATSC 2.0 Standard is adopted to allow mobile, interactive, and hybrid television technologies by connecting screens with Internet services to allow interactive elements into the broadcast stream.
	The Future of Broadcast Television (FOBTV) is established with a mandate to try to harmonize standards to the greatest extent possible, especially in lieu of ever-faster changes in broadcast production and transmission technologies.
	YouTube states that roughly 60 h of new videos are uploaded on their service every single minute and that three-fourths of the material comes from outside the United States.
2010	The ATSC announces a call for proposals for an ATSC 3.0 Standard, which should support a video resolution of 3840 x 2160 at 60 frames per second.
	22% of the world's population now owns an Internet-enabled smart phone that has the ability to stream video content (this figure includes more than two-thirds ofthe U.S. population).
2014	Netflix's OTT service now accounts for 34.2% of all downstream Internet usage during peak hours, including popular original content such as *Orange Is the New Black* and *House of Cards*.
	HDTV penetration reaches 77% of U.S. homes.
	Japan's NHK announces plans to cover upcoming World Cup and Olympics using 8K Super Hi-Vision broadcast.
	Market research forecasts that UHDTV receivers will be in 33% of U.S. homes by 2020, compared with less than 10% today.

of interference issues in surrounding channels (the so-called taboo channels). In the development of high-definition digital television, reduction of this effect therefore became a critical factor. In fact, with the ATSC 1.0 standard in 1995, the interference impact was reduced to co-channel and adjacent channels and 18 MHz of spectrum. That is progress and respect for scarce broadcast spectrum.

Much time is spent on this issue because in broadcasting the "chicken or the egg" question is asked: Which came first; the spectrum or the technology? You can't have one without the other.

Digital television's promise and challenges. Digital television technology may be about to take its next great leap forward. Although the digital evolution is not exactly a Moore's law–like experience of exponential change in television, the pace of change has nevertheless radically altered from the decades it took for higher-resolution images and surround sound to find their way to viewers. Even color took decades to make the transition from black and white.

Digital production, imagery, formats, and distribution mean that broadcast products can appear on every available platform to serve viewers' interests, whenever and however viewers want their television delivered. The Internet is both a challenge and a huge opportunity for television, and digital technology gives broadcasters and content makers the tools to exploit it.

Yet some things remain the same, and one that has not changed is the scarcity of broadcast spectrum. Spectrum is the resource through which broadcasters transmit their signals from one source to many (or all). The combined availability of both spectrum and the technology to transmit and receive signals was what first made radio and then television ubiquitous in communities and countries around the world.

Decades ago, spectrum bands designated for broadcasting were recognized as important by governments to ensure that citizens would benefit from the information, entertainment, and enlightenment made possible by broadcasting services. This allocation allowed for a better-informed civil society, a celebration and expression of culture, and, in times of natural and human-made emergency, a source of vital survival information. This all still holds true today, but there are many other ways that the broadcaster reaches the viewer too (for example, through cable, satellite, the Internet, and mobile). *See* RADIO AND TELEVISION BROADCASTING DURING DISASTERS.

As a consequence, governments in many national jurisdictions are less supportive than they once were of protecting this vital spectrum for broadcasting. Many are proposing to share the broadcast bands with mobile services, specifically International Mobile Telecommunications (IMT) services, a decision with huge consequences for television. In particular, interference with existing signals may place the development of new technology for better resolution and sound at risk: It may constrain over-the-air (OTA) television broadcasts to a lesser quality of picture and sound than would be available by other delivery systems. These systems charge the viewer for what was once freely available, and it is hard to discern the public interest in that. This practice also changes the business models for public and private television services, prompting some of them to say they will be supplying tiered services.

Yet, as noted previously, broadcasters and the engineering and technical community supporting them have so far shown considerable innovation in improving the picture and sound of television while respecting the issue of scarce broadcast spectrum. In the development of HDTV, broadcasters were not given more spectrum. It was only the improvement in digital compression technology that allowed for a full HDTV (1080-line progressive scan) signal to be transmitted in the same 6 MHz of spectrum available to the analog signal in the United States (while offering four times the resolution and twice the width of the analog picture). In the analog broadcasting world that same quality could have taken up to 30 MHz of spectrum.

It is noteworthy that the end result of the conversion to digital television in the United States was a giveback of channels 52–69. The 108-MHz "digital dividend" resulted in the 700-MHz spectrum auction. Broadcasters have a long tradition of using spectrum efficiently and creating better services for the citizens they serve. Beginning with the introduction of monochrome television, then color, and now the recent transition to HD digital, all these huge improvements were achieved within the same amount (or less) spectrum than was used before the improvements were made.

Sadly, several national jurisdictions have not recognized the technical achievements of the past and the hopes for the future by committing to guarantee broadcast spectrum for future broadcasting endeavors. The financial lure and demands of the mobile/Wi-Fi community may well render broadcast spectrum useless for broadcasting. Nonetheless, the broadcast industry continues to innovate and move forward on the assumption that providing freely available OTA services of the highest quality to viewers is still at the core of a contract between governments and citizens concerning the best use of spectrum.

The transition to HDTV in North America took just a little longer than a decade and now is considered largely complete in Canada and the United States, and in progress in Mexico. The North American standard, developed by the Advanced Television System Committee (ATSC), is called ATSC 1.0, and it began to roll out to the mass market at the end of the 1990s. A few years ago, improvements were made and a new standard called ATSC 2.0 was introduced. ATSC 2.0 provides interactive and hybrid television technologies that allow for a "connected TV" linked to the Internet and to some interactive material in the broadcast stream, as well as data storage, targeted advertising, audience measurement, and video on demand.

Emergence of next-generation television. Four centers in the global community are developing digital television standards: the aforementioned ATSC in the United States, Digital Video Broadcasting (DVB) in Europe, Integrated Services Digital Broadcasting (ISDB) in Japan, and work being done in China based on the work of the previous three. The Future of Broadcast Television (FOBTV) is a global organization, formed in 2012, that is made up of all these groups, plus some user organizations such as broadcasters, with its secretariat based in China. The centers' mandate is to try to harmonize standards to the greatest extent possible—a desirable goal in a world of ever-faster change in broadcast production and transmission technologies.

Development is fully underway, with work going on in all of those centers on the next generation of television. Coming so soon after the transition to digital and digital HDTV, many observers of the broadcast industry (and even some members of it) were surprised at the advances in technology that made this possible. In particular, developments in compression and IP (Internet Protocol) technology will allow real advances in picture resolution and sound, multiplatform delivery, interactivity, and mobile services.

In 2012, in response to a call from ATSC for a new standard, ATSC 3.0, the North American Broadcasters Association (NABA) published the Next Generation of Terrestrial Broadcasting requirements. Essentially, it was a user's guide to the next generation of television. The white paper explored the future needs of network broadcasters in North America based on the following important considerations:

1. ATSC 3.0 must be developed within a relatively short time to direct government spectrum policy and repacking decisions.
2. Because ATSC 3.0 is likely to be incompatible with current broadcast systems, it must provide improvements in performance, functionality, and efficiency significant enough to warrant implementation of a system that is not backwards-compatible.
3. Interoperability with broadcast systems and nonbroadcast distribution systems must be considered.
4. Television broadcasting is increasingly evolving to mean multimedia broadcasting, which involves sending multiple signals simultaneously to a variety of devices. The broadcast signal must support a variety of content types, and these content elements may offer a viewing experience either alone or in composite. This change requires support for synchronous and nonsynchronous relationships among disparate content elements, as well as metadata for discovery of content, for scaling the spatial and temporal resolution of images, and for triggering interactivity between screens and between devices.

The NABA paper noted the following elements as central requirements for next-generation television:

1. High definition (HD) is the absolute minimum baseline, and mobile HD service must be a core capability.

2. Integral bi-directional communication is needed for the future health of the broadcast business. The television of the future will depend on two-way connection to provide services including conditional access, audience measurement data, consumer information gathering, interactive content, dynamic ad insertion, audience voting, personalized content, interactive-TV applications, and multicontent synchronization.

3. Interoperability with IP and LTE (Long-Term Evolution) is crucial. Many kinds of interoperability should be considered, but interoperability with IP and LTE networks are crucial to next-generation requirements. Wireless portable and mobile devices will likely be the largest base of receivers for direct reception of over-the-air terrestrial broadcast signals, and they will have IP/Wi-Fi or LTE capabilities. Interoperability with these protocols will offer broadcasters the opportunity to deliver enhanced services that make use of the Internet.

4. The system should provide conditional access capabilities. Broadcasters need flexible conditional access to continually develop new business models, which will evolve with the marketplace.

5. Better compression systems will be required, and the ability to update receivers with compression advances should be part of any new standard. Advances in computing power and multicore processors have made software-based decoding practical in consumer devices. Continued improvements to codec performance for Internet video streaming demonstrate the benefits of being able to update receiver software. Broadcast codecs and receiver performance should similarly be capable of continuing evolutionary improvements.

6. Progressive scan formats are now and will be central to the next generation of television. The underlying technology limitations that were barriers to a 1080/60p format when the ATSC 1.0 standard was established in 1995 have now been overcome.

7. Integer frame rates should be adopted, including higher-than-60-Hz rates. The need for frame-rate compatibility with analog NTSC should be obsolete by the time of ATSC 3.0 deployment. Because all ATSC receivers are capable of both 59.94 and 60.0 families of frame rates, ATSC 3.0 should no longer be constrained to handle legacy 59.94 frame rates.

8. Ultra-high definition (UHD) will characterize the next generation of television. The ability to create content is the first necessary precondition for establishing transmission standards. 4K cameras and production technology are rapidly emerging in commercial products. It is also noteworthy that some feature film masters also currently exist in 4K, because that resolution is currently the high end for the digital cinema specification. 8K camera prototypes have been developed in research laboratories. The ability to mass-manufacture displays is the second necessary precondition for establishing transmission standards. 4K and 8K consumer displays are being offered by many manufacturers. Terrestrial broadcasters need 4K and potentially 8K for retransmission and to remain competitive. Scalable compression and modulation approaches will be essential.

In addition, NABA believes the following topics deserve serious consideration in ongoing discussions about next-generation requirements:

1. Spectrum packing efficiency and frequency reuse.

2. A hierarchical modulation system. NABA stops short of calling this a requirement but rates it as highly relevant to discussions about how to provide a mobile service and a UHD service at the same time, given the various trade-offs in their services and their different effective coverage areas.

3. Planning factors for the receiving antennas appropriate to particular services, such as mobile HD and SHD.

4. Hybrid transmission network possibilities. The traditional "high-power, high-tower" single-transmitter approach for broadcasting may need to be modified. Fully cellular network models are probably cost-prohibitive, but a hybrid model involving a central transmitter with a small number of lower-power macro-cells to complete coverage over a service area deserves consideration. The macro-cells could either be on-channel repeaters or utilize different frequencies. Both possibilities have advantages and disadvantages.

5. Adaptive modulation coding with a return or feedback channel. Adaptive modulation coding has been employed successfully in wireless communications systems for LTE and WIMAX. Adaptive modulation systems can make significant improvements in the rate of transmission and bit error rates by exploiting the channel condition information in wireless propagation environments.

6. The desirability of a voluntary minimum receiver radio-frequency (RF) performance specification (that is not government regulated). Adoption of receiver minimum RF performance specifications should enable more efficient use of the spectrum and create opportunities for new service offerings. Today, although transmitters are required to control out-of-band and spurious emissions to minimize interference, no specifications govern receiver performance. Consideration should be given to how these standards might work. One approach would be to establish specifications for minimum receiver performance characteristics. Another approach would be to define receiver protection limits in which the specifications establish an in- and out-of-band interference environment within which receivers must operate. NABA prefers the establishment of voluntary industry standards to the imposition of standards by regulation.

Although the requirements continue to evolve with both the market and the available technology, there is no question that broadcasting will be launched into a transition to the next generation in the very near future.

For over three-quarters of a century, television broadcasters have demonstrated unprecedented efficiencies and economies in spectrum use and improved services. The next generation of television will be true to this tradition, except that the greater scale of change and improvements will be staggering. They will include system improvements leading to flexibility and scalable broadcast services (from mobile to ultrahigh definition), bit-rate robustness, and access to novel information-based services and connections to the wider world—and all within the same spectrum limits that broadcasters respect today. One has to wonder why there is such fascination with the inefficient one-to-one technology of the mobile and IMT world when the one-to-many signal at the heart of broadcasting is poised for such a huge breakthrough.

For background information *see* BANDWIDTH REQUIREMENTS (COMMUNICATIONS); DATA COMPRESSION; INTERNET; MOBILE COMMUNICATIONS; MULTIMEDIA TECHNOLOGY; RADIO SPECTRUM ALLOCATION; TELECAST; TELECONFERENCING; TELEVISION; TELEVISION CAMERA; TELEVISION RECEIVER; TELEVISION SCANNING; TELEVISION STANDARDS; TELEVISION TRANSMITTER; VIDEOTELEPHONY; WIMAX BROADBAND WIRELESS COMMUNICATIONS; WIRELESS FIDELITY (WI-FI) in the McGraw-Hill Encyclopedia of Science & Technology. Michael McEwen

Bibliography. A. Abramson, *The History of Television, 1880 to 1941*, McFarland, Jefferson, NC, 1987; A. Abramson, *The History of Television, 1942 to 2000*, McFarland, Jefferson, NC, 2003; L. E. Frenzel, *Principles of Electronic Communication Systems*, 3d ed., McGraw-Hill, Boston, 2008; M. S. Richer et al., The ATSC digital television system, *Proc. IEEE*, 94:37–43, 2006, DOI:10.1109/JPROC.2005.861714; M. Starks, *The Digital Revolution: Origins to Outcome*, Palgrave Global Media Policy & Business, 2013; Y. Wu et al., Overview of digital television development worldwide, *Proc. IEEE*, 94:8–21, 2006, DOI:10.1109/JPROC.2006.861000.

Three-dimensional nonvolatile memory scaling

NAND flash memory, which dominates the nonvolatile memory market, has seen exponential growth in the past 15 years as the result of its very rapid scaling. NAND flash has been scaling at a pace of about 18 months per technology node. (A new technology node usually uses minimum features that are approximately 70% of the previous node, resulting in doubling the number of devices, or amount of memory, that can be packed into a specified area.) At such a pace, NAND flash has quickly surpassed all other integrated circuit technologies. In early 2014, NAND flash with a 16-nm feature size was in volume production, compared to DRAM (dynamic random access memory), with feature sizes ranging in the upper 20s of nanometers. Furthermore, NAND flash is capable of storing more than one bit of information per memory cell by adjusting the amount of charge stored to produce multiple logic levels in one memory cell (multilevel logic cell, or MLC) to double or triple its storage density.

Operation of flash memory devices. All flash memory devices are derived from the basic metal-oxide-semiconductor (MOS) transistor (**Fig. 1***a*). In a MOS transistor, the conduction of electricity between source and drain is controlled by the voltage on the gate, which is usually made of polycrystalline silicon (polysilicon or poly-Si). If the voltage on the gate is higher than a threshold voltage (V_t), a conducting channel is established under the gate [which is insulated from the silicon substrate by an oxide, usually silicon dioxide (SiO_2)], and electrons may flow from source to drain.

Currently, almost all NAND flash memory products are made with a conventional two-dimensional (2D) floating-gate memory that uses an extra piece of poly-Si sandwiched between the gate oxide and the control gate of a MOS transistor to store charge (Fig. 1*b*). The floating gate is completely insulated on all sides so that electrons that are injected into it stay there indefinitely (until removed by an intentional erase operation). The charge (electrons) stored in the floating gate increases the voltage, V_t, required to turn on the transistor. Thus, by applying a known voltage to the control gate of the MOS transistor, the logic state of the device can be determined from its output current. For example, a transistor in the erased state (no stored charge) has a lower threshold voltage, $V_{t(\text{erased})}$, and can be switched on by a voltage greater than $V_{t(\text{erased})}$, and one with stored charge has a higher threshold voltage, $V_{t(\text{programmed})}$, and thus remains in the off state when applied the same voltage.

In Fig. 1*b*, the oxide-nitride-oxide (ONO) structure is a triple $SiO_2/SiN/SiO_2$ layer that is draped around the sides of the floating gate (as shown in the side view of Fig. 1*b*) so that it completely encloses the floating gate except for the bottom oxide (the gate oxide, called the tunneling oxide for flash memory devices). Its purpose is to take advantage of the higher K (electric permittivity) value of silicon nitride (SiN) to ensure that the applied voltage will drop largely across the tunneling oxide, through which the programming and erasing actions occur. However, the ONO configuration makes fabrication more complicated, and, more important, it makes scaling more difficult, since the ONO layer on the side of the floating gate takes up room.

An alternative means of storing electrons is to use the SiN itself in the ONO layer instead of the floating gate. SiN has intrinsic defects that act as trapping sites for electrons, thus permitting charge storage (Fig. 1*c*). This type of device (called a charge-trapping SONOS device, where "SONOS" stands for the Si-gate/ONO/Si-channel layering) has a very simple structure and is easier to scale, since the ONO

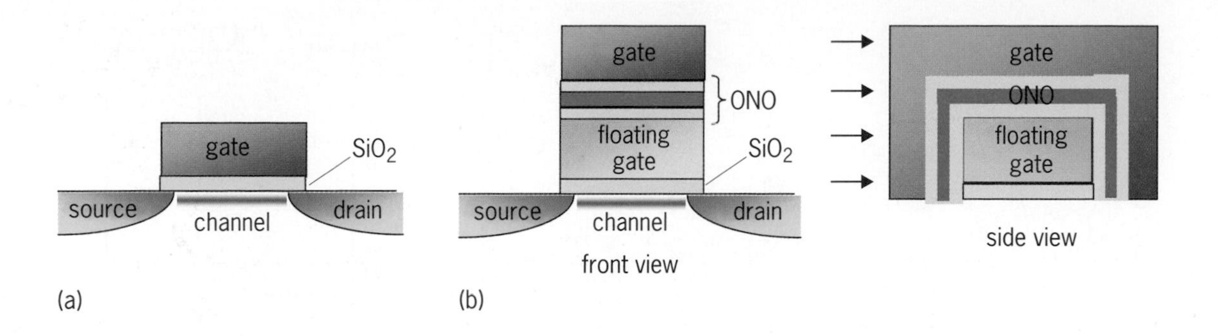

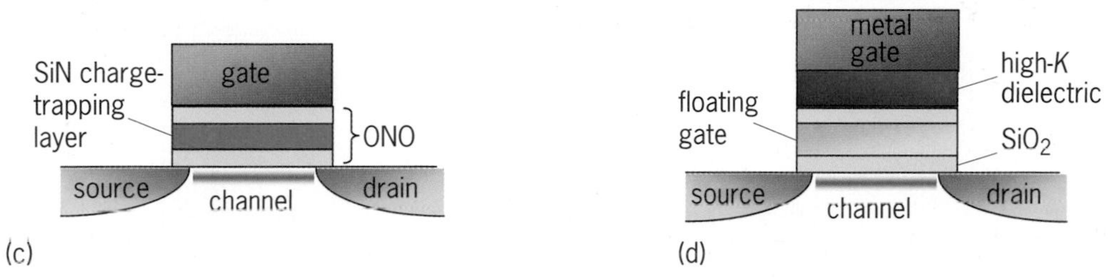

Fig. 1. Various types of flash memory devices. (*a*) MOS (metal-oxide-semiconductor) transistor. (*b*) Floating-gate flash memory device. (*c*) Charge-trapping flash memory device. (*d*) High *K*/metal gate flash memory device.

layer is not on the side of the device and does not occupy additional space. However, deeply trapped electrons are difficult to "de-trap," and thus difficult to erase. Recently, innovations in "barrier engineering" the tunnel oxide have allowed erasing deeply trapped electrons.

The most advanced flash memory technology has adopted a high-*K* dielectric (hafnium oxide, HfO_2) to replace the ONO, and a metal gate to enhance erase performance (Fig. 1*d*). The high-*K* dielectric allows the voltage to drop mostly across the tunneling oxide without having to wrap itself around the side of the floating gate. This technology allows a planar structure similar to the charge-trapping device in Fig. 1*c* but still retains the advantage of a floating gate—it is easy to erase. The metal gate [usually tantalum (Ta) or tantalum nitride (TaN)] provides an additional knob to further enhance the erase performance.

Scaling of 2D flash memory. However, NAND flash is rapidly approaching its scaling limit. The difficulties that limit NAND flash scaling are many, but among the most serious are (1) capacitive coupling interference from the stored charges in neighboring cells that cause misreading of the logic levels, and (2) electric breakdown between neighboring wires. NAND flash requires substantial voltages (15–20 V) to operate, and the space between neighbors is now less than 20 nm, producing high electric fields (up to 10 MV/cm) that are close to the breakdown field in the SiO_2 layer that insulates the adjacent devices. However, the most fundamental difficulty is that (3) the number of stored electrons diminishes by a factor of two at each new technology node (since the area of the memory device is reduced by 50%), thus quickly reducing the total number of stored electrons to less than 100, and these electrons then need to be divided among four logic levels (in the case of a 2-bits-per-cell MLC) or more. Thus, when the number of stored electrons is less than 100, the number of electrons that represents an MLC logic level becomes less than 30, and the number of electrons in each programming pulse (typically several pulses are needed to program a logic level) falls below 10. The statistical fluctuation noise, which is governed by the square root of the number of storage electrons, then makes it difficult to precisely define the logic levels, and adjacent logic levels merge and become indistinguishable. Algorithms enabled by mathematical error-code-correction (ECC) formulas can alleviate this issue, but the complexity of ECC calculation escalates quickly when logic level overlapping increases and can no longer help without drastically increasing the cost.

Once 2D NAND reaches its scaling limit, the only way to increase the memory density is to stack the storage cells vertically—three-dimensional (3D) NAND. There is no consensus as to when 2D NAND will reach its scaling limit. Some NAND vendors extend the 2D NAND roadmap into the low-teen-nanometers regime, and the International Technology Roadmap for Semiconductors (ITRS) forecasts 3D NAND around 2014. Indeed, the first 3D NAND product was introduced in August 2013 with volume production expected in 2014.

Conventional 2D NAND. To understand the 3D NAND architecture, it is best to start with conventional 2D NAND. **Figure 2** shows a 2D NAND array and its principle of operation. The *x*-*y* coordinates of any cell in the array are defined by the bit line (in red) in the *y* direction and word line (in blue) in the *x* direction. In other words, each storage cell in

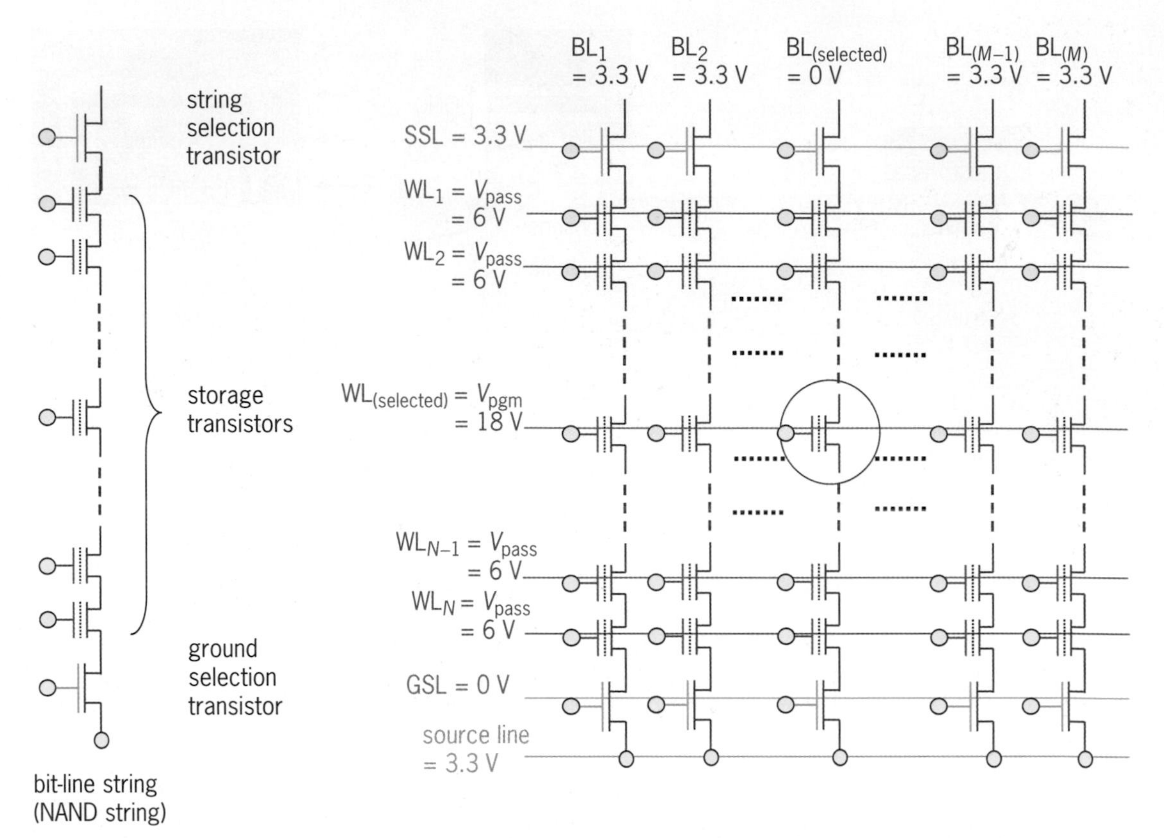

Fig. 2. Operation of a conventional two-dimensional NAND array. The memory block of data shown here consists of *N* rows and *M* columns. *N* is the number of storage devices in each bit-line (BL) string (NAND string), one of which is shown at the left, and *N* = 64 for recent, advanced technology nodes. *M* is the number of NAND strings controlled by a word line (WL), and is about 8–16 kB (kilobytes) in size.

a 2D array is uniquely identified by the bit line and the word line that control it.

The basic building block of the array is the bit-line string (also known as the NAND string), shown at the left of Fig. 2. It consists of 64 storage cells, each a floating-gate MOS transistor, serially connected, with a conventional MOS transistor at each end of the string that is used to control access to the string. The bit-line string resembles a NAND logic gate since the output is pulled down when all the transistors in the string are turned on (conducting).

The word line is connected to the gate that switches the transistor on and off. A word line crosses thousands of bit-line strings, and the cells controlled by the word line form a data page, about 8–16 kB (kilobytes) in size for the most advanced technology, plus some extra bytes for error correction. (1 kB = 1024 × 8 bits.) All the 64 pages that share the same NAND strings form a data block; thus, the data size of a block is about 1 MB (8 Mb). The data are structured in these page and block forms for fast read/write throughput. Programming and reading are carried out one page at a time. However, erasing is applied only to the entire block.

The two control transistors at the ends of a bit-line string are named the string-selecting transistor (SST) and ground-selecting transistor (GST), respectively. Accordingly, the corresponding word lines controlling these transistors are called the string select line (SSL) and ground select line (GSL) [both in green]. This latter convention is so commonly used that the acronyms SSL and GSL are often used for the transistors as well as the word lines feeding them. The top of each bit line is connected to a sensing circuit (not shown in Fig. 2), and the bottom is connected to a common source line (in pink) that supplies the needed current.

To program, a high enough voltage [V_{pass}, approximately 6–9 V, a voltage greater than $V_{t(programmed)}$ for all the storage cells in the NAND string] is applied to all unselected word lines to ensure that all of those devices are turned on and do not constrict the current flow. The voltages shown in Fig. 2 are for programming the cell shown in the circle. The selected bit line is held at 0 V while the others are held at 3.3 V. The selected word line is applied a high voltage (V_{pgm}) of approximately 18 V. The transistor at the intersection then sees 18 V across the gate and the silicon substrate. This voltage drop creates a high electric field across the tunneling oxide, and electrons in the substrate inject into the floating gate through the tunneling oxide by means of a quantum-mechanical tunneling mechanism. (This phenomenon is called Fowler-Nordheim tunneling.) Other transistors on the same word line, however, see a voltage difference of only 14.7 V between the gate and the silicon substrate. Since quantum-mechanical tunneling is very sensitive to the magnitude of the electric field, no electron injection occurs in these devices.

To read data in the page, the selected word line is applied a voltage slightly lower than $V_{t(\mathrm{programmed})}$, and all other cells in the bit-line string are applied a voltage higher than $V_{t(\mathrm{programmed})}$, that is, V_{pass}, to ensure that they are turned on. The output from the bit-line string is then sent to a sensing circuit to determine the logic state of the cell. If the cell has no stored charge (that is, it is in the erased state), then the threshold voltage of the transistor is lower than the applied voltage; in this case the transistor will be turned on and the sensing circuit will detect a current. If the cell has stored charge, and thus has a higher threshold voltage, then the transistor remains in the off state and no current is detected.

To erase, a negative voltage of about the same magnitude as V_{pgm} is applied and electrons stored in the floating gate are injected back into the substrate through the same Fowler-Nordheim tunneling mechanism.

Three-dimensional NAND. Since a MOS transistor is a three-terminal device (with source, drain, and gate electrodes), it can be decoded by three coordinates (x, y, z), and is thus suitable for building a 3D array. 3D NAND may be constructed in a number of ways, each with its own pros and cons. Although a 3D array may be achieved by stacking 2D arrays on top of one another, this approach is very costly because the complicated fabrication processes have to be repeated many times. A number of innovations have been introduced in the last several years that promise cost-effective 3D NAND memories.

In a 2D NAND array, the current always flows in the horizontal direction because the entire array is in the x-y plane. However, 3D NAND has an option to choose the direction of the current flow. Thus, 3D NAND architectures fall into two main categories—vertical channel and horizontal channel. (The channel is where the current flows when the current is turned on.) Because 3D arrays must still be fabricated by 2D monolithic processes, that is, a succession of 2D planar processes, the alternative directions of the channel (current flow) pose different challenges to processing, performance, and scalability.

Most structures are based on charge-trapping SONOS devices (Fig. 1*c*) because their simpler structure is easier to fabricate in 3D, which is naturally more complicated than 2D arrays. Although a 3D array with a conventional floating-gate device has also been proposed, the complex structure restricts the scaling of cell dimensions, and thus has not gained popularity.

Figure 3*a* shows the original bit-cost scalable (BiCS) structure. [Terms such as BiCS and terabit cell array transistor (TCAT) do not provide an intuitive connection to the physical structures that they describe but have been widely adopted.] In this structure, a 3D array is constructed by rotating the conventional bit-line string (in red) 90° from the horizontal to the vertical (z) direction (Fig. 3*b*). The channel of the device is changed from the planar single-crystalline silicon substrate to a vertical poly-Si channel in the shape of a nanowire. The gate surrounds the center channel horizontally. [This type of transistor is commonly referred to as a gate-all-around (GAA) device. In the case of extreme scaling, the center channel is very small, like a nanometer-size wire, and thus the transistor is also called a nanowire device.] This structure is made by stacking many layers of poly-Si and insulator (SiO_2, for example), and then punching a hole

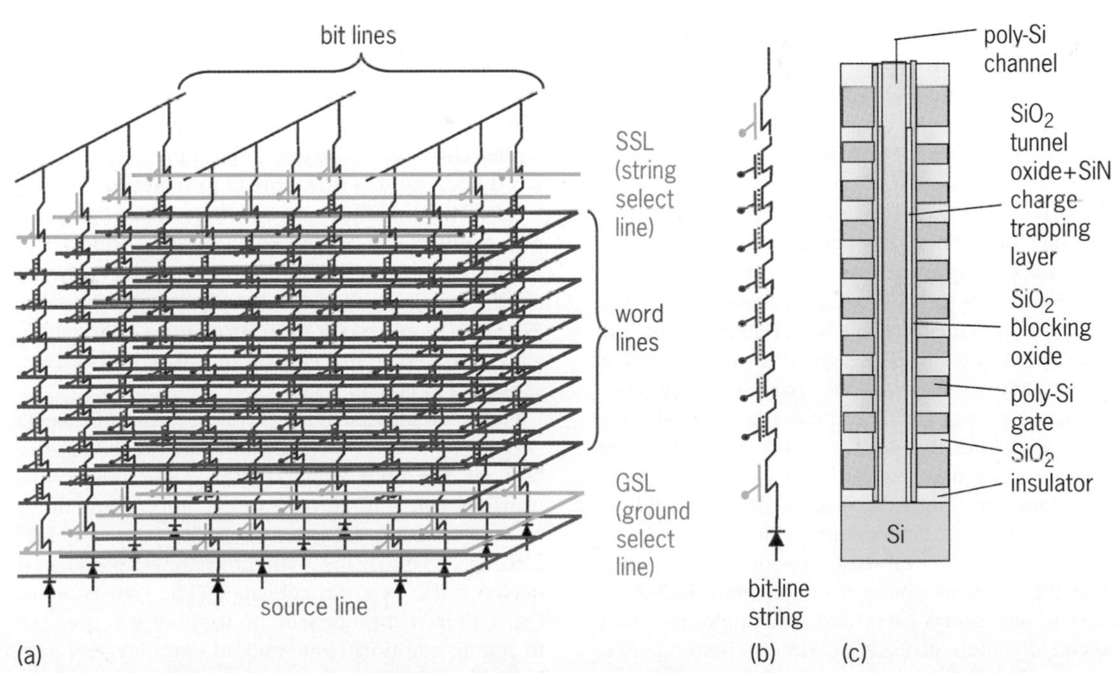

Fig. 3. BiCS (bit-cost scalable) structure. (*a*) Array architecture. (*b*) Equivalent circuit and (*c*) structure of a vertical bit-line string.

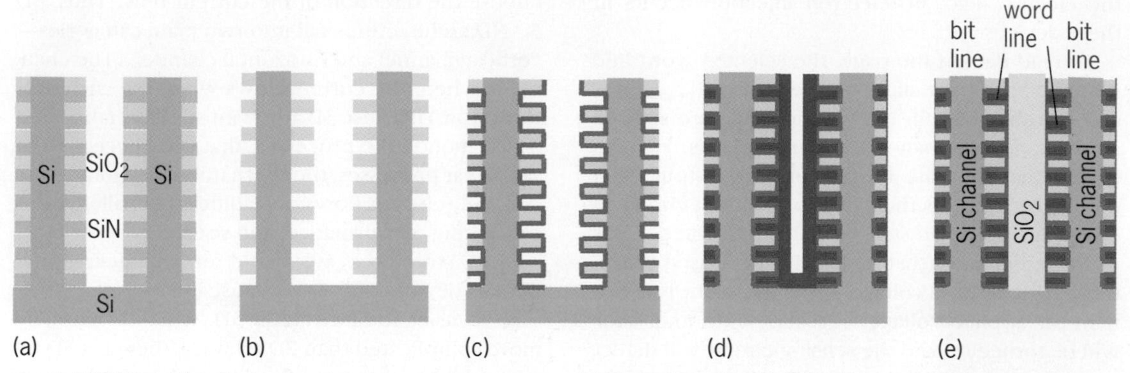

Fig. 4. TCAT (terabit cell array transistor) processing sequence. (*a*) Punch holes in SiO_2/SiN stack, refill with poly-Si, then cut slots (white area in center). (*b*) Selectively remove the sacrificial SiN layers. (*c*) Deposit dielectric SiO_2/SiN/SiO_2 layers. (*d*) Poly-Si (or metal) deposition to form word lines. (*e*) Etch off excess poly-Si (or metal), then refill with SiO_2.

through the entire stack (by one single lithography and etching operation) for the channel (Fig. 3*c*). The ONO dielectric layers in the SONOS nanowire transistor are then deposited conformally inside the hole, followed by poly-Si deposition to form the channel. Thus, many layers of devices may be fabricated in relatively few steps, greatly reducing the fabrication cost.

The word lines (blue in Fig. 3*a* and *b*) stay in the *x-y* planes and control the cells in each plane; this structure automatically groups devices in each plane into pages. As in conventional 2D NAND, the bit-line string is controlled by two MOS transistors, the SSL and GSL, at the two ends. A diode is added below the GSL to isolate individual bit-line strings. An improved version that connects two bit-line strings at the bottom and puts both bit-line string control devices on top is called PBiCS (meaning "pipe-shaped BiCS," not shown in Fig. 3).

Figure 4 illustrates an alternative process, commonly referred to as TCAT. This is a more complex process that may be integrated with high-*K* dielectric and metal gates, thus preserving some of the advantages of the most advanced 2D NAND (Fig. 1*d*). The initial stacking layers consist of only SiO_2 and SiN. After the holes are etched, poly-Si is deposited into them to form channels. (It is also possible to epitaxially grow single-crystalline silicon in the hole, which may give better performance than poly-Si devices.) Then slots along future word-line directions are cut to allow access to subsequent processing (Fig. 4*a*). Next, the sacrificial SiN layers are removed (Fig. 4*b*), and ONO layers are deposited (Fig. 4*c*); the SiN in the ONO layers serves as the charge-trapping layer. Then, the poly-Si (or metal) gates are deposited to form word lines (in blue) [Fig. 4*d*], and the excess material (poly-Si or metal) is removed by an extra lithography-etching step to separate the word lines in different layers. Finally, the cut slot is refilled with SiO_2 (Fig. 4*e*). TCAT fabrication is more complicated than BiCS, but its ability to incorporate high-*K* dielectric and metal gates and use single-crystalline silicon channels gives it potentially better performance, such as higher read speed and faster erase operation.

Figure 5 shows a vertical-gate, horizontal-channel architecture. The most important advantage of maintaining the horizontal channel (as in a 2D array) is its scalability, which will be further discussed. However, this architecture has some important challenges. In the vertical-channel architecture, the string selection devices (labeled "SSL" in Fig. 5*a*, in accordance with the previously discussed convention) are already built into the bit-line strings and are readily contacted outside the array for control signals. In a horizontal-channel array, the vertical gates are shared by all the bit-line strings, including the gates controlling the SSL. This setup disables the SSL function (because all SSLs are turned on and turned off as one, they cannot control each bit-line string individually), and the array cannot be decoded.

Innovative decoding methods have since been proposed, and Fig. 5*a* illustrates the equivalent circuit for one such proposal that embeds an isolated SSL in each bit line to decode the 3D array. For simplicity only short bit-line strings (in red) and only two stacking layers are shown. The word lines (in blue) are now expanded not only to span the in-plane bit-line strings but also to cover all stacked layers. The page (8–16 kB) is still represented by the word line, but it is now spread into all the stacking layers. The common source line (CSL, in light purple) supplies the current to all bit-line strings. The ground select line (GSL, in green) is shared by all bit-line strings in all layers, but the string select line (SSL, also in green) for each string must be individually controlled, in this case by an isolated SSL device, as shown in Fig. 5*b*.

Figure 5*b* illustrates the physical layout of this 3D NAND. To clearly show the SSL and contacts to the bit lines, Fig. 5*b* is rotated by 90° counterclockwise from Fig 5*a*, with three layers, more word lines, and fewer bit lines. Each storage device is a double-gate SONOS transistor, with the channel sandwiched between the two vertical gates. (The transistors in Fig. 1 all have a single gate on top, but it is possible to put an additional gate and an extra layer of gate oxide on the other side of the channel and mirror the top gate. This configuration provides extra control

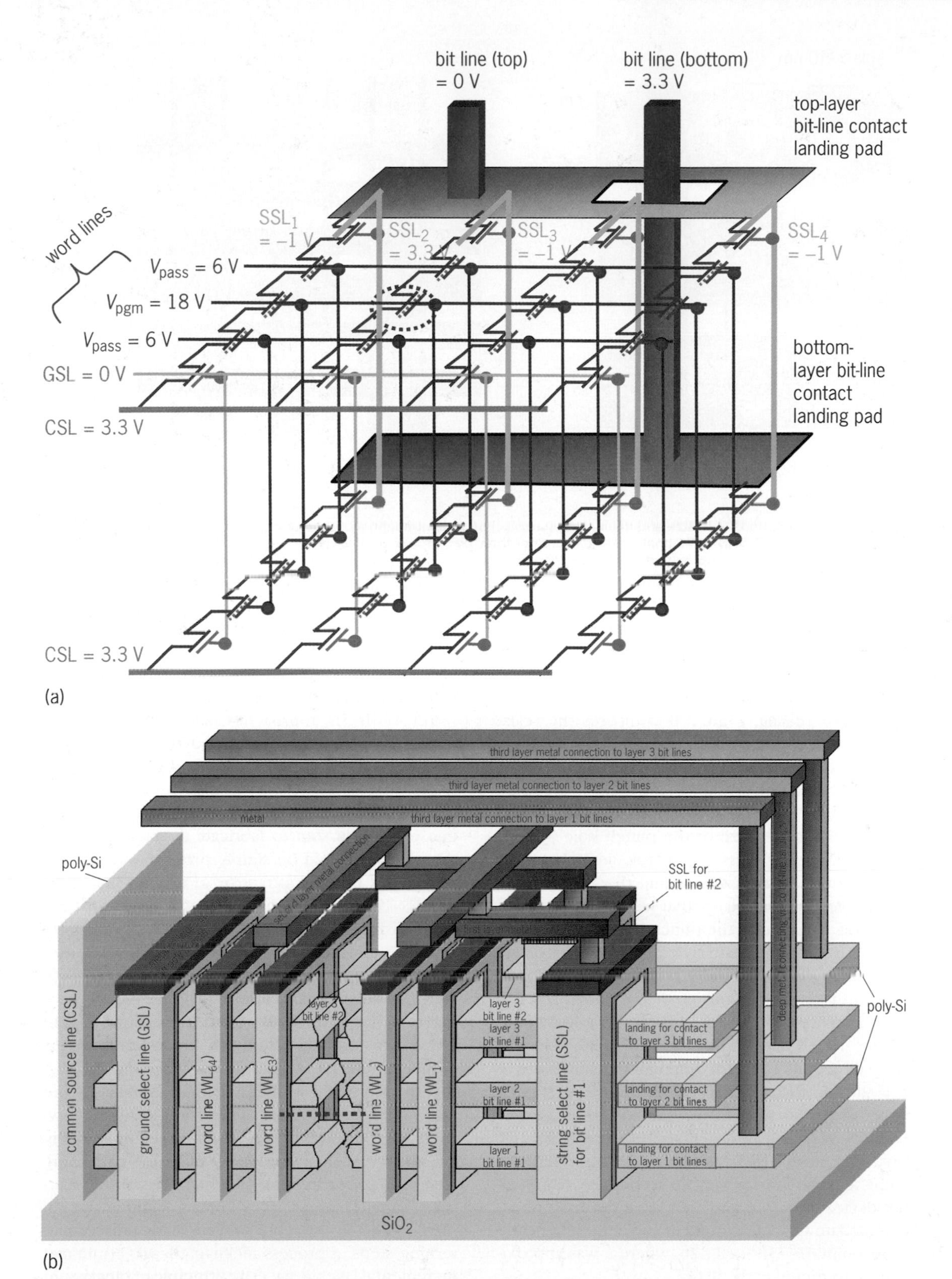

Fig. 5. Horizontal-channel (vertical-gate) 3D NAND array. (*a*) Equivalent circuit. (*b*) Physical layout.

of the channel that benefits very small devices. In the GAA device, previously mentioned in the discussion of the BiCS structure, the gate oxide and the gate around it surround a cylindrical transistor channel to gain even more control.) Bit-line strings that share the same *x*-*y* coordinates are controlled by a unique SSL, which is independently controlled through its own wiring. All bit lines in the same *z* plane are tied together and connected to the same sensing circuit. However, because only one SSL is activated only the corresponding bit-line string is turned on; thus, there is no confusion about which string is

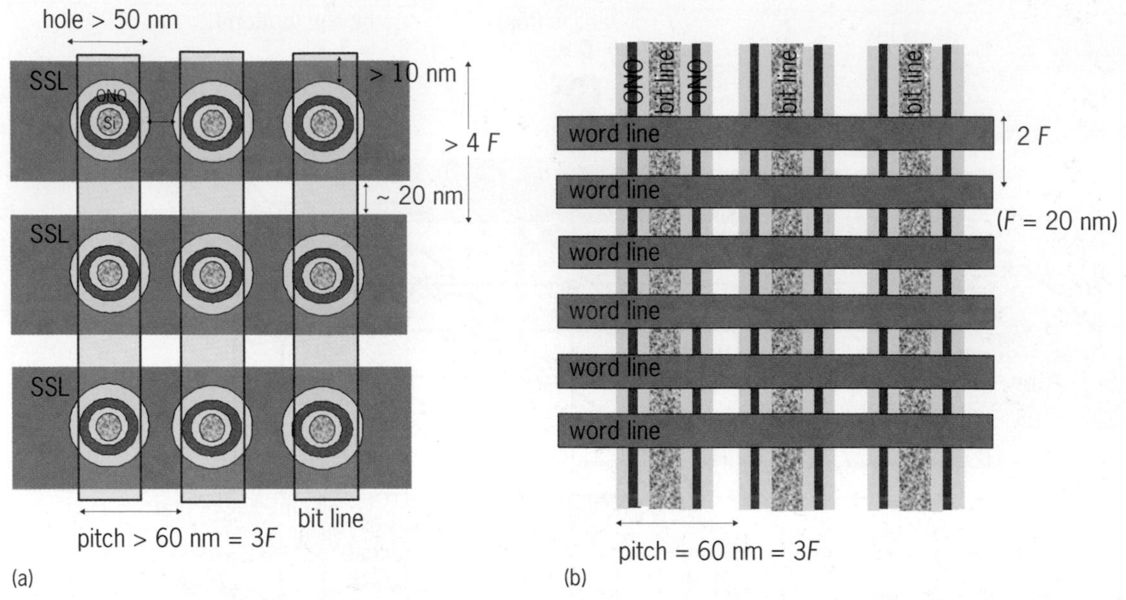

Fig. 6. Scaling limitations (cell size and minimal feature size) of vertical-channel and horizontal-channel 3D NAND arrays. (*a*) Vertical-channel 3D NAND. Minimal cell size is greater than $12F^2$ (F = 20 nm). (*b*) Horizontal-channel (vertical-gate) 3D NAND. Minimal cell size is approximately $6F^2$ (F = 20 nm).

being read. The top of the poly-Si gates is strapped by a low-resistance metal [tungsten silicide (WSi_2) or another metal silicide is commonly used] to reduce the *RC* delay in the word line.

3D NAND scaling. **Figure 6** compares the scalability of vertical-channel and horizontal-channel 3D NAND architectures. Figure 6*a* and *b* look down from the *z* direction onto the *x*-*y* plane. For a vertical-channel structure (Fig. 6*a*), the size of the cell is determined by the size of the punch hole in both *x* and *y* directions, plus overlay tolerances between the nanowire storage device and the edges of the string selecting transistor (that is, the SSL) in the *y* direction. The size of the punch hole must be large enough to accommodate twice the thickness of the ONO layers, plus room for the nanowire channel, and thus is approximately 40–50 nm, or larger. The overlay tolerances from the hole to the edges of the SSL are 10–15 nm each, and thus add 20–30 nm in the *y* direction. Thus, for a minimum feature size of $F = 20$ nm, the *x* dimension of the cell is slightly larger than $3F$ and the *y* dimension is greater than $4F$, resulting in a cell size greater than $12F^2$. Two-dimensional NAND has no such overlay requirement, and the most advanced technology has a planar device structure and thus does not have to accommodate ONO thickness; it can achieve a nearly cross point $4F^2$–$5F^2$ cell size, where *F* was approximately 16 mm in early 2014.

Conversely, the horizontal-channel (vertical-gate) architecture (Fig. 6*b*) has a self-aligned double-gate structure that does not need additional overlay area. Thus, only the *x* dimension of the cell is limited by the ONO thickness, and, with no overlay constraints, there is no penalty in the *y* direction. Therefore, with a minimum feature size of $F = 20$ nm, the *x* dimension is approximately $3F$ and the *y* dimension is only $2F$, resulting in a cell size of approximately $6F^2$, more than two times smaller than that of the vertical-channel structure.

Yet, both 3D approaches have cell sizes larger than the $4F^2$ of 2D NAND, and this intrinsic disadvantage must be compensated by more device layers.

Bit cost for 3D NAND. If 3D NAND truly used only one patterning step to fabricate *N* stacking layers, the bit cost would be simply proportional to $1/N$, and would be very attractive. However, in reality, the need to independently contact all memory layers in order to decode each cell into *x*, *y*, and *z* coordinates adds to both contact area and processing steps. If no innovation is conceived and each new layer requires an extra patterning process and decoding area, the bit cost may be described by **Fig. 7*a***. The bit cost initially decreases, but when the layer number increases it rises again, reflecting the fact that the advantage of higher density is offset by the more complex process. Several innovations, however, have drastically changed this scenario. Many layers may be contacted using only one patterning step by plasma trimming the photoresist into a staircase shape. The principle of binary sum may be implemented to contact 2^N layers with only *N* patterning steps, a process nicknamed MiLC (minimal incremental layer cost). [The principle of binary sum is quite simple. Any integer can be expressed by a sum of the numbers $2^0, 2^1, 2^2, \ldots, 2^N$. For example, $5 = 2^0 + 2^2, 7 = 2^0 + 2^1 + 2^2, 9 = 2^0 + 2^3, \ldots$. Thus, by superimposing the processes, only *N* lithography-etching operations are sufficient to make contacts to 2^N layers. This is a very powerful approach, because, to double the density, 3D NAND needs to double

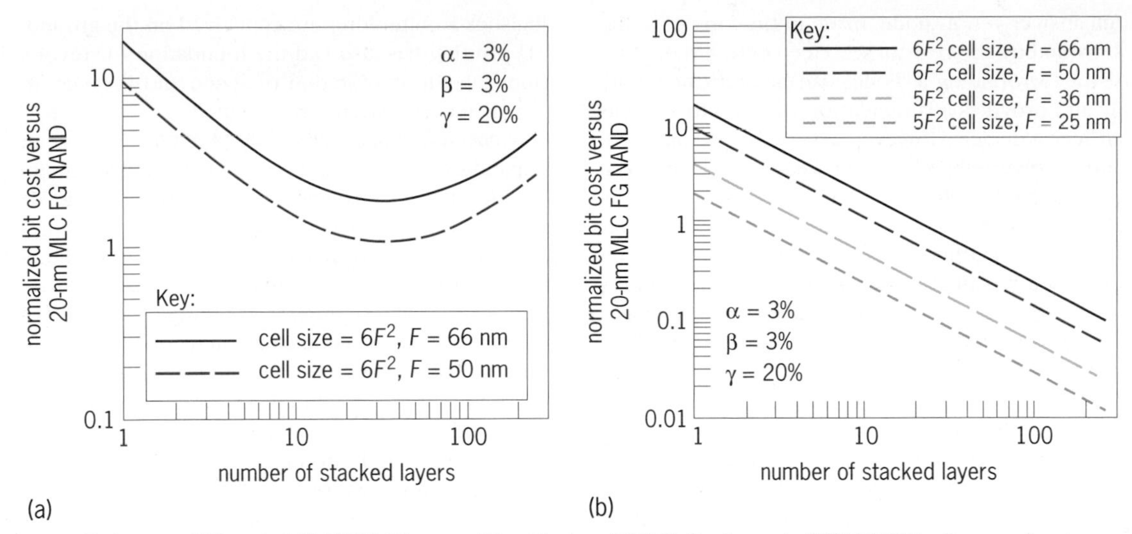

Fig. 7. Estimates of bit cost of 3D NAND [20-nm multilevel logic cell (MLC), floating-gate (FG) NAND] for linear and logarithmic increase of cost and area per added layer. (*a*) If fabrication cost and area increase linearly with layer number, then eventually fabrication cost will surpass density increase because layer number has to double at each new node. (*b*) If fabrication cost and area increase only logarithmically with layer number (through implementation of innovations such as binary sum, which needs only one additional lithography-etching step to double the layers), then bit cost can continue to decrease indefinitely. Alpha (α) and beta (β) are the incremental processing cost and area cost per additional 3D layer, respectively, and gamma (γ) is the peripheral circuit area increase from 2D to 3D. The assumptions of 3% and 20% are arbitrary (although not too far from practical expectations) for purpose of illustration.

the number of layers, and, by using this technique, needs to add only one processing step to accomplish layer doubling.] Finally, by using a tapered surface, contacts may be made edgewise to all layers without adding new processing steps. By implementing one of these innovations, the bit cost may continuously drop when the number of 3D layers increases (Fig. 7*b*).

The biggest challenge in bit cost is the restricted *x-y* scaling because of geometrical limitation. Because 2D NAND has already achieved a node size F less than 20 nm, with a cell size of approximately $4F^2$, the larger cell sizes of 3D NAND must be compensated by more layers. The vertical-gate arrays are more competitive than vertical-channel, because the former suffers geometrical limitation in only one direction (the x direction). At $F = 20$ nm, two layers of vertical-gate devices would surpass the cost of current 2D NAND, but it takes four layers of vertical-channel devices to do the same. Because 2D NAND still continues to scale, but 3D NAND suffers the geometrical limitation and cannot scale below $F = 20$ nm, the number of layers for 3D NAND to be cost competitive is likely to increase.

3D nonvolatile memory using two-terminal devices. Contrary to three-terminal devices such as transistors, two-terminal devices such as resistors and capacitors lack the means to modulate their outputs by a third terminal, and thus have been used mostly as passive elements rather than as active functional devices. However, resistance change is readily detectable and logic states based on resistance differences do not suffer the statistical fluctuation of number of storage electrons that limits the scaling of flash memory. Thus, memories based on resistance change have attracted much attention in the last several years as a potential replacement for NAND flash memory.

Resistance change may be achieved in a number of ways, through changes in magnetization state (in magnetic memory or MRAM), transitions between crystalline and amorphous phases (in phase-change memory or PCRAM), or changes in electrochemical state (in resistive memory or ReRAM). Unlike flash memory, in which the transistor serves as both the storage device (with or without stored electrons) and the memory access device (as an active transistor with high on/off ratio) that shuts off unwanted current flow when not in use, resistors do not have high enough on/off ratio and must be coupled with a separate access device (also known as a cell-selection device) to form a 1T1R (1-transistor, 1-resistor) memory cell. [For reference, flash memory has a simple 1T cell, while DRAM has a 1T1C (1-transistor, 1 capacitor) cell.]

Two-dimensional MRAM and PCRAM using an MOS transistor or diode as the access device have been in volume production recently. To expand into 3D, however, two-terminal devices face a natural barrier, which is how to identify a device by its (x, y, z) coordinates when the device has only two terminals and thus does not allow a third wire to connect to it. A recent innovation is a 3D array in which a MOS transistor that controls a string of resistors in the z direction is placed in the bottom *x-y* plane. The z-direction selection is accomplished by attaching a separate cell-selection device (for example, a diode) with each resistor [to form a 1D1R (1-diode, 1-resistor) cell, for example]. This arrangement solves the decoding issue, but has not provided the full answer yet. A diode, made in the form of a ring surrounding the resistor (or vice versa, a ring of resistor surrounding a diode) can be fabricated readily and provides a sufficient on/off ratio to serve as an access device. However, a diode is a *p-n* junction, and the reverse bias leakage relies on sufficient room for a dopant-depleted region, and thus is not scalable. (The distance required for depletion is greater than 40 nm.) Other cell-selection devices have been proposed recently, but at present the selection device is still a hot research subject and the challenges for 3D two-terminal device arrays have not been fully resolved.

For background information *see* COMPUTER STORAGE TECHNOLOGY; ELECTRICAL BREAKDOWN; ELECTRICAL COMMUNICATIONS; FIELD EMISSION; INTEGRATED CIRCUITS; LOGIC CIRCUITS; MICROLITHOGRAPHY; PERMITTIVITY; SEMICONDUCTOR MEMORIES; TRANSISTOR; TRAPS IN SOLIDS; TUNNELING IN SOLIDS in the McGraw-Hill Encyclopedia of Science & Technology.

Bibliography. S. H. Chen et al., A highly scalable 8-layer vertical gate 3D NAND with split-page bit line layout and efficient binary-sum MiLC (minimal incremental layer cost) staircase contacts, in *2012 IEEE International Electron Devices Meeting*, San Francisco, December 10–13, 2012, pp. 2.3.1–2.3.4, IEEE, Piscataway, NJ, 2012, DOI:10.1109/IEDM.2012.6478963; J. Jang et al., Vertical cell array using TCAT (terabit cell array transistor) technology for ultra high density NAND flash memory, in *2009 IEEE Symposium on VLSI Technology*, June 16–18, 2009, Honolulu, HI, pp. 192–193, IEEE, Piscataway, NJ, 2009; H. T. Lue et al., A highly scalable 8-layer 3D vertical-gate (VG) TFT NAND flash using junction-free buried channel BE-SONOS device, in *2010 Symposium on VLSI Technology*, June 15–17, 2010, pp. 131–132, IEEE, Piscataway, NJ, DOI:10.1109/VLSIT.2010.5556199; H. Tanaka et al., Bit cost scalable technology with punch and plug process for ultra high density flash memory, in *2007 IEEE Symposium on VLSI Technology*, Kyoto, Japan, June 12–14, 2007, pp. 14–15, IEEE, Piscataway, NJ, 2007, DOI:10.1109/VLSIT.2007.4339708.

3D printing applications in space

In the realm of manufacturing, additive processes, also known as 3D printing, are relatively new. Additive manufacturing technology constructs a desired three-dimensional geometry by additively building it layer upon layer. This technology differs completely from traditional manufacturing methods that manufacture destructively by taking away material until a final geometry is realized. The materials that are currently capable of being utilized in 3D printers are extensive, ranging from common household plastics to aerospace-grade metals such as aluminum and titanium. Current terrestrial-based development has opened the door to a paradigm shift in the way logistics architectures are conceived on the ground. 3D printing has also laid the foundation for revolutionizing the exploration of space and the way in which space missions are designed.

Spacecraft applications of 3D printing. There is an expanding interest in applying 3D printing technologies for use in spacecraft applications. Using 3D printing can drastically reduce the time and costs associated with manufacturing and assembly. Components that normally would require multiple parts to realize could be completed as a single part. Rocket engines in particular could benefit from this attribute greatly. Cooling and other fluidic channels that normally would be integrated out of many individual components, often by hand, could be designed into a single part and produced with the same ease as basic geometries. In July 2013, NASA's Marshall Space Flight Center (MSFC) successfully test-fired a rocket engine with an additively manufactured injector (**Fig. 1**), and other components are being investigated for their capability to be manufactured utilizing this technology.

Benefits of 3D printing in space. 3D printing ultimately serves as an expedient way to manufacture on demand and on location. The cost reduction benefits not only ground-based applications but also those in space. Under the current methods, getting anything into space requires it to be launched. This is very costly, and payloads that go up have to survive the extreme forces and vibrations associated with launch. Those payloads are also restricted in size to what can fit inside a launch fairing. If hardware meant for space could be manufactured on location, it would require less mass to produce, as it would not need to survive launch, and require less time to get there, and components could be produced that would be orders of magnitude greater in size than what is possible now.

Launching also requires planning months to years in the future. Sending replacements for noncritical part failures on the *International Space Station* (*ISS*) takes around 6 months, at best. Some of these failures that are very minor in nature may never be fixed. The on-demand capability of 3D printing will benefit these types of scenarios and expedite the logistics process associated with repairs and new parts. More critical items on the *ISS* have spares in case they are needed. This requires a large amount of internal volume for storage and upfront cost to procure the spares. Most of these spares are not used and never will be. If humans are ever to explore and expand further, this type of mission design is not supportable with current resource availability.

Zero-gravity testing. Made In Space, Inc. has been developing 3D printing technology for use in space to solve the logistics issues associated with space flight. This was first done through zero-gravity flights of NASA's Flight Opportunities Program. 3D printers and components were taken on parabolic aircraft flights and tested to determine the difference between manufacturing on the ground and in a simulated zero-gravity environment. More than 500

(a)

(b)

Fig. 1. Additively manufactured rocket injector. (*a*) Manufactured injector. (*b*) Testing of the component in a rocket engine at NASA's Marshall Space Flight Center (MSFC). (*Courtesy of NASA MSFC*)

parabolic flights accumulated around two hours of data about weightlessness with various testing devices. It was shown early that significant deviations, especially in layer thicknesses, occurred between the ground and flight samples on commercially supplied 3D printers. The company then modified and built printing hardware to solve the issues associated with the commercial platforms. This hardware can build nominally through various gravity regimes without manufacturing anomalies (**Figs. 2** and **3**).

3D Print. The company has several contracts to deliver additive manufacturing hardware and their peripheral devices to the *ISS* in the near term. The first machine, scheduled to launch in August 2014, is the first manufacturing device for space. The 3D Printing in Zero-G Experiment (3D Print), installed into the Microgravity Science Glovebox (MSG), can create material specimens that can be down-massed and tested back on Earth. The 3D Print can also manufacture usable parts that can be used by the *ISS* crew, as well as support ongoing science experiments on the *ISS*. This experiment is expected to be a stepping stone to the future of space manufacturing and to validate the capability of off-Earth manufacturing.

Additive Manufacturing Facility. The second facility that is being developed is the Additive Manufacturing Facility (AMF). The AMF is a commercial platform that will enable students, private citizens, companies, and government agencies to manufacture in space. The platform will be open so that a wide variety of applications can be enabled on demand. Entire experiments can be built without launch or existing experiments can be upgraded or repaired in a short period of time. This will allow a vital component to the experiment process that previously has been unavailable in space, rapid iteration. The results from work performed can be analyzed, changes can be produced and installed, and more data can be generated to greatly increase the science being performed.

The AMF will be capable of manufacturing space-rated materials, meaning that they will be able to survive in the space environment. This produces the opportunity for the production of outside components, including the structures of small satellites to be produced, and then utilized in the space environment. Both 3D printing devices will be available in the event that an emergency solution is needed. The crew on the *ISS* has been very resourceful in the past at improvising solutions for events that have occurred. Most recently, a tool was fabricated, primarily out of a toothbrush, to alleviate a potentially dangerous situation that could have led to the abandoning of the station. The manufacturing capability will allow for optimal fixes to be engineered and tested on the ground, then produced in space and used. This will eliminate the jury-rigging solutions that often are only held together with tape, which are solutions that have a substantially less chance of performing the required task.

The future of manufacturing in space is expected to enable new markets and capabilities that have yet to be explored. The limits that have traditionally governed development in space—consigning development to a long, expensive, and logistically difficult process—are on the verge of being overcome by the use of applying additive manufacturing in space.

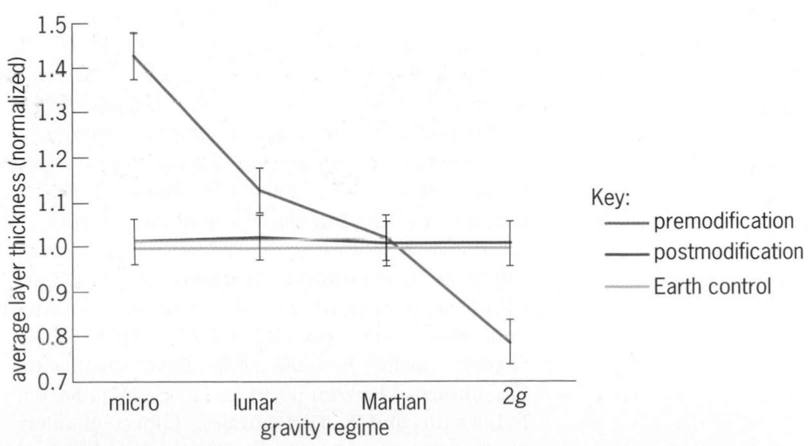

Fig. 2. Average layer thickness of samples across varying gravity regimes. (*Courtesy of Made In Space, Inc.*)

Fig. 3. The first manufacturing device to be utilized in space, the 3D Printing in Zero-G Experiment. (*Courtesy of Made In Space, Inc. and NASA*)

For background information *see* COMPUTER-AIDED DESIGN AND MANUFACTURING; SPACE FLIGHT; SPACE PROCESSING; SPACE STATION; SPACE TECHNOLOGY in the McGraw-Hill Encyclopedia of Science & Technology.

Michael P. Snyder

Bibliography. K. G. Cooper, *Rapid Prototyping Technology: Selection and Application*, Marcel Dekker, New York, 2001; J. J. Dunn et al., 3D printing on the *International Space Station*, in International Astronautical Federation, *64th International Astronautical Congress 2013*, Beijing, China, September 23-27, 2013, Curran Associates, Red Hook, NY, pp. 8976-8980, 2014; M. P. Snyder, Earthly benefits of additive manufacturing, *Environ. Forum*, 30(6):39, 2013; M. P. Snyder, J. J. Dunn, and E. G. Gonzalez, Effects of microgravity on extrusion based additive manufacturing, American Institute of Aeronautics and Astronautics, AIAA 2013-5439, pp. 1988-1993, in *AIAA Space 2013 Conference and Exposition*, San Diego, September 10-12, 2013, Curran Associates, Red Hook, NY, 2013, DOI:10.2514/6.2013-5439.

Three neutrino flavors and their mixing

Prominent among the three thrust areas of particle physics at present are the neutrinos. The three thrust areas are sometimes labeled the intensity frontier, the energy frontier, and the cosmic frontier. The intensity frontier aspires to maximize the event rate for rare processes. Neutrinos, characterized by some as "as close to nothing as something can be," are loath to interact. Fittingly, the study of neutrinos is at the heart of the intensity frontier endeavor.

The energy frontier concerns itself with maximizing the energy of human-made accelerators. This endeavor achieved a major breakthrough in 2012, with the discovery of the mysterious Higgs particle at the world's most energetic accelerator, the Large Hadron Collider (LHC), located near Geneva, Switzerland. The cosmic frontier studies nature's mysteries in the cosmic setting. These mysteries include dark matter, dark energy, cosmic rays, and again, neutrinos, this time at extreme energies. Unlike the cosmic rays, which are electrically charged and therefore bent by the cosmic magnetic field, and unlike photons, which at high energy are absorbed by the ubiquitous cosmic radiation, the charge-neutral neutrinos point back to their sources, and so have the potential to usher in the new field of neutrino astronomy.

This article reviews what is known and not known about neutrino physics. Much of what we know about neutrinos, and hope to learn in the future, is derived from a unique feature of neutrinos—"oscillation" among their "flavor" types. An initial neutrino flavor will in general oscillate into another flavor as the neutrino travels. Oscillations are a quantum mechanical phenomenon. Quantum mechanics is usually observed in the domain of the very small. However, one of the wonders of neutrinos is that their quantum mechanical aspect is observed over large distances, even astronomically large. For example, the explanation of the "anomalous" neutrino flux observed over the Sun-to-Earth distance relies on quantum mechanics. Nobel Prizes in 2008, 2002, 1995, and 1988 were given in whole or part for discoveries in neutrino physics.

Directions and angles. At any instant of time, the position and orientation of a rotating object, such as the Earth, a football, or a top, is describable by three angles (yaw, pitch, and roll, in the language of aerodynamics) measured relative to a fixed axis of $\hat{x}$, $\hat{y}$, and $\hat{z}$ directions. These three orientation angles are called Euler angles.

Why are there three angles, and not more or fewer? Some thought reveals that the number of angles is specified by a rotation in each of the independent planes of the space. Since three-dimensional space has three planes ($\hat{x}$-$\hat{y}$, $\hat{y}$-$\hat{z}$, and $\hat{z}$-$\hat{x}$), there are three independent rotations, each specified by an independent rotation angle. The final outcome of the three rotations depends on the ordering of the individual rotations. One can prove this simply by rotating a (rectangular) book through 90° about two axes in one order, and then in reverse order, and noting that the final outcomes are different. If we lived in two dimensions, say $\hat{x}$ and $\hat{y}$, there would be but a single plane, the $\hat{x}$-$\hat{y}$ plane, and therefore but one rotation angle. In four space dimensions, there would be six rotation angles. Continuing the count, in N space dimensions, there would be a number of planes given by the number of ways two axes may be chosen from the N total axes. Mathematicians denote this count as C_2^N; it is equal to $\frac{1}{2}N(N-1)$.

Neutrino mixing works in a similar fashion. If there are N distinct neutrinos, then their distinctness defines N axes. A vector in this neutrino-space will then have components along each axis, and so describe a linear combination of the "basis" neutrinos that define the axes. What should we take for the "basis" neutrinos? Quantum field theory tells us that it is the distinct mass states that best describe propagation over a distance (equivalently, the evolution of the neutrino in time—think of a movie of the motion). So we define the basis axes to lie along neutrino mass directions, and label the axes by the symbols ν_1, ν_2, and so forth, one axis for each neutrino.

Let us focus on the three known "active" neutrinos. There may be additional neutrinos, called "sterile" neutrinos. However, the evidence for additional neutrinos is not strong, and in some ways, the evidence is contradictory, so it is also possible that there are no more neutrinos to be discovered by our experiments. So we assume three neutrino axes, defined by the three neutrinos, ν_1, ν_2, and ν_3, of differing mass. These axes are the analogs of the fixed $\hat{x}$, $\hat{y}$, and $\hat{z}$ directions of space.

What is the analog of the body-centered orientation axes? It turns out that when neutrinos are produced by the weak interaction, such as occurs in decay of certain particles (charged pions, for example), in the fusion process inside stars such as the Sun, in supernova explosions, or in the big-bang epoch of the early universe, the produced neutrinos lie along another set of axes analogous to the body-centered frame. Our electronics do not actually "see" a neutrino. Commonly, at neutrino production or detection, the event transpires with an associated production or annihilation of a "charged lepton," either the electron/positron ($e^{\pm}$), the muon/antimuon ($\mu^{\pm}$), or the tau/antitau ($\tau^{\pm}$). Hence, these interacting neutrinos are named at production according to their charged lepton partner as the ν_e, the ν_μ, or the ν_τ. It is common to call the three neutrino types neutrino flavors. The new axes are then the direction basis in neutrino flavor space. These neutrino flavor axes are the analog of the body-centered axes for the rigid-body context. The two sets of axes, being three in number, are rotated with respect to each other by three Euler-like angles.

By convention, the neutrinos associated with the negative leptons are true neutrinos, while those associated with the positive antiparticles of the negative leptons are the antiparticles of the neutrino, that is, "antineutrinos."

Antineutrinos and a phase. For neutrinos and antineutrinos, an additional phenomenon may occur. This phenomenon arises purely from quantum mechanics. Quantum mechanics innately requires complex numbers for its description of nature. This means that, unlike the case of rotation matrices of "classical" physics, which are described purely with real numbers (the Euler angles), the rotation matrices of quantum mechanics may have complex phase factors. Recall that complex numbers introduce the definition $i \equiv \sqrt{-1}$. A complex phase factor is a number $e^{i\delta}$, where δ is a real number with a value in the interval $[0, 2\pi]$. Euler's formula helps to understand a phase factor: $e^{i\delta} = \cos\delta + i\sin\delta$. The phase factor has a real part ($\cos\delta$) and an "imaginary" part ($\sin\delta$), but always a unit length, as in Eq. (1).

$$|e^{i\delta}| = |\cos\delta + i\sin\delta| = \sqrt{\cos^2\delta + \sin^2\delta} = 1 \quad (1)$$

One may count how many phases there are in a world with N neutrinos. The mixing matrix is called U. The elements of U, $U^*_{\alpha j} \equiv \langle\nu_\alpha|\nu_j\rangle$, express the amount of overlap of the unit ν_α neutrino flavor axis along the unit ν_j neutrino mass axis. (The Dirac bracket $\langle\nu_\alpha|\nu_j\rangle$ is common notation for the complex-valued generalization of the dot product $\vec{v}_1 \cdot \vec{v}_2$ that describes the overlap of two real-valued vectors.) Since $\alpha = 1, \ldots N$ and $j = 1, \ldots N$, U is an $N \times N$ matrix. An equivalent statement is that the matrix U^* "rotates" the neutrino mass vector $(\nu_1, \ldots, \nu_N)$ into the neutrino flavor vector $(\nu_e, \nu_\mu, \nu_\tau, \ldots, \nu_N)$. ($U^*$ means U but "complex conjugated," that is, with all phases reversed in sign.) The mixing matrix U depends on angles and phases. A physics constraint that the number of neutrinos be the same when referred to either the mass axes or the flavor axes is that the complex-valued matrix U be "unitary," which has a technical definition that is not needed here. What is needed is the result that any $N \times N$ unitary matrix has N^2 free parameters. Furthermore, with N particle pairs, each consisting of a neutrino ν_α and a charged lepton α ($\alpha = 1, 2, \ldots, N$), it can be shown that $(2N - 1)$ relative phases are absorbable into definitions of the complex-valued wave functions describing each particle. So $N^2 - (2N-1) = (N-1)^2$ is the number of physical parameters that are left to describe neutrino mixing, and to be determined by experiments. We used $N_{\text{angles}} \equiv \frac{1}{2}N(N-1)$ of these parameters for the rotation angles in U. That leaves $N_{\text{phases}} \equiv (N-1)^2 - \frac{1}{2}N(N-1) = \frac{1}{2}(N-1)(N-2)$ physical phases in the mixing matrix U.

For a number of reasons, $N_{\text{phases}} = \frac{1}{2}(N-1)(N-2) = 0$, for $N \geq 3$ is a very interesting result. First, it can be shown that the antineutrino mixing matrix is not U, but rather U^*. For a complex-valued U, we have $U^* \neq U$, so the neutrinos and the antineutrinos mix differently. This difference (due to nonzero phases) is being sought in experiments. Second, notice that if there were but one or two neutrino types (flavors), then N_{phases} would equal 0, and the mixing matrix U would be real-valued. Thus, it is first at $N = 3$ that a nonzero phase may "complexify" the neutrino mixing matrix. The 2008 Nobel Prize in physics was awarded for elucidation of this fact.

Ordering the rotations. Because the rotations do not commute, a convention must be adopted for which rotation axes are chosen and in what order. The conventional matrix for mixing among three neutrinos, as established by the Particle Data Group (PDG), is given by Eq. (2). Here, $R_{jk}(\theta_{jk})$ describes

$$U_{\text{PMNS}} = R_{32}(\theta_{32})U_\delta^\dagger R_{13}(\theta_{13})U_\delta R_{21}(\theta_{21})$$

$$= \begin{pmatrix} c_{21}c_{13} & s_{21}c_{13} & s_{13}e^{-i\delta} \\ -s_{21}c_{32} - c_{21}s_{32}s_{13}e^{i\delta} & c_{21}c_{32} - s_{21}s_{32}s_{13}e^{i\delta} & s_{32}c_{13} \\ s_{21}s_{32} - c_{21}c_{32}s_{13}e^{i\delta} & -c_{21}s_{32} - c_{21}c_{32}s_{13}e^{i\delta} & c_{32}c_{13} \end{pmatrix} \quad (2)$$

a rotation in the jk-plane through angle θ_{jk}, $U_\delta = \text{diag}\,(e^{i\delta/2}, 1, e^{-i\delta/2})$, and $s_{jk} \equiv \sin\theta_{jk}$, $c_{jk} \equiv \cos\theta_{jk}$. The acronym PMNS stands for Bruno Pontecorvo, Ziro Maki, Masami Nakagawa, and Shoichi Sakata, all contributors to the early history of neutrino oscillation physics. Note that the phase factor $e^{\pm i\delta}$ is always accompanied by the factor $\sin\theta_{13}$. So if $\theta_{13} = 0$, then the phase δ cancels from U_{PMNS} and U_{PMNS} becomes real-valued (and much simpler).

Two additional phases are omitted that are present only if the neutrinos are of "Majorana" type. These additional Majorana phases do not enter into neutrino oscillations. Whether neutrinos are of Majorana or Dirac types is at present an open question and an active experimental subfield, which will not be pursued here.

What we know. We have a three-active neutrino sector parameterized by three mixing angles (θ_{13}, θ_{21}, and θ_{32}) and one phase (δ) in U, and three neutrino masses (m_1, m_2, and m_3). To date, all three mixing angles have been inferred from neutrino oscillation data. In fact, because of the environments—nuclear reactors, the Sun, and the atmosphere—from which their values were first deduced, these three mixing angles are sometimes referred to as θ_R, $\theta_\odot$, and θ_A. Likewise, the mass-squared difference $\delta m_{21}^2 \equiv m_2^2 - m_1^2$ has been inferred from oscillation data, as has the absolute value of the difference $|\delta m_{32}^2| \equiv |m_3^2 - m_2^2|$.

The neutrino mixing angles are shown geometrically in **Fig. 1**, and their values are presented in the **table**. The generous neutrino mixing angles θ_{12}, θ_{32}, and θ_{13} are 3, 20, and 50 times larger than the corresponding angles from the quark sector. This is one of many surprising features of neutrino physics, and a very beneficial feature for neutrino experimenters. Conversely, the large angles present a challenge to neutrino theorists, who hope to explain them.

How does it come about that we know the absolute sign of δm_{21}^2? It is because this mass-squared

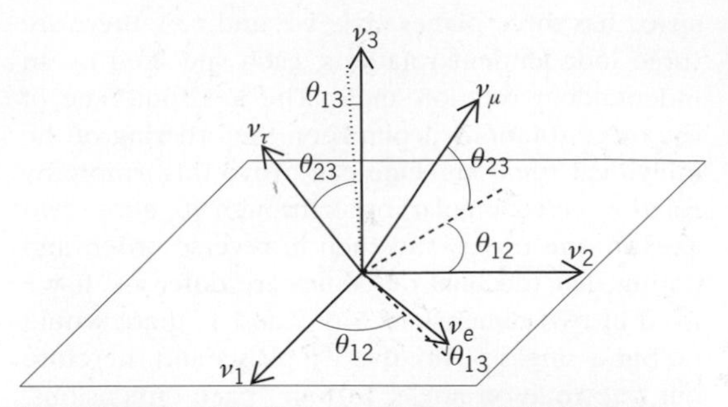

Fig. 1. Display of the three mixing angles that characterize the orientation of the neutrino flavor axes (ν_e, ν_μ, and ν_τ) relative to mass axes (ν_1, ν_2, and ν_3). (*Figure courtesy of S. King*)

Best fit values of neutrino mixing angles compared to quark mixing angles*

	Neutrinos (PMNS)	Quarks (CKM)†
θ_{12}	35°	13°
θ_{32}	43°	2°
θ_{13}	9°	0.2°
δ	unknown	68°

*From oscillation data, $\delta m_{21}^2 = 0.8 \times 10^{-4}$ eV^2 and $|\delta m_{32}^2| = 2.5 \times 10^{-3}$ eV^2 for the neutrino sector.
†CKM denotes the researchers Nicola Cabibbo, Makoto Kobayashi, and Toshihide Maskawa.

difference was inferred from solar neutrino observations; in the Sun, the background of electrons affects ν_e differently than ν_μ and ν_τ. We do not have the space here to explain this subtle effect, but we note that it tells us that the lighter of ν_1 and ν_2 must contain more ν_e in vacuum (free space), while the heavier of the two mass states must contain more ν_e as it emerges from the Sun. These facts fix the mass ordering to be $m_1 < m_2$.

What we do not know yet. Completely unknown at present are the single phase δ, related to any neutrino-antineutrino asymmetry, and the sign of δm_{32}^2. The absolute neutrino masses m_3 and m_2 are, of course, bounded to be at least $\sqrt{\delta m_{32}^2} \sim 0.05$ eV and $\sqrt{\delta m_{21}^2} \sim 0.01$ eV, but the ordering of these two masses is unknown. Whether $m_3 > m_2$ (called the normal mass hierarchy), or $m_3 < m_1$ (called the inverted mass hierarchy) is a central issue in neutrino physics. Direct searches for neutrino masses as kinks in the energy spectrum of the electron from tritium decay yield an upper bound $m_j \lesssim 1$ eV, while arguments from the growth of large-scale structure in the early universe yield the upper bound $\Sigma_j m_j \lesssim 0.5$ eV. Clearly, the three active neutrino masses are very light compared to all other known massive particles, but nonzero (except perhaps for the lightest neutrino). Even the tiny electron mass is 511 keV, a million times or more than that of the neutrino, while the mass of the arguably fundamental top quark is almost another factor of a million larger.

Oscillation phenomenon. Oscillations in time or distance between two states nearby in energy is a common quantum mechanical effect, with a description

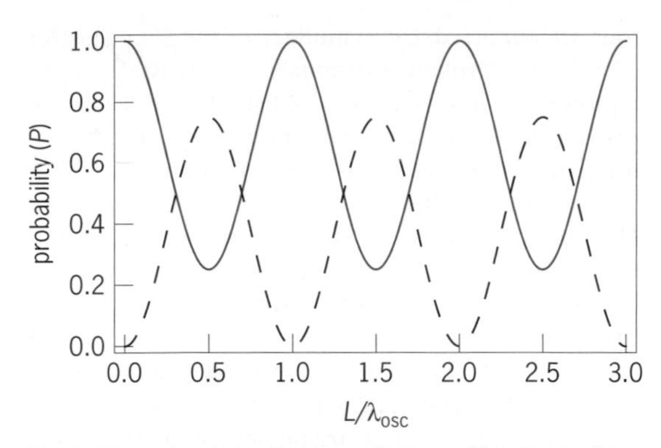

Fig. 2. Example of two-flavor oscillation, with initial ν_μ (blue curve) oscillating to ν_e (broken black curve). The mixing angle is $\theta = 30°$, that is, $\sin^2(2\theta) = \frac{3}{4}$. *(Figure courtesy of Lingjun Fu)*

available in any quantum mechanics textbook. For a two-flavor system, the probability for the initial flavor state α to survive, rather than to oscillate to the second state β, is given by Eq. (3). Here, E is the

$$P_{\alpha\to\alpha} = 1 - \sin^2(2\theta)\sin^2\left(\frac{L\delta m^2}{4E}\right) = 1 - P_{\alpha\to\beta} \quad (3)$$

neutrino energy, δm^2 is the mass-squared difference, and L (or $t = L/c = L$ in our units) is the distance from the source. Notice that the first $\sin^2$ with mixing angle as its argument governs the size of the oscillation, while the argument of the second $\sin^2$ establishes the oscillation length as $\lambda_{osc} = 4\pi E/\delta m^2$ (**Fig. 2**). Thus, a measurement of oscillations can infer both the mixing angle θ and the mass-squared difference $|\delta m^2|$. These results generalize to three-flavor systems, although the formulae become more complicated.

Discovery of large θ_{13} and its implications. For many years, the sparse data on the angle θ_{13} allowed its consistency with zero. However, in the spring of 2012, θ_{13} was definitively measured to be nonzero, and by a large margin. This discovery of a "large" θ_{13} has several consequences. Four that will be discussed briefly are (1) the increased reach for experiments to infer the neutrino mass hierarchy; (2) the significantly larger amount of particle-antiparticle asymmetry (called CP violation) in the theory (which could explain the asymmetric fact that we now live in a matter-dominated universe, rather than a matter-antimatter symmetric universe); (3) the breaking of ν_μ-ν_τ symmetry; and (4) the increased difficulty of constructing models to accommodate the large observed mixing angles.

Point (1) is explained in that oscillations in vacuum and in matter are enhanced by larger mixing angles. Point (2) is more subtle. It turns out that the parameterization-independent measure of CP violation in the three-neutrino system is given by $J \equiv |\text{Im}(U_{\alpha j}U^*_{\beta j}U_{\beta k}U^*_{\alpha k})|$, for any $\alpha \neq \beta$ and any $j \neq k$. With three neutrinos, J has the same value for any choice of these indices, a fact related to the one and only phase in the mixing matrix. In terms of the PDG parameters given in Eq. (2), one finds that J is given by Eq. (4). Clearly, all three angles

$$J = \frac{1}{8}\cos\theta_{13}\sin(2\theta_{13})\sin(2\theta_{12})\sin(2\theta_{32})\sin\delta$$
$$\sim 0.036\sin\delta \quad (4)$$

and the phase must be nonzero for J to be nonzero. We already saw that a zero value for θ_{13} implies a real-valued U_{PMNS}, which in turn implies a vanishing value for J; Eq. (4) generalizes that finding. Notice that J grows as $\sin(2\theta_{13})$, so as θ_{13} moves away from zero, J increases. One may infer the robustness of this result by contrasting J with the analogous quantity in the quark sector, $J_Q \approx 0.30 \times 10^{-4}$. We have $J \approx (1000\, J_Q)\sin\delta$. Of course, $\sin\delta$, completely unknown at present, must be nonzero for J to be nonzero.

Point (3) concerns the breaking of ν_μ-ν_τ symmetry. When θ_{32} is set to the maximal mixing value of 45° and θ_{13} is set to zero, a significant increase in symmetry of the mixing matrix arises. The norms $|U_{\mu j}|$ become equal to the norms $|U_{\tau j}|$, for each $j = 1, 2, 3$. This is the mathematical statement of what is termed ν_μ-ν_τ symmetry. If, in addition, the norms of all second column elements $U_{\alpha 2}$, $\alpha = e, \mu, \tau$, are set to be identical, the symmetry is increased further. This further symmetry requires that $\theta_{21} = \tan^{-1}\left(\frac{1}{\sqrt{2}}\right)$. This version of ν_μ-ν_τ symmetry came to be known as tri-bimaximal mixing (TBM), the unwieldy name reflecting some vestigial history. The TBM mixing matrix is given by Eq. (5). When

$$U_{TBM} = \frac{1}{\sqrt{6}}\begin{pmatrix} 2 & \sqrt{2} & 0 \\ 1 & \sqrt{2} & \sqrt{3} \\ 1 & -\sqrt{2} & \sqrt{3} \end{pmatrix} \quad (5)$$

θ_{13} was thought to be zero or nearly so, U_{TBM} was the natural and popular choice for the zeroth-order matrix about which perturbations could be added to better accommodate data.

Incidentally, the equality of norms in the middle column of U_{TBM} implies an equality of flavors in the mass state ν_2. This equipartition of neutrino flavors among the ν_2 state resolved the "solar anomaly," which had vexed neutrino and solar physicists for more than 20 years. Because of the smoothly varying electron density from the Sun's core to its corona, the solar neutrinos, which are produced in the fusion cycle of the core as pure flavor states ν_e, emerge from the Sun as almost pure mass states ν_2 [via subtle quantum mechanics called the Mikheyev-Smirnov-Wolfenstein (MSW) effect]. Thus, the solar neutrinos are perceived on Earth as very nearly equal mixtures of ν_e, ν_μ, and ν_τ, that is, the measured ν_e flux at Earth is only one-third of the original solar ν_e flux.

Now that θ_{13} is known to be approximately 9°, and 10 standard deviations removed from zero, the ν_μ-ν_τ symmetry of TBM is broken (barring special values of the unknown phase δ). This leaves model building of the neutrino masses and mixing in a much more complicated and confused state, encapsulated in point (4). In particle physics, scientists are accustomed to breaking a symmetrical situation "perturbatively," meaning with small asymmetries. The inferred value of θ_{13}, $\sim 9^\circ$, is not perturbatively small. The ν_μ-ν_τ symmetry of TBM is rather badly violated. One possibility is to introduce large perturbations. The question of whether such a direction will bear any fruit is unresolved, and it is right to ask whether the TBM basis with subsequent large symmetry breaking should be replaced by some other, more symmetric basis that accommodates smaller symmetry breaking than does the TBM model. Again, the fruitfulness of the "new basis" approach is still undetermined.

Neutrino model building and group theory. Generally, the neutrinos' kinetic energy terms and interaction terms are more symmetric than the mass terms; the latter therefore provide symmetry-breaking information. Diagonalization of the mass matrix produces the transformation from the interaction or flavor axes to the mass axes, that is, it, provides the mixing matrix U_{PMNS}. The resulting mixing matrix generally shows some residual symmetry, but also some significant symmetry breaking. For example, the TBM mixing pattern of Eq. (5) follows naturally from diagonalization of the four-parameter mass matrix shown in Eq. (6). This mass matrix has ν_μ-ν_τ

$$M_{\text{TBM}} = \begin{pmatrix} \mu_1 & \mu_2 & \mu_2 \\ \mu_2 & \mu_3 & \mu_4 \\ \mu_2 & \mu_4 & \mu_3 \end{pmatrix} \tag{6}$$

symmetry, as shown by invariance under simultaneous interchange of the second and third rows and second and third columns (the ν_μ and ν_τ rows and columns). This ν_μ-ν_τ symmetry is reflected in the U_{TBM} mixing matrix of Eq. (5).

A discussion of flavor symmetry and symmetry breaking necessarily invokes the mathematics of group theory and group representations. Suppose the neutrinos were massless, or mass-degenerate, meaning that all neutrinos had a common nonzero mass. Then interchange of the neutrinos, or rotation among them, could not change the physics. The collection of all three-dimensional spatial rotation operations is the group named SO(3). The meaning of the "3" is obvious, the "O" stands for "orthogonal," and the "S" signifies unit determinant. However, quantum mechanics is intrinsically complex-valued rather than real-valued, and so the rotations in neutrino space are unitary rather than orthogonal, leading to the "special unitary group" SU(3), characterized by 3×3 unitary matrices having unit determinant. The unit determinant constraint removes one parameter, leaving an $N^2 - 1 \overset{N=3}{\Longrightarrow}$ eight-parameter group. The eight-parameter group SU(3) is then the symmetry group of the three flavors, before any symmetry breaking due to mass differences is admitted.

In the standard model of particle physics, the active neutrinos and their charged lepton partners must rotate together under each SU(3) rotation. When mass differences and off-diagonal terms in the mass matrices of the neutrinos or the charged leptons are admitted, the symmetry of the SU(3) group is said to be "broken." Diagonalization of the charged lepton mass matrix is accomplished with a unitary matrix called $V_{l\pm}$. Diagonalization of the neutral lepton ("neutrino") mass matrix is accomplished with a unitary matrix called V_ν. (If the neutrinos are of the Majorana type, then the mass matrix can be shown to be symmetric, and diagonalization proceeds via $V_\nu M V_\nu^{\text{T}}$, where V_ν^{T} is the transpose of V_ν, rather than via $V_\nu M V_\nu^\dagger$, where $V_\nu^\dagger$ is the conjugate of the transpose of V_ν.) If the two matrices $V_{l\pm}$ and V_ν were identical, then the common unitary rotation would just reflect the SU(3) invariance of the mathematics (what physicists call the "Lagrangian"), and would not correspond to any physical effect. However, if the two matrices are different, then there is physics involved. The invariant mixing matrix (up to some quantum mechanical phases) is the misaligned product $U_{\text{PMNS}} \equiv V_{l\pm} V_\nu^\dagger$. Since overall rotations are not physical, it is convenient, allowable, and common to work in a rotated basis where the charged lepton matrix is already diagonal. But the underlying physics is that symmetries of the massless Lagrangian may be broken in the charged lepton sector via $V_{l\pm}$ or V_ν, or more likely, both.

In the language of group theory, we say that the initial symmetry, here SU(3), is broken by masses to a subgroup $G_{l\pm}$ for the charged leptons, and a subgroup G_ν for the neutrinos. Then U_{PMNS} results from the mismatch of the way nature breaks the large SU(3) symmetry to the two subgroups $G_{l\pm}$ and G_ν. The full continuous group SU(3) offers many breaking patterns that fail to constrain the resulting mixing angles. Recent models constructed to "explain" the two large and one small mixing angles of the neutrino sector use discrete subgroups of SU(3) for $G_{l\pm}$ and G_ν. The smaller discrete subgroups such as S_3, S_4, A_4, A_5, result in an overconstrained system of mixing parameters, and therefore predict mixing angles or relations among mixing angles. Individual particles are assigned to a "group representation," a kind of flavor vector whose members rotate among themselves under general group rotations. Each distinct group has its own unique set of "representations." The small discrete groups contain several singlet representations for particles, and some doublet and triplet representations, perfect for assignments of the three active neutrinos and three charged leptons. Once a particular discrete flavor group is chosen and particles are assigned to the group's representations, the flavor group is broken by assigning large vacuum expectation values ("vevs") to flavor scalars (called flavons or familons), into the neutrino sector group G_ν and into the charged-lepton sector group $G_{l\pm}$. The small groups, including those just mentioned, tend to contain an inherent ν_μ-ν_τ symmetry, which, as we have seen, leads to the prediction that θ_{13} is zero, or in the case of natural choices for symmetry breaking, nearly zero. Consequently, the recent measurement of a rather large value for θ_{13}, 9°, presents a challenge to the implementation of these small discrete subgroups.

An alternative approach is to embrace the larger discrete subgroups of SU(3). Larger groups present larger representations for the particle assignments, and tend at leading order (that is, without perturbations) to get θ_{13} about right. The subgroups $\Delta(96)$ and $\Delta(384)$ are examples in current use. (Here the argument of the discrete group refers to the number of elements in the group.) While these groups may appear "big" compared to the first generation of ansätze, we should not summarily dismiss particular groups as being "too large." Nature, the ultimate arbiter, may have chosen one of them to fit "just right."

A consistency check on the validity of some discrete groups is provided by the model's predictions for rare flavor-changing processes, such as $\mu \rightarrow e\gamma$ and $\mu \rightarrow 3e$. In the standard model, such branching ratios of the muon are proportional to m_ν^2, and so negligibly small. But beyond the standard model, such branching ratios may be observable, due to enhancements from the flavor scalars. The MEG experiment has attained an upper limit of the branching ratio $\mu \rightarrow e\gamma$ of 5×10^{-13}; a reach to 5×10^{-14} is expected within this decade. Also, at Fermilab in the United States, and at J-PARC in Japan, dedicated experiments to search for $\mu \rightarrow e$ transitions in nuclei are presently under construction. Results from these experiments, now and in the future, invalidate or constrain particular neutrino-mixing models, or will do so.

Summary. This article has overviewed some experimental and theoretical aspects of one of the most exciting arenas in present-day particle physics, that of the neutrino. The three angles and one phase that characterize the misalignment of the neutrino "flavors" and neutrino masses were introduced. These parameters, and the neutrino masses themselves, were shown to enter and emerge from neutrino oscillation studies. The phenomenon of neutrino oscillations is quite interesting in its own right, being a macroscopic manifestation of quantum mechanics at work.

The recent inference of the nonzero value for θ_{13} completes our knowledge of the three neutrino mixing angles. The article discussed how the fortuitously large value of θ_{13} increases the reach of experiments to reveal the ordering of the neutrino masses (the mass hierarchy), and to discover matter-antimatter asymmetries in the neutrino sector. Conversely, the large value of θ_{13} also complicates the building of group-theoretic models that are presumed to underlie the neutrino mixing angles and mass values.

Neutrino physics is very much an ongoing enterprise, both experimentally and theoretically. If the past is any guidepost to the future, surprising results can be expected from the continued search into the nature of neutrinos, their masses, mixings, and interactions.

For background information *see* ANTIMATTER; COMPLEX NUMBERS AND COMPLEX VARIABLES; ELECTROWEAK INTERACTION; ELEMENTARY PARTICLE; EULER ANGLES; GROUP THEORY; LEPTON; MATRIX THEORY; NEUTRINO; NEUTRINO ASTRONOMY; QUARKS; RADIOACTIVITY; SOLAR NEUTRINOS; STANDARD MODEL; SYMMETRY BREAKING; SYMMETRY LAWS (PHYSICS); TIME REVERSAL INVARIANCE; WEAK NUCLEAR INTERACTIONS in the McGraw-Hill Encyclopedia of Science & Technology. Thomas J. Weiler

Bibliography. J. K. Ahn et al. (RENO Collaboration), Observation of reactor electron antineutrinos disappearance in the RENO experiment, *Phys. Rev. Lett.*, 108:191802 (6 pp.), 2012, DOI:10.1103/PhysRevLett.108.191802; F. P. An et al. (Daya Bay Collaboration), Improved measurement of electron antineutrino disappearance at Daya Bay, *Chin. Phys. C*, 37:011001 (21 pp.), 2013, DOI:10.1088/1674-1137/37/1/011001; F. P. An et al. (Daya Bay Collaboration), Observation of electron-antineutrino disappearance at Daya Bay, *Phys. Rev. Lett.*, 108:171803 (7 pp.), 2012, DOI:10.1103/PhysRevLett.108.171803; Y. Kuno, A search for muon-to-electron conversion at J-PARC: The COMET experiment, *Progr. Theor. Exp. Phys.*, 2013:022C01 (43 pp.), 2013, DOI:10.1093/ptep/pts089.

Tiger Milk mushroom

The Tiger Milk mushroom (*Lignosus rhinocerotis*; also referred to as *L. rhinocerus*) [**Fig. 1**] is hailed as a national treasure in Malaysia. Local communities have regarded it as an indispensable medicinal mushroom because of its ability to aid in the treatment of numerous ailments. It has been used traditionally for more than 400 years to treat lung and respiratory diseases, asthma, cough, fever, cancer, chronic hepatitis, gastric ulcer, and food poisoning. It is also used as a general tonic.

The Diary of John Evelyn (June 1664) recorded the name of this mushroom as "Lac tygridis" (meaning tiger's milk). According to Malaysian folklore, it was believed that the Tiger Milk mushroom grows

Fig. 1. The Tiger Milk mushroom, *Lignosus rhinocerotis*.

Fig. 2. Sclerotial tissues of the Tiger Milk mushroom are compact and white internally, with a thin, red-brown rind.

at the very spot where a prowling tigress dropped her milk during lactation.

Structurally, the Tiger Milk mushroom (abbreviated TMM) is characterized by a central stipitate pileus [that is, a mushroom cap growing at the end of a stipe (stem)] arising from a distinct buried sclerotium (a food-storage body) [**Fig. 2**], which is the part with the medicinal value. The occurrence of this mushroom in the tropical rainforest is always solitary. Thus, the collection of its underground sclerotia is a tedious and difficult task. As a result, the supply has been limited, hindering the discovery of its full potential to provide health and economic benefits. Recently, however, investigators have reported the successful cultivation of the mushroom with a high production yield, thereby overcoming the supply issue.

Biological properties. There has been a great deal of interest in the TMM in the past few years because of its wide-ranging ethnobotanical uses and its recent successful domestication. Many studies have been initiated to examine its safety and biopharmacological efficacy in order to validate its ethnobotanical claims. Research findings have revealed that the TMM sclerotia contain various biologically active substances, including polysaccharides, polysaccharide-protein complexes (PSPCs), and β-glucans, which demonstrate significant anti-inflammatory, antioxidant, antiproliferative, and immunomodulating effects.

Immunomodulatory activity. Fungal polysaccharides are known as immunopotentiators because they can boost the immune system. The mechanism of immunomodulation by these polysaccharides involves the stimulation of immune cells such as macrophages, natural killer (NK) cells, and T lymphocytes to release cytokines. For example, an in vitro study has reported that the TMM sclerotial polysaccharides stimulated the proliferation of NK cells and macrophages, and another study has confirmed that the TMM sclerotial polysaccharides exhibited immunomodulatory effects by stimulating human innate immune cells and T-helper cells. Currently, various investigations are studying the sclerotial polysaccharides to decipher the receptor-mediated pathways for the immunomodulatory activities of the immune cells.

Antiproliferative activity. The TMM sclerotial polysaccharides have been shown to possess a remarkable degree of antitumor activity in cancer cell–implanted mice, and notable direct cytotoxic effects have been seen in various human leukemic cell lines, human breast carcinoma cell lines, and human lung carcinoma cell lines. A preliminary analysis suggests that the TMM exerts its antiproliferative activity toward cancer cells in a selective manner through the induction of cell cycle arrest and apoptosis (programmed cell death).

Anti-inflammatory activity. Investigations with the extracts of the TMM sclerotia in an in vivo rat model have indicated that the TMM has an acute anti-inflammatory effect. The activity is mainly the result of the presence of the PSPCs. Further research on the mechanism of this property is ongoing.

Nutrient composition and antioxidant activity. A study to examine the nutrient composition and antioxidant properties of the sclerotial powder of the wild-type and cultivated TMM has shown that the sclerotia have a high carbohydrate content but a low fat content. TMM sclerotial extracts have been shown to exhibit strong superoxide anion radical-scavenging activity. Interestingly, the sclerotium of the cultivated TMM is richer in its nutrient composition and has a stronger antioxidant capacity than that of the wild type. These differences are not entirely surprising. The wild-type sclerotium is usually collected only after the formation of the cap because, prior to that, the sclerotium is buried beneath the ground and hence is not noticeable. On the other hand, the sclerotium from the cultivated TMM is harvested just before the formation of the stem and cap; hence, it is at an earlier stage of maturation of the sclerotium and has a higher level of bioactive components.

Antiasthmatic activity. The antiasthmatic effect of the TMM has been investigated in asthmatic rats. Interestingly, the TMM extracts significantly reduced a number of asthmatic parameters in the treated rats, including the levels of total immunoglobulin E (IgE) in serum and the levels of various cytokines in bronchoalveolar lavage fluid (BALF). They also effectively suppressed the numbers of eosinophils (proinflammatory white blood cells) in BALF and attenuated the infiltration of eosinophils in the lung. This finding is the first animal study supporting the ethnobotanical use of the TMM in treating asthmatic problems.

Neurite-stimulating activity. An in vitro study has reported that an aqueous extract of TMM sclerotia contained nerve growth factor (NGF)–like compounds that enhanced neurite outgrowth activity. The aqueous extract stimulated neurite outgrowth in PC-12 cell lines, and a combination of NGF and the extract had additive effects and enhanced neurite outgrowth.

Novel prebiotics. The nondigestible carbohydrates extracted from the TMM sclerotia have been shown to stimulate the growth of *Bifidobacterium longum* and *Lactobacillus brevis*, thus suggesting the potential application of TMM extracts as novel prebiotics for gastrointestinal health.

Genomics and proteomics. In an effort to accelerate the exploitation of TMM extracts, a research group has assembled and annotated the genome sequences of the TMM. The genome-based proteomic analysis of the TMM sclerotia has also revealed a number of proteins with pharmaceutical potential, including putative lectins, immunomodulatory proteins, aegerolysin, and antioxidant proteins. The findings provide imperative resources to facilitate future work on the characterization of the bioactive compounds from the mushroom.

Safety assessment. The TMM has been used extensively and safely for hundreds of years with no known toxicity and side effects, and its safety has also been assessed scientifically. The preclinical toxicological evaluations of cultivated TMM sclerotia have indicated that there are no treatment-related subacute toxicities or chronic toxicities in rats following oral administration of TMM sclerotial powder as measured by weight, general observations, hematological and clinical biochemistry analysis, and histological examinations of the organs. The sclerotial powder also did not have any adverse effects on fertility, and there were no teratogenic (developmental malformation) effects on the offspring of the treated rats. Mutation assays have also shown that the sclerotial powder was not mutagenic. Overall, the results indicate that the TMM sclerotia were devoid of toxicity.

Conclusions. The Tiger Milk mushroom appears to be a promising and relatively untapped source of materials with potential nutritional and medicinal applications. Although the mushroom is far from being thoroughly studied, its biological activities and bioactive components have been clearly demonstrated in the research undertaken so far. Nevertheless, clinical studies will be crucial to further substantiate the ethnopharmacological knowledge scientifically.

For background information *see* ASTHMA; CANCER (MEDICINE); ETHNOMYCOLOGY; FUNGAL BIOTECHNOLOGY; FUNGAL ECOLOGY; FUNGI; IMMUNOLOGY; MUSHROOM; MUSHROOM PHARMACY; MYCOLOGY; NUTRITION; POLYSACCHARIDE in the McGraw-Hill Encyclopedia of Science & Technology. Szu Ting Ng

Bibliography. L. F. Eik et al., *Lignosus rhinocerus* (Cooke) Ryvarden: A medicinal mushroom that stimulates neurite outgrowth in PC 12 cells, *Evid. Based Compl. Alternative Med.*, 2012:320308, 2012, DOI:10.1155/2012/320308; M. L. Lee et al., The antiproliferative activity of sclerotia of *Lignosus rhinocerus* (Tiger Milk mushroom), *Evid. Based Compl. Alternative Med.*, 2012:697603, 2012, DOI:10.1155/2012/697603; S. S. Lee et al., Evaluation of the sub-acute toxicity of the sclerotium of *Lignosus rhinocerus* (Cooke), the Tiger Milk mushroom, *J. Ethnopharmacol.*, 138:192–200, 2011, DOI:10.1016/j.jep.2011.09.004; S. S. Lee et al., Preclinical toxicological evaluations of the sclerotium of *Lignosus rhinocerus* (Cooke), the Tiger Milk mushroom, *J. Ethnopharmacol.*, 147:157–163, 2013, DOI:10.1016/j.jep.2013.02.027; K. H. Wong, C. K. M. Lai, and P. C. K. Cheung, Immunomodulatory activities of mushroom sclerotial polysaccharides, *Food Hydrocolloids*, 25(2):150–158, 2010, DOI:10.1016/j.foodhyd.2010.04.008; K. H. Wong, C. K. M. Lai, and P. C. K. Cheung, Stimulation of human innate immune cells by medicinal mushroom sclerotial polysaccharides, *Int. J. Med. Mushrooms*, 11(3):215–223, 2009, DOI:10.1615/IntJMedMushr.v11.i3.10; Y. H. Yap et al., Nutrient composition, antioxidant properties, and antiproliferative activity of *Lignosus rhinocerus* Cooke sclerotium, *J. Sci. Food Agric.*, 93(12):2945–2952, 2013, DOI:10.1002/jsfa.6121.

Turbulence: When is the Reynolds number high enough?

The Reynolds number is the primary scaling parameter for turbulent flows. For the flow over large vehicles or through large pipelines, the working Reynolds number can be very high, of the order of 10^6 to 10^8 or higher. Experiments and simulations of turbulent flow, however, are often limited to Reynolds numbers of the order of 10^3 to 10^5. It has become evident that turbulence continues to evolve with the Reynolds number, so that low-Reynolds-number data are not a reliable guide to the behavior of high-Reynolds-number flows. The question then becomes, when is the Reynolds number high enough to reveal the asymptotic state of turbulence, if there is one?

Boundary layers. When a body moves through a fluid, the relative velocity between the fluid and the body is zero at its surface. In a frame of reference moving with the body, this so-called no-slip condition causes the fluid velocity to decrease from its "freestream" value far from the body to zero at the body's surface. The region where this decrease in velocity occurs is called a boundary layer (**Fig. 1***a*), and it can be experienced by simply sticking your hand out the window of a car. As you move your hand closer to the car body, you will notice that the speed of the air near the car is less than its speed further away. Another important example is the atmospheric boundary layer. The wind must satisfy the no-slip condition at the Earth's surface, and so a boundary layer forms that can be up to a kilometer thick. The atmospheric boundary layer is an important consideration in the design of, for example, wind turbines, which need to be located well above the Earth's surface to get the benefit of higher wind speeds. *See* OFFSHORE WIND ENERGY CONVERSION.

Almost every object that is exposed to a flow, human-made or not, will be affected by boundary layers, and understanding boundary-layer behavior is a major concern. In order to predict the performance of, for example, large vehicles such as ships and airplanes with confidence, the exact details are needed.

Fig. 1. A turbulent boundary layer flowing over the wall at the bottom of the figure. (*a*) A typical variation of the mean velocity *U* with wall distance *y*. (*b*) Flow visualization using smoke, making the turbulent eddies visible. The yellow and red boxes indicate the inner and outer scaling regions, respectively. (*Photo courtesy of T. Corke, Y. Guezennec, and H. Nagib*)

In these cases, the boundary layer is very thin compared to the length of the vehicle. Therefore, strong velocity gradients occur inside the boundary layer, and this causes a significant drag on the vehicle. This drag is due to viscous friction, since the local stress on the surface is equal to the product of the viscosity of the fluid and the velocity gradient at the surface. Whenever a body, be it an airplane wing, a turbine blade, a fish, a bird, or a car, moves through a fluid, the drag caused by the boundary layer makes a significant contribution to the total resistance experienced by the body. For bulk carriers and tankers, the frictional resistance resulting from the boundary layer is responsible for as much as 70–90% of the ship's total resistance. For a typical cruising airplane, it can be as high as 50%. A similar friction drag is also present in pipe and duct flows. On the inside wall of the pipe or duct, the no-slip condition applies unchanged, and so a substantial resistance is developed, which needs to be overcome by the work done by the pump or the fan.

Reynolds number. In general, boundary layers can be either laminar or turbulent, where laminar flow is smooth and steady, and turbulent flow is erratic, unsteady, and seemingly random (Fig. 1*b*). In turbulent flows, the instantaneous streamwise velocity is often written as the sum of a time-averaged mean value U and an instantaneous fluctuating value u. The state of the boundary layer depends on the ratio of inertial forces to forces due to viscosity. Dimensional analysis of the governing equations—the Navier-Stokes equations—reveals that this ratio, called the Reynolds number, is given by VL/ν, where V and L are the characteristic velocity and length scales of the problem, and ν is the kinematic viscosity of the fluid (the dynamic viscosity divided by the density). The balance between inertial and viscous forces governs the stability of the flow. At low Reynolds numbers, viscous forces are dominant, and they damp instabilities in the flow and promote laminar flow. At higher Reynolds numbers, viscous forces are not large enough to damp the instabilities; this allows small perturbations to grow larger, causing the flow to change from laminar to turbulent. Thus, the existence of turbulence depends on the Reynolds number and the nature of the disturbances that trigger the instabilities in the flow. The disturbances can come from many sources, such as surface roughness, imperfections, or pressure fluctuations. At large enough Reynolds numbers, any small disturbance will trigger turbulence.

Scaling problem. Even after it makes the transition to turbulence, the flow continues to change as the Reynolds number increases. The properties of the turbulent flow, such as its mean velocity distribution, the frictional resistance at the surface, and the statistics of the turbulent velocity fluctuations, all continue to evolve with the Reynolds number. Because the Navier-Stokes equations do not have an analytical solution for turbulent flow, exact predictions are not possible and models need to be developed. The development of such models depends fundamentally on numerical simulations or empirical data. In order to reduce the number of simulations and tests needed to model the effects of an increasing Reynolds number, we look for scaling laws that describe the flow and reduce the complexity. When simulating turbulence in a laboratory or on a computer, tests at lower Reynolds numbers are cheaper and easier, but, unfortunately, many engineering applications operate at very high Reynolds numbers. A vital question in the design of large vehicles and piping systems is how we can use tests and simulations conducted in turbulent flows at relatively low Reynolds numbers to predict the properties of turbulent flows at much higher Reynolds numbers. Uncertainties in the scaling process and the resulting conclusions open up the possibility of very expensive redesigns if the performance is found to be deficient.

High-Reynolds-number turbulence. Almost every existing theory or model for turbulence is based on scaling arguments that are valid only in the limit of infinite Reynolds numbers. We can write the Reynolds number as $L/(\nu/V)$, which we can interpret as the ratio of two length scales: one length scale that characterizes the largest eddies in the flow and one that characterizes the smallest eddies in the flow, where viscous effects are important. The main assumption behind most turbulence theories and models is that the large eddies containing the turbulent energy are many orders of magnitude larger than the small eddies that dissipate the turbulent energy into heat. Between these two limits, there is a continuum of eddy scales where the energy is transferred from larger scales to smaller scales, a process referred to as the energy cascade.

For turbulent boundary layers, this model has a particular interpretation. Near the wall (Fig. 1*b*), the velocity gradients are very large and the dominant eddies are very small, and so we expect the viscosity to be important in this region. The characteristic length scale is usually taken to be $\nu/\sqrt{\tau/\rho}$, where τ is the shear stress at the wall and ρ is the fluid density. Far from the wall (Fig. 1*b*), the velocity gradients are much weaker and the dominant eddies will be much larger, and it is possible that viscosity may be neglected in this region. The characteristic length scale is then δ, the thickness of the boundary layer, and the appropriate Reynolds number is the ratio of the large to the small length scale, that is, $R^+ = \delta\sqrt{\tau/\rho}/\nu$. If there is an intermediate part of the flow where both scalings hold true, it can be shown that the mean velocity will vary logarithmically with the distance from the wall.

This conceptually simple model is arguably the most important theoretical achievement for dealing with turbulent boundary layers. Using similar arguments, several authors have postulated that the variance of the turbulence $\overline{u^2}$ should also behave logarithmically over the same region of the flow (the overbar denotes a time average). The early evidence indicated that the mean velocity followed a logarithmic variation, although it also became clear that there was some dispute about its extent, and in particular at what distance from the wall it was first evident. The log law for turbulence was much more elusive, and there was no clear evidence that it actually existed. It was generally agreed that the problem was that the Reynolds number was insufficiently high, and so there was no clear separation between the inner and outer scaling. Since the 1990s, several laboratories have explored the features of these log laws, in particular to find the minimum Reynolds number needed for them to be present, and to see if their slopes and intercepts display some universal features.

Finding the minimum Reynolds number is important, since it might indicate at what Reynolds number we can expect the dynamics of the turbulent flow to behave as if the Reynolds number were infinite, if it ever will. To achieve this aim, it was necessary to build new facilities that could produce very-high-Reynolds-number flows, and to develop new sensors that could measure turbulence at very small scales.

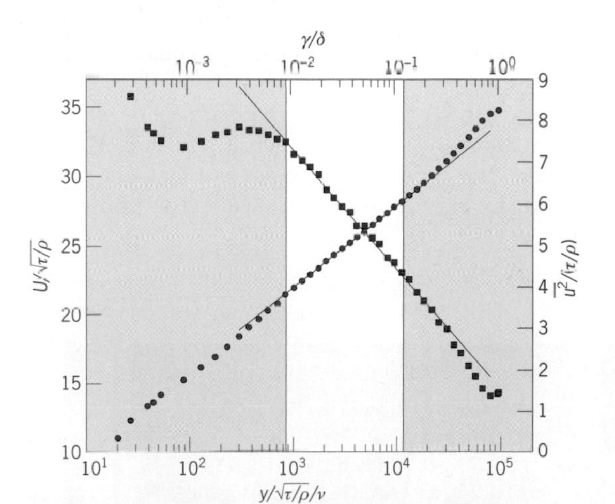

Fig. 2. Dependence of the mean velocity U (red symbols) and the variance of the streamwise velocity fluctuations $\overline{u^2}$ (blue symbols) on the wall distance y in pipe flow with Reynolds number $R^+ = 98{,}000$. The mean and fluctuating velocities have been nondimensionalized by the characteristic velocity scale $\sqrt{\tau/\rho}$, and δ is the pipe radius. Both the mean and the fluctuating velocities are seen to exhibit logarithmic behaviors in the "overlap" region, indicated with a white background.

These new results have revealed that many high-Reynolds-number flows, including boundary layers, pipes, and natural heat convection cells, display this asymptotic behavior. The results are surprisingly consistent; all of these investigations indicate that high-Reynolds-number asymptotes exist and that log laws seem to be present in all of them. For example, in boundary layers, once the Reynolds number R^+ exceeds 20,000, log laws for the mean velocity and the turbulent fluctuations begin to appear in the same physical region of the flow (**Fig. 2**). Most laboratory flows, however, can produce only much smaller values, and so they cannot reveal this apparently asymptotic behavior. Log laws have now also been reported for higher-order moments of the turbulent velocities, such as $\overline{u^4}$, $\overline{u^6}$, and so forth, as well as in the temperature profiles in turbulent heat transfer. These results suggest that there is a universality hidden in turbulence that is apparent only at very high Reynolds numbers, at values typical of large vehicles, pipes, and ducts. The mystery of turbulence is starting to reveal some of its secrets.

For background information *see* BOUNDARY-LAYER FLOW; FLUID-FLOW PRINCIPLES, FLUID MECHANICS; NAVIER-STOKES EQUATION; REYNOLDS NUMBER; SHIP POWERING, MANEUVERING, AND SEAKEEPING; TURBULENT FLOW; VISCOSITY; WIND POWER in the McGraw-Hill Encyclopedia of Science & Technology.

Alexander J. Smits; Marcus Hultmark

Bibliography. G. Ahlers et al., Logarithmic temperature profiles in turbulent Rayleigh-Bénard convection, *Phys. Rev. Lett.*, 109:114501 (5 pp.), 2012, DOI:10.1103/PhysRevLett.109.114501; S. G. Huisman et al., Logarithmic boundary layers in strong Taylor-Couette turbulence, *Phys. Rev. Lett.*, 110:264501 (5 pp.), 2013, DOI:10.1103/PhysRevLett.110.264501; M. Hultmark et al., Turbulent pipe flow at extreme Reynolds numbers, *Phys. Rev. Lett.*, 108:094501 (5 pp.), 2012, DOI:10.1103/PhysRevLett.108.094501; A. J. Smits and I. Marusic, Wall-bounded turbulence, *Phys. Today*, 66(9):25–30, September 2013, DOI:10.1063/PT.3.2114; A. J. Smits, B. J. McKeon, and I. Marusic, High–Reynolds number wall turbulence, *Annu. Rev. Fluid Mech.*, 43:353–375, 2011, DOI:10.1146/annurev-fluid-122109-160753; H. Tennekes and J. L. Lumley, *A First Course in Turbulence*, MIT Press, Cambridge, MA, 1972; A. A. Townsend, *The Structure of Turbulent Shear Flow*, Cambridge University Press, Cambridge, UK, 1976.

Underwater acoustic scattering signature

Since the early twentieth century, SONAR (SOund Navigation And Ranging) has been used in military, commercial, and scientific applications to help find objects submerged in the oceans. Sonar has been very successful for detection (Is there something out there?), localization (Where is it?), some degree of classification (What type of object is it, for example, human-made or marine organism?), and some

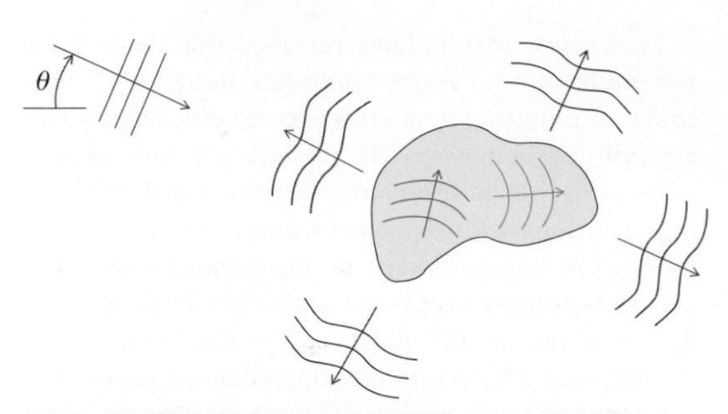

Fig. 1. An object submerged in a fluid is insonified by a plane wave (red arrow), causing the object to vibrate (interior elastic waves) and reradiate scattered waves (black arrows) back into the fluid in all directions.

degree of identification (What is the object?). To more precisely identify the nature of a detected object, an active sonar technique has recently emerged that produces additional information about the object, including not only its size and shape but also its internal composition. The resulting body of information is referred to as the "acoustic signature" of the object.

The physics underlying an acoustic scattering signature. Consider an arbitrary object in free space that is insonified by a monochromatic (single-frequency) sound wave (**Fig. 1**). Real objects are made of solid materials (metals, plastics, bones, flesh, and so forth) that change shape or volume "elastically" when any stress is applied (albeit by microscopic amounts when small acoustic pressures are applied). The incident sound wave is an oscillatory pressure disturbance in the water. As it strikes the object, the oscillating pressure on the surface of the object induces elastic (solid) waves to propagate throughout all parts of the structure. The vibrating object will, in turn, exert an oscillating pressure on the surrounding water, which produces pressure waves in the water that propagate away from the object, so-called scattered waves (also known as echoes). The scattered waves will propagate out in all directions, with different intensities in different directions. As the incident sound wave changes direction relative to the object (the angle θ in the figure), the vibrations in the object will change, which, in turn, will change the scattered waves. And as the frequency is changed, again the sound field will change in all directions.

When an object (called the "target") is insonified by a plane wave (wave fronts are planar, rather than curved, as occurs when the sound source is far from the object), the intensity of the scattered pressure wave back in the same direction that the sound wave came from is expressed by the monostatic (source and observer in same direction) far-field target strength, *TS*, given by Eq. (1), where f is

$$TS(f,\theta) = \underset{r\to\infty}{Lim}\ 20\,\mathrm{Log}_{10}\left(\frac{r\,|p(\mathbf{r})|}{r_0 p_0}\right) \qquad (1)$$

frequency, θ is the aspect angle of the incident sound wave (which is defined for each different problem, but is usually the azimuthal angle, that is, a horizontal angle about the vertical to the ocean bottom), $p(\mathbf{r})$ is the pressure of the scattered wave, $|\,|$ indicates magnitude (since p is a complex-valued function), $\mathbf{r}$ is a position vector from the object to the "observer" (where the target strength is being measured), r is its magnitude, and r_o and p_o are a reference distance and pressure that normalize and make dimensionless the argument of the logarithm. In three-dimensional space, $p(\mathbf{r}) \propto 1/r$ far away from the object, that is, in the "far field," so the numerator, $r\,|p(\mathbf{r})|$, converges to a limiting value as $r \to \infty$. The symbol $\underset{r\to\infty}{Lim}$ means evaluate the scattered pressure far enough away to obtain the limiting value. The units of target strength are decibels (dB).

An acoustic scattering signature is the target strength of an object, Eq. (1), that has been insonified over a broad band of frequencies and, for each frequency, over a broad range of aspect angles. The resulting values of target strength are plotted as a template (the terminology is taken from radar) of *TS* versus f and θ, with *TS* displayed as a color. This is illustrated in **Fig. 2** for a steel cylindrical closed shell, with a color bar on the right to quantify the *TS* values.

One can see that there is a great deal of information about the object in such a template. For example, the geometry of object and incident sound wave in Fig. 2*a* is clearly symmetric about the

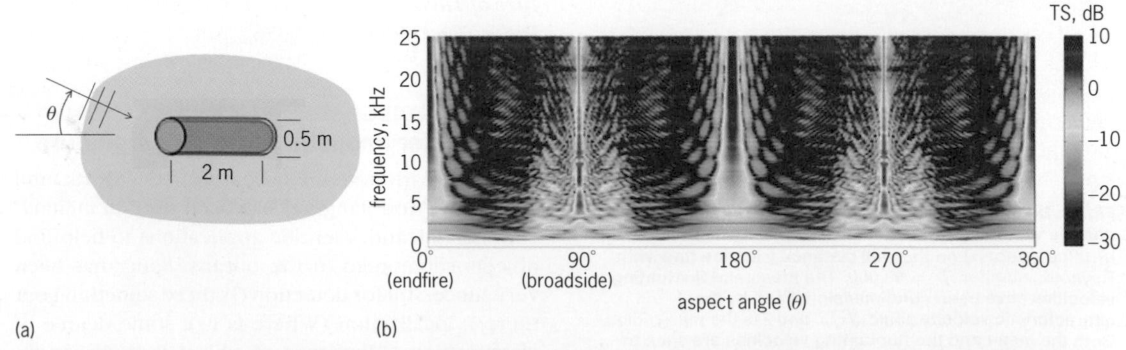

Fig. 2. Example of an acoustic scattering signature. (*a*) Insonification of a steel cylindrical closed shell. (*b*) The resulting acoustic scattering signature template.

incident directions that are parallel to the axis of the cylinder (endfire, $\theta = 0°$) and perpendicular to the axis (broadside, $\theta = 90°$), and this is manifested in corresponding symmetries in the template in Fig. 2*b*. The target strength is also strongest (red color) in these two directions because there are parts of the surfaces of the shell that are oriented perpendicular to the incident wave and therefore reflect part of the energy directly backward; at other angles the surfaces are oblique, reflecting energy in other directions. Other intense areas of the template (yellow, green, light blue) are caused by resonances of various types of internal elastic waves.

Acoustic signature versus imaging. The frequency band of acoustic signatures corresponds to wavelengths that are comparable to the interior and exterior dimensions of the object, sometimes referred to as low to mid frequencies. Sonar classification has traditionally relied on much higher frequencies, where wavelengths are very small relative to the object and therefore can produce a rough image of the object, that is, its external shape, but cannot reveal internal composition because the shorter wavelengths cannot penetrate very deeply before they die out due to attenuation. Consequently acoustic signatures and imaging complement each other in modern sonar systems. A key difference between the two approaches is the amount of information needed to interpret sonar output: Imaging produces a picture of the object so an engineer can identify the object simply by looking at the image, whereas a signature produces only a target strength template, which requires a computer to detect meaningful patterns in the template.

The need for computer modeling. One approach to the computer search for such patterns is to see if there are similarities with the templates of various other objects in a variety of ocean and sediment environments. That requires having a large library of reference acoustic signature templates. To construct such a library, one could perform experiments on actual objects. But experiments are very expensive and time consuming so only a few can be performed, and one cannot perform experiments on unavailable objects or environments. Computers, however, can model virtually any object or environment scenario of interest, including nonexistent scenarios. The cost of computer resources per model is negligible compared to that of a real underwater experiment and often faster by orders of magnitude, sometimes enabling hundreds or thousands of templates to be computed in the same time as performing one underwater experiment. There is clearly a need for a computer modeling system that is both high-fidelity and computationally fast.

Computer modeling. High-fidelity computer simulation is achieved by developing mathematical models, based on the physics of wave propagation, that accurately describe all the vibrations throughout the object and surrounding fluids. To this end, one seeks a solution to two partial differential equations (PDEs)—one for fluids and the other for solids—that describe all the relevant physics. They are reproduced here, not to discuss them, but rather to help readers with modest backgrounds in calculus and physics to appreciate that just these two equations describe not only all conceivable acoustic signature phenomena but also virtually all sound and vibration phenomena in most human experiences (music, traffic noise, human speech, and so forth) throughout the world.

The PDE for describing monochromatic (single-frequency) sound waves in fluids is known as the Helmholtz equation (also called the monochromatic wave equation), Eq. (2), where p is the acoustic pres-

$$-\nabla \bullet \left(\frac{1}{\omega^2 \rho}\nabla p\right) - \frac{1}{B}p = 0 \qquad (2)$$

sure of the scattered waves, $\omega = 2\pi f$, f is frequency, B and ρ are the bulk modulus (stiffness of the fluid resisting compression) and density, respectively, of the fluid, ∇ is the gradient operator (containing derivatives), and $\nabla\bullet$ is the divergence operator (also containing derivatives).

The PDE for describing monochromatic elastic waves in solids is known as the elastodynamic equation, Eq. (3), where $\mathbf{u}$ is the (vector) displacement of

$$-\nabla \bullet (\mathsf{c}\nabla \mathbf{u}) - \omega^2 \rho_S \mathbf{u} = \mathbf{f}_S \qquad (3)$$

a particle of the material as the wave passes by, c is a tensor (sort of a table) of elastic moduli (stiffnesses resisting compression and change of shape), ρ_S is the density of the solid material, and $\mathbf{f}_S$ is an applied force (vector) per unit volume.

Equations (2) and (3) are differential equations that describe all the relevant physics of vibrating matter within a differential, that is, infinitesimal, element of material. It is impossible to obtain an exact solution to these equations for virtually any realistic objects using classical methods of applied mathematics, unless one simplifies the equations a great deal, thereby eliminating much of the physics. Fortunately that is not necessary as the branch of modern mathematics known as finite-element (FE) analysis can produce a solution as close as desired to the exact solution without simplifying any of the physics.

FE analysis is an extension of classical calculus. It began in the mid-twentieth century and has grown rapidly, becoming such a powerful theoretical and numerical technique that it can find solutions to virtually any differential or integral equations that model (simulate) applications of almost any complexity. It has often been described as the most significant revolution in applied mathematics in the twentieth century. The merging of FE analysis with computer technology has created the modern discipline known as computational mechanics. As computers continue to evolve, the power of FE analysis continues to grow apace. This article on acoustic signatures is just one illustration of the power of modern computer simulation.

The essence of FE analysis is to subdivide the problem into a mesh of very small, simply shaped

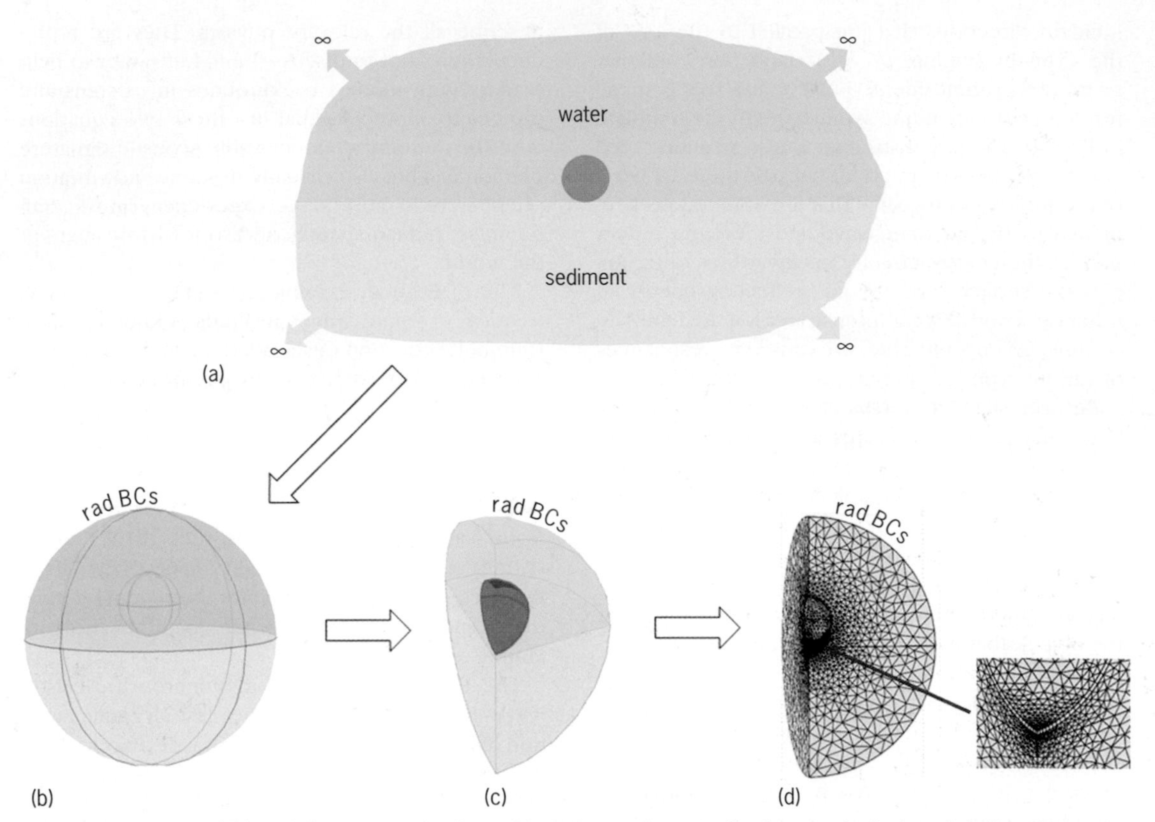

Fig. 3. Finite-element (FE) modeling process for the problem of acoustic scattering from a spherical steel shell resting on the ocean bottom. (*a*) Full model. The infinity symbols (∞) imply the water and sediment each occupy an infinite "half-space." (*b*) Reduced model. (*c*) One quadrant of the reduced model. (*d*) The FE mesh for the quadrant model. rad BC, radiation boundary condition.

"elements" and then to approximately represent Eq. (2) or (3) inside each and every element by transforming the above differential equations into approximately equivalent algebraic equations. The algebraic equations in adjacent elements are interrelated, producing a continuity of physics across all the elements, so all of the element equations are coupled together into a very large system of simultaneous algebraic equations, typically hundreds of thousands or millions, which must then be solved on a computer. As elements in a mesh are made progressively smaller, or the mathematical representation inside each element is enriched, the FE approximate solution becomes progressively more accurate, converging eventually to the exact solution of the original differential equations.

observation point
Helmholtz surface, $\partial\Omega_H$
r
r′

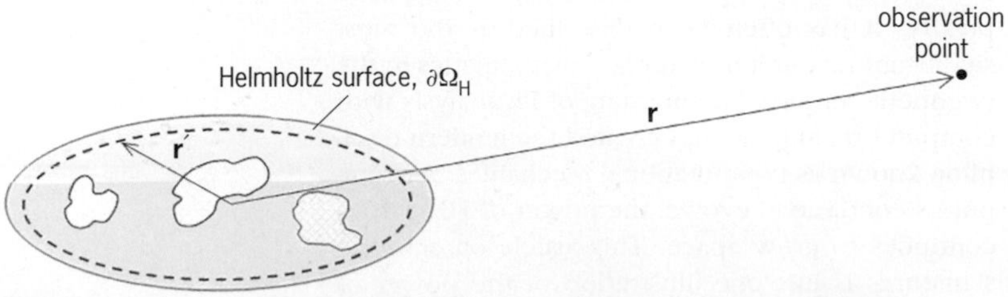

Fig. 4. Geometry of an artificial mathematical surface (broken red line), called the Helmholtz surface, which is used in conjunction with Eq. (4) to evaluate the scattered acoustic pressure at the observation point, which may be anywhere outside the finite-element (FE) model.

The FE modeling process is illustrated in **Fig. 3** for the problem of acoustic scattering from a spherical steel shell resting on (almost touching) a fluid-like sandy ocean bottom, where the very large ocean and sediment regions can be considered mathematically as "infinitely large" regions (Fig. 3*a*). To reduce the computational size of this problem the large water and sediment regions can be replaced by a small surrounding sphere (or ellipsoid) of water and sediment, with all the physics in the removed regions represented instead by mathematical relations applied to the outer boundary of the sphere, known as radiation boundary conditions, or rad BCs (Fig. 3*b*). This reduced model can be reduced further, to just one quadrant, by dividing it by any two perpendicular vertical planes intersecting the center of the spherical shell (Fig. 3*c*). One can then analyze just the one quadrant by decomposing all acoustic fields into components that are symmetric or antisymmetric with respect to those planes, in a way that preserves all the physics in the reduced model. The last step in the modeling process is to create a computationally efficient FE mesh for the quadrant model: larger elements away from the shell to represent long-wavelength propagating waves, smaller elements near to the shell to represent shorter-wavelength evanescent waves, and even smaller elements in the gap between the shell and the ocean bottom (Fig. 3*d*).

The result of the above FE modeling is a solution only within the reduced FE model. However, the solution is usually desired far outside the FE model. Therefore, the last step in the modeling process is to use the FE solution inside the FE model to compute the solution outside the FE model, that is, anywhere in the large ocean space. This is accomplished by using an integral formula called the Helmholtz integral, Eq. (4). The terms inside the big parentheses

$$p(\mathbf{r}) = \iint_{\partial\Omega_H} \left(\frac{\partial G(\mathbf{r}, \mathbf{r}')}{\partial n} p(\mathbf{r}') - G(\mathbf{r}, \mathbf{r}') \frac{\partial p(\mathbf{r}')}{\partial n} \right) d\Gamma \tag{4}$$

are evaluated all over a mathematical surface, called the Helmholtz surface, which circumscribes the object or objects in the target (**Fig. 4**). The symbol $d\Gamma$ represents an infinitesimal element of area on that surface.

In evaluating the integral in Eq. (4), the quantities $p(\mathbf{r}')$ and $\partial p(\mathbf{r}')/\partial n$ are the computed FE solution and its derivative normal (perpendicular) to the surface, respectively. The function $G(\mathbf{r}',\mathbf{r})$ is called a Green's function; it is a powerful concept, fundamental to differential and integral calculus. In this work, it describes how sound waves propagate in the large ocean environment in the absence of the target. That function embodies a lot of physics; it can sometimes be expressed with simple formulas for simple idealized environments but can also be computed numerically in a separate FE modeling exercise for more complicated realistic environments. The result of evaluating Eq. (4) is the value of the scattered pressure, $p(\mathbf{r})$, anywhere outside the FE model. This pressure is also the function $p(\mathbf{r})$ in Eq. (1).

In summary, the computer modeling process is based on a lot of differential and integral calculus. It consists of using FE analysis to find approximate numerical solutions to differential equations, Eqs. (2) and (3), that describe the physics of wave propagation in fluids and solids, and then using integral expressions, Eq. (4), in conjunction with those solutions, to find the final desired solution: the scattered sound pressure anywhere in the ocean. This process is repeated over and over for hundreds of different frequencies, and, for each frequency, hundreds of different directions of the incident sound wave, yielding a different scattered pressure for each frequency and incident direction. Inserting those scattered pressures into Eq. (1) yields the sought-after target strength (*TS*) as a function of frequency and direction, which is the acoustic signature of the object.

Verification and validation. Mathematical modeling is prone to human errors throughout the modeling process, so to achieve reliable solutions it is important to continually subject the models to a process of experimental validation (of the physics) and numerical verification (of the mathematics), called V&V (**Fig. 5**). One can never be certain of computer models; however, as more and more V&V testing is done, it might be said that one's confidence level can approach certainty asymptotically!

Verification is often accomplished by comparing two models using independent modeling techniques and looking for agreement over a broad range of the variables (not just at a few isolated spots). The rationale is this: Two modeling techniques that use very different methods are unlikely to produce the same errors over a broad, continuous span of data; ergo, if both solutions agree, then there are probably no errors, so both have probably found the correct solution. Since verification is testing only for mathematics errors (not physics), then errors can be made as small as desired, and often this means demonstrating accuracies to several significant figures.

Verification of an acoustic signature model is illustrated in **Fig. 6** for the problem in Fig. 3. A plane wave with a pressure magnitude of 1 pascal (Pa) [1 newton (N)/m^2] is incident at an angle of 40° above the sediment. The model computes the scattered pressure at a distance of 10 m from the center of the sphere, back along the same direction that the incident wave came from, also called the backscattered pressure or monostatic response. Since the sphere and its environment are axisymmetric about the vertical (broken line), the backscattered

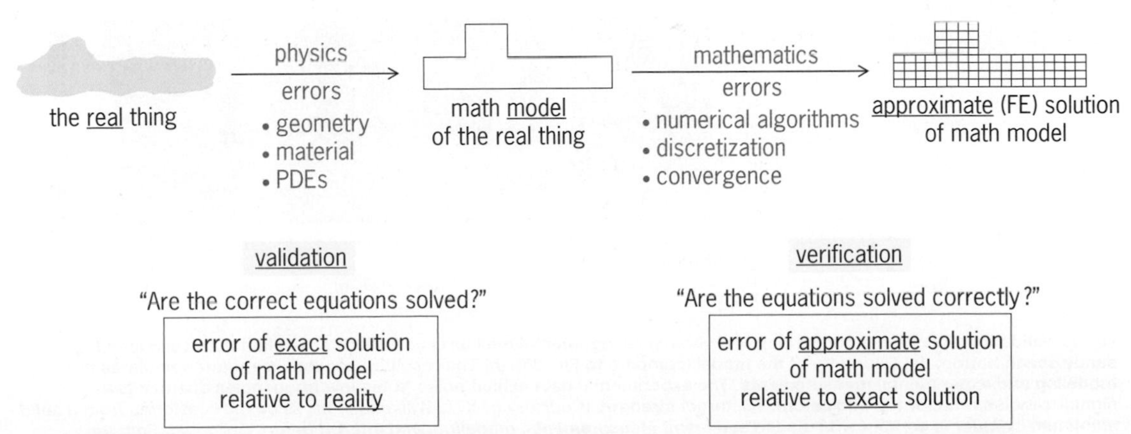

Fig. 5. Validation of physics and verification of mathematics. FE, finite element; PDE, partial differential equation.

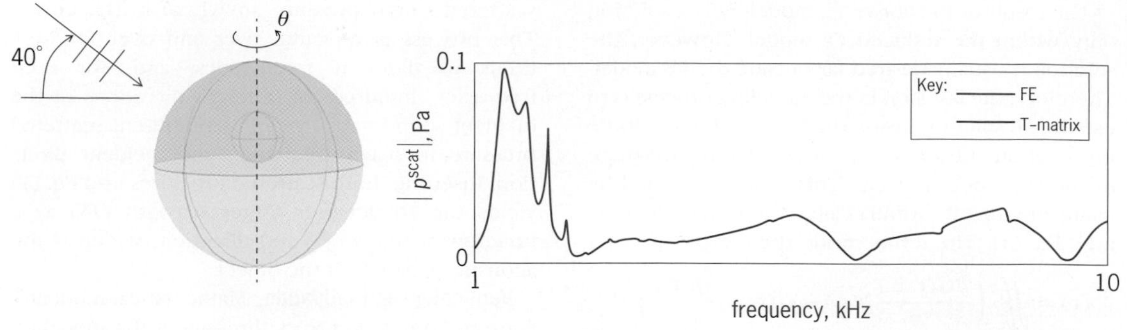

Fig. 6. Verification of model of scattering from spherical shell resting on sandy sediment on ocean bottom (compare to Fig. 3). FE, finite element.

pressure is independent of the aspect angle, θ, which is the azimuthal (horizontal) angle about the vertical.

The geometry of a sphere is very simple, so this problem is amenable to other, non-FE solution techniques, such as the T-matrix method, which is limited to very simple geometric shapes. The mathematical formalism and computer codes are completely different for FE and T-matrix analyses. The plot in Fig. 6 shows two solutions, one computed by the FE method (red curve) and the other by the T-matrix method (black curve), the latter lying directly underneath the former. The curves represent the magnitude of the scattered pressure [in the numerator in Eq. (1)] versus frequency over a range from 1000 to 10,000 Hz. Because of the aforementioned axis symmetry, the curves are independent of θ. Thus, *TS*, which could be computed from these curves using Eq. (1), would be independent of θ and a *TS* template would consist of horizontal bands. The agreement of the two solutions in Fig. 6 is to about three significant figures over the entire broad band of frequencies, including some sharp spikes. Such agreements provide confidence in both mathematical techniques.

Validation is accomplished by comparing computer model predictions with data measured in experiments with real objects. This comparison tests whether the correct physics is being used in the computer models (compare to Fig. 5); for example, are Eqs. (2) and (3) adequate or are additional equations necessary? Or are there physical features in the real object that were intentionally omitted from the model to simplify the modeling, but perhaps should not have been omitted? Has all the

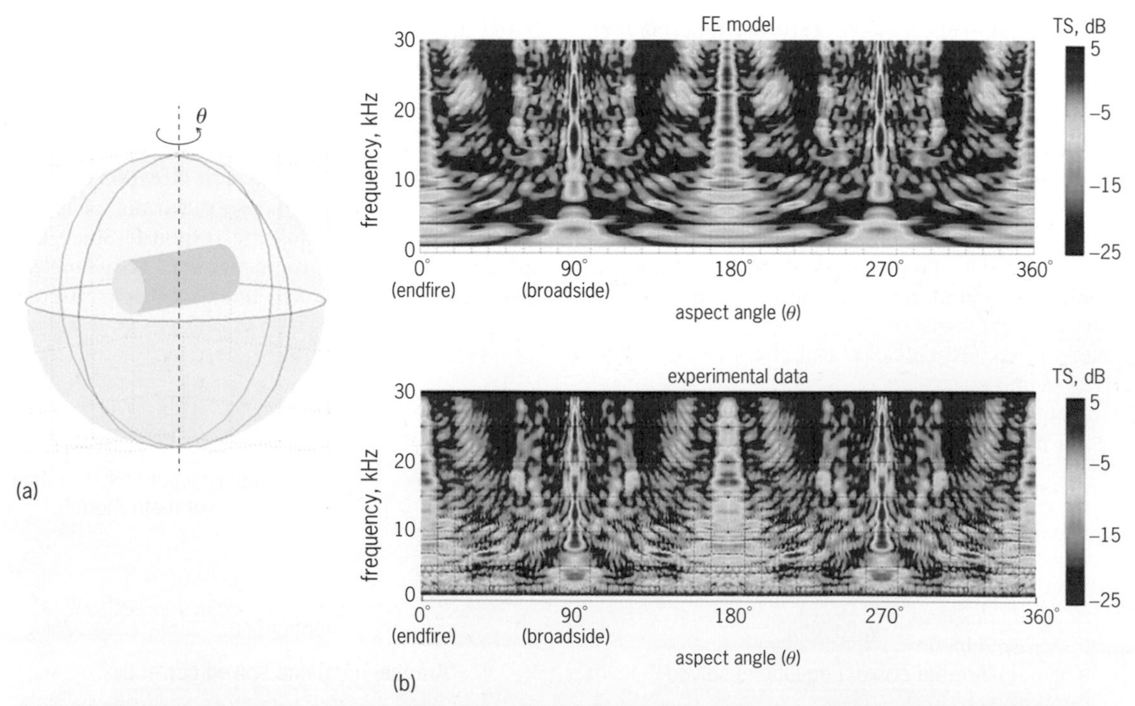

Fig. 7. Validation of finite-element (FE) model vis-à-vis experimental measurements for a solid aluminum cylinder lying on a sandy ocean bottom. (*a*) Geometry of the model (compare to Fig. 3*b*). (*b*) The resulting acoustic signature templates from FE modeling and experimental measurements. The experimental data exhibit noise at the low frequencies due to a low signal-to-noise ratio for the equipment. TS, target strength. (*Courtesy of K. L. Williams et al., Acoustic scattering from a solid aluminum cylinder in contact with a sand sediment: Measurements, modeling, and interpretation, J. Acoust. Soc. Am., 127:3356–3371, 2010, DOI:10.1121/1.3419926*)

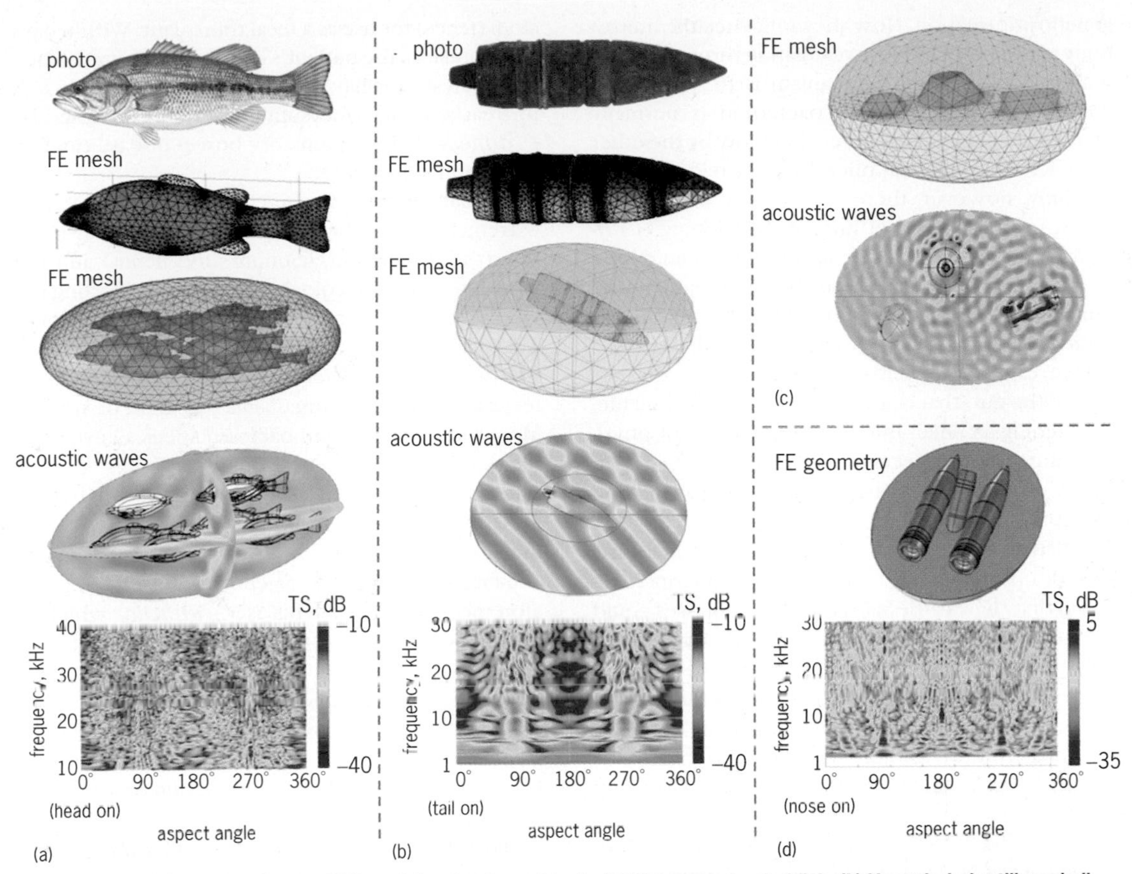

Fig. 8. Images from finite-element (FE) models of various objects. (*a*) Fish and school of fish. (*b*) Unexploded artillery shell partially buried in sediment. (*c*) Rock, mine, and concrete conduit pipe on sediment. (*d*) Unexploded Howitzer shells on sediment and partially buried bullet.

experimental equipment been calibrated properly? Validation tests the physics in the model against the physical features in the real world, which involves ranges of uncertainty in all the physical properties, experimental imprecision, and so forth. Therefore, accuracies are generally much lower than those for verification; agreements to within a few decibels are often considered very good.

Validation of an acoustic signature model is illustrated below for a solid aluminum cylinder lying on a sandy sediment (**Fig. 7**). There is very good agreement (to within about 3 dB) over almost the entire range of frequencies and aspect angles. This provides some confidence that the FE simulation program is based on the correct physics.

Examples. This article has used simple objects—spheres and cylinders—to explain and illustrate the physics, mathematics, and FE concepts for the computer modeling of the acoustic signatures of submerged objects. **Figure 8** shows a variety of images from FE models of more realistic objects, including photos, FE meshes, color contours of acoustic waves during insonification, and templates.

The science and technology of acoustic signatures is not limited to finding objects at the bottom of the ocean but can be applied to other acoustic scattering applications, for example, biomedical ultrasound and nondestructive evaluation, and to other wave-propagation phenomena, for example, radar.

For background information *see* ACOUSTICS; CALCULUS OF VECTORS; DECIBEL; DIFFERENTIAL EQUATION; ECHO; FINITE ELEMENT METHOD; GREEN'S FUNCTION; SONAR; UNDERWATER SOUND in the McGraw-Hill Encyclopedia of Science & Technology.

David S. Burnett

Bibliography. D. S. Burnett, *Finite Element Analysis: From Concepts to Applications*, Addison-Wesley, Reading, MA, 1987; D. S. Burnett, Modeling the acoustic scattering signature of objects at the bottom of the ocean, *Acoustics Today*, vol. 11, no. 1, January 2015; L. E. Kinsler et al., *Fundamentals of Acoustics*, 4th ed., Wiley, New York, 2000; A. D. Pierce, *Acoustics*, Acoustical Society of America, Woodbury, NY, 1989; R. J. Urick, *Principles of Underwater Sound*, 3d ed., McGraw-Hill, New York, 1983, reprint, Peninsula Pub., Los Altos Hills, CA, 1996.

Use of fecal transplants to treat intestinal disorders

The human gastrointestinal (GI) tract is teeming with a rich diversity of microorganisms (viruses, bacteria, and fungi), whose roles in health and disease are under intense scrutiny. These microorganisms outnumber the normal cells of a human body by 10 to 1, and they contain nearly 350 times more genetic information. How they influence the human body is an area of active investigation throughout the world. One such microorganism is the bacterium *Clostridium difficile*. This bacterium is normally held in check and its numbers kept low by the other gut microflora (normal microbiota or microbiome). Recently, however, there has been an alarming increase in the number of humans developing *C. difficile* infections. Often, the application of antibiotic therapy eliminates the "good" bacteria found in the human gut, thereby allowing antibiotic-resistant *C. difficile* to flourish and grow unchecked. This bacterium can cause colitis (inflammation of the last part of the gut, that is, the colon), which can be life-threatening. A small but growing number of physicians are turning to fecal transplants, also called fecal bacteriotherapy, as a last-resort option for treating this colitis.

Historical aspect. How did fecal transplants start? The idea probably began long ago when farmers realized that their cattle with intestinal illnesses could be cured by ingesting the rumen fluid of another, healthy cow. Current human treatment started in the 1980s, when an Australian gastroenterologist, Thomas Borody, was caring for a patient who had returned from a vacation with an incurable colitis caused by an unknown pathogen. After exhausting every conventional treatment, he began looking for alternative therapies. He found one in a 1958 edition of the journal *Surgery* that contained a report of four patients who were cured by fecal transplants. Therefore, the patient's brother was asked to donate his stool (feces) for use as a fecal transplant. Within days after infusion, the patient's colitis disappeared. Since then, physicians have begun to use fecal transplants to treat patients who suffer from colitis caused by *C. difficile* and inflammatory bowel disease (including Crohn's disease).

The human gut. The human gut (intestine) is an extremely long tube consisting of the small intestine (the duodenum, jejunum, and ileum) and the large intestine (the colon), with each supporting the growth of different bacterial species (microbiota). The colon has the largest microbial community in the human body. Microscopic organism counts in feces approach 10^{12} organisms per gram of weight. More than 400 different bacterial species have been identified in human feces. However, it has been estimated from genomic sequencing data that there are approximately 12,000 bacterial genera present in the colon.

Causative agent. The agent responsible for the aforementioned problems is *C. difficile*, which is a gram-positive, obligately anaerobic, fermentative, endospore-forming bacterium (**Fig. 1**) in the phylum Firmicutes. Within this phylum, the class Clostridia contains a diverse group of bacteria. *Clostridium difficile* resides in the human gut in low numbers. Under normal conditions, it is outcompeted by nonpathogenic bacteria.

The disease (pathogenesis). Many *C. difficile* infections arise in people who are undergoing antibiotic chemotherapy. This causes inhibition of the normal gut microflora, enabling the overgrowth of

Fig. 1. Photomicrograph of the endospore-forming bacterium *Clostridium difficile*, which can cause colitis. The cells (approximately 3 μm in length) are stained with crystal violet. The cells appear blue, whereas the spores are colorless and are located inside the cells. (*Photo courtesy of Gilda Jones, Centers for Disease Control and Prevention, Atlanta, GA*)

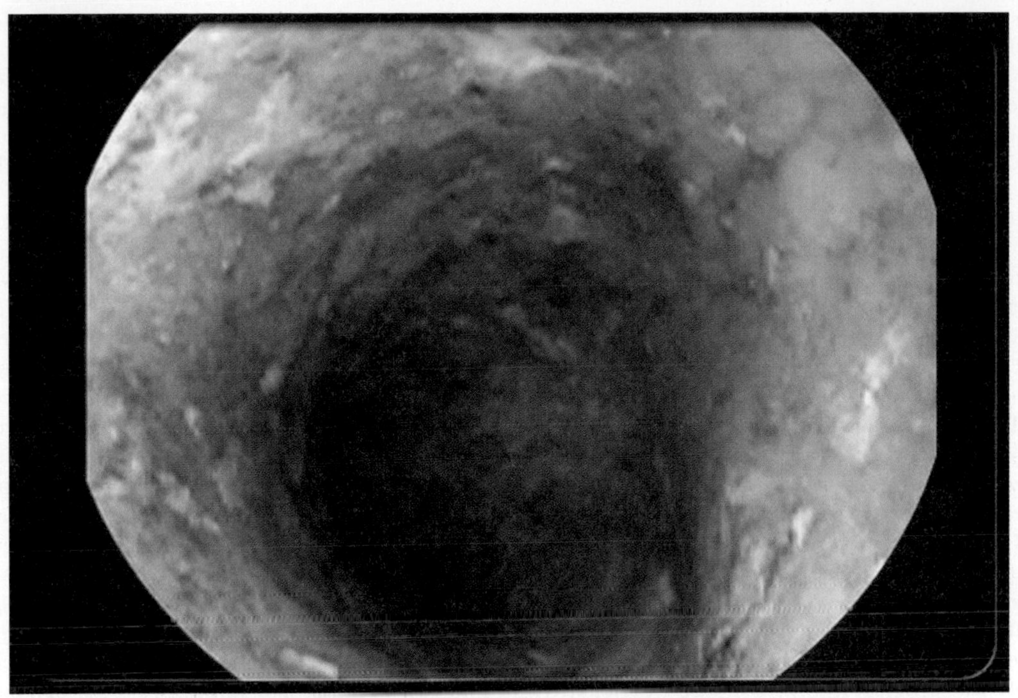

Fig. 2. Pseudomembranous enterocolitis caused by *Clostridium difficile*. A toxin produced by the bacterium growing at the epithelial surface damages and kills cells, causing white exudative plaques to form on the intestinal wall. The small plaques eventually coalesce to form a large pseudomembrane that can slough off into the intestinal contents. (*Photo courtesy of the Centers for Disease Control and Prevention, Atlanta, GA*)

C. difficile and its release of dangerous toxins, which trigger episodes of disabling diarrhea. This is accompanied by sloughing of the gut (colon) epithelium, leading to pseudomembranous colitis (inflammation of the colon) [**Fig. 2**]. This colitis consists of viscous collections of inflammatory cells, dead cells, necrotic tissue, and fibrin that together obstruct the intestine. It is a toxic condition that can lead to sepsis and even death. The most common antibiotics associated with this disease are amoxicillin, ampicillin, clindamycin, and various cephalosporins.

The toxins released by *C. difficile* include an enterotoxin (toxin A) and a cytotoxin (toxin B). Both of these toxins are responsible for the inflammation and diarrhea. The most common symptom of the colitis is a watery diarrhea, consisting of three or more bowel movements per day for two or more days. Other common symptoms include fever, loss of appetite, nausea, and abdominal cramping or tenderness.

Epidemiology. *Clostridium difficile* is most commonly seen in hospitalized patients, but community-acquired infections are on the rise. In the hospital, spores from one patient can be easily transmitted to other patients. *Clostridium difficile* infections can be devastating among the elderly, who often are afflicted with inflammation and diarrhea, and even perforation of the gut in severe cases. Frequently, antibiotic treatment fails. *Clostridium difficile* also often infects infants and young children and has been shown to spread from infants to adults. Recent research carried out in Sweden indicates that *C. difficile* can persist for more than six months in approximately one-third of infected children, and more than half of the long-term colonizing strains are likely to cause relapsing infections in adults. The proportion of toxin-producing strains infecting infants has risen from 20%, as reported in a previous Swedish study conducted in 1980, to more than 71% in the latest Swedish study.

Treatment. Ironically, the first line of treatment for a *C. difficile* infection is the use of oral antibiotics, usually oral metronidazole or oral vancomycin. Newer therapies showing some success include (1) the use of a probiotic "milkshake" containing *Lactobacillus casei*, *L. bulgaricus*, and *Streptococcus thermophilus* to prevent the antibiotic-associated diarrhea caused by *C. difficile* and (2) the transplantation of fecal material to replenish the healthy colonic microbiota.

Fecal transplants as a possible cure. The patient's fecal donor is usually a healthy family member because it is believed that people who have the same environment and diet may have similar gut microflora. The patient's gut is first cleansed by enemas, and the donated stool is screened for pathogenic viruses, bacteria, fungi, and protozoa. The stool is then mixed with some sterile saline and infused into the patient via a nasogastric tube or via a colonoscope. Approximately 85–90% of patients report a complete cure within 48 h to a week.

More recently, Canadian researchers from the University of Calgary have prepared oral pills made from feces from the family members of infected patients. So far, all the patients using the oral pills have been treated successfully and have experienced no recurrences of the disease. To make the pills, researchers processed the feces until it contained only bacteria; then, they encapsulated the bacteria concentrate inside three gelatin capsule layers. This ensured that the pills would not leak or disintegrate before they were passed from the stomach into the small intestine.

Fecal transplant regulation. Currently, the U.S. Food and Drug Administration (FDA) is regulating fecal transplants in the same manner as it regulates unapproved drugs. However, there are many small trials in which fecal transplants have been used safely, without any side effects, and lives are being saved on a daily basis. As a result, it is being recommended that fecal transplants be regulated and overseen like blood transfusions instead of being regulated like drugs. The FDA is also calling for a federally funded registry to compile health and safety data on fecal transplant-treated individuals.

For background information *see* ANTIBIOTIC; CLOSTRIDIUM; COLON; DIARRHEA; DIGESTIVE SYSTEM; GASTROINTESTINAL TRACT DISORDERS; HOSPITAL INFECTIONS; INFLAMMATORY BOWEL DISEASE; INTESTINE; MEDICAL BACTERIOLOGY; MICROBIOTA (HUMAN); TRANSPLANTATION BIOLOGY in the McGraw-Hill Encyclopedia of Science & Technology.

John P. Harley

Bibliography. I. Adlerberth et al., Toxin-producing *Clostridium difficile* strains as long-term gut colonizers in healthy infants, *J. Clin. Microbiol.*, 52: 173–179, 2014, DOI:10.1128/JCM.01701-13; K. C. Carroll and J. G. Bartlett, Biology of *Clostridium difficile*: Implications for epidemiology and diagnosis, *Annu. Rev. Microbiol.*, 65:501–521, 2011, DOI:10.1146/annurev-micro-090110-102824; T. Norén, *Clostridium difficile* and the disease it causes, *Meth. Mol. Biol.*, 646:9–35, 2010, DOI:10.1007/978-1-60327-365-7_2.

Vegetation dynamics

Today, Earth's physical environment is changing dynamically. Changes in climate conditions (such as temperature and precipitation) and land-use patterns (such as forest clearance and coastal reclamation) are of particular importance to vegetation. Whether these changes are gradual or rapid, they often have profound influences on the physiology of plants and on the quality and quantity of the soil resources on which they rely. Because of such dynamism in both abiotic and biotic components on Earth, contemporary science seeks to develop appropriate strategies to conserve (and restore, when necessary) the spatial distribution of vegetation and plant species diversity in various terrestrial ecosystems.

Vegetation is a broad concept; it is not restricted to any particular taxa, life form, or spatial extent. With regard to taxa, vegetation consists of all the plant species that are present in a given area or habitat. Vegetation is characterized by the growth forms of its dominant species, encompassing annual herbs, dwarf shrubs, broadleaf evergreen trees, coastal mangrove stands, sphagnum bogs, and wheat fields. Vegetation can refer to a wide range of spatial scales, such as small roadside weed patches, mixed hardwood-softwood forests across a region, and even the flora of the entire globe.

Definition and measurement. Because of the multifaceted nature of vegetation, its dynamics can be only broadly defined. At its simplest, vegetation dynamics imply changes in the relative abundance or biomass of different plant species at specific temporal and spatial scales, reflecting the net effects of environmental changes, human land use, and biological interactions. Abundance has multiple meanings: typically, for trees, one can measure it as density, indicating how many individuals of a species are present in a certain habitat or as the total diameter of these individual trees measured at breast height. Conventionally, the total diameter value is converted into basal area (m^2 ha^{-1})—the area of the habitat occupied by the cross section of tree trunks and stems at their base. For herbs or forbs, abundance is mostly estimated as percent cover—the proportion of the habitat surface covered by a species.

Ideally, one can quantify the rate of vegetation changes through time by repeatedly recording the above-mentioned abundance indices for each species, or weighing the dry mass of individual plants (that is, biomass). Although these approaches are feasible at multidecadal scales, they are not appropriate for understanding the history of vegetation dynamics over several hundreds or thousands of years. Advances in ecological and sedimentological dating methods have enabled such long-term research. For example, one can reconstruct the past patterns of tree growth in a forest stand and the ambient climate conditions by measuring the width and number of tree rings for individual stems, which is known as dendrochronology. Pollen analysis (palynology) is useful for the same purpose. The pollen grains of plants are highly resistant to decay, and, in many cases, they are preserved for a long time in lake sediments or peat bogs. By dating thin cross sections of a vertical sediment core, one can approximately identify varying types of plant species composition over different periods of time. Such floristic information, in turn, is used to understand changes in climate conditions over the same time periods.

Vegetation dynamics over time and space. The concept of vegetation dynamics is not restricted to the temporal transition of flora; it also refers to spatially varying species composition or biomass productivity either along environmental gradients, or across discrete patches. The former is often called spatial zonation, whereas the latter is known as patch dynamics. Distinct zonal vegetation patterns occur at a variety of spatial scales, encompassing, for example, the sea–land gradient of salinity in a single salt marsh and the equator–pole gradient of temperature in an entire hemisphere. Whereas spatial zonation mostly represents the response of plant species as a whole to a relatively gradual change in physical factors over space, patch dynamics are related to another type of spatial heterogeneity in which diverse patches are created by disturbance events. A patch is a discrete spatial unit with a definite shape, the functional and floristic characteristics of which differ from those of the neighboring patches.

Despite this conceptual difference between gradient zonation and patch dynamics, they are not mutually exclusive because, in many cases, disturbance events occur across environmental gradients. Disturbances can both influence and be influenced by gradients, thereby shaping new plant community patterns that reflect the interplay of these externally driven forces and the preexisting zonal patterns of physical conditions.

Internal and external drivers. In a broad sense, autogenic and allogenic factors drive vegetation dynamics. Autogenic dynamics proceed primarily as a result of biological processes within a plant community, such as nitrogen fixation, production and accumulation of detritus, and interactions among individual plants of different species or of the same species. Through these interactions, the existing plants can facilitate or inhibit the establishment of the other plants. Allogenic changes occur in response to the

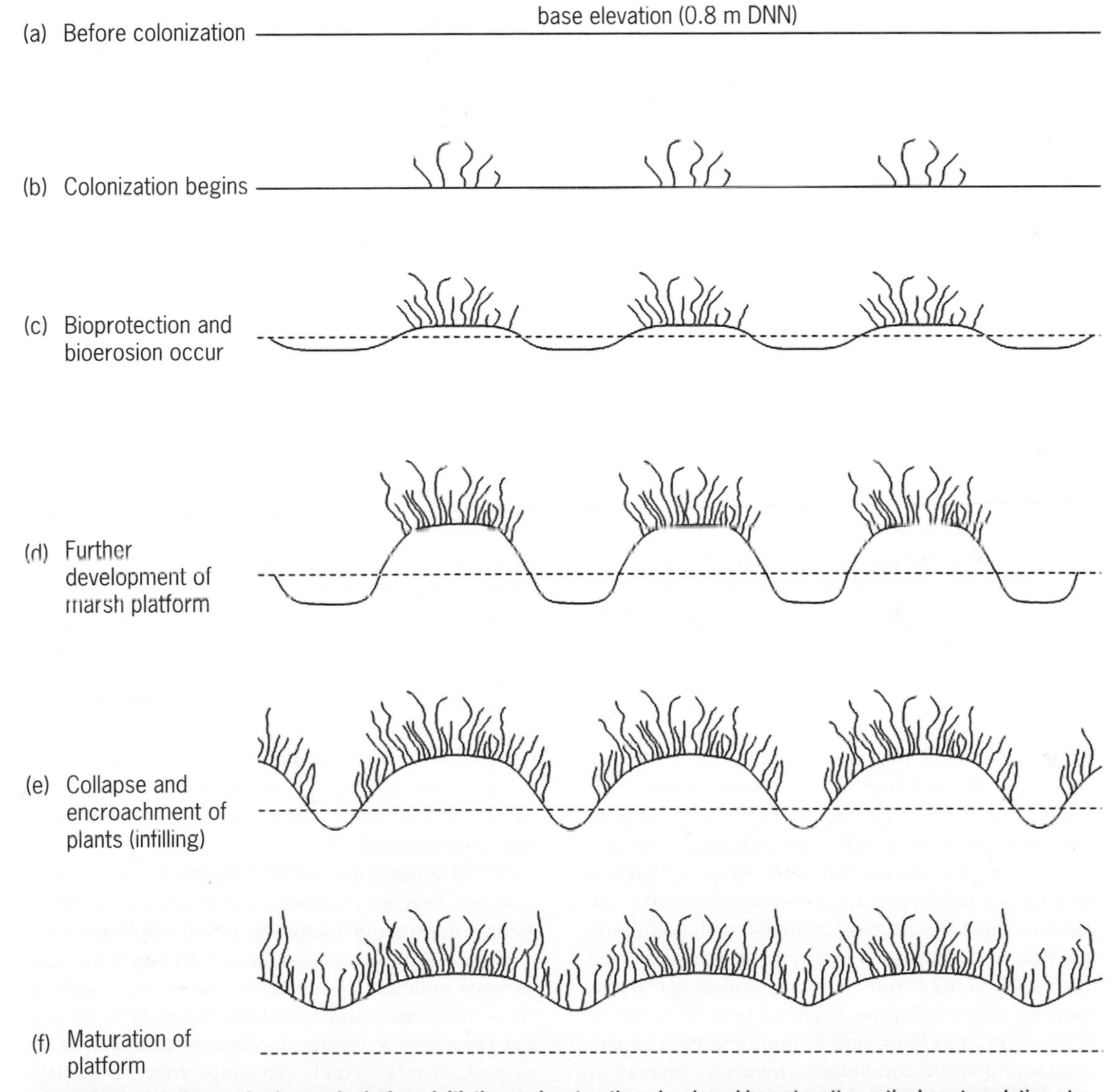

A hypothetical pathway of salt-marsh platform initiation and maturation, developed based on theoretical contemplation at a field site in Denmark. (*a, b*) When a tidal flat surface reaches approximately 0.8 m Danish Ordnance Zero (DNN), pioneer plants begin to colonize bare ground. (*c*) Vegetated patches experience bioprotection and sediment accretion, while unvegetated sites show bioerosion. (*d*) Continuation of bioprotection and bioerosion results in increased elevation difference between patches and bare ground. Note that the vegetation cover has steepened edges. (*e*) Episodic strong waves induce edge collapse, and hence, infilling of bare sites leads to encroachment of plants. Such infilling also occurs because the vegetated patch is now so high that sedimentation is concentrated in the gaps. (*f*) The platform is now mature with a significantly lessened topographic relief and complete vegetation cover. All sites are now above the base elevation.

processes that are not related directly to the biological components mentioned above. Such processes include, but are not limited to, climate change, natural disturbances, human interference, and grazing or browsing by animals.

No generalizable insights exist as to the relative importance of autogenic and allogenic drivers to vegetation dynamics. For example, in riparian (water-bank) zones, allogenic hydrogeomorphic processes are traditionally considered major factors in vegetation succession, whereas the effect of autogenic, plant-induced processes may be increasingly significant as landform stability is achieved. However, recent research indicates that these two types of processes are probably equally—or at least simultaneously—important to the overall vegetation dynamics even under dynamic hydrogeomorphology.

Vegetation as ecosystem engineers. One of the mainstream notions in contemporary ecology and biogeography is the feedback between vegetation and landforms. Until recently, one-way relationships in which topography influences vegetation have been dominantly considered when studying the evolution of a landscape. Today, this unidirectional perspective is increasingly transitioning to an integrated two-way viewpoint in order to address the importance of the reciprocal nature of vegetation–topography interactions. Through this perspective, vegetation influences terrain formations by controlling the mode and path of microbiological, hydrological, and even geologic and climatic processes. Landforms, in turn, constrain vegetation development through topographic controls on these same biophysical processes at a variety of spatial and temporal scales. Such two-way couplings fundamentally create and alter landscape patterns and processes, which is now conceptualized as biogeomorphology, ecogeomorphology, or biomorphodynamics. For example, vegetation–geomorphology interactions in salt marshes have conventionally been studied in the framework of positive feedback (*see* **illustration**). In this mechanism, vegetation cover increases sediment accretion by reducing tidal energy and capturing suspended materials. As the surface elevation gradually increases through such biota-driven enhancement of sedimentation, plant growth is in turn facilitated as a result of reduced seawater inundation and overall physical stresses. Such a process of edaphic (soil) amelioration can lead eventually to the establishment of high-marsh communities that are composed mostly of late-successional plant species. Key to biogeomorphology is the perspective that vegetation is an ecosystem engineer that actively modifies, and is modified by, the ambient topographic conditions. Therefore, vegetation dynamics are perceived not merely as the change of vegetation itself, but more as the evolution of a vegetation–landform complex, which is materialized by the mutual interactions between the two under the combined operation of autogenic and allogenic processes.

For background information *see* BIOGEOGRAPHY; DENDROCHRONOLOGY; ECOSYSTEM; GLOBAL CLIMATE CHANGE; PALYNOLOGY; PLANT GEOGRAPHY; PLANTS, LIFE FORMS OF; SALT MARSH; SEDIMENTOLOGY; SOIL; VEGETATION AND ECOSYSTEM MAPPING in the McGraw-Hill Encyclopedia of Science & Technology.

Daehyun Kim

Bibliography. D. Corenblit et al., Reciprocal interactions and adjustments between fluvial landforms and vegetation dynamics in river corridors: A review of complementary approaches, *Earth-Sci. Rev.*, 84:56–86, 2007, DOI:10.1016/j.earscirev.2007.05.004; C. G. Jones, J. H. Lawton, and M. Shachak, Positive and negative effects of organisms as physical ecosystem engineers, *Ecology*, 78:1946–1957, 1997, DOI:10.1890/0012-9658(1997)078[1946:PANEOO]2.0.CO;2; M. G. Turner, R. H. Gardner, and R. V. O'Neill, *Landscape Ecology in Theory and Practice: Pattern and Process*, Springer-Verlag, New York, 2001.

Venomous crustacean

The convergent evolution of venoms across the animal kingdom is testament to their great adaptive value. Venoms play key roles in predation, defense (including defense against microorganisms), competition, and even communication in a wide variety of terrestrial and aquatic taxa, ranging from jellyfish, cone snails, spiders, scorpions, insects, and centipedes to snakes, fish, and the duck-billed platypus. Venoms are particularly common in arthropods and contribute to the biological success of tens of thousands of species of insects, scorpions, spiders, and centipedes. However, one major group of arthropods represents a glaring exception to this abundance of venomous species: the crustaceans. Although the approximately 70,000 known species of crustaceans exhibit an unparalleled diversity of body plans (including water fleas, crabs, lobsters, barnacles, and brine shrimp), none of them has been known to be venomous. However, current research has now shown that remipedes (members of the class Remipedia, including the species *Speleonectes tulumensis*) are the first venomous crustaceans, adding a new major animal group to the roster of venomous animals.

Animal venoms are complex mixtures. Animal venoms are complex cocktails of proteins, salts, and organic components, including amino acids. They are synthesized in venom glands and delivered through wounds inflicted by specialized structures, such as fangs, stingers, spines, and harpoons. The biological effects of a venom are largely determined by its protein mix, which can range from small peptides to large multiunit enzymes. Each distinct protein or peptide is referred to as a toxin, and venoms can contain an enormous diversity of toxins. Individual species of cone snails, for example, can have venom repertoires of hundreds of distinct toxins, with some species possibly even having more than a

Fig. 1. Head and front end of the trunk of the remipede *Speleonectes tulumensis*. (*Copyright © and courtesy of Bjoern von Reumont*)

thousand distinct toxins. Not surprisingly, complex venoms can interfere with many physiological processes, causing rapid and often irreversible changes. For this reason, venoms have become an invaluable resource for the development of pharmaceuticals that can be used in many different contexts, from the control of agricultural pests to the treatment of human disease.

Venomous remipede crustaceans? The two dozen known species of remipedes are ghostly, unpigmented, and blind crustaceans that live exclusively in submerged marine caves. When they were first described in 1981, no mention was made of the possibility that they could be venomous. Superficially, remipedes resemble centipedes. They have long, homonomously (relatively similar) segmented bodies, with most segments being equipped with paddlelike swimming appendages. However, the remipede head features several pairs of prehensile limbs (**Figs. 1** and **2**). The pair that follows the mandibles, known as the maxillules, is especially robust. Later investigations revealed that the remipede maxillules may function as an injection apparatus. They contain a reservoir that opens via a pore on the sharp tip of the maxillule, and the reservoir is connected via ducts to large glands located in the body of the animal (Fig. 2). Field and laboratory observations have shown that remipedes use their maxillules to stab putative prey, suggesting that remipedes could perhaps inject a noxious substance. However, the nature of this noxious substance, if any, had previously remained unknown.

Remipede venom glands express venom toxins. Research has revealed that remipede venom glands express a mixture of toxin genes, and a detailed three-dimensional reconstruction of the venom injection apparatus shows that the animal is able to inject venom in a precise and controlled manner (Fig. 2). The arrangement of muscles suggests that the animal can separate the stabbing motion of the maxillules from the expulsion of venom from the venom reservoir.

About ten different types of toxins are expressed in the remipede's venom glands. The two most abundantly expressed types of toxins are enzymes: proteinases and chitinases. The activities of these enzymes probably help the animal to digest its prey. Proteinases break down proteins, whereas

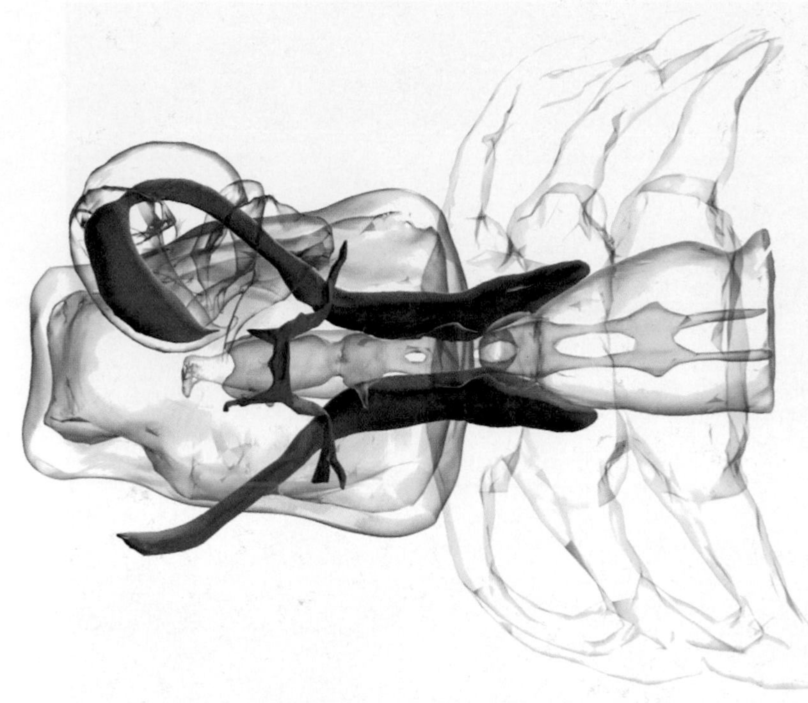

Fig. 2. Three-dimensional reconstruction of the head region of *Speleonectes tulumensis* in ventral view. The venom delivery system is indicated in purple. The left maxillule is shown, containing a venom reservoir connected via a duct to a gland in the trunk. (*Copyright © and courtesy of Alexander Blanke*)

chitinases break down chitin (a polysaccharide material), of which the cuticle of the remipede's crustacean prey is largely composed. Contraction of the remipede's muscular esophagus allows the remipede to suck up the prey's liquefying tissues, much in the way that spiders feed. Indeed, in the field, remipedes have been observed to consume the internal tissues of their prey, after which they release the empty husk of the prey's cuticle.

However, the action of such enzymes is probably not fast enough to prevent the prey from escaping this blind predator into the dark depths of the cave. Interestingly, studies suggest that the remipede is able to prevent the escape of its victims by injecting a paralyzing neurotoxin. Although the remipede evolved its venom independently from chelicerates (including the arachnids), its neurotoxin shows striking similarities to a group of neurotoxins known from some spider species. The spider neurotoxins force sodium ion channels to open at resting membrane potentials. This generates a continuous stream of action potentials along the nerves of insect prey, leading to spastic paralysis. It is likely that remipedes use their neurotoxin in the same way to paralyze crustaceans, on which they are known to feed.

The remipede toxin mix is unique. Other venomous arthropods that eat other arthropods, such as spiders, scorpions, and centipedes, have venoms that are much richer in different types of neurotoxins compared to remipede venom. Remipede venom is unique among those of venomous arthropods because it is dominated by a mix of large enzymes. In this respect, especially the predominance of proteinases, remipede venom is superficially more similar to that of viperid snakes than to that of other venomous arthropods. Thus, remipede venom nicely illustrates both the hand of history and the constraints that drive convergent evolution. Every independently evolved venom has a unique history and is therefore a unique toxin cocktail. However, poorly understood constraints on the evolutionary recruitment of genes to become toxins can cause striking similarities of composition and action in venoms of distantly related taxa. In addition, very closely related species of snakes or cone snails, for example, can have remarkably different venoms. It would thus be very interesting to study the venom of other species of remipedes.

Need for further work. A caveat of these investigations is that, so far, only the expression of toxin genes in remipede venom glands has been profiled. To confirm that the toxin proteins are actually present in the venom, it will be necessary to collect crude venom. Venom from animals such as snakes, spiders, centipedes, and even cone snails can be milked by researchers in order to analyze its protein composition. However, given remipedes' inaccessible habitat, aquatic habits, low population densities, and small size, it will be very challenging to collect crude remipede venom to complement the genetic work.

For background information *see* ARTHROPODA; CRUSTACEA; ENZYME; GLAND; POPULATION ECOLOGY; PREDATOR-PREY INTERACTIONS; PROTEIN; REMIPEDIA; TOXIN in the McGraw-Hill Encyclopedia of Science & Technology.

Ronald A. Jenner

Bibliography. N. R. Casewell et al., Complex cocktails: The evolutionary novelty of venoms, *Trends Ecol. Evol.*, 28:219–229, 2013, DOI:10.1016/j.tree.2012.10.020; B. Fry et al., The toxicogenomic multiverse: Convergent recruitment of proteins into animal venoms, *Annu. Rev. Genom. Hum. Genet.*, 10:483–511, 2009, DOI:10.1146/annurev.genom.9.081307.164356; G. F. King, Venoms as a platform for human drugs: Translating toxins into therapeutics, *Expert Opin. Biol. Ther.*, 11:1469–1484, 2011, DOI:10.1517/14712598.2011.621940; B. M. von Reumont et al., The first venomous crustacean revealed by transcriptomics and functional morphology: Remipede venom glands express a unique toxin cocktail dominated by enzymes and a neurotoxin, *Mol. Biol. Evol.*, 31(1):48–58, 2014, DOI:10.1093/molbev/mst199.

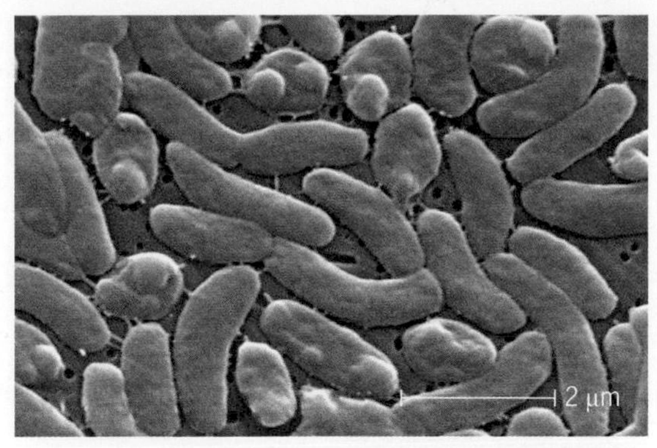

Fig. 1. *Vibrio vulnificus*, **the opportunistic bacterial species that causes necrotizing fasciitis (flesh-eating disease). Note the curved shape of the rods. [*Courtesy of Janice Haney Carr, U.S. Centers for Disease Control and Prevention (CDC); http://www.cdc.gov/vibrio/index.html*]**

Vibrio vulnificus

In 2013, 31 people in Florida were infected by a saltwater bacterial pathogen known as *Vibrio vulnificus*. This species is a rare cause of flesh eating disease, which in medical terms is called necrotizing fasciitis. Of the 31 people infected in 2013, nine died from their disease.

Background. *Vibrio vulnificus* is an opportunistic bacterial species that causes highly lethal infections in exposed individuals. This pathogen infects an individual either through the consumption of raw seafood that is contaminated with the bacteria or through direct contact of an open wound with saltwater containing the bacteria. It is closely related to *Vibrio cholerae*, which is the bacterial species responsible for the very contagious and dangerous disease known as cholera. *Vibrio vulnificus* is a gram-negative rod-shaped organism that has a single curve (**Fig. 1**). The term "vibrio" refers specifically to this curved rod shape, which is a diagnostic tool in identifying this pathogen. *Vibrio* species require the presence of salt in their environment, which is why they are known as halophiles, or species that require salt. Therefore, they are usually found in warm saltwater, such as ocean waters. *Vibrio vulnificus* is found worldwide as part of the normal microbiota in coastal marine environments and has been isolated from both water and a variety of shellfish species, including shrimp, oysters, and clams.

Vibrio vulnificus is classified into three different biotypes (that is, various strains of a species of microorganisms having differentiable physiologic characteristics). Biotype 1 strains cause the majority of human infections, whereas biotype 2 strains primarily cause disease in eels. A third group, biotype 3, has characteristics from both types 1 and 2 and was responsible for an outbreak of human wound infections in Israel during the 1990s. However, this biotype was not implicated in the Florida outbreak in 2013.

Disease. *Vibrio vulnificus* causes severe disease in individuals who are exposed to the pathogen through the consumption of contaminated shellfish or in those who are exposed to contaminated seawater through open wounds. Most people are resistant to infection. However, individuals with compromised immune systems or other underlying illnesses are much more likely to develop severe disease from this pathogen. Cases that occur after consuming shellfish usually present as a severe, fulminant (rapid) systemic infection, with symptoms that include fever, chills, nausea, hypotensive septic shock (life-threatening low blood pressure), and secondary lesions forming on the patients' extremities. Patients who develop septicemia (blood poisoning) caused by *V. vulnificus* have a greater than 50% chance of dying from their disease.

Wound infections caused by *V. vulnificus* are typically the result of exposing an open wound to contaminated seawater through recreational water sports (such as swimming, boating, or waterskiing), fishing injuries, or seafood preparation. These wound infections progress rapidly to cellulitis (inflammation of the skin and connective tissues), ecchymoses (the plural form of ecchymosis: the escape of blood into the tissues from ruptured blood vessels, causing a livid black-and-blue or purple spot or area), and bullae (fluid-containing blisters). In some cases, these infections can proceed to necrotizing fasciitis or flesh-eating disease, which is characterized by infection of the deeper layers of skin and subcutaneous tissues. This infection can spread easily across the fascial plane [that is, the plane of connective tissue (fascia)] within the subcutaneous tissue, causing necrosis (tissue death) of the subcutaneous tissue and fascia, while usually leaving the underlying muscle alone (**Fig. 2**).

Symptoms of necrotizing fasciitis begin rapidly with the onset of severe pain, which often seems excessive in relationship to the external appearance of the wound. Frequent early symptoms include inflammation, fever, and tachycardia (a heartbeat that exceeds the normal range; generally, a resting heart rate of 100 beats per minute or higher is considered to be tachycardic). Within hours, the tissue becomes

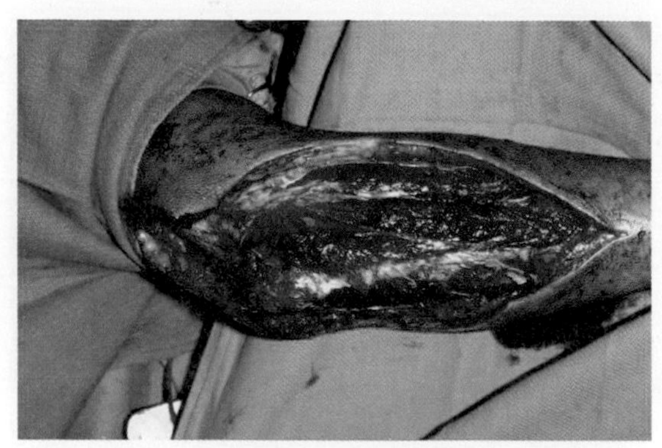

Fig. 2. An example of necrotizing fasciitis (flesh-eating disease). (*Courtesy of M. K. Cowan, Microbiology: A Systems Approach, 3d ed., McGraw-Hill, New York, 2012*)

swollen and the skin becomes discolored and purple. Then, blisters develop and necrosis begins to occur. Untreated cases of necrotizing fasciitis have a very high mortality rate, typically 70% or greater. Even with treatment, victims can still die of the disease. Treatment consists of aggressive surgical removal of necrotic infected tissues and intravenous antibiotic therapy. However, the use of antibiotics may not be effective, depending on the rapidity of the spread of infection. Multiple surgeries are frequently required because this disease spreads very quickly through the tissues, and one surgery may not remove sufficient tissue to stop its spread. Amputation may be required in some cases. In those cases where multiple surgeries are required, skin grafts may be necessary to close the wounds after the disease is cured. In general, cases are treated similarly to a severe burn because open wounds are likely to become infected by opportunistic pathogens found in the hospital environment.

Outbreak in Florida. In the summer and early fall months of 2013, an outbreak of *V. vulnificus* infections in Florida gained national attention after 31 people became infected with the pathogen. One victim was a 59-year-old male, who died of his infection just 28 h after being exposed. He became exposed while boating on the Halifax River in Volusia County, Florida, setting out crab traps. At first, his wound looked like a bug bite. However, the wound soon festered, after which the pathogen traveled to his kidneys, causing systemic disease and eventual death. A second case was linked to the same area, leading local health officials to notify the public of the risks of exposure to water in that area.

Overall, *V. vulnificus* caused a total of 31 cases of flesh-eating disease in Florida over the course of several months in 2013. Of those cases, 9 died from their illness, which is a mortality rate of 29%. These cases included septicemic infections resulting from exposure through seafood and instances of necrotizing fasciitis that occurred as a result of wound exposure. Because Florida's waters are naturally warm throughout the year, the state sees the largest number of these cases annually in the United States. Avoidance of the disease is straightforward, primarily through proper cooking and handling of shellfish before consumption, as well as preventing water from coming in contact with an open wound. Individuals at higher risk for this pathogen, including diabetics or HIV/AIDS patients, should also pay close attention to their potential for exposure because they are more likely to develop severe disease as a result of *V. vulnificus*.

For background information *see* BACTERIA; BIVALVIA; CHOLERA; CLINICAL MICROBIOLOGY; DISEASE ECOLOGY; EPIDEMIOLOGY; FOOD MICROBIOLOGY; FOOD POISONING; INFECTIOUS DISEASE; MEDICAL BACTERIOLOGY; OPPORTUNISTIC INFECTIONS; PATHOGEN; SKIN; WATER-BORNE DISEASE in the McGraw-Hill Encyclopedia of Science & Technology.

Marcia M. Pierce

Bibliography. M. K. Cowan, *Microbiology: A Systems Approach*, 4th ed., McGraw-Hill Education, New York, 2015; B. A. Froelich et al., Apparent loss of *Vibrio vulnificus* from North Carolina oysters coincides with a drought-induced increase in salinity, *Appl. Environ. Microbiol.*, 78(11):3885–3889, 2012, DOI:10.1128/AEM.07855-11; M. K. Jones and J. D. Oliver, *Vibrio vulnificus*: Disease and pathogenesis, *Infect. Immun.*, 77(5):1723–1733, 2009, DOI:10.1128/IAI.01046-08; P. R. Murray, K. S. Rosenthal, and M. A. Pfaller, *Medical Microbiology*, 7th ed., Mosby, St. Louis, 2013; E. Nester, D. Anderson, and C. E. Roberts, Jr., *Microbiology: A Human Perspective*, 7th ed., McGraw-Hill, New York, 2012; Z. Tao et al., Prevalence and population structure of *Vibrio vulnificus* on fishes from the northern Gulf of Mexico, *Appl. Environ. Microbiol.*, 78(21):7611–7618, 2012, DOI:10.1128/AEM.01646-12.

Visual attention

Humans cannot process every stimulus that reaches their senses simultaneously, nor can they think about a great number of different problems and interests at the same time. Attention refers to a family of mental and neural tools that allow cognitive processing to be limited to a subset (sometimes a very small subset) of available stimuli, thoughts, and behaviors at any one time. Even though we talk about "attention" as if it were a single thing, the brain has a large number of different attentional processes. The discussion in this article will be restricted to visual attention, although it can be applied to other senses or thought processes.

Serial and parallel processing. Even within the visual modality, attention has many aspects. For instance, you could focus object-based attention on THIS word. You could also devote feature-based attention to all red regions in your current field of view. Visual selective attention is necessary for some tasks; just try reading two sentences at the same time. However, selection is not absolute. As you attend to these words that you are reading, it is probably clear that the rest of the world does not completely vanish. This suggests that there might be two paths to visual awareness. Some aspects of the visual input are processed nonselectively, in parallel at many or all locations at the same time. Other aspects require selective processing, where massively parallel processing runs into a bottleneck that can only handle a limited number of selected items in series. One of the goals of attention research has been to determine when stimuli are handled by parallel processing and when serial processing is needed.

Given that there are tasks that must be done one at a time, it is important to be able to control the deployment of attention. Visual attention can be guided to specific items in the visual field by a limited set of basic features, including color, size, and orientation, which can be processed in parallel. Thus, if you are looking for bananas, you cannot identify bananas in parallel across the entire visual field; however, you can guide your attention to yellow, curved items, thereby devoting attention to the items most likely to be bananas. Attention can also be given endogenous commands (for example, attend to the left), and it can be "captured" by events in the world, notably the abrupt appearance of a new object.

Studying attention. Much information about attention can be gleaned from behavioral studies in which an observer looks at a display and produces a response, often pressing a key on a computer keyboard. For example, consider a version of the banana search. Observers might be asked to find an image of bananas on a computer screen displaying a variable number of distractor objects. The amount of time (the reaction time, RT) required to make a "yes" response to indicate that a target is present, or a "no" response, can be recorded, along with the accuracy of the response. In a search for bananas among oranges, RTs will be independent of the number of oranges because the properties of color and shape can be processed in parallel. However, if observers needed to find the bunch with a specific brand-name sticker, the RT would increase linearly with the number of items. This is consistent with deployment of serial attention from item to item until the target is found. It also is consistent with a variety of parallel models in which all items are processed at once, but where increased numbers of distractors make that parallel processing less efficient. If observers were searching for the correct bunch among a mixture of oranges and bananas, the RTs would be consistent with a search guided to only the yellow items.

There are many other behavioral assessments of attention. To measure attentional capacity, one might ask how many moving, identical items can be tracked at the same time (about 4). To measure attentional control, one might flash a light on the right, but instruct observers to move their eyes in the opposite direction. This assesses the ability to countermand the attention-capturing power of the flash. To measure the temporal resolution of attention, one might present multiple items at the point of fixation, asking observers to monitor for targets (for example, digits) among distractors (for example, letters) at a rate of 10 per second. Eye tracking can be very useful in assessing the spatial deployment because the mechanisms for deploying gaze and attention are closely related. It is important to remember, however, that the point of fixation and the locus (location) of attention need not be the same. One can deploy attention to one location while the eyes fixate on another. Moreover, the maximum rate of attentional deployment appears to be significantly greater than the limit of 3 or 4 saccades (small, rapid movements of the eye when changing focus) per second for voluntary eye movements.

Behavioral measures are limited in what they can reveal about how and where attentional functions are implemented in the physical brain. On these topics, information again comes from a range of methods. Clues about normal attentional function can be obtained from studies of the damaged brain. These studies can involve surgical lesions in animals or accidental lesions (from stroke or injury) in human patients. For example, damage to specific areas (such as the parietal lobe) on one side of the brain can lead to "neglect" of the opposite side of the world or of the body. Thus, patients with damage on the right side of the brain might fail to notice people to their left. They might leave untouched the food on the left half of a plate or even forget to dress the left half of the body.

Electrophysiological data have been recorded from single neurons (brain cells) of animals trained to do attentional tasks. Analyses of these data have determined that attention can change the sensitivity of cells to specific stimulus properties, and attention can also change the spatial receptive fields of those cells. Functional neuroimaging studies have allowed a noninvasive look into the biological processes underlying attention in humans. Functional magnetic resonance imaging (fMRI) relies on the fact that blood flow and blood oxygenation levels increase locally in the vicinity of active neurons. There are different patches of the brain's cortex that become active when observers view different types of stimuli (for example, faces and houses). By having observers attend to one type of stimulus or another, it is possible to use fMRI to watch observers "change their minds." Using methods [including magneto-encephalography (MEG)] that have greater temporal resolution, it is possible to map out some of the neural networks underlying such attentional switching. Recordings of event-related potentials (ERPs) from the scalp also can be used to answer questions about the time course of attention.

For background information *see* BRAIN; COGNITION; ELECTRODIAGNOSIS; ELECTROENCEPHALOGRAPHY; EYE (VERTEBRATE); HEMISPHERIC LATERALITY; MEDICAL IMAGING; NEUROBIOLOGY; PERCEPTION; PSYCHOLOGY; VISION in the McGraw-Hill Encyclopedia of Science & Technology. Jeremy M. Wolfe

Bibliography. M. M. Chun, J. D. Golomb, and N. B. Turk-Browne, A taxonomy of external and internal attention, *Annu. Rev. Psychol.*, 62:73–101, 2011, DOI:10.1146/annurev.psych.093008.100427; M. S. Gazzaniga and G. R. Mangun, *The Cognitive Neurosciences V*, MIT Press, Cambridge, MA, 2014; A. C. Nobre and S. Kastner (eds.), *Oxford Handbook of Attention*, Oxford University Press, New York, 2014; J. H. Reynolds and L. Chelazzi, Attentional modulation of visual processing, *Annu. Rev. Neurosci.*, 27:611–647, 2004, DOI:10.1146/annurev.neuro.26.041002.131039; O. Sacks, *The Man Who Mistook His Wife for a Hat and Other Clinical Tales*, Simon & Schuster, New York, 1985.

Vitamin D–enhanced mushrooms

Although the mushroom is thought of as a culinary vegetable, it comes from the biological kingdom of Fungi. It has a different nutritional profile from vegetables, and it has the ability to generate vitamin D when it is exposed to ultraviolet (UV) light. This ability is something that plant foods cannot do.

By 1935, it was known that vitamin D existed in mushrooms. However, not a lot of interest in this fact was shown until the 1990s, when researchers at the University of Helsinki reported that wild mushrooms were naturally high in vitamin D as a result of sun exposure. All mushrooms contain ergosterol, which is a type of sterol that is not found in plants or animals. When mushrooms receive sunlight or are placed under a UV light source, the ergosterol is converted to ergocalciferol, also known as vitamin D_2.

Importance of vitamin D. Vitamin D is essential for activating calcium absorption from the intestines and for the cells that form new bone from calcium and phosphorus. Without sufficient vitamin D, bones fail to grow normally in children or become brittle and soft in adults. Most countries recommend a daily consumption of 5–20 μg [200–800 international units (IU)] of vitamin D. This amount can be difficult to obtain solely from the diet because vitamin D is found in very few foods. Many people will consume only 2–4 μg (80–160 IU) of vitamin D daily from a typical Western diet, unless they specifically choose vitamin D-fortified foods. However, not all countries permit fortification, or the level of fortification may be limited.

The risk of vitamin D deficiency depends on many factors other than food. It is known that humans naturally make vitamin D when they expose their skin to sunlight, especially in the middle of the day. However, the risk of vitamin D deficiency rises as the exposure to sunlight decreases. This may be a matter of geography and season, with a greater risk being correlated with the distance from the equator at which one lives and during wintertime. Skin color also has a bearing, with people with dark skin having a greater risk because they require longer sunlight exposure to generate the same amount of vitamin D as light-colored people. In addition, age and obesity can influence the risk because aging diminishes the ability to generate vitamin D from sunlight, and fat cells concentrate vitamin D, leaving less vitamin D in the blood.

Approximately 1 billion people worldwide are deficient in vitamin D or have insufficient levels of vitamin D. No country appears to be immune from vitamin D deficiency. Higher blood levels of vitamin D are seen in Norway and Sweden, probably as a result of a high intake of fatty fish and cod liver oil, whereas the low average levels seen in Greece, Spain, and Italy may be the result of darker skin and sun avoidance. Even in sunny Australia, about one-third of the population is deficient in vitamin D.

Although the incidence of rickets (a disorder of calcium and phosphorus metabolism that affects bony structures, resulting from vitamin D deficiency) has dropped since the 1950s, it is common in children of African, Southeast Asian, and Middle Eastern immigrants who have moved to countries with colder climates. There still appear to be high rates of rickets in certain areas of Africa (for example, Ethiopia and Yemen) and Asia (for example, Tibet and Mongolia).

The consumption of foods that are high in vitamin D has a simple and clear role in reducing the incidence of vitamin D deficiency. Dietary vitamin D is present in foods in predominantly two forms, D_2 and D_3. Vitamin D_2 is found in mushrooms and yeast. Wild mushrooms in Europe commonly have 2–40 μg (80–1600 IU) of vitamin D per 100 g (3.5 oz). The most commonly consumed mushroom in Western cuisine is *Agaricus bisporus* (**Fig. 1**), for example, button, portobello, and field mushrooms. Hence,

Fig. 1. ***Agaricus bisporus*** **button mushrooms. (*Courtesy of the Australian Mushroom Growers Association*)**

most of the research on vitamin D_2 has been undertaken with these mushrooms. Commercial mushrooms, after a short exposure to UV light, can generate 10–30 μg (400–1200 IU) of vitamin D in a 100-g serving.

Vitamin D_3 can be found naturally in oily fish (for example, salmon), fish oil, liver, and egg yolks, and it is added to vitamin D-fortified foods and powdered drink bases. Both vitamins D_2 and D_3 are equally effective in raising the blood levels of vitamin D in the body.

Methods of increasing vitamin D in mushrooms. There are a number of methods employed to increase the levels of vitamin D in mushrooms.

Sunlight. Researchers at the University of Sydney have observed vitamin D production after mushrooms were exposed to the gentle midday winter sun. Within 2 h, the larger mushrooms had produced around 10 μg (400 IU) of vitamin D in 100 g. Smaller mushrooms (buttons) could generate 10 μg/100 g after only 1 h in the sun because they have a higher ratio of surface area to volume than cup mushrooms. The time required in the sun to generate 10 μg or more of vitamin D will be less in the spring and autumn. In a Californian study, mushrooms were able to generate more than 20 μg (800 IU) per serving after being placed in the midday autumn sunlight for a couple of hours.

UV light exposure. A more efficient way to stimulate the production of vitamin D in commercial mushrooms is to use controlled amounts of UV light. Normal UV lamps are effective. However, they may require some minutes to generate sufficient amounts of vitamin D, possibly causing significant browning of the mushrooms' skin and making them impractical for commercial production.

Many mushroom farmers have chosen to use a more intense UV source called a pulsed UV light (**Fig. 2**). Proof of the effectiveness of pulsed UV light on vitamin D levels in mushrooms came from studies by the University of Western Sydney and Pennsylvania State University. Just 1 s of pulsed UV light was able to produce in excess of the day's needs of vitamin D without any discoloration, making the process commercially viable.

With either UV or sunlight exposure, more vitamin D_2 is produced under the following conditions: when the light exposure is longer; when the light is more intense; when the surface area exposed to light is greater (hence, sliced mushrooms generate vitamin D_2 more quickly than whole mushrooms); and when the mushrooms are exposed to UV light at any time during growth, after harvest, during transport, or at wholesale or retail markets. In addition, more vitamin D_2 is produced in the top layer of mushrooms in multilayer packs because the top

Fig. 2. After 1–2 s of pulsed UV light, mushrooms have sufficient vitamin D to meet daily nutritional needs. Note that the UV light machine is kept behind the plastic curtain because of occupational and health concerns. (*Courtesy of M. Fensom, White Prince Mushrooms, Australia*)

mushrooms shield the mushrooms underneath them from the light.

Fluorescent light on retail mushrooms. Although mushrooms are generally grown in the dark, there is some vitamin D in store-bought mushrooms, probably because any fluorescent light includes part of the UV spectrum responsible for stimulating vitamin D production in mushrooms. Samples from major Australian retailers had 4–6 μg (160–240 IU) of vitamin D/100 g, which is more than the 1 μg (40 IU)/100 g found in mushrooms straight from the farm. This is similar to the 0–5 μg (0–200 IU) of vitamin D/100 g found in retail mushrooms in the United States.

The vitamin D in mushrooms is easy to absorb. There is at least an 85% retention rate of vitamin D in mushrooms after frying for 5 minutes. Furthermore, there is very little loss of vitamin D_2 when mushrooms are refrigerated for 8 days or even frozen for 3 months.

Researchers at Boston University Medical Center found that mushrooms given UV light generated vitamins D_2, D_3, and D_4 as well as other compounds, such as lumisterol. Vitamin D_2 was by far the most dominant form of vitamin D. The role of vitamin D_4 and the photogenerated compounds is not clear, but it does demonstrate that the mushroom is a remarkable food and is likely to reveal a few more surprises with further research.

Outlook. Mushrooms are the only fresh food with the potential to provide daily vitamin D needs in a single serving (100 g). Vitamin D-enhanced mushrooms are a simple and delicious way for people to get 100% of their daily nutritional needs for vitamin D, especially if they are unable to get adequate sun exposure. Commercial vitamin D-enhanced mushrooms can be found in supermarkets in the United States, Canada, Australia, and New Zealand.

For background information *see* BONE; BONE DISORDERS; ERGOSTEROL; ETHNOMYCOLOGY; FUNGAL BIOTECHNOLOGY; FUNGAL ECOLOGY; FUNGI; MALNUTRITION; MUSHROOM; MUSHROOM PHARMACY; MYCOLOGY; NUTRITION; RADIATION BIOLOGY; ULTRAVIOLET RADIATION; ULTRAVIOLET RADIATION (BIOLOGY); VITAMIN; VITAMIN D in the McGraw-Hill Encyclopedia of Science & Technology.

Glenn Cardwell

Bibliography. M. K. Holick et al., Vitamin D_2 is as effective as vitamin D_3 in maintaining circulating concentrations of 25-hydroxyvitamin D, *J. Clin. Endocrinol. Metabol.*, 93:677–681, 2008, DOI:10.1210/jc.2007-2308; M. D. Kalaras, R. B. Beelman, and R. J. Elias, Effects of postharvest pulsed UV light treatment of white button mushrooms (*Agaricus bisporus*) on vitamin D_2 content and quality attributes, *J. Agr. Food Chem.*, 60:220–225, 2012, DOI:10.1021/jf203825e; S. R. Koyyalamudi et al., Vitamin D2 formation and bioavailability from *Agaricus bisporus* button mushrooms treated with ultraviolet irradiation, *J. Agr. Food Chem.*, 57:3351–3355, 2009, DOI:10.1021/jf803908q; P. Mattila et al., Sterol and vitamin D_2 contents in some wild and cultivated mushrooms, *Food Chem.*, 76:293–298, 2002, DOI:10.1016/S0308-8146(01)00275-8; R. R. Simon et al., Vitamin D mushrooms: Comparison of the composition of button mushrooms (*Agaricus bisporus*) treated postharvest with UVB light or sunlight, *J. Agr. Food Chem.*, 59:8724–8732, 2011, DOI:10.1021/jf201255b.

Whale theft of fish from longline fisheries

Sperm whales (*Physeter macrocephalus*) remove fish from longline fishing gear. This behavior is called depredation. Whales will directly approach a vessel, sometimes after following it for days and deftly remove the catch from a longline (**Fig. 1**). In 2003, Gulf of Alaska (GOA) fishermen, looking for solutions to minimize their economic loss and reduce the potential entanglement of sperm whales in their gear, contacted scientists and formed the Southeast Alaska Sperm Whale Avoidance Project (SEASWAP).

(a)

(b)

Fig. 1. Sperm whales have learned that an easy meal can be found by following (*a*) fishing vessels and removing the catch from the gear. This is a global problem and researchers in the Gulf of Alaska are looking for solutions. (*b*) Sperm whales sometimes investigate the vessel visually with a spy hop and acoustically, whereby as the whales come toward the vessel, echolocation clicks are heard through the hull or inside the aluminum bait shed. (*Images copyright © SEASWAP*)

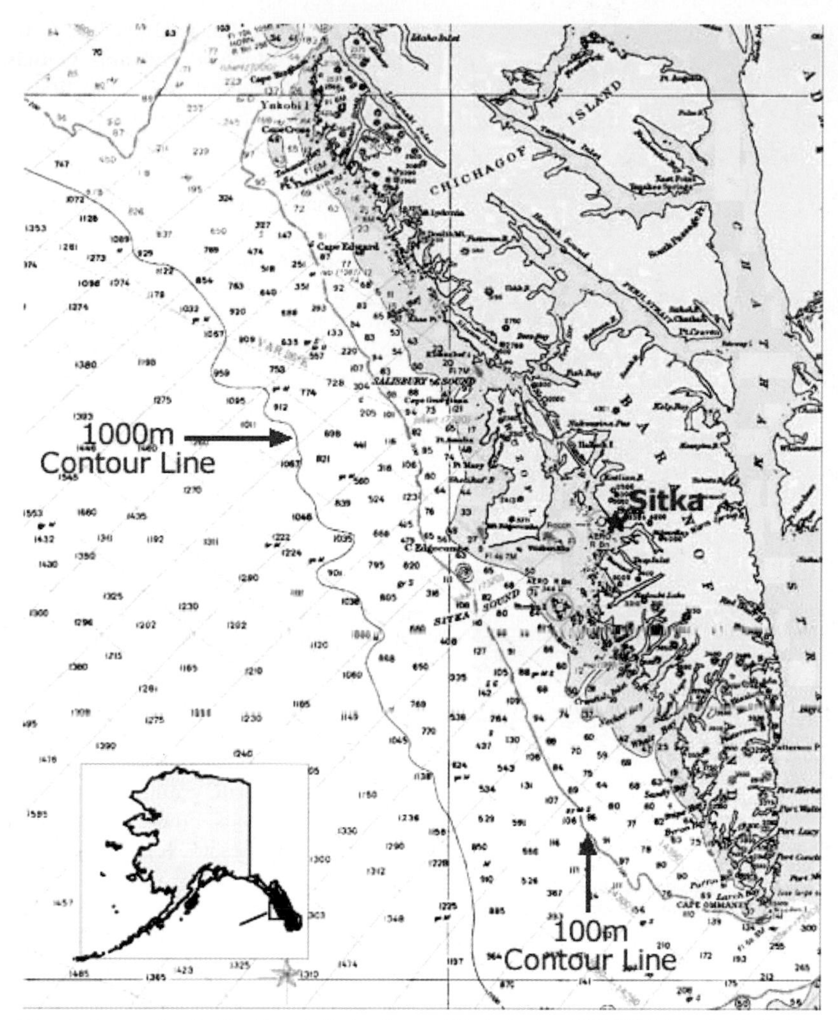

Fig. 2. Map of the study area for the Southeast Alaska Sperm Whale Project (SEASWAP) in the Gulf of Alaska.

This unusual mix of scientists, fishermen, and managers have come together to understand the whale theft problem and find solutions to minimize interactions. The study area is focused along the eastern side of the GOA, where the community of Sitka, Alaska, is located (**Fig. 2**). Sitka is conveniently close to the shelf break, allowing relatively easy access to where sperm whales are waiting for fishermen to retrieve their catch.

In 1980, T. Kawakami reviewed sperm-whale diets worldwide and found fish to be the predominant component of their diet. In 1982, D. E. Gaskin reported that the diet of pelagic (deep ocean) populations of sperm whales includes more squids (cephalopods), while coastal populations predominantly forage on fish. Stomach samples from specimens examined at whaling stations from whales caught in Alaska revealed that cephalopods were important food in the Western Aleutians and Bering Sea but that fish became progressively more significant toward the eastern Aleutians and into the GOA. Sablefish is the target catch for many fishermen in the GOA, and is a natural prey for sperm whales roaming these waters. In the Bering Sea, sperm whales with a diet composed of squid do not remove fish from the longline fishery. In direct contrast is the GOA, where depredation occurs, and sablefish and other deep-water fishes, such as halibut and grenadiers, are natural prey for sperm whales.

We now know from discussions with GOA fishermen that sperm whales primarily remove sablefish and to a lesser extent halibut and sometimes lingcod. They are selective thieves. They leave rockfish on the longline and are even capable of removing a single sablefish between two rockfish. Fishermen know this because a whale may leave jaws, lips, or a shredded body on the hook as evidence (**Fig. 3**); occasionally these remains are on a hook with rockfish on either side. This tells us that a sperm whale can essentially pluck a targeted fish off a hook.

Sablefish fishery. Sablefish are a commercially important fish species in the northeastern Pacific Ocean. The Alaska sablefish fishery primarily is hook-and-line, the gear is fished on-bottom, and 30–40 million circle-type hooks are deployed each year. Catches in the Alaskan exclusive economic zone (EEZ) have averaged 15,000 million tonnes in recent

Fig. 3. Shredded fish are left behind on hooks by sperm whales. This is evidence of depredation but cannot be used solely to document depredation because a whale can pluck a fish off a line and leave an empty hook. (*Image copyright © SEASWAP*)

years with an annual value of about $100 million. In Alaska, sperm-whale depredation on commercial sablefish longline sets has increased since a change in management was implemented in 1995, which extended the fishing season from two weeks to nine months. This change has allowed more opportunity for whales to participate in this behavior. Coupled with the end of commercial whaling a few decades ago, there likely has been an increase in the numbers of sperm whales, as well.

Sablefish live where the water is deep off the shelf break; more than 400 fathoms (800 m). This is where the fishermen set 2-3 nautical miles (4-5 km) of ground line on the bottom of the ocean, anchored and then marked with a vertical line attached to large inflatable buoys. Sometimes a flagpole is attached to mark each end of the set for greater visibility. Hooks are spaced about 1 m apart and attached to the longline with a short line called a gangion. Lured by the opportunity to catch an easy meal, sperm whales will follow vessels for days, sometimes sleeping next to the vessel or waiting near the ends of the longline set for greater visibility. Once the haul or recovery of the catch starts, the whales will actively remove fish from the longline at varying depths below the vessel.

Depredation studies. SEASWAP has been able to use fishing vessels of less than 18 m as platforms for understanding this problem and to test potential deterrents to reduce depredation. Vessels participating in SEASWAP record their catch, deploy recording instruments, and take opportunistic photographs to identify each whale (**Fig. 4**).

SEASWAP research began by looking at the big picture; that is, assessing which individual whales were in the study area and how many were removing fish, identifying the cues whales use to find a fishing vessel, and then bringing it into a narrower focus by studying the behavior directly at the longline. SEASWAP has documented 115 individual whales and estimated that 136 (124, 153; 95% Bayesian credible interval) male sperm whales were involved in depredation of sablefish in the eastern GOA between 2003 and 2014. The whales are solitary, although on occasion, they travel as pairs.

Sperm whales are present in the GOA year round, as detected acoustically by D. K. Mellinger and colleagues, with fewer detected in winter than in summer. However, the residency of individual sperm whales in the GOA and movements from higher to lower latitudes are unclear. Satellite-tracked tag data has shown that whales depart the GOA in summer and move to the waters offshore of Mexico and to the Gulf of California (**Fig. 5**). Genetic analysis of tissue samples has determined that these whales are part of a mixed stock, coming from multiple populations. Sperm whales are vocally active and acoustics was the primary tool in unraveling their behavior while near fishing vessels. SEASWAP found engine cycling was the acoustic cue alerting the whales to fishing activity and whales can hear the vessels on average at 5 mi (8 km) away. From 1-10 whales have been seen interacting with a fishing vessel while recovering its catch.

It is difficult to estimate the loss of fish due to depredation. Estimates of fish loss are generally conservative because it is not possible to attribute an empty hook (because the bait could have been

Fig. 4. The flukes of a sperm whale rise above the ocean during a dive. A sperm whale will be at depth for 40 min or so when normally feeding, but if the whale is feeding near a vessel and presumably off the longline, the dive time is about 15 min. Photographs document that each whale is unique and are used to estimate the numbers of whales in the study area of the eastern Gulf of Alaska. (*Image copyright © SEASWAP*)

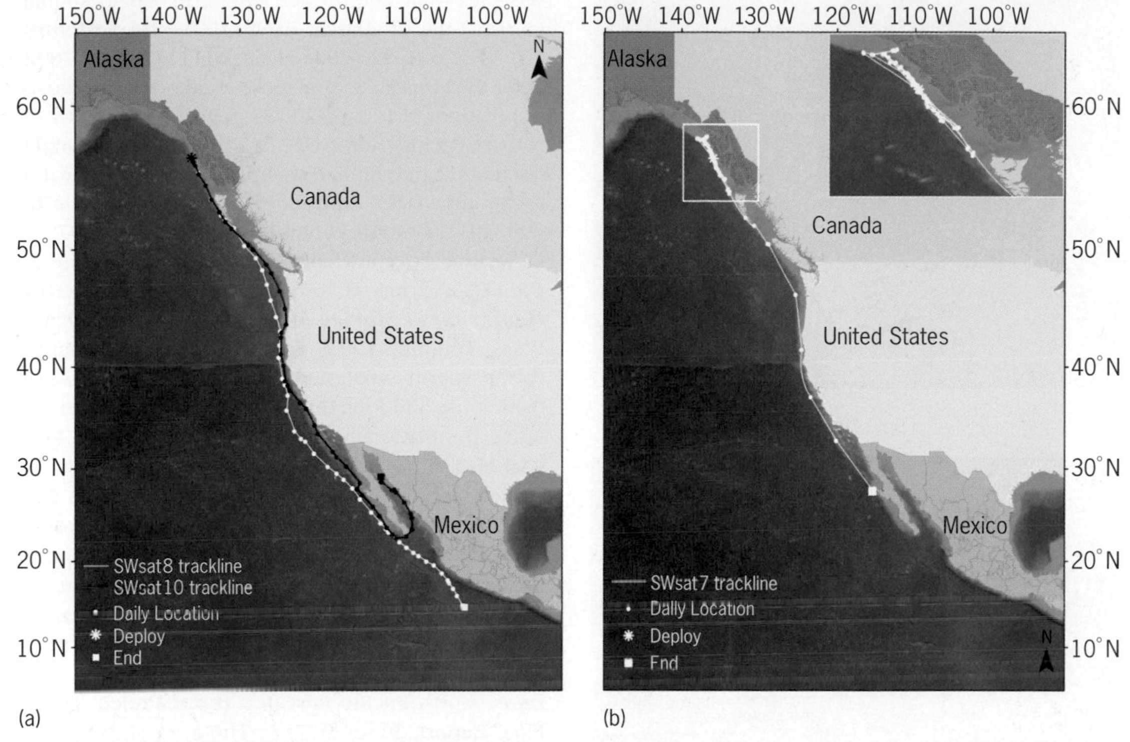

Fig. 5. Maps of satellite-tracked movements of three sperm whales tagged in the eastern Gulf of Alaska (GOA) in 2009 (*a*) SWsat8 and SWsat10 headed south almost immediately after tagging, with the last transmissions from Mexican waters. (Tag durations were 45 and 52 days, respectively). Light blue depicts the oceanic shelf and dark blue represents the oceanic basin with the continental slope the transition from light to dark blue. Locations connected by lines represent the most direct path between sequential position estimates, not the actual path of the whales. One degree of latitude is equivalent to 60 nautical miles. (*b*) SWsat7 was tagged off Sitka in June 2009 and remained in the GOA until October 8 when he headed south (tag duration 158 days); zoomed area for SWSat7 corresponds with the boxed area showing the whale locations prior to October 8. See J. Straley et al., 2014 for more details on the movements of tagged sperm whales.

removed, disintegrated, or fallen off the hook) to depredation. Additionally it can be difficult to distinguish whether other species, such as sharks, have contributed to the damage or loss of hooked fish. Damage and loss of fish has significant economic and management implications for both fishermen and fishery biologists tasked with assessing fish stocks. In general, depredation by sperm whales seems to be low to moderate, but it is highly variable in extent both among and within fishing areas. A SEASWAP goal is to determine how many fish are lost to depredation, and is finalizing an assessment of an acoustic metric. The variability became evident the first few years of SEASWAP when fishermen recorded details of fishing effort in logbooks. These data showed the catch was higher when whales were present. This finding mystified SEASWAP. When investigated further, it was found that sperm whales and fishermen both knew locations where the good fishing grounds were located. Locations with no whales present were not as productive for fishermen as areas where whales were present, even if depredation occurred. SEASWAP also captured a whale on video camera at depth fishing off the longline (**Fig. 6**). In doing so, the whale approached the line, grabbed it with its mouth, and created tension by moving up the line. The tension created by these actions popped a fish off the line and the whale then extracted itself from the line and swam after the fish. These brief insights were of immeasurable importance in providing guidance to SEASWAP. Foremost, these amazing animals have many ways of removing fish from a line and can delicately and deftly maneuver around a longline.

Outlook. The story of SEASWAP is still unfolding. SEASWAP is currently evaluating a deterrent for the fishermen to use that may be successful in delaying or deterring whales coming to the vessel. This is accomplished by setting a decoy vertical line a noted distance from the real set. This decoy line looks identical to real longline sets and is equipped with a playback of engine cycling, an acoustic cue the whales use to locate vessels. Of course, these are intelligent mammals and they may outsmart our every step. But if we have a few tools to use in different situations, or on a rotation, we hope to keep the whales away from the gear or even delay their arrival, which will be to everyone's benefit.

[Acknowledgement: Co-investigators and partners include Aaron Thode (acoustics for understanding northern latitude sperm-whale vocalizations); Victoria O'Connell; Chris Lunsford (NOAA federal sablefish survey); Russ Andrews, John Calambokidis, and Greg Shorr (tagging); Linda Behnken and

(a)

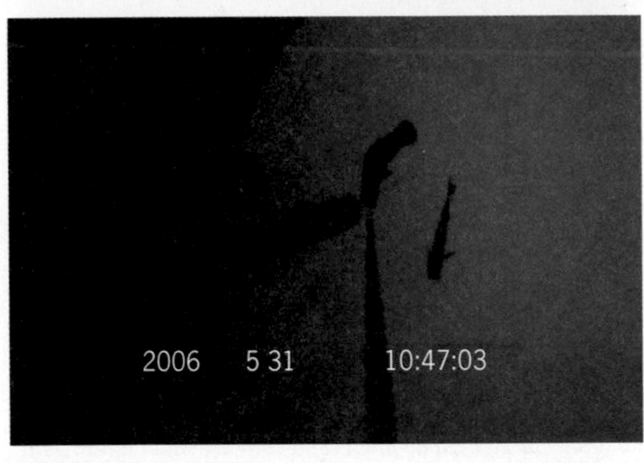

(b)

Fig. 6. This sequence of images was captured from video footage from a camera that was placed on the longline with fish attached 3 m up the line. (*a*) The whale grabbed the line and created tension, popping one fish off the hook. (*b*) When the fish popped off the hook the whale delicately released itself from the line and pursued the fish. It was never actually seen if the whale captured the fish because it went out of the view range of the camera. All acoustic indications are that the whale pursued the fish, however. A description of this footage and the behavior of the whale can be found in D. Mathias et al., 2009. (*Images copyright © SEASWAP*)

Dan Falvey; and the local longline fleet partners. SEASWAP has been supported by state, federal, private and industry funding.]

For background information *see* BERING SEA; CETACEA; GADIFORMES; MARINE FISHERIES; PACIFIC OCEAN; PLEURONECTIFORMES in the McGraw-Hill Encyclopedia of Science & Technology. Jan Straley

Bibliography. D. E. Gaskin, *The Ecology of Whales and Dolphins*, Heinemann, Exeter, NH, 1982; T. Kawakami, A review of sperm whale food, *Sci. Rep. Whales Res. Inst.*, 32:199–218, 1980; D. Mathias et al., Acoustic and diving behavior of sperm whales (*Physeter macrocephalus*) during natural and depredation foraging in the Gulf of Alaska, *J. Acoust. Soc. Am.*, 132:518–532, 2012, DOI:10.1121/1.4726005; D. Mathias et al., Relationship between sperm whale (*Physeter macrocephalus*) click structure and size derived from videocamera images of a depredating whale (sperm whale prey acquisition), *J. Acoust. Soc. Am.*, 125(5):3444–3453, 2009, DOI:10.1121/1.3097758; D. K. Mellinger., K. M. Stafford, and C. G. Fox, Seasonal occurrence of sperm whale sounds in the Gulf of Alaska, 1999–2001, *Mar. Mamm. Sci.*, 20(1):48–62, 2004, DOI:10.1111/j.1748-7692.2004.tb01140.x; S. Mesnick et al., Sperm whale population structure in the eastern and central North Pacific inferred by the use of single-nucleotide polymorphisms, microsatellites and mitochondrial DNA, *Mol. Ecol. Res.*, 11 (suppl. 1), 278–298, 2011, DOI:10.1111/j.1755-0998.2010.02973.x; D. W. Rice, Sperm whale (*Physeter macrocephalus*), pp. 177–233, in S. H. Ridgway and R. Harrison (eds.), *Handbook of Marine Mammals*, vol. 4, Academic Press, London, 1989; J. M. Straley et al., Depredating sperm whales in the Gulf of Alaska: Local habitat use and long distance movements across putative population boundaries, *Endang. Spec. Res.*, 24:125–135, 2014, DOI:10.3354/esr00595; J. M. Straley, Sperm whales & fisheries: An Alaskan perspective of a global problem, *J. Am. Cetacean Soc.*, 41(1):38–41, 2012; M. F. Sigler et al., Sperm whale depredation of sablefish longline gear in the northeast Pacific Ocean, *Mar. Mamm. Sci.*, 24:16–27, 2008, DOI:10.1111/j.1748-7692.2007.00149.x; A. Thode et al., *Evaluation of Sperm Whale Deterrents*, North Pacific Research Board Project F0527 Final Report, June 2007; A. Thode et al., Observations of potential acoustic cues that attract sperm whales (*Physeter macrocephalus*) to longline fishing activities in the Gulf of Alaska, *J. Acoust. Soc. Am.*, 122(2):1265–1277, 2007, DOI:10.1121/1.2749450; H. Whitehead, *Sperm Whales Social Evolution in the Ocean*, University of Chicago Press, 2003.

WISE spacecraft

The NASA *Wide-field Infrared Survey Explorer* (*WISE*) is a space-based telescope that imaged the entire celestial sky at infrared wavelengths. Launched in December 2009, *WISE* spent more than a year taking images to construct two all-sky mosaics at infrared wavelengths of 3.4, 4.6, 12, and 22 μm. Traveling in an Earth-circling orbit, *WISE* pointed at a different celestial position every 11 s, using movable mirrors within the telescope. The telescope's imaging pattern and cadence allowed astronomers to use these overlapping, time-separated exposures to create two highly detailed mosaics of the entire sky. In addition to identifying galaxies, black holes, stars, and nebulae, *WISE* detected a number of nearby asteroids and comets. The mission generated a number of publicly accessible data catalogs for continued characterization of celestial objects, and additional catalogs based on *WISE* data are still being developed by the astronomical community.

Background. Astronomers observe light energy from the sky over a range of wavelengths known as the electromagnetic spectrum. The shortest-wavelength, highest-energy emission comes from gamma rays, and the longest-wavelength, lowest-energy radiation derives from radio waves. The human eye is capable of observing energy in the middle of this range, as visible light. Infrared light, which is undetectable by the human eye, is observed at longer wavelengths than visible light. Infrared energy is an important indicator of heat production, whether by a distant galaxy or a human being.

Data from different wavelength regions of the spectrum provide unique astronomical information regarding the physical structure, chemical composition, origin, and other properties of celestial objects. Infrared telescopes are particularly useful for investigating optically opaque, dust-shrouded regions of the sky, since these instruments detect longer-wavelength energy, which is capable of escaping from a dusty region (**Fig. 1**). Infrared observations provide valuable data about a variety of sources that are relatively warm against the cold background of space, including asteroids and comets, brown dwarf stars, and ultraluminous infrared galaxies (ULIRGs). At the time *WISE* was launched, two infrared, space-based telescopes were in operation, NASA's *Spitzer Space Telescope* and the European Space Agency's *Herschel Space Observatory*. These two observatories, among others, have followed up the *WISE* survey with detailed studies of a number of sources.

Fig. 1. *WISE* image of the nebula IRAS 12116-6001 in the Southern Cross constellation. False-color images taken in all four passbands have been superimposed. Warmer objects detected at 3.4 and 4.6 μm, such as the bright star on the right side of the image, are colored blue. The adjacent star to the lower right is cooler. It is observed at 12 μm and colored green. The extended green and red regions are dust and gas clouds that glow brightly at 12 and 22 μm, respectively. These infrared-bright gas clouds cannot be seen in visible light. (*Courtesy of NASA/JPL-Caltech/UCLA*)

Spitzer, *Herschel*, the *Hubble Space Telescope*, and many ground-based telescopes implement observing programs that target individual objects. Observatories such as *WISE* are designed, instead, to take images at many celestial positions for assembly into a mosaic. These mosaics, or all-sky surveys, can be assembled using data from ground- or space-based telescopes, and they are capable of identifying time variable phenomena and discovering new targets for future study. All-sky surveys provide a legacy of observations of astronomical objects throughout the universe, from near-Earth asteroids (NEAs) to brown dwarf stars in our Milky Way Galaxy to very bright galaxies that lie billions of light-years away.

The *WISE* survey at 3.4, 4.6, 12, and 22 μm (**Fig. 2**) builds upon previous infrared-wavelength surveys, including 2MASS, a ground-based study at 1.25, 1.65, and 2.17 μm; the NASA *Cosmic Background Explorer* satellite, which surveyed the sky in wide-wavelength bands from 1.25 to 240 μm; the *Infrared Astronomical Satellite*, a joint mission among NASA, the Netherlands Agency for Aerospace Programs, and the UK Science and Engineering Research Council, which studied the sky at 12, 25, 60, and 100 μm; and *Akari*, a Japan Aerospace Exploration Agency satellite, which took wideband images between 1.7 and 180 μm. Spatial resolution of images from previous all-sky surveys was significantly poorer than that achieved with *WISE*. Results from the *WISE* all-sky survey, like those of its predecessors, serve as important reference data for future ground- and space-based astronomical observations.

Mission. When a spacecraft is launched into space, the exact launch time must be carefully determined to ensure that the vehicle reaches its planned orbit. When a mission is designed to orbit or fly by another planet or solar system object, the launch window is tightly constrained to times when the destination body is at the right location relative to Earth. For Earth orbiting spacecraft, the window is more flexible, although timing is still important to ensure the correct final orbit. The *WISE* telescope was launched into low Earth orbit on a Delta II rocket from Vandenberg Air Force Base on December 14, 2009, at 6:09 a.m. local time. To fulfill its role as

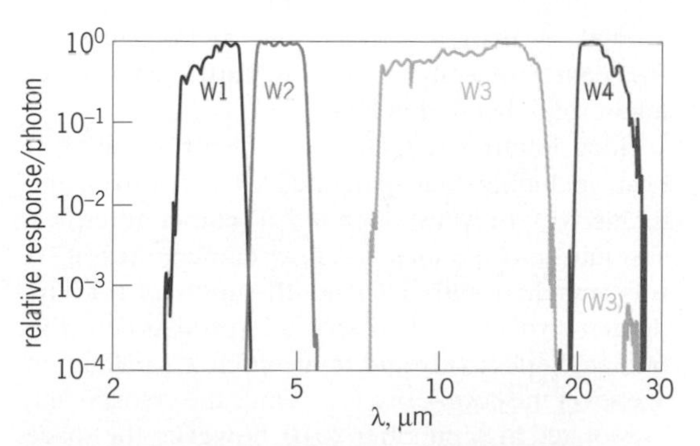

Fig. 2. The four wavelength regions where the *WISE* telescope observed the celestial sky from space. Charge-coupled detector (CCD) response per photon, normalized to a value of unity, is plotted against wavelength. The central wavelengths of the four detectors, W1, W2, W3, and W4, lie at 3.4, 4.6, 12, and 22 μm, respectively. (*Courtesy of NASA/JPL-Caltech/UCLA*)

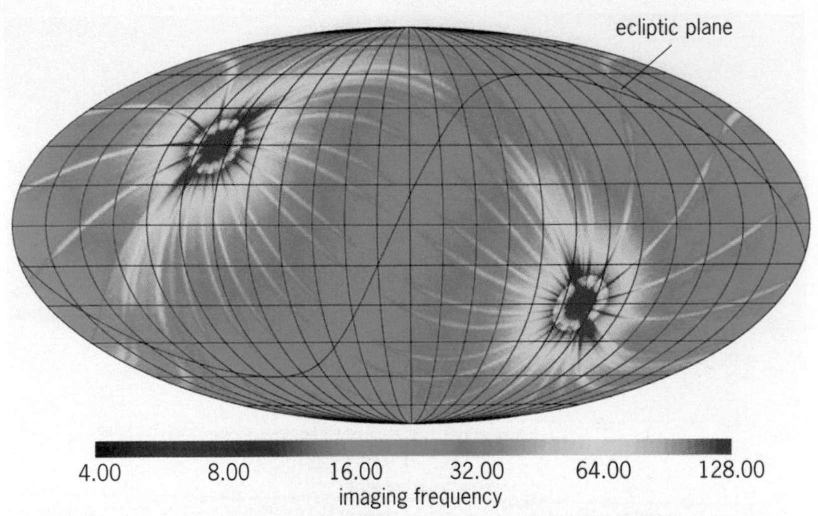

Fig. 3. Celestial sky coverage by the *WISE* spacecraft, in galactic coordinates, over the course of a year. Red patches at the ecliptic poles were imaged the maximum number of times. Blue areas lie in the ecliptic plane (which is a curved line in these coordinates) and were imaged the fewest number of times because of the mission's Earth-circling orbit along the day-night terminator. A total of 2,784,184 frames were transmitted through the end of the mission. (*Courtesy of NASA/JPL-Caltech/UCLA*)

an all-sky surveyor, *WISE* moved into a polar orbit 525 km (326 mi) above the Earth, moving along the day-night terminator. The spacecraft orbited the earth every 95 min, or 15 times each day. It was positioned with solar panels facing the Sun to provide continuous power, and the telescope and detectors were pointed away from the Sun to minimize the effects of stray or reflected sunlight. To maintain a Sun-synchronous orbit along the day-night terminator, the *WISE* orbit was shifted, or precessed, regularly.

Mosaics of the sky at infrared wavelengths of 3.4, 4.6, 12, and 22 μm, at angular resolutions of 6.1, 6.4, 6.5, and 12.0 arcseconds/pixel, respectively, were pieced together using individual exposures taken every 11 s. Each image covered a 47 × 47-arcminute piece of sky, and the telescope's scan pattern resulted in image overlap of 12 or more exposures on the ecliptic plane and several hundred at the ecliptic poles (**Fig. 3**). The spacecraft transmitted these images, more than 15 terabytes of data, to the ground for analysis through the Tracking and Data Relay Satellite System. The low-Earth orbit helped minimize transmission loss.

Since infrared detectors are highly sensitive to heat, including that generated by the instruments themselves, they are operated at temperatures near absolute zero. To maintain low temperatures, *WISE* was launched with a 10-month supply of solid hydrogen cryogen, and the spacecraft and its detectors operated effectively and transmitted 2.7 million images over the course of a year. Once the cryogen fully evaporated in September 2010, however, the spacecraft and its instruments warmed to 75 K (−198°C), and only the shorter-wavelength 3.4- and 4.6-μm detectors were able to provide viable data. From September 2010 to February 2011, the spacecraft continued operation as the NEOWISE mission and surveyed 70% of the sky for warmer solar system objects such as asteroids and comets.

NEOWISE. The NEOWISE mission extended WISE's science goals to include exhaustive searches for nearby asteroids and comets, and completion of a survey of the inner Main Asteroid Belt. Using data from both the cryogenic and postcryogenic periods, NEOWISE employed a software system to detect moving foreground sources in all exposures. Moving sources at nearby positions in the sky were reported to the International Astronomical Union Minor Planet Center (MPC) as candidate asteroids or comets. The MPC computed orbits for these objects, compared them to orbits of known asteroids and comets, and posted the results on its website to allow confirmation through follow-up observations. Particular attention was given to identifying smaller near-Earth objects (NEOs), which are near-Earth asteroids and comets, since their sizes, numbers, and locations had not been well characterized previously.

NEOWISE reactivated mission. Because of the success of the NEOWISE program, NASA resumed operation of the *WISE* spacecraft in September 2013. The telescope, once turned back away from the Earth, resumed its post-cryogenic operating temperature of 75 K and continued to map the celestial sky at 3.4 and 4.6 mm. The extended mission was expected to identify more than 150 new NEOs and to characterize approximately 2000 known NEOs, some of which could be candidate landing sites for President Obama's asteroid initiative.

Instruments. Major components of the *WISE* spacecraft included the telescope, the detectors, and the cryostat. The spacecraft weighed 661 kg (1457 lb) and was 2.85 m (9.35 ft) tall, 2 m (6.56 ft) wide, and 1.73 m (5.68 ft) deep (**Fig. 4**). The telescope used a 40-cm (16-in.) mirror to focus an infrared image onto four charge-coupled detectors (CCDs) simultaneously. For the first 10 months of operation, the 3.4- and 4.6-μm detectors were cryogenically cooled to 32 K (−241°C) and the longer-wavelength 12- and 22-μm detectors operated at 8 K (−265°C), while the telescope itself was cooled to 12 K (−261°C). The detectors and telescope were cooled by a cryostat that consisted of two doughnut-like tanks of frozen hydrogen.

Discoveries. Infrared observations yield important information that cannot be obtained using visible light. Infrared telescopes can peer through gas and dust clouds to reveal the interior structure of nebulae, make observations of very cool objects, and obtain Doppler-shifted data on objects from the early universe. Infrared telescopes are also well suited for identifying asteroids and comets in the solar system (**Fig. 5**). Space-based telescopes such as *WISE* detect objects at more infrared wavelengths than their ground-based counterparts because they operate outside the Earth's atmosphere and consequently are not limited by atmospheric energy absorption. Ground-based observations are restricted to very specific infrared wavelength bands because of energy absorption, primarily resulting from water vapor. Space-based telescopes have the added advantage of operating at cooler temperatures, where heat from the telescope and its components does not interfere with infrared observations.

WISE mapped the entire sky at infrared wavelengths and detected many faint, previously unknown celestial objects. The observatory made a number of significant discoveries, with more important results anticipated as astronomers continue working with the mission's data catalogs.

Inside the solar system. Among its major discoveries, *WISE* found that the number of midsize NEOs was significantly lower than previous estimates and that more than 90% of the largest NEOs had been identified to date. The mission also found that 20–30% of potentially hazardous asteroids (PHAs), those that orbit within 8 million km (5 million mi) of Earth, had been discovered to date. Analysis of the *WISE* data indicates that there are 3200–5200 PHAs with diameters greater than 100 m (328 ft) and that they lie near the Earth's orbital plane.

WISE also characterized the origin and composition of asteroids that orbit the Sun farther away. Trojan asteroids move in orbits near Jupiter or Saturn, and they appear to have been created in their local neighborhood rather than moving there from other locations in the solar system. Trojan asteroids are different in composition from main-belt asteroids and Kuiper Belt objects, which lie between the orbits of Mars and Jupiter and outside the orbit of Neptune, respectively. *WISE* also unexpectedly discovered a Trojan asteroid that had been transported close to the Earth's orbit. By the end of spacecraft operations, *WISE* had detected more than 157,000 asteroids in the main belt, 500 NEOs, and 120 comets.

Asteroids and comets provide clues to the history of the solar system through their orbits and compositions. These bodies contain pristine material from the early ages of the solar system, and some of them have migrated to new locations because of collisions and gravitational interactions. NEOs, objects that move into or near the Earth's orbit, are also of public interest as "Earth crashers" that represent a potential hazard to our planet.

Beyond the solar system. *WISE* has found the closest star system discovered in the last century, a binary system of brown dwarf stars at a distance of 6.5 light-years. It is the first observatory to identify the coldest stellar objects in the universe, cold brown dwarf stars known as Y dwarfs. WISE observations, combined with data from NASA's *Spitzer Space Telescope*, initially measured Y dwarf surface temperatures around 423 K (150°C). WISE subsequently surpassed its own record findings by identifying an even colder star at a distance of 7.2 light-years. This supercool star has surface temperatures in the range 225–260 K (−48 to −13°C).

WISE has observed stellar evolution inside interstellar dust clouds. In addition to viewing newly formed stars, the observatory has identified massive stars that release gas and radiation to create cavities. The telescope has also helped to characterize a white dwarf star that exploded as a famous supernova and lies 9000 light-years away, RCW86. The explosion that created RCW86 was observed by the ancient Chinese 2000 years ago. Modern-day observations of the supernova remnant confirm that this star absorbed nearby material to create a cavity before its explosion was observed in the year 185 CE.

Farther away, *WISE* has identified millions of supermassive black hole candidates as well as a number of dust-laden galaxies that are extremely bright in infrared bands, yet remain dust-obscured galaxies in visible light ("hot DOGs"). The telescope has also made detailed observations of millions of more common galaxies, such as our neighbor Andromeda,

Fig. 4. ***WISE*** **flight system in the payload processing facility at Vandenberg Air Force Base, California. The Earth-orbiting telescope was launched on a Delta II rocket on December 9, 2009. The domed cover was ejected two weeks after launch. (*Courtesy of NASA/JPL-Caltech/UCLA*)**

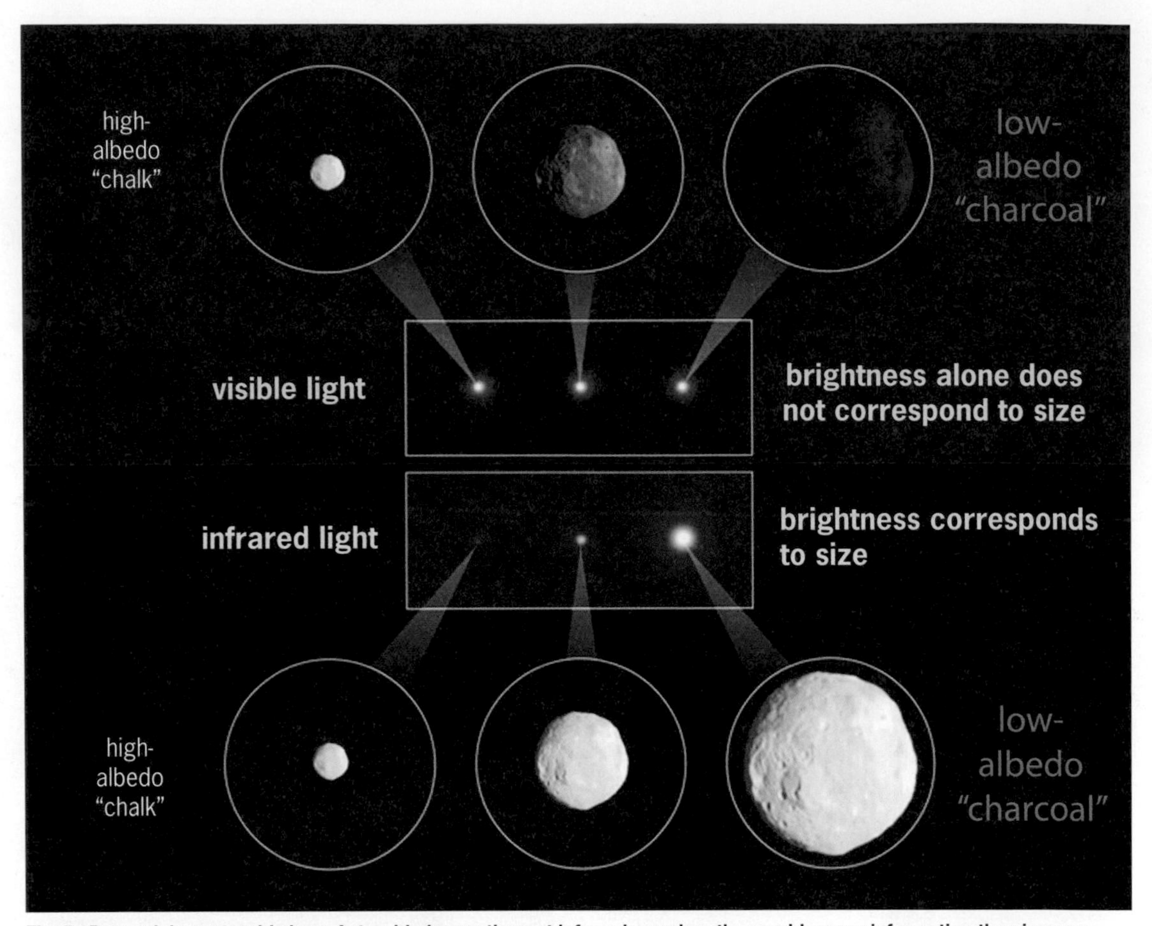

Fig. 5. Determining asteroid sizes. Asteroid observations at infrared wavelengths provide more information than images taken at visible wavelengths. In visible light, a small bright object can appear as bright as a large dark object. Since infrared telescopes measure the amount of heat emitted by an object, a small object will appear less bright than a large one. (*Courtesy of NASA/JPL-Caltech/UCLA*)

also known as M31. *WISE* has also identified a cluster of thousands of galaxies at a distance of 7.7 billion light-years from Earth.

Using the data from *WISE* in conjunction with observations at other wavelengths from different ground- and space-based telescopes, astronomers have produced a number of findings, including the discovery of a galaxy in a very unusual state of evolution. A gravitationally collapsing gas cloud in the center of this galaxy, known as SDSSJ1506+54, is being processed into stars with nearly 100% efficiency at the theoretical maximum rate of star formation. As galaxies evolve, new stars formed within the collapsed gas will disperse the gas itself by the production of radiation and charged particles, leading to a slowdown in stellar production. The fact that new stars in SDSSJ1506+54 are being produced at a highly efficient rate suggests that their activity has not progressed to the stage where the gas cloud has stopped collapsing.

Data and image catalogs. Science and engineering teams on the WISE mission combined raw data from the telescope with positional and instrument calibration data to generate individual images. These images were analyzed carefully, either individually or in groups, to identify asteroids, comets, stars, nebulae, galaxies, and other astronomical sources. The mission generated a number of catalogs, and analysis efforts continue to produce new data sets. These catalogs are available to the public.

WISE All-Sky Data Release. The *WISE* mission team processed the more than 15 terabytes of data transmitted from the spacecraft to generate a set of all-sky data at all four *WISE* wavelengths; this is publicly accessible through the Infrared Science Archive (IRSA) of the NASA Infrared Processing and Analysis Center (IPAC). The data set was released in March 2012, and the records included a catalog of celestial coordinates, brightness, and other information for more than 560 million individual astronomical objects as well as more than 18,000 images of the entire celestial sky. The majority of the cataloged entries were of stars and galaxies.

NEOWISE Post-Cryo Preliminary Data Release. Based on observations taken from September 29, 2010 to February 1, 2011, IRSA produced a public data release of WISE images and data tables. In June 2012, a preliminary catalog of single-frame images and a database of detected asteroids and comets were released. As these observations were made after the on-board coolant was depleted, the longer-wavelength detectors were inoperable, and only the

3.4- and 4.6-mm detectors were active. Database records included a table of coordinates, brightness, and other information for nearly 5 million individual astronomical objects and more than 900,000 images of portions of the celestial sky. This archive also included a database of all known asteroids and comets as of June 2012. On May 22, 2013, an updated catalog was made available to the public. This version used updated calibration parameters to reprocess all NEOWISE data, based on changes in the spacecraft during the post-cryogenic operations phase. NEOWISE Post-Cryo Data products included observed brightness and coordinate values, not only for solar system objects, but also for more distant celestial sources. In addition to yielding a wealth of information on the inner Main Asteroid Belt, this release also provided brightness comparisons for stars, galaxies, and other distant objects that were observed during both the primary and post-cryogenic periods.

AllWISE. This project will add data from both the cryogenic WISE mission and the post-cryogenic NEOWISE mission to generate more detailed images in all four passbands. A database of multi-epoch observations, brightness values, and derived proper motions is planned for release in November 2013.

WISE Enhanced Resolution Galaxy Atlas and WISE Extended Source Catalog. An international team of astronomers will use image processing techniques to produce high-resolution images of large, bright galaxies as well as small, faint ones.

For background information *see* ALBEDO; ANDROMEDA GALAXY; ASTEROID; ASTRONOMICAL CATALOGS; BLACK HOLE; BROWN DWARF; CHARGE-COUPLED DEVICES; COMET; GALAXY, EXTERNAL; IMAGE PROCESSING; INFRARED ASTRONOMY; INFRARED RADIATION; INTERSTELLAR MATTER; NEBULA; SPITZER SPACE TELESCOPE; STAR; STELLAR EVOLUTION; SUPERNOVA; TELESCOPE; TROJAN ASTEROIDS in the McGraw-Hill Encyclopedia of Science & Technology.

Bidushi Bhattacharya

Bibliography. A. Mainzer et al., Preliminary results from NEOWISE: An enhancement to the *Wide-field Infrared Survey Explorer* for solar system science, *Astrophys. J.*, 731:53 (13 pp.), 2011, DOI:10.1088/0004-637X/731/1/53; E. L. Wright et al., The Wide-field Infrared Survey Explorer (WISE): Mission description and initial on-orbit performance, *Astron. J.*, 140:1868–1881, 2010, DOI:10.1088/0004-6256/140/6/1868.

XML (Extensible Markup Language)

Ever since its release almost 20 years ago, XML has continuously gained popularity and has become an internationally accepted and implemented standard for electronic publishing. XML is a fundamental component of Web development today. It has been widely adopted by academic journals, e-commerce, news sites, Web services, and social media. Communities in any domain can take advantage of the XML structure for sharing their resources. The use of XML for online publishing as well as offline storage can assist with information exchange and extraction.

History. Like the name suggests, markup languages are used to mark up data inside documents. They represent a set of formal conventions for adding extra information about the content. Markup languages use annotations (tags) to describe the structure and attributes of the data and ultimately aim to separate the content of documents from their textual form. This annotation concept emerged in the late 1960s, from IBM, as the Generalized Markup Language (GML). Gradually, it evolved into the Standard Generalized Markup Language (SGML) and was recognized in 1986 by the International Organization for Standardization (ISO) as a standard for information processing (ISO 8879). SGML is conceptually very powerful and complex. However, it was not well-received by the general public who considered SGML to be unnecessarily complicated and expensive to implement, especially in a Web environment. Specifically designed for the creation of Web documents, a simpler version called HyperText Markup Language (HTML) surfaced in the early 1990s. HTML was regarded as a divergent subset of SGML and became the core language of almost all Internet content today. The tags and attributes used in HTML mainly address the presentation of the page. Web browsers interpret and use the tags and attributes to properly format and display text for the human eye. One drawback of HTML is that its set of tags is predetermined and cannot be extended.

The information technology (IT) world needed a language that was simpler than SGML, but still powerful enough to accommodate various domains. A new dialect targeting Web applications materialized, under the name of XML (eXtensible Markup Language). The specifications were released in 1998 by the World Wide Web Consortium (W3C), which is the major international standards organization for regulating the World Wide Web technologies. Ten years later, the fifth and latest edition was published, which is currently in use.

XML is not a language per se, but a meta-language and therefore extensible. The set of rules are adaptable to specific annotations, allowing users to customize their own tags, to define new orderings of tags, and ultimately to design novel markup languages. Both HTML and XML share similarities with their predecessor technology, SGML. As opposed to HTML, which can be considered an application of SGML, XML is a less complex subset of SGML, while still maintaining the power and value of SGML and allowing richer structures than HTML. One application of XML is XHTML (also a recommendation by the W3C).

Structure. The XML syntax is both human readable and machine readable. The protocol of markup languages is platform independent, hence portable to all operating systems and generic enough to be processed by many applications. Using tags in XML is very similar to the way they are used in HTML; they are mainly information fields about the data. Tags are indicated by surrounding angle brackets, which are reserved characters especially used for identifying the start and end of a markup. The end tag also contains a forward slash. The text contained inside the tags is referred to as an element. For example, in HTML, the following line specifies that the contained text must be displayed using italicized letters:

```
<em>Breakfast at Tiffany's</em>
```

Aside from the font specification, the reader cannot deduce if the phrase refers to the movie, the book, the song, or the time and place for a meeting. On the other hand, the markups in XML deliver more information about the elements they enclose. The tags clearly indicate that in this document, "Breakfast at Tiffany's" is the title of a novel:

```
<book>Breakfast at Tiffany's</book>
```

In HTML, the tags are fixed and used by the browser to display the data, whereas in XML users can define their own set of tags to organize the data. Moreover, unlike HTML where the closing tag can sometimes be omitted, in XML all tags must be balanced. Consider the following description of a book in XML:

```
<book id="2117">
    <author>Capote, Truman</author>
    <title>Breakfast at Tiffany's</title>
    <type>novella</type>
    <year>1958</year>
    <publisher>Random House</publisher>
</book>
```

Web crawlers or any parsing application can exploit the fact that all books are uniquely identified by the attribute id and that all book authors are enclosed in <author> tags. The attribute values are always quoted (book id = "2117"). There exist many open-source tools for processing XML documents (such as editors, validation tools, and parsers). The verification of XML documents to guarantee they are valid [having a correct syntax and conforming to a Document Type Definition (DTD)] ensures data reusability and allows automated programs to capture more easily large collections of data for analyses.

XML is suitable for offline document storage and management, with the caveat that it should not be used as a traditional database because of its repetitive tag structure. A substantial collection of documents in XML (raw text) will occupy more disk space than the same data would if stored in a reasonable database. Typically, XML databases map XML documents into more traditional database structures. However, the verbosity of XML is desirable in communication between servers, because it adds safety and the receiver can easily verify the integrity of the file. It is also worth noting that XML is not a programming language, it is a definition for expressing data, not computer programs.

Many flavors of XML have been developed. Usually, the tags represent terminologies of different domains. Controlled vocabularies are critical in the creation of a robust and time-withstanding markup language. The set of legal tags, attributes, and rules must be predefined for any specialized XML. They are specified using DTDs, or XML Schemas. Any newly developed markup language will have a domain-specific semantic, but a regularized XML syntax. Some of the notable specialized markup languages include MathML (created by W3C and the American Mathematical Society), Chemical Markup Language (CML) developed by Peter Murray-Rust and Henry Rzepa, Security Assertion Markup Language (SAML) advanced by the Security Services Technical Committee, Financial Products Markup Language (FpML), and Geography Markup Language (GML) defined by the Open Geospatial Consortium (OGC). The Keyhole Markup Language (KML) was submitted by Google to the OGC and is complementary to most of the key existing OGC standards; it is focused on geographic visualization, including annotation of maps and images. Other notable markup languages are BioPAX for visualization and analysis of biological pathways, MusicXML for representing musical notation, NeuroML used in computational neuroscience, and EPUB (a standard for electronic publishing of e-books) developed by the International Digital Publishing Forum.

There are other markup languages that transcend domains and can be used in many disciplines. For example, Scalable Vector Graphics (SVG) is the XML format for defining graphics; it is a format developed and maintained by the SVG Working Group at W3C, and its capability was enhanced to accommodate mobile devices. Interoperability between markup languages such as HTML and SVG allows graphics to be easily displayed in Web browsers. For more complex information, there are specialized tools available. For example, Jmol is a Java-written high-performance 3D rendering tool that accepts a variety of XML-based files. Among many others, Jmol can display objects from CML files, PBD files (Protein Data Bank), and CIF (Crystallographic Information File).

Another example of an interdisciplinary markup standard is the CALS (Continuous Acquisition and Life-cycle Support) Table Model, used for tabular data. OASIS (Organization for the Advancement of Structured Information Standards, a consortium formerly known as SGML Open) has only recommended a subset of the original full CALS, citing higher probability for interoperability and exchange among various SGML products.

Outlook. There are many ongoing activities related to XML technologies and standards on W3C and OASIS. Their collaboration with the scientific communities is essential, and so far their efforts have been successful. S. Grijzenhout and M. Mar recently reported the condition of XML data on the Web. Their study revealed the surprisingly good quality of XML documents, which facilitates the direct application of XML technologies. Information retrieval is one discipline that can gain enormously from metadata. For example, distinguishing ambiguous words by the context they appear in can ultimately improve precision in document search.

Since its implementation, XML has become a universal standard that has revolutionized online scientific publishing and will continue to grow alongside the World Wide Web.

For background information *see* DATA STRUCTURE; DATABASE MANAGEMENT SYSTEM; INTERNET; METADATA; WORLD WIDE WEB in the McGraw-Hill Encyclopedia of Science & Technology.

Ana Stanescu

Bibliography. S. Grijzenhout and M. Mar, The quality of the XML web, *Web Semant. Sci. Serv. Agents World Wide Web*, 19:59–68, 2013, DOI:10.1016/j.websem.2012.12.001; P. Murray-Rust and H. S. Rzepa, Scientific publications in XML—towards a global knowledge base, *Data Sci. J.*, 1(1):84–98, 2002, DOI:10.2481/dsj.1.84.

Nobel Prizes for 2014

The Nobel Prizes for 2014 included the following awards for scientific disciplines.

Chemistry. The chemistry prize was awarded to Erik Betzig of the Howard Hughes Medical Institute, Ashburn, Virginia, Stefan W. Hell of the Max Planck Institute for Biophysical Chemistry, Göttingen, and German Cancer Research Center, Heidelberg, Germany, and W. E. Moerner of Stanford University, Stanford, California, for developing techniques that extend light (optical) microscopy to the nanoscale, called super-resolution fluorescence microscopy. These techniques allow researchers to see things as small as individual viruses and proteins and to observe biological processes in living cells in real time. Super-resolution fluorescence microscopy promises to improve our understanding of how cells work in normal physiology and malfunction in disease.

In optical microscopy, resolution—the smallest distance between two distinctly visualized points on an object—is limited by diffraction to about half the wavelength used. This constraint is commonly called Abbe's limit, after the German physicist and optical researcher Ernst Abbe (1840–1905) who discovered it. For visible light (with wavelengths of 400–700 nanometers), the resolution limit is no less than about 200 nm. Electron microscopy has much higher resolving power because electrons have a wavelength of about 0.01 nm, but it requires very thin samples and cannot image live cells.

Hell, Moerner, and Betzig found a pair of ways to surpass the diffraction limit of optical microscopy without breaking the laws of physics. To do so, they manipulated the output of fluorescent molecules (fluorophores) and used a healthy dose of computing power to construct high-resolution images for objects smaller than 200 nm.

In 2000, Hell imaged a fluorophore-tagged *E. coli* bacterium at a resolution three times as high as previously seen with optical microscopy using a technique called stimulated emission depletion (STED) microscopy. He discovered if he excited the fluorophores with a blue light beam surrounded by a ring of red light (the STED beam), the red light would quench the yellow fluorescence from the sample except for a small area in the center. If he increased the intensity of the red light, the size of the fluorescence spot became smaller and the image resolution became higher. By scanning the beams through the sample, a point-by-point image was obtained. A sample can be repeatedly and rapidly scanned for time-lapsed microscopy. In 2012, Hell and coworkers reported STED images of neurons in a living mouse at less than 70-nm resolution.

Moerner and Betzig worked independently on a common concept to develop the method known as single-molecule microscopy. In 1997, Moerner discovered fluorescent proteins that would turn on (fluoresce) using one wavelength (488 nm) of light and then slowly bleach (turn off). However, to turn on the fluorescent proteins again once they had faded required a second, different wavelength (405 nm) of light. In 2006 using similar fluorescent proteins, Betzig and coworkers reported that by photoactivating a small number of these fluorescent molecules attached to the membrane of the lysosome, they could record their positions very accurately. After repeating the procedure numerous times, they were able to combine all the data sets to produce a composite superhigh resolution image of the lysosome membrane.

For background information *see* BIOPHYSICS; CELL (BIOLOGY); DIFFRACTION; FLUORESCENCE; FLUORESCENCE MICROSCOPE; GREEN FLUORESCENT PROTEIN; LASER; MICROSCOPE; OPTICAL MICROSCOPE; RESOLVING POWER (OPTICS); VIDEO MICROSCOPY in the McGraw-Hill Encyclopedia of Science & Technology.

Physics. The physics prize was awarded to Isamu Akasaki of Meijo University and Nagoya University in Nagoya, Japan, Hiroshi Amano of Nagoya University, and Shuji Nakamura of the University of California at Santa Barbara, United States, for their invention of efficient blue-light-emitting diodes (LEDs), which has enabled the fabrication of bright and energy-saving white-light sources along with other improvements to a variety of optoelectrical devices.

Light-emitting diodes consume much less power than conventional light sources because they convert electricity directly into visible photons. In contrast, an incandescent lamp produces light by electrically heating a metallic filament until it glows, wasting much energy as heat in the process. White LED lamps, which are based on blue LEDs, are rapidly replacing incandescent lamps, and it has been said that the twenty-first century will be lit by LED lamps, as the twentieth century was lit by incandescent bulbs. Moreover, LED lamps are constantly being improved, and now can generate over 300 lumens per watt of electric power, compared to 16 for incandescent lamps and approximately 70 for fluorescent lamps. This advance in energy efficiency can significantly affect the world economy since over 20% of the world's electric power is used for lighting. Furthermore, LED lamps can last for 100,000 hours on average, compared to 1000 for incandescent lamps and 10,000 for fluorescents, so their use of material resources is also much more efficient. While it has been argued that the availability of inexpensive power may actually result in a net increase in world energy consumption, rather than a saving, there is no doubt that LED lighting technology can enhance the well-being of most of the world population, especially in developing countries. LEDs can be particularly beneficial for the over 1.5 billion people who lack access to electric power grids since the low power requirements of LEDs allow them to be powered cheaply by local solar power.

Red and green LEDs have been known since the 1960s, but until they were joined by a blue light source the white lamps needed to illuminate homes and offices could not be produced. However, the production of a blue LED was a challenge that defied the efforts of researchers and industry for 30 years. Akasaki, Amano, and Nakamura challenged conventional wisdom, took considerable risks, conducted thousands of experiments, and remained undaunted when most of them failed, finally accomplishing what large research laboratories had been unable to achieve. Akasaki collaborated with Amano at Nagoya University, while Nakamura worked at the Nichia Corporation in Tokushima, Japan.

The creation of an LED that could emit blue light required a semiconductor with a large band gap (the energy difference between the conduction band and the valence band, which are the sources, respectively, of the electrons and "holes" that combine, resulting in the emission of photons). Also, this structure would preferably be a direct band gap, which would allow the recombination of electrons and holes without the mediation of phonons. Gallium nitride (GaN) was a material that fulfilled these requirements, but it posed enormous practical difficulties since no one was able to grow GaN crystals of sufficient quality, despite repeated attempts to do so. However, in 1986, Akasaki (who had been working with GaN since the late 1960s) and his then-doctoral student Amano were able to create high-quality crystals by placing a layer of aluminum nitride (AlN) on a sapphire substrate and then growing the GaN on top of it. Nakamura began his research to develop a blue LED in 1988, and in 1991 he was also able to create high-quality GaN, but by a different method that involved first growing a thin GaN layer at a relatively low temperature, and then growing subsequent layers at higher temperatures.

Meanwhile, in 1988, Akasaki and Amano overcame what had been considered the second major obstacle to the fabrication of GaN-based LEDs: the seeming impossibility of creating a *p*-type layer of this compound. (Such a layer, which in effect provides positively charged "holes" for excited electrons to fill, is a basic component of an LED.) Remarkably, their breakthrough came about accidentally when they noticed that their material was glowing more intensely when it was studied in a scanning electron microscope, suggesting that the microscope's electron beam was making the *p*-type layer more efficient. In 1992, Nakamura was able to explain this effect: Hydrogen forms complexes with the acceptor atoms in the *p*-type layer, which makes them passive, but electron beams dissociate these complexes and thereby activate the acceptors. Nakamura also was able to find a simpler and cheaper way of activating the acceptors that did not require an electron beam and involved only heating the material.

Previous development of LEDs and laser diodes had shown that high efficiency required the use of heterojunctions and quantum wells, in which electrons and holes are injected into a small volume so that recombination occurs more efficiently and with minimal losses. Both the Akasaki and Nakamura groups worked on the growth and doping of various alloys to realize such structures, which became increasingly complex. In 1995–1996, both groups developed a blue laser, in which a blue LED, the size of a sand grain, is a critical component. Since blue light has a shorter wavelength than red light, its use allows the storage of more information in a given area, and this increase in storage capacity has led to the development of high-density digital video disks (DVDs), including Blu-ray Discs, and better laser printers.

White LEDs used for lighting can be based either on blue LEDs that excite a phosphor so that blue light is converted to white light, or on a combination of red, green, and blue LEDs, enabling the development of highly flexible light sources with a great variety of possible applications. In addition to the lighting of homes and offices, blue LEDs already provide the dominant technology for back-illuminated liquid-crystal displays in television screens, computer monitors, and mobile devices, particularly mobile phones, where they also provide a lamp and flash for the camera. A future application may be sterilization of polluted water by an ultraviolet LED, a further development of the blue LED.

For background information *see* ACCEPTOR ATOM; LASER; LIGHT-EMITTING DIODE; LIGHT PANEL; QUANTIZED ELECTRONIC STRUCTURE (QUEST); SEMICONDUCTOR; SEMICONDUCTOR HETEROSTRUCTURES; VIDEO DISK in the McGraw-Hill Encyclopedia of Science & Technology.

Physiology or medicine. The prize in physiology or medicine honored the discovery and study of cells that constitute a positioning system in the brain. Half of the prize was awarded to John O'Keefe of University College, London, United Kingdom, and the other half jointly to May-Britt Moser of the Centre for Neural Computation in Trondheim, Norway, and Edvard I. Moser of the Kavli Institute for Systems Neuroscience in Trondheim, Norway.

The ability of animals to identify spatial and temporal relationships among stimuli in their environment is well established. For instance, a digger wasp will remember the location of its nest relative to arbitrary landmarks, and will fly to the wrong place if those landmarks are moved. Although this navigational ability has been documented since the late 1940s, an explanation as to how and where this ability arises was lacking until the groundbreaking discoveries of O'Keefe, Moser, and Moser.

Building on the concept of a cognitive map system originally proposed by experimental psychologist Edward Tolman in 1948, John O'Keefe was the first to study physiological mechanisms and correlates of behavior in freely moving animals. In the early 1970s, he laid the groundwork for positioning research by recording cellular activity in nerve cells within the hippocampus—a major structure deep within the vertebrate brain that is part of the limbic system and is associated with memory. While collaborating with fellow neuroscientist Jonathan Dostrovsky, O'Keefe discovered so-called place cells in the hippocampus of rats, which become active whenever the animal occupies a specific position in its environment. O'Keefe later deduced that, when an animal is on the move, place cells fire at different positions (place fields) along its route, forming a spatial map of the animal's surroundings.

By the late 1970s, O'Keefe had elucidated the cognitive mapping theory of hippocampal function, with place cells at its crux. At first, the scientific community met this theory with skepticism. However, through a series of meticulous experiments, O'Keefe demonstrated that combinations of place cell activity elicited in different environments could create and preserve multiple maps within the hippocampus simultaneously, allowing an animal to learn and remember various spatial and temporal positions within its territory. Subsequent studies by other researchers confirmed these seminal findings.

The prevailing theory among neuroscientists throughout the 1980s and 1990s was that the formation of place fields originated within the hippocampus. Then in the mid-1990s, married collaborators May-Britt Moser and Edvard Moser became interested in how the brain computes spatial location and spatial memory. They wanted to know whether activity outside of the hippocampus could cause place cells to fire.

While recording brain activity in freely moving rats in 2005, the Mosers observed a surprising pattern of activity near the hippocampus in an area called the entorhinal cortex, the main site of nerve entry into the limbic system. The Mosers had discovered a new type of nerve cell—the grid cell—that can create a coordinate system in the medial entorhinal cortex. Within this part of the brain, the animal's surrounding environment is spatially mapped by dividing it into hexagonal grids of positions. Each grid cell is associated with one of these hexagonal grids, and it fires as the animal moves from place to place within that grid. Grids of neighboring cells have the same spacing and orientation, but are slightly offset from each other. Collectively, these overlapping grids cover every point in the environment. Thus, grid cells constantly update the animal's awareness of its location in external space.

The discovery of grid cells provided the first clues about how the brain uses neural firing patterns to create spatial maps, and in subsequent research, the Mosers elucidated the relationship between place cells and grid cells. While place cells mark an animal's position in the environment, grid cells allow the brain to compute distances. Along with other cells of the entorhinal cortex that recognize the direction in which the head is pointed and the border of the environment, grid cells form circuits with place cells in the hippocampus.

In effect, the limbic system operates somewhat like a neurological Global Positioning System (GPS). Whereas the GPS receives position, velocity, time, and related data from an assembly of satellites, the limbic system receives such input from all of the body's sensory organs via the entorhinal cortex. From there, sensory information goes to the hippocampus, where the data are integrated over time into neural representations of position, which establish an internal coordinate system. Hippocampal output returns to the entorhinal cortex, which distributes the integrated positional data to all of the sensory organs and prepares them to receive new input.

The laureates' findings have prompted examination of the hippocampal-entorhinal structure in other mammals, including humans. Results of such studies suggest that this system of highly specialized brain cells is robust and has been conserved in vertebrate evolution. The findings are also opening new doors for a better understanding of Alzheimer's and other neurological diseases that impair cognition. In the early stages of Alzheimer's disease, the hippocampus and entorhinal cortex are often damaged, resulting in the loss of spatial recognition. Ultimately, the laureates' findings could translate into treatments for Alzheimer's patients, or others with spatial memory problems, to ensure that navigational cells are functioning properly.

For background information *see* ALZHEIMER'S DISEASE; BRAIN; CELL (BIOLOGY); CELL BIOLOGY; COGNITION; MEMORY; NERVOUS SYSTEM (VERTEBRATE); NEUROBIOLOGY; PSYCHOLOGY; VERTEBRATE BRAIN (EVOLUTION) in the McGraw-Hill Encyclopedia of Science & Technology.

Contributors

Contributors

The affiliation of each Yearbook contributor is given, followed by the title of his or her article. An article title with the notation "coauthored" indicates that two or more authors jointly prepared an article or section.

A

Abergel, Dr. Chantal. *Information Génomique et Structurale, Centre National de la Recherche Scientifique, Marseille, France.* PANDORAVIRUS—coauthored.

Aboutabl, Dr. Mohamed S. *Department of Computer Science, James Madison University, Harrisonburg, Virginia.* SECURE INTERNET COMMUNICATION.

Ackerman, Laura K. G. *Department of Chemistry, University of Rochester, New York.* TANDEM CATALYSIS—coauthored.

Agarlita, Dr. Sorin Cristian. *Interior Body and Security, Continental Automotive Romania, Timisoara, Romania.* SENSORLESS ELECTRIC DRIVES—coauthored.

Agarwal, Prof. Ramesh K. *Department of Mechanical Engineering and Materials Science, Washington University in St. Louis, St. Louis, Missouri.* GREEN AVIATION.

Åkerman, Dr. Johan. *Department of Physics, University of Gothenburg, Sweden.* SPIN TORQUE.

Alarcón, Daniela I. *Indiana University of Pennsylvania.* RESTORING REGENERATION—coauthored.

Allaei, Dr. Daryoush. *Chief Executive Officer, SheerWind Inc. Chaska, Minnesota.* NEW WIND POWER TECHNOLOGY: INVELOX—coauthored.

Andreopoulos, Prof. Yiannis. *Department of Mechanical Engineering, The City College of the City University of New York.* NEW WIND POWER TECHNOLOGY: INVELOX—coauthored.

Asmar, Dr. Sami W. *Jet Propulsion Laboratory, NASA, Pasadena, California.* GRAVITY RECOVERY AND INTERIOR LABORATORY (GRAIL) MISSION—coauthored.

Averitt, Dr. Richard D. *Physics Department, Boston University, Boston, Massachusetts.* TERAHERTZ TIME-DOMAIN SPECTROSCOPY AND MATERIAL DYNAMICS.

B

Bailey, Dr. Caleb M. *Stowers Microscopy Center, Stowers Institute for Medical Research, Kansas City, Missouri.* IMAGING OF DEVELOPMENTAL BIOLOGICAL PROCESSES—coauthored.

Bakac, Dr. Andreja. *Department of Chemistry, Iowa State University, Ames.* OXYGEN ACTIVATION BY METAL COMPLEXES.

Balakrishnan, Dr. Kaushik. *College of Optical Sciences, University of Arizona, Tucson.* SELF-ASSEMBLED ORGANIC OPTICAL MATERIALS—coauthored.

Bare, Dr. Jane C. *National Risk Management Research Laboratory, U.S. Environmental Protection Agency, Cincinnati, Ohio.* INCORPORATING ENVIRONMENTAL EXPOSURE SCIENCE INTO LIFE-CYCLE ANALYSIS—coauthored.

Batlle, Prof. Xavier. *Department of Fundamental Physics and Institute of Nanoscience and Nanotechnology (IN2UB), University of Barcelona, Catalonia, Spain.* INFLUENCE OF NANOSTRUCTURAL FEATURES ON THE PROPERTIES OF MAGNETIC NANOPARTICLES—coauthored.

Beard, Dr. K. Christopher. *Department of Ecology and Evolutionary Biology, Biodiversity Institute, University of Kansas, Lawrence.* ANTHROPOID ORIGINS.

Bernacchioni, Dr. Caterina. *CERM, University of Florence, Sesto, Fiorentino, Italy.* FERRITIN: IRON STORAGE IN BIOLOGICAL SYSTEMS—coauthored.

Bhandari, Dr. Mohit. *Division of Orthopaedic Surgery, Centre for Evidence-Based Orthopaedics, McMaster University, Hamilton, Ontario, Canada.* SPRING UNLOADER FOR KNEE OSTEOARTHRITIS—coauthored.

Bhattacharya, Dr. Bidushi. *W. M. Keck Science Department, Claremont McKenna, Pitzer, and Scripps Colleges, Claremont, California.* *WISE* SPACECRAFT.

Block, Dr. Jon E. *The Jon Block Group, San Francisco, California.* SPRING UNLOADER FOR KNEE OSTEOARTHRITIS—coauthored.

Boldea, Prof. Ion. *Electrical Engineering, Politehnica University of Timisoara, Romania.* Sensorless electric drives—coauthored.

Bowden, Dr. Anton E. *Department of Mechanical Engineering, Brigham Young University, Provo, Utah.* Artificial spinal disc replacement.

Bradley, Dr. Marty K. *Propulsion and Environmental Analysis, Boeing Research and Technology, The Boeing Company, Huntington Beach, California.* Options for future methane- and LNG-fueled aircraft—coauthored.

Bragg-Sitton, Dr. Shannon M. *Fuel Performance and Design, Idaho National Laboratory, Idaho Falls, Idaho.* Accident-tolerant fuels.

Bringmann, Prof. Oliver. *Department of Computer Science, Faculty of Science, University of Tübingen, Germany.* Self-adaptive design techniques—coauthored.

Brown, Dr. Brian V. *Entomology Section, Natural History Museum of Los Angeles County, California.* Phorid flies.

Brown, Dr. Peter J. *Department of Physics and Astronomy, Texas A&M University, College Station, Texas.* Nearby supernova SN 2014J.

Bruce, Prof. Duncan. *Department of Chemistry, University of York, York, United Kingdom.* Luminescent liquid crystals for OLED display technology—coauthored.

Bryant, Dr. Vaughn M. *Department of Anthropology, Texas A&M University, College Station.* Forensic palynology.

Burnett, Dr. David S. *Naval Surface Warfare Center, Panama City Division, Panama City, Florida.* Underwater acoustic scattering signature.

C

Cameron, Dr. Amy Yule. *Department of Psychiatry and Human Behavior, Warren Alpert Medical School of Brown University, Providence, Rhode Island; and Department of Psychology, Clark University, Worcester, Massachusetts.* Dialectical behavior therapy.

Cao, Dr. Ke. *College of Civil Engineering, Tongji University, Shanghai, China.* Application and research of modular buildings—coauthored.

Cardwell, Glenn. *Accredited Practicing Dietitian, Nutrition Impact Proprietary Limited, Bentley DC, Western Australia.* Vitamin D-enhanced mushrooms.

Chen, Prof. Gang. *Department of Mechanical Engineering, Massachusetts Institute of Technology, Cambridge.* Nanostructured thermoelectric energy scavenging.

Chen, Dr. Wenbiao. *Department of Molecular Physiology and Biophysics, Vanderbilt University School of Medicine, Nashville, Tennessee.* CRISPR/Cas9 gene editing—coauthored.

Clapham, Dr. Matthew E. *Department of Earth and Planetary Sciences, University of California, Santa Cruz.* History of insect body size.

Claudy, Lynn. *National Association of Broadcasters, Washington, DC.* Radio and television broadcasting during disasters.

Claverie, Dr. Jean-Michel. *Structural and Genomic Information Laboratory, Aix-Marseille Université, Marseille, France.* Pandoravirus—coauthored.

Conley, Dr. Matthew P. *Department of Chemistry, ETH Zürich, Zürich, Switzerland.* Alkane metathesis—coauthored.

Copéret, Prof. Christophe. *Department of Chemistry, ETH Zürich, Zürich, Switzerland.* Alkane metathesis—coauthored.

Cox, Kelly. *The Boeing Company, Boeing Research and Technology-Australia, Brisbane, Queensland, Australia.* Environmental life-cycle impact of alternative aviation fuels.

D

Daly, Dr. Norelle L. *School of Pharmacy and Molecular Sciences, James Cook University, Smithfield, Cairns, Queensland, Australia.* Macrocyclic peptides used as drugs.

Darr, Charles V. *Cruise Lines International Association, Arlington, Virginia.* Cruise ships—coauthored.

De Gaudenzi, Dr. Riccardo. *European Space Agency, Noordwijk, The Netherlands.* S-MIM radio interface for mobile satellite services—coauthored.

Del Valle, Kierstin M. *Cruise Lines International Association, Arlington, Virginia.* Cruise ships—coauthored.

Diebold, Prof. Gerald J. *Department of Chemistry, Brown University, Providence, Rhode Island.* Photoacoustic imaging in biomedicine.

Diehl, Dr. H. Thomas. *Fermi National Accelerator Laboratory, Batavia, Illinois.* The Dark Energy Survey and Camera—coauthored.

Dixon, Dr. Kingsley W. *Kings Park and Botanic Garden, West Perth; and School of Plant Biology, University of Western Australia, Crawley.* Karrikins—coauthored.

Dorn, Dr. Gerald W., II. *Center for Pharmacogenomics, Division of Biology and Biomedical Sciences, Washington University in St. Louis, Missouri.* PINK1-Parkin pathway.

Doyle, Dr. Michael P. *Center for Food Safety, Department of Food Science and Technology, University of Georgia, Griffin.* Foodborne disease.

E

El-Hawary, Prof. Mohamed E. *Associate Dean of Engineering, Dalhousie University, Halifax, Nova Scotia, Canada.* Offshore wind energy conversion—coauthored.

El-Metwally, Prof. Salah El-Din E. *Structural Engineering Department, El Mansoura University, Mansoura, Egypt.* Fundamental design of reinforced concrete water structures.

Elliott, Dr. David K. *School of Earth Sciences and Environmental Sustainability, Northern Arizona University, Flagstaff.* Origin of vertebrates.

Engel, Dr. Michael S. *Division of Entomology, University of Kansas, Lawrence.* Early Cretaceous insect camouflage.

Eronen, Dr. Tommi. *Department of Physics, University of Jyväskylä, Finland.* Nuclear structure studies with ion traps—coauthored.

Eskin, Dr. N. A. Michael. *Department of Human Nutritional Sciences, University of Manitoba, Winnipeg, Canada.* Nutraceuticals.

F

Fassett, Dr. Caleb I. *Department of Astronomy, Mt. Holyoke College, South Hadley, Massachusetts.* MESSENGER mission results.

Flaugher, Dr. Brenna L. *Fermi National Accelerator Laboratory, Batavia, Illinois.* The Dark Energy Survey and Camera—coauthored.

Flematti, Dr. Gavin R. *School of Chemistry and Biochemistry, University of Western Australia, Crawley.* Karrikins—coauthored.

Fox, Dr. Christopher. *Department of Computer Science, James Madison University, Harrisonburg, Virginia.* Agile methods in software engineering.

Fraile Rodriguez, Dr. Arantxa. *Department of Fundamental Physics and Institute of Nanoscience and Nanotechnology (IN2UB), University of Barcelona, Catalonia, Spain.* Influence of nanostructural features on the properties of magnetic nanoparticles—coauthored.

Freese, Dr. Curtis H. *Freese Consulting, Westport, Massachusetts.* Prairie restoration on the American Prairie Reserve.

G

Gage, Dr. Fred H. *Laboratory of Genetics, Salk Institute for Biological Studies, La Jolla, California.* Aneuploidy in neurons—coauthored.

Giazotto, Dr. Francesco. *NEST: National Enterprise for nanoScience and nanoTechnology, Istituto Nanoscienze-Consiglio Nazionale delle Ricerche and Scuola Normale Superiore, Pisa, Italy.* Superconducting quantum interference proximity transistor (SQUIPT).

Glass, Keely. *Department of Chemistry, Duke University, Durham, North Carolina.* Color of ancient cephalopod ink—coauthored.

Gu, Dr. Jiang. *Department of Pathology, Shantou University Medical College, Guangdong, China.* Big data and personalized medicine: the role of pathology—coauthored.

Gupta, Dr. Pranshu. *Department of Mathematics and Computer Science, DeSales University, Center Valley, Pennsylvania.* Big data—coauthored.

H

Hamel, Jeffrey. *Advanced Products Team, GE Aviation, Cincinnati, Ohio.* Options for future methane- and LNG-fueled aircraft—coauthored.

Hamilton, Prof. Joseph H. *Landon C. Garland Distinguished Professor of Physics, Department of Physics and Astronomy, Vanderbilt University, Nashville, Tennessee.* Prolate-oblate shape transitions in nuclei—coauthored.

Harley, Dr. John P. *Department of Biological Sciences, Eastern Kentucky University, Richmond.* Use of fecal transplants to treat intestinal disorders.

Hassett, Dr. Daniel E. *Department of Veterinary Pathobiology, and Laboratory for Infectious Disease Research, University of Missouri, Columbia.* DNA vaccination.

Haugen, Dr. Linda. *Forest Health Protection, State and Private Forestry, U.S. Forest Service, St. Paul, Minnesota.* Dutch elm disease.

Hayton, Dr. Trevor W. *Department of Chemistry, University of California, Santa Barbara.* Coordination chemistry of actinide elements—coauthored.

Holmes, Dr. Stephen M. *Center for Nanoscience, University of Missouri-St. Louis.* Molecular magnets.

Howell, Dr. Miya D. *Monsanto Company, Chesterfield, Missouri.* Evolutionary transition from C_3 to C_4 photosynthesis in plants—coauthored.

Hugenholtz, Dr. Christopher. *Department of Geography, University of Calgary, Alberta, Canada.* Remote sensing with small unmanned aircraft systems.

Hultmark, Dr. Marcus. *Department of Mechanical and Aerospace Engineering, Princeton University, Princeton, New Jersey.* Turbulence: When is the Reynolds number high enough?—coauthored.

I

Imre, Prof. Sándor. *Department of Networked Systems and Services, Budapest University of Technology and Economics (BME), Budapest, Hungary.* Quantum communications.

Inomata, Dr. Takeshi. *School of Anthropology, University of Arizona, Tucson.* Maya civilization.

Irle, Dr. Mark A. *Research Director, École Supérieure du Bois, Nantes, France.* Measuring formaldehyde emissions from wood-based panels.

Ito, Dr. Shosuke. *Department of Chemistry, Fujita Health University, Toyoake, Aichi, Japan.* COLOR OF ANCIENT CEPHALOPOD INK—coauthored.

J

Janssen, Dr. Jan-Theodoor. *Time, Quantum and Electromagnetics Division, National Physical Laboratory, Teddington, Middlesex, United Kingdom.* GRAPHENE DEVICES FOR QUANTUM METROLOGY.

Jeffery, Dr. William R. *Department of Biology, University of Maryland, College Park.* EYE DEVELOPMENT (VERTEBRATE)—coauthored.

Jenner, Dr. Ronald A. *Department of Life Sciences, The Natural History Museum, London, United Kingdom.* VENOMOUS CRUSTACEAN.

Jiménez, Dr. Víctor M. *CIGRAS (Centro para Investigaciones en Granos y Semillas), Universidad de Costa Rica, San Pedro, Montes de Oca, San José.* PAPAYA.

Jokinen, Dr. Ari. *Department of Physics, University of Jyväskylä, Finland.* NUCLEAR STRUCTURE STUDIES WITH ION TRAPS—coauthored.

Jones, Dr. Christopher W. *School of Chemical & Biomolecular Engineering, Georgia Institute of Technology, Atlanta.* SUPPORTED AMINE MATERIALS FOR CO_2 CAPTURE—coauthored.

Jones, Dr. Warren. *Marcus Autism Center, Children's Healthcare of Atlanta, and Emory University School of Medicine, Atlanta, Georgia.* AUTISM AND EYE GAZE ABNORMALITIES—coauthored.

K

Karafiath, Gabor. *Naval Surface Warfare Center, Carderock Division, West Bethesda, Maryland.* SEATRAIN—coauthored.

Kelemen, Rachel E. *Department of Chemistry, University of Rochester, New York.* TANDEM CATALYSIS—coauthored.

Kim, Dr. Daehyun. *Department of Geography, University of Kentucky, Lexington.* VEGETATION DYNAMICS.

Kim, Dr. Na Young, *Ginzton Laboratory, Stanford University, Stanford, California.* POLARITON LASER.

Klin, Dr. Ami. *Marcus Autism Center, Children's Healthcare of Atlanta, and Emory University School of Medicine, Atlanta, Georgia.* AUTISM AND EYE GAZE ABNORMALITIES—coauthored.

Kozhevnikov, Dr. Valery N. *Department of Applied Sciences, University of Northumbria, Newcastle-Upon-Tyne, United Kingdom.* LUMINESCENT LIQUID CRYSTALS FOR OLED DISPLAY TECHNOLOGY—coauthored.

Kulesa, Dr. Paul M. *Director of Imaging Center, Stowers Microscopy Center, Stowers Institute for Medical Research, Kansas City, Missouri.* IMAGING OF DEVELOPMENTAL BIOLOGICAL PROCESSES—coauthored.

L

Labarta, Prof. Amilcar. *Department of Fundamental Physics and Institute of Nanoscience and Nanotechnology (IN2UB), University of Barcelona, Catalonia, Spain.* INFLUENCE OF NANOSTRUCTURAL FEATURES ON THE PROPERTIES OF MAGNETIC NANOPARTICLES—coauthored.

Larson, Dr. Eric B. *Executive Director and Senior Investigator, Group Health Research Institute, Seattle, Washington.* DEMENTIA.

Laursen, Wendy. *Maritime journalist, Batemans Bay, New South Wales, Australia.* LIQUEFIED NATURAL GAS AS A MARINE FUEL—coauthored.

Letcher, Prof. Trevor M. *Stratton on the Fosse, United Kingdom.* EMERGING ENERGY TECHNOLOGIES.

Li, Prof. Guo-Qiang. *State Key Laboratory for Disaster Reduction in Civil Engineering, Tongji University, Shanghai, China.* APPLICATION AND RESEARCH OF MODULAR BUILDINGS—coauthored.

Lu, Dr. Chih-Yuan. *President, Macronix International Co., Ltd., Hsinchu, Taiwan, Republic of China.* THREE-DIMENSIONAL NONVOLATILE MEMORY SCALING.

Lu, Dr. Ye. *College of Civil Engineering, Tongji University, Shanghai, China.* APPLICATION AND RESEARCH OF MODULAR BUILDINGS—coauthored.

Lui, Dr. Eric M. *Department of Civil & Environmental Engineering, Syracuse University, New York.* LONG-SPAN HYBRID SUSPENSION AND CABLE-STAYED BRIDGES—coauthored.

Luo, Dr. Yixiao. *Lawrence Berkeley National Laboratory, Berkeley, California.* PROLATE-OBLATE SHAPE TRANSITIONS IN NUCLEI—coauthored.

Lynch, Rebecca M. *Philadelphia College of Osteopathic Medicine, Suwanee, Georgia.* MICROBIAL FORENSICS—coauthored.

Lysak, Dr. Martin A. *Laboratory of Plant Cytogenomics, Central European Institute of Technology (CEITEC), Masaryk University, Brno, Czech Republic.* CHROMOSOME "FUSIONS" IN KARYOTYPE EVOLUTION.

M

Ma, Dr. Li. *Department of Biology, University of Maryland, College Park.* EYE DEVELOPMENT (VERTEBRATE)—coauthored.

Maisey, Dr. John G. *Division of Paleontology, American Museum of Natural History, New York.* SHARK HISTORY.

Major, Dr. Robert J. *Department of Biology, Indiana University of Pennsylvania.* RESTORING REGENERATION—coauthored.

Malik, Dr. Praveen. *National Research Center on Equines, Hisar, Haryana, India.* EQUINE GLANDERS—coauthored.

Mata-Toledo, Dr. Ramon A. *Professor of Computer Science, James Madison University, Harrisonburg, Virginia.* Big data—coauthored.

McConnell, Dr. Michael J. *Department of Biochemistry and Molecular Genetics, University of Virginia School of Medicine, Charlottesville.* Aneuploidy in neurons—coauthored.

McDonald, Dr. John W. *International Center for Spinal Cord Injury (ICSCI), Kennedy Krieger Institute, Baltimore, Maryland.* Functional electrical stimulation therapy—coauthored.

McEwen, Michael. *Director-General, North American Broadcasters Association, Toronto, Ontario, Canada.* The next generation of television.

Mio, Dr. Jeffery Scott. *Department of Psychology and Sociology, California State Polytechnic University, Pomona.* Social class and poverty: missing elements in psychological research.

Moore, Dr. Chester G. *Department of Microbiology, Immunology, and Pathology, College of Veterinary Medicine and Biomedical Sciences, Colorado State University, Fort Collins.* Encephalitis (arboviral).

Moore, Mark D. *NASA Langley Research Center, Hampton, Virginia.* Technologies for a new generation of small electric aircraft.

Moriuchi, Jennifer. *Department of Psychology, Emory University, Atlanta, Georgia.* Autism and eye gaze abnormalities—coauthored.

Muhlfelder, Dr. Barry. *W. W. Hansen Experimental Physics Laboratory, Stanford University, Stanford, California.* Gravity Probe B: final results.

Muir, Jeffrey M. *The Jon Block Group, San Francisco, California.* Spring unloader for knee osteoarthritis—coauthored.

Mutuku, Dr. Josiah Musembi. *Plant Immunity Research Group, RIKEN Center for Sustainable Resource Science, Yokohama, Japan.* *Striga*: physiology, effects, and genomics—coauthored.

N

Nemunaitis, Dr. John. *Department of Oncology, Mary Crowley Cancer Research Centers; Gradalis, Incorporated; Texas Oncology, Professional Association; and Medical City Dallas Hospital; Dallas, Texas.* Gene therapy (cancer).

Ng, Dr. Szu Ting. *Ligno Biotech, Balakong Jaya, Selangor, Malaysia.* Tiger Milk mushroom.

Noffke, Dr. Nora. *Department of Ocean, Earth, and Atmospheric Sciences, Old Dominion University, Norfolk, Virginia.* Ancient microbial ecosystem.

P

Page-McCaw, Dr. Patrick S. *Department of Molecular Physiology and Biophysics, Vanderbilt University School of Medicine, Nashville, Tennessee.* CRISPR/Cas9 gene editing—coauthored.

Palmer, Dr. A. Richard. *Department of Biological Sciences, University of Alberta, Edmonton, Canada.* Barnacle reproduction.

Park, Prof. Nam-Gyu. *School of Chemical Engineering, Sungkyunkwan University, Jangan-gu, Suwon, Republic of Korea.* Sensitized mesoscopic solar cells.

Pask, Dr. Andrew J. *Department of Zoology, University of Melbourne, Victoria, Australia.* Sex chromosomes.

Pau, Prof. Stanley. *College of Optical Sciences, University of Arizona, Tucson.* Self assembled organic optical materials—coauthored.

Paxson, Dr. Daniel E. *Controls and Dynamics Branch, NASA Glenn Research Center, Cleveland, Ohio.* Pressure-gain combustion.

Pierce, Dr. Marcia M. *Department of Biological Sciences, Eastern Kentucky University, Richmond.* Obesity as a factor of gut microbiota; *Vibrio vulnificus*.

Platt, Dr. Donald. *Micro Aerospace Solutions, Inc., Melbourne, Florida.* Space flight, 2013.

Popa, Adriana. *Department of Chemistry, Case Western Reserve University, Cleveland, Ohio.* Functional inorganic nanomaterials—coauthored.

Pornet, Clément. *Visionary Aircraft Concepts, Bauhaus Luftfahrt, e.V., Munich, Germany.* Hybrid electric and universally electric aircraft concepts.

R

Rasmussen, Prof. John O. *Lawrence Berkeley National Laboratory, Berkeley, California.* Prolate-oblate shape transitions in nuclei—coauthored.

Rayfield, Dr. Emily J. *School of Earth Sciences, University of Bristol, Clifton, United Kingdom.* Finite element analysis in paleontology.

Recio, Dr. Albert. *International Center for Spinal Cord Injury (ICSCI), Kennedy Krieger Institute, Baltimore, Maryland.* Functional electrical stimulation therapy—coauthored.

Rowell, Dr. Roger M. *Professor Emeritus, Department of Biological Systems Engineering, University of Wisconsin, Madison.* Blue stain in wood.

Rudion, Prof. Krzysztof. *Institute of Power Transmission and High-Voltage Technology, University of Stuttgart, Germany.* Offshore wind energy conversion—coauthored.

S

Sadegh, Prof. Ali M. *Department of Mechanical Engineering, The City College of the City University of New York.* New wind power technology: INVELOX—coauthored.

Sakwa-Novak, Miles A. *School of Chemical & Biomolecular Engineering, Georgia Institute of Technology, Atlanta.* SUPPORTED AMINE MATERIALS FOR CO_2 CAPTURE—coauthored.

Samia, Dr. Anna. *Department of Chemistry, Case Western Reserve University, Cleveland, Ohio.* FUNCTIONAL INORGANIC NANOMATERIALS—coauthored.

Scalise, Dr. Sandro. *Institute of Communications and Navigation, German Aerospace Center (DLR), Oberpfaffenhofen, Germany.* S-MIM RADIO INTERFACE FOR MOBILE SATELLITE SERVICES—coauthored.

Schwartz, Dr. Bennett L. *Department of Psychology, Florida International University, Miami.* METAMEMORY.

Schweizer, Dr. Thomas. *Department of Computer Science, Faculty of Science, University of Tübingen, Germany.* SELF-ADAPTIVE DESIGN TECHNIQUES—coauthored.

Schwiegerling, Prof. Jim. *College of Optical Sciences, University of Arizona, Tucson.* PLENOPTIC IMAGING.

Shirasu, Dr. Ken. *Plant Immunity Research Group, RIKEN Center for Sustainable Resource Science, Yokohama, Japan.* *STRIGA*: PHYSIOLOGY, EFFECTS, AND GENOMICS—coauthored.

Shors, Dr. Teri. *Department of Biology and Microbiology, University of Wisconsin-Oshkosh.* MIDDLE EAST RESPIRATORY SYNDROME (MERS); SCHMALLENBERG VIRUS.

Siegwart, Dr. Daniel J. *Simmons Comprehensive Cancer Center, University of Texas Southwestern Medical Center, Dallas.* POLYMERS FOR DRUG DELIVERY.

Simon, Dr. John D. *Department of Chemistry, University of Virginia, Charlottesville.* COLOR OF ANCIENT CEPHALOPOD INK—coauthored.

Singh, Dr. Raj Kumar. *Indian Veterinary Research Institute, Bareilly, Uttar Pradesh, India.* EQUINE GLANDERS—coauthored.

Singha, Dr. Harisankar. *National Research Center on Equines, Hisar, Haryana, India.* EQUINE GLANDERS—coauthored.

Slewinski, Dr. Thomas L. *ELS Research Scientist, Monsanto Company, Chesterfield, Missouri.* EVOLUTIONARY TRANSITION FROM C_3 TO C_4 PHOTOSYNTHESIS IN PLANTS—coauthored.

Slutsky, Jonathan. *Naval Surface Warfare Center, Carderock Division, West Bethesda, Maryland.* SEATRAIN—coauthored; SHIP MODERNIZATION AND CONVERSION.

Smiles, Danil E. *Department of Chemistry, University of California, Santa Barbara.* COORDINATION CHEMISTRY OF ACTINIDE ELEMENTS—coauthored.

Smith, Dr. Steven M. *School of Chemistry and Biochemistry, ARC Centre of Excellence in Plant Energy Biology, University of Western Australia, Crawley.* KARRIKINS—coauthored.

Smits, Prof. Alexander J. *Department of Mechanical and Aerospace Engineering, Princeton University, Princeton, New Jersey.* TURBULENCE: WHEN IS THE REYNOLDS NUMBER HIGH ENOUGH?—coauthored.

Snyder, Michael P. *Director of Research and Development, Made In Space, Inc., NASA Ames Research Park, Moffett Field, California.* 3D PRINTING APPLICATIONS IN SPACE.

Soltis, Dr. Douglas E. *Department of Biology, University of Florida, Gainesville.* *AMBORELLA* GENOMICS.

Spallek, Dr. Thomas. *Plant Immunity Research Group, RIKEN Center for Sustainable Resource Science, Yokohama, Japan.* *STRIGA*: PHYSIOLOGY, EFFECTS, AND GENOMICS—coauthored.

Stanescu, Ana. *Department of Computing and Information Sciences, Kansas State University, Manhattan.* XML (EXTENSIBLE MARKUP LANGUAGE).

Stanley, Dr. George D., Jr. *Department of Geosciences, University of Montana, Missoula.* GEOLOGIC HISTORY OF REEFS.

Straley, Prof. Jan. *Department of Biology, University of Alaska Southeast, Sitka, Alaska.* WHALE THEFT OF FISH FROM LONG-LINE FISHERIES.

Styczynski, Prof. Zbigniew A. *Institute of Electric Power Systems, Otto-von-Guericke University, Magdeburg, Germany.* OFFSHORE WIND ENERGY CONVERSION—coauthored.

Subrahmanyam, Dr. Kaveri. *Department of Psychology, College of Natural and Social Sciences, California State University, Los Angeles.* ELECTRONIC COMMUNICATION AND IDENTITY DEVELOPMENT.

T

Taylor, Dr. Clive R. *Department of Pathology, Keck School of Medicine, University of Southern California, Los Angeles.* BIG DATA AND PERSONALIZED MEDICINE: THE ROLE OF PATHOLOGY—coauthored.

Teo, Tony. *Technology and Business Director, Region Americas, DNV GL - Maritime, Katy, Texas.* LIQUEFIED NATURAL GAS AS A MARINE FUEL—coauthored.

Todd, Dr. Nancy E. *Department of Biology and Environmental Studies, Manhattanville College, Purchase, New York.* ELEPHANT PHYLOGENY.

Traub, Dr. Wesley A. *Exoplanet Exploration Program, Jet Propulsion Laboratory, NASA, Pasadena, California.* EXOPLANET RESEARCH.

Turano, Dr. Paola. *CERM and Department of Chemistry, University of Florence, Sesto, Fiorentino, Italy.* FERRITIN: IRON STORAGE IN BIOLOGICAL SYSTEMS—coauthored.

V

Vacariu, Kim. *Western Director, Wildlands Network, Portal, Arizona.* CONNECTED WILDLANDS: CORRIDORS FOR SURVIVAL.

Valdastri, Dr. Pietro. *Department of Mechanical Engineering, Vanderbilt University, Nashville, Tennessee.* CAPSULE ROBOTS FOR ENDOSCOPY—coauthored.

Valentine, Vanessa Nicole. *Department of Mechanical Engineering, Vanderbilt University, Nashville, Tennessee.* CAPSULE ROBOTS FOR ENDOSCOPY—coauthored.

Vallero, Dr. Daniel A. *National Exposure Research Laboratory, U.S. Environmental Protection Agency, Research Triangle Park, North Carolina.* INCORPORATING ENVIRONMENTAL EXPOSURE SCIENCE INTO LIFE-CYCLE ANALYSIS—coauthored.

van Leeuwen, Dr. Floor. *Institute of Astronomy, University of Cambridge, United Kingdom.* GAIA MISSION.

Vergari, Fabrizio. *Selex ES, Avionics LOB, Pomezia, Italy.* APPLICATION OF SOFTWARE-DEFINED RADIO.

W

Waikel, Dr. Rebekah L. *Department of Biological Sciences, Eastern Kentucky University, Richmond.* MICROBIAL FORENSICS—coauthored.

Wang, Dr. Kevin G. *Department of Aerospace and Ocean Engineering, Virginia Polytechnic Institute and State University, Blacksburg, Virginia.* HETEROGENEOUS COMPUTING.

Wanninger, Dr. Andreas. *Department of Integrative Zoology, Faculty of Life Sciences, University of Vienna, Austria.* EVOLUTION OF APLACOPHORAN MOLLUSKS.

Weiler, Prof. Thomas J. *Department of Physics and Astronomy, Vanderbilt University, Nashville, Tennessee.* THREE NEUTRINO FLAVORS AND THEIR MIXING.

Weix, Dr. Daniel J. *Department of Chemistry, University of Rochester, New York.* TANDEM CATALYSIS—coauthored.

Whitehorn, Dr. Nathan. *Department of Physics, University of California, Berkeley.* HIGH-ENERGY ASTROPHYSICAL NEUTRINOS AT ICECUBE.

Wilby, Dr. Philip. *British Geological Survey, Keyworth, Nottingham, United Kingdom.* COLOR OF ANCIENT CEPHALOPOD INK—coauthored.

Williams, Dr. Scott A. *Department of Anthropology, New York University, New York.* EARLIEST HOMININS.

Woldegebriel, Zekarias Tadesse. *Department of Civil & Environmental Engineering, Syracuse University, New York.* LONG-SPAN HYBRID SUSPENSION AND CABLE-STAYED BRIDGES—coauthored.

Wolfe, Dr. Jeremy M. *Visual Attention Lab, Harvard Medical School and Brigham and Women's Hospital, Cambridge, Massachusetts.* VISUAL ATTENTION.

Wultsch, Dr. Claudia. *Sackler Institute for Comparative Genomics, American Museum of Natural History, New York.* PREDATORY FELINE GENES.

Z

Zuber, Prof. Maria T. *Vice President for Research and E. A. Griswold Professor of Geophysics, Department of Earth, Atmospheric and Planetary Sciences, Massachusetts Institute of Technology, Cambridge.* GRAVITY RECOVERY AND INTERIOR LABORATORY (GRAIL) MISSION—coauthored.

Index

Index

Asterisks indicate page references to article titles.

A

E

F

H

I

M

N

P

R

S

W

X